T0319343

Functionally Graded Materials 2000

Related titles published by The American Ceramic Society:

Ceramic Processing Science (Ceramic Transactions, Volume 112)
Edited by Shin-ichi Hirano, Gary L. Messing, and Nils Claussen
©2001, ISBN 1-57498-104-8

Advances in Ceramic Matrix Composites VI (Ceramic Transactions Volume 124)
Edited by J.P. Singh, Narottam P. Bansal, and Ersan Ustundag
©2001, ISBN 1-57498-123-4

Boing-Boing the Bionic Cat and the Jewel Thief
By Larry L. Hench
©2001, ISBN 1-57498-129-3

Boing-Boing the Bionic Cat
By Larry L. Hench
©2000, ISBN 1-57498-109-9

The Magic of Ceramics
By David W. Richerson
©2000, ISBN 1-57498-050-5

Ceramic Innovations in the 20th Century
Edited by John B. Wachtman Jr.
©1999, ISBN 1-57498-093-9

Advances in Ceramic Matrix Composites V (Ceramic Transactions Volume 103)
Edited by Narottam P. Bansal, J.P. Singh, and Ersan Ustundag
©2000, ISBN 1-57498-089-0

Advances in Ceramic Matrix Composites IV (Ceramic Transactions Volume 96)
Edited by J.P. Singh and Narottam P. Bansal
©1999, ISBN 1-57498-059-9

*Innovative Processing and Synthesis of Advanced Ceramics, Glasses, and
Composites (Ceramic Transactions Volume 85)*
Edidted by Narottam P. Bansal, Kathryn V. Logan, and J.P. Singh
©1998, ISBN 1-57498-030-0

For information on ordering titles published by The American Ceramic Society, or
to request a publications catalog, please contact our Customer Service
Department at 614-794-5890 (phone), 614-794-5892
(fax), <customersrvc@acers.org> (e-mail), or write to Customer Service
Department, 735 Ceramic Place, Westerville, OH 43081, USA.

Visit our on-line book catalog at <www.ceramics.org>.

Volume 114

Functionally Graded Materials 2000

Proceedings of the 6th International Symposium on Functionally Graded Materials, Stanley Hotel, Estes Park, Colorado, USA September 10–14, 2000.

Edited by

Kevin Trumble
Purdue University

Keith Bowman
Purdue University

Ivar Reimanis
Colorado School of Mines

Sanjay Sampath
State University of New York—Stony Brook

Published by
The American Ceramic Society
735 Ceramic Place
Westerville, Ohio 43081
www.ceramics.org

Proceedings of the 6th International Symposium on Functionally Graded Materials, Stanley Hotel, Estes Park, Colorado, USA September 10–14, 2000.

Cover photo: "SEM micrographs of bonding interface in FGMs," is courtesy of H. Kobayashi and appears as figure 4 in the paper "Fabrication of PSZ-Al$_2$O$_3$ Functionally Graded Disks and Plates," which begins on page 299.

Library of Congress Cataloging-in-Publication Data
A CIP record for this book is available from the Library of Congress.

For information on ordering titles published by The American Ceramic Society, or to request a publications catalog, please call 614-794-5890.

ISSN 1042-1122
ISBN 978-1-57498-110-2

Contents

Biomedical Applications

Tribological Coatings

Thermal Barrier Coatings

Powder Processing

Infiltration Processing

Deposition and Casting

Properties Modeling

Properties Characterization

Fracture Mechanics Modeling

Fracture Characterization

Preface

Functionally Graded Materials (FGM) has served as a unifying theme for interdisciplinary research for more than a decade. The biannual International Symposium on Functionally Graded Materials has provided a forum for research on materials with spatial variations in microstructures or chemistries. Meetings in Sendai 1990, San Francisco 1992, Lausanne 1994, Tsukuba 1996, and Dresden 1998 have engendered a small, but richly interactive, community of FGM researchers from university, industry, and government labs all around the world. In this tradition, FGM 2000 was held at the historic Stanley Hotel in Estes Park, Colorado, September 10–14, 2000.

More than 150 researchers from about 20 countries participated in FGM 2000, presenting 120 oral presentations and 18 posters over four days. All presentations were very well attended and the discussions were lively and productive. As the papers in this volume attest, FGM continues to be a vigorous topic stimulating new materials research.

The papers of these proceedings are divided into sections corresponding approximately to the sessions in which they were presented. The initial sections represent the main applications of functional materials, biomedical applications, tribological coatings, and thermal barrier coatings. The largest group of papers is on processing, subdivided into sections on powder processing (reactive and non-reactive routes), infiltration processing, and casting and depositions. Another large group of papers—mechanical, thermal, and electrical properties of FGM—are divided into two sections, properties modeling and properties characterization. Fracture mechanics modeling and fracture characterization papers complete the proceedings.

All papers were peer reviewed. The session chairs served as associate editors for reviewing all papers. Their efforts and those of the reviewers and authors are gratefully acknowledged.

Financial support of National Science Foundation, Dr. Lise Schioler, Army Research Office, Dr. David Stepp, and Office of Naval Research, Dr. Asuri Vasudevan is gratefully acknowledged.

Kevin P. Trumble
Keith J. Bowman
Sanjay Sampath
Ivar Reimanis

Associate Editors

Sami El-Borgi (Mechanics Modeling)
Polytechnic School of Tunisia

Keith Bowman (Properties Characterization III—Impact)
Purdue University

Rowland Cannon (Coatings II—Tribological)
University of California – Berkeley

Edwin Fuller (Coating IV—Thermal Barrier)
NIST

Michael Gasik (Processing II—Powder [Reactive])
Helsinki University of Technology

Changchun Ge (Processing V—Deposition and Casting)
Beijing University of Science and Technology

Leonard Gray (Biomedical Applications I)
Oak Ridge National Laboratory

Mark Hoffman (Properties Characterization IV—Fracture)
University of New South Wales

E. A. Levashov (Processing I—Powder [Reactive])
Moscow Institute of Steel and Alloys

Yoshinari Miyamoto (Processing IV—Powder)
Osaka University

Yoshinari Miyamoto (Functional Materials II)
Osaka University

Dan Mumm (Coatings I—Tribological)
Princeton University

Dietrich Munz (Mechanics Modeling II)
University Karlsruhe

Achim Neubrand (Properties and Performance)
Darmstadt University of Technology

Glaucino Paulino (Properties Characterization V—Fracture)
University of Illinois

Manfred Peters
(Properties Characterization II—Mechanical and Thermal)
DLR German Aerospace Center

QingJie Zhang (Properties Modeling II)
Wuhan University

Maria Peters (Processing VI—Powder)
Los Alamos National Laboratory

Wolfgang Pompe (Biomedical Applications II)
Dresden University of Technology

Juergen Roedel (Processing III—Powder)
Technische Universitat Darmstadt

Sanjay Sampath (Coatings III—Thermal Barrier)
State University of New York – Stony Brook

E. D. Steffler
(Properties Characterization I—Mechanical and Thermal)
Idaho National Engineering and Environment Laboratory

Kevin Trumble (Processing VII and VIII—Infiltration)
Purdue University

Will Windes (Properties Modeling I and Functional Materials I)
Idaho National Engineering and Environment Laboratory

Functional Materials

FABRICATION OF CERAMIC/EPOXY PHOTONIC CRYSTALS WITH GRADED LATTICE SPACINGS BY STEREOLITHOGRAPHY

Soshu Kirihara and Yoshinari Miyamoto
Joining and Welding Research Institute, Osaka University
11-1 Mihogaoka Ibaraki, Osaka 567-0047, Japan

Kenji Kajiyama
Ion Engineering Research Institute Corporation
2-8-1 Tuda-Yamate Hirakata, Osaka 573-0128, Japan

ABSTRACT

Three-dimensional photonic crystals with periodic variations in dielectric constant which can totally reflect electromagnetic waves were fabricated by a method of stereolithographic rapid prototyping. The structures are composed of millimeter-order epoxy lattices including 10 vol.% TiO_2 particles with a diamond structure. The attenuation of transmission amplitude through the photonic crystals clearly showed the formation of bandgaps in a microwave range. The diamond photonic crystals with graded lattice spacings which can form wide bandgaps were also fabricated. Development of engineering these photonic crystals and lattice modifications has a large potential of applications to high performance filters, antennas, and other devices in GHz range for telecommunication systems.

INTRODUCTION

Three-dimensional photonic crystals with periodic variations in dielectric constant can form bandgaps against electromagnetic waves[1,2]. These artificial crystals can totally reflect a light or a microwave. It is well known that the periodic variation of crystal potential in semiconductors forms electronic bandgaps by three dimensional scattering of electron waves associated with Bragg reflection.

We have fabricated photonic crystals composed of millimeter order epoxy FCC lattices by using stereolithographic rapid prototyping and examined the bandgap formation in 5~20 GHz range[3]. Firstly, the structure of a photonic crystal is designed with a CAD program using a personal computer. The structure design is sliced to a set of thin sections and converted to a numerical code (STL data) which is transferred to the stereolithographic machine. Then, it forms a two-dimensional layer on photo-sensitive liquid resin by UV-laser scanning, and builds up to a three-dimensional structure by repeating layer formations based on the STL data[4]. Silica or titania fine particles with high dielectric constants were dispersed into the photo-polymer epoxy resin in order to control the bandgap.

Recently, we succeeded in fabrication of a diamond structure of the titania/epoxy lattice. It is believed that the photonic crystal with the symmetry of diamond structure can form a perfect bandgap in all directions[5]. We also tried to make photonic crystals with graded lattice spacings in order to extend the ability of the bandgap control. The wide bandgap is expected to be formed with graded lattices. The engineering of photonic crystals and lattice modifications is expected to lead to many applications in telecommunication systems; new high frequency filters, compact wave guides, high performance directional antennas for future intelligent transportation systems, and barriers against electromagnetic interference.

In this paper, the attenuation of transmission amplitude of microwaves through the titania/epoxy diamond photonic crystals with or without graded lattice spacings will be shown, and relationships between the lattice arrangements of the crystal and the form of the bandgap will be discussed.

EXPERIMENTAL PROCEDURE

The photonic crystals were fabricated from the photopolymer epoxy resin (D-MEC LTD. SCR-730). Titania ceramic particles were dispersed into this liquid resin with 10% in volume fraction. The particle size is about 10 μm in average. The dielectric constants of the polymerized epoxy and titania are about 2.8 and 100, respectively. The three-dimensional lattice structures were processed by the stereolithographic system (D-MEC LTD. SCS-300P). The titania dispersed epoxy resin was photo-polymerized through the UV laser scanning. The laser spot was about 100μm in beam diameter. The layers of titania/epoxy composite with 100μm in thickness were formed and stacked to 3D structures with 0.15% part accuracy through the CAD/CAM processes. The dispersion of the titania particles in the crystal lattice was observed by using SEM.

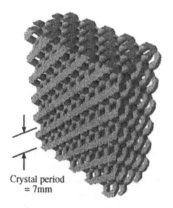
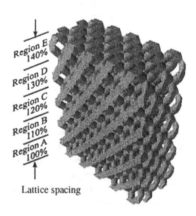

Fig. 1 A three-dimensional model of a diamond photonic crystal. This lattice structure imitated to a diamond structure.

Fig. 2 The three-dimensional model of the diamond photonic crystal with a graded lattice spacing.

The lattice model of a photonic crystal designed is shown in Fig. 1. The crystal with 15 ×34×42mm in dimension was composed of dielectric lattices with 2×3mm rods in diameter. The periodic variation in dielectric constant was realized by the repeated arrangement of lattice and air gap. The epoxy/titania lattice was designed to have the symmetry of a diamond structure to exhibit the bandgap perfectly in all directions[6]. In this diamond photonic crystal, the lattice spacing for the <100> direction was designed to be 7 mm. A graded lattice was designed along the <100> direction, which is shown in Fig. 2. The lattice spacing was increased stepwise from A to E uniform regions by 100, 110, 120, 130 and 140%, respectively. The crystal size was 15×34×42mm.

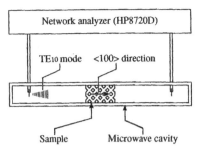

Fig. 3 An experimental configuration to measure the transmission attenuation in a microwave range through the diamond photonic crystals.

The attenuation of transmission amplitude of microwaves through the photonic crystals was measured by using a network analyzer (HP-8720D). Figure 3 shows the measurement system. In a metallic cavity, electromagnetic waves with TE10 mode was transmitted through the photonic crystals for <100> direction. The attenuations for <110> and <111> directions were measured for uniform crystals without graded lattices. A titania/epoxy bulk sample with the same size as the photonic crystal was fabricated to compare the microwave attenuation. The dielectric constant of the bulk sample was measured by using a dielectric probe kit (HP-85070B).

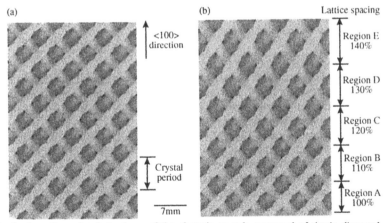

Fig. 4 three-dimensional structures of the photonic crystals composed of titania dispersed epoxy lattices with similar arrangement to diamond structure. (a) and (b) are diamond photonic crystal with or without graded lattice spacing, respectively.

RESULTS AND DISCUSSION
Structure of photonic crystals

Figure 4 (a) and (b) shows the uniform and graded lattice structures of the titania/epoxy photonic crystals fabricated by stereolithography. The millimeter order diamond lattice structures could be formed with high precision using the three-dimensional CAD data. The microstructre of the lattice is shown in Fig. 5. The titania particles were dispersed uniformly without pores in the epoxy matrix. The dielectric constants of the epoxy bulk samples with and without titania dispersion are shown as a function of frequency in Fig. 6. The titania dispersions increased the dielectric constant of the bulk. The increasing ratio in dielectric constant was roughly proportionate to the volume ratio of titana particles.

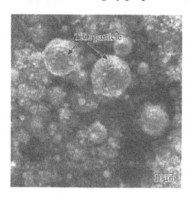

Fig. 5 A microstructre of the ceramic particles dispersed epoxy lattice in the diamond photonic crystal. Fine particles are broken pieces of epoxy, when the lattices were fractured.

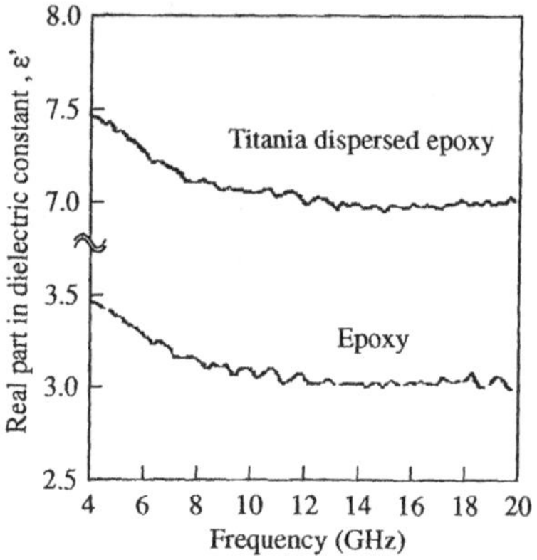

Fig. 6 Real parts in dielectric constant of epoxy bulk samples with or without the titania particles dispersion.

Microwave properties

The attenuation of transmission amplitude through a diamond photonic crystal and a bulk sample are shown in Fig. 7. The bulk sample almost transmits electromagnetic waves without absorption in the measured frequency range. On the other hand, the photonic crystal exhibits the narrow and deep forbidden gap in 12.4 to 12.9 GHz range. The maximum attenuation is quite large reaching about 55 dB. Such large attenuations were observed against other directions of <110> and <111>. These evidences suggest that the narrow forbidden band was formed in every direction.

Figure 8 shows the attenuation of transmission amplitude through the graded photonic crystal. In this case, a broad forbidden gap was formed in 8.3 to 13.5 GHz range. This result shows that the forbidden bandgap can be extended effectively. Though the attenuation of transmission amplitude, that is, the reflection in this experiment is slightly reduced and the attenuation curve is oscillated as seen in Fig.8. The maximum attenuation was about 25 dB.

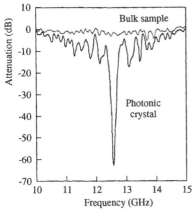

Fig. 7 Attenuations of transmission amplitude through the titania/epoxy diamond pahotonic crystal and the bulk sample.

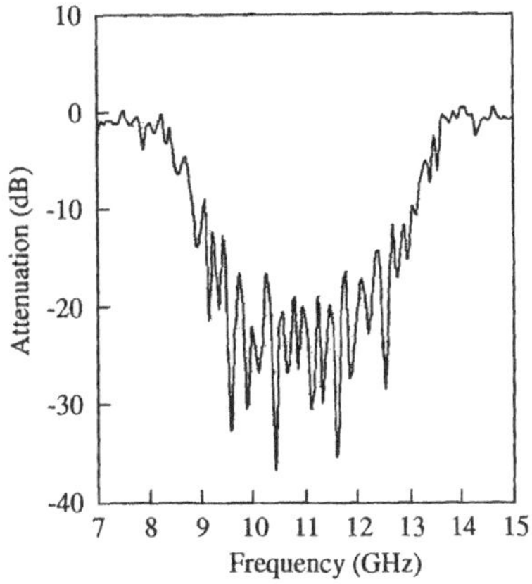

Fig. 8 The transmission amplitude through the diamond pahotonic crystal with the graded lattice spacing.

Effect of lattice defect in graded photonic crystal

It is possible to produce energy levels in the photonic bandgap like donor or acceptor levels in semiconductors[1]. These energy levels can be formed by introducing lattice defects such as vacancy or interstitial types. A large potential in application of photonic crystals are expected by designing lattice defects[5]. For example, the curved channels in a photonic crystals can form the waveguide. An advantage of stereolithography exists in easiness of design and fabrication of lattice defects[4]. Figure 9 shows a lattice structure of

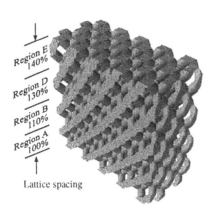

Fig. 9 The model of the diamond photonic crystal with the graded lattice spacing including layer defects.

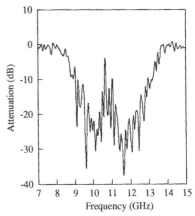

Fig. 10 The transmission amplitude through the diamond photonic crystal with the graded lattice spacing including the layer defects.

the graded photonic crystal which is missing a layer of the region C. The attenuation of transmission amplitude through this photonic crystal is shown in Fig. 10. A broad bound-mode appears for 10.3 to 11.1 GHz in the forbidden gap ranging from 8.3 to 13.5 GHz

CONCLUSIONS

Three-dimensional titania/epoxy photonic crystals with a diamond lattice structure were fabricated by a stereolithographic rapid prototyping method. Periodic variations in dielectric constant were realized in millimeter-order lattice arrangements. The large attenuation of transmission amplitude of electromagnetic waves through these photonic crystals confirmed the formation of the photonic bandgaps. Subsequently, the diamond photonic crystals with the graded lattice spacing were processed. These graded photonic crystals extended the bandgap. The graded crystal with a large gap of lattice spacing formed a broad bound-mode.

ACKNOWLEDGEMENTS

This study is financially supported by a Grant-in-Aid for Scientific Research (A) (2) No. 11305046 of the Ministry of Education, Science, Sports and Culture, Japan. The authors wish to thank D-MEC LTD. for their assistance in fabrication of photonic crystals by stereolithographic techniques, Institute of Laser Engineering, Osaka University in electromagnetic wave measurement, and Mr. Shoichi Terashima of Joining and Welding Research Institute, Osaka University, in three-dimensional graphic design of photonic crystals.

REFERENCES

[1] J. D. Joannopoulos, R. D. Mead and J. N. Winn: "Photonic crystals - Modeling the flow of light.", Princeton Univ. Pres, New Jersey, (1995)1.
[2] E. R. Brown, C. D. Parker and E. Yablonovichii: J. Opt. Soc. Am. B., 10(1993)404-407.
[3] S. Kirihara, Y. Miyamoto and K. Kajiyama: J. Japan Soc. Powder and Powder Metallurgy, 47(2000)239-243.
[4] J. D. Joannopoulos, P. R. Villeneuve and S. Fan: Nature, 386(1997)143-149.
[5] P. F. Jacobs: "Rapid Prototyping and Manufacturing.", Nikkei BP Publishing Center Inc., Tokyo, (1993)1..
[6] E. Yablonovitchii: J. Morden Optics, 41(1994)173-194.

OPTICAL CHARACTERIZATIONS OF A 42-LAYER SiO_2-ZrO_2 SYSTEM MULTILAYER FILM WITH STEPWISE GRADED REFRACTIVE INDEX PROFILES

Xinrong Wang, Lidong Chen and Toshio Hirai
Institute for Materials Research, Tohoku University
2-1-1 Katahira, Aoba-Ku, Sendai 980-8577
JAPAN

Yoshihiro Someno
ALPS Electric Co. Ltd., 6-1 Aza,
Nishida Kakuda, Kakuda, Miyagi
981-1505, JAPAN.

ABSTRACT

A new SiO_2-ZrO_2 system multilayer reflection filter with stepwise graded refractive index profiles was designed and prepared by helicon plasma sputtering. This filter has a 42-layer configuration that can be written as BK7|LABC(2D)(2E)CFBGAH(LH)^{9}LGAFB(2D) (2E)(2D)CBAL|Air. H, L, A, B, C, D, E, F and G are ZrO_2, SiO_2 layers and SiO_2-ZrO_2 composite layers with quarter-wave optical thickness and different refractive indices. The prepared filter exhibited a good optical performance characterized by a sharp cutoff reflection band around the reference wavelength of 770 nm.

INTRODUCTION

Oxide multilayer films are widely used as optical filters due to their high transparency, low absorption and good reliability. SiO_2-TiO_2 system multilayer filters have widely been developed.[1-5] Previously, we have also successfully designed and fabricated some novel SiO_2-TiO_2 system reflection filters with stepwise graded refractive index profiles.[6,7] It was revealed that stepwise graded refractive index profiles are effective in improving the optical property and simplifying the design and manufacture of optical filters. However, structural analysis of prepared SiO_2-TiO_2 multilayer films showed that columnar crystalline TiO_2 became easier to form in the amorphous TiO_2 layers as the layer number increased. Such crystal inclusions have some absorption and degrade the optical characteristics compared with design. On the other hand, it was reported that ZrO_2 film has a higher crystallization temperature (about 300°C-400°C) than TiO_2 film (about 100°C-300°C).[8-10] It is expected that a high-performance filter can be developed by using SiO_2-ZrO_2 material system. Although SiO_2-ZrO_2 system multilayer films have been fabricated by several researchers,[11-12] the optical characters are unsatisfactory because of the large sidelobes and the small refractive index difference between SiO_2 and ZrO_2 which makes the optical design difficult. In the present study, we applied the graded index concept to the SiO_2-ZrO_2 system film and designed a new reflection filter with high optical performance.

DESIGN OF OPTICAL MULTILAYER FILTER

A 31-layer SiO_2-TiO_2 system reflection filter with stepwise graded refractive index profiles we previously designed, is used as the initial configuration of SiO_2-ZrO_2 system multilayer filter. The central wavelength used in the calculation is 770 nm. This configuration of the multilayer filter can be written as follows:

BK7|LABCDECFBGAH(LH)⁴LGAFBEDCBAL|Air, where H, L, A, B, C, D, E, F and G are ZrO_2, SiO_2 layers and SiO_2-ZrO_2 composite layers with quarter-wave optical thickness and different refractive indices of 1.986, 1.470, 1.535, 1.603, 1.666, 1.728, 1.802 and 1.853, respectively. These data of the refractive indices are obtained from experimental results of SiO_2, ZrO_2 and SiO_2-ZrO_2 composite films deposited by helicon plasma sputtering. The transmittance spectrum is calculated by means of optical multilayer equations.

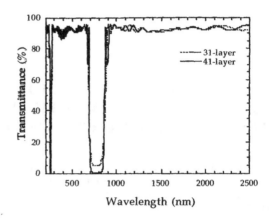

Fig. 1 Transmittance spectra of a 31-layer (dotted line) and a 41-layer filter (solid line) with stepwise graded refractive index profiles.
31-layer: BK7|LABCDECFBGAH(LH)⁴LGAFBEDCBAL|Air
41-layer: BK7|LABCDECFBGAH(LH)⁹LGAFBEDCBAL|Air

[13] For calculation of transmittance spectrum, the incident wavelengths at an incident angle of 0° and the refractive indices of SiO_2, ZrO_2 and SiO_2-ZrO_2 composite films as well as reflectance on interface between substrate and air are used. The calculations for the designed multilayer films are performed using Kidger Optics FILM-2000 thin film design software. Transmittance spectrum of this filter is shown as a dotted line in Fig. 1. It shows high transmittances outside the stopband except for two small reflection peaks near the stopband. However, reflectance of the stopband is less than 97.3%. To obtain higher reflectance of the stopband, it is usually necessary to increase the number of layers because the ratio of the refractive indices of ZrO_2 and SiO_2 is small. Therefore, a 41-layer filter with the following configuration was designed:

BK7|LABCDECFBGAH(LH)⁹LGAFBEDCBAL|Air.

The optical property is shown as a solid line in Fig. 1. It shows that the reflectance of the stopband is greater than 99% in the range from 720 to 828 nm for the 41-layer filter. But two peaks near the stopband become larger and their reflectance increases to about 30%.

It was previously reported that sidelobes are sensitive to the patterns of the stepwise graded refractive index. We adjusted the configuration patterns of the graded refractive index parts (such as LABCDE and EDCBA) and calculated the transmittance spectra. As a result, the design of a 42-layer filter with the following configuration and a good optical transmittance spectrum was obtained:

42-layer: BK7|LABC(2D)(2E)CFBGAH(LH)⁴LGAFB(2D)(2E)(2D)CBAL|Air.

The relationship between the refractive index and thickness for the above formula is shown in Fig. 2. In the 42-FGM configuration, the graded index portion of the surface was modified unsymmetrically with a different shape compared to the portion closed to the filter substrate.

The transmittance spectrum of this filter is shown in Fig. 3. For comparison, a transmittance spectrum of an alternating 41-layer filter H(LH)²⁰ is also shown. It is seen that the stopband from 718 to 830 nm has high reflectance greater than 99.0%, in which the reflectance at the reference wavelength of 770 nm is 99.7%. The full width at half maximum (FWHMs) of stopband is 174 nm from 696 to 862 nm. In the transmission regions from 300 to 690 nm and from 870 to 2500 nm high transmittances (between 90% and 95%) were obtained and the sidelobe was greatly suppressed by stepwise graded refractive index profiles.

PREPARATION AND STRUCTURE

The multilayer film is deposited on BK7 glass and Si (100) substrates by using a helicon plasma sputtering system with a pair of helicon cathodes. The chamber is evacuated below a baseline vacuum level of 5×10^{-5} Pa before the deposition. The ambient gas pressure of the system is 1.8×10^{-1} Pa. Ar and O_2 are used as the sputtering gas and the reactive gas with a partial pressure of 1.2×10^{-1} Pa and 6.0×10^{-2} Pa, respectively.

SiO_2-ZrO_2 composite films with various refractive indices are synthesized by changing the radio frequency powers of the SiO_2 and ZrO_2 (containing 10 mol% CeO_2) targets in the respective cathodes. Their thickness and refractive indices are measured at 633 nm using a Gaertner L116-B ellipsometer.

Figure 4 shows a SEM micrograph of a cross-sectional 42-layer SiO_2-ZrO_2 film (the relationship between refractive index and thickness for this filter is shown in Fig. 2) on a Si (100) substrate. The multilayer film has a total designed thickness of 5287 nm. It is seen that the prepared film has the same multilayer structure as the designed filter. In the SEM image, ZrO_2 layers or ZrO_2-rich layers are darker, and SiO_2 layers or SiO_2-rich layers are brighter. The smooth surface of a multilayer film can be seen. It indicates there are smooth interfaces between SiO_2 and ZrO_2 as well as ZrO_2-SiO_2 composite films.

The transmittance spectrum of the prepared 42-layer SiO_2-ZrO_2 multilayer film with stepwise graded refractive index profiles is shown as a solid line in Fig. 5. For comparison, the calculated transmittance spectrum of the designed 42-

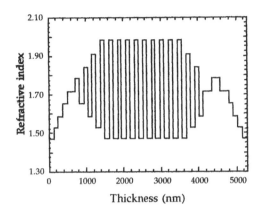

Fig. 2 Relationship between refractive index and thickness of a 42-layer SiO_2-ZrO_2 system filter with stepwise graded refractive index profiles.

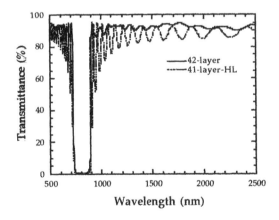

Fig. 3 Transmittance spectra of a 42-layer and a 41-layer-HL filters.
42-layer: BK7|LABC(2D)(2E)CFBGAH(LH)^{4}LGAFB(2D)(2E) (2D)CBAL|Air.
41-layer-HL: BK7| H(LH)^{20} |Air.

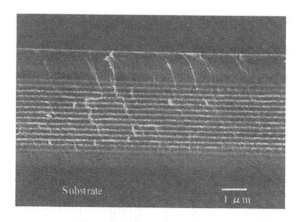

Fig. 4 SEM microphotograph of a 42-layer SiO$_2$-ZrO$_2$
system filter with stepwise graded refractive
index profiles.

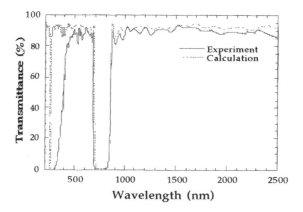

Fig. 5 Measured and calculated transmittance spectra of a
42-layer SiO$_2$-ZrO$_2$ system filter with stepwise
graded refractive index profiles.

layer SiO$_2$-ZrO$_2$ filter is also shown as a dotted line in Fig. 5. In the measured transmittance spectrum, the reflectance of the reflection band of the multilayer film is more than 99.0% in the wavelength region from 704 to 828 nm, in which the transmittance is up to 99.9% at the central wavelength of 770 nm. The full width at a half maximum is about 178 nm from 684 to 862 nm, which is consistent with the designed values. There are also high transmittances in the transmission wave-length region (468~660 nm and 874~2500 nm) outside the reflection band due to the suppression of the sidelobes in the experimental transmittance spectrum as in the case of the calculated spectrum. There is, however, a significant discrepancy between the experimental and calculated values below 500 nm. This may be due to the absorption of the multilayer film and substrate in the wavelength region. In general, it is indicated that the measured transmittance spectrum corresponds well with the calculated results described above.

SUMMARY

We designed and prepared a new SiO_2-ZrO_2 system multilayer reflection filter with stepwise graded refractive index profiles. This 42-layer reflection filter has a configuration of: LABC(2D)ECFBGAH(HL)^{9}LGAFB(2D)(2E)(2D)CBAL, where, A, B, C, D, E, F, G, H and L are the SiO_2-ZrO_2 composite films with different refractive indices and quarter-wave optical thickness. The transmittance spectrum of this filter exhibits sharp cutoff characteristics with few sidelobes, and has high reflectances of greater than 99% in the stopband and high transmission in the pass fields. The prepared 42-layer film has a high optical performance that is in good agreement with the design.

ACKNOWLEDGMENTS

A part of this work was supported by a Grant-in-Aid for Encouragement of Young Scientists (A) (No. 11750578) and a Grant-in-Aid for Scientific Research (A) (No. 11355027) from the Japan Society for the Promotion of Science. We wish to thank Mr. E. Aoyagi and Mr. U. Hayasaka at the Institute for Materials Research, Tohoku University for the SEM observations.

REFERENCES

[1]H. Sankur and W. Gunning, " Crystallization and Diffusion in Composite TiO_2-SiO_2 Thin Films," *J. Appl. Phys.*, **66**, 4747-53 (1989).

[2]M. F. Ouellette, R. V. Lang, K. L. Yan, R. W. Bertram, R. S. Owles and D. Vincent, "Experimental Studies of Inhomogeneous Coating for Optical Applications," *J. Vac. Sci. Technol.*, **A9**, 1188-92 (1991).

[3]H. G. Lotz, " Computer-Aided Multilayer Design of Optical Filters with Wide Transmittance Bands using SiO_2 and TiO_2," *Appl. Opt.*, **26**, 4487-93 (1987).

[4]H. Demiryont, "Optical Properties of SiO_2-TiO_2 Composite Films," *Appl. Opt.*, **24**, 2647-53 (1985).

[5]J. Chen, S. Chao, J. Kao, H. Niu and C. Chen, " Mixed Films of TiO_2-SiO_2 Deposited by Double Electron-Beam Coevaporation," *Appl. Opt.*, **35**, 90-96 (1996).

[6]X. Wang, H. Masumoto, Y. Someno and T. Hirai, "Helicon Plasma Deposition of TiO_2/SiO_2 Multilayer Optical Filter with Gradient Refractive Index Profiles," *Appl. Phys. Lett.*, **72** (25), 3264-67 (1998).

[7]X. Wang, H. Masumoto, Y. Someno, L. Chen and T. Hirai, "Fabrication of a 33-layer Optical Reflection Filter with Step-Wise Graded Refractive Index Profiles," *J. Mater. Res.*, **15**, 274-79 (2000).

[8]H. K. Pulker, "Characterization of Optical Thin Films," *Appl. Opt.*, **18**, 1968-74 (1979).

[9]P. J. Martin, " Ion-Based Methods for Optical Thin Film Deposition," *J. Mater. Sci.*, **21**, 1-9 (1986).

[10]N. Albertinetti and H. T. Minden, "Granularity in Ion-Beam-Sputtered TiO_2 Films," *Appl. Opt.*, **35**, 5620-26 (1996).

[11]P. J. Martin, H. A. Macleod, R. P. Netterfield, C. G. Pacey, and W. G. Sainty, " Ion-Beam-Assisted Deposition of Thin Films," *Appl. Opt.*, **22**, 178-84(1983).

[12]A. Feldman, E. N. Farabaugh, W. K. Haller, D. M. Sanders, and R. A. Stempniak, "Modifying Structure and Properties of Optical Films by Co-Evaporation," *J. Vac. Sci. Technol.*, **A4**, 2969-75 (1986).

[13]H. A. Macleod, Thin-Film Optical Filters, 2nd, Edited by Adam Hilger, Techno House, England, Chap. 8, p. 334. 1986.

LAUNCH INTO THE SPACE WITH FGM

A. Kumakawa, M. Niino, S. Moriya, and A. Moro
Kakuda Research Center, National Aerospace Laboratory
Kimigaya, Kakuda, Miyagi 981-1525, Japan

ABSTRACT

The authors suggested several applications of FGMs (Functionally Graded Materials) which might be contributed to the design and construction of new types of hypersonic space planes and/or reusable launch vehicles (RLVs). The next-generation of space plane must be developed within a relatively short period of time to be competitive in the world market, while we strive to reduce the cost and establish reliability. In the development of such vehicles, it is a good opportunity to demonstrate the superiority of the FGM concept. We have accumulated the findings and expertise gained though numerous projects in Japan. These results are not only limited to the field of thermal stress relaxation, but also are extended in the fields of energy converter, the application of which has a good potential for various usages such as health monitoring sensors of RLVs.

INTRODUCTION

The concept of FGM was conceived about fifteen years ago its potential as ultra thermo-resistant material for TPS (thermal protection system) which was considered the key component to develop space planes.

In 1987, the Science and Technology Agency of Japan organized a project under a national budget to develop functionally graded materials for the relaxation of thermal stress. Later, since 1993, a new research program "Study in the Development of Energy Conversion Materials through the Formation of Gradient Structures" promoted the study of energy converting FGM that remarkably improved energy efficiency.[1] The way to incorporate positive aspects of results obtained through these projects into the design of new space planes will be considered in this study.

WHY THE NEED TO REDUCE COSTS TO A HUNDREDTH

The international R&D trend to develop the next generation of space transportation systems seems to be in the development of RLVs. They have rocket engines as the main propulsion system. In Japan, the controversy continues between the different groups favoring two different types of RLV, Single Stage to Orbit (SSTO) and Two Stages to Orbit (TSTO). The SSTO has single body to reach the space while the TSTO has two separated ones. Figure 1 shows an image of the TSTO vehicle. The final goal is to reuse the bodies of the vehicle for several hundred times. The engine will also be reused hundreds of times, and thus the purpose of the reusability is to reduce the cost to one-hundredth. The development of the TSTO is expected completed in the late of 2010's.

However, the launch cost reduction towards one-hundredth means that large economical investment for this project is required. The most promising approach among the potential demands for realization of such

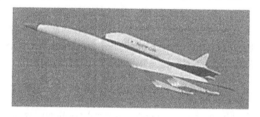

Figure 1   An image of the TSTO.

reusable space planes is the formation of international consolidations. Also promising is the demand for new structural materials for space power plants.

The scheme for solar energy plants in space was previously formed in response to the global need for the alternative energy source that can replace fossil and nuclear power plants.[3] To construct a solar power plant in space, twenty to thirty thousand tons of construction materials must be launched from the earth to the orbit. Thus it will be necessary to reduce the cost to one hundredth of the present cost of launching space vehicles, because the energy must be provided for the same price at the current value. This will be an another requirement of a momentous reformation of our technology.

THE PRINCIPAL ACCOMPLISHMENTS OF FGM RESEARCH PROGRAMS.

We have tried the development of an engine for the HOPE-X, the experimental reusable plane, applying the FGM concept with a view to higher durability and performance. So far we have developed two engines, namely, the Orbiter and Maneuvering System (OMS) Engine [4] to obtain propulsion on orbit, and the Reaction Control System (RCS) Engine [5] to control the attitude of the plane.

The OMS Engine for HOPE-X Plane

To cool the engine wall, the combination of a regenerative cooling system (in which coolant passes through the passage machined axially along the combustion chamber) and film cooling system (in which a part of liquid fuel is injected along the internal wall of the combustion chamber to cool it off). The fuel as filmed coolant does not contribute to propulsion, so to achieve the higher performance of the engine, the amount of filmed coolant must be reduced. Therefore, a partially stabilized zirconia (PSZ) coating method has been applied. PSZ is the material used for the thermal barrier, and by putting the layer of PSZ on the internal wall surface of the combustion chamber, filmed coolant is unnecessary in this case, and at the same time, the temperature of the coolant is maintained below the thermo-decomposition temperature. To relax thermal stress in this thermal barrier coating system, PSZ was gradually applied to the substrate consisting of Nickel and Chromium to form PSZ/NiCr FGM.

Figure 2 shows the concept of an OMS engine and its firing test, in which thermal cycles were repeated 250 times, up to 1,230 seconds in total. The specific impulse that shows the combustion efficiency was 315 seconds, the best in the world among this type of engine. After durability tests, de-lamination of the PSZ layer was observed at the nozzle throat where the largest thermal loading occurs. However, there was no extraordinary increase of the temperature of the regenerative coolant during the combustion test, thus proving that heat resistance and safety can be guaranteed.

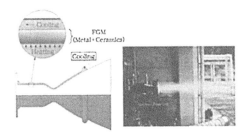

Figure 2   The construction of the rocket engine and combustion test.

RCS Engine for HOPE-X Plane

Engines of space planes are required to have high heat resistance and minimum weight. Carbon Composite Materials (hereafter referred to as the C/C material) are thought to satisfy these criteria. To use the C/C materials in the oxidizing environment, it must have the oxidation resistant protective coatings such as SiC (Silicon Carbide). However, due to the difference of thermal expansion coefficients of SiC and C/C materials, conventional coating method often results in delamination of the coating material from the C/C layer in high temperature such as in the rocket combustion chamber, thus causing problems with reliability and durability. We used SiC/C FGM for the RCS engines of HOPE with a view to improve durability and relax thermal stress. For the substrate C/C material to be airtight, the procedure of pitch infiltration, hot isostatic pressing and high temperature sintering was repeated eight times to densify it. The 0.04-mm thick graded layer of SiC/C was made by chemical vapor infiltration (CVI). Within compositionally controlled source gas mixture of tetrachlosilane-methane-hydrogen ($SiCl_4$-CH_4-H_2), the $SiCl_4$/CH_4-gas ratio is changed with time by the computer, thus changing the composition rate of the layer from C to SiC continuously. Finally, SiC layer was formed by chemical vapor deposition to make the whole layer 0.1mm thick. Figure 3 shows the structure of the engine.

The RCS engines are used in orbit to control the vehicle position, burning the fuel for a short time at intervals. Therefore, the evaluation of this engine was carried out in the pulse combustion test with short combustion time and many combustion cycles, namely, 7,465 cycles of fuel combustion in the total of 7465 seconds. When FGM was not used, cracks between C/C layer & SiC layers occurred after a single cycle, but when used, it withstood 900 cycles before delamination was observed. This shows a good prospect for the engine's practical applications.

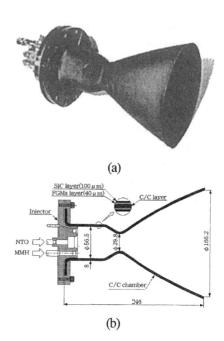

(a)

(b)

Figure 3   Structure of the engine.
(a) The RCS thruster used in this test.
(b) The designed thruster uses a steel injector body and a film cooled C/C-SiC FGM chamber.

FGM Returned from Space

HIPMEX (High Performance Material Experiment) is a part of NEDO (New Energy and Industrial Technology Development Organization)'s "Research Project of Super Refractory Materials" in which SiC-C/C FGM is actually used as the thermo-protective material of the capsule and is exposed to the atmosphere when the capsule reenters into it on its return from space.[6] On January 15th, 1995, M-SII launched this type of capsule in Kagoshima Space Observation Center but failed to orbit it, and the capsule was got back from space in July 1996. Detailed data of this experiment are not available, so we must speak in terms of estimation: Supposedly the greatest thermal loading was given at 1720 K for approximately 10 min. As a comparison four other materials made of SiC and C/C materials (which were not FGM) were launched with the capsule, and after being got back, their condition was compared with that of SiC-C/C FGM.   The soundness of the material was best preserved in the FGM of SiC-C/C, and the microscopic examination shows the SiC layer to be almost sound and reusable. There is damage at the base, but other specimens also bear the same visible damage so external forces produced on earth after it was collected probably caused them.

ULTRA-LIGHT TPS AND FGM

The demand for high insulating efficiency is greater for TSTOs because of the longer length of time spent in the atmosphere, therefore a particularly large amount of heat insulating material has to be used as shown in Figure 4 by using conventional TPS design.

By applying FGM technology, TPS is expected to have ultra-high insulating efficiency, so as to resist temperatures up to 2,300 K. This technological innovation will greatly reduce the thickness of TPS and thus significantly lighten the body weight of the vehicle. Among the "candidate" TPS ma-

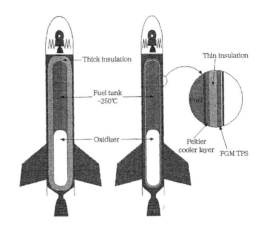

Figure 4   Super lightweight body construction.

terials to be used in the place of the conventional ceramic tile is that of thermo-resistant FGM coatings on the surface of the vehicle body, the invention of Rocketdyne Ind. in the U.S.

Furthermore, Kawasaki et al. have worked on improvement of oxidation resistance by applying the FGM concept, and they have $PSZ/Al_2O_3/NiCrAlY$ as an example of functionally graded oxidation resistant TBC. They got the result that durability was considerably improved with FGM coatings compared to the conventional two-layer coatings.

LONG LIFE ENGINES AND FGM

RLVs require long life engines because they are not disposed of, as are conventional engines. Here again, radically new technologies and ideas are needed. The essential elements in designing these long life engines are the design of TBC (thermal barrier coatings) that block the heat flow from the 3,300 K combustion flame into the combustion chamber wall and the design of compliant structure that relieves the thermal strain of the combustion chamber.[7]

The temperature of the wall surface of the inner cylinder made of copper alloy, which faces the combustion flame, must be cooled down sufficiently in order to withstand a few hundred firings of the engine. To achieve this under the heat flux that reaches up to 100 MW/m^2, it is considered that the cylinder wall needs 30μm thick ZrO_2 coating layer.   In addition, it is considered that the coarse interface is the primary cause of delamination of the coating layer from the substrate, and that the layer should be processed on the smooth surface. Also required of the coating layer is high thermal shock resistance, because vertical cracks are induced by the thermal shock during the shut down of the engine. For the above requirements to be met, i.e. the thinness of the layer and its tight adherence to the substrate, the research program began on the processing method using the Electron Beam-Physical Vapor Deposition (EB-PVD) technique. Shown here is one of the advantageous features of the coatings produced by EB-PVD, namely, a columnar microstructure. Since each column-shaped crystal is independent of other crystals, these can separately follow either the expansion or contraction of the substrate material, supposedly resulting in efficient relaxation of thermal shock. Figure 5 shows a prototype coating from the EB-PVD method. The cup test to measure the mechanical adherence of the coating to the substrate is being carried out in cooperation with the Hirai laboratory of Tohoku Univ. Also scheduled is to evaluate the thermal shock resistivity, using the high power YAG pulse laser to simulate the high heat flux of 100 MW/m^2 in rocket engines. The results of these evaluation experiments will be a basis on which we will continue to improve the resistance of the materials to the combustion gases of the high pressure rocket engines.

At the National Aerospace Laboratory cooperating with MHI we also started investigation of the structure with the outer shell made of C/C composites that is expected to have effects of the low stiffness structure because C/C composites are light and thermal expansion coefficient is very low. As a result of model analysis, by controlling the stiffness of the outer cylinder made of C/C composites, the longevity of the inner cylinder made of Cu alloy was doubled compared to the metal shell made of conventional Inconel-alloy.

Prolongation effects of life expectancy of the engine by TBC and compliant structure can be multiplied.

Figure 5   Prototype coating by the EB-PVD method.

HEALTH MONITORING SYSTEM AND FGM

Both the engine and the body of the reusable vehicle must have health monitors implanted in the critical parts so as to have safety guaranteed after hundreds and thousands of recycles.

In order to maintain soundness in the structures, RLVs must be examined by quantitative damage detection systems, and need to be repaired or have the parts replaced according to the extent of the damage incurred. This health monitoring system is another important aspect in our goal to establish reusable vehicles. Sensors to monitor the condition of the vehicle parts are under development. Among them is the thermal sensor made from the thinned thermoelectric converter which should be able to detect temperature rises due to damage in parts such as the thermal barrier layer and bearings of the turbo pump. FGM is thought to be effective in relaxation of thermal stress around the spots where the sensors are inserted, so that the sensors are protected from thermal expansion of the surrounding material.

In the United States, the CFRP-produced extremely low temperature tank of X33 is monitored by a network of fiber optic materials all over the structure, so as to detect any local fracture or stress concentration. Again the FGM concept could be applied to fiber optic materials in order to produce higher efficiency.

OTHER POSSIBLE APPLICATIONS OF FGM CONCEPT

The FGM concept is favorable for the component control of solid lubricant materials so as to bind the body structure and the surface lubricant as tightly as possible.

The bodies and engines have many extremely hot spots with coolant adjacent to them. At such spots where heating and enforced cooling occur at the same time, large temperature differences can be used as a power source by inserting FGM thermoelectric converters.

Another example is debris protection materials, which are usually made from one component. But if the material can have functional varieties, namely, both shock absorbing properties and toughness as an integrated part of the structure, damage caused by debris in space may be protected to a large extent.

SUMMARY

The authors have tried to make a few suggestions that we think may contribute to the design and construction of new types of hypersonic space planes, RLVs. To develop RLVs, many more approaches to adopt the FGM concept must be discussed. The second-generation space planes must be developed within a relatively short period of time to be competitive in the world market, while we strive to reduce the cost and establish reliability. Fortunately, we in Japan have accumulated the findings and expertise gained through numerous projects; namely, the project that started in 1987 under the Science and Technology Agency, the other FGM project from 1995 under the Ministry of Education, NEDO's "Research Project of Super Refractory Material," and finally, the large FGM fabrication project under the leadership of Mechanical Engineering Laboratory of Industrial Science and Technology. These research projects are not limited to the field of thermal stress relaxation, but are extended to the functional materials such as energy converter, the application of which has a good potential for various usages such as health monitoring sensors. A wide range of technology must be applied for the RLV to realize so the co-operation beyond the borders from governmental, academic or industrial organizations is hoped for. This fiscal year the Functionally Graded Materials Forum will start investigating further research areas and potential applications of this field, in response to the request of the National Space Development Agency of Japan.

REFERENCES

[1]M. Niino and M. Koizumi, "Overview of FGM Research in Japan", *MRS Bulletin*,20, [1] 14-23 (1995).

[2]ANON, Proceedings of Conference on Future Space Transportation System, Research and Development Bureau of Science and Technology Agency, (1999) (in Japanese).

[3]ANON, Proceedings of Forum of Visionary Laser Energy Network, Foundation for Promotion of Japanese Aerospace Technology, (1999) (in Japanese).

[4]Y. Kuroda, *et al.*, "Durability and High Altitude Performance Tests of Regenerativelly-Cooled Thrust Engine made of ZrO2/Ni Functionally Graded materials," Proceedings of the 4th International Symposium on Functionally Graded Materials, 469-474, (1996).

[5]N. Kiuchi, *et al.*, "Application of C/C Composite with Sic/C Functionally Graded Material (FGM) Coating to Rocket Combustion Chambers," Proceedings of the 5th Japan International SAMPE Symposium, 1165-1172, (1997).

[6]N. Kiuchi, *et al.*, "Results of High Performance Material Experiment (HIPMEX) on C/C Composites with SiC/C-FGM Coating," Proceedings of Symposium on Functionally Graded Materials, 95-100, (1997) (in Japanese).

[7]S. Moriya, *et al.*, "Discussion on Life Prolongation of Rocket Combustion Chamber", Proceedings of the 40th Conference on Aerospace propulsion, 187-191, (2000) (in Japanese).

INTERDEPENDENCE OF COUPLED FUNCTIONAL PROPERTIES IN THERMOELECTRIC FeSi$_2$ BASED FGM

E. Müller, K. Schackenberg, H. Ernst, H. T. Kaibe, L. Rauscher, C. Reinhard, and W. A. Kaysser
German Aerospace Center (DLR)
Institute of Materials Research
D-51170 Cologne, Germany

ABSTRACT

Highly sensitive thermal sensors can be developed when replacing metallic alloys in high temperature thermoelectric (TE) detectors by semiconducting materials. FGM can be used to control their temperature characteristics. Segmented TE FGM were made of FeSi$_2$:Al, FeSi$_2$:Co, FeSi$_2$:Mn,Al. The correlated variation of TE properties was evaluated. The doping content of the material was varied to control the temperature dependence of the Seebeck coefficient S. Consequently, the electrical (σ) and thermal conductivity (κ) are shifted. For highly responsive heat flux sensors, high S and low κ are desired. Significant κ reduction was achieved by Al+Mn double doping, as well as by increased porosity and strong alloy scattering (Co doping). Increased S values have been found for high Mn content. The functional properties are essentially influenced by doping content and microstructural features such as conductive and non-conductive second phase inclusions. Formation of metallic ε-FeSi could be suppressed by adding excess Si to the powdermetallurgical preparation route. The systematic variation of the thermal properties in dependence on the Al, Mn, and Co content reveals several distinct processes of phonon scattering. Evidence of significant reduction of thermal conduction by oxidic inclusions was found. The slope with T of the electrical conductivity is changed from increasing to decreasing type by increase of the Al content. The $S(T)$ curve shows an opposite tendency for Al,Mn double doping above room temperature. A numerical calculation shows that the temperature dependence of the integral Seebeck coefficient of a segmented FGM can be modified in a controlled manner over the relevant application temperature range.

INTRODUCTION

Doped iron disilicide has been suggested for use in thermal sensors [1]. Linearization of the responsivity characteristics and widening of the usable temperature range of measurement can be achieved by the FGM principle when applied to TE thermal sensors. If the signal responsivity of a sensor can be made independent of temperature, then easy direct measurement of thermal properties will be enabled without costly electronic linearization and without a simultaneous temperature measurement and processing. It avoids switching of the measurement range and allows for the amplifier to work in the range of its best accuracy. Thus, a low-cost measuring system for variable temperature conditions would be available. Long-term stability of the functional properties is required as well as suitability for harsh conditions, e.g. for thermal monitoring in aircraft engine test facilities.

Powdermetallurgically processed FeSi$_2$ is well known as a TE generator material [2]. The material is advantageous due to its chemical high temperature stability (up to about 900 °C). Detailed studies on the high temperature aging of the functional properties have been published [3]. It has been shown that the functional properties of FeSi$_2$ can be widely tuned by alloying with a few percent of a doping element (e.g. Al) [4]. This allows preparation of FGM without changing the semiconducting base material, thus maintaining mechanical integrity and a good functional stability even during long-term high temperature operation.

Thin film preparation of $FeSi_2$ by sputtering has been proposed for sensor preparation and has been thoroughly studied [5]. Restrictions of the film thickness to a few micrometers maximum is imposed by the thermomechanical stability of the sputtered films. In contrast, for functional stability reasons, thicker layers of TE material are proposed here, where bulk-like stability behavior of the functional properties at elevated temperature can be expected. Consequently, other preparation techniques have to be employed and specific layered composites have to be developed. A schematic of a layered heat flux sensor is given in Fig. 1.

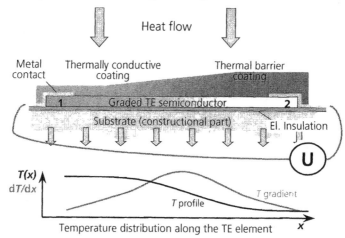

Fig. 1: Graded layered TE thermal sensor (schematic)

As a prerequisite for the graded sensor concept proposed, a well-defined control of the functional properties has to be accomplished. The need for extending the number of studied doping elements with known properties in the relevant concentration range appeared. Co as an n-type dopant was suggested. Furthermore, it has been shown that Al is of significant influence on the thermal properties [6,7] and Al+Mn double doping is leading to a further reduction of thermal conductivity [8]. Moreover Al+Mn may enable a stronger variation of the temperature characteristics of the TE properties than pure Al doping.

One technological aim of this investigation is to supplement the established hot uniaxial pressing (HUP) technique by pressure-less sintering [9], whereas HUP is limiting the sample geometry to voluminous bulk. Since pressure-less sintering is yielding higher material porosity, the comparison to HUP allows to conclude on the porosity influence on the functional properties. From literature, no harmful influence of porosity up to 10% on the properties of thermoelectric $FeSi_2$ was reported [10].

A previous study on oxide influence [7] has shown that material with increased oxide content does, under certain circumstances, behave no worse than dense hot pressed material. So, an increased oxide portion in the material, which is inavoidable in an easy preparation route with pressure-less sintering of milled powder, can be accepted. Nevertheless, the formation of oxides (SiO_2 or Al_2O_3) leads to a Si loss, and is consequently shifting the stoichiometric Fe:Si balance. Iron excess causes the formation of metallic ε-FeSi phase inclusions which deteriorate the TE properties (short-circuiting the internal thermo-voltage). From a stoichiometric starting ratio of the components, after gas-atomizing and ball milling an overall Si deficiency is obtained [4]. So, excess elemental Si has to be added. A suitable preparation mode has been evaluated.

EXPERIMENTAL

To study the homogenization of additional Si, undoped gas-atomized $FeSi_2$ was milled together with 2 / 4 / 5 / 9 at% of powdered Si in a planetary ball mill for 60 min at 300 rpm. Samples have been hot pressed for 30 min at 50 MPa, 920 °C (2 / 9 at%) or sintered for 5 h at 1150 °C (4 / 5 at%) for consolidation. XRD plots and electron back-scattering (EBS) images have been evaluated to analyze the phase constitution. EBS was found to be more sensitive to the phase detection at small percentage as far as the second phase is agglomerated to particles of more than 100 nm in size. ε phase (FeSi) provides a light contrast and elemental Si a clear dark one against the semiconducting β phase ($FeSi_2$). Metallic ε phase could be identified in the 2 at% Si excess sample from XRD plots, and it was also visible in the EBS micrographs as small light phase inclusions of globular shape (Fig. 2). These inclusions disappeared for the hot pressed 9% sample simultaneously to the peaks in the XRD in favor of a larger portion of dark phase (Si). For the pressure-less sintered samples, both the 4% and 5% samples did not show any Si- or ε-peak in the XRD plots. The EBS image of the 5 at% Si sintered material shows (besides a larger porosity of about 20 %) a very homogeneous microstructure containing a few small dark, sharply limited globular inclusions, probably consisting of Si. Since elemental Si as a semiconducting inclusion if present in excess is of minor influence on the TE properties, an amount of 5 at% additional Si was chosen as a technological baseline for the $FeSi_2$:Co sample preparation.

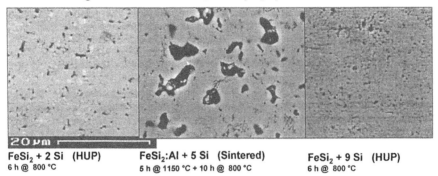

FeSi₂ + 2 Si   (HUP)   FeSi₂:Al + 5 Si   (Sintered)   FeSi₂ + 9 Si   (HUP)
6 h @ 800 °C   5 h @ 1150 °C + 10 h @ 800 °C   6 h @ 800 °C

Fig. 2: Microstructure of $FeSi_2$ in dependence on elemental Si addition (2 at%, 5 at%, 9 at%)

The phase development during the sintering preparation route and a succeeding short-term aging at 800 °C (up to 20 h) was recorded by XRD (Fig. 3). Annealing for 10 hours was found to be sufficient for completion of the α → β phase transition, with no more change in the XRD plot during further annealing. Remarkably, SiO_2 peaks were appearing still after the first 800 °C annealing step (10 h), whereas the oxygen incorporation is mainly taking place during milling before sintering. Thus, the oxides seem to be dissolved or finely distributed in an amorphous state during sintering and begin to agglomerate still at lower temperature.

Preparation of material with varying doping and of segmented FGMs started from pre-alloyed Al, Co, Mn doped and non-doped powders. $FeSi_{1.95}Al_{0.06}$, $Fe_{0.95}Co_{0.05}Si_2$, and $FeSi_2$ powders were made by gas-atomizing. The $Fe_{0.915}Mn_{0.085}Si_2$ powder used has been crushed from the melt. Si powder for stoichiometric compensation was crushed too, and ball-milled for 20 min before mixing. The final composition of homogeneous samples as well as of the FGM segments was obtained by mixing two of these powders in small number ratios (1:3, 2:2, 3:1), accordingly referred to as "0%" for undoped powder, "100%" for fully doped powder, and "25%", "50%",

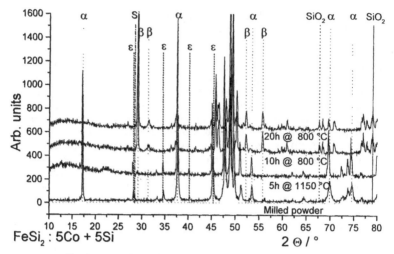

Fig. 3: XRD plots of pressure-less sintered $Fe_{0.95}Co_{0.05}Si_2$ + 5 at% Si

"75%", resp., indicating the weight fraction of doped powder in the mixture. For double doping (Al+Mn), the fraction of the Mn doped powder is indicated. The powder mixtures were comminuted in the planetary ball mill (1 h @ 300 rpm) to a sub-micron particle size. From the powders, series of samples were prepared by hot pressing ($FeSi_2$:Al and $FeSi_2$:Mn,Al, "0%" to "100%" each) for 15 min at 940 °C [7,8] and then annealed for 2 h at 800 °C. A $FeSi_2$:Co series ("0%" to "100%") was after cold compression consolidated by pressure-less sintering (5 h @ 1150 °C) and annealed to complete the α→β phase transition (10 h @ 800 °C).

The electrical conductivity σ was monitored in dependence on temperature and doping content: Whereas $FeSi_2$:Co shows a very regular behavior (σ proportional to the doping concentration, similarity of the curves for all doping concentrations; Fig. 4, right), this was not the case for Al doping. From the plot of σ versus the total Al concentration in the material (Fig. 4, left), a linearity is found indicating a large portion of Al which was incorporated into the material in an electrically inactive chemical state, very probably as oxide. This was confirmed by EPMA studies [7]. A change of the electrical conduction mechanism of $FeSi_2$:Al with increasing doping content was proved by a principal change of the temperature slope of σ(T).

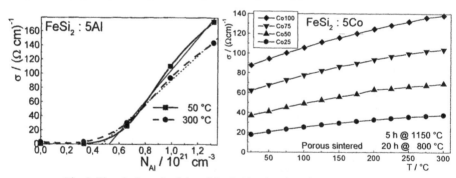

Fig. 4: Electrical conductivity of the $FeSi_2$:Al and $FeSi_2$:Co sample series

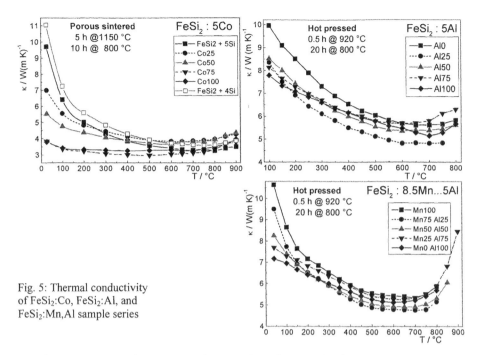

Fig. 5: Thermal conductivity
of FeSi₂:Co, FeSi₂:Al, and
FeSi₂:Mn,Al sample series

The thermal conductivity was monitored in dependence on the doping content (Fig. 5). A number of striking effects become obvious, with alloy scattering of phonons on Co substituting Fe as one of the strongest. In the temperature range below 500 °C the $\kappa(T)$ curve changes significantly from falling for non-doped FeSi₂ to nearly constant with temperature for 5% of the Fe atoms in the lattice replaced by Co. At higher temperature the order of the κ-values of the samples is reversed because of an increasing carrier contribution to the thermal conduction. On the contrary, Mn doped FeSi₂ does not show any similar reduction like Co doped, even with as much as 8.5% Mn content. So a lattice substitution of Mn atoms at Fe lattice positions can hardly be assumed. For Al, a gradual alloy scattering effect has to be stated. Almost independently of temperature, a drastic reduction of the thermal conductivity (i.e. improvement of the TE figure of merit) is caused by material porosity. Non-doped sintered FeSi₂ (with a porosity up to 20 %) was reduced to less than 3.5 W/(m·K) (Fig. 5, left) compared to more than 5.5 W/(m·K) for hot pressed material (Fig. 5, top right) at 600-700 °C. At the highest temperature measured, the thermal conductivity is strongly increasing with temperature due to the bipolar electronic contribution, preferentially for the highest Al and Co concentration, resp.

Earlier it was shown that for FeSi₂:Al the slope of the Seebeck coefficient $S(T)$ is changing, with its magnitude crossing a maximum. This effect is linked to the change of the electrical conduction mechanism with increasing Al concentration. A similar behavior has been found even more pronounced for the FeSi₂:Mn,Al sample series (Fig. 6), with nearly linear curves over a range of several hundred K in width.

Five-layer segmented FGMs have been prepared from FeSi₂ doped with Co, Al, and Al+Mn, with segment concentration according to the sample series described. For comparison of the techniques, hot pressing and pressure-less sintering have been applied. The flattest segment interfaces

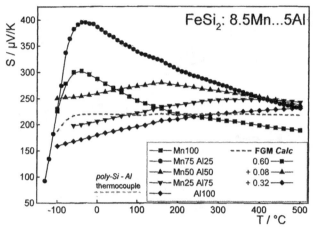

Fig. 6: Temperature dependence of the Seebeck coefficient of the FeSi$_2$:Mn,Al sample series. Dotted lines: Calculated $S(T)_{FGM}$ function (constant with temperature)

and better mechanical stability were achieved for hot pressed FGM (Fig. 7, right), whereas different shrinkage of adjacent segments leads to an interface bending of the sintered samples (Fig. 7, top). Although the compositional difference between neighboring segments would be too small for visual distinction, the different milling and densification behavior leads to a different porosity structure with good contrast in the optical microscope. After sintering in a cylindrical matrix, up to 3% difference in diameter between the segments of a stack have been found. For a sample diameter of several millimeters this imposes the danger of cracks in the outer segments.

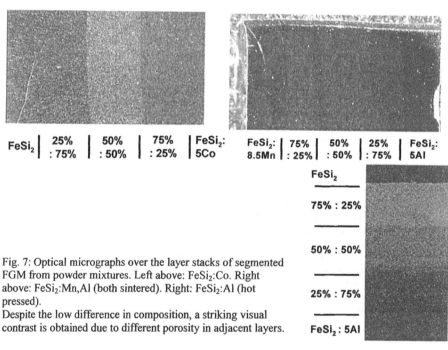

Fig. 7: Optical micrographs over the layer stacks of segmented FGM from powder mixtures. Left above: FeSi$_2$:Co. Right above: FeSi$_2$:Mn,Al (both sintered). Right: FeSi$_2$:Al (hot pressed).
Despite the low difference in composition, a striking visual contrast is obtained due to different porosity in adjacent layers.

DISCUSSION

Preliminary results on a doping controlled tuning of TE properties considering the Seebeck coefficient of Al-doped $FeSi_2$ [7] have shown that in the region above room temperature the electrical conductivity changes its slope in dependence on the doping content from descending (5 at% Al) to ascending (undoped $FeSi_2$) characteristics, as a consequence of the change of the electric conduction mechanism from band conduction to hopping. Accordingly, a slope change of the Seebeck coefficient with very high values at an intermediate doping content has been observed. A similar behavior was found, even more pronounced, for $FeSi_2$:Al,Mn double doped material when gradually changing from fully Al doped (5 at% Al) to fully Mn doped material (8.5 at% Mn; Fig. 6). Increasing dopant level resulted in left-handed curvature and vice versa. This provides ideal pre-conditions for FGM design.

An example of the linearization of sensor characteristics by the FGM principle shall be given: Analytically, the output voltage of a graded sensor element is obtained by a linear combination of the Seebeck values of the FGM segments, weighted with the inverse thermal conduction along these segments. The latter may be either determined

a)   by the thermal conductivity and geometry of the TE segment itself (self-supporting structure),

b)   by the thermal conductance and geometry of the supporting structure (TE layer on a carrier).

For the easiest case, a one-dimensional set-up was considered based on a supporting structure which thermal properties were assumed to be independent on temperature. Then, a Seebeck curve of the FGM has to be tuned which is independent of temperature, as well. A numerical calculation revealed that a one-dimensional FGM consisting of three segments, built up in length out of 60% $FeSi_2$:Mn (Mn100), 8% $FeSi_2$:Mn,Al (Mn50), and 32% $FeSi_2$:Al would yield a constant Seebeck coefficient between –50 °C and +500 °C with a comparably high value of 216 μV/K and a deviation from this value below the measurement accuracy (dashed line in Fig. 6). This magnitude is about three times higher than that of commercially used poly-Si/Al thin film thermocouples, which thermo-power is kept intentionally low for linearization by p-n combination to achieve a flat temperature characteristics.

Similarly, within the limits of the strongest decline and incline of the $S(T)$ curves, other functions of a synthetic temperature dependence can be obtained e.g. for a TE layer on a ceramic carrier, where the relative temperature dependence of the thermal conductance of the ceramic carrier has to be reproduced by $S(T)_{FGM}$. From the given sample series, such a tune is possible with convenience, since $S(T)$ changes its slope with doping change, whereas thermal conductivity maintains its principal temperature dependence (undergoing merely some change in magnitude).

CONCLUSIONS

$FeSi_2$ has turned out to be a very suitable material for heat flux sensor development, since it is high temperature resistant and enables a high signal responsivity. $FeSi_2$ based sample series and FGMs with several dopings (Al, Mn, Co) yielding p- and n-type thermoelectrics have been prepared by milling and hot pressing or pressure-less sintering.

The material properties can not only be tuned by the concentration of the doping element, but also significantly by the doping species, the oxide content, the Si:Fe ratio, and the porosity of the material. An easy preparation route has been developed for addition of elemental Si which allows to suppress metallic second phase inclusions.

A remarkable material improvement (reduction of thermal conductivity) has been achieved by high Co doping content and high porosity, causing amplified alloy scattering and boundary scattering of phonons.

Al/Mn double doping was identified as an efficient way to control the TE material properties. When proceeding from Al to Mn doping, a change of the conduction type is observed, linked to a change of the slope of $S(T)$ over a broad temperature range. The Seebeck coefficient as a function of the Mn and Al concentration exhibits a pronounced maximum. Under these conditions, the $S(T)$ slope of segmented FGM becomes tunable by segment thickness control. A calculation from experimental data yields a value of $S(T)_{FGM} = 216\pm2$ μV/K which is constant over a temperature range from -50 up to 500 °C. Still higher values are possible for a smaller temperature range. Refining of the doping concentration grid in the region of highest Seebeck coefficient is planned.

Supplementing the successful hot pressing technique, FGM preparation by pressure-less sintering has been tested. The shrinkage behavior was found to depend sensitively on doping additives, so a deformation of the FGM outer shape and the segment interfaces was caused.

To make the obtained functional effects exploitable to a sensor system, layered systems with suitable carriers have to be prepared. Covering $FeSi_2$ with TBC and thermally conductive coatings is under study e.g. by EB-PVD technique.

ACKNOWLEDGMENT

Part of this work was funded by the Deutsche Forschungsgemeinschaft (DFG) under the project number Ka 664/9 which is grateful acknowledged.

REFERENCES

[1] E. Müller, K. Schackenberg, and W. A. Kaysser, „Iron Disilicide for High Temperature Thermal Sensors“, *Proc. 5th European Workshop on Thermoelectrics, 20./21. 9. 1999, Pardubice, CZ*, 70-75 (1999).

[2] U. Birkholz, E. Groß, and U. Stöhrer, „Polycrystalline Iron Disilicide as a Thermoelectric Generator Material“, pp. 287-298 in *CRC Handbook of Thermoelectrics*, Ed. by D. M. Rowe, CRC Press, Boca Raton, 1995.

[3] K. Schackenberg, E. Müller, J. Schilz, H. Ernst, and W. A. Kaysser, „Carrier Density Behavior During Aging of Doped Plasma Spray Formed Iron Disilicide“, *Proc. XVII. Int. Conf. on Thermoelectrics, Nagoya, Japan*, IEEE Piscataway, NY 422-425 (1998).

[4] E. Müller, K. Schackenberg, F. Arenz, J. Schilz, and W. A. Kaysser, „Graded $FeSi_2$ Thermoelectric Elements - Preparation by Intermixed Additives Sintering“, *Proc. 5th Int. Symp. on Functionally Graded Materials, Dresden, Germany*, Trans Tech Switzerland 681-686 (1999).

[5] H. Griessmann, A. Heinrich, J. Schumann, D. Elefant, W. Pitschke, and J. Thomas, „Thermoelectric Transport Properties, Structure Investigations and Application of Doped β-$FeSi_2$ Thin Films“, *Proc. XVIIIth Int. Conf. on Thermoelectrics, Baltimore, MD*, IEEE Piscataway, NY, 662-666 (2000).

[6] E. Müller, K. Schackenberg, J. Schilz, „Improvement of the figure of merit of thermally sprayed iron disilicide in a strongly disordered microstructure“, *XIXth Int. Conf. on Thermoelectrics, 20-24/08/2000, Cardiff, U.K.*

[7] E. Müller, K. Schackenberg, H. Ernst, E. de Groote, W. A. Kaysser, „$FeSi_2$ for Sensor Application - Control of Functional Properties by Composition“, pp. 409-414 in *Proc. Euromat 99, Vol. 13, Functional Materials*, Ed. K. Grassie, E. Teuckhoff, G. Wegner, J. Haußelt, H. Hanselka, Weinheim 2000..

[8] H. T. Kaibe, H. Ernst, L. Rauscher, K. Schackenberg, E. Müller, Y. Isoda, I. A. Nishida: "Electrical and Thermal Properties of p-Type $FeSi_2$ with Mn and Al Double Doping", *Proc. XVIIIth Int. Conf. on Thermoelectrics, Baltimore, MD*, IEEE Piscataway, NY, 133-136, (2000).

[9] I. A. Nishida, "Fabrication and Thermoelectric Properties of Semiconducting Iron Disilicide", *Tetsu to Hagane (Iron and Steel)* **81** [10] 454-460 (1995).

[10]E. A. Groß, "Pulvermetallurgie von Höherem Mangansilizid und Eisendisilizid zur thermoelektrischen Energiekonversion" (Powder Metallurgy of Higher Manganese Silicide and Iron Disilicide for thermoelectric energy conversion), *PhD Thesis, University of Karlsruhe, Germany*, 1993.

FERMI LEVEL PINNING OVER A WIDE TEMPERATURE RANGE IN FUNCTIONALLY GRADED IV-VI SEMICONDUCTORS

Z. Dashevsky, S. Shusterman, and M.P. Dariel
Department of Materials Engineering, Ben-Gurion University of the Negev,
P.O.B. 653, Beer-Sheva 84105, Israel

ABSTRACT

The elevated electron mobility of the $A^{IV}B^{VI}$ semiconductor compounds in the heavily doped state and their low lattice thermal conductivity make them useful materials for thermoelectric energy conversion. High efficiency thermoelectric conversion is achieved by using materials with a maximum figure merit, $Z = S^2 \times \sigma / k$ (S – thermoelectric power, σ and k -- electrical and thermal conductivity) over a wide temperature range. High quality homogeneous thermoelectric materials usually display elevated values of Z only over a narrow temperature range. A maximum value of Z is attained only for one specific position of the Fermi level, E_F with respect to the conduction band edge, E_C. In order to maintain the optimal Z value, namely maintain a constant location of the Fermi level, N_D must increase with increasing temperature. The objective of the study was to develop a dopant concentration profile, along the temperature gradient, that would give rise to a constant location of the Fermi level, hence optimal Z, over a wide temperature range (100 – 600°C) in PbTe single crystals.

Indium has a high solubility up to 5 at. % and displays unique donor properties in PbTe (each In atom occupies a Pb lattice site, contributes one free electron and two states). By diffusion into PbTe crystals from a surface source, In concentration profiles were generated that allowed keeping the Fermi level practically in coincidence with the impurity level (pinning of the Fermi level) over the operating temperature range.

INTRODUCTION

Thermoelectric materials operate in a temperature gradient between a heat source and a constant, low temperature (sink). The efficiency of heat conversion into electric energy is one of the main factors that determine the performance of thermoelectric materials. As a first approximation, it may be evaluated from the following expression [1]:

$$\eta = \frac{\Delta T(\sqrt{1 + ZT_{av}} - 1)}{(\sqrt{1 + ZT_{av}})T_h + T_c} \quad . \tag{1}$$

where T_h and T_c are the absolute temperatures of the hot and cold sides of the thermoelement, respectively, T_{av} – average temperature in the operating temperature range = $(T_h + T_c)/2$, and Z is the figure of merit that depends on the properties of the material $Z=\sigma S^2/k$, where S is the Seebeck coefficient, σ, the electrical conductivity, and k, the thermal conductivity. Unfortunately, the three material parameters that determine ZT_{av} are not independent. In general, as the electrical conductivity increases, the Seebeck coefficient decreases with increasing carrier concentration. The best compromise seems to lie in using heavily doped semiconductors to produce a carrier concentration of about 10^{19} cm^{-3}. The thermal conductivity k is the sum of two contributions, that due to the charge carriers, k_e, and that to the lattice vibrations (phonons), k_L. Although k_e is proportional to electrical conductivity σ, in many semiconductors $k_L > k_e$, and thence it is important to minimize k_L.

At a carrier concentration of about 10^{19} cm^{-3}, the Fermi level E_F is close to the edge of the conduction band E_C (for n-type conductivity). The figure of merit reaches a maximum value at a temperature that varies with the charge carrier concentration. The performance of a thermoelectric material can be improved significantly if the charge carrier concentration is adjusted so that the actual temperature distribution along the sample coincides with the temperature distribution of the maxima of the figure of merit. In monolithic thermoelectric materials, the charge concentration is usually constant and, consequently, the figure of merit does not reach its optimum value. Three approaches can be taken in order to attain a maximal figure of merit over a wide temperature range, these are: (a) varying the carrier concentration (i.e., doping with shallow level impurities) [2]; (b) tailoring the width of the band-gap (i.e., by changing the composition) [3]; (c) producing a concentration gradient of impurities that have deep lying levels (i.e., doping with In) [4]. An even more promising approach is to take advantage of two of the above degrees of freedom in order to tailor an optimal figure of merit.

$A^{IV}B^{VI}$ Semiconductors For Thermoelectric Energy Application

$A^{IV}B^{VI}$ (chalcogenides of IV-group elements) semiconductors are well-known materials that have found widespread applications in thermoelectric energy devices. The advantages of these materials are their:

The low thermal conductivity: The phonon part of PbTe, k_L is $\approx 2\times10^{-2}$ W/cm-K at room temperature. The small value of k_L in PbTe is determined by the large mass of the atoms in the compound.

The high mobility of the majority carriers: The best n-type PbTe crystals have a carrier mobility $\mu \geq 10^3$ cm^2/V-s at room temperature.

The large density of states. Conduction and valence bands consist of four equivalent sub-bands (valleys).

The large value of dielectric permitivity ε_o. In PbTe, $\varepsilon_o \approx 400$ (T = 300 K). The elevated value of ε_o stands behind the sharply decreased impurity scattering of the charge carriers in the heavily doped state.

The ability to dope the material to a high carrier concentration (10^{19} cm^{-3}). This generally gives optimal $Z=\sigma S^2/k$, ratios.

The ability to control and vary the width of the band gap by adding a third component to PbTe. For instance, the addition of Sn ($Pb_xSn_{1-x}Te$) decreases and the addition of Mn ($Pb_yMn_{1-y}Te$) increases the band gap.

Doping by Group III Impurities

Indium has been used as doping element in $A^{IV}B^{VI}$ (PbTe, $Pb_{1-x-y}Sn_xMn_yTe$) semiconductors. In PbTe, it displays donor-like properties and generates a deep-lying level, namely a narrow impurity band. In spite of the high In solubility in PbTe, the electron concentration does not exceed n = $(3-5)\times10^{18}$ cm^{-3}, corresponding to 0.05 % In. In contrast to shallow lying states, deep lying ones consist of wave functions that belong to several bands. Consequently, their energy levels are not connected to any particular band edge, E_C and E_V, and they may lie within a band as well as within a band gap. For an In impurity on a substitutional site, the s-shell does not participate in the bond formation and the In orbitals hybridize into an excited sp^2-state. The corresponding *s*-orbital is, therefore, of non-bonding nature and is strongly localized on the impurity atom. As a result, the corresponding *s*-level should be considered as the deep-lying one [6]. Such a level not only releases but also accepts electrons, i.e. it displays both donor and acceptor properties (an amphoteric level). The behavior of indium in $A^{IV}B^{VI}$ has been subject to several investigations [5-7]. It was established that:

- At low temperature the In energy level is close to E_C. Its position doesn't depend on the In concentration up to $N_{In} \approx 3$ at %.

- Each In atom occupies one Pb lattice site, contributes one free electron and two states. Thus, the impurity level is half- filled.

- At low temperature, the chemical potential is determined practically by the impurity level (pinning of the Fermi level).

- The position of the impurity level with respect to E_C can be varied by changing the alloy composition e.g. in $Pb_{1-x}Sn_xTe$ [5].

CALCULATION

In the framework of the above-described model, the position of the Fermi level as a function of temperature was calculated for a given In concentration. When impurity atoms are introduced, the Fermi level must adjust itself to preserve charge neutrality [8]:

$$n = N_D^+ + p, \tag{2}$$

where n is the electron density in the conduction band, p, the hole density in the valence band, and N_D^+ is the number of the ionized donors (indium).

The electron density n is

$$n = 4\pi(2m_{de} k_B T/ h^2)^{3/2}F_{1/2}(\mu^*), \tag{3}$$

where k_B is the Boltzmann constant, h is the Planck constant, m_{de} is the density-of-state- effective mass for electrons in the conduction band, in PbTe, $m_{de} = (M_c)^{2/3} (m_l \times m_t^2)^{1/3}$ (M_c is the number of equivalent minima in the conduction band, m_l and m_t are the longitudinal and transverse effective mass in the conduction band); μ^*, is the reduced Fermi level equal $(E_F - E_C)/kT$; and $F_{1/2}(\mu^*)$ is the Fermi integral.

The hole density p (minority carriers) is

$$p = 2 \, (2\pi m_{dh}kT/\,h^2)^{3/2}\exp[(E_V - E_F)/kT], \tag{4}$$

where m_{dh} is the density-of-state- effective mass for holes in the valence band, in PbTe $m_{dh} = (M_V)^{2/3} (m_{lp} \times m_{tp}^2)^{1/3}$ (M_V is the number of equivalent minima in the valence band. m_{lp} , m_{tp} are the longitudinal and transverse effective mass in the valence band).

The number of ionized donors N_D^+, is given by [7]:

$$N_D^+ = N_D \left[1 - \frac{2}{exp\left(\dfrac{E_D - E_F}{kT}\right)} \right], \tag{5}$$

where N_D is the indium concentration, E_D is the ionization energy for an In impurity.

Plots of the Fermi level, the carrier concentration and the Seebeck coefficient as a function of the In impurity concentration are shown in Figs. 1-3. Note in Fig.1 that as the temperature increases, the indium concentration has to increase in order for the Fermi level to approach the impurity level close to the conduction edge E_C, (the optimal position for the achievement of maximal value of figure of merit over a wide temperature range). For an In concentration of 5×10^{18} cm^{-3} at the low operating temperature (300 K) edge of the sample and up to a 2×10^{20}cm^{-3} In concentration at the high operating temperature (900 K) edge, the Fermi level is pinned with respect to the In impurity level. In such a sample with a graded In profile, the average figure of merit $Z \sim 1.5\times10^{-3}$ K^{-1} and the efficiency (eq. 1), $\eta \sim 12$ %. This calculated value is by a factor of two higher than the efficiency of the homogeneous n-type PbTe commonly used in commercial thermoelectric devices [9].

The optimization of the thermoelectric properties was made in the framework of a model based on one type of carriers (electrons). At high temperature $T \geq 700$ K, the negative influence of minority carriers (holes) on the thermoelectric properties cannot be neglected. However in the present case, with an electron concentration on the impurity level higher than the carrier concentration in the conduction band, (Fig. 2), electrons dropping from the impurity level to the valence band annihilate most minority carriers.

EXPERIMENTAL
Single Crystals

Single crystals of $A^{IV}B^{VI}$ chalcogenides were grown in a Czochralski crystal puller. The crystals were grown from a PbTe melt with an excess Te (up to 1 at %) composition that generates Pb vacancies = V_{Pb}^{++} (acceptors).

To prevent decomposition of the melt during the crystal growth, a liquid encapsulation method is employed. The liquid encapsulant was molten boron trioxide (B_2O_3). A single crystal seed was used in the growth process, so that direction of growth was approximately parallel to the [001] axis. The temperature gradient at the crystallization front and the crystal pull rate were of the order of 20–25 Kcm^{-1} and 5-10 mmh^{-1}, respectively. In order to enhance the equalization of the heat flux through the lateral surface of the crystal, the crucible and the crystal were rotated in opposite directions at an angular rate of 1 s⁻¹. The crucible rotation provides an effective and rapid method for ensuring the homogeneity of the melt.

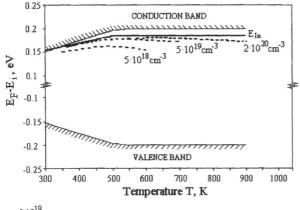

Figure 1. The Fermi level of PbTe doped with indium as a function of temperature and indium concentration.

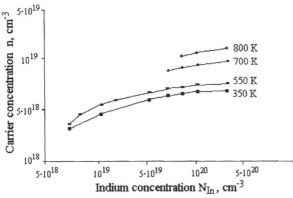

Figure 2. Electron density for PbTe doped with indium as a function of indium concentration and temperature.

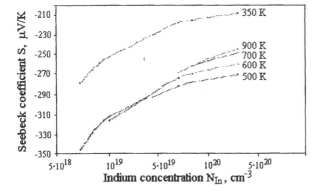

Figure 3. Seebeck coefficient for PbTe doped with indium as a function of indium concentration and temperature.

Doping

Diffusion of In from a gaseous source (In$_4$Te$_3$) was used for doping the specimens, allowing to maintain a constant level of surface In concentration during the diffusion anneal. The concentration profile of the In dopant lies in the 0.1 - 2 at % range and can be varied by changing the diffusion length and the annealing temperature (600 to 750°C). The concentration profile of In within the crystal was consistent with the solution of the diffusion equation for a constant surface source. The surface concentration (C^S) of In was determined by EDS (C^S $_{700°C}$ = 2.0 ± 0.7 at %). The concentration at a second point C(x) was determined from Seebeck coefficient measurements [4]. The variation of the Seebeck coefficient S, measured in a direction parallel to the penetration of In by diffusion in a p-type PbTe (grown from a melt with excess Te) crystal is shown in Fig. 4. The Seebeck coefficient for a p-type PbTe crystal in which no In penetration took place has a constant positive value (line 1). The S values for all crystals in which In had diffused (curves 2 to 5), drop at the front-end of the crystal to a large negative value reflecting the effect of the large In addition. The presence of In over-compensates the V_{Pb++} acceptors and transforms the initial p-type PbTe into n-type PbTe. The values of S, as one proceeds into the crystal, increase (algebraically), according to the penetration profile of In, change sign and revert to the positive value, characteristic of In-free PbTe. We can estimate the In concentration N_{In} at a location x at which the Seebeck coefficient reverses its sign from positive (sample with V_{Pb++} acceptors) to negative (sample doped with In). At point x, both donor and acceptor impurities are present simultaneously and their concentrations are approximately equal one to other.

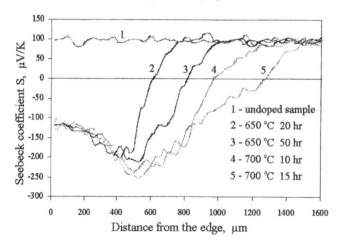

Figure 4. The variation of the Seebeck coefficient, measured at room temperature along the penetration profile of In into a p-type PbTe single crystal.

The diffusion coefficient of indium in PbTe was deduced from best fits of the experimentally deduced concentrations to the appropriate solution of the diffusion equation.

$$C(x,T) = C_s erfc(x / 2\sqrt{Dt})$$ (6)

Diffusion profiles of indium into PbTe as a function of the distance for different diffusion times and temperatures are shown in Fig. 5. The diffusion coefficient of indium in PbTe crystals at 700 °C that was deduced is high, D ≈ 5×10^{-9} cm^2/sec. The process could, therefore, be used to dope the PbTe crystal with In up to a depth of 5 mm.

Functionally Graded Materials 2000

RESULTS

A 4 mm thick slice was cut from the *p*-type PbTe sample that had been doped with In originating from the gaseous phase. The portion that was cut coincided with the region in which an In gradient had been set up, as determined by the Seebeck coefficient measurements. The two end-surfaces of the graded crystal were polished and positioned between two flat surfaces. The lower surface was kept at constant $T_c = 50°C$ and the temperature of the upper surface was increased up to $T_h = 600°C$. The results indicate that the dependence of V on $\Delta T = T_h - T_c$ is linear within a close approximation (Fig.6). Thus, the Seebeck coefficient $S = V/\Delta T$ is practically constant ($S \approx -200$ µV/K) over a wide temperature range. This behavior of $S(T)$ is very different from that observed in homogeneous *n*-type PbTe (doped only with iodine) [10]. In our case, as previously mentioned, electrons in the In-generated level minimize the effect of the minority carriers.

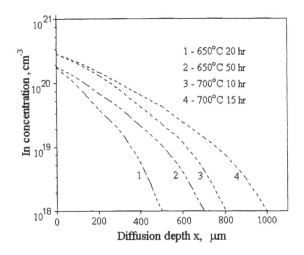

Figure 5. Diffusion profile of indium into PbTe is a function of the distance for different diffusion times and temperatures.

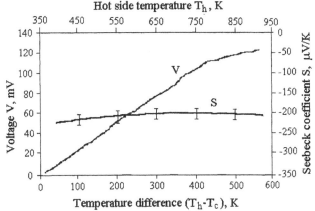

Figure 6. The thermovoltage and Seebeck coefficient of a graded PbTe<In> crystal as a function of the temperature of the hot side ($T_c = 350$ K).

SUMMARY

Single p-type PbTe crystals were grown by the Czochralski technique from a melt containing excess Te. Indium that generates donor levels was diffused from a gaseous source into the single p-type PbTe crystal, transforming its conductivity from p- to n-type over the whole penetration depth (5 mm). Over the region in which the concentration of In varied from 2×10^{20} cm^{-3} (near the surface) to 5×10^{18} cm^{-3}, the location of the Fermi level with respect to the edge of the conduction band does not depend on the operating temperature (pinning of the Fermi level). This material provides a practically constant, high figure of merit, $Z \approx 1.5 \times 10^{-3}$ K^{-1}, over a wide temperature range (50-600°C). Thus, our study demonstrated the feasibility of the FGM approach for preparing a thermoelectric material with improved efficiency for energy conversion.

REFERENCES

[1]G.D. Mahan, "Good Thermoelectrics", *Solid State Physics*, **51** 81-158 (1997).

[2]T. Kajikawa, "Research on Enhancement of Thermoelectric Figure of Merit through Functionally Graded Materials Processing Technology in Japan"; pp. 475-482 in *Functionally Graded Material*, Edited by I. Shiota and Y. Miyamoto. Elsevier Science Publishers, Amsterdam, 1997.

[3]J. Singh, *Semiconductor Devices*, McGraw-Hill, New York, 1994, pp. 124-127.

[4]Z. Dashevsky, S. Shusterman, A. Horowitz, and M.P. Dariel, "The Effect of a Graded In Profile on the Figure of Merit of PbTe"; *in Proceedings of Material Research Symposium MRS 1998 Fall Meeting*, Boston, USA, **545** 150-159 (1999).

[5]B.A. Akimov, A.V. Dmitriev, D.R. Khokhlov, ad L.I. Ryabova, "Carrier Transport and Non-Equilibrium Phenomena in Doped PbTe and Related Materials", Phys. Stat. Sol. (a) **137** 9-55 (1993).

[6]D.E. Onopko, and A.I. Ryskin, "Structure of Metastable Column-III Centers in IV-VI Compounds"; Th-P186, pp. 1-4 in *Proceedings of 24th Int. Conf. on the Physics of Semiconductors*, Jerusalem, 1998.

[7]V.I. Kaidanov, and Yu. I. Ravich, "Deep and Resonance States in $A^{IV}B^{VI}$ Semiconductors, *Soviet Physics Uspekhi*, **28** 31-52 (1985).

[8]S.M. Sze, *Physics of Semiconductor Devices*, J. Willey, New York, 1981, p. 23.

[9]B.C. LeSage, G. Meszaros and T. Nyston, "On the Suitability of Low Cost PbTe Devices for High Grade Waste Heat Power Generation Applications"; pp. 426-428, *in Proceedings of the 17h International Conference on Thermoelectrics;* Edited by Kunihito Kuamoto, Nagoya, 1998.

[10]Yu.I. Ravich, B.A.Efimova, I.A. Smirnov, *Semiconducting Lead Chalcogenides*, Plenum Press, New York, 1970.

DOPANT DIFFUSION PROCESS IN THERMOELECTRIC MATERIAL PbTe

Yoshikazu Shinohara, Yoshio Imai
and Yukihiro Isoda
National Research Institute for Metals, STA
1-2-1, Sengen, Tsukuba 305-0047, Japan

Hiromasa T. Kaibe
Tokyo Metropolitan University
1-1, Minamiosawa, Hachioji 192-0364,
Japan

ABSTRACT

Diffusion of PbI_2 and $ZrCl_4$ dopants into PbTe has been investigated as a fundamental method to form continuous gradient carrier concentrations. High purity of non-doped PbTe, which was p-type, was prepared using rocking furnace, and heat treated in PbI_2 and $ZrCl_4$ vapors. Diffusion thickness was evaluated by resistance measurement of one-probe and four-probe methods. All of the treated PbTe in PbI_2 vapor changed to n-type independently of heat treatment conditions. This is not due to PbI_2 diffusion but to an intrinsic transport-property change of PbTe. The treated PbTe in $ZrCl_4$ vapor showed diffusion thickness of 0.3mm at 973K for 50h. $ZrCl_4$ diffusion in PbTe is as slow as Pb self-diffusion.

INTRODUCTION

Lead telluride PbTe is typical n-type thermoelectric material applied in the temperature range from 300K to 700K[1]. It has been already used for power generating devices at special locations, *i.e.* space, deep sea, desert, pole, *etc*[2]. Its more common applications need even higher thermoelectric energy conversion efficiency η. The higher η is achieved by the higher figure-of-merit Z ($=\alpha^2/(\rho\kappa)$, α: Seebeck coefficient, ρ:electrical resistivity, κ:thermal conductivity). PbTe has variations of Z with temperature and the Z has a maximum at the corresponding temperature T_m, as shown in Fig.1. The T_m can be increased by increasing its carrier concentration n[3]. When the n changes gradually in PbTe, the Z can be kept high in the wide temperature range up to 700K, as indicated by a dotted line in Fig.1. The graded carrier concentration is effective in improving the η of thermoelectric material[4].

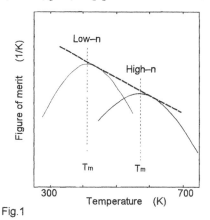

Fig.1

Variations of figure–of–merit with temperature for low–n and high–n PbTe

There are two types of carrier concentration gradient; step-like gradient and continuous gradient. The former can be formed by joining segments with different n, while the latter is by sintering mixed powders, which gradually change the mixing ratio of powders with different n [4,5]. Continuous carrier gradient is more desirable for high Z than step-like one. Thermoelectric properties of PbTe, however, are difficult to control by mixing PbTe powders with different n[5]. Generally, the diffusion process forms a continuous gradient. When the dopant can diffuse from one side of the non-doped PbTe ingot, which is called as directional diffusion control, continuous carrier gradient is successfully formed. However, there are few reports on dopant diffusion in PbTe, especially for n-type dopants.

In the present paper, a high purity of PbTe ingot was prepared and the diffusion of two kinds of n-type dopants into PbTe was investigated as a fundamental research of the directional diffusion control.

EXPERIMENTAL

Pb(6N) and Te(6N) were weighed at a composition of PbTe, and capsulated in an evacuated quartz tube under 1×10^{-3}Pa. They were melted at 1300K by the rocking furnace with a rocking cycle of 1 Hz to obtain homogeneous ingots, and subsequently solidified at a cooling rate of 10K/h. The obtained ingot was p-type with n of $1.6 \times 10^{24}/m^3$. The Hall mobility μ_H of 0.09Vs/m^2 achieved is not less than the literature data[6,7], which indicates an ideal purity of PbTe ingot. The non-doped ingot was cut into specimens of $3 \times 3 \times 20$mm.

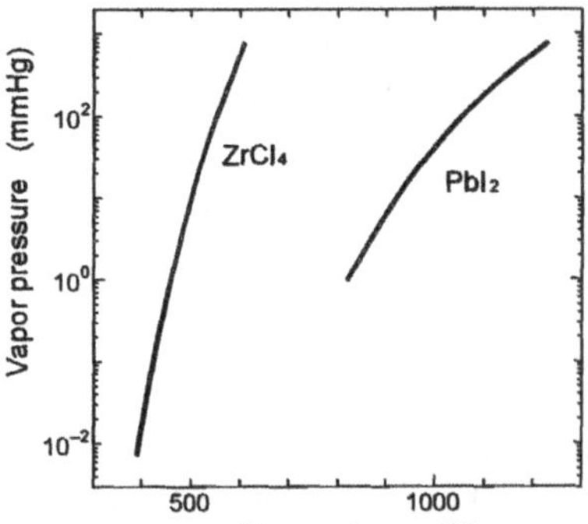

PbI$_2$ and ZrCl$_4$ were used as n-type dopants for the present study, because PbI$_2$ is a typical dopant for practical applications and ZrCl$_4$ is a dopant free from Pb element. The vapor pressures of dopants are shown in Fig.2[8]. The PbTe specimens and dopants were capsulated 5cm apart in evacuated quarts tubes under 1×10^{-3}Pa and heat treated at 823, 873, 909, 950, 973 and 1003K for 24, 50 and 100h. Both the dopants had vapor pressures high enough at the heat treatment temperature. The dopants evaporated and diffused into the specimens during heat treatment, as shown in Fig.3.

Fig.2

Vapor pressure variations of dopoants as a function of temperature

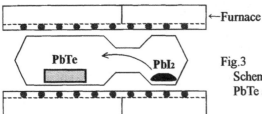

Fig.3
Schematic view of heat treatment of PbTe specimens in the dopant vapors

Diffusion of the dopants into the specimens was evaluated by the electrical resistivity ρ. We applied two kinds of typical ρ measurements; one-probe method and four-probe method as shown in Fig.4. When the diffusion distance is long enough to ignore the effect of electrodes, the one-probe method is useful for the diffusion study. The four-probe method, which needs the flat surface of specimen, has an advantage that the ρ of a thin diffusion layer can be measured.

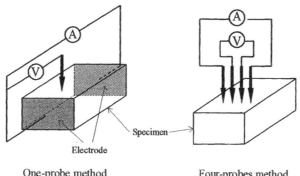

Fig.4   One-probe method   Four-probes method

PbTe ingots doped with PbI_2 and $ZrCl_4$, especially low-n ingots, were also prepared to evaluate the thermoelectric properties as a basis of the above diffusion test. Pb(6N) and Te(6N) were weighed at a composition of PbTe, and capsulated with some amount of dopant in an evacuated quarts tube under 1×10^{-3} Pa. The fabrication process was similar to that of non-doped PbTe. The ρ and Hall coefficient R_H were measured by the dc method with high speed and high resolution, and thermal conductivity κ was by static comparative method[9]. Seebeck coefficient α was obtained from a slope of (thermo-electromotive force, E_0)-(given temperature difference, ΔT) curve.

RESULTS AND DISCUSSIONS
Thermoelectric properties of PbTe ingots doped with PbI_2 and $ZrCl_4$ are shown in Table 1. Both the dopants give almost the same high values of μ_H, though the other properties have some difference because of the different n. This result suggests that both the dopants act on PbTe similarly at room temperature and that the n of diffusion tested specimens can be estimated from the measured values of ρ.

Table 1   Thermoelectric properties of PbTe ingots doped with PbI_2 and $ZrCl_4$
           at room temperature

Dopant	n (/m³)	μ_H(m²/Vs)	ρ (μΩm)	α (μV/K)	κ (W/Km)	Z (/K)
PbI_2	3.2×10^{24}	0.15	13.3	-184	2.0	1.3×10^{-3}
$ZrCl_4$	1.8×10^{24}	0.13	28.5	-242	2.2	0.93×10^{-3}

Figure 5 and 6 show the temperature dependence of ρ and α for PbTe ingots doped with PbI$_2$ and ZrCl$_4$, respectively. The PbI$_2$ doped ingot has an increase in ρ with temperature, whereas the ZrCl$_4$ doped ingot stops the increase above 550K. The ZrCl$_4$ doped ingot has lower n than the PbI$_2$ doped ingot as shown in Table 1. Thus the so-called intrinsic region appears above 550K. The α of both the ingots has a similar tendency of parabolic change with temperature. The difference is that the α curve of the ZrCl$_4$ doped one shifts to a lower temperature, which is also caused by the lower n.

The non-doped stoichiometric PbTe, which is p-type at room temperature, changes to n-type above 500K, and come back to p-type below 500K. On the other hand, both the doped ingots, which have low n, remained n-type during heat treatment up to 800K. This result suggests that although the conduction type changes from p-type to n-type during diffusion treatment, the specimens are n-type after cooled down to room temperature.

Surface melting during diffusion treatment changed morphology of the PbTe specimens. The results are summarized in Table 2. In PbI$_2$ vapor, the specimens were free from surface melting at 823K, while partially surface melted at 873K and surface melted or melted down at 909K or more. According to the Pb-Te phase diagram[10], PbTe exists only in a very narrow compositional range of Pb:Te=1:1 and the non-stoichiometry causes the partial melting below 700K. When PbI$_2$ diffuses into the surface of PbTe specimens, the surface comes to be rich in Pb and melts locally. Table 2 indicates that the temperatures above 873K activated PbI$_2$ diffusion into PbTe.

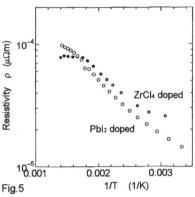

Fig.5 Variations of resistivity as a function of temperature

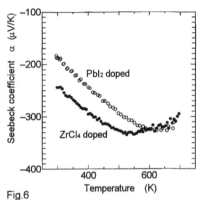

Fig.6 Variations of Seebeck coefficient as a function of temperature

Table 2   Morphology change of PbTe specimens after PbI$_2$ and ZrCl$_4$ diffusion treatment

	823K	873K	909K	950K	973K	1003K
PbI$_2$	○	△	×	×	×	×
ZrCl$_4$	-	○	-	-	△	-

○: No change   △: Partially surface melted   ×: Surface melted or melted down

All of the specimens treated in PbI$_2$ vapor showed n-type conduction. If PbI$_2$ deeply diffuses into the specimen to the extent that does not cause surface melting, a high temperature may change the specimen wholly from p-type to n-type. In this assumption, the n should be changed with treatment temperature. However, all of the specimens showed the same n of 2~3 × 10^{24}/m^3 independently of treatment temperature and time. From this result, the change of conduction type is not due to PbI$_2$ diffusion. The μ_H of all the treated specimens was 0.07~0.09m^2/Vs, which is different from the value of PbI$_2$ doped PbTe shown in Table 1. Heat treatment in PbI$_2$ vapor causes the intrinsic change in transport properties of PbTe. Resistance measurement by one-probe and four-probe methods was applied to the treated specimens, but the distinct evidence of dopant diffusion was not obtained.

In ZrCl$_4$ vapor, the specimens revealed the different behavior from in PbI$_2$ vapor; not surface melted at 873K, though partially melted at 973K. Since ZrCl$_4$ does not contain Pb, the dopant diffusion has less effect on the melting point change of PbTe. The reason of the partial melting at 973K is not clarified, but Zr in diffused ZrCl$_4$ may react with Te in PbTe to increase the Pb content at the surface, resulting in local melting.

The inside of all the specimens treated in ZrCl$_4$ vapor still remained p-type. The n and μ_H were 1.6 × 10^{24}/m^3 and 0.09Vs/m^2, respectively, which are the same values as the non-doped original specimen. ZrCl$_4$ vapor has no effect on the transport properties of PbTe.

Figure 7 shows resistance variations of the specimen treated in ZrCl$_4$ vapor as a function of distance from an edge by one-probe method. Slope of the plots gives the resistivity at the distance. The measurement was performed for the specimen before(△) and after(▲) surface polishing by 0.3mm. The absolute values are different between before and after polishing, which is due to contact with the electrodes. The before-polished specimen has a steeper slope than the after-polished one. The measured values by one-probe method include the resistance change at the surface to some extent. This result indicates that the ZrCl$_4$ diffusion layer has higher ρ than the inside. The diffusion thickness of ZrCl$_4$ is near 0.3mm.

Figure 8 shows resistance variations as a function of distance from an edge by the four-probe method. The specimen is the same in Fig.7. The measurement was performed before and after surface polishing by 0.1, 0.2, 0.4 and 0.6mm. The four-probe measurement was also applied to the non-doped original specimen(○). As is seen in the dots of ○, the measured values slightly increase toward a specimen edge by the edge effect.

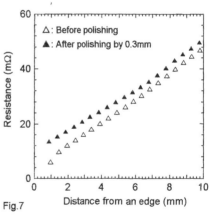

Fig.7

Resistance measurement by one-probe method for diffusion-treated PbTe in ZrCl$_4$ vapor at 973K for 50h

The before-polished specimen (indicated as 0mm in Fig.8) has a similar tendency to the original specimen, whereas the polished specimens suddenly increase resistance at a distance of 0.4mm toward a specimen edge. This increase corresponds to $ZrCl_4$ diffusion, which indicates that the diffusion layer has higher ρ than the inside. The diffusion thickness of $ZrCl_4$ is about 0.4mm.

The diffusion layer was not directly determined whether it was p-type or n-type. We found the fact that the resistance rapidly changed only at a distance between 0.3 and 0.4mm, as if resistivity of an open circuit was measured by voltmeter. This fact indicates that there is a thin region with almost zero n between 0.3 and 0.4mm. Conduction type probably changes between 0.3 and 0.4mm by $ZrCl_4$ diffusion; n-type at a distance <0.3mm and p-type at >0.4mm. The n of the diffusion layer 0.3mm thick is estimated to be less than $1 \times 10^{24}/m^3$.

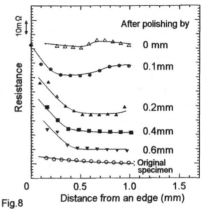

Fig.8 Resistance measurement by four–probes method for diffusion–treated PbTe in $ZrCl_4$ vapor at 973K for 50h

The diffusion layer of 0.3mm was formed by treatment at 973K for 50h. The $ZrCl_4$ diffusion into PbTe is as slow as Pb self-diffusion. It is probable that $ZrCl_4$ diffuses in ionic PbTe by similar mechanism to Pb self-diffusion[11]. If the grade carrier concentration 5mm thick is formed by $ZrCl_4$ diffusion, more than one year will be taken at 973K. The dopants without surface melting of PbTe are required for the directional diffusion control.

ACKNOWLEDGEMENT

This study was granted as research sponsored by special coordination funds of STA for promoting science and technology. The authors acknowledge Mr.M.Hashimoto in Niigata Univ. for preparation of the non-doped PbTe specimens.

REFERENCES

1)N.B.Elsner,J.Chen and G.H.Reynods, "Fabrication of selenide segented elements", *Proc. 3rd International Conference on Thermoelectric Energy Conversion(ICTEC)*, 105-108 (1980).

2)A.G.McNaughton, "Commercially Available Generators"; pp.459-469 in *CRC Handbook of THERMOELECTRICS*, Edited by D.M.Rowe. CRC Press, New York, 1994.

3)J.F.Goff and J.R.Lowney, "The thermoelectric figure-of-merit", *Proc.1st ICTEC*, 47-49 (1976).

4)Y.Shinohara, Y.Imai, Y.Isoda, I.A.Nishida, H.T.Kaibe and I.Shiota, "Thermoelectric properties of Segmented Pb-Te Systems with Graded Carrier Concentrations", *Proc. 16th International Conference on Thermoelectrics(ICT)*, 386-389 (1997).

5)R.Watanabe,M.Miyajima,A.Kawasaki and H.Okumura, "Percolation Design of Graded Composite of Powder Metallurgically Prepared SiGe and PbTe", *Proc. 4th International Conference on Functionally Graded Materials(FGM)*, 515-519 (1996).

6)R.S.Allgaier and W.W.Scanlon, "Mobility of Electrons and Holes in PbS, PbSe and PbTe between Room Temperature and 4.2K", *Phys.Rev.*, **111** 1029-1037 (1958).

7)A.J.Crocker and L.M.Rogers, "Interpretation of the Hall Coefficient, Electrical Resistivity and Seebeck Coefficient of P-type Lead Telluride", *J.Appl.Phys.*, **18** 563-573 (1967)

8)D.R.Lide; *CRC Handbook of CHMISTRY and PHYSICS 81st Edition*, pp.665-668, CRC Press, Washington,D.C., 2000.

9)K.Uemura and I.A.Nishida; *Thermoelectric Semiconductors and Applications* (in Japanese), pp.459-469, Nikkan Kogyo Shinbunsha,Tokyo,1988.

10) T.B.Massalski ed.; *Binary Alloy Phase Diagrams*, [3] pp.3401, ASMI, USA, 1990.

11)M.P.Gomez, D.A.Stevenson and R.A.Huggins, "Self-diffusion of Pb and Te in Lead Telluride", *J.Phys.Chem.Solids*, **32** 335-344 (1971).

FUNCTIONALLY GRADED BISMUTH ANTIMONY TELLURIDE CRYSTALS FOR LOW-TEMPERATURE PELTIER COOLERS GROWN BY ZONE MELTING

M. Ueltzen, W. Heiliger, W. Seifert, P. Reinshaus
Fachbereich Physik, Martin-Luther-Universität Halle-Wittenberg,
D-06099 Halle (Saale), Germany

ABSTRACT

A growth process for functionally graded thermoelectric materials based on modified zone melting has been developed. The influence of the material parameters and the limits of the technique are calculated from a simple material balance model and compared with practical results. The grown crystals are characterized by electron probe micro-analysis (EPMA) and by SEEBECK micro-thermoprobe measurements, which yield a locally resolving micro-imaging of the thermoelectric power. The characterization measurements show the capability of the growth technique to control the axial composition (Bi:Sb ratio, Te excess) of the ingot. Even a series of graded regions within a crystal of a few centimeters in length can be produced. The FGM concept will be applied to the enhancement of the efficiency of PELTIER coolers below room temperature. A mathematical description of thermoelectric cooling and heat pumping in functionally graded materials will show advantages and the limits of the FGM concept in this application.

INTRODUCTION

The concept of Functionally Graded Materials (FGM) for thermoelectric applications is well introduced. There is much activity, especially in Japan, in Functionally Graded thermoelectrics with the main focus on energy conversion [1, 2]. A locally maximized figure of merit for the working conditions, i. e. the temperature profile for the thermoelectric application of the material, should result in a higher effectiveness of thermoelectric generators, for instance [3, 4].

In the case of PELTIER cooling, we have to discuss two very different working conditions: the maximum temperature difference at zero heat transport and the maximum heat pumping at isothermal conditions. The usual operation of a PELTIER cooling device is between these extrema.

PHYSICAL BACKGROUND OF FGM PELTIER COOLERS
Thermoelectrics

A PELTIER cooling device uses the thermoelectric coupling between charge transport and heat transport. Under isotropic conditions, the constitutive equations are

$$\mathbf{j_{el}} = \sigma \mathbf{E} - \sigma S \nabla T \tag{1}$$
$$\mathbf{j_q} = -\kappa \nabla T + S T \mathbf{j_{el}} \tag{2}$$

with j_{el} - current density, j_q - heat flux, σ - electric conductivity, S - SEEBECK coefficient, κ - heat conductivity, T - temperature, and ∇ - grad.

Assuming steady state conditions, the governing equations are known from the principles of conservation of charge and energy

$$\nabla \cdot \mathbf{j_{el}} = 0 \tag{3}$$
$$\nabla \cdot \mathbf{j_q} = \mathbf{j_{el}} \cdot \mathbf{E} \tag{4}$$

Applying a 1-D model with a constant electric current density j_{el}^0 in the x-direction (see Fig. 1), equation (3) is automatically satisfied; from equation (4) one gets with E from equation (1) the basic equation of our 1-D model

$$\partial/\partial x \left(-\kappa \, \partial T/\partial x + S \, T \, j_{el}^0 \right) = (j_{el}^0)^2/\sigma + S \, j_{el}^0 \, \partial T/\partial x \tag{5}$$

which can be rewritten in an equation for the unknown temperature distribution in the PELTIER cooler:

$$\kappa \, \partial^2 T/\partial x^2 - j_{el}^0 \, \partial(ST)/\partial x + S \, j_{el}^0 \, \partial T/\partial x = -(j_{el}^0)^2/\sigma \tag{6}$$

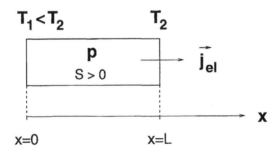

Fig. 1: 1-D model of the PELTIER effect

With S = const. and introducing $c_0 = (j_{el}^0)^2/\sigma\kappa$, the temperature distribution is

$$T(x) = -c_0/2\, x^2 + c_1\, x + c_2 \qquad (7)$$

In the same manner follows for the potential (with $E = -\nabla\varphi_{el}$)

$$\varphi_{el}(x) = S\, c_0/2\, x^2 - x\, (j_{el}^0/\sigma + S\, c_1) + c_3 \qquad (8)$$

The constants c_1, c_2, and c_3 are determined by the boundary conditions. Here, two cases are discussed:
(i) the temperatures at both ends are given.
(ii) the temperature of only one side and the heat flow are given.

Heat flow perpendicular to the x-axis is not allowed in this one-dimensional model. This means adiabatic conditions at the side of the PELTIER leg. This restriction is in good accordance to the practical conditions of PELTIER cooling devices, because many legs are arranged thermally parallel which suppresses energy exchange by radiation.

A straightforward calculation gives the following results:
(i) boundary conditions of the 1. type

$$T(x=0) = T_1 \qquad (9)$$
$$T(x=L) = T_2 \qquad (10)$$
$$\varphi_{el}(0) = 0 \qquad (11)$$

with the resulting constants

$$c_1 = 1/L\, (T_2 - T_1 + c_0/2\, L^2) \qquad (12)$$
$$c_2 = T_1 \qquad (13)$$
$$c_3 = 0 \qquad (14)$$

(ii) mixed boundary conditions

$$j_q(x=0) = j_q^0 \qquad (15)$$
$$T(x=L) = T_2 \qquad (16)$$
$$\varphi_{el}(0) = 0 \qquad (17)$$

with the resulting constants

$$c_1 = 1/(\kappa + j_{el}^0 SL)\, (j_{el}^0 S\, (c_0/2\, L^2 + T_2) - j_q^0) \qquad (18)$$
$$c_2 = 1/(\kappa + j_{el}^0 SL)\, (\kappa\, (c_0/2\, L^2 + T_2) + j_q^0 L) \qquad (19)$$
$$c_3 = 0 \qquad (20)$$

and the temperature distributions $T(x)$ for both cases discussed below.

Mathematical modeling

The results are presented using MATHEMATICA software, for which purpose the following material parameters (oriented at realistic values for bismuth antimony telluride in the low temperature region) in SI values were chosen: $S = 2*10^{-4}$ V/K, $\kappa = 2$ W/(m·K), and $\sigma = 10^5$ A/(V·m). With the length of a typical PELTIER leg of 5 mm (L = 0.005 m) the heat transport and the temperature distribution can be discussed. Typical practical values of the electric current are 50 A through a typical area of 0.5 cm^2, corresponding to 10^6 A/m^2.

Under these requirements, the well known linear dependency between heat flow and temperature difference over the PELTIER cooler with the maximum heat flow at zero temperature difference, as well as the relation between heat transport and electric current, showing the optimum conditions of operation of a PELTIER cooler, can be discussed. Fig. 2 shows a MATHEMATICA plot of the parameters. The calculated data are in good agreement with typical experimental data of bismuth antimony telluride based PELTIER coolers.

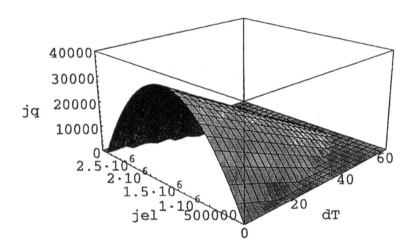

Fig. 2: Calculated parameters of a PELTIER cooler (all data in SI)

Temperature distribution in a PELTIER cooler

Furthermore, we can discuss the local temperature distribution in the PELTIER leg. Fig. 3 shows the calculated temperature distributions for optimum current density for the two most important cases of operation: a) isothermal conditions $T_1 = T_2$, that means maximum heat transport of the device, and b) maximum temperature difference of the cooler, that means no heat transport (cf. boundary conditions above). The temperature inside the PELTIER cooler follows CLEMENS-ALBAN's temperature distribution [5]. This is characterized by a deviation from the linear law. The maximum overheating in the cooler is calculated to about 25 K with the realistic material parameters and realistic conditions of operation given above.

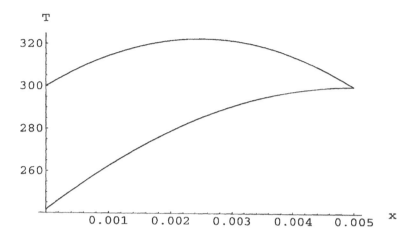

Fig. 3: CLEMENS-ALBAN's temperature distribution in a PELTIER cooler (all data in SI)

The practical conditions of operation of a PELTIER cooling device are between those two described cases. It can be assumed that the temperature distribution in the material varies with the ratio of temperature difference and pumped heat flow between the two calculated curves. An optimization of thermoelectric devices by the FGM concept has to take into account these special temperature profiles inside the PELTIER device.

CRYSTAL GROWTH AND CHARACTERIZATION

A technology for the growth of functionally graded thermoelectric materials based on zone melting was developed. Pieces of homogeneous material (e.g. 8 mm in length) are stacked into a quartz ampoule. This stack is processed by zone melting as usual with growth velocities of 3 mm/h. Further technological details are described elsewhere [6].

The maximum material gradient produced by this technique with an abrupt change from bismuth telluride to antimony telluride in the initial stack is given by the following formula (see [6]):

$$c_s(z') = c_{i0} \cdot \left[1 - \exp\left[-\frac{\rho_s}{\rho_l} \cdot \frac{k}{\delta} \cdot (z' - (z_0' - \delta))\right]\right] \qquad (21)$$

where c and ρ are the concentration and the density of the solid phase (s) or the liquid phase (l), respectively, z' is the coordinate fixed to the growing crystal, z_0' is the position of the abrupt change of bismuth telluride to antimony telluride in the initial stack, $k = c_s/c_l$ is the segregation coefficient and δ is the height of the molten zone. This method enables maximum material variations in the order of "zone height divided by the segregation coefficient".

Figure 4 shows the concentration profiles (atomic fraction) along the crystal and the locally measured SEEBECK coefficient. Concentration profiles were measured using a CAMEBAX microanalyzer at 20 kV electron beam and in a wavelength dispersive analysis mode. The measurement technique for the locally resolved SEEBECK coefficient was described in an earlier publication [7].

The concentration profile in growth direction of the crystal shows a complementary behaviour of antimony and bismuth. They occupy the same sites in the crystal structure. The starting composition of the crystal was a piece of 50:50 mixed crystal followed by alternating pieces of antimony telluride and bismuth telluride. During zone melting the abrupt material transitions are smeared into continuous transitions as discussed above. The dependence from the material related parameters k and δ explains the different behaviour of the gradients in the different directions as shown in Fig. 4.

OUTLOOK

There are some open questions in the scientific understanding of FGM thermoelectrics as well as in materials science of manufacturing thermoelectric FGM.

The CLEMENS-ALBAN's law of the temperature distribution in a PELTIER device was discussed for homogeneous material from the viewpoint of a theoretical description of the physical effects. This modeling will be extended to inhomogeneous material, esp. to FGMs, including the implementation of the THOMSON effect. By means of the physical model, the advantages of the FGM concept for the different working conditions of a PELTIER cooler can be discussed. To ensure the correlation of the model with the practical behaviour of a PELTIER cooler, a measurement system for a comprehensive characterization of cooling devices in different working conditions (between maximum temperature difference and maximum pumped heat) is under construction. With those measurements, the correlation of the theoretical CLEMENS-ALBAN's law of the temperature distribution with the real behaviour can be proved. Furthermore, the

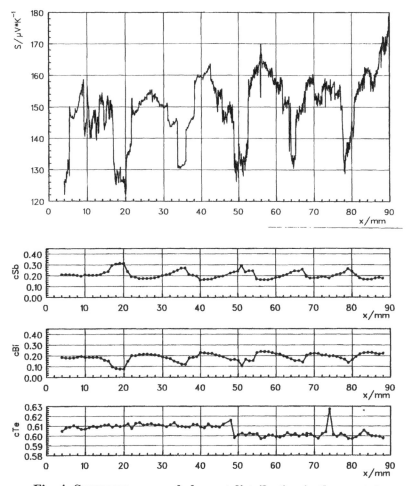

Fig. 4: SEEBECK scan and element distribution in the crystal

impact of technical problems like additional electrical and heat flow resistance at the contacts in the device can be studied.

A further way for manufacturing bismuth antimony telluride based FGMs is the controlling of the tellurium content of the grown crystal by variations of the growth velocity. By means of this technique, the optimum tellurium concentration according to the bismuth antimony ratio realized while crystalizing should improve the local thermoelectric parameters.

In order to ensure a positive impact of the FGM concept in low-temperature thermoelectrics, the further search for materials suitable for the temperature region below room temperature is necessary. The great scientific and technical interest in this question is shown by a recent publication in "Science" [8]. The material discussed there seems to be compatible to the standard room temperature bismuth antimony telluride material in order to develop FGM based materials.

ACKNOWLEDGMENTS

This work was sponsored by the Deutsche Forschungsgemeinschaft DFG.

REFERENCES

[1] I. Shiota and M.Y. Miyamoto (Editors): Functionally Graded Materials 1996, Elsevier Science B.V. 1997.

[2] W.A. Kaysser (Editor): Functionally Graded Materials 1998, Materials Science Forum Vols. 308-311, Trans Tech Publications 1999.

[3] L. Helmers, E. Müller, J. Schilz, W.A. Kaysser: Graded and stacked thermoelectric generators - numerical description and maximisation of output power, Mater. Sci. Eng. B56 (1998) 60-68.

[4] J. Schilz, L. Helmers, W.E. Müller, M. Niino: A local selection criterion for the composition of graded thermoelectric generators, J. Appl. Phys. 83 (1998) 1150 - 1152.

[5] W.E. Müller, DLR Colonia: private communication (1999).

[6] M. Ueltzen, W. Heiliger, P. Reinshaus: On the potential of zone melting for the growth of functionally graded $(Bi,Sb)_2Te_3$-mixed crystals, in preparation for Cryst. Res. Technol. (2000).

[7] P. Reinshaus, H. Süßmann, G. Erz, U. Kramer, W. Heiliger: Scanning Thermo Probe Technique - a Method for High Resolution Characterization of Graded Semiconductors and Metals, Materials Science Forum Vols. 308-311 (1999) 890-895.

[8] D.-Y. Chung, T. Hogan, P. Brazis, M. Rocci-Lane, C. Kannewurf, M. Bastea, C. Uher, M.G. Kanatzidis: $CsBi_4Te_6$: A high-performance thermoelectric material for low-temperature applications, Science 287 (2000) 1024-1027.

Biomedical Applications

FABRICATION OF BIOACTIVE FGM FROM Ti AND HYDROXYAPATITE

Mamoru Omori, Akira Okubo and Toshio Hirai
Institute for Materials Research, Tohoku University, 2-1-1 Katahira, Aoba-ku, Sendai 980-8577
Rika Miyao and Fumio Watari
School of Dentistry, Hokkaido University, 13-7 Kita, Kita-ku, Sapporo 060-8586

ABSTRACT

Composites of hydroxyapatite (HA) and titanium (Ti) were decomposed for several days. The composite prepared from titanium hydride (TiH) and HA was stable for more than two years, and its toughness was very low. HA was decomposed by the catalytic action of Ti in spite of the use of TiH. Functionally Graded Materials (FGMs) fabricated from TiH and HA were cracked in the HA layer by residual stress due to a thermal expansion mismatch between Ti and HA, as the density of the HA layer was high. Partially nitrided titanium (Ti(N)) was synthesized in N_2 gas at 800°C for 180 min, resulting in the formation of stable composites with HA. HA was decomposed by a catalytic action of Ti(N). The composite of titanium nitride (TiN) and HA was stable more than 1000°C. A Ti/TiN-HA FGM consisted of a dense Ti layer and a stable HA-TiN composite of which HA was not decomposed. This FGM was composed of layers of Ti, Ti(N), 80vol%Ti(N)-20vol%HA, 60vol%Ti(N)-40vol%HA and 40vol%TiN-60vol%HA. This FGM maintained the ductility of Ti and crystal structure of HA and avoided cracking, using the hard-to-sinter and inactive compound such as TiN.

INTRODUCTION

Hydroxyapatite (HA) is well established as a biocompatible ceramic. This ceramic is not a load-bearing material [1 – 3] and is useful only in applications where mechanical forces do not exist or are primarily compressive. Hydroxyapatite-coated titanium is characterized by both bioactivity and mechanical reliability. Various techniques of coating have been attempted. With plasma-spray coating, coated materials applicable in manufacturing as well as in bone and tooth implants have been produced. The plasma spraying technique does not result in dense films on substrates. The HA coating contains many pores [4, 5], and porosity is associated with increased degradation. The temperature of plasma coating is uncertain, and may sometimes be more than the decomposition temperature. The HA coating undergoes changes in crystal structure, phase composition, specific surface and morphology [6 – 10]. Pure HA is decomposed more than 1000°C even in a vacuum, but the decomposition of HA is facilitated near 800°C by the catalytic action of Ti and TiO_2 [11, 12]. Mechanical failure occurs mainly at the interface of HA and Ti [7]. This failure may be due to the compounds resulting from the decomposition of HA by the catalytic action. The difference in thermal expansion of HA and Ti is large and produces residual stress that can influence the stability of the bonding of HA and Ti. Stress is relaxed in plasma-sprayed HA films by lower moduli and density due to pores.

This paper is concerned with the stability of HA and Ti composites and the fabrication of stable functionally graded materials (FGMs). Composites were prepared from Ti, partially nitrided Ti (Ti(N)), or titanium hydride (TiH) and HA, and their properties were measured. FGMs were fabricated from HA and Ti(N) or TiN, residual stress being reduced by dispersed pores in the HA layers.

EXPERIMENTAL PROCEDURE

Materials were hydroxyapatite (Sumitomo Osaka Cement Co. Ltd., 99.5%), titanium (Sumitomo Sitix Co. Ltd., 99.5%), titanium hydride (Sumitomo Sitix Co. Ltd., 99.5%), and titanium nitride (Japan New Metals Co. Ltd., 1 µm) powders. HA powder was blended with Ti , TiH and TiN powders using an agate mortar and pestle for 20 min. The blended material was sintered between 1100 and 1150°C at 20 MPa in a vacuum by the spark plasma system (SPS) (Izumi Technology Co. Ltd., SPS1050).

Two kinds of Ti powder, that under 124 µm and that under 44 µm, were partially nitrided in N_2 gas (99.99%) at 800 – 950°C for 15 – 180 min to prepare partially nitrided titanium (Ti(N)). The Ti(N) powder was mixed with HA in a planetary ball mill (Thinky Co. Ltd. Japan, MX-201), using only a polyethylene container to avoid fracture of nitride film.

The blended powders of HA-Ti and HA-TiH were sintered between 1100 and 1150°C at 20 MPa in a vacuum by SPS. The mixture of HA and Ti(N) was sintered in a vacuum at 1000°C

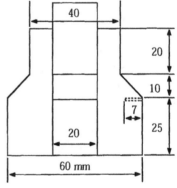

Fig. 1 Graphite die for temperatutre gradiant.

for 15 min by SPS.

FGMs were fabricated under a temperature gradient in Ar gas, using a graphite die with the upper and lower parts having different diameters (Fig. 1) [13]. Powders consisting of different compositions were layered in the graphite die. The FGM fabrication was carried out by SPS. The heating program was as follows: from 20 to 850°C for 10 min, from 850 to 900°C for 10 min and at 900°C for 15 min. The sintering temperature was obtained near the bottom of the powder samples, and the metal layer was heated to more than 900°C.

The density of composites was determined by Archimedes' principle using water immersion. Elastic moduli of the bulk, 5 - 10 mm in thickness and 20 mm in diameter, were measured by the pulse-echo overlap ultrasonic technique using

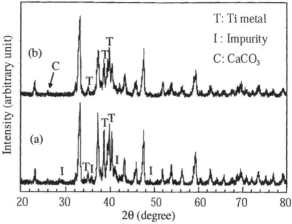

Fig. 2 XRD patterns of composites: (a) 50vol%Ti-50vol%HA and (b) 50vol%TiH-50vol%HA.

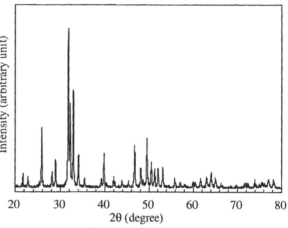

Fig. 3 XRD pattern of hydroxyapatite.

Table1 Properties of TiH-hydroxyapatite composites

TiH (vol%)	HAP (vol%)	Bulk density (g/cm³)	Hardness Hv (GPa)	Young's modulus (GPa)	Toughness (MPa·m$^{1/2}$)	Poisson's ratio
0	100	3.15	5.9±0.4	121	0.62±0.21	0.28
10	90	3.31	5.8±0.3	128	1.0±0.3	0.27
20	70	3.83	7.9±0.7	191	2.9±0.4	0.24
40	60	4.12	9.2±0.5	167	2.9±0.3	0.25
50	50	4.31	9.3±0.5	234	3.8±0.4	0.27
70	30	4.51	9.0±0.6	209	2.2±0.5	0.27

an ultrasonic detector (Hitachi Kenki Co. Ltd. Japan, ATS-100) and a storage oscilloscope (Iwasaki Tsushinki Co. Ltd. Japan, DS6411). The polished surface was observed with an optical microscope (Nicon,). A Vickers hardness tester (Akashi Co. Ltd. Japan, AVK) was used for hardness and toughness measurements of composites under a load of 98 N. Fracture toughness was calculated from Vickers hardness, Young's modulus, and crack length at indents [14].

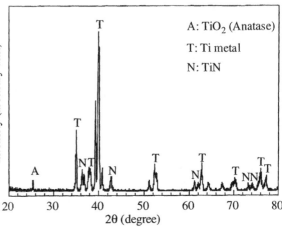

Fig. 4 XRD pattern of partially nitrided Ti.

RESULTS AND DISCUSSION
COMPOSITES

Composites prepared from HA and Ti were decomposed into small particles after a few days. The XRD pattern of the compact before decomposition is shown in Fig. 2. There was no hydroxyapatite, the pattern of which is indicated in Fig. 3. There were diffraction peaks due to Ti. The composite prepared from HA and TiH was stable for more than 2 years. The XRD shown in Fig. 2 was different from those of the HA-Ti composite with regard to weak diffraction peaks. The small peaks may be responsible for the decomposition. Since hydrogen gas evolved from 600 to 800°C, TiH decomposed to form Ti and was not detected in the pattern. HA is stable in air up to 1250°C [15] and to 1050°C in a vacuum [16], but it is decomposed at temperatures lower than 1000°C by the catalytic action of HA and TiO_2 [11, 12]. The composites were

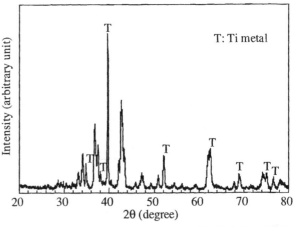

Fig. 5 XRD pattern of composite of 40vol%Ti(N)-60vol%HA.

sintered in a vacuum to remove hydrogen gas from TiH. HA of the HA-TiH composites was decomposed by the catalytic action and the vacuum treatment.

Mechanical properties of HA-TiH composites are listed in Table 1. The toughness of 50vol%HA-50vol% TiH (50HA-50Ti) was the greatest of all composites but far less than that of Ti (80 − 12 MPa·m$^{1/2}$). The high Young's modulus of 50HA-50TiH was not the same as those of Ti metal and HA. Ti in the composite changed from metal to a brittle material. When $CaCO_3$ is produced by the decomposition of HA, a phosphorus element is released. A weak diffraction peak due to $CaCO_3$ was detected as shown in Fig. 2. The free P penetrated Ti and decreased the toughness of the composites.

Ti metal was partially nitrided to decrease the catalytic action on HA. Partially nitrided titanium (Ti(N)) was prepared under various conditions, and sintered with HA at 1000°C, as shown in Table 2. The XRD pattern of Ti(N) is shown in Fig. 4. A diffraction peak due to TiO_2 (anatase) was detected, because oxygen in N_2 gas was not completely removed. Composites prepared from a HA powder and Ti(N) powders except for those heated at 800°C for 180 min cracked after being exposed to air for a few days. The TiN film prepared at a low temperature of 800°C was protective against this circumstance. However, as shown by the XRD pattern of the stable composite in Fig. 5 , HA ceased to exist with its decomposition into other compounds, except for CaP_2O_9 and $Ca_3(PO_4)_2$.

Table 2 Preparation of partially nitrided titanium Ti(N) and its composite with HA

Temp. (°C)	Time (min)	Weight Gain (wt%)	Crack in composite***
900*	15	6.6	X
900*	30	7.5	X
950*	15	11.4	X
800**	60	3.8	X
800**	120	5.0	X
800**	180	6.3	O

*Ti<150 mesh, ** Ti<45 mesh and ***composite with HA at 1000°C for 15 min

FUNCTIONALLY GRADED MATERIALS

Functionally Graded Materials (FGMs) fabricated from TiH and HA were cracked in the HA layer by residual stress due to a thermal expansion mismatch between Ti and HA, as the density of the HA layer was high. It was difficult to fabricate FGMs with a dense Ti layer. Even if the surface layer was changed from the HA to a 40vol%TiH -60vol%HA layers, the densification of the Ti layer was not achieved.

Table 3 Fabrication of Ti(N)/HA FGM

Ti(N)	Ti	HA (vol%)	Temp. (°C)	Crack on HA	Density of Ti(N) or Ti
100	-	100	850	X	X
100	-	100	800	X	X
100	-	100	750	X	X
100	-	90	850	X	X
100	-	80	850	X	X
100	-	60	850	O	X
100	-	60	900	O	X
-	100	60	900	O	O

The composite of HA and Ti(N) does not decompose and can be used as intermediate layers in FGMs. Functionally graded materials were fabricated from HA and Ti(N). Results of fabrication of FGMs are listed in Table 3. The graded layers consisted of Ti(N)80vol%-HA20vol% (Ti(N)80-HA20, Ti(N)60-HA40, Ti(N)40-HA60 and Ti(N)20-HA80. The sintering temperature was measured near the HA layer and was insufficient to achieve the densification of Ti(N). The Ti(N) powder was sintered at 1100°C. Having an insufficient number of pores to relax residual stress, the HA surface layer was changed from HA to a composite of HA and Ti(N). The Ti(N)40-HA60 surface was not cracked. However, the Ti(N) layer was in the course of being sintered, and its density was lower than that of the dense Ti(N). It is necessary to use Ti for the preparation of a dense metal layer. When the temperature of the Ti(N)40-HA60 layer was 900°C, the Ti layer was sintered at a graded temperature of about 1000°C. A Ti(N) layer was inserted between the Ti and Ti(N)80-HA20 layers to avoid the formation of unstable compounds due to the catalytic action of Ti on HA. A problem with this dense FGM is that the HA composite layer does not have the crystal structure of HA.

Titanium nitride was sintered at 1600°C by SPS. The composite of TiN and HA prepared at 1000°C contained many pores. The decomposition of HA by TiN has not been studied. FGMs with various graded layers are shown in Table 4. The Ti layer of F-1 did not contain pores. Layers from HA to TiN80 (TiN80vol%-HA20vol%) were separated from Ti(N) and Ti layers, because the TiN80 layer was not sintered at about 1000°C and was susceptible to stress. The densification of the Ti layer was achieved at 900°C. The layers from TiN20 to TiN80 of the F-2 sample cracked and separated from the Ti(N) and Ti layers. The cracking and separation of F-3 was similar to that of F-2. The sintered composite from HA and TiN was composed of pores, the number of which increased with increasing the ratio of TiN. The surface layers of HA and TiN20 of F-4 and F-5 cracked and separated from the other sintered layers of Ti(N) and Ti. The F-6 sample having a 40TiN surface layer appropriate for reducing residual stress, was the only FGM free from cracks. The TiN layer was a mixture of TiN and HA crystals and was not too brittle.

CONCLUSION

Table 4 Fabrication of Ti/TiN-HA FGM

	HA	TiN20*	TiN40	TiN60	TiN80	Ti(N)	Ti
F-1	o**	o	o	o	o	o	o
F-2	-	o	o	o	o	o	o
F-3	-	-	o	o	o	o	o
		TiN20	Ti(N)40	Ti(N)60	Ti(N)80		
F-4	o	o	o	o	o	o	o
F-5	-	o	o	o	o	o	o
			TiN40				
F-6	-	-	o	o	o	o	o

*TiN20 = TiN20vol%-HA80vol%

**o indicats the comoposition of a layer.

Composites of hydroxyapatite (HA) and titanium decomposed in air. The composite prepared from titanium hydride and HA was stable, but its toughness was very low. Nitrided titanium (Ti(N)) was synthesized in N_2 gas at 800°C for 180 min, resulting in the formation of stable composites with HA. HA was decomposed by the catalytic action of Ti and Ti(N). Ti(N)/HA FGMs were fabricated at 900°C. The FGM prepared from Ti, Ti(N) and HA consisted of a dense Ti layer and a stable Ti(N)-HA layer, the HA of which HA decomposed. Titanium nitride did not decompose HA near 1000°C. The Ti/TiN-HA FGM fabricated at 900°C was composed of dense Ti and stable HA-TiN layers.

ACKNOWLEDGMENT

We acknowledge the support of the Ministry of Education, Science and Culture under a Grant-in Aid for Scientific Research on Priority Areas (B) (No. 11221203, "Harmonic Material Design of Multi-Functional Composites"). We appreciate being permitted to use the spark plasma system (SPS) managed by the Laboratory for Developmental Research of Advanced Materials, Institute for Materials Research, Tohoku University.

REFERENCES

[1]G. De With, H. J. A. Van Dijk, N. Hattu and K. Prijs, "Preparation, Microstructure and Mechanical Properties of Dense Polycrystalline Hydroxyapatite", J. Mater. Sci., 16 1592-98 (1981).

[2]P. Van Landuyt, F. Li, J. P. Keustermans, J. M. Streydio, F. Delannay and E. Munting, "The Influence of High Sintering Temperatures on the Mechanical Structures", J. Mater. Sci. Mater. Med., 6 8-13 (1995).

[3]D. S. Metsger, M. R. Rieger and D. W. Foreman, "Mechanical Properties of Sintered Hydroxyapatite and Tricalcium Phosphate Ceramic" J. Mater. Sci. Mater. Med., 10 (1999) 9-17.

[4]T. W. Bauer, B. N. Stulberg, J. Ming and R. G. T. Geesink, "Uncemented Acetabular Components: Histologic Analysis of Retrieved Hydroxyapatite-Coated and Porous Implants", J. Arthroplasty, 8 167-77 (1993).

[5]J. E. Dalton and S. D. Cook, "In Vivo Mechanical and Histological Characteristics of HA-Coated Implants Vary with Coating Vendor", J. Biomed. Mater. Res., 29 239-45 (1995).

[6]D. P. Rivero, J. Fox, A. K. Skipor, R. M. Urban and J. O. Galante, "Calcium Phosphate-Coated Porous Titanium Implants for Enhanced Skeletal Fixation", J. Biomed. Mater. Res., 22 191-201 (1988).

[7]S. R. Radin and P. Ducheyne, "Plasma Spaying Induced Changes of Calcium Phosphate Ceramic Characteristics and the Effect on in Vitro Stability", J. Mater. Sci.

Mater. Med., 3 33-42 (1992).

[8]H. Ji, C. B. Ponton and P. M. Marquis, "Microstructural Characterization of Hydroxyapatite Coating on Titanium", J. Mater. Sci. Mater. Med., 3 283-87 (1992).

[9]M. Weinlaender, J. Beumer III, E. B. Kenney and P. K. Moy, "Raman Microprobe Investigation of the Calcium Phosphate Phases of Three Commercially Available Plasma-Flame-Sprayed Hydroxyapatite-Coated Dental Implants", J. Mater. Sci. Mater. Med., 3 397-401 (1992).

[10]R. McPherson, N. Gane and T. J. Bastow, "Structural Characterization of Plasma-Sprayed Hydroxyapatite Coatings", J. Mater. Sci. Mater. Med., 6 327-34 (1995).

[11]P. Ducheyne. S. Radin, M. Heughebaert and J. C. Heughebaert, "Calcium Phosphate Ceramic coatings on Metallic Porous Surfaces. The Effect of Structure and Composition on the Electrophoretic Deposition, Vacuum Sintering, and In Vitro Dissolution", Biomaterials, 11 [3] 244-54 (1990).

[12]J. Weng, X. Liu, X. Zhang and X. Ji, "Thermal Decomposition of Hydroxyapatite Structure Induced by Titanium and Its Dioxide", J. Mater. Sci. Lett., 13 159-61 (1994).

[13]M. Omori, A. Okubo, G. H. Kan and T. Hirai, "Preparation and Properties of Polyimide/Cu Functionally Graded Material", pp. 767-772 in 4th. Int. Symp. Functionally Graded Mater., Tsukuba, Ed by I. Shiota and Y. Miyamoto, Elsevier, Amsterdam, 1996.

[14]G. R. Anstis, P. Chantikul, B. R. Lawn and D. B. Marshall, "A Critical Evaluation of Indentation Techniques for Measuring Fracture Toughness: I, Direct Crack Measurements", J. Am. Ceram. Soc., 64 [9] 533-38 (1981).

[15]M. Jarcho, C. H. Bolen, M. B. Thomas, J. Bobick, J. F. Kay, and R. H. Doremus, "Hydroxyapatite Synthesis and Characterization in Dense Polycrystalline Form", J. Mater. Sci., 11 2027-35 (1976).

[16]J. C. Tronbe and G. Montel, "Some Features of The Incorporation of Oxygen in Different Oxidation States in The Apatitic Lattice-I", J. Inorg. Nucl. Chem., 40 15-21 (1978).

FUNCTIONALLY GRADED COLLAGEN-HYDROXYAPATITE MATERIALS FOR BONE REPLACEMENT

Wolfgang Pompe, Michael Gelinsky, Ines Hofinger and Birgit Knepper-Nicolai
Dresden University of Technology, Department of Materials Science
Hallwachsstrasse 3, 01069 Dresden, Germany

ABSTRACT

Two complementary processing routes for the formation of functionally graded collagen/ hydroxyapatite (HAP)-composites for bone replacement are described: one approach is based on the preparation of laminated materials composed from collagen/HAP-tapes manufactured by vacuum filtration of liquid precursors. The mechanical properties of these tapes are outstanding. A second group of materials concerns monolithic composites of mineralized collagen and HAP. In a sintered porous HAP material a liquid collagen/HAP-precursor has been infiltrated. Depending on the process parameters graded profiles of chemical composition and porosity have been prepared. The effects of osteocalcin and hyaluronic acid on the structure, the mechanical and biological behavior have been investigated. It has been shown that osteocalcin influences the dissolution/ precipitation kinetics of HAP. The chemotactic activity and the adhesion promoting effect of osteocalcin for osteoblasts and osteoclasts can be used for the initiation of the remodeling of the bone replacement. Alternatively, functionalization of surface near regions with hyaluronic acid allows a controlled limitation of the cell interaction.

MOTIVATION

Recently, an ambiguous activity can be observed in materials science to develop a new generation of biomaterials for bone replacement which combines bone-like mechanical properties with characteristic features of the living bone: the capability of remodeling by cellular activity. These two goals open the avenue for application of basic concepts of functionally graded materials to new bone replacement materials. As pointed out in the work of numerous groups in the last years, there is a natural approach for preparation of bone replacements starting from HAP-bioceramics[1,2,3,4]. Obviously, HAP-cements combine good mechanical properties with a certain potential for resorption depending on the particular chemical composition and structure[1,2]. Thus, first successful applications in medicine have been reported. However, their load bearing capability as well as remodeling behavior is yet far away from that of living bone. Therefore, stepwise further features of hard tissue have to be implemented in the new biomaterial. The first natural step is the preparation of a collagen/HAP-composite[2,4]. where, the two main components of the extracellular matrix of bone are combined. However, to approach the most important hierarchical structure of hard tissue, further processing steps are necessary, for this two alternatives, or better complementary approaches, exist – the manufacturing of the bone-like structure during the in vitro-processing of the material, or the use of cellular processes for an in vivo-remodeling of a structurally non-perfect precursor structure leading to the final bone replacement. Whereas the first process needs a high flexibility of the in vitro processing, for instance by assembling of appropriate preformed building units different in porosity, chemical composition etc., or by use of hierarchical structure forming elements in the manufacturing process. The second approach only works when the surface near region of the biomaterial can be functionalized to activate bone resorbing as well as bone matrix producing cells, the osteoclasts and the osteoblasts, respectively, and to protect particular other regions of the implant against cellular interaction completely. In the following we will present examples for both approaches, the *in vitro* formation of a hierarchical structure of collagen/HAP-composites, and the functionalization of surface near regions collagen/HAP-composites for controlled cellular activity. The combination of both approaches yields a functionally graded material in which the optimization of the mechanical reliability and the activation of remodeling processes can be realized in the same material.

IN VITRO FORMATION OF A GRADED STRUCTURE OF COLLAGEN/HAP-COMPOSITES

Collagen I/HAP-composites can be prepared from liquid precursors using various processing routes as controlled drying, freeze-drying, centrifugation, tape casting, phase separation, or press filtration. As discussed in ref.[5], these methods offer the possibility to generate graded pore structures as well. However, for realistic processing times there are limitations when the gradation of the structure should be tailored in a sufficiently large parameter range. In this case, preformed building units of different chemical composition and pore distribution should be used to create the intended macroscopically graded component. In the following, we will give two examples of well established processing techniques in polymer technology and powder metallurgy.

Functionally graded structuring with collagen/HAP-tapes

In order to generate a graded collagen/HAP-composite with reasonable mechanical properties, a preferred alignment of the collagen fibrils would be favorable. Therefore, it should be useful to build up a laminated structure from tapes where the collagen fibrils are distributed randomly in plane of the tape. A suitable technique for tape manufacturing is the vacuum press filtration. As reported already in ref.[3], such tapes can be prepared from mineralized collagen fibrils[4] which are chemically crosslinked after the densification process. The alignment of the collagen fibrils and the porosity of the tape can be varied in a certain range by the process conditions (vacuum pressure, initial concentration of the solution). The remarkable mechanical properties of the tape (in the dry state fracture strain of about 4 %, and strength of about 12 MPa, in the wet state fracture strain > 20 %, strength of about 1.5 MPa) make it a feasible building block. With the same method (press filtration) exclusively tapes of different chemical composition can be prepared as well. Thus, we have manufactured tapes composed from collagen I and collagen III, as well as HAP/collagen I, HAP/collagen III and HAP/collagen I/hyaluronic acid (see scanning electron microscopy (SEM) images in Fig. 1).

In order to build up laminated structures composed of the various tapes, it is possible to use a thin HAP-cement layer as an interconnecting material. Due to the different shrinkage of the collagen containing tape and the solidifying HAP-interlayer, residual stresses can be critical. Therefore, the lamination process has to be performed under pressure in controlled humidity. As shown in the laser scanning microscopy (LSM) images in Fig. 2 laminates with dense interfaces can be prepared.

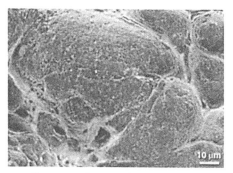
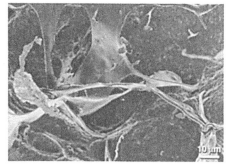

Fig. 1: SEM images of HAP/collagen I tape (left) and HAP/hyaluronic acid tape (right)

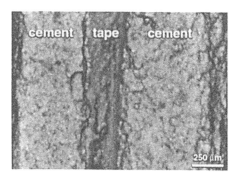
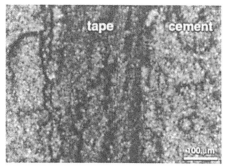

Fig. 2: LSM images of laminates of thin HAP-cement layers

Vacuum infiltration of liquid collagen/HAP-precursors in porous HAP ceramics

An alternative way of preparing graded structures exists when we start from a porous sintered HAP ceramic. As shown by Greil and coworkers[6], HAP ceramics with defined pore size can be manufactured starting from HAP green bodies with pore forming additives as thermally unstable polymer spheres. The sintered porous HAP body can be used as the framestructure to create a graded collagen/HAP-material. A suspension of mineralized collagen fibrils can be infiltrated in the pore channels by vacuum infiltration. The infiltration process yields a structure with a graded macroprofile of large pores filled with the mineralized collagen. However, depending on the processing conditions, different microstructures can be formed as visible in the comparison of Fig. 3 and 4. During the infiltration two competitive processes are acting: the dissolution-precipitation of nanocrystalline or amorphous HAP at the collagen fibrils, and the deposition of HAP at the well formed crystalline surfaces of the large grains of the sintered body. At longer reaction times the second process should dominate. Consequently, we observe a structure mainly composed of large HAP grains and pores filled with minor mineralized collagen networks (Fig. 3). In the opposite case of short reaction time, the pores are filled with a dense composite of nanocrystalline HAP connected by mineralized collagen fibrils (Fig. 4). Therefore, with variation of the process time via changing flow resistance during the filtration the final microstructure can be influenced. Furthermore, by addition of a second biomolecule (osteocalcin) the dissolution-precipitation kinetics of the HAP can be slowed down leading to a higher nanocrystalline content of HAP (see Fig. 5).

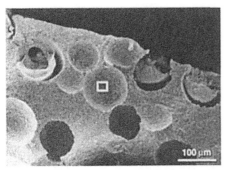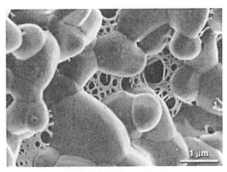

Fig. 3: SEM images of porous HAP ceramics after vacuum infiltration of liquid collagen/ HAP-precursors with low filtration rate

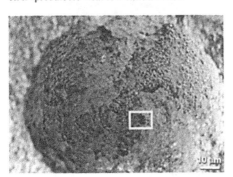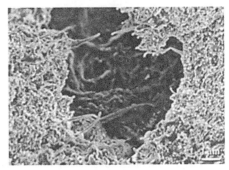

Fig. 4: SEM images of porous HAP ceramics after vacuum infiltration of liquid collagen/ HAP-precursors with high filtration rate

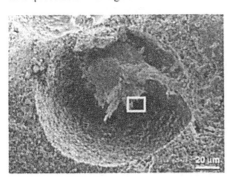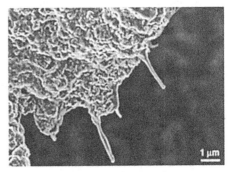

Fig. 5: SEM images of porous HAP ceramics after vacuum infiltration of liquid collagen/ HAP-precursors with low filtration rate in presence of osteocalcin

FUNCTIONALIZATION OF SURFACE NEAR REGIONS OF HAP/COLLAGEN-
COMPOSITES FOR CONTROLLED CELLULAR ACTIVITY

The integration strategy of an implant in the living tissue should follow one of the two possible paradigms:

(i) the surface region of the implant should be a carrier for signals, which direct the specific cellular activity to promoting the ingrowth or the remodeling of the implant in the tissue;

(ii) the surface region of the implant should suppress intensive cell interaction in order to keep a maximum of long term structural integrity in the surrounding tissue.

For hard tissue implants both approaches can be relevant. Thus, an optimum replacement of hard tissue always should be intended to create a structure with maximum adaptation to the living tissue. The study of natural processes in living bone has given promising ideas to develop an appropriate functionalized surface region to control cellular activity.

Activation of bone cell interaction by osteocalcin functionalized surface regions

Besides collagen, which is the main biopolymer part of the extracellular matrix of hard tissue, there are further non-collagenous proteins found in bone. Among these is osteocalcin, also called Bone Gla Protein (BGP), the most abundant. BGP is secreted by osteoblasts, the bone forming cells, and incorporated into the extracellular matrix. It consists of a single chain of 46-50 amino acids including two to three gamma-carboxy glutamic acids residues (GLA), which are essential for the calcium and hydroxyapatite binding properties of osteocalcin[7].

As shown by Romberg et al.[8], osteocalcin inhibits the growth of HAP crystals, whereby the presence of GLA residues is crucial for this effect. In regard on the influence of osteocalcin on cellular activity, it has been shown that osteocalcin is chemotactic for human osteoclast-like cells[9]. Additionally, initial adhesion of these cells was promoted on osteocalcin-coated substrate. As discussed by Hauschka et al.[10], the cell-binding property of osteocalcin seems to be related to the amino- and carboxyterminal ends of the folded osteocalcin molecule (see Fig. 6). Most important for the intended cellular remodeling is the observation that the initial chemotactic activity of osteocalcin towards human osteoclast-like cells starts a sequence of intracellular processes in the osteoclast, leading to secretion of osteopontin, bone sialoprotein and fibronectin[9]. These proteins induce the formation of specific focal adhesion contacts, which implies the involvement of integrins. Subsequently, the osteoclast will be able to polarize and to organize the so called sealing zone and ruffled border, where the resorption process takes place. In conclusion, osteocalcin offers two options simultaneously – the limitation of the dissolution/precipitation process of the HAP, and the acceleration of the osteoclasts resorption activity at the implant surface.

The influence of osteocalcin on the initial adherence of bone cells of osteocalcin-impregnated surface regions of collagen/HAP-composites has been investigated using collagen/HAP-cement samples (with 2.5% collagen). Figs. 7 and 8 show the adhesion of osteoblasts (SAOS-2: human osteosarcoma cell line) on these samples in absence and presence of osteocalcin. As seen in Figs. 7 and 8, osteoblasts adhere to the same extent to both surfaces, but cells are already more flattened on osteocalcin containing cements, which hints to a promotion of initial adherence. This finding is confirmed by studies using the confocal laser scanning microscope: osteoblasts on composites, on which osteocalcin was chemisorbed in surface near regions, are well spread after 2 hours and display prominent lamellipodae, whereas osteoblast on collagen/HAP-composites are round. Anyway, osteoblasts proliferated on both cements to the same extend (not shown). The adhesion-promoting activity of osteocalcin in HAP-composites might be due to the osteocalcin content itself, providing an adhesive ligand which is recognized by specific receptors. On the other hand, osteocalcin influences the microstructure of the HAP-composite, a fact which might also be involved in the adhesion-promoting properties.

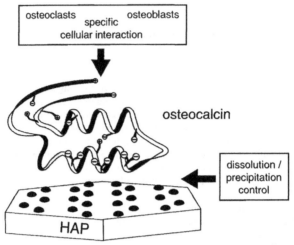

Fig. 6: The two structure forming effects of osteocalcin leading to remodeling

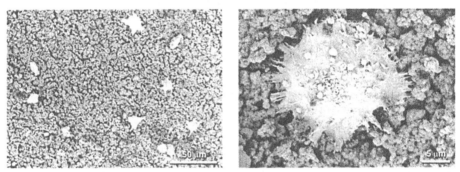

Fig. 7: SEM images of the adhesion of osteoblasts on collagen/HAP-cement samples in absence of osteocalcin

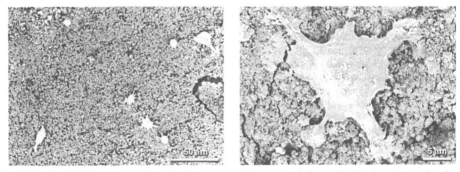

Fig. 8: SEM images of the adhesion of osteoblasts on collagen/HAP-cement samples in presence of osteocalcin

The observation, that osteocalcin promotes the adhesion of osteoblasts is a new interesting finding which could open additional possibilities for accelerated in-growth. Additionally, osteocalcin is expected to be chemotactic for (pre)osteoclasts, thus inducing recruitment and first adherence, which is the prerequisite for tight adherence via integrins, polarization and resorption. Investigation of the osteoclast response to osteocalcin-containing composites are currently performed.

Inhibition of cellular processes by hyaluronic acid functionalized surface regions

As mentioned above, there are also situations that in the nearest neighborhood of an implant or at particular surfaces cellular adhesion and proliferation should be avoided. Cartilage gives an exceptional example in nature that such material can exist. Similar concepts can be followed when we again use the possibility of functionalization of surface near regions with an appropriate biomolecular component. For implants with contact to hard and soft tissue the structure and biochemistry of cartilage can be used as an orientation for the right choice. Hyaluronic acid, one of the main molecular components of cartilage, is one candidate for such a design, Hyaluronic acid fulfills a structure building function in cartilage as it acts as a backbone for assembling proteglycans. However, it is also known that hyaluronic acid undergoes increasing hydrophobicity with increasing crosslinking via the formation of ester-groups[11]. Macroscopically, this behavior can be observed by measuring the effect of crosslinking on the contact angle of a water droplet at hyaluronic acid film surfaces (see Fig. 9). Such a change of the wetting behavior of a material should be reflected in the cell adhesion as well. Seeding osteoblasts on titanium or glass substrates partially coated with a film of highly crosslinked hyaluronic acid, complete cellular depletion of the coated areas has been observed (Fig. 9). A different behavior has been found when collagen/HAP-tapes were modified with hyaluronic acid. MTT-tests evaluating the cell vitality show no significant difference between both materials. The hydrophobic, crosslinked hyaluronic acid seems not to influence the total hydrophobicity of the material which is dominated by the polar mineralized collagen.

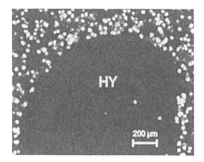
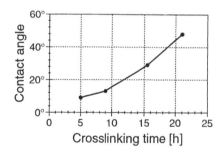

Fig. 9: Cells do not adhere on a highly crosslinked hyaluronic acid (HY) coating on a titanium substrate (left). Increase of the contact angle of a HY-coating with crosslinking time (right)

CONCLUSIONS
The developed design principle for collagen/HAP-composites combines two aspects of functional gradation which are relevant for the successful application of the bone replacement material. Firstly, the initial mechanical response of the composite can be adapted to the given loading situation in the particular part of the body by tailoring the kind and orientation of the biopolymer reinforcement of the brittle HAP matrix. Secondly, the porosity can be varied as well as the pore size in the surface near region of the implant. By this way, a first rough layout of a

hierarchical bone structure can be built. With the intention to achieve a real bone replacement, the second kind of functionalization – the directed remodeling of the artifical replacement by cellular processes, seems to be even more important. It has been shown that the chemisorption of osteocalcin in the surface near region of the collagen/HAP-composite is one promising option to switch on the remodeling cascade. The chemotactic effect on osteoclasts as well as on osteoblasts can be used for a controlled and accelerated remodeling and incorporation of the new formed material as living hard tissue. In case osteoclast adhesion and activity can be promoted on the artifical collagen/HAP-material, pronounced resorption of the composite is expected. Due to the natural controlled interaction between osteoclasts and osteoblasts, subsequent formation of new bone by osteoblasts is probable in the surface near region, leading to replacement of the composite by natural bone.

ACKNOWLEDGEMENTS
   The authors would like to thank the Deutsche Forschungsgemeinschaft supporting this work in the program „Functionally Graded Materials". Furthermore we thank our colleagues M. Witt, R. Funk, from the Institute of Anatomy, as well as U. Hempel, and K. Wenzel from the Institute of Physiological Chemistry, Dresden University for essential contributions.

REFERENCES
   [1] F.C. Driessens, J.A. Planell, M.G. Boltong, I. Khairoun and M.P. Ginebra, "Osteotransductive Bone Cements", *Proc. of the Institution of Mechanical Engineers,* **H212** 427–435 (1998).
   [2] K.S. TenHuisen, R.I. Martin, M. Klimkiewicz and P.W. Brown, "Formation and Properties of a Synthetic Bone Composite: Hydroxyapatite-Collagen", *Journal of Biomedical Materials Research,* **29** 803–810 (1995).
   [3] W.Pompe, S. Lampenscherf, S. Rößler, D. Scharnweber, K. Weis, H. Worch and J. Hofinger, "Functionally Graded Bioceramics", *Materials Science Forum,* **308-311** 325–330 (1999).
   [4] J.-H. Bradt, M. Mertig, A. Teresiak and W. Pompe, "Biomimetic Mineralization of Collagen by Combined Fibril Assembly and Calcium Phosphate Formation", *Chemistry of Materials,* **11** 2694–2701 (1999).
   [5] S. Lampenscherf, K. Weis and W. Pompe, "Pore and Structure Formation in Collagen-Hydroxyapatite Precursors: Experiment-Model", *Materials Science Forum,* **308-311** 362–367 (1999).
   [6] J.-P. Werner, C. Lathe, P. Greil and W. Frieß, "Pore-Graded Hydroxyapatite Materials for Implantation", *6th Conference of the European Ceramic Society* Vol. 2, Brighton (UK), 509–510 (1999).
   [7] P.A. Price, A.A. Otsuka, J.W. Poser, J.K.Kristaponis and N. Raman, "Characterization of a Gamma-Carboxyglutamic Acid Containg Protein from Bone", *Proc. of the National Academy of Science of the USA,* **73** 1447–1451 (1976).
   [8] R.W. Romberg, P.G. Werness, B.L. Riggs and K.G. Mann, "Inhibition of Hydroxypatite Crystal Growth by Bone - Specific an other Calcium-Binding Proteins", *Biochemistry,* **25** 1176–1180 (1986).
   [9] C.Chemu, S.Colucci, M. Grano, P. Zigrino, R. Barattolo, G. Zambonin, N. Baldini, P. Vergnaud, P.D. Delmas and A.Z. Zallone, "Osteocalcine Induces Chemotaxis, Secretion of Matrix Proteins, and Calcium-Mediated Intracellular Signaling in Human Osteoclast like Cells", *Journal of Cell Biology* **127** 1149–1158 (1994).
   [10] P.V. Hauschka, J.B.Lian, D.E.C. Cole and C.M.Gundberg, "Osteocalcin and Matrix Gla Protein: Vitamin K-Dependent Proteins in Bone", *Physiological Reviews,* **69**, 990–1047 (1989)
   [11] K. Tomihata and Y. Ikada, "Crosslinking of Hyaluronic Acid with Water-Soluble Carbodiimide", *Journal of Biomedical Materials Research,* **37** 243–251 (1997)

GRADIENT TISSUE REACTION INDUCED BY FUNCTIONALLY GRADED IMPLANT

F.Watari, A.Yokoyama, H.Matsuno, R.Miyao, M.Uo, Y.Tamura and T.Kawasaki
Hokkaido University Graduate School of Dental Medicine, Sapporo 060-8586, JAPAN

M.Omori and T.Hirai
Institute for Metal Research, Tohoku University, Sendai 980-8577, JAPAN

ABSTRACT

An animal implantation test was done to investigate the influence of FGM implant on the response of tissue. The Ti/Co FGM of cylindrical shape with the gradually changed Co concentration from pure Ti to pure Co in the longitudinal direction was implanted in soft tissue of rats. The thickness of the fibrous connective tissue formed around implant was increased gradiently with Co concentration. The FGM with the graded composition from pure Ti to Ti-20%hydroxyapatite (Ti/20HAP) implanted in bone marrow of rabbits showed that new bone formation in direct contact to the implant surface increased with HAP content. The Ti/100HAP FGM implanted in bone marrow of rats showed the similar results. These results demonstrated that the tissue reaction, either biocompatibility in soft tissue or osteogenesis in hard tissue, occurred over a gradient on the order of millimeters in response to the graded structure of FGM.

INTRODUCTION

The implants are used as artificial bone in reconstruction of bone and teeth. The materials need to satisfy the all-round properties[1,2] of biocompatibility, strength and corrosion resistance[3]. The functions necessary for each part of implant may differ from part to part. For example a dental implant is fixed through from the inside of jaw bone where bone affinity is important, to the outside, that is, in oral cavity where the sufficient strength is necessary for occlusion and mastication. FGM structure would be suitable for this kind of applications[4-7].

The function of implant may develop further in future to control various bioresponse for the use as a small artificial internal organ. For the control by materials it is necessary first to confirm whether the tissue reacts gradiently in response to the graded structure of FGM in a small scale of μm to mm range. For this purpose the Ti based FGM implants with the components slightly different in biocompatibility were prepared. These are the FGMs with the graded composition in the longitudinal direction from pure Ti to pure Co (Ti/100Co), from pure Ti to Ti-20%HAP (hydroxyapatite) (Ti/20HAP) and from pure Ti to pure HAP (Ti/100HAP).

Ti is one of the best biocompatible metals and used most widely as implant. Co is used for an orthodontic wires and denture materials as a main component of Co-Cr alloys. HAP, main component of bone and teeth, has bioactive properties for new bone formation.

EXPERIMENTAL PROCEDURE
1) Specimen

Mixed powders with the varying concentration of 99.98% Ti (SUMITOMO SITIX) and HAP(SUMITOMO OSAKA CEMENT) or 99% Co(NILACO) were packed into the thermo-contractive tube with 0.25mm thickness, changing gradually the concentration along the longitudinal direction. After the heat treatment of a tube at 60°C, Ti/Co and Ti/20HAP were compressed by CIP at 800-1000MPa. The implants of the miniature cylindrical shape 2ϕx7mm were made finally by sintering in vacuum of 10^{-3}Pa at 1100°C for Ti/Co and at 1300°C for pure Ti, Ti/20HAP FGM. For Ti/100HAP FGM mixed powders of Ti hydrate(SUMITOMO SITIX) and HAP were packed in graphite mold of 20ϕx(7-14)mm with the gradient composition ranged from pure Ti to 100%HAP in the height direction and sintered at 950°C under 40MPa by SPS(Spark Plasma Sintering/IZUMITECH). After sintering square rods of 1x1x10mm were cut out and mechanically polished up to #2000 on the surface for implants.
2) Animal implantation test

Ti/Co FGM was implanted into subcutaneous soft tissue in the dorsal thoracic region of rats for 2 weeks. Ti/20HAP FGM was implanted into bone marrow of tibia of rabbits for 4 weeks. Ti/100HAP FGM was implanted for 2-8 weeks in the bone marrow of femora of rats. Pure Ti was implanted for every case for comparison.

3) Observation

The FGM before implantation was observed by optical microscopy(OM: OLYMPUS VANOX-S, ZEISS AXIOSKOP 50), SEM (HITACHI S2380N), and elemental mapping was done by XSAM(X-ray Scanning Analytical Microscope/ HORIBA XGT2000V)[8] and EPMA(JEOL JXA8900).

After implantation the histological observation by OM was performed in reflection and transmission mode after the tissue around implant was fixed, dehydrated, embedded in resin or paraffin, sectioned and stained. The unsliced other part of tissue blocks embedded in resin was served for elemental mapping by XSAM and EPMA as unstained.

RESULTS

Fig.1 shows the histological observation of the fibrous connective tissue formed around the Ti/Co FGM implant in the soft tissue of rat after 2 week implantation. The implant which was removed from soft tissue had been situated in the upper side of the photograph. The left side is the part of pure Ti. The concentration of Co increases from left toward the right along the longitudinal direction and the right end is pure Co.. The thickness of the fibrous tissue layer was small in the left side and increased toward the right direction.

Fig.2 is the enlarged view of each region in the left(a) and right(b) of Fig.1. The thickness of fibrous tissue layer is about several fibroblasts in the region where tissue was contact to pure Ti(Fig.2a). In the Ti-Co alloyed

Fig.1 Soft tissue around Ti/Co FGM implant inserted in rat for 2 weeks.

region the tissue layer thickness became larger. In the region where tissue was contact to Co rich or pure Co implant surface(Fig.2b) the layer thickness is several times of that in the pure Ti region and necrosis occurred in the direct contact area to Co. The inflammatory granulation tissue with expanded capillaries and round cells of lymphocytes was observed.

The elemental mapping of S, P, Ca, Ti and Co was done for the same

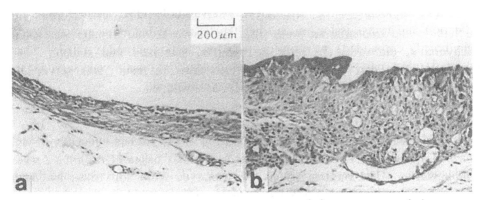

Fig.2 Enlargement of Fig.1 around Ti side(a) and Co side(b).

Fig.3 Co mapping by XSAM for the surrounding tissue of Ti/Co FGM implant of Fig.1.

specimen as Fig.1 by XSAM. S, Ca and P whose concentration was low enough, of the order 100ppm, in soft tissue, were detected by XSAM and showed the whole area of soft tissue. The Ti mapping showed no trace of Ti dissolution. The mapping of Co(Fig.3) revealed that the dissolution of Co into the surrounding soft tissue was prominent around the right part of implant. The changing concentration of dissolved Co along the implant

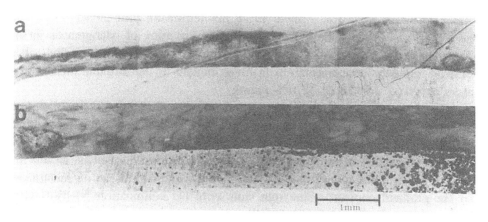

Fig.4　New bone formation around　Ti(a) and Ti/20HAP FGM(b) implants after 4 week insertion in cavitas medullaris of tibia of rabbit.

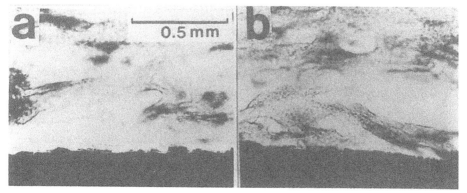

Fig.5　Enlargement around Ti(a) and Ti-20HAP(b) regions of Fig.4b.

toward the right is well coincided with the increase of thickness of tissue layer and the degree of inflammatory response in Figs.1 and 2.

Fig.4 shows the new bone formation around Ti(a) and Ti/20HAP FGM (b) implant in bone marrow of tibia of rabbits after 4 week insertion. Observation was done by optical microscopy in reflection mode. The increase of HAP particles inside the Ti matrix from left to right could be recognized by the distribution of black dots inside the implant(b). With the increase of HAP content the thickness of newly formed bone observed with black contrast was increased along the surface of implant toward the right, while little new bone was formed yet at this period for pure Ti(a).

Fig.5 shows the enlarged histological observation of osteogenesis in the Ti part(a) and around Ti-20HAP part(b) of Ti/20HAP FGM(Fig.4b). It is evident that new bone formation in direct contact to the implant surface was more progressed in the HAP rich region.

Fig.6 is the combined photograph of the Ti/100HAP FGM before implantation taken in reflection mode by OM (lower half) and histological image of the corresponding part after 4 week implantation in femora of rat taken in transmission mode (upper). The left side of implant is Ti and the right is HAP in the lower image. The graded composition can be recognized by the gradient contrast. The right side with the composition 90-100%HAP looks white. In the upper image the contrast of implant is black since the material of implant with about 100 μm thickness is not transparent, while the tissue is observable due to its transparency. The formation of new bone surrounding the implant looks more advanced in direct contact along the surface of HAP-rich part on the right.

Fig.6   Ti/HAP FGM implant observed in reflection mode by OM (lower half) and histological image after insertion for 4 weeks in femora of rat observed in transmission mode (upper).

DISCUSSION

The difference in bioreaction was relatively easy to recognize for the combination of biocompatible and biomalignant materials as in the case of Ti/Co FGM implant in soft tissue. It was difficult, however, to show the difference for the combination of biocompatible materials. The components of Ti/20HAP FGM, Ti and HAP, were both biocompatible. In such a case the influence of other factors are often dominant. The new bone formation is easily affected by the shape, porosity, surface roughness of implants, the inserted position, distance from cortical bone and perforated region, and individual difference of animals other than implant materials. In the relatively ideal conditions of Figs.4 and 5 the implant was situated at a constant distance from cortical bone in the bone marrow of tibia of rabbits. It was shown that the coexistence of the HAP component seemed to induce the effect to accelerate the formation of new bone from the earlier stage and this effect worked gradiently to the graded composition of Ti/20HAP FGM implant. The case for Ti/100HAP FGM of Fig.6 confirmed this tendency. The both gradient reactions to FGM, formation of fibrous connective tissue in soft tissue and osteogenesis in hard tissue, imply thus the possibility to control the tissue response over millimeter scales in each part through the designed gradient composition and structure of FGM implant. This would contribute to the application of biomaterials for use as artificial internal organs.

REFERENCES

[1] T.Imai, F.Watari, S.Yamagata, M.Kobayasi, K.Nagayama and S.Nakamura, "Mechanical properties and estheticity of FRP orthodontic wire fabricated by hot drawing", *Biomaterials*, **19**[23], 2195-2200 (1998).

[2] F.Watari, M.Takahashi and K.Yada, "X-ray Diffraction Study of Carburization Transformation from Tantalum to Tantalum Carbide"; pp.317-320 in *Proc.Int.Conf.Solid-Solid Phase Transformations'99 (JIMIC-3)*, Edited by M.KOIWA, K.OTSUKA and T.MIYAZAKI, The Japan Institute of Metals, Sendai, 1999.

[3] F.Watari, "In situ etching observation of human teeth in acid agent by atomic force microscopy", *Journal of Electron Microscopy*, **48**[5], 537-544 (1999).

[4] F.Watari, A.Yokoyama, F.Saso, M.Uo and T.Kawasaki, "Functionally Gradient Dental Implant Composed of Titanium and Hydroxyapatite";

pp.703-708 in *Proc.3rd Int.Symp.on Structural & Functional Gradient Materials*, Edited by B.Ilschner and N.Cherradi, Presse Polytechniques et Universitaires Romandes, Lausanne, 1995.

[5] F.Watari, A.Yokoyama, F.Saso, M.Uo and T.Kawasaki, "Fabrication and Properties of Functionally Graded Dental Implant", *Composites Part B*, **28B**, 5-11 (1997).

[6] F.Watari, A.Yokoyama, F.Saso, M.Uo and T.Kawasaki, "Elemental mapping of functionally graded dental implant in biocompatibility test"; pp.749-754 in *Functionally Graded Materials 1996*, Edited by I.Shiota and Y.Miyamoto, Elsevier, Amsterdam, 1997.

[7] F.Watari, A.Yokoyama, F.Saso, M.Uo, H.Matsuno and T.Kawasaki, "Biocompatibility of Titanium/hydroxyapatite and Titanium/Cobalt Functionally Graded Implants"; pp.356-361 in *Functionally Graded Materials 1998*, Edited by W.A.Kayser, Trans Tech Publications, Zurich, 1999.

[8] M.Uo, F.Watari, A.Yokoyama, H.Matsuno and T.Kawasaki, "Dissolution of nickel and tissue response observed by X-ray analytical microscopy", *Biomaterials*, **20**[8], 747-755 (1999).

CHARACTERIZATION AND OPTIMIZED DESIGN OF HA-Ti/Ti/HA-Ti SYMMETRICAL
FUNCTIONALLY GRADED BIOMATERIAL

Chenglin Chu
Department of Mechanical Engineering
Southeast University
Nanjing, 210018, China
Email: chenglin-chu@263.net

Zhongda Yin and Jingchuan Zhu
BOX 433, School of Materials Science and
Engineering, Harbin Institute of Technology
Harbin, 150001, China

Shidong Wang and Pinghua Lin
Department of Mechanical Engineering
Southeast University
Nanjing, 210018, China

ABSTRACT
    The optimized distribution function of components of the hydroxyapatite (HA)-Ti symmetrical
functionally graded biomaterial (FGM) has been obtained based on the classical lamination theory
and thermo-elastic mechanics firstly. The FGM was fabricated by hot pressing successfully whose
actual distributions of compositions and thermal residual stress have been analyzed by electron
probe microanalyser (EPMA) and X-ray diffraction method, which are consistent with those
designed theoretically. Its integral apparent bending strength can reach to 159MPa, which is higher
than that of human bone. HA-Ti symmetrical FGM behaved integral-fracture without falling apart
or delaminating under all loading modes. As a result, the pure titanium layer could provide an
effective support of mechanical properties for the whole FGM.

INTRODUCTION
    Hydroxyapatite/Ti system functionally graded biomaterials have been developed successfully
by powder metallurgical method, which make full use of the excellent bioactivity of HA and the
high strength and toughness of Ti metal [1-5]. At the same time, the mechanical properties,
fracture behaviors and the relaxing characteristics of thermal residual stress of this FGM have also
been reported correspondingly [2,6,7].
    In this paper, the thermal stress relaxation design and structure optimization of
HA-Ti/Ti/HA-Ti symmetrical FGM were made to acquire the optimal relaxation of the residual
thermal stress during the fabrication process. Then HA-Ti/Ti/HA-Ti symmetrical FGM with the
optimum graded composition was fabricated by hot pressing. The magnitude and distribution of
the residual thermal stress in the FGM were tested by X-ray stress analyzer. Moreover, the failure
mechanism and the integral apparent bending strength of the FGM were studied in detail.

OPTIMIZED DESIGN MODEL OF FUNCTIONALLY GRADED BIOMATERIAL
    The ambient temperature is also a important factor that can lead to the deformation of FGM
besides the mechanical load. FGM often consists of some graded layers with different
thermo-elastic behaviors. The change of ambient temperature could lead to the existence of
additional residual thermal macroscopic stress and strain in each graded layer besides the

thermo-elastic deformation of the whole FGM due to the heterogeneous characteristic along the thickness direction of FGM[8-10]. The sintering temperatures are far higher than the utilizing ones (or the body temperatures) of HA-Ti/Ti/Ti-HA symmetrical FGM. As a result, the yielding of residual thermal macroscopic stress in the FGM could not be avoided, which will play a important role on the preparation and the integral strength of FGM[6,7]. So the thermal stress relaxation design and structure optimization of HA-Ti/Ti/HA-Ti symmetrical FGM should be made to acquire the optimal relaxation of the residual thermal stress during the fabrication process.

Fig. 1 shows the analysis model for FGM plate with a thickness of h, which consists of n layers of infinite macroscopic boards. To analyze the elastic properties of FGM plate simply and conveniently, the following hypotheses should be made: (1) the hypothesis of consistent deformation between layers: each layer bonds tightly and is deformed consistently with its contiguous layers. (2) the hypothesis of invariable straight normal: the normal on the median plane of FGM plate keeps vertical before and after the deformation. (3) the hypothesis of $\sigma_z = 0$: the positive stress in the direction of thickness is too little to be taken into account. (4) the hypothesis of the plane stress state: each layer in the FGM may be considered in the plane stress state approximatively. The HA/Ti system materials with very homogeneous chemical compositions and microstructure stand in linear elastic stage and presents intergranular fracture without macroscopic plastic deformation[1,2]. At the same time, the dimension of thickness in the FGM is far less than those of the plate surface.

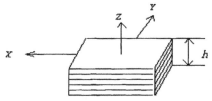

Fig. 1 The analysis model for FGM plate

Based on the above analysis model, the elastic properties of FGM were analyzed by thermo-elastic mechanics. The following analytic expression of the residual thermal macroscopic stress during the fabrication process at an arbitrary point in FGM was gained. The detailed theoretical discussion is available elsewhere [1,6,7].

$$\{\sigma^R\} = [Q](\{\varepsilon^{0T}\} + z\{\kappa^T\} - \{\alpha\}\Delta T) \tag{1}$$

where $\{\sigma^R\}$ is the matrix of residual thermal stress, $[Q]$ the matrix of the stiffness of the positive axis, $\{\varepsilon^{0T}\}$ the matrix of the thermal strain of the median plane, $\{\kappa^T\}$ the matrix of thermal deformation curvature, $\{\alpha\}$ the matrix of thermal expansion coefficient. ΔT the difference between the fabricating temperature and the room one.

EXPERIMENTAL

The raw materials used were titanium powders and HA powders. The chemical composition of the titanium was(wt.%): Ti 99.3, Fe 0.039, O 0.35, N 0.035, C 0.025, CL 0.034, H 0.024 and Si 0.0018. Hydroxyapatite was prepared by the reaction between $Ca(NO_3)_2$ and $(NH_4)_2HPO_4$. Its Ca/P ratio was 1.67±2.0%. Sizing by means of Laser Particle Sizer(OMEC LS-POP(III)) showed the Ti particles had a average size of 45.2μm (93.64wt% of Ti particles were in the range 37.0-60.0μm), whereas the average size of HA particles is 1.75μm (82.12wt% of HA particles were found to be between 0.35-3.70μm). There are significant agglomerations of HA powders shown by scanning electron microscopy (SEM). The starting powders with different HA/Ti mixing ratios were first blended by ball milling for 12 hours. Then the mixed powders were stacked layer by layer in a steel die according to the optimized compositional profile and compacted at 200MPa. Thus the green compacts were hot-pressed at 1100 °C under a pressure of 20MPa in nitrogen atmosphere for 30-90min with a heating rate of 10 °C/min and a cooling rate of 6 °C/min.

Cross sections of FGM sample were polished and covered with a thin film of carbon by vacuum-deposition. Then the distribution and change of each element along the graded direction in FGM was analyzed by electron probe microanalyser (EPMA, JEOL JCXA-733 type). The residual thermal macroscopic stress during the fabrication process in FGM was tested layer by layer using 0°-45° method by X-ray stress analyzer(XSA, AST-X2001 type). The testing conditions include the X-ray power of $30KV/6.6mA$, the 2θ range of 125°-162°, the angle resolving power of 0.029°/point. The tested samples were first mechanical ground and polished, then corroded by a solution of $HF+HNO_3$, which could reduce the influence of grinding stress. Three-point bending tests were performed to determine integral bending strength of FGM. The fracture surface of the bending samples was covered with a thin film of gold by vacuum-deposition and then examined by scanning electron microscope (SEM).

RESULTS AND DISCUSSION
1. Optimized Design of HA-Ti/Ti/HA-Ti Symmetrical FGM
    The graded compositional distribution function of HA-Ti/Ti/HA-Ti symmetrical FGM is shown as the following,

$$f_{HA}(\xi) = \begin{cases} 0 & |\xi| < L_0 \\ \left(\dfrac{|\xi| - L_0}{1 - L_0 - L_1}\right)^n \times 40\% & L_0 < |\xi| < 1 - L_1 \\ 40\% & |\xi| > 1 - L_1 \end{cases} \qquad (2)$$

where $f_{HA}(\xi)$ is the volume fraction of HA ceramic which changes with the distance ξ in the FGM, n the compositional distribution exponent, L_0 the half thickness of the pure titanium layer in the middle of FGM, L_1 the thickness of Ti-40vol%HA surface layer. The introducing dimensionless variable ξ is defined as,

$$\xi = z/d \qquad (3)$$

where z is the coordinate along the thickness direction of FGM as shown in Fig. 1. The origin $z_0 = 0$ corresponds to the middle position of the pure titanium graded layer in FGM. d is the half thickness of FGM along the graded direction. If the coordinate system z is translated into the coordinate system ξ, the origin $\xi_0 = 0$ corresponds to the middle position of the pure titanium graded layer in FGM. the situation $\xi = 1$ corresponds to the outer side of the surface layer Ti-40vol%HA. L_0 and L_1 are also dimensionless variables. Figure 2 shows the compositional continuous distributions of HA-Ti/Ti/Ti-HA symmetrical FGM dependent on n. The optimal design of this FGM is to determine the graded compositional distribution function when the residual thermal macroscopic stress during the fabrication process is minimum. The magnitude of L_0 and L_1 in the compositional distribution function should be preestablished according to the actual requirements of hard tissue replacement towards the FGM. So the optimized design of

HA-Ti/Ti/Ti-HA symmetrical FGM is to determine the compositional distribution exponent n as done previously for other types of FGM[1].

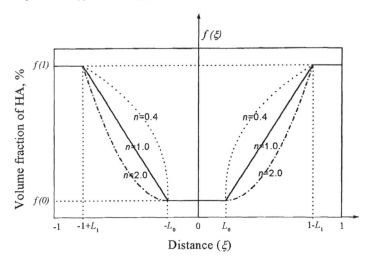

Fig. 2 Compositional continuous distributions of HA-Ti/Ti/Ti-HA symmetrical FGM dependent on n

The difference between the sintering temperature of FGM(1100℃) and the room one(25℃) is -1075℃. The compositional distribution exponent n is optimized between 0.1 and 10 with a optimized step of 0.1. The necessary thermo-elastic properties of HA/Ti composite materials were listed in Table 1[1,6]. Among them, Poisson's ratio was estimated by the mixing law. Figure 3 shows the changes of the maximum tensile stress in HA-Ti/Ti/Ti-HA symmetrical FGM with the compositional distribution exponent n. It is obvious that the maximum tensile stress increases with the rise of n for different L_0 and L_1. The larger the thickness of pure titanium layer L_0 is, the larger the maximum tensile stress is. Moreover, all the maximum tensile stresses arise in the surface layer. As a result, it is unnecessary to optimize the situation where the maximum tensile stress appears. To sum up, the compositional distribution exponent $n = 0.1$ is the optimal choice for HA-Ti/Ti/Ti-HA symmetrical FGM.

Table 1 Elastic modulus, thermal expansion coefficient(20-900℃) and Poisson's ratio of HA/Ti composite materials[1,6]

Materials	Elastic modulus (GPa)	Thermal expansion coefficient $(10^{-6}℃^{-1})$	Poisson's ratio
Pure Ti	107.95	10.9	0.34
Ti-20%HA	102.64	11.72	0.328
Ti-40%HA	87.71	12.94	0.316

2. Fabrication and Residual Thermal Stress Measurement of HA-Ti/Ti/HA-Ti Symmetrical FGM

The FGM with the optimum graded composition was fabricated by hot pressing under the optimal sintering conditions. Its thickness along the graded direction, L_0, L_1 and n is 5mm,

0.20(1/5), 0.40(2/5), and 0.1 respectively. The sintered sample shows no bending deformation and microcracks on the surfaces, which suggests that the optimized design is successful.

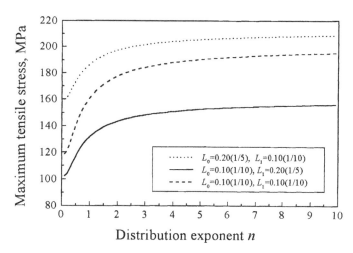

Fig. 3 Changes of the maximum tensile stress in HA-Ti/Ti/Ti-HA symmetrical FGM
with the compositional distribution exponent n

To realize the actual distributions of Ca, P and Ti elements in the FGM with the optimum graded composition, the change of each element along the graded direction was analyzed by EPMA. The elemental distribution actual profile of HA-Ti/Ti/Ti-HA symmetrical FGM with the optimum graded composition is shown in Figure 4. It could be found that each element in HA-Ti/Ti/Ti-HA symmetrical FGM exhibits a symmetrical graded distribution. The content of Ti is high in the middle and low at both sides, while the contents of Ca and P change inversely. Their varying tends are basically consistent with those of the designed compositional distribution.

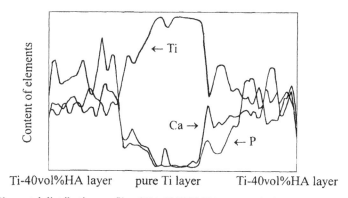

Fig. 4 Elemental distribution profile of HA-Ti/Ti/Ti-HA symmetrical FGM with the optimum
graded composition by EPMA line analysis

The residual thermal stresses at different situations along the graded direction in the HA-Ti/Ti/Ti-HA symmetrical FGM with the optimum graded composition were tested by X-ray stress analyzer(XSA) as shown in Figure 5. The residual thermal stress in HA-Ti/Ti/Ti-HA symmetrical FGM also exhibits a symmetrical graded distribution as the compositions. The surface layer of Ti-40vol%HA is acted by a residual thermal tensile stress about 80-100MPa, while the large residual thermal compressive stress acts on the middle pure titanium layer. The transition layer is acted by the little residual thermal compressive stress. The above tested values of residual thermal stress are basically consistent with the theoretically calculated ones.

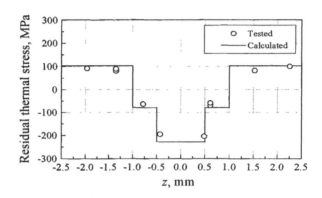

Fig. 5 Residual thermal stress in HA-Ti/Ti/Ti-HA symmetrical FGM
with the optimum graded composition

3. Integral Apparent Strength and Failure Mechanism of HA-Ti/Ti/HA-Ti Symmetrical FGM

The compositional gradient could lead to a new characteristic for the propagation of microcracks in HA-Ti/Ti/Ti-HA symmetrical FGM under the loads besides its influence on the residual thermal stress. Although the influence of the graded characteristic on the propagation of microcracks has not been made clear, some literatures have found that the cracks perpendicular to the graded layers in FGM are prone to be deflected, which could cause the flaking of the top ceramic layer[10]. The mechanical properties of hydroxyapatite ceramic or HA/Ti composites could not meet the requirement for use as hard tissue replacement implants[1,2]. Thus whether the pure titanium layer in HA-Ti/Ti/Ti-HA symmetrical FGM could provide an effective support of mechanical properties is very important. If the hydroxyapatite ceramic or HA/Ti composite layers in FGM with the poor mechanical properties are flaked during the loading process, the destroy of FGM will be shifted to an earlier time. As a result, the pure titanium graded layer with the highest toughness could not support the whole FGM from the point of view of mechanical properties. So the integral-fracture mode under different loading modes is anticipant. At present, it is short of the experimental and theoretical analyses for the failure mode of FGM under load [10]. The HA-Ti/Ti/Ti-HA symmetrical FGM is mainly acted by the transverse loads in the human bodies, so the integral apparent bending strength and the failure mode of this FGM were studied by three-point bending method. The span of the tested sample is larger than five times of its thickness in order to eliminate size effects [1].

The integral apparent bending strength and elastic modulus of HA-Ti/Ti/Ti-HA symmetrical FGM with the optimum graded composition were listed in Table 2. It is found that HA-Ti/Ti/Ti-HA symmetrical FGM always has a high integral apparent bending strength under any a loading mode. The integral apparent bending strength of FGM under Mode-1 is similar to that of FGM under Mode-2, which are 164.5MPa and 153.3MPa respectively. The average value(158.9MPa) is higher than that of human bone(121-149MPa)[11,12]. Fig. 6 shows macroscopic fracture surface characteristics of HA-Ti/Ti/Ti-HA symmetrical FGM with the optimum graded composition under different loading modes. It should be noted that the FGM performed without falling apart or delaminating under both loading modes. Under the loading Mode-1, when the propagating cracks arrived at the middle of the graded layer of Ti-20vol%HA, they were transversely deflected along the graded layer of Ti-20vol%HA due to the great resistance of the pure titanium graded layer, as shown in Fig. 6(a). The cracks still entered into the pure titanium layer and resulted in the integral-fracture of FGM finally. In this way, the pure Ti layer could provide an effective support of mechanical properties for the whole FGM.

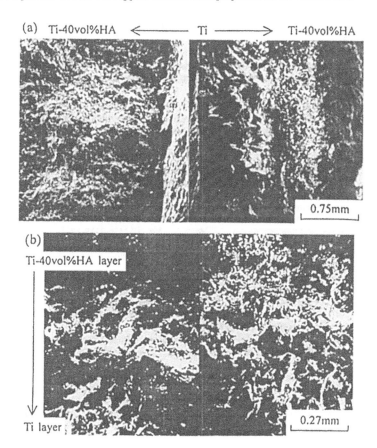

Fig. 6 SEM fractographs shows macroscopic fracture surface characteristics of HA-Ti/Ti/Ti-HA symmetrical FGM with the optimum graded composition under different loading modes: (a)Mode-1; (b)Mode-2

Table 2 Integral apparent bending strength and elastic modulus of HA-Ti/Ti/Ti-HA symmetrical FGM with the optimum graded composition

Code of loading modes	Loading modes	Integral apparent bending strength (MPa)	Apparent elastic modulus (GPa)
1	Surface of Ti-40vol%HA layer is tensed	164.5	91.8
2	The side vertical to surface layer is tensed	153.3	100.3

CONCLUSIONS

(1)The optimal compositional distribution exponent n for HA-Ti/Ti/Ti-HA symmetrical FGM is acquired by the thermal stress relaxation design and structure optimization, which is 0.1. The tested values of residual thermal stress in the FGM are basically consistent with the theoretically calculated ones.

(2)HA-Ti/Ti/Ti-HA symmetrical FGM with the optimum graded composition has been fabricated by hot pressing successfully, which has no any microcracks on the surfaces. Its integral apparent bending strength can reach to 158.9MPa, which is higher than the bending strength of human bone.

(3)HA-Ti/Ti/Ti-HA symmetrical FGM behaved integral-fracture without falling apart or delaminating under all loading modes. As a result, the pure titanium layer could provide an effective support of mechanical properties for the whole FGM.

REFERENCES

[1]C.L. Chu, *Fabrication and Microstructure-Properties of Hydroxyapatite/Ti Functionally Graded Biomaterial*, PhD Dissertation, Harbin Institute of Technology, 2000.

[2]C.L. Chu, J.C. Zhu, Z.D. Yin and S.D. Wang, "Hydroxyapatite-Ti Functionally Graded Biomaterial Fabricated by Powder Metallurgy," *Materials Science & Engineering*, **A271** 95-100(1999).

[3]A. Bishop, C.Y. Lin, M. Navaratnam, R.D. Rawlings, H.B. Mcshane, "A functionally gradient material produced by a powder metallurgical process," *J. Mater. Sci. lett.*, **12** 1516-1518(1993)

[4]C.L. Chu, Z.D. Yin, J.C Zhu, M.W. Li and S.D. Wang, "The sintering of hydroxyapatite-Ti system materials by hot pressing and the structural stability of hydroxyapatite," *Transactions of Nonferrous Metals Society of China*, **10**[4] Accepted for publication (2000)

[5] F. Watari, A Yokoyama, F. Saso, M. Uo, H. Matsuno and T. Kawasaki, "Imaging of gradient structure of titanium/apatite functionally graded dental implant," *J. Japan Inst. Metals*, **62**[11] 1095-1101(1998)

[6]C.L. Chu, J.C. Zhu, Z.D. Yin and S.D. Wang, "Preparation and Thermal Stress Relaxation Characteristics of HA-Ti/Ti/HA-Ti Axial Symmetrical Functionally Graded Biomaterial," *Transactions of Nonferrous Metals Society of China*, **9** [S1] 57-62(1999)

[7]C.L. Chu, Z.D. Yin, J.C Zhu and S.D. Wang, "Preparation and Thermal Stress Relaxation Characteristics of Hydroxyapatite/Ti Functionally Graded Biomaterial," *Journal of Inorganic Materials*, **14** [5] 775-782(1999)

[8]S. Suresh and A. Mortensen, "Functionally graded metals and metal-ceramic composites: Part 2 Thermo-mechanical behavior," *International Materials Reviews*, **42** [3] 85-116(1997)

[9]X.F. Tang, L.M. Zhang and R.Z. Yuan, "Thermal Stress Relaxation Design and Fabrication of PSZ-Mo Functionally Gradient Materials," *Journal of The Chinese Ceramic Society*, **22** [1] 44-49(1994)

[10]A. Neubrand and J. Rodel, "Gradient Materials: An Overview of a Novel Concept," *Z. Metallkd.*, **88** 358-371(1997)

[11]L.L. Hench, "Bioceramics: From Concept to Clinic," *J. Am. Ceram. Soc.*, **74**[7] 1487-1510(1991)

[12]W. Suchanek and M. Yoshimura, "Processing and properties of hydroxyapatite-based biomaterials for use as hard tissue replacement implants," *J. Mater. Res.*, **13** [1] 94-117(1998)

TENSILE BEHAVIOR OF FUNCTIONALLY GRADED BRAIDED CARBON FIBER/EPOXY
COMPOSITE MATERIAL

Qiongan Wang, Zhengming Huang and S. Ramakrishna
Dept. of Mechanical and Production Engineering
The National University of Singapore,
Engineering Drive 1 Singapore 117576

ABSTRACT
    Primary objective of this research work is experimentally to investigate the tensile behavior of
functionally graded braided composite rods. Composite rods were made using tubular braided
carbon fiber fabrics and epoxy resin. Functional gradient was achieved by changing the braiding
angle continuously along the length direction of the rod. For comparison and analysis purpose,
both functionally-graded-braiding-angle composite rods (FG rods) and constant-braiding-angle
composite rods (CBA rods, non-FGM) were fabricated. Mechanical properties of these composite
rods were determined through tensile tests. From the test results of CBA rods, the effect of
braiding angle on the composite properties was established. The data were also used to analyze the
mechanical properties of FG rods. It has been demonstrated that the elastic tensile modulus of a
FG rod can be estimated from the elastic tensile moduli of a series of CBA rods. It has also been
found that the tensile strength of a FG rod is always higher than the tensile strength of the CBA
rod whose braiding angle is equal to the largest braiding angle of this FG rod, although they were
originally supposed to be equal based on the results of CBA rods.

1. INTRODUCTION
    Functionally Graded Material (FGM) is one of the most innovative concepts introduced into
composites world in the last two decades [1, 2]. It is a class of material that possesses a continuous
variation in composition and/or microstructure that gives rise to a smooth and spatially controlled
change in properties. Previously this group of material was mainly fabricated from a combination
of metal and ceramic materials, with either a constructive process or a transport based process [2].
Only recently the concept has been applied to the area of fiber reinforced polymer composites [3].
In the study of Jang and Lee [4, 5], and Jang and Han [6], functional gradient of the glass fiber
(GF)/carbon fiber (CF) mixed mat was realized by changing the feeding ratio of chopped GF and
CF. Lee et al. [7] employed a centrifugal separation method to obtain a compositional gradient of
short carbon fibers in epoxy resin. Reinforcement in the FGMs investigated in these studies was
short fibers. There is little documented research that was carried out on continuous fiber reinforced
FGMs.
    However, it has been recognized that continuous fiber reinforced polymer composites play a
very important part in the modern composites world [8, 9]. The excellent mechanical properties
that they provide are generally attractive. As a result, it is wise and promising to take them as one
of the alternative solutions for FGM. As a distinctive branch of continuous fiber reinforced
polymer composites, braided composites especially deserve our attention. It is known that braiding
angle has a significant and decisive effect on the mechanical properties of braided composites [10,
11]. Therefore functional gradient may be achieved in braided composites by changing the
braiding angle. Recently, this idea has been proposed by Ramakrishna et al. [12, 13] to prepare

FGM to substitute non-FGM for dental post application. It was pointed out that the FGM dental post could reduce stress concentration at the root region of restored tooth and thus improve its survivability [14, 15]. However, a necessary step is to understand thoroughly its mechanical properties in order to put this type of FGM into final use.

In this project, continuous carbon fibers were braided into tubular preforms with continuous braiding angle variation along the longitudinal direction. They were then made into composite rods (FG rods) with epoxy resin. Tensile behavior of these composites was mainly investigated. For the purpose of comparison and analysis, properties of constant-braiding-angle composite rods (CBA rods, non-FGM) were also studied. The effect of braiding angle on the composite properties based on the data of CBA rods and the property relationship between FG rods and CBA rods were presented in this paper.

2. EXPERIMENTAL

2.1 Materials

PAN type T300B carbon fibers from Toray Company (Japan) were used in this study. Filament number in a single yarn is 3K (i.e., 3000 filaments). Epoxy resin R50 with a hardener of H64 from Chemicrete (Singapore) Private Limited was used as matrix material.

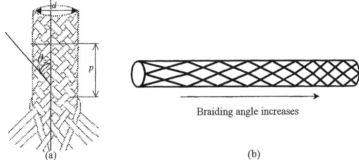

Braiding angle increases

(a)   (b)

Fig. 1 Schematic view of (a) diamond-braided tubular preform and its geometric parameters, and (b) functionally graded preform with changing braiding angle.

2.2 Preparation of Specimens

10 carbon yarns were diamond-braided into tubular preforms using Kokubun braiding machine (model 102-C13). The preform was then impregnated with epoxy resin with the aid of vacuum, followed by pultrusion operation through a straight polyethylene tube with a diameter of 2.0 mm. Composite rod was obtained by removing the tube after 24 hours. By changing the braiding angle θ (Fig. 1) during the preform fabrication stage, FG rods were prepared. The photographs of the preform and the composite rod were shown in Fig. 2.

2.3 Preparation of End Tabs

To ensure that specimens are gripped effectively in the tensile test, end tabs are needed. Two aluminum alloy tubes were used to form end tabs on a specimen. The tubes were aligned on the appropriate position of the specimen, as illustrated in Fig. 3. They were fixed uprightly and epoxy resin was poured into the upper tube. Resin-impregnated carbon fiber rods with approximate length to the tube were then inserted into the tube for the purpose of reinforcement. After epoxy is cured, the other side was formed by the same method.

2.4 Measurement of Braiding Angles

It is difficult to measure the braiding angle directly because the specimen is round. In this study, the braiding angle is calculated from the pitch p and diameter d (Fig. 1 (a)) of the preform:

$$\theta = Tan^{-1}(d / p) \tag{1}$$

One carbon yarn was marked in a preform so that yarn trace could be identified and pitch length could be measured. Diameters of all of the preforms were approximated to the diameter of the polyethylene tube used in the pultrusion process.

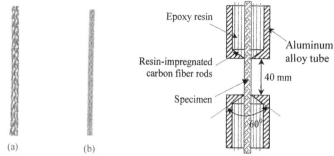

(a)　　　　　(b)

Fig. 2 Photographs of (a) tubular braided preform, and (b) braided composite rod.

Fig. 3 Illustration of end tab preparation

2.5 Measurement of Fiber Volume Fractions

Fiber volume fraction was measured by matrix digestion technique according to ASTM standard D3171-67 (Reapproved 1990). Volume percent fiber in the composite was calculated as follows:

$$\upsilon_f = \frac{W_f / \rho_f}{\left(W - W_f\right)/ \rho_m + W_f / \rho_f} \tag{2}$$

where, W = weight of composite specimen, W_f = weight of fiber in the composite, ρ_f = fiber density, and ρ_m = matrix density. Prior to this measurement, the quality of specimens fabricated through above-illustrated method was examined under microscope. There are no observable voids or porosity in the composites. Fig. 4 shows a cross-section micrograph of the composites.

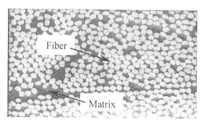

Fig. 4: Cross-section view of specimen under microscope

2.6 Measurement of Mechanical Properties

Tensile tests were carried out on INSTRON 8516 machine according to ASTM standard D3039/D3039M-95a. However, necessary modification was made with our experience because the specimen is not standard. Gauge length of the specimen was 40 mm and crosshead speed was set to 0.5 mm/min. The specimens have an average diameter of 2.04 mm.

For CBA specimens, seven groups were prepared. Their braiding angles were 5.2^0, 8.4^0, 9.7^0, 11.3^0, 12.6^0, 16.6^0, 17.2^0, respectively. Three or more specimens were tested for each group. For FG specimens, two groups were prepared. In the first group, each specimen has braiding angle changed only once, from 5.2^0 to 17.2^0. In the second group, each specimen has braiding angle changed twice, from 5.2^0 to 11.3^0 finally to 17.2^0. These two groups were divided into 3 and 4 subgroups, respectively, as demonstrated in Fig. 5 and Fig. 6. Three or more specimens were tested for each sub-group.

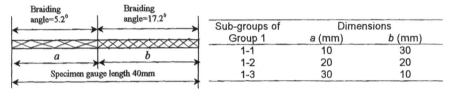

Sub-groups of	Dimensions	
Group 1	a (mm)	b (mm)
1-1	10	30
1-2	20	20
1-3	30	10

Fig. 5 Schematic illustration of Group1 FG specimens and their dimensions in the experiment.

Sub-groups of	Dimensions		
Group 2	a (mm)	b (mm)	c (mm)
2-1	10	10	20
2-2	13.3	13.3	13.3
2-3	10	20	10
2-4	20	10	10

Fig. 6 Schematic illustration of Group2 FG specimens and their dimensions in the experiment.

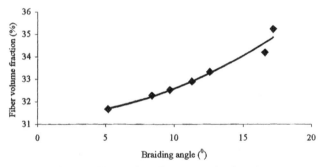

Fig. 7 Effect of braiding angle on fiber volume fraction of CBA rods.

3. RESULTS AND DISCUSSION

3.1 Effect of Braiding Angle on Fiber Volume Fraction of CBA Rods

Fiber volume fraction against braiding angle was plotted in Fig. 7 for CBA rods. Composite materials fabricated like this have a narrow range of fiber volume fraction. Fiber volume fraction varied from 31.66% to 35.22% when braiding angle varied from 5.2^0 to 17.2^0.

3.2 Tensile Behavior of Constant-braiding-angle Composite Rods (CBA rods)

For CBA Rods, the effect of braiding angle on tensile strength and elastic tensile modulus was determined, and illustrated in Fig. 8 and Fig. 9, respectively. Both the strength and modulus increase with the decrease of the braiding angle. Moreover, the rates of increase decrease as the braiding angle decreases. It is believed that with the decrease of the braiding angle, the alignment

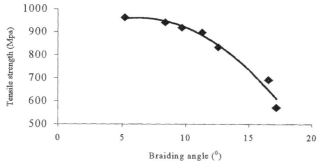

Fig. 8 Effect of braiding angle on tensile strength of CBA rods

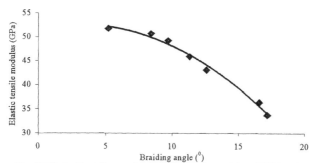

Fig. 9 Effect of braiding angle on elastic tensile modulus of CBA rods.

of fiber increases and thus tensile strength and elastic tensile modulus increase. The influence of fiber alignment on the strength and modulus is not linear. It is less significant when the braiding angle is smaller. Hence, the rate of increase decreases as the braiding angle decreases.

3.3 Tensile Behavior of Functionally-graded-braiding-angle Composite Rods (FG Rods)

Tensile strengths and elastic tensile moduli of Group 1 FG rods are plotted against the length fractions of the contained braiding angles in Fig. 10. Data of CBA rods are introduced into the chart by assuming that either of the angles occupies 100% length fraction.

The study of CBA rods has shown that the larger the braiding angle, the lower is the strength. Based on this, it was originally supposed that the strengths of Group1 FG rods should be equal to the strength of specimen group CBA 17.2^0 because the largest braiding angle of all these rods is 17.2^0. However, it can be seen from the chart that they are higher than that. Moreover, the strength increases as the length fraction of 5.2^0 braiding angle increases, but decreases as the length fraction of 17.2^0 braiding angle increases. These results can be properly characterized by a term - dimension-dependent strength enhancement. It is believed that the strength enhancement is caused by tension-induced microstructure readjustment within a FG rod. When tension is imposed, larger braiding angles of a FG rod will decrease because of the existence of smaller braiding angles. As a result, the strength of the FG rod will increase. Furthermore, the less are the length fractions of larger braiding angles, or the more are the length fractions of smaller braiding angles, the more is the decrease of larger braiding angles. Hence, the strength enhancement will be greater. From the test results of Group 2 FG rods (Table I), this kind of strength enhancement is also observed: all the strengths of Group 2 FG rods are higher than the strength of CBA rod 17.2⁰. However, the regularity of the strength enhancement is more complex for this group because there are more braiding angles.

However, it can still be determined that the tensile strength of a FG rod is dependent on its largest braiding angle since fracture did take place in the section of largest braiding angle in the tests for both Group 1 and Group 2 FG rods.

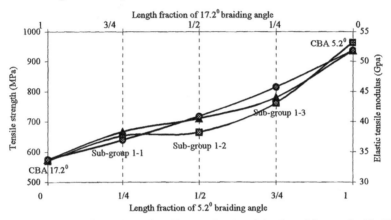

Fig. 10 Elastic tensile modulus and tensile strength vs length fractions of the contained braiding angles based on the test results of Group 1 FG rods▦ tensile strength▲ elastic tensile modulus, and ✸ calculated modulus.

It is found from Fig. 10 that the modulus of a FG rod is also dependent on the length fractions of the contained braiding angles. To understand the relationship further, a simple theoretical formula was derived for calculating the elastic tensile modulus of a FG rod. Suppose that the length of θ_i braiding angle section of a FG rod (with total length L) is L_i (i=1, 2, 3...). The modulus of a CBA rod with braiding angle θ_i is E_i. Assume that the modulus of θ_i braiding angle section of the FG rod is also E_i and the sections of the FG rod are connected in series in tensile test. Then, we have:

$$\sigma = \sigma_1 = \sigma_2 = \sigma_3 = ..., \text{ and } \sigma_1 = E_1\varepsilon_1, \sigma_2 = E_2\varepsilon_2, \sigma_3 = E_3\varepsilon_3 = ... \tag{3}$$

where, σ and ε represents stress and strain, respectively, and the subscripts stand for different sections of the FG rod. Based on the definition of Young's Modulus, the overall modulus of the FG rod is:

$$E = \frac{\sigma}{\varepsilon} = \frac{\sigma}{\left(L_1\varepsilon_1 + L_2\varepsilon_2 + L_3\varepsilon_3 + ...\right)/L} \tag{4}$$

Substituting (3) into (4) yields:

$$\frac{1}{E} = \frac{l_1}{E_1} + \frac{l_2}{E_2} + \frac{l_3}{E_3} + \dots \tag{5}$$

where, $l_1 = L_1 / L$, $l_2 = L_2 / L$, and $l_3 = L_3 / L \cdots$

The elastic tensile moduli calculated with Equation 5 for Group 1 FG rods are also shown in Fig. 10. A good accordance was observed between these calculated moduli and the experimentally obtained ones. For Group 2 FG rods, this kind of accordance was also observed. It was indicated in Fig. 11 by a negligible difference between the calculated moduli and the experimentally obtained ones.

Table I Elastic tensile moduli and tensile strengths for Group 2 FG rods (data of specimen group CBA 17.2^0 is listed here for reference).

Specimen group	Elastic tensile modulus (GPa)	Calculated elastic tensile modulus (GPa)	Tensile strength (MPa)
2-1	41.95	39.845	680.78
2-2	42.73	42.402	643.43
2-3	43.72	43.221	685.06
2-4	44.11	44.411	697.84
CBA 17.2^0	33.74	/	570.73

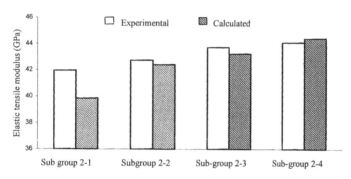

Fig. 11 Comparison between experimentally obtained moduli and the calculated ones for Group 2 FG rods.

3.4 Failure Mechanism Scanning Electron Microscopy (SEM) Analysis

To understand the failure mechanism, fracture surfaces of both CBA specimens and FG specimens were examined with SEM. No significant difference was found between them. Both fracture and pullout of fibers were observed on the surface. The failure mechanism can be characterized as rupture of fibers, followed by pullout of some fibers.

4. CONCLUSION

This research work investigated the tensile properties of braided carbon fiber/epoxy composite rods, which included constant-braiding-angle composite rods (CBA rods) and functionally-graded-braiding-angle composite rods (FG rods). Both the elastic tensile modulus and tensile strength of a CBA rod increase as the braiding angle decreases. This behavior can be utilized to make functionally graded material. By changing the braiding angle continuously along the length direction of the rod, the FG rod was prepared. Since each section (each with different braiding angle) of a FG rod can be separately regarded as a CBA rod, the elastic tensile modulus of a FG rod can be calculated using the data of CBA rods. A simple theoretical formula was derived for the calculation. It was shown that the experimentally obtained moduli kept in good accordance with the calculated ones. The tensile strength of a FG rod was proven to be higher than that of the section containing the largest braiding angle although they were originally supposed to be equal. Moreover, the strength of a FG rod is dependent on the length fractions of the contained braiding angles. Generally, the less are the length fractions of larger braiding angles, the higher is the strength.

REFERENCE

[1] Akira Kawasaki and Ryuzo Watanabe, "Concept and P/M Fabrication of Functionally Gradient Materials," *Ceramics International* **23** 73-83 (1997).

[2] A. Mortensen and S. Suresh, "Functionally Graded Metals and Metal-Ceramic Composites: Part 1 Processing," *International Materials Reviews*, **40** [6] 239-265 (1995).

[3] M. Funabashi, in *Proceedings of the Fourth Japan International SAMPE Symposium*, September (1995).

[4] Jyongsik Jang and Cholho Lee, "Performance improvement of GF/CF Functionally Gradient Hybrid Composite," *Polymer testing*, **17** 383-394 (1998).

[5] Jyongsik Jang and Cholho Lee, "Fabrication and Mechanical Properties of Glass Fiber-Carbon Fiber Polypropylene Functionally Gradient Materials," *Journal of Materials Science*, **33** 5445-5450, (1998).

[6] J. Jang and S. Han, "Mechanical Properties of Glass-fiber Mat/PMMA Functionally Gradient Composite," *Composites: Part A* **30** 1045-1053 (1999).

[7] N.J. Lee, J. Jang, M. Park, and C.R.Choe, "Characterization of Functionally Gradient Epoxy/Carbon Fiber Composite Prepared under Centrifugal Force," *Journal of Material Science*, **32**, 2013-2020, (1997).

[8] Ryuta Kamiya, Bryan A. Cheeseman, Peter Popper, and Tsu-Wei Chou, " Some Recent Advances in the Fabrication and Design of Three-dimensional Textile Preforms: a Review," *Composite Science and Technology* **60** 33-47 (2000).

[9] J.-H. Byun and T.-W. Chou, "Modelling and Characterization of Textile Structural Composites: a Review," *Journal of Strain Analysis*, **24** [4] 253-262 (1989).

[10] Rajiv A. Naik, Perter G. Ifju, and John E. Masters, "Effect of Fiber Architecture Parameters on Deformation Fields and Elastic Moduli of 2-D Braided Composites," *Journal of Composite Materials*, **28** [7] 656-681 (1994).

[11] Hui-Yu Sun and Xin Qiao, "Prediction of the Mechanical Properties of Three-dimensionally Braided Composites," *Composites Science and Technology* **57** 623-629 (1997).

[12] S. Ramakrishna, V.K. Ganesh, S.H. Teoh, P.L. Loh, and C.L. Chew, "Fiber Reinforced Composite Product with Graded Stiffness," Singapore Patent Application No. 9800874-1, 1998.

[13] S. Ramakrishna, V.K. Ganesh, S.H. Teoh, P.L. Loh, and C.L. Chew, "Fiber Reinforced Composite Product with Graded Stiffness," U.S. Patent Application No. 09/291,698, April 1999.

[14] V.K. Ganesh and S. Ramakrishna, "Textile Composites for Biomedical Applications," *Proceedings of Techtextile Symposium '98 –* Vol 2: Health and Protective Textiles, Lyon, France, pp.49-54 (1998).

[15] V.K. Ganesh, S. Ramakrishna, P.L. Loh, C.L. Chew and S.H. Teoh, "Functionally Graded Textile Composites for Dental Application," *Proceedings of the 1st Asian-Australasian Conference on Composite Materials* (ACCM-1), Osaka, Japan, 1998.

FUNCTIONALLY GRADED MATERIALS OF BIODEGRADABLE POLYESTERS AND BONE-LIKE CALCIUM PHOSPHATES FOR BONE REPLACEMENT

Carsten Schiller, Michael Siedler, Fabian Peters, Matthias Epple*
Solid State Chemistry
Faculty of Chemistry
University of Bochum
Universitätsstrasse 150
D-44780 Bochum
Germany

ABSTRACT

The concept of functionally graded materials was transposed to biomaterials designed for filling bone defects. Ideally an implant should have chemical, biological, mechanical and morphological properties that resemble the defect site. We report on composite materials of polyesters (polyglycolide and polylactide) with bone-like amorphous carbonated calcium phosphate that exhibit variable porosities (from compact to 80 vol%). The mechanical properties show the expected decrease with increasing porosity but are still in a sufficient range. Enhanced degradation at 70 °C was used to study the bulk erosion behavior. The erosion rate was higher for macroporous polyglycolide (400 μm pore diameter), compact polyglycolide and lower for microporous polyglycolide (<1 μm pore diameter) and poly-L-lactide. It can be fine-tuned in a functionally graded way to meet the specific requirements at the operation site.

INTRODUCTION

There is a strong demand for bone substitution materials in clinical medicine. Bone defects remain, e.g., after tumor extractions, comminuted fractures or inflammations. Due to this considerable importance, many materials have been proposed for filling of bone defects, among them calcium phosphate ceramics, calcium phosphate glasses (bioglasses), metals, and (often biodegradable) polymers.[1,2] Unfortunately, no presently available material is able to fulfill all requirements: biocompatibility, mechanical strength, and unlimited availability.

Autologous spongiosa from the patient, which is still the best solution today, the "golden standard" for the surgeon, suffers from the facts of a limited availability and the necessity for an often painful secondary operation. Consequently, the search for an ideal bone substitute (or bone graft) is still ongoing.

When a complex biological material as bone has to be replaced, the first consideration should be composition and properties. Mammalian bone is a highly ordered composite material consisting of about 30 wt% organic components (e.g. collagen type I) and 70 wt% inorganic bone mineral (nanocrystalline carbonated hydroxyapatite). It is organized on different structural levels, ranging from the sub-micrometer scale to the centimeter scale.[3] It is important to note that its structure is graded from a compact exterior (cortical bone) to a highly porous interior (cancellous or spongy bone). The graded nature is responsible for an optimal use of material to fulfill the necessary mechanical function.[4]

We believe that an optimal bone substitution material should be adapted to a defect in terms of its chemical composition and to its graded structure. We are furthermore convinced that a degradable implant that is finally replaced by bone is advantageous over a permanent one that would remain in the body forever (like a metal). Here we present results on a tailor-made material that mimics both features of bone: composition and structure.

CHEMICAL COMPOSITION

Conceptually, biodegradable polyester allows the temporary replacement of the bone until the implant is resorbed and replaced again by new bone.[5] In the ideal case, the degrading implant is replaced continuously by new bone. However, during the degradation of a polyester acidic monomers (carboxylic acid) are released and may decrease the local pH-value to a cell toxic level (4 or below).[6,7] This may cause inflammation and osteolysis (dissolution of the surrounding bone mineral). To stabilize the pH-value on a physiological level, the polyester can be combined with a basic filling material.[6,8-11] Our approach was to combine biodegradable polyesters (polyglycolide and polylactide) with a basic bone-like ceramic phase (carbonated amorphous calcium phosphate) to a highly biocompatible composite material.

The Polymer: Microporous Polyglycolide

Polyglycolide as a non-toxic biodegradable polyester has a wide use in medicine. Conventionally, polyglycolide is prepared by ring-opening polymerization. We developed a preparation method for polyglycolide by a solid-state reaction from sodium chloroacetate.[12,13]

$$n \ X-\underset{H_2}{\overset{O}{\underset{|}{C}}}-\overset{\overset{O}{\|}}{C}-O^- \ M^+ \longrightarrow X-\underset{H_2}{\overset{|}{C}}-\overset{\overset{O}{\|}}{C}-O\left[\underset{H_2}{\overset{|}{C}}-\overset{\overset{O}{\|}}{C}-O\right]_{n\text{-}2}\underset{H_2}{\overset{|}{C}}-\overset{\overset{O}{\|}}{C}-O^- \ M^+ \ + \ \text{n-1 MX}$$

During the reaction sodium chloride is precipitated as interconnected micron-sized cubic crystals within a polymeric matrix. Hot-pressing of the salt-loaded powder permits the preparation of larger objects in the cm scale. Due to the interconnection of the crystals the salt can be easily leached out with water and the porous polyglycolide structure remains (see Figure 1).

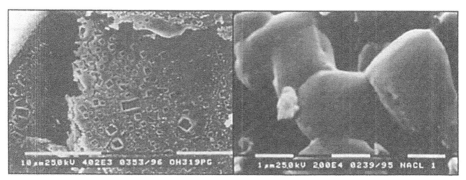

Figure 1: SEM-images of porous polyglycolide after leaching out the sodium chloride (left, magnification 4000x) and interconnected sodium chloride crystals after burning up the polymer (right, magnification 20000x)

In addition to the micron-sized pores (micropores) it is possible to introduce larger pores (macropores) by simply adding a water soluble spacer ("porogen"; e.g. sugar or sodium chloride), followed by pressing and leaching with water.[14] This reaction can also be transposed to 2-chloropropionates, yielding polylactide.

The Ceramics: An Amorphous Carbonated Calcium Phosphate

Due to their chemical similarity to the bone mineral, calcium phosphates are "known" by the body and therefore exhibit an excellent biocompatibility (see, e.g.,[15,16]). Upon implantation, they are well accepted by the body. The most prominent ones are hydroxyapatite (the mineral phase of bone; $Ca_{10}(PO_4)_6(OH)_2$) and tricalcium phosphate (TCP; $Ca_3(PO_4)_2$). However, some of these ceramics participate only insufficiently in the biological remodeling process of the body. Our skeletal system is continuously dissolved by bone-resorbing cells (osteoclasts) and rebuilt by bone-forming cells (osteoblasts). While bone with its nanocrystals of hydroxyapatite can be well resorbed by osteoclasts, they appear to have difficulties with highly crystalline calcium phosphates as they are usually obtained from chemical synthesis, followed by sintering,[17-19] a fact that can be ascribed to the higher thermodynamic stability and resulting lower solubility of sintered material in comparison to nanocrystalline (disordered) material.[20]

Following these thoughts, we have prepared an amorphous calcium phosphate phase that is thermodynamically even less stable than bone mineral. As in bone, a fraction of the phosphate groups was substituted by carbonate. Osteoclasts should be able to dissolve this material better. The resulting calcium and phosphate ions can be used for the formation of new bone mineral. Figure 2 shows three illustrative X-ray powder diffraction patterns that highlight the differences in crystallinity.

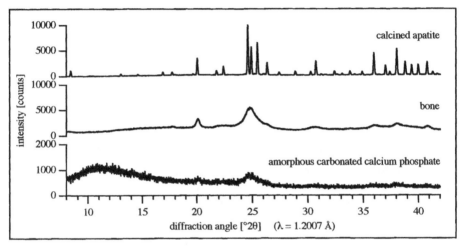

Figure 2: X-ray powder diffraction patterns of crystalline hydroxyapatite (top), bone and amorphous carbonated calcium phosphate

The Composite Material

By mechanically mixing polyglycolide and amorphous carbonated calcium phosphate in a ratio of about 5:2, a composite material can be obtained which shows a physiological pH-value (7.4) during storage in water as opposed to pure polyglycolide (pH 4.2 after 16 h) or the pure calcium phosphate (pH>9).

The biocompatibility of these composite material was tested *in-vitro* with bone forming cells (osteoblasts). The results demonstrated the excellent biocompatibility of the composite material.[7]

OPTIMIZED MORPHOLOGY AND COMPOSITION: A FUNCTIONALLY GRADED MATERIAL

An ideal bone substitution material should mimic the graded bone morphology. To perform the specific requirements at the implantation site, a spatially variable morphology, spatially variable mechanical properties and spatially variable degradation times are necessary.

To allow the ingrowth of bone forming cells, the implant should have interconnected pores with at least 200 μm diameter.[5] In contrast, the surface in contact with the soft tissue should prevent an immigration of fibroblasts that inhibit the bone formation by forming fibrous tissue. To achieve that, the average pore size at the surface must be below 10 μm. The microporosity is still advantageous to permit a free exchange of fluids for nutrition and to remove degradation products. Additionally, it may be desired that the implant degrades faster from the interior (where it meets the cancellous bone and where the osteoblasts are) than from the outside (where no osteoblasts are and where the ingrowth of fibrous tissue has to be prevented). Furthermore, the main load would rest on the cortical exterior rather than on the spongious interior, therefore we need more and longer stability on the exterior (see Figure 3).

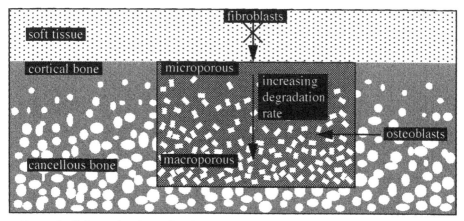

Figure 3: Schematic view of an implant with gradual transitions in porosity and degradation rate filling a bone defect

These requirements can be fulfilled by variations in
- porosity (micropores form solid-state reaction vs. macropores from porogens)
- nature of the polyester (fast degrading polyglycolide vs. longer stable polylactide)

- mixing ratio between the polymer and the calcium phosphate (influences degradation time and mechanical properties)

A graded variation in these parameters is necessary to permit a smooth variation of properties and also to prevent a possible loss of mechanical function at a sharp border between two implant types.

To demonstrate this concept we produced different objects with gradual transitions in these three properties by melt-pressing of spatially mixed powders. The experimental procedure was analogous as described in refs.[7,12,14]. The dimensions were $5 \cdot 5 \cdot 30$ mm^3 in all cases.

A Gradient in Porosity

Objects of polyglycolide (polyglycolic acid; PGA) with a gradual variation in the macroporosity (addition of different amounts of sugar (ca. 400 μm) to the polyglycolide-NaCl mixture (NaCl: <1 μm) obtained from solid-state reaction) were prepared. This means that an underlying microporosity (<1 μm) is gradually supplemented by macroporosity (ca. 400 μm).This is illustrated in Figure 4.

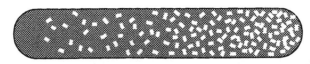

Figure 4: Schematic view of an object with a gradual transition in the volume-fraction of macropores (white) in a microporous matrix (black)

The porosity distribution over the geometry of the implant can be conveniently determined by measuring the locally variable density of the porous object. The results are compiled in Figure 5, demonstrating the graded porosity.

The mechanical properties depend - of course - strongly on the porosity. Clearly, the highly porous part of the object has inferior mechanical stability. These properties were determined on five objects without gradient that correspond to the different parts of the graded object. The results of stress-pressure diagrams (Zwick Z 2.5/TN1S; 1 mm min^{-1}; $5 \cdot 5$ mm^2 plane parallel faces of a 14.5 mm long object; according to DIN EN ISO 604) are shown in Figure 6. It can be seen that both compression strength and elastic modulus strongly decrease with increasing porosity. However, despite its high porosity, the material still has acceptable properties when compared to an elastic modulus of 6.5 GPa of compact polyglycolide.[21]

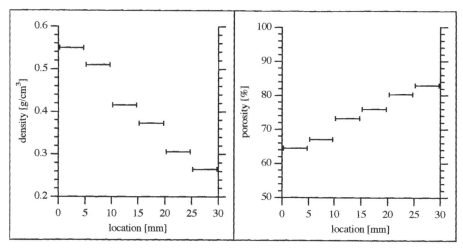

Figure 5: Analysis of an object with micropores and a gradual transition in the volume-fraction of additional macropores. The object was cut into six pieces and the density of each piece was measured by weighing (left). The porosity was calculated from density (right).

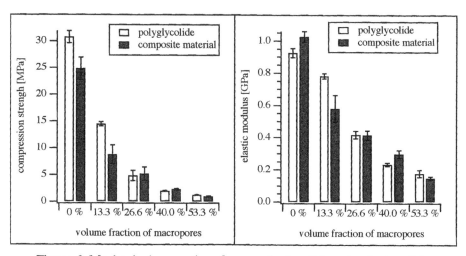

Figure 6: Mechanical properties of pure polyglycolide and polyglycolide-amorphous calcium phosphate-composite material, both with micropores and defined volume-fractions of macropores (left: compression strengh; right: elastic modulus)

The degradation of the implant was simulated at 70 °C in phosphate-buffered aqueous solution (modified Sørensen buffer at pH 7.4), in accordance to ISO 13781. Note that the buffer capacity was rapidly exceeded and that the pH went down to about 3 to 3.5. Figure 7 shows optical micrographs of compact polyglycolide (left), microporous polyglycolide (<1 µm) (center) and of a functionally graded object with micropores only on the left and micropores+macropores on the right. The figure shows the objects after 17 and 24 h. It can be noted that the compact material erodes rapidly, the microporous material is fairly stable and the graded object starts to erode on the macroporous side. The fact that the microporous material is more stable than the compact one can be explained by the specific bulk-erosion mechanism of polyesters that tend to accumulate acidic degradation products inside, leading to enhanced autocatalytic ester hydrolysis at low pH.[22] This is possible only to a minor degree in a microporous material in which the solvent always penetrates the interior and reduces gradients in pH. After four days, all implants were fully eroded to a dispersion.

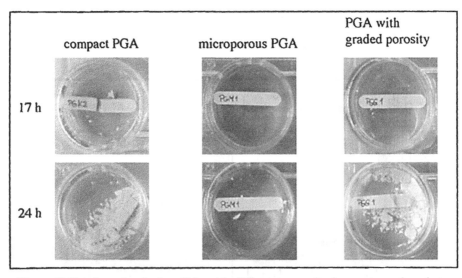

Figure 7: Erosion of polyglycolide with different porosity in aqueous phosphate buffer solution at 70 °C (the graded object has a gradual transition from micropores on the left to micropores+macropores on the right)

A Gradient in the Polymer Composition

A gradient in degradation rate can also be realized by a polymer-polymer-gradient of polymers with different biodegradation lifetimes. To test this idea, we prepared a compact functionally graded object with polyglycolide on the one side and poly-L-lactide (polylactic acid; PLLA) on the other side (no pores in either case). It is known that polyglycolide implants degrade within weeks or months whereas poly-L-lactide needs about one year to degrade.[23,24]

The results of a degradation study at 70 °C are shown in Figure 8. The object starts to degrade on the polyglycolide-rich side whereas the polylactide-rich side is stable up to weeks. This demonstrates that a gradual change in degradation properties can be achieved by a functionally graded material.

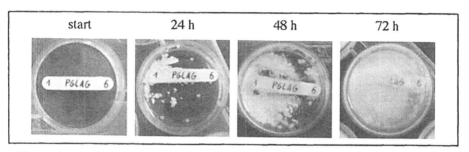

Figure 8: Erosion of an object with a gradual transition from polyglycolide (left side) to poly-L-lactide (right side) in aqueous phosphate buffer solution at 70 °C

If even longer degradation times are needed, one can use higher homologues of polylactide, namely poly-(R)-3-hydroxybutyrate (P3HB) and poly-(R)-3-hydroxybutyrate-co-(R)-3-hydroxyvalerate (P3HBHV). If subjected to enhanced degradation at 70 °C, these materials show no observable erosion after 5 months

Note that these degradation experiments do not necessarily reflect the conditions in the body as enzymatic degradation is excluded in this setup. However, this mechanism is of considerable importance *in vivo*.[25]

A Gradient in the Mixing Ratio of Polymer and Ceramic

Although the requirement of internal pH stabilization necessitates a given ratio between polymer and ceramic, it is of interest to investigate materials with a gradient in composition, as they should permit varying degradation rates and mechanical properties. We have prepared samples of carbonated amorphous calcium phosphate and polyglycolide by melt pressing. The composition can be assessed by infrared spectroscopy at different points. For quantification, we used the IR bands at 1745 cm^{-1} (polyglycolide) and 565 cm^{-1} (calcium phosphate). The results are shown in Figure 9, demonstrating that a gradient can be established by simple mixing and melt-pressing.

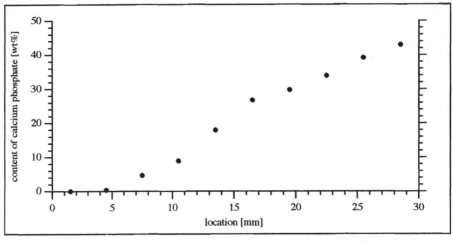

Figure 9: Analysis of an object of composite material (polyglycolide, amorphous calcium phosphate) with a gradual transition in the mixing ratio (results of quantitative IR spectroscopy)

CONCLUSIONS

Functionally graded materials consisting of polyesters and calcium phosphate ceramics were prepared. Gradients in porosity, nature of the polymer and mixing ratio of polymer and ceramics were realized. It could be shown that the degradation rate *in vitro* can be controlled by choice of these parameters. Likewise, the mechanical properties strongly depend on the porosity. The results should help in design of tailor-made biomaterials for appropriate filling of bone defects, that incorporates chemical and morphological optimization of a prospective implant.

REFERENCES

[1]S. I. Stupp, P. V. Braun, "Molecular manipulation of microstructures: biomaterials, ceramics, and semiconductors," *Science* **277** 1242 (1997).

[2]J. M. Rueger, "Bone replacement materials - state of the art and the way ahead," *Orthopäde* **27** 72 (1998).

[3]S. Weiner, H. D. Wagner, "The material bone: structure-mechanical function relations," *Annu. Rev. Mater. Sci.* **28** 271 (1998).

[4]W. Pompe, S. Lampenscherf, S. Rößler, D. Scharnweber, K. Weis, H. Worch, J. Hofinger, "Functionally graded bioceramics," *Mat. Sci. Forum* **308-311** 325 (1999).

[5]S. L. Ishaug, G. M. Crane, M. J. Miller, A. W. Yasko, M. J. Yaszemski, A. G. Mikos, "Bone formation by three-dimensional stromal osteoblast culture in biodegradable polymer scaffolds," *J. Biomed. Mater. Res* **36** 17 (1997).

[6]C. M. Agrawal, K. A. Athanasiou, "Technique to control pH in vicinity of biodegrading PLA-PGA implants," *J. Biomed. Mater. Res.* **38** 105 (1997).

[7]W. Linhart, F. Peters, W. Lehmann, K. Schwarz, A. F. Schilling, M. Amling, J. M. Rueger, M. Epple, "Biologically and chemically optimized composites of carbonated apatite and polyglycolide as bone substitution materials," *J. Biomed. Mater. Appl. Biomater.* **54** 162 (2001).

[8]C. C. P. M. Verheyen, J. R. de Wijn, C. A. van Blitterswijk, K. de Groot, "Evaluation of hydroxylapatite/poly(L-lactide) composites: mechanical behaviour," *J. Biomed. Mater. Res.* **26** 1277 (1992).

[9]R. Zhang, P. X. Ma, "Porous poly(L-lactic acid)/apatite composites created by biomimetic process," *J. Biomed. Mater. Res.* **45** 285 (1999).

[10]N. Ignjatovic, S. Tomic, M. Dakic, M. Miljkovic, M. Plavsic, D. Uskokovic, "Synthesis and properties of hydroxyapatite/poly-L-lactide composite biomaterials," *Biomaterials* **20** 809 (1999).

[11]O. Betz, A. Ignatius, D. Reif, B. Leuner, L. Claes, *25th Annual Meeting Transactions, Society for Biomaterials (Providence, Rhode Island)*, "New resorbable composites made of TCP/polylactide and B14N/polylactide as bone graft substitutes in a loaded implant model: a histological analysis," pp. 453 1999.

[12]M. Epple, O. Herzberg, "Polyglycolide with controlled porosity - an improved biomaterial," *J. Mater. Chem.* **7** 1037 (1997).

[13]M. Epple, O. Herzberg, "Porous polyglycolide," *J. Biomed. Mater. Res. Appl. Biomat.* **43** 83 (1998).

[14]K. Schwarz, M. Epple, "Hierarchically structured polyglycolide - a biomaterial mimicking natural bone," *Macromol. Rapid Commun.* **19** 613 (1998).

[15]J. C. Elliot, *Structure and chemistry of the apatites and other calcium orthophosphates*, Vol. 18, Elsevier, Amsterdam, 1994.

[16]R. Z. LeGeros, "Biological and synthetic apatites"; p. 3 in *Hydroxyapatite and related materials*, edited by P. W. Brown and B. Constantz, CRC Press, Boca Raton 1994.

[17]J. D. de Bruijn, Y. P. Bovell, J. E. Davies, C. A. van Blitterswijk, "Osteoclastic resorption of calcium phosphates is potentiated in postosteigenic culture conditions," *J. Biomed. Mater. Res.* **28** 105 (1994).

[18]S. Yamada, D. Heymann, J. M. Bouler, G. Daculsi, "Osteoclastic resorption of calcium phosphate ceramics with different hydroxyapatite/β-tricalcium phosphate ratios," *Biomaterials* **18** 1037 (1997).

[19]Y. Doi, T. Shibutani, Y. Moriwaki, T. Kajimoto, Y. Iwayama, "Sintered carbonate apatites as bioresorbable bone substitutes," *J. Biomed. Mater. Res.* **39** 603 (1998).

[20]R. P. Shellis, A. R. Lee, R. M. Wilson,"Observations on the apparent solubility of carbonate-apatites," *J. Coll. Interface Sci.* **218** 351 (1999).

[21]P. Törmälä, "Biodegradable self-reinforced composite materials; manufacturing structure and mechanical properties," *Clin. Mater.* **10** 29 (1992).

[22]A. Gopferich, "Erosion of composite polymer matrices," *Biomaterials* **18** 397 (1997).

[23]R. Ginde, R. Gupta, "In vitro chemical degradation of polyglycolic acid pellets and fibers," *J. Appl. Polym. Sci.* **33** 2411 (1987).

[24]E. A. R. Duek, C. A. C. Zavaglia, W. D. Belangero, "In vitro study of poly(lactic acid) pin degradation," *Polymer* 6465 (1999).

[25]R. A. Kenley, M. O. Lee, T. R. Mahoney, L. M. Sanders, "Poly(lactide-co-glycolide) decomposition kinetics in vivo and in vitro," *Macromolecules* **20** 2398 (1987).

Tribological Coatings

DAMAGE TOLERANT TRIBOLOGICAL COATINGS BASED ON THERMAL SPRAYED FGMS

L. Prchlik, A. Vaidya, S. Sampath
Center for Thermal Spray Research
Department of Materials Science and Engineering
State University of New York
Stony Brook, NY 11794-2275

ABSTRACT

A wide variety of refractory carbide based coatings are used in industrial components to lower friction and improve wear resistance. Traditional thermal sprayed hard coatings have deficiencies in these applications due to their high hardness, brittleness and inherent stresses. FGM concepts offer a means for tuning the tribological performance, allow for processing of thick deposits and enable meeting other system requirements. In this study, steel/cermet based FGMs have been prepared through plasma spray and high velocity oxy-fuel spray processes (HVOF). HVOF WC-Co/stainless steel and plasma Mo-Mo$_2$C/stainless steel FGM were fabricated after optimising the processing conditions. Basic mechanical properties, wear and frictional responses have been evaluated, and their implications for the FGM real-component performance were discussed.

INTRODUCTION

Surface engineering by thermal spray is a pragmatic and highly cost effective method for applying protective coatings and functional surfaces. Such coatings find extensive wear applications in industrial machinery. Thermal spray has traditionally been the process of choice for many industrial sectors: aerospace, pulp and paper, marine, diesel engine and oil industries. These coatings, usually deposited through high velocity oxy-fuel or plasma spray processes are based on cermet compositions such as WC-Co, Cr$_3$C$_2$-Mo, and Mo$_2$C-Mo and perform well under stringent conditions [1-3]. In recent years, there has been an increased demand for design-based implementation of coatings. Here the surface hardness, sliding friction and compatibility between mating surfaces play a key role in system performance.

In service, the coatings are subjected to a variety of abrasive, adhesive and erosive wear conditions. Hence, an understanding of the tribological behavior of the deposits in relation to properties is critical in evaluating their performance. The tribological behavior of thermal spray materials is closely linked to the layered microstructure. Inhomogeneities in mechanical properties such as hardness, anisotropy in toughness and varying surface profiles all lead to wide variations in wear behavior.

The flexibility of thermal spray allows for the rapid synthesis of FGMs. FGM-based engineered surfaces offer versatility in terms of materials selection, design and application [4-5]. In order to fully utilize the beneficial effects of design-based graded coatings, it is critical to understand the relationships among processing, microstructure, properties and performance of such systems. In the area of tribological coatings, the change in composition as a function of depth adds a further dimension in understanding and predicting contact damage response of the materials. The goal of this study is to examine fundamentally, dry sliding friction, contact damage and wear behavior of selected carbide reinforced metallic composite and graded materials. Specifically, Mo-Mo$_2$C/stainless steel (SS) and WC-Co/SS composites have been produced and studied. The choice of these materials is based on their relevance to sliding wear applications in automotive and earth moving machinery [6-8]. The Mo-based alloy coatings offer excellent scuff resistance during dry unlubricated contact. However, they are prone to brittle breakout of the coating and spallation [9]. The ferrous materials, when used as a matrix, can reduce this limitation. In the case of WC-Co stainless steel FGMs, improved adhesion due to increased material compatibility is expected.

EXPERIMENTAL METHODS

Processing

Two systems of graded coatings were fabricated: i) HVOF deposited WC-Co/stainless steel; ii) plasma sprayed Mo-Mo$_2$C/stainless steel. In the first case, commercial grade stainless steel powder (ANVAL 316, particle size +18/-42μm) and a WC-Co (Diamalloy 2004, particle size +5/-45μm) were used to prepare the graded and composite coatings. These coatings were deposited on steel substrates (1040) with the Praxair HV2000 HVOF system. Spraying was carried out by using propylene as the fuel gas and nitrogen as the carrier gas for powders. The gun was kept at a fixed spray distance of 230 mm and traversed at 6mm/s during the coating process. The substrates were mounted on a carousel of 189mm diameter, rotating at 400rpm. The FGM consisted of six uniform layers of approximately the same thickness, graded in 20% composition steps. Prior to the deposition of graded materials, deposition rates of both single phases were measured. Based on this measurement, powder feed-rates were adjusted appropriately to achieve the required content of each phase in the composites. The full FGM coating was graded from 100% stainless steel (at the substrate) to 100% WC/Co (at the surface). Table I lists the spraying parameters used for each step of the graded coating.

Table I. Spray parameters used for HVOF processing of WC-Co/stainless steel coatings.

Composition Steel/ WC-Co	H$_2$ Flow [scfh / slm]	O$_2$ [scfh / slm]	Feed rate [g/min] SS/WC-Co	Deposition Rate [mm/pass]	Number of Passes (for FGM)
100% / 0%	1400 / 658	350 / 165	40/0	0.0127	14
80% / 20%	1400 / 658	350 / 165	32/8	0.0111	16
60% / 40%	1400 / 658	400 / 188	24/16	0.0096	18
40% / 60%	1400 / 658	400 / 188	16/24	0.086	20
20% / 80%	1485 / 698	450 / 212	8/32	0.0074	24
0% / 100%	1485 / 698	450 / 212	0/40	0.0065	28

Table II. Spraying parameters for Mo-Mo$_2$C composites and FGM.

Parameter	Description/Value
Gun	F4-MB
Gun Nozzle	8 mm
Gun Power	500 A
Gun Voltage	69 V
Primary (Ar) gas flow	40-48 slpm
Secondary (H2) gas flow	10-8 slpm
Carrier (Ar) gas flow	2.5 slpm
Spraying distance	110 mm
Vertical traverse speed	30 mm/s
Spindle speed/diameter	160 rpm /⌀ 140 mm
Total feed rate (Carbide + Stainless Steel)	8 gm/min.

The Mo-Mo$_2$C/SS coating was sprayed with the same stainless steel powder (ANVAL 316, particle size +18/-42 μm) and Mo-Mo$_2$C powder from OSRAM Sylvania SX274 (particle size +49/-

108µm). The coatings were prepared on steel substrates with a Plasma Gun F4-MB (Plasma Technik Switzerland). The plasma spray parameters are given in Table II. Graded coatings were prepared with five composition steps (0%, 25%, 50%, 75% and 100% Mo-Mo_2C content). As in the case of HVOF deposits, deposition rates were measured and feed rates and number of passes were adjusted to achieve the desired linear grading profile. The Mo-Mo_2C/SS consisted of five layers.

Testing and Characterization

Coating microstructures and properties were evaluated using metallographic techniques, microhardness and indentation fracture toughness [10] measurements. For WC-Co/SS graded coatings, the composition evaluation was performed both by image analysis of back-scattered EM images and energy dispersion microanalysis (EDAX) analysis. In the case of molybdenum carbide-stainless steel coatings, EDAX and quantitative X-ray analysis were used.

Room temperature friction and wear tests were performed for a ball-on-disc configuration shown in Fig.1. Prior to dry sliding tests, coatings surfaces were polished using the following procedure: i) fine wet grinding on 45 µm diamond abrasive wheel; ii) polishing by diamond suspension emulsions 9, 3 and 1 µm; iii) ultrasonic cleaning in methanol for 15 s and iv) drying at 100°C for 10 min. The final surface roughness of specimens was measured, and the R_a values were typically lower than 0.1 µm.

Unidirectional sliding friction tests were carried out in air with stationary silicon nitride balls (Si_3N_4) at 50 N load on a UMT-CETR friction/wear tester. The eccentric position of balls produced wear scar of 7 mm diameter, which for a rotational speed 27.4 rpm, and 120 min of sliding resulted in 72 m of sliding at a velocity of 0.01 m/s. Profiles of wear scars were measured with an optical 3D Zygo profilometer (a scanning white light interferometer), which had the lateral resolution of 0.7 µm and an effective vertical resolution of 0.01 µm.

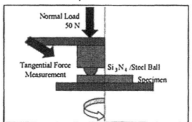

Figure 1 Configuration of the ball-on-disc friction-wear tester.

RESULTS AND DISCUSSION

Coating deposition and characterization

Mo_2C/stainless steel coatings form a complex system consisting of molybdenum, hexagonal molybdenum carbide, metallic elements (Fe, Ni, Cr) and their oxides. Quantitative X-ray analysis showed that the content of Mo_2C in Mo-Mo_2C regions was approximately 19wt.% compared to 77% in feedstock powder SX 274. This corresponds to an intense decarburization of the powder during the spraying process [8]. The deviation between targeted and achieved compositions of graded coatings was estimated by EDAX. The results are listed in Table III for three composite coatings. A very good agreement was achieved for 50%/50% mixture. For other composites, an adjustment of feed rates during FGM spraying was necessary.

Composition of Mo₂C/SS mixture coatings	
Targeted	Measured (EDAX)
25% Mo/Mo₂C + 75% SS	22%/78%
50% Mo/Mo₂C + 50% SS	51%/49%
75% Mo/Mo₂C + 25% SS	68%/32%

Similar analysis for WC-Co/stainless steel composites/FGMs revealed that the deviation of achieved and targeted compositions was smaller than 5%.Fig. 2 shows optical micrographs of the two prepared FGMs.

Figure 2 Microstructures of (a)HVOF WC-Co/SS and (b)Plasma Mo-Mo₂C/SS

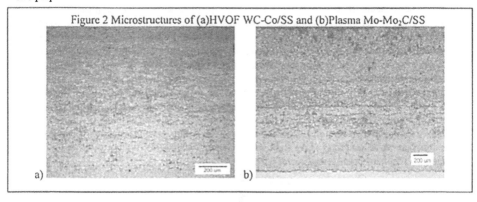

Figure 3 Microstructures of (a) 100% and (b) 68% Mo₂C-Mo coatings.

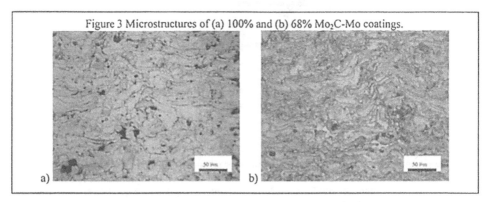

The graded coatings are formed by layered structures on both macro-and micro-scale. Macro-layers, corresponding to spraying steps, as well as localized microlayers of one phase within each of the 'uniform' layers, can be observed. The phase distribution appears to be somewhat more uniform in case of the HVOF deposit. This is mainly due to higher particle velocities, axial feed to the flame and smaller difference in particle size of feedstock powders for the HVOF process as compared to plasma. Thus, this reduces the difference in particle trajectories and resulting phase segregation in the deposit. Differences between flame velocities and feeding techniques also lead to lower splat thicknesses in HVOF coatings as compared with plasma. Fig. 3 shows the cross-sections of plasma sprayed coatings with 100% and 68% content of Mo-Mo_2C constituent, respectively. The addition of stainless steel reduced the open porosity and made the structure more compact. Similar effects of porosity reduction were not observed in the case of HVOF deposit.

Hardness and fracture toughness

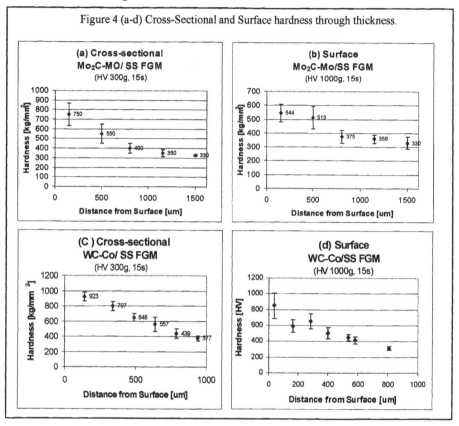

Figure 4 (a-d) Cross-Sectional and Surface hardness through thickness.

Fig. 4 illustrates the fact that in spite of a linear composition gradient for both FGMs, their hardness profiles were different. In the case of WC-Co/SS coating, the hardness follows very closely the rule-of-mixtures for the cross section and surface measurements. This can be understood on the basis of the fact that the WC-Co/SS coating was fabricated by the HVOF process, which results in a fairly dense microstructure. Hence, properties of the coating are less affected by defects – such as porosity – in the coating. A small variation of surface hardness from the linear profile can be noticed at a depth of 170 µm. This is believed to be due to the variation of coating properties close to the interface for two steps.

Surface hardness of the Mo-Mo_2C/SS coating decreases only slightly when a small amount of stainless steel is added. Since the response of 100% Mo-Mo_2C to a surface indentation is determined by the cracking and intersplat sliding in the plane of the coating (HV_{25} of Mo-Mo_2C is approximately 1320 kg/mm^2), the addition of steel makes the structure less prone to cracking due to better intersplat bonding, reduction of porosity and greater ductility of steel splats as compared to Mo-Mo$_2$C. For the same material, a decrease of cross-sectional hardness with depth/composition is more rapid due to an uneven distribution of Mo-Mo$_2$C splats. For larger contents of stainless steel (50, 75%), hardness is determined mainly by properties of the steel constituent.

Indentation fracture toughness measurement on 100% and 75% Mo-Mo$_2$C, performed according to [10] at 20 kg load, yielded values 1.4 and 2.4 MPa.m$^{0.5}$, respectively. The fracture toughness of WC-Co obtained by the same technique was 3.9 MPa.m$^{0.5}$. In the layer with larger stainless steel content, the measurements could not be evaluated since the indentation did not produce observable cracks.

Friction and wear behavior

The results of unlubricated friction and wear experiments with Si$_3$N$_4$ balls are presented in Fig. 5. This accelerated laboratory experiment was utilized to examine the mechanisms of material failure, rather than simulate actual component situation. The reported steady-state values of coefficient of friction (COF) were reached usually in less than 20 minutes of sliding. For the WC-Co coatings, the COF values were roughly proportional to the content of stainless steel constituent present in the tested layer. The wear of coating rapidly increased with an increase in stainless steel content. The wear of the balls showed a similar trend. Due to the sliding setup with the sample face-up, a significant portion of the hard loose debris generated during the test was entrapped between the sliding surfaces. The debris was acting as a third body, causing accelerated wear of the counterbody. When the amount of stainless steel exceeded 75% percent, the COF was somewhat lower as well as the wear of the ball, since smaller amounts of hard WC-Co debris were generated. Nevertheless, the wear of this coating was the highest, as was to be expected from hardness measurements.

The wear/friction behavior of Mo-Mo$_2$C/stainless steel FGM appeared to be more complex. Neither the COF, nor the wear were lowest for the layer with the 100% Mo-Mo$_2$C content. In fact, wear of the counterbody was highest from all FGM layers tested. This was rather surprising, but a close examination of the wear tracks explained this behavior. Large amount of open porosity can be observed on the surface of 100% Mo-Mo$_2$C layer. This porosity corresponds to limited interlocking of the splats. After a sufficiently large sliding distance, localized delamination from the surface occurs. No regular scar was formed on 100% Mo/Mo$_2$C surface during sliding. Instead, entire splats were pulled locally and large pits were formed in the wear scar. Since the hardness of the Mo-Mo$_2$C splats on the microscale is large (1320 HV$_{25}$), the edges of the pits abraded the ball and caused the high wear. A small addition of stainless steel improved the support of the splats, and localized splat delamination did not take place so excessively. Therefore, reduction of wear of the coating itself and a large decrease of the ball wear were noted. This is in a good agreement with the surface hardness measurements and fracture toughness values. As discussed earlier, the surface hardness of Mo-Mo$_2$C with 25% stainless steel addition did not decrease significantly, and, at the same time, the fracture toughness of the mixed layer was considerably higher. A further increase of steel content did not improve the properties of plasma sprayed coatings, since the brittle damage mechanism was replaced by adhesive wear.

Comparison of the wear tests for the two FGMs shows that neither of them perform better in the entire thickness range. The wear rates of WC-Co layers with low steel content were much lower than those of Mo-Mo$_2$C layers with the same steel content. On the other hand, with increasing depth/steel content, the molybdenum-based FGM exhibited lower wear rates. It is believed that high temperature oxidation of stainless steel powder in plasma and inferior intersplat binding in the HVOF deposit are responsible for the observed behavior. An investigation revealing the wear mechanism in multi-constituent thermal sprayed coatings sprayed by different processes is currently underway.

Figure 5 (a – f) Wear of Coatings, Counterbodies and COF for HVOF and Plasma Sprayed FGMs.

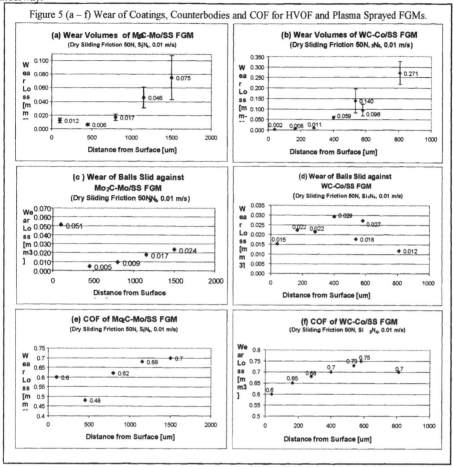

SUMMARY AND CONCLUSIONS

The graded coatings composed of stainless steel/Mo-Mo$_2$C and WC-Co were successfully prepared by plasma and HVOF spraying. Feed rates required for the deposition of mixture/graded coatings with a specific composition were calculated from deposition efficiencies measured during deposition processes of both powders used. Targeted and measured compositions of mixture coatings appeared to be in a reasonable agreement, considering the complexity of the process.

The friction/wear experiments have shown that composite coatings can have a higher resistance against a surface damage that would correspond to the simple rule of mixtures. The high porosity of 100% Mo/Mo$_2$C and low fracture toughness adversely affects friction and wear characteristics. Addition of stainless steel is effective in reducing the coating porosity, increasing fracture toughness and improving friction/wear response.

Therefore, coatings toughened by a minor addition of stainless steel appear to have interesting tribological properties and their behavior needs to be further investigated. It is also suggested that in some case only partial grading of FGM may be employed in order to achieve an optimum tribological response. Similar effects have not been observed for the HVOF FGM. The WC-Co coating is sufficiently dense in itself and any additions of stainless steel to this carbide phase do not improve the density. With steel additions, the resistance against crack propagation increases, but with no positive effect on wear resistance.

Comparison of both FGMs shows that in dry sliding friction, WC-Co/SS exhibits lower wear rates, but its wear rates are increasing more rapidly with depth (than those of Mo-Mo$_2$C/SS). Studies investigating the abrasion response of both materials are currently underway.

REFERENCES

[1] F. Rastegar and A.E. Craft, "Piston ring coatings for high horsepower diesel engines," *Surface coatings and Technology*, **61** 36-42 (1993).

[2] J. Nertz, B. Kuschner and A. Rotolico, "Microstructural Evaluation of Tungsten Carbide Coatings," *J. Thermal spray Tech.*, **1**[2] 147 (1992).

[3] D. R. Sielsky and P. Sahoo, "Chromium Carbide coatings for High Temperature Erosion Resistance," *Thermal Spray: Practical Solutions for Engineering Problems*, Ed: C.C. Berndt, ASM, 159-167 (1996).

[4] Proceedings of the Second International Symposium on Functionally Gradient Materials, San Francisco, Ed: J. Birch, M. Koizumi, T. Hirai and Z.A. Munir, The American Ceramic Society, *Ceramic Transactions*, **34**, 1992.

[5] Proceedings of the Third International Symposium on Structural and Functionally Gradient Materials, Ed: B. Ilschner and N. Cherradi, Lausanne, 1994.

[6] S.F. Wayne, J.G. Baldoni and S.T. Buljan, "Abrasion and Erosion of WC-Co with Controlled Microstructures", *Tribol. Trans.*, **33**[4] 611-617 (1990).

[7] S. F.Wayne and S. Sampath, "Structure/Property Relationships in Sintered and Thermally Sprayed WC-Co," *J. Thermal spray Tech.*, **1**[4] 307-315 (1992).

[8] S. Sampath and S.F. Wayne, "Microstructure and Properties of Plasma-Sprayed Mo-Mo$_2$C Composites," *J. Thermal Spray Tech.*, **3**[3] 282-288 (1994).

[9] S.F. Wayne, S. Sampath and S. Anand, "Wear mechanism in Thermally Sprayed Mo-Based Coatings," *Tribol. Trans.*, **37**[3] 636-640 (1994).

[10] A.G. Evans and T.R. Wilshaw: "Quasi-static solid damage in brittle solids, I. Observations, Analysis and Implications", *Acta Metall.*, **24** 939-956 (1976).

INTERNAL STRESS DISTRIBUTION IN THE FUNCTIONALLY GRADED DIAMOND/SILICON NITRIDE COATINGS

M. Kamiya, R. Sasai, S. S. Lee and H. Itoh
Research Center for Advanced Waste and Emission Management, Nagoya University, Furo-cho, Chikusa-ku, Nagoya 464-8603, Japan

K. Tanaka
Department of Mechanical Engineering, Graduate School of Nagoya University, Furo-cho, Chikusa-ku, Nagoya 464-8603, Japan

ABSTRACT
Strongly adherent diamond coating on silicon nitride substrate was prepared by two-stage microwave plasma CVD in the $CO-H_2$ system. The substrate was pretreated in a hot and strong acid and then microflawed ultrasonically with diamond grains. An anchored deposition of diamond into the micropores and the subsequent long-term diamond CVD resulted in a graded microstructure from substrate to diamond film. Internal stress distribution was analyzed by the X-ray stress measurement method to determine the intrinsic/thermal stress relationship in the coatings. The residual stress value of diamond film gradually approached that of the pretreated substrate with increasing CVD treatment time.

INTRODUCTION
For the application of chemically vapor deposited (CVD) diamond coatings to cutting tool, the improvement of adherence between diamond and substrate is one of the crucial problems to be solved. Diamond film is easily peeled from the substrate before achieving normal wear, when it is used for milling high-silicon aluminum alloy work material. The major cause for low adherence strength is that internal stress develops and causes the film decohesion. Residual stress in the film, which is composed of intrinsic stress and thermal stress, is usually stronger than the adherence strength of diamond film to substrate. Intrinsic stress results from the lattice mismatch between diamond and substrate in addition to the various defects formed during

deposition. On the other hand, thermal stress results from the difference in the thermal expansion coefficient between diamond and substrate, which is induced during the cooling process of the specimen after deposition. For resolving this problem, Itoh et al.[1,2] have selected silicon nitride substrate with a relatively similar thermal expansion coefficient to diamond, and pretreated the substrate in a hot and strong acid. Then they microflawed it ultrasonically with diamond grains and finally deposited diamond on the pretreated substrate with the two-stage CVD. The graded texture and structural variation of C-C bonding from the substrate side to the diamond film were characterized by SEM/XMA and TEM analyses or micro-Raman spectroscopy. The tool life of the diamond coated specimen increased significantly compared with that produced by ordinary CVD method.

Residual stress in the diamond film has been studied by various techniques, such as X-ray stress measurement method[3], Raman spectroscopy[4], etc. However, the interaction between the adherence strength of diamond film to substrate and the residual stress generated in the coatings has not been described in detail. In this paper, strongly adherent diamond coating on silicon nitride substrate is demonstrated by a sophisticated pretreatment of substrate and the subsequent two-stage microwave plasma CVD in the $CO-H_2$ system. Internal stress distribution in the coatings is analyzed by X-ray stress measurement method, and then we discussed the relationship between the adherent strength and internal stress.

EXPERIMENTAL PROCEDURE

Commercially available sintered silicon nitride (NGK Spark Plug, SX8) with the shape of square chip ($10 \times 10 \times 3$mm) was used as a substrate for diamond coating. In order to form micropores in the surface region by removing SiAlON phase, the substrate was pretreated at 40℃ for 60 min in a strong acid of HNO_3 (60%) and HF (47%) in a volume ratio of 1:1. Then it was microflawed in ethanol suspended with diamond grains (grain size <0.25μm) in an ultrasonic field[5]. Diamond coatings were prepared by microwave plasma CVD in the $CO-H_2$ system. The two-stage CVD conditions are shown in Table 1. The first-step CVD was performed under the conditions with higher microwave power, lower pressure, lower CO concentration and shorter reaction time to deposit fine-grained diamond. On the other hand the second-stage CVD was performed with lower microwave power, higher pressure, higher CO concentration and longer reaction time for increasing growth rate of diamond film. Total flow rate was kept constant at 200 mL/min in both CVD stages.

The deposited film was identified by using X-ray diffraction (XRD; Rigaku, RADIII) and micro-Raman spectroscope (JASCO, NR-MPS-21). Surface morphology and texture of specimens before and after the pretreatment and CVD treatment were

Table 1　Two-stage CVD conditions for diamond coating.

	First-stage	Second-stage
Microwave power (W)	750	550
Total pressure (kPa)	2	4
CO concentration (%)	10	25
Total flow rate (mL/min)	200	200
Reaction time (h)	1	1-29

observed by scanning electron microscope (SEM; Hitachi, S-2400).

Internal stresses of non-pretreated, pretreated, and CVD treated (1 h - 30 h) specimens were analyzed by X-ray stress measurement apparatus (Mac Science, MXP18) using chromium as a target material.　Conditions for internal stress measurement of diamond and silicon nitride are shown in Table 2.

Table 2　Conditions for X-ray stress measurement and X-ray penetration depth calculation of diamond and Si_3N_4

	Diamond	Si_3N_4
Characteristic X-ray	Cr-Kα	Cr-Kα
Diffraction line	220	411
Theoretical $2\theta_0$ (deg)	130.424	125.403
Young's modulus (GPa)	1210	418
Poisson's ratio	0.104	0.395
X-ray absorption coefficient (m^{-1})	508	4130

RESULTS AND DISCUSSION

Figure 1 shows X-ray diffraction patterns of non-pretreated substrate and diamond coated specimens. Here, the total CVD treatment time includes the first-stage CVD time of 1 h and the subsequent second-stage CVD time of 29 h. Two peaks at nearly 44° and 75° corresponding to diamond 111 and 220 peaks were identified. Other peaks are from diffraction lines of β-Si_3N_4. The preferred orientation of diamond 110 direction was found to increase with increasing treatment time. Itoh et al.[1] reported that actual cutting test by milling the aluminum-20 wt% silicon alloy work revealed a significantly long tool life in using the thick diamond coated specimen with a long-term CVD.

SEM photographs of the specimen surfaces at various stages of treatments are shown in Fig. 2. The surface of non-pretreated substrate shows some grinded streaks that are induced in the grinding work, as shown in Fig.2(a). Micropores are formed on

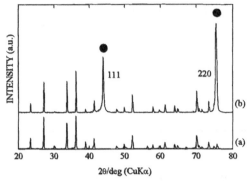

Fig.1 X-ray diffraction patterns of the non-treated and diamond coated specimens. (a) non-treated substrate, (b) diamond coated specimen (CVD time: 30 h).

the substrate surface by removing the SiAlON phase from the grain boundary of sintered β-Si$_3$N$_4$ after the strong acid pretreatment. The grinded streak pattern disappeared after this pretreatment. At 1 h of CVD treatment time by the first-stage CVD, dense and fine diamond grains were deposited. After the subsequent long time CVD, fine-grained diamond crystals grow to the crystallites with diameter about

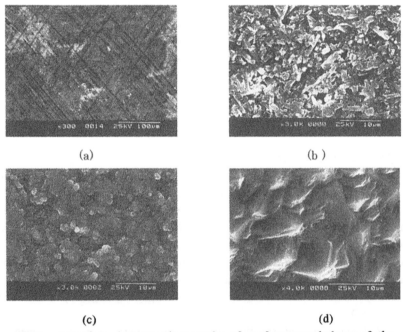

Fig.2. Scanning electron micrographs of surface morphology of the specimens after and before two-stage CVD. (a) non-pretreated substrate, (b) after acid pretreatment, (c) CVD time, 1 h, (d) CVD time, 30 h.

10μm.

According to the previous work on X-ray stress measurement[6], the depth through which characteristic X-ray of Cr-Kα penetrates into the film and substrate, was calculated by using the equation:

$$T = \frac{\sin^2\theta - \sin^2\psi}{2\mu \cdot \sin\theta \cdot \cos\psi}$$ (1)

where T is the penetration depth of X-ray, and θ, μ and ψ are, respectively, the theoretical diffraction angle, the absorption coefficient of the characteristic X-ray for material and the angle between normals of lattice plane and specimen surface. Conditions and results of calculation for X-ray penetration depth through diamond and silicon nitride substrate are shown in Table 2 and Fig. 3, respectively.

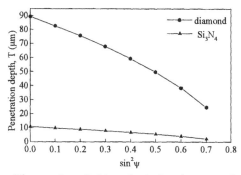

Fig.3. Penetration depth of X-ray through diamond and silicon nitride substrate.

The results of this calculation in case of short time CVD suggest that X-ray can penetrate into the substrate through the diamond coated film in the whole ψ angle range. However, it is too thick to penetrate the diamond film in higher ψ angle range in case of long-time CVD. It should be noted that the measured stress value for silicon nitride is the weighted average stress from the substrate surface up to 10μm depth.

The 2θ-sin²ψ diagrams for silicon nitride are shown in Fig. 4. The 2θ-sin²ψ relationship for non-treated silicon nitride is apparently non-linear, as seen in Fig. 4(a), because a strong compressive stress will be introduced in the vicinity of the surface due to the grinding work. It is found that differentials against sin²ψ increase with increasing sin²ψ. However, 2θ-sin²ψ diagrams for the pretreated and CVD treated (30h) substrates shows a linear relationship, as shown in Fig. 4 (b). This would be caused by the fact that the grinding streaks disappeared and the stress was released by the acid pretreatment.

Internal stresses measured for silicon nitride treated under various conditions of pretreatment and CVD treatment times are plotted in Fig. 5. Strong compressive stress due to grinding work of non-pretreated substrate is released from about 600 MPa to

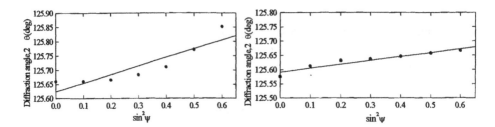

Fig.4 2θ-sin²ψ diagrams for silicon nitride substrate.
(a) non-treated, (b) CVD time, 30 h.

200 MPa by acid pretreatment. For the specimens with short CVD time (1 h and 2 h), compressive stress increased compared with that of pretreated substrate. This is attributed to the tensile stress which generates on the free surface of diamond and compensates the compressive stress at the interface between diamond and silicon nitride. Therefore, the internal stress value of silicon nitride gradually approached that of the pretreated substrate with increasing CVD treatment time.

Figure 6 shows the 2θ-$\sin^2\psi$ diagrams for diamond films obtained with the reaction times of 10, 20 and 30 h. Non-linear relationship is found in all the diagrams. This is responsible for the influences of stress distribution that exists in the diamond film and/or the preferred orientation of diamond film to 110 direction, which corresponds to the measured diffraction line. However, if the plots in all the range of $\sin^2\psi$ can be approximated as linear, the gradients of these lines would be negative. This indicates that the strong tensile intrinsic stress is compensated by a large compressive thermal stress that generates in the diamond film. It is noted by more detailed observation of these diagrams that each 2θ-$\sin^2\psi$ curve has a positive slope in the low ψ angle range, while a negative slope is confirmed in the high ψ angle range. In the middle range of

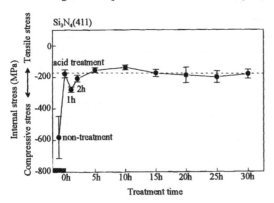

Fig.5 Internal stress measured for various conditions of pretreatment and CVD treatment times.

ψ angle, the positive slopes gradually decrease and the negative slopes shift to higher value of ψ angle. This results in the formation of a balanced region with the slopes being nearly equal to zero with increasing CVD time.

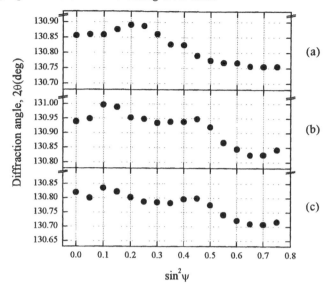

Fig.6 2θ-$\sin^2\psi$ diagrams for diamond films obtained with reaction times of (a)10 h, (b) 20 h, and (d)30 h.

Qualitative analysis was made for these diagrams considering the difference in penetration depth of X-ray at each $\sin^2\psi$. The diagram in the low ψ angle range reflects much stress information in the interfacial region between diamond and silicon nitride because the penetration of X-ray is deep. On the other hand, the diagram in the high ψ angle range reflects more information from the surface region of diamond film on account of the shallow X-ray penetration depth. This analysis concludes that thermal compressive stress works in the interface region between diamond and silicon nitride, while the intrinsic tensile stress works in the free surface of diamond film, which results in the formation of a stress balanced region where the thermal stress balances with the intrinsic stress in the diamond film. This stress balanced region expands by increasing CVD treatment time. The result agrees with the decreased compressive stress of silicon nitride, as shown in Fig. 5. In case of shorter CVD time, strong tensile stress affects the stress relationship at the interfacial region, where the intrinsic stress in thinner diamond film is predominant, so that the compressive stress in silicon nitride increases.

Diamond film has a poor adherence strength in case of shorter CVD time because of the great difference in stress between the film and the substrate. On the other hand, in case of longer CVD time, a gradual stress distribution in the coatings is formed with compressive stress mode in the substrate and the stress balanced region in the thick diamond film that compensates the intrinsic stress. The difference in stress in the interfacial region is smaller, when compressive thermal stress in the interfacial region is gradually relaxed by the stress balanced region in the diamond film. This is the reason why the adherence strength of the diamond film to substrate is improved by long-term CVD.

REFERENCES
[1]H. Itoh, S. Shimura, K. Sugiyama, H. Iwahara, "Improvement of Cutting Performance of Silicon Nitride Tool by Adherent Coating of Thick Diamond Film," *J. Am. Ceram. Soc.*, **80**[1], 189-96 (1997).

[2]H. Itoh, S. S. Lee, K. Sugiyama, H. Iwahara, T. Tsutsumoto, "Adhesion Improvement of Diamond Coating on Silicon Nitride Substrate," Surface and Coatings Technology, **112**, 199-203 (1999).

[3]M. A. Taher, W. F. Schmidt, H. A. Naseen, W. D. Brown, A. P. Malshe, "Effect of Methane Concentration on Physical Properties of Diamond-coated Cemented Carbide Tool Inserts obtained by Hot-filament Chemical Vapor Deposition," *J. Mater. Sci.*, **33**, 173-182 (1998).

[4]S. K. Choi, D. Y. Jung, H. M. Choi, "Intrinsic Stress and its Relaxation in Diamond Film Deposited by Hot Filament Chemical Vapor Deposition," *J. Vac. Sci. Technol.*, A**14**(1),165-169 (1996).

[5]H. Itoh, T. Osaki, H. Iwahara, and H. Sakamoto, "Nucleation Control of Diamond Synthesized by Microwave Plasma CVD on Cemented Carbide Substrate," *J. Mater. Sci.*, **26** [15] 3763-68 (1991).

[6]T. Sasaki, M. Kuramoto, Y. Yoshioka, "Determination of stress Distribution in the Thickness Direction of Thin Films by Means of X-Rays," *Journal of the Japanese Society for Non-Destructive Inspection*, **42**(1), 37-43(1993).

MICROSTRUCTURE AND PROPERTIES OF GRADED METAL-CARBIDE CLADDINGS

Carmen Theiler, Thomas Seefeld, Gerd Sepold
BIAS, Bremen Institute of Applied Beam Technology
Klagenfurter Str. 2
D-28359 Bremen

ABSTRACT
The powder-fed laser beam cladding process is an attractive processing method to produce a graded metal-carbide multilayer structure by varying the chemical composition starting from a metal-rich composite up to a carbide-rich outer region. Detailed investigations have been carried out using NiBSi as matrix alloy and chromium carbide (Cr_3C_2) as hard and wear resistant phase. The particle shape and the distribution of the reinforced phase were determined for cladded graded structures using chromium carbide with two different particle sizes. The dependence of the microstructure and the wear resistance on the carbide content and the laser process parameters were determined.

INTRODUCTION
Surface engineering has been defined as "the application of surface technologies to engineering components in order to produce a composite material with properties unattainable in either the base or surface material".[1] For instance, the life cycle of components may be extended by application of wear resistant claddings consisting of a high content of hard carbide particles embedded in a ductile metal matrix. These metal matrix composite materials, however, have physical and mechanical properties different from the base material. Under service load, stress related failure may occur due to the mismatch at the interface between substrate and surface layer; hence the application of a graded layer that has a high carbide content at the surface in order to meet the functional requirements and a high metal content in the vicinity of the interface to match the base metal properties.[2,3] Functionally graded materials (FGM) and graded surface layers can be produced using a laser beam cladding technique that allows a locally limited heat input and a good control of the microstructure.[4,5]

EXPERIMENTAL
In laser beam cladding, a consumable powder is fed continuously into a melt pool on the surface of the substrate. The melt pool is generated and maintained through the interaction with the laser beam which forms a high intensity heat source. As the substrate is scanned relative to the beam, the material solidifies and a clad track is deposited on the surface (Figure 1).

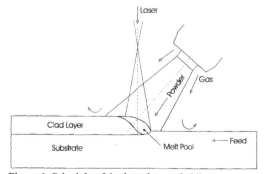

Figure 1. Principle of the laser beam cladding process

A layer is formed successively by tracks deposited side by side and multilayer structures are generated by depositing multiple layers on top of each other.

For dosing the powder mixture in the desired composition two separate pneumatic powder hoppers – one filled with the matrix alloy powder, the other filled with the carbide powder – are used. Each of these powder hoppers can be controlled separately. The two powder streams are fed together and the ratio of the metal and hard phase powder can be adjusted exactly.[4,5]

A nickel-base alloy (NiBSi) as matrix alloy and two chromium carbide powders (Cr_3C_2) with different particle sizes as hard phase are used, where in the following text the finer chromium carbide is denominated as "I" and the coarser one as "II". Table I shows the chemical composition, the grain size and their morphologies.

Starting with the nickel alloy, the amount of chromium carbide in the consumable powder mixture is increased steadily to generate multilayer structures with a compositional gradient.

A fibre-coupled 600 W cw Nd:YAG laser with 200 mm focussing optics has been used. Working in the focus with a beam diameter of 0.6 mm, the graded structures were generated. For this study, only the relative velocity of substrate and laser beam (20 – 40 mm/s) are varied. All other process parameters (laser power, the offset between two subsequent clad tracks and layers, powder feedrate) were kept constant.

Table I. Particle size, morphology and chemical composition of the used powders

Powder	NiBSi	Cr_3C_2 (I)	Cr_3C_2 (II)
Particle size	-105 +45 µm	-45 +05 µm	-106 +45 µm
Morphology			
80 µm			
Chemical composition	Si 2,30 %; B 1,39 %; Fe 0,06 %; Ni balance	Fe 0,29 %; Cr 86,0 %; C 13,43 %	Fe 0,21 %; Cr 87,06 %; C 12,97 %

For microstructural investigations and the determination of the mechanical properties, multilayer structures with a step-wise gradient in the chemical composition were produced. The samples allow a direct correlation between composition and microstructure or material properties, respectively. The wear tests are carried out using samples with a constant composition.

The microstructural examinations were carried out on cross sections with quantitative image analysis (Image C). The size and the shape of the carbides for both material combinations, depending on the chromium carbide content (0 – 70 vol% Cr_3C_2) and on the velocity, were determined. The wear investigations are carried out using a wear tribometer. An aluminium oxide ball, which is mounted on a pin, loads the cladded samples with a specified force (F = 10.7 N). The ball is moved linearly over the surface of the sample with a fixed number of cycles. Because of the friction between ball and sample a linear track of wear arises, and the cross section area of the track was measured afterwards. In dependence on the chemical composition and the relative velocity, the rate of wear for different numbers of cycles (10,000 up to 100,000) was determined.[6]

MICROSTRUCTURE

A longitudional section of a graded metal- carbide particle cladding is given in Figure 2. The image covers carbide contents from 30 vol% (at the bottom) to 70 vol%, and the compositional gradient is clearly evident.

Figure 3 shows the typical microstructure of a metal-carbide particle composite. In general, with an increase of the added carbide the microstructure is characterised by an increasing number and size of undissolved or partly dissolved carbide powder particles (1). The carbide particles dissolve from outside to inside, consequently the particles may lose their blocky shape, become round and their size is reduced. At a total dissolution they reprecipitate in needle-shape (2) mainly in the modification Cr_3C_2.[7]

2 mm

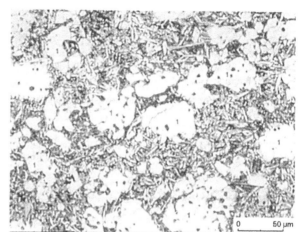

0   50 µm

Figure 2.
Longitudional section of a gradient material with
30 – 70 vol% Cr_3C_2 (I)

Figure 3. NiBSi + 50 vol% Cr_3C_2 (II), v_s = 20 mm/s

(1) blocky, primary chromium carbides
(2) needle-shaped, secondary chromium carbides

The kinetics of carbide dissolution and consequently the size, number and shape of the undissolved carbides depend on the process parameters, particularly on the heat input. Temperature and interaction time determine the degree of dissolution of the carbide powder particles and thus the chemical composition of the matrix as well as the precipitations of carbides from the melt.[4,8,9]

Generally, the number of undissolved carbides increases with increasing velocity, whilst the number of precipitated secondary carbides is reduced, Figure 4. The porosity, however, is particularly present at higher velocities (Figure 4c), when the heat input is low and gases are trapped in the cladding. The higher porosity may occur as a result of the decomposition of the carbides if process parameters are detrimental. The best result is obtained using a velocity of 30 mm/s. The embedded chromium carbide is mainly undissolved and homogeneously distributed. Increasing the interaction time (v_s = 20 mm/s), a high fraction of the carbide is dissolved and reprecipitated.

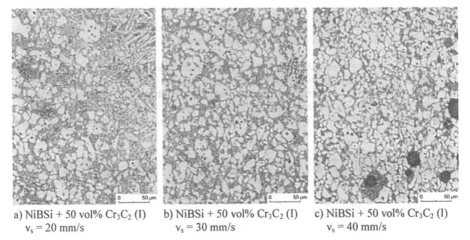

a) NiBSi + 50 vol% Cr_3C_2 (I)
   $v_s = 20$ mm/s

b) NiBSi + 50 vol% Cr_3C_2 (I)
   $v_s = 30$ mm/s

c) NiBSi + 50 vol% Cr_3C_2 (I)
   $v_s = 40$ mm/s

Figure 4. Effect of the relative deposition velocity on the microstructure of NiBSi + 50 vol% Cr_3C_2 (I)

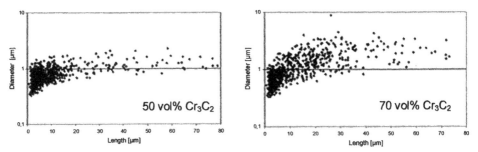

Figure 5. Diameter and length of the precipitations depending on the carbide content for NiBSi + Cr_3C_2 (I), $v_s = 20$ mm/s

Quantitative micorstructural analysis was carried out to explain and quantify the influence of process parameters and chemical composition on the microstructure and wear properties. As the dissolution behaviour of the added carbide particles is of special interest, their number, shape and area fraction was determined. The dissolution process or behaviour is independent on the powder combination. With increasing interaction time (i.e. decreasing velocity) and increasing carbide content the number of secondary precipitations increases and they become longer and thicker (Figure 5).

Indeed, the dissolution starts earlier using the coarser chromium carbide powder in contrast to the carbide I (Figure 6). At a velocity of 20 mm/s the first needle-shaped precipitations occur at 40 vol% of the chromium carbide II (Figure 6b), whereas undissolved carbides of the finer particle powder occur using the same process parameters (Figure 6a). The diagrams illustrate that the added carbide content, which is fed into the melt pool, is in good agreement with the total carbide content detected by the analysis software. By X-ray diffraction phase analysis mainly, Cr_3C_2 type carbides and only small quantities of the chromium rich Cr_7C_3 type carbides have been found.[6]

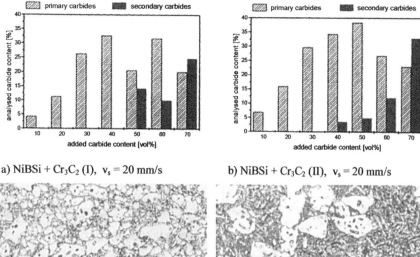

a) NiBSi + Cr_3C_2 (I), v_s = 20 mm/s

b) NiBSi + Cr_3C_2 (II), v_s = 20 mm/s

c) NBSi + 40 vol% Cr_3C_2 (I), v_s = 20 mm/s

d) NiBSi + 40 vol% Cr_3C_2 (II), v_s = 20 mm/s

Figure 6. Microstructure of NiBSi + Cr_3C_2 depending on the chromium carbide size

WEAR PROPERTIES

Independent of the particle size of the chromium carbide, the total wear [μm^2] increases with increased number of cycles (Figure 7a + Figure 7b) and decreasing carbide content (Figure 7c + 7d). Varying the relative velocity the best wear result is obtained at v_s = 30 mm/s (Figure 7a + Figure 7b). Remarkable is the jump of the total wear at 100,000 cycles using the coarser chromium carbide (Figure 7b).

The results of the wear tests can be correlated with the microstructure. Both the wear resistance and the microstructure show the best results at a velocity of 30 mm/s. The microstructure is nearly defect-free and most of the added chromium carbide is embedded undissolved in the metal matrix (Figure 4). The good quality of the metal-carbide composite is responsible for the low rate of wear. At v_s = 20 mm/s the hard phases dissolve; at a higher velocity (v_s = 40 mm/s) defects occurred like bonding problems, cracks or pores.

At a number of cycles lower than 75,000 the wear resistance using the coarser carbide (II) is more favourable comparable to the finer one (I). At higher number of cycles it is vice versa. Figure 8 shows the failure behaviour of the metal-hard phase composite. Generally the hard phases break into pieces under the load. The interface strength between metal matrix and hard particles is excellent, no detachment of whole particles is observed. Beyond that, only primary carbides fail, needle-shaped precipitated chromium carbides show no failure. Using a finer chromium carbide

(I), the hard particles are very small and homogeneously distributed, the load is distributed on a higher number of particles. The failure of the coarser Cr_3C_2 particles significantly weakens the microstructure and the wear resistance falls rapidly off. Only the primary chromium carbides break into pieces, the needle-shaped precipitations don't fail during the wear testing and are well embedded in the metal matrix.

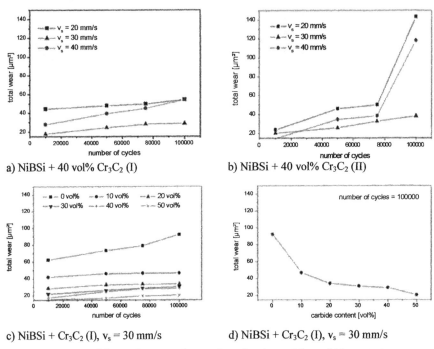

a) NiBSi + 40 vol% Cr_3C_2 (I)

b) NiBSi + 40 vol% Cr_3C_2 (II)

c) NiBSi + Cr_3C_2 (I), v_s = 30 mm/s

d) NiBSi + Cr_3C_2 (I), v_s = 30 mm/s

Figure 7. Dependence of the total wear on the number of cycles and the carbide content

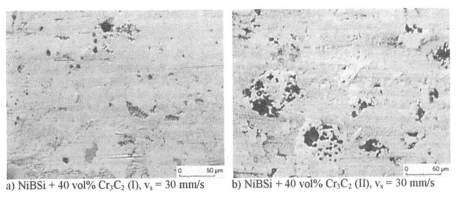

a) NiBSi + 40 vol% Cr_3C_2 (I), v_s = 30 mm/s   b) NiBSi + 40 vol% Cr_3C_2 (II), v_s = 30 mm/s

Figure 8. Images from NiBSi + Cr_3C_2 after the wear testing (number of cycles 100,000)

CONCLUSIONS

Multilayer coatings with a gradient in the chemical composition can be produced using the powder-fed laser beam process. The concentration gradient allows the generation of metal-carbide composites with a high content of hard particles.

Independent of the particle size ratio between the metal and the carbide powder, crack-free samples can be produced at similar process parameters. The dissolution behaviour of the added hard phases happens in the same manner.

With decreasing relative velocity and/or increasing carbide content the added chromium carbides (Cr_3C_2) dissolve and precipitate mainly in the same modification. Using a coarser chromium carbide the dissolution process starts earlier in comparison to a finer one.

The wear test results are excellent, if cladded multilayer structures are defect-free and the hard particles are well embedded in the metal matrix. Increasing the carbide content the capacity of resistance to wear increases. Using the coarser chromium carbide (II) a failure is observed at higher number of cycles. The embedded hard particles break into pieces and quarry out, whereas the bonding strength between metal matrix and hard phases remains strong.

ACKNOWLEDGEMENTS

The authors would like thank the German Research Association (DFG) for their beneficial support of the research described in this paper.

REFERENCES

[1] T. Bell, "Surface engineering: past, present, and future," *Surface Engineering*, 6 [1] 31-40 (1990).

[2] S. Suresh and A. Mortensen, "Fundamentals of Functionally Graded Materials"; IOM Communications Ltd, London (1998).

[3] R. Rabin and I. Shiota, "Functionally Gradient Materials," *MRS Bulletin*, XX [1] 14-16 (1995).

[4] M. Gremaud, J.D. Wagnière, A. Zryd and W. Kurz, "Laser Metal Froming: Process Fundamentals," *Surface Engineering*, 12 [3] 251-259 (1996).

[5] T. Seefeld, C. Theiler, E. Schubert and G. Sepold, "Laser Generation of Graded Metal-Carbide Components"; pp. 459-466 in *Functionally Graded Materials 1998*, Edited by W.A. Kaysser. ttp trans tech publications Ltd, Switzerland (1999).

[6] "Verschleiß-Meßgrößen," DIN 50321, Dez., 1979.

[7] C. Theiler, T. Seefeld, E. Schubert and G. Sepold, "Laser Beam Cladding of Graded Layers and Freeform Components"; pp. 455-460 in *ECLAT 1998*, Edited by B.L. Mordike. Werkstoff-Informationsgesellschaft mbH, Frankfurt (1998).

[8] U. Draugelates, B. Bouaifi and B. Ouaissa, "Einfluss der Carbidauflösung auf die Eigenschaften hartstoffverstärkter Schutzschichten," *Schweissen & Schneiden*, 52 [1] 12-17 (2000).

[9] A. Luft, A. Techel, S. Nowotny and W. Reitzenstein, "Microstructures and Dissolution of Carbides Occuring during the Laser Cladding on Steel with Tungsten Carbide Reinforced Ni- and C-Hard Alloys," *Prakt. Metallogr.*, 32 [5] 235-247 (1995).

ROLE OF IMPERFECTIONS AND INTRINSIC STRESSES ON THE THERMO-MECHANICAL AND TRIBOLOGICAL PROPERTIES OF THERMAL SPRAYED METAL-CERAMIC FGMS

Lubos Prchlik, Jiri Matejicek, Anirudha Vaidya and Sanjay Sampath
Center for Thermal Spray Research
Department of Materials Science
State University of New York at Stony Brook
NY 11794-2275

ABSTRACT

Functionally graded oxide-metal thermal spray coatings are finding increased applications in industrial components. The versatility, economics and inherent flexibility of thermal spray processes offer unique advantages for the synthesis and application of such coatings. Thermal spray processes produce layered microstructures with some degree of porosity and microcracking especially in the ceramic component. The coatings also contain process induced quenching and thermal stresses, which can affect both the fabrication of thick coatings and their performance. The spray method and process conditions directly affect these characteristics. In this study, oxide-metal FGM – NiCrAlY/Yttria Partially Stabilized Zirconia - coatings have been prepared by the plasma spray process. Graded as well as composite deposits have been prepared and their properties investigated. The intrinsic stresses, microstructures and elastic modulii of the coatings were investigated. The in-situ curvature technique applied to FGM allowed predictions of the distribution of stresses through the thickness of the FGM, and represents a powerful tool for the prediction of contact response of such coatings.

INTRODUCTION

Using plasma spray, layered and continuous FGMs can be prepared [1]. Layered FGMs have been, in fact, used for a long time in the form of so-called bond coats improving the adhesion between a substrate and a ceramic coating. Preparing graded materials by thermal spraying is based on the simultaneous deposition of multiple constituents, whose ratio can be changed during the deposition process. One can, therefore, relatively easily prepare continuously graded coatings, whose continuity or discreetness is, nevertheless, to a certain extent limited by the intrinsic process characteristics. Since the thickness of a single splat ranges from 1 to 10 μm, a 'continually' graded coating consists of more or less uniform layers of a thickness of the order of tens of microns. The main advantage of the plasma spraying is that the process enables the build-up composite materials consisting of components of very different properties.

In situations where a coating is exposed to more than one type of aggressive environment, composite and/or graded materials can be designed to best satisfy the requirements of a given application. In some applications, such as abradable seals for turbine shrouds, the coating thickness requirements are rather high (several hundreds of microns). In such situations, it is difficult, and often impossible, to fabricate thick deposits without using a grading approach to minimize process induced thermal stresses and delamination.

The use of graded microstructures allows tailoring of the abradability and thermal barrier effectiveness [2]. For instance FGMs of M-CrAlY alloys with Yttria Partially Stabilized Zirconia (PSZ) will allow for enhanced adhesion, reduced interfacial stresses and enhanced reliability for relatively low temperature applications (< 1000°C), where metal oxidation is not severe. Applications for graded coatings include high temperature abradable coatings on turbine shrouds (both aero-engines and land-based turbines), thick thermal barriers on valves and pistons in diesel engines, etc. Important issues, such as high temperature deformation, thermal fracture and delamination, have already been addressed from experimental [3-5] and theoretical points of view [6]. In order to fully describe and understand the delamination process within the thermal spray coating and at the interface, a knowledge of residual stress distribution is important. The main objective of the present work was to synthesize graded PSZ/NiCrAlY coatings and to evaluate the residual stress and elastic modulus by straight-forward and industrially relevant techniques.

Residual stresses in coatings can be measured by various techniques [7]. Here, *in-situ* curvature technique was used, and the elastic modulus and quenching stress were obtained from curvature/temperature spraying records. 'Quenching stress' is a result of rapid quenching of the molten droplet upon impact on the substrate. Thermal mismatch stress is generated due to different thermal expansivities of the coating and the substrate. Residual stress is then a superposition of both contributions [8].

EXPERIMENTAL METHODS

Synthesis

Prior to FGM processing, a set of six uniform PSZ/NiCrAlY coatings with different compositions was prepared. This set consisted of 100% PSZ, 100% NiCrAlY coatings and four mixture coatings graded in 20% steps (20, 40, 60 and 80 percent of NiCrAlY). Commercially available NiCrAlY and PSZ powders (Praxair 346/1 and Metco 204NS) were sprayed using a Metco 3MB plasma spray system consisting of independently controlled powder feeders for the metal and ceramic. The powders were introduced at specified ratios through a single external port perpendicular to the flame line using a Y-connection. Monitoring of deposition efficiencies allowed preparation of composites of the desired compositions with an error less than 5%. Details of the synthesis and characterization of more complex grading profiles are provided in [9]. Optimized spraying conditions are listed in Table I.

Table I: Plasma spray parameters.

Parameter	Value
Gun power	600 – 650 Amps
Primary gas flow	40 lpm
Secondary gas flow	10 lpm
Carrier gas flow	2.5 lpm (divided between powders)
Spray distance	100 mm
Total feed rate	40 g/min

FGMs were synthesized under similar spraying conditions. Graded coatings consisted of six layers varying in 20% steps.

Testing methods

Coating samples, prepared by standard metallographic techniques, were used to evaluate the microstructure and cross-sectional hardness. All indentation measurements were carried out on Wilson/Instron tester TU 230-C2550. The duration of hardness tests was 15 seconds, and reported values were averaged from a minimum of ten measurements.

The calculation of quenching stress and elastic modulus from in-situ temperature/curvature records is a well-established method for a relatively fast evaluation of important properties of thermal spray coatings [10]. The method is based on a simultaneous monitoring of temperature and curvature of the sample during coating deposition. The curvature is measured by non-contact sensors; the temperature on back-and-front of the sample was monitored by a thermocouple and pyrometer, respectively.

Quenching stress in a newly deposited thermal sprayed layer can be calculated from curvature change before and after the layer deposition. When deposition of a new layer with small thickness h_d onto the substrate with thickness h_s ($h_d \ll h_s$) is considered, one obtains from moment equilibrium [10,11]:

$$\sigma_q = \Delta\kappa \frac{E_s}{(1 - v_s)} \frac{h_s^2}{6h_d} \qquad (1)$$

where σ_q is the quenching stress, $\Delta\kappa$ the observed curvature change, h_d is the deposit thickness and v_s, E_s, h_s are Poisson ratio, elastic modulus and thickness of the substrate. The constant $(1-v_s)$ in the denominator represents the assumption that coating is sprayed onto a wide beam is assumed. Similarly, the elastic modulus can be calculated for the post-deposition cooling record as [11]:

$$E_d = \frac{\Delta\kappa}{\Delta T} \frac{h_s^2 . E_s}{(\alpha_s - \alpha_d)6h_d} \frac{(1 - v_d)}{(1 - v_s)} \qquad (2)$$

where v_d is the Poisson ratio of the deposit; α_s and α_d are the thermal expansion coefficients of substrate and deposit, respectively. Evaluation of quenching stress and elastic modulus directly from an FGM spraying record becomes somewhat more complex, since the position of neutral axis and the stiffness of substrate are gradually changing during FGM build-up. The details of the derivation based mainly on relationships of linear solid mechanics for layered materials [11,12] will be published elsewhere [13].

The method of stress-modulus determination through thickness can be summarized as follows.
1) Temperature/curvature record is obtained from FGM spraying.
2) Thickness of each layer is evaluated from knowledge of final thickness of the FGM and relative deposition efficiencies of each layer. The thickness can also be measured *in-situ* by an additional sensor, or post processing on a cross-section sample.
3) Cooling curves for each layer are separated, $\Delta\kappa_i/\Delta T$ values are evaluated for layers with known values of the coefficient of thermal expansion α (CTE). Using the least square fit method, elastic modulus E_i of these layers is calculated.
4) For remaining layers, CTE is measured or estimated by the rule of mixtures (constant stress or constant strain model), and E_i are obtained as in step 3.
5) Using curvature change from each spraying cycle, the quenching stress in each layer is calculated.
6) Based on the knowledge of σ_q, E and v for each layer, the resulting stress distribution in FGM can be predicted [11].

RESULTS AND DISCUSSION

Microstructure and hardness
        Fig. 1 shows the microstructures of both mixture coatings and the FGM. Table II presents the values of Vickers microhardness measured on the cross-sections of the mixture coatings. The highest values were not observed for the pure ceramic layer, but rather for layers with a addition of NiCrAlY (40%), which corresponds to the reduction of porosity and overall toughening of the structure.

Table II: Hardness variation through thickness.

Composition (%PSZ)	0	20	40	60	80	100
HV_{300} [kg/mm^2]	291±10	327±22	368±40	487±42	452±38	439±82

Measurement of elastic modulus and residual stress by in-situ curvature technique
        A typical record from the FGM spraying is shown in Fig.2. It can be seen that the sections for each spraying cycle are separated by the cooling intervals. These cooling curves are used for elastic modulus calculation. Curvature changes after deposition of each layer are used to estimate the quenching stress in each of the layers. Composition, thickness, position of neutral axis (z_0) and stiffness (Σ_n) are summarized for each spraying step in Table III. Coefficients of thermal

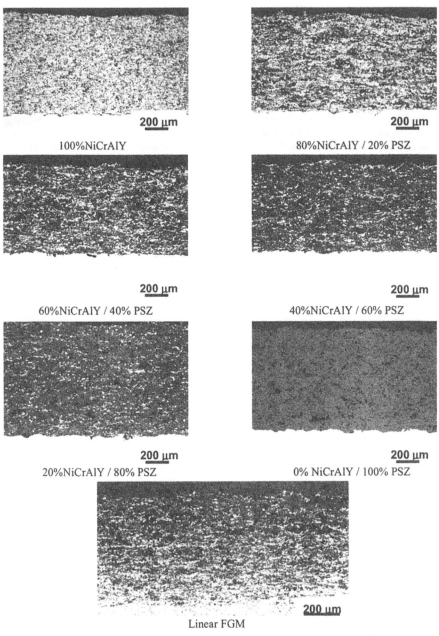

Fig. 1 NiCrAlY/PSZ composites and FGM prepared by plasma spraying.

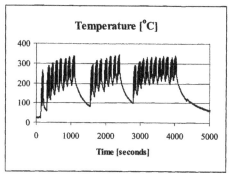

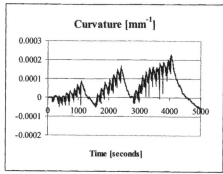

Fig. 2 Typical curvature and temperature records from FGM spraying.

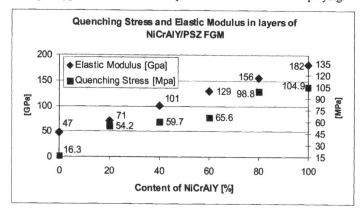

Fig. 3 Plot showing the dependence of elastic modulus and quenching stress on the composition of FGM layers.

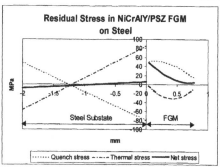

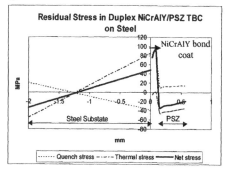

Fig. 4 Through thickness profiles for graded and duplex coatings with approximately equivalent thermal resistivity.

expansion were measured for 100%NiCrAlY and 100% PSZ coatings. For the composites, following constant strain approximation was used ($V_{NiCrAlY}$ is the volume fraction of metallic phase):

$$\alpha = \frac{\alpha_{NiCrAlY}E_{NiCrAlY}V_{NiCrAlY} + \alpha_{PSZ}E_{PSZ}(1 - V_{NiCrAlY})}{E_{NiCrAlY}V_{NiCrAlY} + E_{PSZ}(1 - V_{NiCrAlY})} \qquad (3)$$

Calculated vales of coefficient of thermal expansion (α), elastic modulus (E) and quenching stress are presented on right side of the same table.

The chart in Fig. 3 shows that the values of E and QS approximately follow the rule-of-mixtures. It is worth mentioning that the errors in the calculation of modulus and stress for top layers are much larger than the errors for the layer closer to the substrate. This is caused by lower values of quenching stress and modulus of subsequent layers and also by the fact that these layers are deposited onto a thicker and stiffer 'substrate'. This causes reduction of curvature change and increases the relative error of its measurement.

Table III. Details of FGM synthesis and el. modulus/quenching stress calculation from *in-situ* curvature/temperature records.

Layer (step)	Compo-sition	Thickness [mm]	z_0 [mm]	Σ_n [Nm²]	α [10^{-6}K^{-1}]	E [Gpa]	$\Delta\kappa$ [m^{-1}]	QS [Mpa]
Substrate	Steel 4040	1.95	0.98	130	13	(210)	-	-
1	0% PSZ	0.19	1.06	166	12.41	182	0.12	104.9
2	20% PSZ	0.16	1.12	197	12.27	156	0.09	98.8
3	40% PSZ	0.14	1.16	224	12.08	129	0.05	65.6
4	60% PSZ	0.13	1.20	248	11.78	101	0.04	59.7
5	80% PSZ	0.13	1.23	266	11.26	71	0.04	54.2
6	100% PSZ	0.11	1.24	278	10.13	47	0.01	16.3

Experimentally obtained values of elastic modulus and quenching stress were used to predict the residual stress profile through thickness. Fig 4a shows the residual stress profile for the graded coating fabricated in this study. Fig. 4b, on the other hand, shows the residual stress profile in a 'traditional' duplex coating, whose thickness was adjusted to have the same thermal resistance as the graded coating. Material composition grading approach reduces processing induced stresses in the steel substrate. The stress gradients at the substrate/bond coat and bond coat/coating interfaces are smaller as well. At the same time, the surface PSZ layer stress shifts from a compressive in duplex coating to a near-zero value in the FGM. The effects of this stress-level change on spalling, thermal cycling behaviour and contact damage response need to be evaluated.

SUMMARY AND CONCLUSIONS

NiCrAlY-PSZ FGM and composites were successfully synthesized with plasma spray technique. The coatings prepared were characterized by traditional techniques as well as an *in-situ* curvature method that has not been commonly used for graded coatings earlier. The method yielded consistent results of elastic modulus and quenching stress, even though the accuracy of analysis may be lower when a layer with low quenching stress and elastic modulus is deposited onto a thick substrate. Various experimental and analytical aspects of this method are currently being evaluated.

In the system studied, both the elastic modulus and quenching stress approximately followed the rule-of-mixtures. Elastic modulus and the quenching stress were gradually decreasing with increasing zirconia content. The method yields the overall stress in each layer and is not capable of distinguishing the stress levels in each of the phases present. The deconvolution of stress in each phase could be achieved either by diffraction techniques or estimated using thermal and mechanical properties of each constituent. Nevertheless, a knowledge of quenching stress and elastic modulus enables prediction of stress distribution through the whole FGM thickness. Significant reduction of stress at the coating/substrate interface was shown. The effect of this

140

residual stress variation on coating properties – including contact response and thermal cycling - is currently being studied.

The presented method in conjunction with other characterization techniques can serve as a powerful tool for the prediction of coating delamination and tribological response.

REFERENCES

[1]S. Sampath, H. Herman, N. Shimoda, and T. Saito, "Thermal Spray Processing of FGMs," *MRS Bulletin*, **20**[1] 27-34 (1995).

[2]S. Sampath, L. Prchlik and K. Kishi, Synthesis, "Thermo-mechanical Properties and Frictional Response of NiCrAlY-Zirconia Composite and Graded Coatings", Thermal Spray – Surface Engineering via Applied Research, Ed. C.C. Berndt, Proc. of the First International Thermal Spray Conference, ASM International, Montreal, Canada, 1227-32 (2000).

[3]S. Kuroda, T. Fukushima and S. Kitahara, "Significance of the Quenching Stress in the Cohesion and Adhesion of Thermally Sprayed Coatings", Proc. of Int. Thermal Spray Conference, Orlando-USA, 903-909 (1992).

[4]Y.R. Takeuchi and K.Kokini, "Thermal Fracture of Multilayer Ceramic Thermal Barrier Coatings", *Transactions of the ASME*, **116** 266-71 (1994).

[5]A. Kawasaki, R. Watanabe, Materials Science Forum Vols. 308-311, 402-9 (1999).

[6]U. Leushake, Y.Y. Yang, W. Shaller, Materials Science Forum Vols. 308-311, 936-41 (1999)

[7]J. Matejicek, Processing Effects on Residual Stress and Related properties of Thermally Sprayed Coatings, PhD Thesis, State University of New York, 1999.

[8]J. Matejicek, S. Sampath, P.C. Brand and H.J. Prask, "Quenching, Thermal and Residual Stress in Plasma Sprayed Deposits: NiCrAlY and YSZ Coatings", *Acta mater.*, **2** 407-617 (1999).

[9]S. Sampath, W.C.Smith, T. Jewett, and H. Kim, Materials Science Forum Vols. 308-311, 383-88, (1999).

[10]S. Kuroda, T. Dendo and S. Kitahara, Quenching Stress in Plasma Sprayed Coatings and its Correlation with the Deposit Microstructure, *J. of Thermal Spray Technology*, **4**[1] 75-84 (1995).

[11]Y.C. Tsui, T.W. Clyne, "An Analytical Model for Predicting Residual Stresses in Progressively Deposited Coatings," *Thin Solid Films*, **306** 23-33 (1997).

[12]O. Kesler, M. Finot, S. Suresh and S. Sampath, Determination of Processing-Induced Stresses and Properties of Layered and Graded Coatings: Experimental Method and Results for Plasma-Sprayed Ni-Al$_2$O$_3$, *Acta mater.*, **43**[8] 3123-34 (1997).

[13]L. Prchlik, A. Vaidya, S.Sampath, Direct Evaluation of elastic modulus and quenching stress from FGMs spraying records, in preparation.

OXIDATION RESISTANCE OF CARBON/CARBON COMPOSITES COATED WITH A Si-MoSi₂ BY SLURRY DIPPING PROCESS

Jae-Ho Jeon, Yoo-Dong Hahn
Department of Materials Engineering, Korea Institute of Machinery and Materials, 66 Sangnam-Dong, Changwon 641-010, Korea

Hai-Tao Fang, Zhong-Da Yin
Materials Science and Engineering School, Harbin Institute of Technology, Harbin 150001, P.R.China

ABSTRACT

Carbon-carbon (C/C) composites have received increased attention as structural materials for aerospace application due to their attractive physical properties. Because all the carbon forms oxidize above 400°C causing catastrophic damage, the development of reliable oxidation protection is crucial to utilizing the full potential of C/C composites. Pure Si inner layer and Si-Mo outer layer was coated to a 2-D C/C composite by a slurry dipping process, and then it was heat-treated at 1420°C for 10min in a vacuum. The oxidation resistance of the C/C composite with coated layers was examined and discussed in terms of microstructure and phase.

INTRODUCTION

Carbon fiber reinforced carbon matrix (C/C) composites are promising structural materials for use in hypersonic aircraft, re-entry vehicles, rocket nozzles, and nuclear fusion reactor due to their outstanding thermomechanical properties such as high specific strength at elevated temperature and thermal shock resistance.[1] However, a serious drawback of C/C composites is the poor oxidation resistance at the temperature over 450°C in environments containing oxygen.[1-4] Utilization of the full potential of C/C composites requires the development of a reliable oxidation protection system for C/C composites to prevent oxygen attack.[2-4] One of important techniques to enhance the oxidation resistance of C/C composites is to introduce protective ceramic coatings on the surface of the composites. The coating must be oxidation resistant, must have low oxygen permeability, and must be compatible with the composite substrate.

The chemical vapor deposition (CVD) process is widely used for producing a dense and uniform ceramic coating on C/C composites. But ceramic coatings prepared by a conventional CVD process suffer from high residual thermal stress caused by large mismatch in coefficient of thermal expansion (CTE) between the coatings and C/C composites. Thus, spallation of coating layer occurred frequently during thermal cycling oxidation test, or even during the cooling from the high CVD deposition temperature to room temperature. This phenomenon is responsible for the breakdown of the oxidation resistance.[5-7]

Slurry dipping process is attractive, since it is simple, inexpensive, and offers good processibility and compositional flexibility.[8-9] Furthermore, due to the infiltration of the coating materials into pores of the substrates and the formation of carbide interface layers, the coatings applied by this approach exhibited excellent adhesion to the substrate, thereby achieving good thermal shock resistance. In this work, a Si-MoSi$_2$ coating was fabricated on the surface of a two-dimensional C/C composite by a slurry dipping process followed by sintering. The relationship between microstructure and the oxidation behavior of the coated C/C composite were investigated.

EXPERIMENTAL

Specimens (3×4×20mm) used as the substrate were cut from a C/C composite bulk with a density of 1.85g/cm^3 using a diamond saw. Before coating, the specimens were hand-polished through 600 grit SiC paper and ultrasonically cleaned with acetone. The raw materials for the coating are high purity Si and Mo powder with average particle size of 7μm and 5μm, respectively. Pure Si slurry for the inner layer and a 60Si-40Mo (weight ratio) slurry for outer layer were prepared by mixing with MEK solvent and 2wt% PVB binder using magnetic stirrer for 4 h. A pre-coating with pure Si slurry inner layer of 110μm and 60Si-40Mo slurry outer layer of 550μm was applied to the C/C substrate by a dipping method. After drying, the as-coated samples were placed in a corundum crucible, and then sintered at 1420°C for 10 minutes in a vacuum of 1.33 Pa to perform metallurgical reaction, liquid phase sintering, and interfacial reaction with the substrate.

The cyclic oxidation test was performed in a corundum tube furnace at 1370°C in air flowing by natural convection. The samples placed upon a corundum support were put inside or taken out of the tube furnace directly in about thirty seconds. Several cycles with different holding time at 1370° and 1400°C were carried out to evaluate the effect of the coating on the oxidation protection of C/C substrate, and to test thermal shocks resistance of the coating simultaneously. Cumulative weight change of the samples after every thermal cycle was measured by a precision balance and was reported as a function of holding time. Surfaces and cross sections of coated C/C composites were examined using scanning electron microscopy (SEM). X-ray diffraction (XRD) was employed to provide phase identification. In addition, energy dispersive spectrometer (EDS) was used for chemical composition analysis.

RESULTS AND DISCUSSION

Microstructure and Phase Analysis

Microstructures of polished cross section of the fused slurry coating and the top region of the C/C substrate imaged by SEM are shown in Fig. 1. Cracks were found to form in the coating due to the mismatch in coefficient of thermal expansion between the coating and the substrate. The coating layer consists of

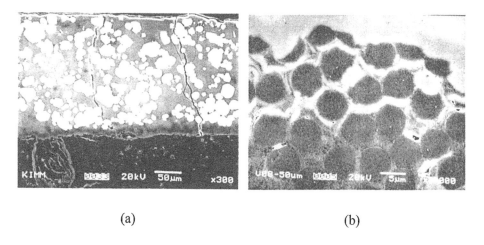

(a)   (b)

Fig. 1. SEM micrograph of the polished cross section of (a) the coating layer and (b) the top region of the C/C substrate.

the white disperse phase and the matrix phase. A continuous interface layer with the thickness of about 6μm was observed between the coating and the C/C substrate. Carbon fibers in the top region of C/C substrate are surrounded by a gray phase with the similar contrast of the interface layer, and the content of the gray phase appeared to gradually decrease from surface to the deeper region.

The surface of coating layer was directly exposed to X-ray in order to identify the phases in the coating. In the case of the phase identification in the interface layer, on the other hand, the C/C substrate was ground carefully to expose the interface layer before XRD examination. XRD patterns, as shown in Fig. 2, reveal the major phases in the coating are $MoSi_2$ and Si. The $MoSi_2$ phase is the product of the reaction between Si and Mo during sintering. SiC peaks are obvious in the interface layer despite C peaks coming from residual C/C substrate, Si and $MoSi_2$ peaks coming from the coating. The SiC interface layer was formed by the reaction between liquid Si and the C/C substrate during the sintering. Based on the XRD patterns, the X-ray mapping, and EDS analysis, the white disperse phase and the matrix phase in the coating are $MoSi_2$ and Si, respectively. It was also confirmed that the gray phase surrounding carbon fibers in the top region of the C/C composite is SiC produced by reaction between infiltrated Si and carbon in the substrate.

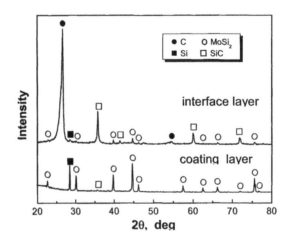

Fig. 2. XRD patterns of the coating layer and the interface layer.

The top region of the C/C composite can be stated as a functionally graded coating (FGC) because the volume ratio of SiC/C is gradually changed from surface to the deeper region. This FGC would reduce mismatch in CTE and result in the reduction of the cracking frequency. It is also expected that SiC/C FGC acted as a thermal stress relaxation layer for the outer Si-MoSi$_2$ coating.[10]

Since liquid Si is able to wet carbon materials easily, liquid Si infiltrated into the substrate under the sintering condition mentioned above as seen in Fig. 1. The infiltration of liquid Si leads to decrease the content of Si in the Si-Mo coating hence damaging the sinterability of the Si-Mo coating. It should be stressed that Si-40wt%Mo coating can not be sintered densely on the C/C substrate without Si inner layer. Therefore, it is necessary to apply pure Si slurry inner layer in order to keep the content of Si in the Si-Mo coating.

Cyclic Oxidation Test

Fig. 3 shows the results of cyclic oxidation test at 1370°C. The bare C/C composite reached 50% burn-off within 1h. In the case of the coated samples, no weight loss, but weight gain was measured after each cycle. The weight gain represents the oxide layer grown on the coating. A polished cross section of the coated C/C composite after the oxidation test for at 1370°C for 115

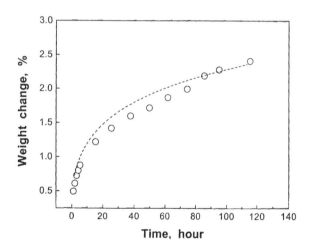

Fig. 3. Weight changes during cyclic oxidation test at 1370°C of the coated C/C composite.

hours is shown in Fig. 4. The oxide layer as inferred from the gray areas covering the coating surface integrally adheres well to the underlying coating. The formation of the integral surface oxide layer indicates that the coating oxidized in passive mode. On the other hand, no spallation of the coating was observed after the oxidation test corresponding to severe thermal shocks 14 times. It is believed that the continuous SiC reaction layer and the infiltration of Si into the substrate provide strong adhesion for the coating to prevent the spallation.

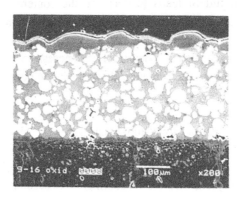

Fig. 4. Polished cross section of the coated C/C composite after the oxidation test at 1370°C for 115 hours.

Crack Healing during Oxidation Test

Without additional information given by microexamination, however, we can not conclude that this coating is able to protect the C/C composite perfectly at 1370° and 1400°C because weight gain coming from the oxidation of the coating might cover weight loss caused by the oxidation of the C/C composite substrate through the cracks. To this end, the cracks present on the surface of the coating before and after oxidation tests were investigated using SEM. Fig. 5 shows the surface morphologies on the same location of as-fused, and the first and second cycle oxidation tested samples. The oxidation test was carried out at 1370°C. It can be noticed that the position of the cracks changed after each cycle. Based on this observation, it is deemed that previous cracks healed and then disappeared in holding stage of each cycle, and that new cracks were formed during the cooling from 1370°C to room temperature. Therefore, no oxidation of the C/C substrate under the Si-MoSi$_2$ coating could take place at 1370°C due to complete healing of the cracks.

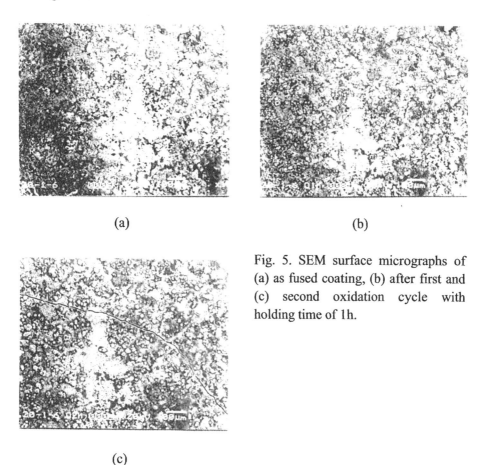

(a)

(b)

Fig. 5. SEM surface micrographs of (a) as fused coating, (b) after first and (c) second oxidation cycle with holding time of 1h.

(c)

CONCLUSIONS

The Si-MoSi$_2$ coating on the C/C composite for oxidation protection was developed by a slurry dipping method. Pure Si slurry inner layer in the percoating was necessary to fabricate a dense coating. Based on good integrity of the coating, the continuous SiC interface layer and the SiC/C FGC, the coating exhibited excellent oxidation resistance and thermal shock resistance.

REFERENCES

[1] J. D. Buckley, "Carbon-Carbon, An Overview," *Ceram. Bull.*, **67** [2] 364-68 (1988).

[2] J. E. Sheehan, K. W. Buesking and B. J. Ssullivan, "Carbon-Carbon Composites," *Annu. Rev. Mater. Sci.*, **24** 19-44 (1994).

[3] J. R. Srief and J. E. Sheehan, "Ceramic Coatings for Carbon-Carbon Composites," *Ceram. Bull.*, **67** [2] 369-74 (1988).

[4] M. E. Westwood, J. D. Webster, R. J. Day, F. H. Hayes and R. Taylor, "Oxidation Protection for Carbon Fibre Composites," *J. Mater. Sci.*, **31** 1389-97 (1996).

[5] H. Fritze, J. Jojic, T. Witke, C. Ruscher, S. Weber, S. Scherrer, R. Weib, B. Schultrich and G. Borchardt, *J. Eur. Ceram. Soc.*, **18** 2351-64 (1998).

[6] A. Sskai, N. Kitamori, K. Nishi and S. Motojima, "Preparation and Characterization of SiC-coated C/C Composites using Pulse Chemical Vapor Deposition (Pulse-CVD)," *Mater. Lett.*, **25** 61-64 (1995).

[7] H. T. Tsou and W. Kowbel, "A Hybrid PACVD SiC/CVD Si3N4/SiC Multilayer Coating for Oxidation Protection of Composites," *Carbon*, **33** [9] 1279-88 (1995).

[8] A. Joshi and J. S. Lee, "Coating with Particulate Dispersion for High Temperature Oxidation Protection of Carbon and C/C Composites," *Composites Part A*, **28A** 181-89 (1997).

[9] H. S. Hu, A. Joshi and J. S. Lee, "Microstructural Evolution of a Si-Hf-Cr fused Slurry Coating on Graphite for Oxidation Protection," *J. Vac. Sci. Technol.*, **A 9** [3] 1535-38 (1991).

[10] Y.-C. Zhu. S. Ohtani, Y. Sato and N. Iwamoto, "Formation of a Functionally Gradient (Si_3N_4+SiC)/C Layer for The Oxidation Protection of Carbon-Carbon Composites," Carbon, **37** 1417-23 (1999).

PROCESSING AND CHARACTERIZATION OF GRADED ALUMINIUM COMPONENTS WITH HIGH HARDNESS AND IMPROVED WEAR BEHAVIOR USING PLASMA TRANSFERRED ARC WELDING (PTA) PROCESSES

U. Dilthey, B. Balachov and L. Kabatnik
ISF - Welding Institute, Aachen University
Pontstrasse 49
D-52062 Aachen, Germany

ABSTRACT
Aluminium alloys are widely used materials due to their advantageous properties like excellent corrosion behavior and low density. However, applications are limited when hard surfaces with a good wear resistance are needed. The Plasma Transferred Arc (PTA) welding process has been used to modify aluminium alloys to obtain materials of improved wear behavior and high hardness exceeding HV400. Due to the differences in the structural material properties of the base material and the alloyed surface layer, a graded material composition is needed. Both different welding powders mainly based on the system Al-Cu-Ni and the influence of different process parameters, e. g. weld current, powder feeding rate and single/multi-layer welds, on the graded structure and specific properties has been investigated. Possible applications are discussed.

INTRODUCTION
Aluminium alloys show attractive physical properties such as ductility, excellent electrical conductivity, good corrosion resistance and a high strength-weight ratio. Therefore aluminium is the most widely used non-ferrous metal for several industrial applications. The performance of aluminium is limited when high hardness and/or good wear resistance are needed. Different approaches have been taken to produce hard and wear resistant coatings on aluminium surfaces using different processes, e. g., ion-plating of hard ceramics or hard anodizing of surfaces. Another way for increasing the wear resistance of aluminium alloys is the addition of ceramic particles or carbides by using PTA welding techniques.[1,2]

For most surfacing processes the thickness of coatings or affected zones in the metal is typically in the range of micrometer. However, some applications need either thick coatings or alloyed zones in the range of several millimetres. Especially thick coatings containing single phases or whole areas with different physical properties, e. g., thermal expansion, might develop cracks which can lead to the failure of complete assemblies. Hence this study is performed to investigate the possibilities of producing thick, wear-resistant coatings with a high hardness, while avoiding any cracking of the coating system. The approach of producing graded structures is one way of solution for this problem which will be discussed in this paper. As a coating method a modified Plasma Transferred Arc (PTA) welding process has been used to obtain these thick graded coatings on aluminium based materials with an increased hardness and improved wear resistance.

EXPERIMENTAL PROCEDURES

The PTA process uses a concentrated plasma arc to melt the surface of the base metal while a metal based powder is being fed into the arc, melted and supplied to the molten pool. The use of powders as filler materials has the advantage of providing a wide choice of weldable materials. While the torch is being moved above the base metal surface a weld overlay is being formed. The formation of this layer and its specific properties are dependent on the composition of the filler metal powder, the powder feed rate, the weld current, the welding speed and the torch oscillation, respectively. The heat input into the base material and consequently the resulting dilution is mainly influenced by the weld current and the movement of the torch.

Whenever aluminium alloys are welded, it is essential to remove the oxide layer. Otherwise defects in the weld overlay or the metallurgical joint between filler and base metal can result from the incomplete removal of these oxides. The melting point of aluminium oxide is 2050°C, much higher than the melting temperature of aluminium alloys which is about 660°C dependent on the alloy composition. At the ISF - Welding Institute a specially designed welding torch has been developed which allows to use the plasma nozzle as the welding electrode. The polarity of the nozzle is permanently positive, which is realized by the DCCP-connection of the welding torch (Direct Current Combined Polarity). By the means of this DCCP-connection the oxide layer is removed completely using the so-called cathodic cleaning effect. The heat input to the plasma nozzle is relatively low due to its direct cooling system. A scheme of the DCCP - PTA welding process is shown in Figure 1.[3]

As illustrated, two electric arcs are used in the process: the pilot arc and the welding arc. While the pilot arc is mainly necessary to ignite the welding arc and stabilize the welding process, the latter is being used for melting the powder as well as the base material surface. The pilot arc itself is ignited by an H.F. arc starter. Usually inert gases, e. g. argon or mixtures of argon and helium, are used. In this study argon was taken as plasma, conveying and shielding gas.

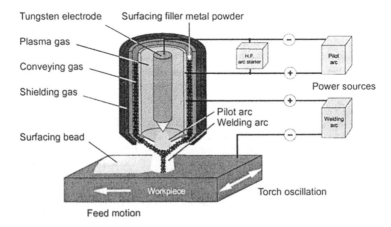

Figure 1: Principle of DCCP - PTA welding process

MATERIALS

Base materials

Base materials were aluminium alloy plates with a dimension of about 150 mm in length, 50 mm in width and a thickness between 10 and 20 mm. The respective chemical compositions are shown in Table I. Before any cladding was performed the surface had been cleaned with acetone to remove oil and other contaminations.

Table I. Chemical composition of base alloys (wt.%)[4]

Alloy	Si	Fe	Cu	Mn	Mg	Cr	Zn	Ti	Other
AlMg4.5Mn 5083	0.40	0.40	0.10	0.40-1.0	4.0-4.9	0.05-0.25	0.25	0.15	0.15
AlSi7Mg 356.0	6.5-7.5	0.6	0.25	0.35	0.20-0.45		0.35	0.25	0.15

Powders

Based on the solubility of different elements in aluminium in rapid solidification conditions, as well as ternary phase systems, weld powder compositions according to the system Al-Cu-Ni has been chosen to achieve a high hardness, a sufficient toughness of the coatings and crack-free layers.[5] The range of the different powder compositions is shown in Table II. The powders Al and a master alloy AlNi₃ were mixed mechanically before welding. Copper was fed separately. Two powder feeding systems were used to avoid a probable decomposition of the powder mixture before melting in the plasma arc due to a different density of the elements.

Table II. Weld powder alloy composition range

Alloy (wt.%)	Minimum	Maximum
Al	bal.	bal.
Cu	35	53
Ni	13	22

Welding procedures

The influence of the welding process on the coating had been studied by variation of the main welding parameters weld current, powder feeding rate and the effect of single or multi-layer welds.

Graded structures form by developing large diffusion zones between the base material and the weld overlay and by a different range of dilution between the single layers. Hence, the graded structure is more distinct over a range of several millimetres when the coating consists of more than one layer.

Coatings with up to three layers have been welded. The influence of the specific heat input has been studied by variation of the weld current for each layer between 90 A and 160 A. However, the optimum range could be found between 100 A and 140 A. If the weld current is too low the energy input is not sufficient to melt the powder completely, while weld currents exceeding 150 A lead to turbulences in the molten pool. Although the welding speed as well as the oscillation of the torch also have an influence, these two parameters were kept on a constant level.

The powder feed rate was varied in order to influence the dilution and the composition of the resulting layer. Generally, dilution is less with an increase of the powder feeding rate.

In Table III the main welding parameters used for the welding experiments are listed.

Table III. Welding parameters

Parameter	Range
Welding current (A)	100 - 160
Pilot arc current (A)	40
Welding speed (cm/min)	10
Torch oscillation width (mm)	8
Oscillation frequency (min^{-1})	100
Plasma gas (l/min) - Argon	0.5
Carrier gas 1 (l/min) - Argon	2
Carrier gas 2 (l/min) - Argon	2
Shielding gas (l/min) - Argon	10
Powder feeding rate 1,2 (g/min)	4.9 - 5.2

TESTING OF COATINGS

The coatings have been characterized by the aid of cross-sections of the weld. Vickers hardness (HV0.5) of the alloy was tested in a vertical line distribution in the centre of the weld stretching from the base material to the surface of the weld. Hardness is dependent on the material composition and the weld structure and was taken to identify gradations within the coating.

EDX analysis was additionally used to relate the phase compositions with the specific hardness level. For each measurement an area of 0.3 mm by 3.0 mm was scanned. The distance between two of those areas was 0.1 mm.

Adhesive wear testing was carried out for a duration of 10 hours using a block-on-ring testing device against 100Cr6 steel. Stainless steel samples were used as a reference. The test was done with graded aluminium specimens coated with three layers. The pressure applied was 1 MPa and wear velocity was 0.5 m/s.

For testing the wear behavior in abrasive environments a pin-on-disc configuration with abrasive paper (SiC, 180 grain size) was used. The paper was replaced after a wear distance of 25 m. The test was done with a force of 5 N and a wear testing speed of 0.63 m/s.

RESULTS AND DISCUSSION

The modified PTA welding process has proved to be a suitable method for applying thick wear resistant coatings on aluminium substrates. Besides an improved wear behavior, the modified surface of coated aluminium substrates has an increased hardness of more than 400 HV. The composition and structure of the coatings can be influenced by various factors. In addition to the selected combination of alloys for surfacing filler metal powders, the welding process itself is essential for creating a graded structure in the coating. Grading of these coatings is necessary to avoid cracks which occur due to different thermal expansion rates and brittleness within the coating as shown in Figure 2. The gradation is based on two effects. Diffusion processes can take place easily with the selected alloys mainly at the fusion line of base metal and weld material and at the interface between two respective layers. Additional grading effects are caused by different levels of dilution of the materials. The ternary alloy system Al-Cu-Ni showed a good ability of achieving a hardening effect of the materials while the composition maintains a minimum level of toughness to prevent cracking. Due to additional diffusion between two adjacent layers the effective increase/decrease of alloying elements is almost continuous.

EDX-linescans have significant differences between graded and ungraded coatings. In the ungraded sample shown in Figure 3 the respective element level reached its maximum within a distance of about 0.7 mm while the structure of a graded coating stretches from a distance up to 4.4 mm. Maximum levels of copper reached up to 60 % while the nickel content reached between

Functionally Graded Materials 2000

18 % and 28 %. The maximum numbers exceed the alloy content in the premixed powder, which can be explained by the tolerances of the EDX analysis. It can be assumed that the resulting maximum level is equivalent to the filler metal content. Corresponding micrographs show the interface of the base-material, centre alloyed area and the top part of the coating, Figure 2. Phase analysis was taken to identify the intermetallic phase Al_2Cu near the fusion line in addition to the aluminium matrix. With a higher nickel content additional phases consisting of Al-Cu-Ni were identified as shown in Figures 2 and 4. One typical phase could be identified as Cu_4NiAl_7. Cracking suggests that the toughness was reduced.

The graded coating microsections are shown in Figure 4. EDX-analysis of a graded coating is presented in Figure 5. In this case cracks could not be found either close to the fusion line or near to the surface. With increasing level of nickel and copper more intermetallic phases develop, covering most of the area near to the top of the coating.

The hardness of the alloyed matrix is dependent on different factors, e. g. content of copper - mainly forming Al_2Cu phases - as well as nickel, where additional intermetallic phases can be identified. The maximum level reached about 700 HV mostly in the top layer. In Figure 6 the hardness distribution within a coating consisting of three layers is shown. The total thickness was about 7 mm. Porosity, which is likely to occur whenever aluminium alloys are welded, was mainly found close to the surface. After machining of components, pores at the surface are usually removed.

Especially close to the fusion line between base material and weld metal, the effect of diffusion can be seen with Al_2Cu phases stretching into the aluminium matrix of the base material, which can be observed both in graded as well as ungraded coatings. This observation is more distinct in graded structures.

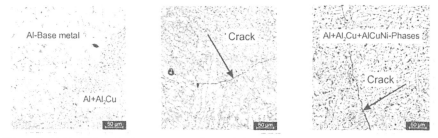

Figure 2: Al-Cu-Ni based microstructure of ungraded coating

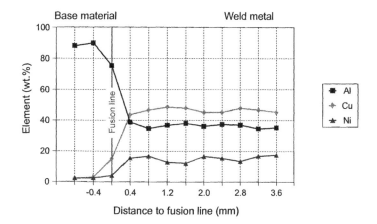

Figure 3: Al-Cu-Ni distribution in ungraded coating

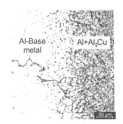 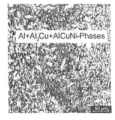 

Al-Base metal Al+Al₂Cu Al+Al₂Cu+AlCuNi-Phases Intermetallic phase

Figure 4: Al-Cu-Ni based microstructure of graded coating

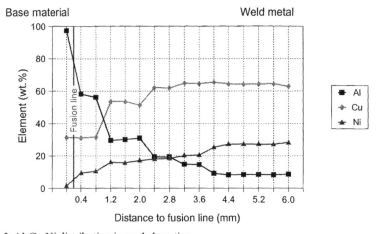

Figure 5: Al-Cu-Ni distribution in graded coating

A significant increase in wear resistance could be observed both under adhesive and abrasive wear conditions. For the modified aluminium alloy the wear volume loss as an indicator for adhesive wear resistance was 0.96×10^{-4} mm³/(N·m) which was about 33 % less compared to stainless steel.

In Figure 6 the wear testing results of a graded weld metal structure (AlCu50Ni22) is compared to the behaviour of the base material (AlSi7Mg). Two characteristic observations can be made. The composition of the base material is not changing. This is the reason why the material loss within a defined period of time is constant. As a consequence the increase of the accumulated wear loss is linear. The graded structure of the welded specimen causes a non-linear increase of the wear loss. Moreover the average material loss within a coating thickness of about 7 mm was 1.29 mg/mm², about 85 % less than the loss of the base material alloy. There was a relation between the resulting hardness, shown in Figure 7, and the corresponding abrasive wear resistance. That is the reason for an excellent wear resistant behaviour at the beginning of the test which was about 7 times less than the original alloy. Both levels of material loss of both alloys equalize when the fusion line of the welded structure is reached.

Functionally Graded Materials 2000

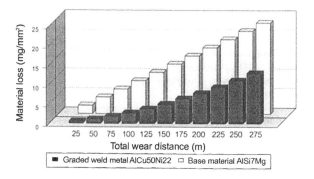

Figure 6: Wear testing results (Pin on disc), Weld metal - Base material

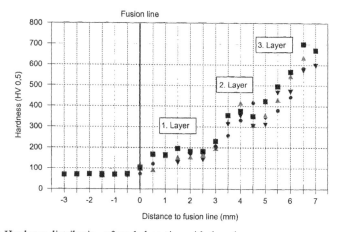

Figure 7: Hardness distribution of graded coating with three layers

CONCLUSIONS

PTA welding using an electrode with a permanent positive polarity is suitable for producing wear resistant coatings based on aluminium alloys with a high hardness. Cracks can be avoided by a graded structure stretching from the fusion line between base metal and weld material to the surface of the coating. This grading is more distinct when two or three layers are welded. The structure of the coating can be mainly influenced by the welding parameters weld current and powder feeding rate.

As materials, the ternary system Al-Cu-Ni proved to be suitable for providing high hardness as well as sufficient toughness. The content of copper was up to about 50 wt.% while a nickel content of about 20 wt.% could be achieved. The hardness reaches up to 700HV. The value is related to the material composition and dependent on the respective Ni and Cu content in the coating forming intermetallic phases like Al_2Cu or $AlNi_4Cu_7$.

The coated samples showed an improved wear behaviour. Material loss could be reduced by 85 % compared to the base material in an abrasive wear testing environment.

APPLICATIONS

Many applications can be identified for aluminium alloys with high hardness and a good wear resistance. Tools for blow moulding processes for various plastic material parts, e. g. bottles in different sizes or components for the automotive industry. For many moulds a combination of aluminium and steel alloys are used for areas with a high compressive load or increased temperature level. Production costs of these moulds are relatively high compared of mould forms entirely made of aluminium alloys.

The life span of aluminium pistons could be extended by strengthening the area of the piston ring groove. Therefore materials with high hardness as well as an excellent strength characteristic in high temperature levels are essential. Steel rings are often used to serve as a suitable base for the ring groove which could be replaced by a welded coating layer.

Wherever materials with a good wear characteristic, high hardness and low density are needed, aluminium alloys which are modified by using the PTA welding process might play an expanded role.

ACKNOWLEDGEMENTS

The authors would like to thank the Deutsche Forschungsgemeinschaft (DFG) for the financial support under grant no. DI 434 / 43 - 1. Wear testing was done at Materials Science Institute of Aachen University and at the Institute for Composite Materials Ltd., Kaiserslautern.

REFERENCES

[1] T. Hashimoto, "Formation of Hardened Layers on Aluminium Alloy Surfaces by the Plasma Transferred Arc Process", *Welding International*, **14** (2000) 3, pp. 173-177.

[2] F. Matsuda, K. Nakata, S. Shimuzu, K. Nagai, "Carbide Addition on Aluminium Alloy Surface by Plasma Transferred Arc Welding Process", *Transactions of the JWRI*, **19** (1990) 2, pp. 81-87.

[3] E. Lugscheider, G. Langer, K. Schlimbach, U. Dilthey and L. Kabatnik, " Possibilities for Improving Wear-Properties of Aluminium-Alloys by Plasma Powder Welding Process", *Proc. of United Thermal Spray Conference (UTSC '99)*, Düsseldorf, Mar 17-19, 1999.

[4] N.N., "Aluminium-Taschenbuch", Aluminium-Verlag, Düsseldorf, 1998.

[5] C. Subramaniam, "Some Considerations Towards the Design of a Wear Resistant Aluminium Alloy", *Wear*, **155** (1992), pp. 193-205.

CRACKING BEHAVIOR OF GRADED CHROMIUM NITRIDE COATINGS ON BRASS
FOR WEAR RESISTANT APPLICATIONS.

Saki Krishnamurthy and Ivar E. Reimanis
Department of Metallurgical and Materials Engineering,
Colorado School of Mines,
Golden, CO-80228

ABSTRACT
        A test designed to assess the fracture behavior of brittle coatings on ductile substrates was
applied to Cr_xN coatings on brass. Brass was chosen so that the substrate yield stress could be
systematically varied with the intent of evaluating the range of applicability of the current model
used to describe fracture. The substrate was pulled in tension, and the initiation, evolution and
morphology of cracks formed was examined using optical or scanning electron microscopy.
Single composition coatings (Cr_2N) and bilayer coatings (Cr_2N/CrN) were prepared by sputter
deposition. Initial results show that the bilayer coating system exhibited higher fracture resistance
than the single composition coatings. Potential reasons for this improved fracture performance are
discussed.

INTRODUCTION

        Surface modifications in the form of metal, ceramic or polymer coatings have been used
for a long time by aerospace and microelectronics industries to improve wear resistance and
prevent corrosion. Ceramic coatings on metals have been mainly used for improving wear
resistance. The simplest configuration is to apply a single composition coating on the metal
substrate. However, since performance requirements may vary spatially within the coating, a
functionally graded structure should provide better overall performance. Brittle ceramic coatings
on stressed ductile metal substrates frequently fail by the formation of a series of cracks
perpendicular to the imposed tensile stress. The formation of these cracks has been described
using a shear lag approximation which relates the substrate yield stress to the spacing of cracks
that form [12,14]. In the present study the substrate yield stress is varied systematically and its
influence on the crack initiation and saturation in single layer and bilayer coatings is studied.
These results are discussed in the context of existing thin film cracking models and mechanics
analyses.

BACKGROUND

        Ceramic thin films are generally used because of their excellent hardness and wear
resistance. However, their service life largely depends on their ability to remain adhered and resist
failure when subjected to mechanical and thermal stresses. Scratch tests are limited to qualitative
results, and typically do not provide information required to understand the failure mechanisms.

More controlled film cracking techniques have been developed to understand the failure and fracture toughness of the films.

The technique applied here employs uniaxial loading of the film and substrate such that isostrain conditions are maintained. Other experiments [1-4] have shown that a critical level of strain results in the formation of periodic cracks in the film perpendicular to the loading direction, as illustrated in Figure 1. As the strain increases the number of cracks increase and, ultimately, a saturation crack spacing is achieved beyond which no new cracks are observed [5-9]. The shear lag theory [10-13] has been used to describe the evolution of film cracking when the adhesion between the film and substrate is high. High adhesion ensures strain continuity at the interface, so that the substrate deformation is completely transferred to the film. The stresses around the tip of each of the periodic cracks induce plastic flow in the substrate. Below each crack at the interface of film/substrate, shear stresses exist over a characteristic length (L) which is related to the yield stress of the substrate [14-16]. The saturation crack spacing is given by,

$$\lambda = \frac{\pi \delta \sigma^f}{\tau^i} \qquad (1)$$

where σ^f is the tensile strength of the coating, δ is the thickness of the film, λ is the saturation crack spacing and τ^i is the interfacial shear stress.

Periodic cracking has also been modeled in functionally graded brittle coatings on ductile substrates [17,18]. Erdogan et al. [18, 19] have shown that a gradation in the elastic modulus of the coating influences the crack driving force. Bao and Wang [17] have also shown through finite element modeling that the strain energy release rate depends on the saturation crack spacing and this effect is influenced by the architecture of the functionally graded coating. The present work examines the influence of a graded architecture on cracking behavior.

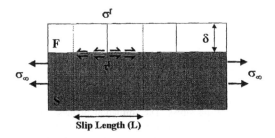

Fig.1: Schematic of the periodic cracks formed in the coating.

EXPERIMENTAL METHODS

The experiments consisted of three steps: 1) annealing brass samples, 2) sputter coating chromium nitride coated on the brass and 3) uniaxial tensile deformation of the coated samples. Cold rolled sheets of brass were cut and machined into tensile specimens whose gauge length dimensions are 1' x 0.5' x 0.08' according to ASTM (E8-99) standards. These specimens were annealed at temperatures ranging from 400-800°C with the intent of changing grain size. This resulted in a difference in the yield strength (Figure 2) of the brass which was measured by tensile tests of the uncoated brass samples. All the samples had identical work hardening behavior, for which the work hardening exponent $n \approx 1$.

An unbalanced magnetron sputtering (UBMS) device was used, as described in more detail elsewhere [20-23]. Two types of coatings were deposited: a single layer coating of Cr_2N and bilayer coating of Cr_2N and CrN. Table 1 shows the parameters used for the sputtering process. The samples were then mechanically polished to 1 μm to obtain a fine surface finish, suitable for deposition.

In-situ tensile tests were carried out in a MTS model tensile testing machine. Coated samples were observed during the test using a long distance optical microscope. The initiation of cracks and their subsequent saturation was recorded by a video recorder. Frames were then grabbed from the video from which the crack initiation strain and crack saturation strain were determined (Figure 3).

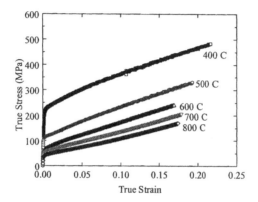

Fig.2: Plot of the true stress vs true strain of brass samples resulting in different yield stresses of the substrate.

Table 1: Conditions under which coatings were deposited in the UBMS.

Type of Sample	Layers	Argon/Nitrogen Ratio	Power (kW)	Thickness (μm)
Single Layer Coatings	Pure Cr	100/0	1	~4.5
	Cr_2N	45/55	2	
Bilayer Coatings	Pure Cr	100/0	1	~7.5
	CrN	25/75	2	
	Cr_2N	45/55	2	

These tensile tests were carried out for the single coat samples and the bilayer coated samples. The crack initiation strain was plotted as a function of the yield stress and saturation crack spacing was plotted as a function of strain.

RESULTS AND DISCUSSION

Crack Initiation

Uniaxial deformation of the coated specimens resulted in an array of periodic cracks perpendicular to the tensile axis. Crack initiation strain was the strain at which the very first cracks appeared on the video. The crack initiation strain does not depend on yield stress as shown in Figure 4. Figure 4 also shows that the crack initiation strain for the bilayer coating is slightly higher than that for the single layer coating.

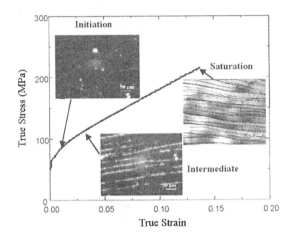

Fig.3: Shows the diagram of a typical stress-strain curve obtained from uniaxial deformation tests and the representative micrographs for crack initiation and saturation.

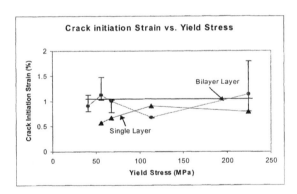

Fig.4: Shows the plot of crack initiation strain vs yield stress of the substrate.

It is unclear whether the cracks nucleate at the surface of the coating or the interface of coating and substrate. Figure 5 shows the scanning electron micrographs of the coating after saturation crack spacing is achieved. The cracks in the single layer coating are straight and do not exhibit any evidence of deflection. However, cracks in the bilayer coating deflect near the internal interface of the coating. The apparent difference in cracking morphology between the two samples is not yet understood, but is believed to be caused by residual stress differences. A mechanics analysis [19] indicates that a crack should accelerate upon approaching a more compliant material under isostrain conditions. In the present case, if it is assumed that the crack nucleates at the coating surface, the observation of crack deflection indicates that the driving force

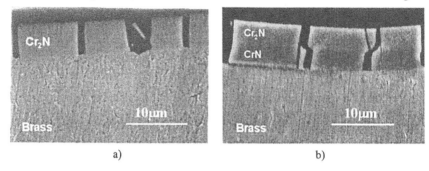

a)   b)

Fig.5: Micrographs showing the saturation cracking of single layer (a) and bilayer coating (b), both with a substrate yield stress of 113MPa.

decreases as the crack approaches the compliant material, opposite to what the mechanics analysis predicts. Crack deflection would be expected to occur if the second layer (CrN) exists in a residual state of biaxial compression, such that the effect of the compressive stress field overcame the effect of the modulus mismatch. Very high levels of compressive residual stress have been previously reported in these types of coatings.

Saturation Crack Spacing

As the strain on the samples is increased the number of cracks increase until they reach a constant value, viz., saturation. The plot of average crack spacing and strain (Figure 6a) shows that for the single layers, as the yield stress
increases, the saturation crack spacing decreases, in accordance with the shear lag theory.
On the other hand, the saturation crack spacing of the bilayer coating does not vary with the yield stress of the substrate. There is no obvious reason for this, and future experiments will provide an insight to the factors that control the crack spacing in bilayer and functionally graded coatings.

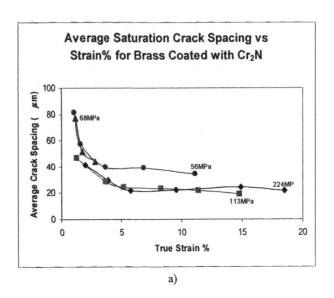

a)

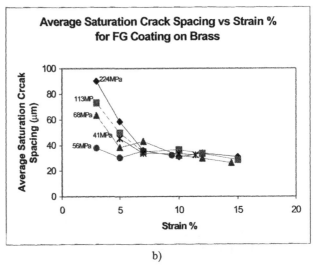

b)

Fig.6: Plot of average saturation crack spacing and strain % for a) single layer and b) bilayer coatings.

CONCLUSIONS

The above results indicate that bilayer coatings behave quite differently from single composition coatings when uniaxially strained. Future experiments will attempt to determine the origin of these differences. Part of the challenge lies in establishing the origin of the cracks. The possibility of deflecting the cracks or arresting them within the coatings should have important implications for wear and corrosion resistance. Modifications of the shear lag theory may be required to understand the crack saturation spacing of functionally graded coatings.

REFERENCES

1. Ignat M., Marieb T., Fujimoto H. and Flinn P.A., "Mechanical Behavior of Submicron Multilayers Submitted to Microtensile Experiments", Thin Solid Films, **353**, 201-207, (1999).
2. Shieu F.S. and Shiao M.H., "Measurement of the Interfacial Mechanical Properties of a Thin Ceramic Coating on Ductile Substrates", Thin Solid Films, **306**, 124-129, (1997).
3. Evans A.G., Crumley G.B. and Demaray R.E., "On the Mechanical Behavior of Brittle Coatings and Layers", Oxidation of Metals, **20**, 193-216, (1983).
4. Kuper A., Clissold R., Martin P.J. and Swain M.V., "A Comparative Assessment of Three Approaches for ranking the Adhesion of TiN Coatings onto Two Steels", Thin Solid Films, **308-309**, 329-333, (1997).
5. Agrawal D.C. and Raj R., "Measurement of The Ultimate Shear Strength of a Metal-Ceramic Interface", Acta Metall., **37**, 1265-1270, (1989).
6. Jeong J.H. and Kwon D., "Evaluation of the Adhesion Strength in DLC Film-Coated Systems Using the Film-Cracking Technique", J. Adhesion Sci. Tech., **12**, 29-46, (1998).
7. Kirk P.B. and Pilliar R.M., "The Deformation Response of Sol-Gel-Derived Zirconia Thin Films on 316L Stainless Steel Substrates Using a Substrate Straining Test", J. Mater. Sci., **34**, 3967-3975, (1999).
8. Chen B.F., Hwang J., Yu G.P. and Huang J.H., "In Situ Observation of the Cracking Behavior of TiN Coating on 304 Stainless Steel Subjected to Tensile Strain", Thin Solid Films, **352**, 173-178, (1999).
9. Hu M.S. and Evans A.G., "The Cracking and Decohesion of Thin Films on Ductile Substrates", Acta Metall., **37**, 917-925, (1989).
10. Aveston J., and Kelly A., "Theory of Multiple Fracture of Fibrous Composites", J. Mater. Sci., **8**, 352-362, (1973).
11. Hsueh C.H., "Pull-out of a Ductile Fiber from a Brittle Matrix-Part 1- Shear lag Model", J. of Mater. Sci., **29**, 4793-4801, (1994).
12. Yanaka M., Tsukahara Y., Nakaso N. and Takeda N., "Cracking Phenomena of Brittle Films in Nanostructure Composites Analysed by a Modified Shear lag Model with Residual Strain", J. Mater. Sci., **33**, 2111-2119, (1998).
13. Marshall D.B., Cox B.N. and Evans A.G., "The Mechanics of matrix Cracking in Brittle Matrix Fiber Composites", Acta Metall., **33**, 2013-2021, (1985).
14. Scafidi P. and Ignat M., "Cracking and Loss of Adhesion of Si3N4 and SiO2:P Films Deposited on Al Substrates", J. Adhesion Sci. Tech., **12**, 1219-1242, (1998).
15. Wang J.S., Sugimura Y., Evans A.G. and Tredway W.K., "The Mechanical Performance of DLC Films on Steel Substrates", Thin Solid Films, **325**, 163-174, (1998).
16. Ramsey P.M., Chandler H.W. and Page T.F., "Bending Tests to Estimate the Through-Thickness Strength and Interfacial Shear Strength in Coated Systems", Thin Solid Films, **201**, 81-89, (1991).

17. Bao G. and Wang L., "Multiple Crcaking in Functionally Graded Ceramic/Metal Coatings", Int. J. Solids Struct., **32**, 2853-2871, (1995).
18. Schulze G.W. and Erdogan F., "Periodic Cracking of Elastic Coatings", Inter. J. Solids Struct., **28-29**, 3615-3634, (1998).
19. Erdogan F. and Ozturk M., "Periodic Cracking of Functionally Graded Coatings", Int. J. Engg. Sci., **33**, 2179-2195, (1995).
20. Reichelt K. and Jiang X., "The Preparation of Thin Films by Physical Vapour Deposition Methods", Thin Solid Films, **191**, 91-126, (1990).
21. Window B. and Harding G.L., "Ion-Assisted magnetron Sources: Principles and Uses", J. Vac. Sci. Tech., **A8**, 1277-1282, (1990).
22. Kelly P.J. and Arnell R.D., "Characterization Studies of the Structure of Al, Zr and W Coatings Deposited by Closed-Field Unbalanced Magnetron Sputtering", Univ. of Salford, UK.
23. Rossnagel S.M. and Hopwood J., "Metal Ion Deposition from Ionized Magnetron Sputtering Discharge", J. Vac. Sci. Tech., **B12**, 449-453, (1994).

TERMOREACTIVE ELECTROSPARK SURFACE STRENGTHENING (TRESS)

E.A. Levashov*, E.I. Kharlamov   M. Ohyanagi, M. Koizumi
SHS-Center of Moscow Steel and Alloys   Ryukoku University, Ohtsu, Japan
Inst.
Leninsky prospect, 4, Moscow 117936

S. Hosomi
Tomei Diamond Co., Ltd., Joto, Japan

ABSTRACT
    The mechanism of mass transfer and structure formation in coatings based on titanium borides, including diamond-containing ones, in the Thermoreactive Electrospark Surface Strengthening (TRESS) and Electrospark Alloying (ESA) processes are studied. It is shown that impulse discharge energy essentially affect the structure and properties of coatings and the degree of graphitization.

INTRODUCTION
    This work is devoted to investigation of coatings produced with the use of charge electrodes, Ti+B, Ti+B+diamond, and SHS-electrode SHIM-4 (TiB_2-Ti) by Thermoreactive Electrospark Surface Strengthening (TRESS) and Electrospark Alloying (ESA) techniques [1,2].
    The thickness and properties of coatings can be altered by raising the impulse discharge energy raise (increase in consumption of electric energy) [3], decreasing in melting temperature of one or several components in the electrode composition, decrease in structural strength, some growth of porosity, changing the sizes of structural components in the electrode material, and change of substrate composition.
    Despite such a wide diversity of factors that determine the structure and properties of TRESS-coatings, a search of new promising compositions of charge electrodes is now in progress, allowing us to improve the wear resistance of the coatings.

1. EXPERIMENTAL PROCEDURE
    Thermoreactive electrospark strengthening was carried out in air or in nitrogen environment in the Elitron-22A and Elitron-52B plants at different energies of impulse discharges. Charge electrodes 3-4 mm in diameter and 30-60 mm long were produced by drawing in tubes of exothermic mixtures of titanium powders (PTM grade) with boron (brown, amorphous). (The composition is designed for TiB-Ti (SHIM-4) and diamond (size fraction of 8/16 μm, production of Tomei Diamond, Japan.) The electrodes were produced by cold drawing in several passes in the air in different tubes (Cu, Fe, Al) [4].
    The mass transfer kinetics (erosion of anode ΔAi and cathode weight increment ΔKi) were determined by gravimetric method under total alloying time of 10 min according to the procedure described elsewhere [5,6]. Titanium alloy VT3-1 and stainless steel 12Kh18N9T were used as substrates.

The X-ray studies were carried out with the use of DRON-3 and DRON-4 diffractometers in Co-K_α radiation. X-ray photography was performed point by point with a step angle of 0.05 degree and exposure for 3 sec in every point.

The metallographic analysis of specimens was carried out in the Neophot-32 light microscope and the Tesla BS-540 electron microscope.

Resistance to abrasive wear was measured in the friction machine. The test wear contact is the following: coated flat over rotating cylindrical surface of a diamond-containing roller. Sliding friction rate was 1.45 m/sec and the load on the friction couple was 1 kg.

2. RESULTS AND DISCUSSION

2.1. Study of anode-electrode structure

The Ti-B system pertains to the category of relatively high-exothermic compositions. It is known from work related to studying the burning processes in the Ti-B system [1,2] that the reactive surface is formed by capillary spread of titanium melt over firm branched surface of amorphous boron. In this case, being independent of stoichiometric titanium-to-boron ratio, the phase of TiB_2 titanium diboride with hexagonal lattice is the primary reaction product. The primary product formed in the reaction zone is characterized by grain size of 1-2 μm irrespective of the weight of the initial reagents.

In the case of charge with a great surplus of titanium relative to the stoichiometry of TiB_2 (as occurred in this particular case), secondary reaction can occur, for example, $TiB_2 + Ti \rightarrow 2TiB$, by recrystallization of diboride grains into titanium boride having the orthorhombic lattice through the titanium melt.

Fig. 1 represents the working part of the end of a Ti+B charge electrode placed in a steel sheath after performing the process at the impulse discharge energy P=0.156 J. Large buildups, which are evidence of the appearance of large amount of liquid phase due to the chemical reaction with high rate of heat generation, can be observed on the electrode end. Moreover, the electron microscopy showed that the width of the zone of reaction products is essentially dependent on of the impulse discharge energy. This can be seen from Fig. 1 where the structure of transverse cross section of the charge electrode is shown. The reacted zone is seen on the right hand and the initial compacted charge is seen on the left hand. The zone width is about 20 μm at P=0.416 J (Fig.1(a)) and approximately a hundred microns at P=2.08 J. (Fig.1(b)). The further increase in the impulse discharge energy leads to the interacting charge growth, resulting in extension of the reaction product zone. The X-ray phase analysis of the electrode end upon carrying out of the TRESS process and the coatings themselves showed that the impulse discharge energy further increases the titanium boride content. All these factors result in extensive transfer on the substrate of reaction products together with the envelope material of charge electrode.

2.2. Composition and properties of coatings produced using Ti+B charge electrode

The results of quantitative X-ray phase analysis of the coatings produced in the Elitron-52B plant showed, in the case of Ti+B electrode put into steel sheath, the presence of Ti_2B_5 phase, the content of which varies from 5 wt. % up to 11 wt. % when impulse discharge energy raises from 0.129 J up to 0.416 J. The remaining phases in the coating are Fe and $FeTiO_3$.

In addition, the coatings produced with the use of electrodes of the same composition (Ti+B) but in different sheaths have been studied. Processing of 12Kh18N9T steel by TRESS-electrode (Ti+B) in copper sheath leads to formation of different modifications of titanium boride (TiB, Ti_2B_5, TiB_2) in the surface layer. Moreover, the exothermal chemical reaction comes more extensively, and the proportion of borides in the coating increases when P raises. The content of TiB_2 increases from 5 % up to 10 %. Replacement of copper sheath with aluminum one results in formation of titanium borides TiB_2, TiB, $TiB_{(x)}$ in the surface layer (Tab. 1). When P raises, the

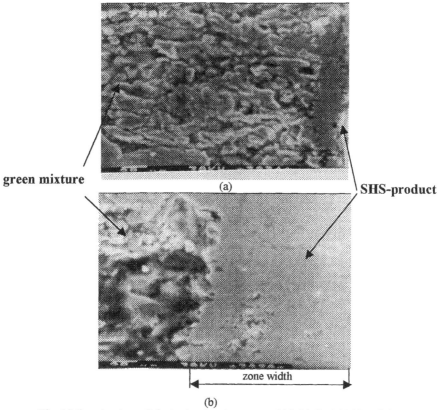

green mixture (a) **SHS-product**

zone width

(b)

Fig. 1 Microstructure of electrodes showing: zone width (a) P=0.416 J and (b) P=2.08 J

Table 1. Results of X-ray phase analysis of coatings in the case of aluminum sheathed Ti+B electrode

Phase	Impulse discharge energy (P), J		
	0.129	0.416	2.08
Fe	0.20	15.02	24.14
TiB_2	11.81	19.02	
Al	29.02	6.92	
$TiAl_3$	15.52	8.72	
TiAl	7.92	8.12	
Fe_4Al_{13}	22.72		
Al_2O_3	12.82	27.64	
TiB		14.82	14.82
$TiB_{(x)}$			4.62
$Ti(Al,Fe)_3$			20.12
$Fe_{23}Ti_2$			36.33

content of iron appreciably increases (0.2 – 15.0 – 24.1 %). A portion of free aluminum in the coating decreases. Apparently, this is connected with increasing the liquid bath volume and more extensive mixing of elements.

The kinetics of mass transfer of coatings produced with the use of SHS and charge electrodes in different sheaths in the Ti – B system has been studied in the Elitron-52B plant.

When using the SHS-electrode of SHIM-4 grade (TiB-Ti), a decrease in cathode weight, as connected with both anode erosion that occurs predominantly in solid phase, and discharge channel action on the steel surface. When alloying over the applied layer, embrittlement of the surface and detachment of the particles on the substrate takes place.

An appreciable cathode weight increment being dependent on energy P can be observed in the dependence of ΔK on alloying time t by TRESS-electrode of titanium with boron in copper sheath (Ti+B). An extremum, brittle fracture threshold (t_b), appears at P = 0.129 – 0.416 J. The ΔK value decreases during the further processing.

Grasping of an electrode to the substrate is observed during the first minutes when strengthening steel by a charge electrode in steel sheath. It is follows from the analysis of the kinetic curves that the processing by electrodes in steel sheath is more preferable as compared to electrodes in aluminum and copper sheaths, since the transport coefficient ($K_t = \Delta K/\Delta A$) runs to maximum 48 % at P = 0.129 J and 44 % at P = 0.416 J.

Application of an electrode in aluminum sheath at low conditions provides a negligible cathode weight increment only. The cathode weight increment appreciably increases when impulse discharge energy P is raised. At the same time, the increase of P results in heating the electrode and deformation of the aluminum sheath. This adverse factor restricts application of electrodes in the aluminum sheath. Thus, compared to SHS-electrodes, the charge TRESS-electrodes allow the coating thickness to be increased. Energy being equal to P=0.416 J is more optimal, since the process is unstable at lower conditions and the coating roughness increase at higher conditions.

Tab. 2 represents generalized results of studies as of coating properties produced with the use of charge electrodes Ti+B in different sheaths as compared to the coatings applied by SHS-electrodes of SHIM-4 grade of the same element composition and consisting of two phases: α-Ti and TiB.

The coatings applied by TRESS-electrodes are predominantly thicker. So, the coatings made of (Ti+B) in aluminum sheath are 70–80 µm thick at P=0.129 J and 270–290 µm thick at P=0.416 J. The coatings made of (Ti+B) in copper sheath are 30–50 µm thick (P=0.129 J) and 70–80 µm thick (P=0.416 J). The coatings made of steel sheathed electrode are 110–120 µm thick (P=0.416 J).

Table 2. Properties of coating on steel 12Kh18N9T

Electrode, ESA mode (P), J		Microhardness, GPa		Continuity of surface layer, %
		Substrate	Coating	
SHIM-4	0.144	2.28±0.11	6.12±0.48	80–85
	0.416	2.34±0.10	5.84±0.35	80–85
Cu (Ti – B)	0.144	2.40±0.18	4.60±0.68	50–60
	0.416	2.69±0.11	3.24±0.28	80–85
Al (Ti – B)	0.144	2.30±0.09	5.25±0.26	85–90
	0.416	2.88±0.10	3.10±0.30	90–95
Steel (Ti – B)	0.129	2.50±0.12	4.89±0.51	70–75
	0.416	2.44±0.12	5.41±0.50	65–70

2.3. Diamond-containing coatings in the Ti-B system

Charge electrodes in steel and copper tubes were used in the work. Diamond, produced by Tomei Diamond, Japan, was introduced in drawing with size fraction of 8/16 μm in amount of 60 vol. % with addition of TiH_2 (5%). All experiments were carried out on titanium substrate VT3–1. It is related to the fact that metals of iron group and their intermetallics do not allow the diamond phase to be unambiguously identified by X-ray, so the major lines of diamond (A4) coincide with the lines of these metals and their alloys. It is evident that the general behavior of the curves for electrodes in different tubes is similar. Titanium content in the coating decreases at the expense of coating thickness growth. The amount of the carbides and carbonitrides increases with the impulse discharge energy as a result of interaction with the graphitized diamond. The difference is in the diamond content. When using copper sheath, the amount of diamond in the coating is little. In the case of steel sheath, a maximum of diamond content (40 %) at energy 0.04 J is a achieved. X-ray phase analysis identified titanium boride only in the coatings made of the steel sheathed electrode. In this case, its content smoothly grows from 0 up to 18 % with the impulse discharge energy from 0.02 J up to 0.16 J.

The kinetics of electrode (anode) mass transfer to the substrate (cathode) was studied in the Elitron-22A plant as in previous work [1,2]. A cumulative weight increment ΔK for diamond-containing electrode in steel sheath at impulse discharge energy P=0.015–0.091J has negative values at the expense of predominant erosion of the substrate. Fig. 2 shows microstructure of diamond-containing coating for the steel-sheathed electrode at impulse discharge energy P=0.04 J.

2.4. Mechanical wear tests

The final answer to the question of what coating is better–thin one with hard matrix or thicker one with plastic bonding and also the comparison of wear-resistant diamond-containing coatings and diamond-less coatings can be obtained in particular tests of coatings in the friction machine. Fig. 3 represents the results for wear testing of coatings applied on steel R6M5.

The coatings were produced at P=0.043 J using charge electrodes Ti+B+diamond (8/16μm), Ti+B in copper sheath, the same composition but in steel sheath, Ti+B in aluminum sheath. In addition, the coatings produced by electrode SHIM-4 were used for comparison. It can be seen from the figure that diamond-containing coatings possess great wear resistance as compared to coatings without diamond and the coating produced by synthesized electrode SHIM-4. It should be noted that the coating made of SHIM-4 has higher wear resistance as compared to the coatings made of charge electrodes Ti+B in copper and steel sheaths without diamond.

Judging from the run of the roller (a bend on the curve H(L)), the coatings made of the electrode containing diamond and the electrodes Ti+B in copper sheath and SHIM-4 are thinner. For the first two coatings this is related to great mass of diamond portion, i.e up to 60 vol. %. For the most part, diamond does not participate in the chemical reaction and is transferred mechanically. It can be graphitized, therefore formation of wear-resistant carbides and carbonitrides is possible. With raising discharge energy, the substrate erosion effect is smoothed out by extensive transfer of synthesis products together with melting copper on the substrate. A similar picture is observed in the coating produced from SHIM-4 electrode. Hence diamond-containing coatings have improved wear resistance owing to diamond, as well as titanium carbides and carbonitrides formed as a result of partial graphitization of diamond.

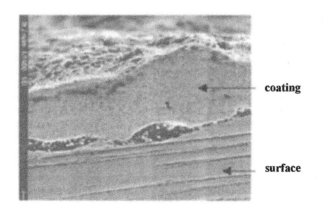

Fig. 2 Microstructure of diamond containing coating

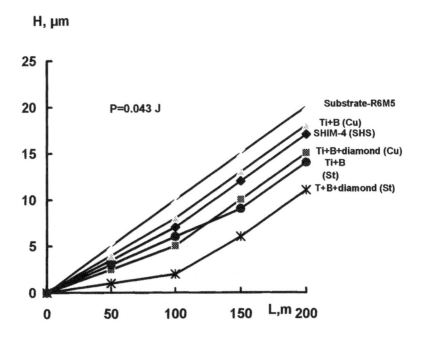

Fig. 3 Dependence of depth of worn layer (H) on roller run length (L)

Functionally Graded Materials 2000

CONCLUSIONS

1. The work studies the mechanism of mass transfer and structure formation in coatings based on titanium borides, including diamond-containing ones, in the TRESS and ESA processes.
2. It is shown that electrode sheath material and impulse discharge energy essentially affect the structure and properties of coatings and the degree of diamond graphitization.
3. Kinetics, X-ray phase analysis, microhardness, and wear resistance were measured on coatings produced by Ti+B electrodes in steel, copper, and aluminum sheaths, by the same electrodes but with diamond content 60 vol.% and SHIM-4 electrode. The coating made of SHIM-4 electrode has the greatest microhardness (6.12 GPa). The most wear-resistant coating with diamond content up to 40 % in the above-mentioned systems was produced from steel-sheathed Ti+B-diamond electrode.

REFERENCES

[1] E.A. Levashov, E.I. Kharlamov, A.E. Kudryashov, "Thermoreactive Electrospark Surface Strengthening with the Use of Charge Electrodes". *Izvestiya Vuzov.Tcvetnaya Metallurgia*, [2] 39-45 (1998).
[2] E.A. Levashov, E.I. Kharlamov, A.E. Kudryashov, A.S. Rogachev, M. Ohyanagi, M. Kaizumi, S. Hosomi, "About the Method of Thermoreactive Electrospark Surface Strengthening", *Journal of Material Synthesis and Processing*, 7 [1] 25-35 (1999).
[3] E.A. Levashov, E.I. Kharlamov, "Kinetics and Mechanism of Diamond-Containing Coatings Formation by Thermoreactive Electrospark Surface Strengthening Process". Book of Abstracts. 5th International Symposium on Self-Propagating High-Temperature Synthesis (SHS-99). Moscow, August 16-19 90-91 (1999).
[4] E.A. Levashov, E.I. Kharlamov, A.A. Korostelin, "Production and Application of Charge Electrodes for Thermoreactive Electrospark Surface Strengthening (TRESS)", *Journal of Material Synthesis and Processing*, [5] 39-46 (1999).
[5] A.D. Verkhoturov, I.A Podchernyaeva, L.F. Pryadko, *"Electrode Materials for Electrospark Alloying"*, Nauka, Moscow 1988
[6] A.E. Gitlevich, V.V. Mikhailov, N.Y. Parkansky, et el. *"Electrospark Alloying of Metal Surfaces"*, Shtiintsa, . Kishinev 1985

Thermal Barrier Coatings

CRACKING BEHAVIOR OF NiCrAlY/YSZ THERMAL BARRIER COATINGS UNDER FOUR POINT BEND LOADS

Ahmet Kucuk, Christopher G. Dambra, Christopher C. Berndt, Ufuk Senturk and Rogerio S. Lima
Center for Thermal Spray Research, Materials Science and Engineering Department
State University of New York at Stony Brook, Stony Brook, NY-11794, USA

ABSTRACT
Thermal barrier coatings with a NiCrAlY bond coat and a yttria partially stabilized zirconia overlayer were air plasma sprayed using various process parameters including different bond and top coat thickness, substrate temperature and stand off distance. The coatings were four point bend tested and coupled with simultaneous acoustic emission (AE) detection via a transducer. Cracks which occurred during the tests were examined using an optical microscope after the tests. The crack patterns obtained were correlated to the mechanical properties of coatings and the recorded AE signals during the tests.
        The thermal spray processing route for the formation of these coatings can be interpreted in terms of a 3-D structure that consists of variably-stressed and compositionally identical (yet phase-differentiated) microstructural units of splats. Therefore, these structures are very much akin to FGMs and their properties are interpreted with respect to their processing schedules to highlight how the cracking behavior can be altered.

1. INTRODUCTION

Functionally graded materials (FGMs) intended for applications including electronic, biomedical, structural, and thermal environments, hold the promise of superior properties over their conventional, non-graded counterparts [1]. The first reported work on thermal spraying graded cermets was reported in the classic textbook by Gerdeman and Hecht [2]. In this compilation of Arc Plasma Technology in Materials Science, reference is made to the manufacturing of $NiCr/ZrO_2$ cermets in the early 1960s. The dramatic results achieved is that ceramic contents of upto 70 wt. % can increase the thermal shock resistance from several cycles to greater than 19 cycles and up to 60 cycles. Work is also reported on composites (or what is now referred to as "FGMs") of $NiAl$-Al_2O_3, Ni-$SrTiO_3$, and Ni-$SiZrO_4$ [2].
        Thermal barrier coatings (TBCs) used in gas turbine and diesel engines, with their duplex composition and variably stressed microstructure, play a significant role in the FGM field; i.e., material systems consisting of a Ni based alloy bond coat (usually NiCrAlY) and a yttria partially stabilized zirconia (YSZ) top coat. The top coat permits higher engine working temperature and increased fuel efficiency by insulating the base material from heat generated in the combustion chamber. The bond coat provides oxidation resistance and enhanced adhesion between the base metal and top coat. For better TBCs, effort has been made to generate true functionally graded TBCs by gradually varying the NiCrAlY/YSZ ratio throughout the thickness of the coating [3-7]. The attributes of such a compositional gradient is that it provides a lower stress concentration [8], greater amount of cycles to failure in thermal cyclic tests [4, 6], and lower energy release rate for delamination under service conditions [3]. However, these microstructures require complex

processing procedures [1] and higher coating thicknesses to achieve the same insulative characteristics due to their higher thermal conductivity and lower porosity levels [3].

It is common practice that TBCs are air plasma sprayed. It has been shown that variations in the process variables significantly alter the characteristics of the coating [9-12]. Therefore, it is necessary to optimize process conditions for the duplex TBC systems with improved characteristics before attempting to design TBCs with gradual composition variation throughout the thickness. In the current study, cracking behavior under tensile loads for variably processed duplex TBC systems, which could mimic the high temperature loading conditions, was examined using acoustic emission analysis and microscopy. In this fashion, a fundamental understanding of damage mechanisms in these coatings was sought.

2. EXPERIMENTAL PROCEDURE

2.1. Sample Preparation

Commercially available 8 wt% yttria stabilized zirconia (YSZ) (Metco 204NS, Sulzer-Metco, Westbury, NY, USA) as a top coat and NiCrAlY (Praxair NI-346-1, Praxair Surface Technologies, Indianapolis, IN, USA) as a bond coat, were air plasma sprayed onto mild-carbon steel substrates (90 x 26.5 x 2.6 mm^3). Spray parameters were altered with respect to (i) top coat thickness (either 300 or 500 μm), (ii) bond coat thickness (either 100 or 250 μm), (iii) the standoff distance (either 80 or 100 mm) and (iv) the substrate pre-heat temperature (either at 273 or 393 K).

The substrate temperature was measured using a hand held infrared temperature detector. The substrates were grit blasted and cleaned with ethyl alcohol before spraying. The average roughness of the grit blasted substrate was measured as 4.0 ±0.5 μm using a Hommel T1000 mechanical profilometer (Hommel America, New Britain, CT, USA). Coatings were sprayed to dimensions of 30 x 26.5 mm, leaving approximately 20 and 40 mm of uncoated substrate on both sides since these areas made contact with the bend testing device. Air cooling on the back side of the substrates avoided overheating during the spray process.

Table 1 lists the samples sprayed in the current study. Six samples from each group in Table 1 were sprayed (see Table 2 for the spray parameters) using a Metco 3MB plasma torch with Metco GH nozzle (Sulzer-Metco, Westbury, NY, USA) mounted on a six-axis articulated robot (Model S400, GMF Fanuc, Charlotteville, VA-USA).

2.2. Four Point Bend Tests and Acoustic Emission Analysis

The four point bend tests were performed at a crosshead displacement rate of 10 μm/sec using an Instron universal test machine (Model 8502, Instron, Canton, MA, USA). Further details of the four point bend tests were previously given [11].

During the bend tests, acoustic emission activity was monitored using MITRAS 2001 AE system (Physical Acoustics Corp, Princeton, NJ) controlled through an IBM compatible PC. Details of AE data analysis were given elsewhere [12].

2.3. Inspection of Cracks

After the four point bend tests, the visual inspection of surface cracks, on a randomly selected coating from each group, was carried out using a liquid dye penetrant (Spotcheck, Magnaflux, Glenview, IL-USA) according to ASTM E165 [13]. Representative samples from each group were also impregnated into epoxy under vacuum and then cut into two in the tensile strain direction; i.e., the cross section perpendicular to the loading axis and support bars. The polished cross sectional areas were examined using a reflected light microscope (Nikon-Epiphot, Nicon Inc., Melville, NY-USA). Two different samples from each group were used for visual inspection and microscopic examination.

Table 1. Samples sprayed according to experimental design.

Sample	Bond Coat (μm)	Top Coat (μm)	Substrate Temp. (K)	Stand off Distance (mm)
S1	100	300	393	80
S2	100	300	393	100
S3	250	300	393	80
S4	250	500	393	100
S5	250	500	393	80
S6	100	500	273	80
S7	100	500	273	100
S8	250	300	273	80
S9	100	500	393	80
S10	100	300	273	80
S11	250	300	393	100
S12	250	300	273	100
S13	250	500	273	80
S14	250	500	273	100
S15	100	300	273	100
S16	100	500	393	100
S17	175	400	333	90

Table 2. Spray parameters for the coating system.

	YSZ	NiCrAlY
Torch Type	Metco 3MB	Metco 3MB
Nozzle	Metco GH	Metco GH
Current (A)	600	500
Voltage (V)	70	70
Primary Gas, Ar (l/min)	40	40
Secondary Gas, H_2 (l/min)	11	8
Powder Carrier Gas, N_2 (l/min)	3.5	3.65

3. RESULTS

3.1. Four Point Bend Test

The yield strength and bending modulus for each sample (Table 3) were measured and reported previously [11]. In summary, it was found that coatings sprayed with a thinner bond coat on a cold substrate exhibited a higher yield strength and stiffness; while changes in top coat thickness and stand off distance did not statistically influence the yield strength and bending modulus values [11].

3.2. Acoustic Emission

The acoustic emission (AE) response of each coating that was previously analyzed [12] and is listed in Table 3. In summary, coatings with thicker top and bond coats, sprayed on a pre-heated substrate, at a shorter stand off distance, exhibited more AE activity and released higher energy during the four point bend tests. In addition, samples were grouped as "low", "medium", and "high" response categories depending on the amount of cumulative AE energy released during the tests. The samples from groups S1, S11, S12, S15, and S16 were in the " low" category, while the samples from groups S2, S6, S7, S8, S10, S14 and S17 were in the "high" category. Moreover, the samples from groups S3, S4, S5, S9, and S13 were included in the "medium" category.

3.3. Crack Examination

The typical sample views for dye penetrant inspection and microscopic examination are given in Figs. 1 and 2, respectively. As seen in Fig.1, a series of parallel cracks appeared along the length of the samples and were oriented perpendicular to the tensile stress field. In general, the cracks were either localized around support bars or continuous throughout the mid-span of the sample. The cross-sectional microscopic examination (at around 100X magnification) also revealed similar crack formation patterns in the top coat and bond coat layers; i.e., either continuous or localized vertical cracks around the support bars. The micrographs also revealed

Table 3. Summary of AE events observed during the four point bend test along with the average bending yield stress and modulus value for each group ("a.u." indicates arbitrary units).

Sample	$N_{AE} \times 10^{-3}$ Elastic	$E_{AE} \times 10^{-3}$ Elastic (a.u.)	$N_{AE} \times 10^{-3}$ Total	$E_{AE} \times 10^{-3}$ Total (a.u.)	σ_{YB} (MPa)	E_B (GPa)
S1	1.1±0.4	0.6±0.4	2.5±1.2	1.9±0.9	301±23	114±11
S2	2.8±0.8	3.9±2.2	10.1±1.4	12±2	260±26	89±11
S3	4.7±0.5	3.8±2.2	19.6±4.4	52±24	92±38	29±14
S4	1.9±0.7	1.1±1.0	16.2±2.4	76±20	14±18	4±4
S5	4.2±1.6	5.8±2.8	20.9±5.0	40±17	225±41	72±17
S6	1.6±0.3	2.1±0.3	11.3±1.0	16±3	160±53	51±19
S7	2.8±0.8	3.6±1.7	10.1±1.4	12±2	260±26	89±11
S8	2.2±0.7	1.5±0.7	8.6±5.1	5±2	328±43	114±7
S9	2.7±0.5	4.6±1.7	22.0± 6.7	52±23	62±43	21±14
S10	2.7±0.8	3.4±1.9	12.3±2.0	15±3	206±88	78±36
S11	1.8±1.5	1.8±1.7	4.1±1.6	6±3.8	27±50	8±14
S12	1.1±0.3	0.7±0.5	2.9±1.1	3.7±2.5	73±99	21±29
S13	3.6±1.2	4.1±2.4	16.5±2.8	21±8	250±30	86±15
S14	1.8±0.7	1.0±0.5	11.4±2.7	42±14	1±25	1±6
S15	0.44±0.06	0.49±0.06	0.95±0.15	1.2±0.3	440±61	171±17
S16	0.8±0.4	1.1±0.5	2.8±0.9	2.9±1.1	371±59	144±19
S17	1.0±0.3	0.7±0.6	11.7±4.9	30±20	11±44	4±13

that the ceramic top coat delaminated from the metallic bond coat or the bond coat delaminated from the substrate for some samples, such as samples S4 and S5. Note that the cross sectional microscopic examination permitted crack analysis in the bond coat and the interfaces while dye penetrant examination was restricted to the top coat. The number of cracks detected using dye penetrant varied from 8 to 40 while 3 to 30 vertical cracks in the top coat and 7 to 34 vertical cracks in the bond coat were detected via microscopic examination; depending on the process conditions. The results on the cracking behavior of each sample that were visually analyzed using dye penetrant, and also examined in greater detail using an optical microscope on cross sectional areas, are summarized in Table 4. As seen, the average crack separation (i.e., average distance between the vertical cracks) measured from the micrographs, taken at about 100X magnification, in the top and bond coats are also included in Table 4. One should note that only the vertical macrocracks that were visible at 100X magnification were taken into consideration in this work. The average crack separation in the top coats varied from 100 μm to 2.5 mm while it varied from 150 μm to 1 mm for bond coats. The coefficient of variance in the average crack separation varied from 7 to 80% and 13 to 65% for top coat and bond coat layers, respectively.

The total delamination amount (horizontal length) in top coat/bond coat and bond coat/substrate interfaces were also measured (Table 4 and Fig. 3). Some coatings such as S1 and S11 (Fig. 3a) exhibited no delamination whereas others such as S3 and S4 (Fig. 3c) yielded severe delamination at the top/bond coat interface or bond coat/substrate interface during the four point bend tests. The total delamination length at the top coat/bond coat interface varied from 0 to 13 mm. The coating which exhibited severe top coat/bond coat delamination, also yielded less vertical cracks in the top coat while there are relatively more in the bond coat. The top coat/ bond coat delamination was near the interface in the top coat rather than at the exact interface.

The optical micrographs and the stress distribution results were used to explain the formation and propagation of the macro-cracking sequence in the coating. The cracks started at the surface of the top coat, where the tensile four point bend stresses are the highest, and then

Functionally Graded Materials 2000

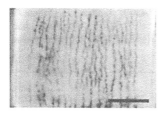 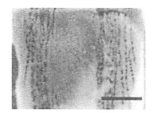

Fig 1. Dye penentrant inspection showing the localized (a) or homogeneously (b) distributed parallel cracks (Bars=1 cm).

propagated to form vertical cracks. These vertical cracks were usually inner-splat cracks rather than inter-splat cracks; i.e., they passed through the splats rather than going through the splat boundaries. After propagating through the top coat, they either (i) deflected to form delamination within the top coat or (ii) went directly to the bond coat to form vertical cracks in the bond coat. In case (i), vertical cracks in the bond coat originated from the delaminated top coat layer and then bond coat delamination occurred in some coatings. In case (ii), some of the vertical bond coat cracks led to bond coat delamination in some of the coatings. The vertical cracks in the bond coat propagated through the splat boundaries (Fig. 2).

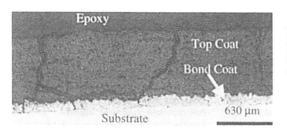

Fig. 2. Typical optical micrograph showing vertical cracks in top and bond coat along with delamination in top and bond coat.

4. DISCUSSION
4.1. AE vs. Cracking

The number of cracks along with the average crack separation and total delamination length is compared with the number of AE events and the cumulative AE energy exhibited during the tests in Fig. 3. The samples were categorized as "low", "medium", and "high" depending on the cumulative AE energy response during the tests. One striking fact is that the samples in the "low" category (Fig. 3a) exhibit almost no delamination, whereas the samples in the "high" category severely delaminated (Fig. 3c). It was not surprising that the delamination generated more AE energy when the microstructure of thermally sprayed coating was considered. The thermally sprayed coatings evolve from cohered splats, which form from the spreading of molten/semi-molten particles on the substrate (or underlying layer) at a high impact velocity [9]. The cohesion between splats at the direction perpendicular to the spray direction is better than the cohesion in the direction parallel to the spray direction. This would be expected because both the mechanical anchoring and the chemical binding between splats, and the total contact area per two splats, are higher in the perpendicular direction. The hardness measurements carried out on the top surface and cross sectional areas show that the cross sectional hardness was higher; thereby indicating that the cohesion was enhanced in the direction perpendicular to spray direction [14]. Therefore, delamination would require more energy and generate a higher AE response. Voyer et

Table 4. Summary of the data collected and the observations made during this study. N is equal to the number of cracks observed and is separated into right and left sides, (R:L), when applicable. The heading d refers to the average distance between cracks. A brief description of the cracking behavior is also given.

Sample	Top Coat		Bond Coat		Dye	Comments
	N	d(μm)	N	d (μm)	N	
S1	30	530 ± 183	34	394 ± 249	R: 8 L: 4	Cracking well distributed, high relative strength, no delamination
S2	R: 7 L: 6	867 ± 296	R: 14 L: 18	373 ± 180	25	Localized cracking, delamination of top and bond coats
S3	R: 9 L: 7	537 ± 318	R: 8 L: 6	390 ± 249	R: 7 L: 7	Localized cracking, delamination of top and bond coats, irregular cracking pattern
S4	8	2216 ± 476	R: 4 L: 9	894 ± 517	10	Cracking is distributed, severe delamination of top coat, large average crack separation
S5	R: 7 L: 5	942 ± 363	R: 7 L: 9	666 ± 237	R: 6 L: 7	Localized cracking, severe delamination of top and bond coats
S6	R: 8 L: 5	1135 ± 527	R: 10 L: 3	680 ± 178	R: 5 L: 6	Localized cracking, delamination of top coat
S7	R: 5 L: 3	545 ± 429	R: 6 L: 11	314 ± 78	R: 7 L: 4	Localized cracking, severe delamination of top coat
S8	R: 3 L: 5	766 ± 262	R: 2 L: 6	395 ± 115	R: 4 L: 5	Localized cracking, delamination of bond coat, good adhesion at top/bond coat interface
S9	R: 3 L: 4	793 ± 534	R: 8 L: 13	328 ± 73	R: 6 L: 3	Localized cracking, large thru-cracks, major delamination of top coat at both supports
S10	R: 3 L: 6	1004 ± 85	R: 5 L: 9	443 ± 148	R: 7 L: 7	Localized cracking, beginnings of delamination in top coat at one support
S11	R: 4 L: 3	1448 ± 107	R: 3 L: 4	612 ± 82	20	Localized cracking, start of branching at bond coat suggesting delamination about to occur
S12	R: 2 L: 4	663 ± 504	R: 3 L: 4	729 ± 145	R: 5 L: 6	Localized cracking, delamination of top and bond coat both at the same support
S13	R: 4 L: 6	733 ± 153	R: 4 L: 6	749 ± 134	R: 4 L: 4	Localized cracking, delamination of bond coat only suggesting better adhesion between top and bond
S14	R: 2 L: 1	464	R: 5 L: 6	459 ± 114	R: 3 L: 5	Localized cracking, large thru-cracks, delamination of top coat, least amount of cracks
S15	R:19 L: 8	535 ± 165	R: 13 L: 4	465 ± 88	40	Cracking relatively distributed, strongest sample in group, least amount of AE avtivity
S16	R: 5 L: 9	702 ± 160	R: 3 L: 5	745 ± 205	40	Localized cracking, no delamination
S17	R: 5 L: 9	979 ± 511	R: 6 L: 10	641 ± 395	R: 4 L: 5	Localized cracking, delamination of top coat

al. [15] also reported that top coat delamination which occurred during thermal cycling of TBCs, generated a larger number of AE signals than reported for vertical cracks.

In addition, the coatings that yielded cracks with larger openings (larger crack width) generated higher AE responses during the tests. No simple relationship between the number of macrocracks, the average crack separation and the AE response of the coating during the tests, has been established.

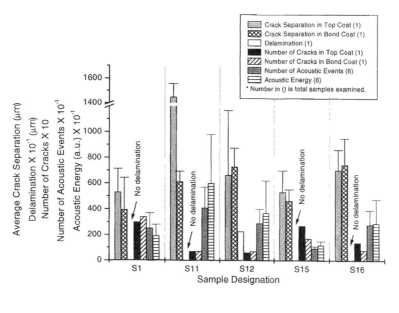

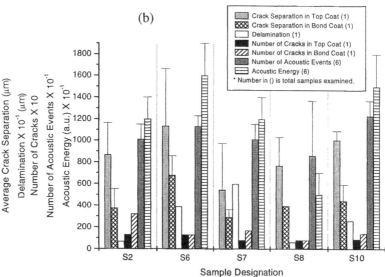

Fig. 3. Comparison of cracking and the AE response of coatings during the four point bend tests. Samples were classified as (a) "low" (b) "medium", and (c) "high" (next page) response categories depending on the cumulative AE released during the tests. Note that the scales are different.

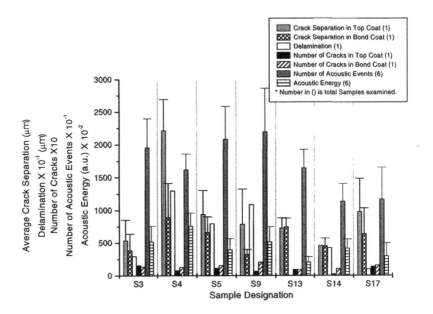

Fig. 3 (cont.). Comparison of cracking and the AE response of coatings during the four point bend tests. Samples were classified as (a) "low" (previous page) (b) "medium" (previous page), and (c) "high" (next page) response categories depending on the cumulative AE released during the tests. Note that the scales are different.

4.2. Influence of Process Parameters on Cracking Patterns

The influence of the processing parameters on the crack formation during the four point bend tests was reported in detail elsewhere [16]. In summary, coatings with thicker top and bond coat showed more susceptibility for delamination.

5. CONCLUSIONS

Plasma sprayed thermal barrier coatings, with varied processing parameters, were tested using a four point bend test arrangement coupled with an *in-situ* acoustic emission monitoring system. After the tests, the formation of macro-cracks were analyzed using dye penetrant inspection and optical microscopy. The dye penetrant visual inspection revealed that a series of parallel cracks formed throughout the length of the samples and were perpendicular to the tensile stress field. These cracks were either localized around support bars or continuous throughout the mid-span of the sample. Microscopic examination on the cross-section of the coatings confirmed these observations. In addition, the microscopy revealed that there exist vertical cracks in the top and bond coat as well as delamination at the top coat/bond coat and bond coat/substrate interfaces. The vertical cracks propagated through the splats in the top coat while they traveled through the splat boundaries in the bond coat. The top coat layer delaminated near the top coat/bond coat interface, rather than at the exact interface. Horizontal delamination propagated through the splat boundaries.

The coatings with higher delamination and wider vertical crack openings, exhibited higher AE energy during the four point bend tests. Coatings with thicker top coats and bond coats were more susceptible to delamination.

6. FUTURE WORK

In the current study, the focus was on the macro-cracks formed during a four point bend test. Micro-cracks formed during processing and deformation are as important as, if not more important than, macro-cracks. A detailed study on the quantification of micro-cracks and their morphology is in progress. In addition, the relationships between AE response, mechanical properties and micro-cracking are also being analyzed. Once all of these studies are completed, this information will be used to optimize the design of functionally graded coating systems.

Acknowledgement
One of the authors (CGD) would like to thank the URECA Summer Fellowship program at SUNY at Stony Brook for their financial support. This work was sponsored under NSF-MRSEC DMR grant number 9632570 and ONR grant number N00014-97-0843.

REFERENCES

1. W. A. Kaysser, (Ed.)"Functionally Graded Materials 1998", *Mater. Sci. Forum,* **308-11** (1999).
2. D. A. Gerdeman and N. L. Hecht, *Arc Plasma Technology in Materials Science,* Springer-Verlag, New York, NY (1972).
3. I. Hofinger, H.-A. Bahr, H. Balke, G. Kirchhoff, C. Hausler, and H. J. Weiss, "Fracture Mechanical Modeling and Damage Characterization of Functionally Graded Thermal Barrier Coatings by Means of Laser Irradiation", *Mater. Sci. Forum,* **308-311** 450-456 (1999).
4. A. Kawasaki and R. Watanabe, "Cyclic Thermal Fracture Behavior and Spallation Life of PSZ/NiCrAlY Functionally Graded Thermal Barrier Coatings", *Mater. Sci. Forum,* **308-311** 402-409 (1999).
5. T. Laux, A. Killinger, M. Auweter-Kurtz, R. Gadow, and H. Willhelm, "Functionally Graded Ceramic Materials for High Temperature Applications for Space Planes", *Mater. Sci. Forum,* **308-311** 428-33 (1999).
6. K. A. Khor, Y. W. Gu, and Z. L. Dong, "Plasma Spraying of Functionally Graded Yttria Stabilized Zirconia/NiCoCrAlY Coating System Using Composite Powder", *J. Thermal Spray Technol.,* **9** [2] 245-49 (2000).
7. K. A. Khor, Y. W. Gu, and Z. L. Dong, "Properties of Plasma Sprayed Functionally Graded YSZ/NiCoCrAlY Composite Coatings"; pp. 1241-48 in *Thermal Spray: Surface Engineering via Applied Research.* Edited by C.C. Berndt. ASM Internal., Materials Park, OH, 2000.
8. O. Kesler, J. Matejicek, S. Sampath, and S. S, "Measurement of Residual Stress in Plasma-Sprayed Composite Coatings with Graded and Uniform Compositions", *Mater. Sci. Forum,* **308-311** 389-395 (1999).
9. L. Pawlowski, *The Science and Engineering of Thermal Spray Coating.* John Wiley & Sons, New York (1995).
10. C. Funke, B. Siebert, D. Stover, and R. Vassen, "Properties of ZrO2-7wt% Y2O3 Thermal Barrier Coatings in Relation to Plasma Spraying Conditions", in *Thermal Spray: A United Forum for Scientific and Technological Advances.* Edited by C. C. Berndt. ASM Internal., Materials Park, OH, 1998.

11. A. Kucuk, C. C. Berndt, U. Senturk, R. S. Lima, and C. R. C. Lima, "Influence of Plasma Spray Parameters on Mechanical Properties of Yttria Partially Stabilized Zirconia Coatings I: Four Point Bend Test", *Mater. Sci Eng. A*, **284** 29-40 (2000).

12. A. Kucuk, C. C. Berndt, U. Senturk, and R. S. Lima, "Influence of Plasma Spray Parameters on Mechanical Properties of Yttria Stabilized Zirconia Coatings II: Acoustic Emission Activities", *Mater. Sci. Eng. A*, **284** 41-50 (2000).

13. "ASTM E 165-91 Standard Test Method for Liquid Penetrant Examination"; pp. 55-71 in ASTM Annual Book of Standards vol. 03.03, ASTM, Philadelphia, PA 1994.

14. R. S. Lima, A. Kucuk, and C. C. Berndt, "Evaluation of Microhardness and Elastic Modulus of Thermally Sprayed Nanostructured Zirconia Coatings", *Surf. Coat. Technol.*, Submitted (2000).

15. J. Voyer, F. Gitzhofer, and M. I. Boulos, "Study of the Performance of TBC under Thermal Cycling Conditions using an Acoustic Emission Rig", *J. Thermal Spray Technol.*, **7** [2] 181-190 (1998).

16. A. Kucuk, C. G. Dambra, and C. C. Berndt, "Influence of Plasma Spray Parameters on Cracking Behavior of Yttria Stabilized Zirconia Coatings", *Practical Failure Analysis*, **1** [1] 55-64 (2001).

DEVELOPMENT OF FUNCTIONALLY-GRADED, NZP-BASED THERMAL BARRIER COATINGS

R. Nageswaran
COI Ceramics, Inc.
181 W. 1700 South
Salt Lake City, UT 84115

S. Sampath
State University of New York, Stony Brook
Dept. of Materials Science & Engg.
Stony Brook, NY 11794

ABSTRACT

Next generation aerospace propulsion and power generation systems have to meet requirements of higher efficiencies, greater fuel economy, and longer lifetimes. Higher firing temperatures and better insulation are required to improve engine efficiency and lifetimes of components in such systems. State-of-the art thermal barrier coatings (TBCs) based on yttria partially-stabilized zirconia (YSZ) have deficiencies for higher temperature applications due to: (i) oxidation related problems, (ii) low thermo-chemical stability, (iii) in-service microstructural changes, and (iv) low inherent strain tolerance.

This (ongoing) program is aimed at developing an advanced TBC technology that is based on low thermal expansion NZP ceramics with very low thermal and oxygen ion conductivity, and high thermal cycling stability. A functionally graded approach is being pursued to overlay NZP on YSZ to reduce the expansion mismatch between NZP and YSZ. Results, so far, indicate that the use of NZPs in conjunction with YSZ has considerable potential to advance the state of the art with respect to TBCs.

INTRODUCTION

Overlay ceramic thermal barrier coatings (TBCs) on metallic components of turbine and heat engines have not only resulted in reduced component surface temperatures but also facilitated an increase in turbine inlet temperatures in supersonic[1], land-based, and marine gas turbines; in turn, increasing the efficiency of these engines. A typical TBC is a two layer coating system consisting of an oxidation resistant bond coat (eg. MCrAlY) on top of a superalloy substrate and a ZrO_2 based ceramic top coat on the bond coat. The bond coat serves to minimize oxidation of the substrate as well as to provide a mechanism for anchoring the ceramic coating.

The minimum general requirements for an effective ceramic TBC were recognized in the 1970's to be[2-6]: (i) low thermal conductivity, (ii) high temperature viability (>1000°C), (iii) phase stability, (iv) resistance to thermal fatigue and thermal shock, (v) reasonably matching thermal expansion with the substrate, and (v) ease of application. To date, the most commonly employed TBC has, in fact, been partially-stabilized zirconia (PSZ), where the stabilizer is generally yttria. Intrinsically, YSZ has low thermal conductivity and is thermally stable to high temperatures.

Several researchers have identified the major life limiting step during extended cyclic thermal exposure of the TBCs to be the oxidation of the bond coat[7-11]. ZrO_2 being an oxygen ion conductor, readily allows oxygen migration from the surface of the coating to the metallic interlayer leading to oxidation. As the oxide (principally Al_2O_3) grows, stresses are introduced due to volumetric dilation which can result in spallation. The oxidation effect is compounded by the cyclic thermal fatigue experienced by the component while in service. Improved oxidation resistance has also been shown to result in increased ceramic-bond coat interface strength[12].

The current bill-of-materials for ZrO_2-based TBCs, dates essentially back to the 1970's. It is apparent that significant advances are not expected to arise from extensions of these conventional coatings, and major innovations in materials/design are essential in order to achieve the next tier of TBCs. A relatively new family of low thermal expansion materials known as NZPs seems to have considerable potential for improving the performance of TBCs in particular. The [NZP] family of materials[13-14] is rich in compositional variations, as it permits flexible ionic substitutions. They also have a very attractive set of properties[15].

Of the various NZP compositions, two new NZP compositions viz. $Ba_{1.25}Zr_4P_{5.5}Si_{0.5}O_{24}$ (BS-25) and $Ca_{0.5}Sr_{0.5}Zr_4P_6O_{24}$ (CS-50) have been developed and extensively characterized by COI Ceramics, Inc. (formerly LoTEC, Inc.). The attractive properties of these materials (see Table 1), including hot corrosion resistance, render them useful for TBC applications. NZP is more attractive for TBC applications as a secondary coating in addition to YSZ; its purpose being to provide a better oxygen and thermal barrier. (It has to noted that the NZPs and the bond coat are thermo-chemically incompatible.) However, the CTE mismatch between NZP and YSZ needs to be overcome by functional grading of the TBC system[16] involving NZP and YSZ.

The overall objective of this research program was to develop commercially attractive, functionally graded TBC coatings based on NZP materials to improve life-time and performance of heat engine and power generation systems.

EXPERIMENTAL APPROACH AND RESULTS

In order to accomplish the main objectives of this research, systematic development of the materials, processing, and design steps was pursued. A detailed discussion of the various research and development activities, including findings from the testing and analysis, is provided in the following.

(A) Feedstock Development

NZP (BS-25 and CS-50) powders were first prepared using the solid state mixing and reaction synthesis route. Calcination was done at 1200°C to obtain fully crystalline, single phase NZP powders. The calcined powders were crushed and screened through an 18 mesh screen. The screened NZP powders were mixed with requisite amounts of water and dispersant, and milled in a vibratory mill for 8 hrs. in order to obtain a low viscosity slurry. (These conditions were based on ECI's vast experience with preparation of NZP slips for the slip casting process). The low viscosity aqueous slurry was shipped to a commercial spray drying facility in order to obtain flowable NZP powders.

Plasma spray trials were then conducted with spray-dried BS-25 and CS-50 (NZP) powders. The first few trials resulted in reasonably good coatings but the deposition efficiency of the coatings was less than desirable, likely due to crumbling of the powders in the plasma flame which led to excessive melting and vaporization. Subsequently, a suitable heating schedule was devised for the spray-dried powders in order to improve the crush strength. Figure 1 depicts the size distributions (typically, 50 to 60 microns median size) and the spherical shape of the particles, respectively, of the spray dried and heat treated BS-25 powders. The powder characteristics of the CS-50 spray dried and heat treated powders were very similar to those of the BS-25 powders.

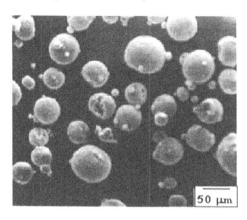

Figure 1. Typical spray dried & heat treated NZP feedstock for plasma spray.

(B) Development of NZP-Based Thermal Barrier Coatings

The development of the plasma spray process for attaining functionally graded NZP-based TBCs consisted of four stages. All plasma spray processing development was done by State University of New York (SUNY) at Stony Brook.

In the first stage, NZP powders were air-plasma sprayed (APS) directly on mild steel (dummy) substrates. Feedstock characteristics were optimized by monitoring the deposition efficiency during the spray process as well as by observing surface characteristics using SEM analysis. In the second stage, the APS process was used to deposit NZP coatings on grit-blasted, mild steel substrates for characterization of physical and thermal properties of the free standing coatings obtained by chemical etching. Results of the characterization were also used for initial refinement of plasma-spray process variables such as feed rate, gun parameters etc.

The third stage of the development was centered on obtaining multi-layered coating configurations involving bond coat, 3mol% (8w%) YSZ[3] (trade named 204 NS), and the NZPs. These coatings, also referred to as Type A, were ~125 μms. thick overlay coatings of BS-25 or CS-50 deposited on top of a 500 microns thick YSZ coating, as

---

[3] Sulzer Metco (US) Inc., Downey, CA 90241.

shown in Figure 2. The final stage of thermal spray process development in Phase I comprised of deposition of quasi-continuous, or discrete functionally graded coatings involving YSZ and NZP mixtures. The quasi-graded TBC samples enabled the identification of key processing issues that need to be addressed in Phase II in order to successfully process continuously graded coatings.

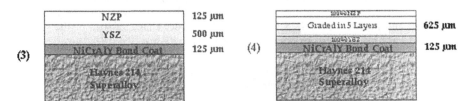

Figure 2. Schematic of the NZP-based overlay and discretely graded coating designs.

For the plasma spray forming of the multi-layered and functionally graded TBCs, a Sulzer-Metco plasma gun was used. The metallic substrates were machined to size, grit blasted, cleaned, chamfered on the edges, and then sprayed. A 125 μm thick bond coat of NiCrAlY was then low-pressure plasma sprayed (LPPS) onto the substrate surface. The 8w%YSZ and NZP oxide coatings were deposited using air plasma-spraying (APS) to the required thicknesses.

(D) Characterization of NZP-Based TBCs

(a) Surface Microstructures: A few of the substrate samples coated with spray-dried and heat treated NZP powders were analyzed for deposit microstructures using SEM. Observations of these micrographs clearly showed the formation of melt-formed "splats". There were no unmelted particles in the microstructure, and the splats were interlocked with one another. In both BS-25 and CS-50 coatings, very few microcracks were observed to be running parallel to the substrate. Figure 3 is a micrograph showing the typical surface microstructure of the APS NZP deposits, which clearly show "splats".

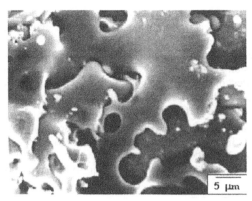

Figure 3. SEM microstructure of air plasma sprayed (APS) NZP TBC showing "splats".

(b) Phase Analysis: Having confirmed the formation of "splats" from molten BS-25 or CS-50 particles, it was imperative that X-ray analysis be conducted to determine if there were any changes in phase composition. Phase analysis of NZP deposits revealed that the BS-25 and CS-50 compositions were intact after plasma spray forming, although a small amount of monoclinic zirconia (m-ZrO_2) had formed.

High resolution X-ray analysis was conducted at High Temperature Materials Laboratory (HTML) of Oak Ridge National Laboratory (ORNL) to examine if the formation of monoclinic zirconia was due to decomposition of a small amount of the NZP during melting. The lattice parameters were extracted from this analysis by curve-fitting and using the Reitveld refining program. Table 1 summarizes the lattice parameters of the BS-25 and CS-50 plasma-spray coatings. Comparison of this data with standards clearly indicates that there is little or no change in the stoichiometry of either BS-25 or CS-50 upon melting.

Table 1. Lattice Parameters of the NZP-TBC Samples As Compared to Standards.

Sample I.D.	a,	b,	Sample I.D.	a,	b,
BS-25 (Standard)[5]	8.68	24.01	CS-50 (Standard)[6]	8.72	23.15
BS-TBC (Type 1)	8.69	24.03	CS-TBC (Type 1)	8.71	23.13
BS-TBC (Type 2)	8.68	23.99	CS-TBC (Type 2)	8.71	23.15
BS-TBC (Type A)	8.68	24.02	CS-TBC (Type A)	8.71	23.14
BS-TBC (Type B)	8.68	24.03	CS-TBC (Type B)	8.71	23.11

(c) Density and Porosity: Density and porosity measurements were conducted using two techniques: (i) using free standing coatings and (ii) combined SEM with image analysis. Free-standing coating samples were coated with wax and the densities determined using Archimedes method.

Porosity measurements of the NZP coatings were conducted using an SEM equipped with image analysis capabilities. The density and porosity obtained using the two different methods were in excellent agreement with each other. Densities of the BS-25 and CS-50 free standing coatings were measured to be 2.65 gm/cc (75.7% of theoretical) and 2.26 gm/cc (68.1% of theoretical), respectively. The porosity levels in the BS-25 TBC are comparable to that routinely observed in low density YSZ TBCs[7].

(d) Microstructural Analysis: SEM micrographs of the sectioned and polished TBC samples are shown in Figures 4(a) and (b). In both cases, the microstructure shows a smeared interface between the YSZ and the NZP coating. No discontinuity could be observed at the interfaces. In both pictures, the NZP shows up as the darker phase in back-scattered SEM. Figure 4(b) shows clearly the graded configuration, with light (YSZ) and dark (BS-25) regions, illustrating the success of the functionally graded TBC processing approach. The thickness of the different layers in these coatings conforms to the desired thickness, as has been shown in Figure 4. For instance, in Fig. 4(a), the NZP top coat is ~125 µms thick and the YSZ main coat is ~ 500 microns. In the case of 4(b), the 5 graded layers of BS-25 and YSZ have a combined thickness of ~ 625 microns.

Figure 4. SEM micrographs of: (a) Type A and (b) Type B (graded) NZPs.

(e) Thermal Conductivity: Thermal conductivity was computed from thermal diffusivity (α), specific heat capacity (c_p), and density (ρ) data, obtained at ORNL, Oak Ridge, for freestanding NZP coatings using the following:

$$\alpha = \frac{\kappa}{\rho \, c_p}$$

The plot in Figure 5 shows a comparison of the thermal conductivities of the APS BS-25 and CS-50 free standing coatings with that of YSZ obtained from Reference 6. In Figure 5, the data for YSZ is for a relatively low density APS coating, 4.7 gm/cc (~77.5% of theoretical). For comparable coating densities, it is interesting to note that the thermal conductivity of BS-25 is significantly lower than that of YSZ even up to 1000°C. The thermal conductivity of the CS-50 coating is also much lower than that of YSZ, however, its density is only 68%.

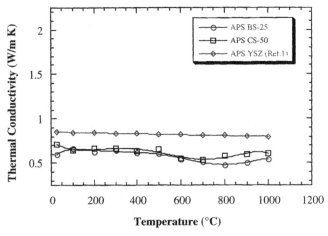

Figure 5. Temperature dependent thermal conductivity of BS-25, CS-50, & YSZ TBCs.

(f) Thermal Cycling Tests: The Haynes 214 superalloy disks were first chamfered on the edges to minimize spurious edge effects and premature spalling of the coatings. The samples with the simple overlay NZP coating (on top of YSZ) and with the functionally graded NZP + YSZ based coating architecture were subjected to cyclic up and down thermal shock testing in a tube furnace held at 1100°C, initially, and at 1200°C, subsequently. For the up-shock, all samples that were at room temperature (25°C) were rapidly moved inside the hot zone of the tube furnace (within 10 secs.) Once inside, the samples were allowed to stay there for 25 minutes for thermal equilibrium. For the down-shock testing, the hot samples were removed very rapidly from the furnace and placed under a fan, to facilitate forced cooling where the surface temperature dropped to < 500°C from ~1000°C in ~20 secs.

All TBC coated samples were initially cycled for 25 heat and cool cycles between R.T. and 1100°C. At the initiation of failure in any of the samples, photographic evidence was gathered. Only those samples that survived the 1100°C cycling test were subjected to further cycling between R.T. and 1200°C. Twenty (20) such heat and cool cycles were conducted. Once again, all samples were examined for failure and photographs were taken using a digital camera to record the onset and progress of spalling or other coating failures. Results of thermal cycling tests yielded the following:

- *Irrespective of the coating configuration (Type A or Type B), BS-25 based TBCs survived 25 thermal cycles between 1100 ℃ and R.T., and 20 additional cycles between 1200and R.T. (Same was true for the YSZ-based TBCs as well.)*
- *In the case of CS-50 based TBCs, spalling failure was observed even after few cycles to 1100 ℃ both in overlay and graded TBC designs. Superalloy samples coated with an overlay coating of CS-50 on top of YSZ seemed to survive a few more thermal shock cycles to 1100 ℃. (See Fig. 6)*

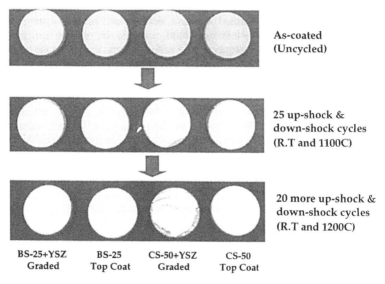

Figure 6. Illustration of the survival / failure of NZP-based TBCs after thermal cycling.

(g) Oxidation Resistance: For isothermal oxidation resistance testing, the 1.75" dia. HAYNES 214 superalloy buttons with Type A and Type B TBC coatings were placed in an alumina crucible and placed in a tube furnace in air at 1100°C for 100 hrs. After oxidation exposure, the disks were sectioned with a CBN wafering blade. The cross section of each sample was then ground and polished using a multi-step process involving abrasive paper and polishing compounds. After final polishing using 0.05 µm silica suspension, the polished sample was cleaned, dried, and etched in a lactic-acid based etch for very short times (5 –10secs). The etched samples were analyzed using SEM with EDS in order to quantify the thickness of the thermally grown oxide (TGO) layer above the bond coat.

After the testing, it was found that the CS-50 TBCs (whether graded or not) did not survive the 100 hrs. exposure at 1100°C. The entire coating spalled off. However, the superalloy samples coated with BS-25 TBCs survived the oxidation testing. The interfacial oxide scale thickness was approximately ~ 6 µm after the 1100°C isothermal exposure, however, the scale thickness was comparable to that of pure YSZ based TBC. This suggests that the oxidation mechanism might be controlled by porosity in the coating rather ionic conduction or diffusion through it. Therefore, there is potential to significantly reduce the oxide scale thickness by making the coatings denser and by grading the NZP with YSZ. Functional grading could significantly minimize the tensile mismatch stresses in the NZP coating (tensile stresses lead to opening of microcracks and pores, thereby, providing easier paths for oxygen permeation).

SUMMARY AND CONCLUSIONS

Powder feedstock with high crush strength and good thermal spray efficiency was successfully developed for both BS-25 and CS-50. The air plasma spray (APS) route to processing of NZP-based TBCs has been successfully developed. Free standing coatings of NZP were subjected to phase, density, and thermal testing. Except for the formation of a small amount of monoclinic zirconia, the plasma sprayed NZP coatings exhibited very good phase retention, good adhesion and densification (85 to 88% dense), and low thermal conductivity (~ 0.50 to 0.60 W/mK versus 0.80 to 0.85 W/mK for YSZ of comparable density).

Notably, overlay coatings (Type A), on top of YSZ, and quasi functionally-graded NZP + YSZ coatings (Type B) were developed. Feasibility of functionally graded coatings involving NZP-based TBCs has been successfully demonstrated. Both NZP overlay coatings and NZP + YSZ compositionally graded coatings were deposited on superalloy buttons using plasma spray, and samples tested for thermal cycling durability and isothermal oxidation resistance. The BS-25-based TBCs exhibited good thermal cycling resistance as the samples endured 25 rapid heat and cool thermal shock cycles between R.T. and 1200°C, and survived oxidation testing at 1100°C for 100 hrs. On the other hand, CS-50 based TBCs did not survive both thermal cycling (failed after 8 cycles) and the oxidation testing.

Future work is necessary to further develop and test overlay / continuously graded BS-25 + YSZ coatings in order to realize all the benefits of an NZP-based TBC and, thereof, improve the state-of-the-art.

6.0 REFERENCES

1. R. A. Miller, *NASA-TBC Workshop Proceedings*, 17 (1995).
2. W. J. Brindley and R.A. Miller, *Adv. Mat. & Proc.* **8**, 29-33, (1989).
3. R. A. Miller, *Surf. & Coat. Tech.*, **30**, 1-11(1987).
4. D. J. Wortman et al., *Mat.Sci. & Eng.*, A-121, 433-440 (1989).
5. F. O. Soechting, *NASA-TBC Workshop Proceedings*, **1**, (1995).
6. S. M. Meier et al., *JOM,* **43**[3] 50-53 (1991).
7. R. A. Miller, *J. Amer. Ceram. Soc.,* **67**[8], 517-521, (1984).
8. Kh. Schmidt-Thomas, *Surf. & Coat. Tech.*, **68/69**, 113-115, (1994).
9. R. Roy et al., *Mat. Res. Bull.*, **19**[2], 471 (1984).
10. S. Y. Limaye et al., *J. Mat. Sci.,* **26**[1], 93 (1991).
11. W. Y. Lee et al., *J. Amer. Ceram. Soc.,* **79**[10], 2759-62 (1996).
12. A. H. Bartlett and R. D. Maschio, *J. Amer. Ceram. Soc.,* **78**[4], 1018-24 (1995).
13. G. M. Ingo and G. Padeletti, *Surf. & Interface Analysis*, **21**, 450-454 (1994).
14. C. Y. Jian et. al, *Composites Engg*, **5**[7], 879-889, (1995).
15. R. Nageswaran et al., *Final Report*, Contract # DE-FG03-97ER82418, May 1998.
16. S. Sampath et al, *MRS Bulletin*, **20**[1], 27-31, (1995).

7.0 ACKNOWLEDGEMENTS

The authors would like to gratefully acknowledge the Phase I SBIR grant provided by NSF under contract number DMI-9961056. We appreciate the efforts of Glenn Bancke at State University of New York at Stony Brook for conducting all the thermal spray processing and Lien Diep for materials synthesis and testing of the coatings at COI Ceramics in Salt Lake City.

EB-PVD ZIRCONIA THERMAL BARRIER COATINGS WITH GRADED Al$_2$O$_3$-PYSZ INTERLAYERS

T. Krell
WIWEB
Wehrwissenschaftliches Institut für Werk-, Explosiv- und Betriebsstoffe
85424 Erding, Germany

U. Schulz, M. Peters, W.A. Kaysser
DLR
Deutsches Zentrum für Luft- und Raumfahrt e.v.
Institut für Werkstoff-Forschung
51147 Köln, Germany

ABSTRACT
    Thermal barrier coatings (TBCs) with graded Al$_2$O$_3$-PYSZ interlayers were processed by means of electron beam-physical vapor deposition (EB-PVD) in a dual source process. Three alternative coating routes with intermediate annealing were carried out. Interruption of processing and carrying out annealing treatments at 1100 °C were necessary in order to get a stable phase and grain structure of the graded Al$_2$O$_3$-PYSZ interlayer. Investigation of graded coatings with SEM and TEM show a dense and coarse columnar microstructure. Single columns consist of nanocrystalline alumina and zirconia grains. Thermal cycling and isothermal oxidation behavior at 1100 °C of the different graded TBC systems were investigated in comparison to the standard duplex zirconia TBC. For one graded coating system slight advantages in lifetime were found.

INTRODUCTION
    There is an increasing demand for the use of EB-PVD thermal barrier coatings in aircraft and stationary gas turbo engines for the thermal insulation of internally cooled metallic turbine components. The use of 200 μm thick thermal barrier coatings (TBCs) allows to increase the lifetime of those turbine components and should enable the operation of modern gas turbines at significantly higher gas temperatures for higher efficiency. Therefore, the development of coating systems

with increased lifetime and reduced thermal conductivity is a major focus of investigation worldwide [1].

The conventional TBC system is a duplex system and consists of a ceramic top coat for thermal insulation and a metallic bond coat. State-of-the-art material for the ceramic layer is partially yttria stabilized zirconia (PYSZ) with a yttria content of 6-8 wt.-% which consists of the non-transformable tetragonal t´-phase [2, 3]. The introduction of the FGM concept to a TBC design is aimed at achieving improvements of TBC reliability and longevity by reducing stresses. A recently presented approach is an EB-PVD TBC system with a chemically graded transition from Al_2O_3 to PYSZ [4-6]. This concept is intended to combine the good thermal insulation of stabilized zirconia with the low oxygen diffusivity of an alumina barrier while minimizing mismatch stresses at the metal-ceramic interface. Such graded coatings are produced by controlled simultaneous evaporation of alumina and zirconia.

EXPERIMENTAL PROCEDURE

The processing of chemically graded Al_2O_3-PYSZ thermal barrier coating systems by a dual source deposition in dependence of different coating parameters is described elsewhere [5, 6]. Three different graded TBC systems were produced. Processing of the three routes including several annealing treatments is given by (|| marks an interruption in processing):

1. Route:
   EB-PVD of Al_2O_3 || vacuum annealing 9 h / 1100 °C || EB-PVD of the graded interlayer and the PYSZ thermal barrier coating
2. Route:
   EB-PVD of the Al_2O_3 and the graded interlayer || vacuum annealing 9 h / 1100 °C || EB-PVD of the PYSZ thermal barrier coating
3. Route:
   EB-PVD of Al_2O_3 || vacuum annealing 9 h / 1100 °C || EB-PVD of the graded interlayer || vacuum annealing 9 h / 1100 °C || EB-PVD of the PYSZ thermal barrier coating

Annealing at 1100 °C for 9 h in vacuum ($<10^{-5}$ mbar) was carried out in order to establish formation of the stable Al_2O_3-corundum phase in the graded interlayers. Without annealing, amorphous alumina was formed which transformed to metastable alumina phases during subsequent PYSZ deposition or thermal loading thus leading to premature coating failure. Investigation of coating morphology was carried out by SEM and TEM.

Thermal cycling behavior of the different graded TBC systems was investigated in a resistance-heated furnace. The cycle period included 50 min heating at a temperature of 1100°C in air and 10 min cooling to nearly room temperature by help of ventilators. As substrates we used IN 100 rods with a standard EB-PVD NiCoCrAlY bond coat with 80 to 100 μm thickness.

Isothermal oxidation with an oxidation time of 300 h at 1100 °C was carried out in air in a Setaram thermobalance, recording the time-dependent oxidative mass gain. Substrates were cylindrical bond-coated IN 100 coupons (15 mm Ø, 1.5 mm thick).

RESULTS AND DISCUSSION
Morphology and phase composition

The microstructure of the graded alumina-zirconia coatings is shown by the cross sections of Fig. 1 and 2. Essentially it is identical for all three coating routes. Fig. 1 presents the cross section of a complete graded TBC combining a graded Al_2O_3-PYSZ interlayer of 20 μm and a PYSZ top coat of 200 μm thickness. In Fig. 2 the discrete graded interlayer consisting of eight different composite zones with varying content of both species is visible.

The microstructure of different Al_2O_3-PYSZ composite zones was already presented in a former article [5] showing the results of SEM investigation and phase analysis by XRD. We observed a dense and coarse columnar morphology for the gradient steps. Only the zones with predominant zirconia content showed the typical highly columnar microstructure. Homogeneous Al_2O_3-PYSZ composite coatings with a wide range of alumina composition reveal the same dense and coarse morphology in contrast to standard zirconia coatings with a feather-like microstructure [5]. On a far smaller scale we found a multilayer structure for homogeneous and graded Al_2O_3-PYSZ coatings which resulted from sample rotation

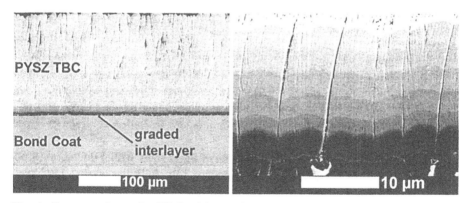

Fig. 1: Cross section of a TBC with chemically graded interlayer

Fig. 2: Discrete chemically graded Al_2O_3-PYSZ interlayer

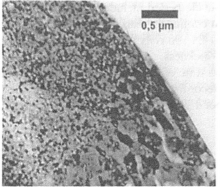 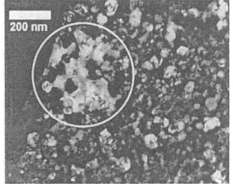

Fig. 3: Nanocrystalline grain structure   Fig. 4: α-alumina grain

during the deposition process. Additionally, a nano-grain structure was indicated by peak-broadening of strong zirconia peaks. These results could be confirmed by TEM analysis on graded alumina-zirconia interlayers. Fig. 3 shows a bright-image picture of a transition between two zones with a balanced content of alumina and zirconia. These zones are a mixture of alumina and zirconia grains with diameters of several ten to hundred nanometers depending on the content of each phase. Fig. 4 presents a dark-image picture of a large alumina grain imbedding smaller zirconia grains (white circle). While the zirconia grains are in general crystalline in the t'-phase, small alumina grains are amorphous and only a minority of large alumina grains with a diameter of several hundred nanometers are crystalline in the stable α–phase. In general dark and bright grains cannot be assigned to a single phase because differences of brightness in these pictures is related to diffraction contrast and not to phase contrast. Therefore the multilayer structure is not visible in these pictures.

Thermal cycling

The results of the thermal cycling test at 1100 °C are presented in the bar diagram of Fig. 5. Thermal cycling lifetimes of the duplex TBC system and the three different graded TBC systems are shown as average values of three different samples for each coating system. Additionally, standard deviation is given as error bars.

For the standard duplex system a lifetime of 580 cycles with a deviation of 60 cycles was found. In comparison to the standard TBC only the graded TBC system of the second deposition route exhibits a slightly increased lifetime of 670 cycles with a clearly higher standard deviation. In contrast, the two other graded TBC systems of the 1. and 3. coating routes show a premature failure.

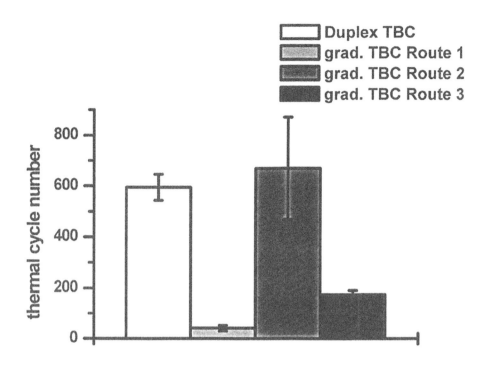

Fig. 5: Thermal cycling numbers for duplex and chemically graded TBC systems

After thermal cycling cross sections of the specimens were prepared in order to investigate the failure region of each coating system by SEM. In Fig. 6 and 7 the results are shown for the duplex TBC system and a graded TBC system, respectively. For the duplex system failure took place close to the zirconia interface in an upper part of the TGO (thermally grown oxide), consisting of small-grained α–alumina with high porosity. The lower part of the TGO is a dense α–alumina with small yttria-rich inclusions which are visible as bright spots in a dark matrix. Presumably these inclusions are (Y, Al)O-phases like $YAlO_3$ (YAP-phase) or $Y_3Al_5O_{12}$ (YAG-phase). Exact determination by EDX was impossible due to their small grain size. Their role in coating failure is not clear yet.

Surprisingly, coating failure behavior of all graded TBC systems was nearly identical. Crack formation and coating delamination was only observed in the first zone of the graded interlayer consisting of pure α–alumina or at the gradient-TGO interface. Instead of the originally columnar microstructure the first gradient zone of pure α–alumina shows a strongly sintered and highly porous morphology. Additionally, vertical cracks reaching down to the TGO and bond coat are visible.

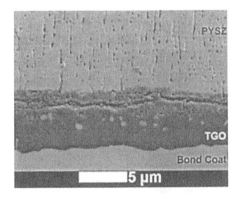
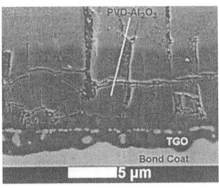

Fig. 6: Duplex system after thermal
cycling

Fig. 7: Chemically graded system after
thermal cycling

Similar to the duplex TBC the TGO consists of dense α–alumina with small
(Y, Al)O-phases.

Crack formation and delamination in duplex and graded TBCs was mainly observed in alumina layers with a highly porous microstructure. Those regions represent the weakest link of a layer compound and advancing of cracks is easily
possible from one pore to another. Alumina is the phase with the lowest coefficient of thermal expansion (CTE) of all coating layers while the CTE of the metallic bond coat is much bigger. This difference in CTEs of neighboring coating
parts should result in substantial compressive stresses in the alumina layers during
cooling of specimens. Therefore, it can be assumed that compressive stresses are
the prime reason for coating failure of the TBC systems investigated here.

Premature failure of the 1. graded TBC system can be explained by formation
of metastable alumina phases in the graded interlayer, which are not transformed
by the preceding annealing treatment but during the thermal cycling test. The reduced lifetime of the 3. graded TBC system is probably caused by degradation
due to pre-oxidation during the double annealing treatment before and after deposition of the graded interlayer. Therefore the 2. graded TBC system is the optimum choice of processing and annealing parameters.

Isothermal oxidation

The oxidation behavior of the graded and conventional duplex coating system
is shown in Fig. 8. All oxidation curves show small steps of mass increase along
their curvature. This could be an indication for local crack formation at the interface leading to small areas of bond coat which are unprotected, thus enabling
rapid oxidation.

Surprisingly, only the graded TBC system of the 2. coating route has a slightly

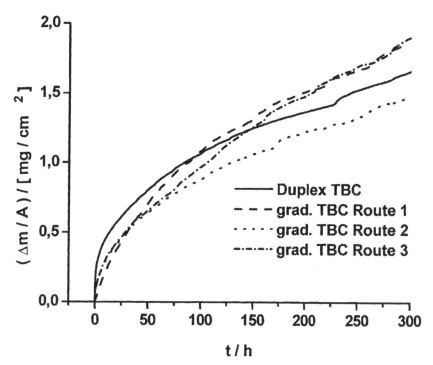

Fig. 8: Oxidation curves of duplex and chemically graded TBC systems

lower normalized mass gain than the duplex TBC with similar oxidation rate (= curve slope). For the 1. and 3. route TBC systems only up to 100 h and 150 h, respectively, the oxidative weight gain is lower than for the duplex TBC but the oxidation rate is higher during the whole 300 h. However, it must be considered that pre-oxidation due to annealing treatments took place in the graded TBC systems. Therefore, the samples are already submitted to oxidative mass gain which is not registered by the thermobalance. Thus the oxidation behavior of the graded TBC is poorer in comparison to the duplex system as shown in Fig. 8.

Concerning the special microstructure of graded interlayers it seems to be clear so far that the graded TBC systems represent no oxidation protection for the metallic bond coat. EB-PVD alumina layers with high open porosity and vertical cracks and Al_2O_3-PYSZ composite zones with nano-crystalline microstructure and therefore increased density of grain boundaries facilitate oxygen diffusion into the bond coat in comparison to the duplex TBC.

SUMMARY

EB-PVD alumina-zirconia thermal barrier coatings with graded interlayers of three different processing routes were produced by means of dual source evaporation. The gradient zone of each coating route consisted of different nanocrystalline zones with constant Al_2O_3-PYSZ content. Thermal cycling and isothermal oxidation behavior of the graded coating systems were compared to the standard duplex coating system. For all TBC systems the thermally grown oxide consisted mainly of α–alumina with small Y-rich inclusions. All coating systems failed during thermal cycling in the thermally grown alumina or in the EB-PVD alumina region. This result can be explained by thermal mismatch stresses due to differences of thermal expansion between the alumina and the metallic bond coat. Only for one graded coating system slight advantages concerning thermal cycling behavior and isothermal oxidation was found. Further research is necessary to fully exploit the potential advantages of a graded coating design for EB-PVD systems.

REFERENCES

[1] W.A. Kaysser, M. Bartsch, T. Krell, K. Fritscher, C. Leyens, U. Schulz, M. Peters, "Ceramic Thermal Barriers for Demanding Turbine Applications" ceramic forum international / Ber. DKG 77(2000) No. 6, pp 32-36

[2] H.G. Scott, "Phase relationships in the zirconia-yttria system", Journal of Material Science 10(1975) pp 1527-1535

[3] L. Lelait, S. Alperine, C. Diot, "Microstructural investigation of EBPVD thermal barrier coatings", Journal de Physique IV, Colloque C9, 3(1993) pp 645-654

[4] T. Krell, U. Schulz, M. Peters, W.A. Kaysser, „Influence of various process parameters on morphology and phase content of EB-PVD thermal barrier coatings, Proc. Euromat 97, Vol. 3, pp. 3/29-3/32, Netherlands Society for Materials Science, Zwijndrecht, The Netherlands, 1997

[5] T. Krell, U. Schulz, M. Peters, W.A. Kaysser, "Graded EB-PVD Alumina-Zirconia Thermal Barrier Coatings - An Experimental Approach", Materials Science Forum Vols. 308-311, pp 396-401, Proc. 5th International Symposium on Functionally Graded Materials, Dresden, Germany, 1998

[6] T. Krell, Thermische und thermophysikalische Eigenschaften von elektronenstrahlgedampften chemisch gradierten Al_2O_3/PYSZ-Wärmedämmschichten, Ph.D. thesis, VDI-Verlag, Düsseldorf, 2001

GRADED THERMAL BARRIER COATINGS: CRACKING DUE TO LASER IRRADIATION AND DETERMINING OF INTERFACE TOUGHNESS

H. Balke, H.-A. Bahr, A.S. Semenov,
I. Hofinger and C. Häusler
Dresden University of Technology
Department of Mechanical Engineering
01062 Dresden, Germany

G. Kirchhoff and H.-J. Weiss
Fraunhofer Institute for Material
and Beam Technology.
Winterbergstrasse 28
01277 Dresden, Germany

ABSTRACT
    The damage resistance of (graded) thermal barrier coatings (TBCs) can be quantified by cyclic surface heating with laser beam. Damage was found to develop in several stages: vertical cracking → delamination → blistering → spalling. This sequence can be understood as an effect of progressive shrinkage due to sintering during thermal cycling, which increases the energy release rate for vertical cracks which subsequently turn into delamination cracks. Taking into account TBC shrinkage, delamination crack propagation is modelled by means of FEM. An increase of interface fracture toughness due to grading and a decrease due to aging has been measured in a 4-point bending test modified by a stiffening layer. Correlation with the observed damage in cyclic heating is discussed. It is explained in which way grading is able to reduce the damage.

INTRODUCTION
    Thermal barrier coatings (TBCs) on gas turbine engine blades are applied with the purposes to increase the energetic efficiency by running at higher temperature. The service life of TBCs is limited by spalling fracture [1]. It had been demonstrated previously that cyclic surface heating by means of laser irradiation is a suitable test for the performance of non-graded and graded TBCs [2-4]. This presentation compares the behaviour of non-graded and graded TBCs under cyclic laser heating for a wide range of temperatures and numbers of cycles.

DAMAGE DUE TO LASER HEATING
    The investigations have been conducted on samples prepared by DLR Köln by means of electron beam physical vapour deposition (EB-PVD). The non-graded samples are flat Nimonic plates with about 100 µm NiCoCrAlY bond coat and 250 µm partially stabilised ZrO_2. The graded samples contain a 50 µm intermediate layer between bond and ZrO_2, with a graded transition from Al_2O_3 to ZrO_2 [5].
    The sample surface is locally heated by means of a 1kW Nd-YAG laser beam, as seen in Fig. 1. The beam is deflected into a circular path on the sample surface by two synchronized oscillating mirrors. The temperature is measured by high speed pyrometer combined with another oscillating mirror. The start of crack can be detected by an acoustic sensor. The typical cycle consists of two phases: active heating for 1 s and 10 s pause (see Fig. 2). The number of cycles

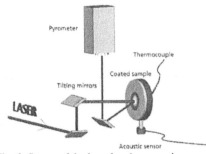

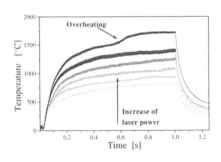

Fig. 1. Set-up of the laser heating experiment.　　**Fig. 2.** Thermal cycling: part of an 11s cycle.

per sample has been chosen between several tens to several thousands. The maximum surface temperature reached 1600°C.

Failure has been found to proceed in several consecutive stages, as seen in Fig. 3. Note that N and T_{max} have been varied. The visible changes and stages of damage were investigated by microscopy and metallographic cuts. There is no visible damage after 11 cycles for all TBCs. The short vertical cracks are observed after 30 cycles with T_{max}=1500°C in the non-graded TBC (Fig. 3a) and after 300 cycles with T_{max}=1530°C in the graded TBC (see Fig. 3d). The long vertical cracks and delamination arise after 1400 cycles with 1500°C in the non-graded TBC (see Fig. 3b) and after 300 cycles with 1600°C in the graded TBC (see Fig. 3f). Under the above loading conditions, local spalling is observed only for non-graded TBC.

Non-graded EB-PVD TBC (homogeneous ZrO_2):

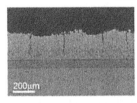 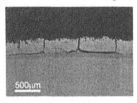 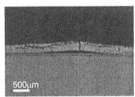

a)　N=30, T_{max}=1500°C　　　b)　N=1400, T_{max}=1500°C　　　c)　N=900, T_{max}=1600°C

Graded EB-PVD TBC (graded transition $Al_2O_3 \rightarrow ZrO_2$):

 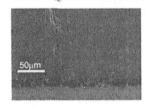 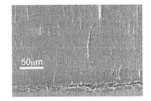

d)　N=300, T_{max}=1530°C　　　e)　N=1000, T_{max}=1530°C　　　f)　N=300, T_{max}=1600°C

Fig. 3. Progress of damage in non-graded and graded TBCs.
N and T_{max} denote the number of cycles and maximum temperature.

It can be concluded that TBCs respond to thermal cycling of increasing severity with the following sequence of phenomena,

vertical cracks → delamination → blistering → spalling,

with the onset of the stages of damage being delayed by grading. This experimental evidence can serve as a guidance for the following analytical and numerical fracture mechanical analysis with the aim to optimize the effect of grading.

Since, in present experiments, damage depends on only two parameters, N and T_{max}, it can be conveniently represented by damage maps as in Fig. 4. Apparently grading delays delamination and subsequent further damage. The blistering and spalling has been identified visually. Delamination can be detected by pyrometry. It reveals itself by a sudden rise of the surface temperature. The kink in the upper curve of Fig. 2 is related to a sudden increase of thermal flow resistance due to delamination.

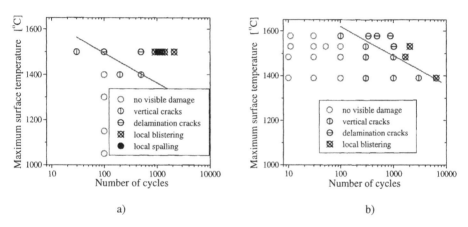

a)   b)

Fig. 4. Damage maps of cyclically heated homogeneous TBC (a) and graded TBC (b).

FRACTURE MECHANICAL MODELLING
Shrinkage Stress as a Driving Force

The considerable residual crack opening after cyclic heating indicates the presence of average shrinkage strain of about 1%, most probably due to sintering of the porous ceramic (Fig. 5). Sintering of TBCs has been discussed in [1,4]. Prolonged heating of TBC samples detached from the substrate by chemically dissolving the NiCoCrAlY bond coat resulted in shrinkage strains $\varepsilon_S \approx 1.4\%$ for 100h at 1200°C and $\varepsilon_S \approx 0.6\%$ for 250h at 1050°C (Fig. 6, curvature due to inhomogeneity of TBC). The progressive shrinkage due to sintering in thermal cycling increases the energy release rate for vertical cracks which subsequently turn into delamination cracks. (Note that the temperature needed for TBC damage in Fig. 3 is much higher because of the smaller total duration of laser heating with N cycles).

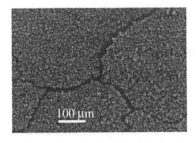

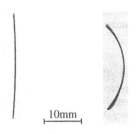

Fig. 5. Shrinkage cracks in thermally cycled TBC.

Fig. 6. Detached TBC before (left) and after (right) deformation due to sintering for 100h at 1200°C.

<u>Simplified Model of TBC Specimen</u>

Let us consider a two-dimensional model for the analysis of delamination cracks after thermal shock (Fig. 7a), with a representative element containing only one crack (Fig. 7b). This model implies plane strain. Elastic isotropy is assumed, with E_1, v_1, E_2 and v_2 as the elastic constants for the TBC and substrate, respectively. The sintering effect is characterized by two parameters: average shrinkage strain ε_S and depth δ. Delamination crack propagation has been successively computed by means of the finite element method (FEM). The finite element model corresponding to the problem depicted in Fig. 7b is shown in Fig. 7c for the deformed state. A highly refined mesh with special elements is used in the vicinity of the crack path.

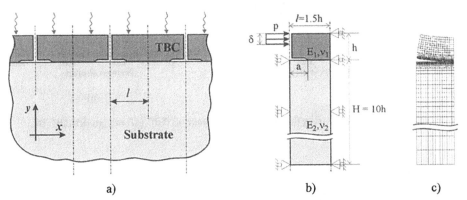

a)   b)   c)

Fig. 7. Model for the analysis of delamination crack propagation due to sintering.

<u>The Finite Element Procedures</u>

The MARC finite element program package [6] was applied for the numerical calculation. The 2D mesh of 8-noded isoparametric finite elements with 4 integration points for stiffness matrix computation is used. The number of elements varies with relative thickness h/H and with dimensionless crack length a/l and reaches about 1500. A highly refined self-similar focused mesh is used in the vicinity of the crack tip. The size of elements around the crack tip is about h/680. The depth of substrate is taken as 10h, which is a sufficiently good approximation to infinity. The right side of the model (Fig. 7b) is fixed in horizontal direction, which corresponds to the semi-plane condition. The deviation of G due to this simplification does not exceed 8%.

The energy release rate G is calculated as J-integral via differential stiffness technique [6]. The mode mixity angle ψ^* is derived by means of the crack surface displacement method [7]. A check of the accuracy of the model and methods is provided by comparison with the analytical result for homogeneous material in the limit of large l/h and $\delta = h$: $G_{SS} = \left(1 - v_1^2\right) p^2 h / \left(2E_1\right)$.

In this model, the effect of sintering is simulated by assuming a homogeneous initial strain ε_S in an upper stratum of depth δ of the TBC. As another check of the numerical computations, the equivalent problem of external pressure $p = E_1 \varepsilon_S / (1-v_1)$ applied on the boundary was also calculated. The coincidence of the results turned out to be better than < 0.5 %, proving consistency of the calculations.

<u>Finite Element Analysis of Delamination</u>
Delamination is analysed on the base of elastic finite element solutions for variable crack lengths and loading conditions. The crack configurations for different crack lengths are shown in Fig. 8.

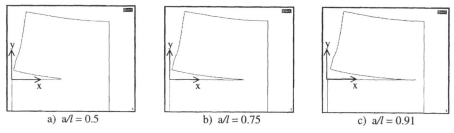

a) $a/l = 0.5$ b) $a/l = 0.75$ c) $a/l = 0.91$

Fig. 8. Crack configurations for different crack lengths (ε_S=1%, δ/h=0.8).

The deformations seen in Fig. 8 (10 times magnified) show that the upper layer experiences both opening and sliding relative to the substrate, indicating "mixed mode" conditions. As seen in Fig. 9 and 10, the two parameters, energy release rate G and mode mixity angle ψ^*, depend strongly on the loading parameter δ while ε_S affects only G. With δ/h = 0.2 and δ/h = 0.4 the vertical crack flanks get into contact, which has been taken into account in the results of Fig. 9.

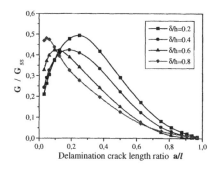

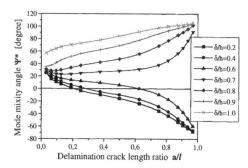

Fig. 9. Energy release rate vs. crack length.   **Fig. 10.** Mode mixity angle vs. crack length.

The effect of this contact is small.

The prominent feature of the G(a/l)-curves in Fig. 9 is the maximum, which is about half the steady-state value G_{SS} for an equivalent specimen with infinite width l defined by equation (1) as follows:

$$G_{SS} = \frac{1+\nu_1}{1-\nu_1}\, \frac{E_1 \varepsilon_s^2 h}{2}\, \left(\frac{\delta}{h}\right)^2 \left[1+3\left(1-\frac{\delta}{h}\right)^2\right]. \tag{1}$$

The mode mixity angle ψ^*, introduced in accordance with the approach proposed by Rice [8] and Charalambides et al. [9], represents the phase angle of the complex quantity $K^* = KL^{i\varepsilon} = |K|e^{i\psi^*}$, where $\varepsilon = 1/2\pi \ln\{[(3-4\nu_1)\mu_2 + \mu_1]/[(3-4\nu_2)\mu_1 + \mu_2]\}$ is a bimaterial constant [8], μ_1 and μ_2 are the shear moduli. The arbitrary reference length L is set equal to TBC thickness h = 300 μm. The mode mixity angle ψ^* can be expressed as

$$\Psi^* = \arctan\left[\frac{\mathrm{Im}(Kh^{i\varepsilon})}{\mathrm{Re}(Kh^{i\varepsilon})}\right]. \tag{2}$$

The crack surface relative displacement distributions $\Delta u_x(r)$ and $\Delta u_y(r)$ are used in numerical analysis of ψ^* computation by

$$\Psi^* = \arctan\left[\frac{\Delta u_x(r)}{\Delta u_y(r)}\right] + \varepsilon \ln\left(\frac{h}{r}\right) + \arctan(2\varepsilon). \tag{3}$$

The optional value of r for accurate ψ^* computation is defined in accordance with the crack surface displacement method [7]. Fig. 10 shows ψ^* vs. crack length. Within the technologically relevant range of elastic moduli and TBC thickness, ψ^* does not much vary.

Delamination Conditions in Non-Graded and Graded TBC

The developed model for the calculation of the energy release rate G due to the measured shrinkage strain allows to predict delamination on the base of the criterion G > $G_C(\psi^*)$. The interface toughness G_C has been measured in a four-point bending test [9] modified by a stiffening layer [10] (Fig. 11). Results of non-graded and graded TBCs are listed in the table in Fig. 11 for three kinds of samples differing by the degree of aging: as received, 100h aged at 1000°C, 10h aged at 1100°C in the furnace. It is seen that graded TBC has higher G_C in all cases, and aging at high temperature reduces G_C.

G_C [N/m]	Non-graded EB-PVD-TBC	Graded EB-PVD-TBC
Without aging	>81	>81
100h at 1000°C	63	>81
10 h at 1100°C	37	45

Fig. 11. TBC delamination testing equipment and results ($\psi^* \approx 70°$).

As another effect of heating during laser cycling, the energy release rate G rises as a result of shrinkage strain ε_S. The results of Fig. 12 have been obtained with the assumption of $\delta/h = 0.4$ and $\varepsilon_S \approx 0.4\%$ for 100h at 1000°C ($G_{SS} = 158$ N/m) and $\varepsilon_S \approx 0.7\%$ for 10h at 1100°C ($G_{SS} = 483$ N/m), as suggested by the shrinkage evident in Figs. 5 and 6.

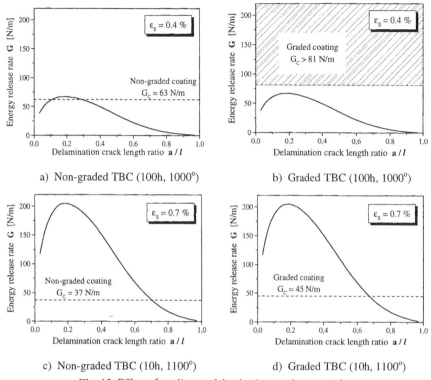

a) Non-graded TBC (100h, 1000°)   b) Graded TBC (100h, 1000°)

c) Non-graded TBC (10h, 1100°)   d) Graded TBC (10h, 1100°)

Fig. 12. Effect of grading on delamination crack propagation.

After 100h at 1000°C, G keeps below G_c in the case of graded TBC (Fig. 12b), but G exceeds G_c in the non-graded TBC so that the delamination crack could be driven to $a/l \approx 0.3$ in Fig. 12a. (This, in turn, would make higher local temperature above the crack, hence more shrinkage stress which would drive the crack even further.) At 1100°C G increases and G_c decreases which leads to large delamination crack lengths $a/l \approx 0.7$ according to Fig. 12c,d. This could explain the observed behaviour of non-graded and particular graded TBCs under laser heating. (It must be mentioned that the experimental G_c in Fig. 12 is related to $\psi^* \approx 70°$ while ψ^* is near zero for $\delta/h = 0.4$ according to Fig. 10, so the real G_c might be smaller. Note that the real shrinkage strain could also be smaller than the assumed value so that the effects visualized in Fig. 12 essentially remain there).

Discussion

The proposed failure model of TBC based on delamination crack propagation with shrinkage stress due to sintering as a crack driving force is compatible with experimentally observed phenomena of progressive cracking due to laser heating in Figs. 3 and 4. This justifies our belief to have found a viable fracture mechanical approach to the damage phenomena of TBCs, which indicates how to modify TBCs with the aim to delay failure as far as possible, as by an additional graded layer which can increase the delamination toughness G_c, for example.

SUMMARY AND CONCLUSIONS

TBC damage resistance can be tested by means of cyclic surface heating by laser irradiation. In this way the influence of aging and grading on the performance of TBCs can be quantified. The observed sequence of TBC damage involves several stages: vertical cracking $\rightarrow$ delamination $\rightarrow$ blistering $\rightarrow$ spalling. Progressive shrinkage in thermal cycling increases the energy release rate for the vertical cracks, which subsequently turn into delamination cracks. It has been shown that grading is able to improve the performance of TBCs by an increase of interface toughness. Experimental evidence for this increase is obtained from four-point bending test modified by a stiffening layer. With this experimental technique, also the effect of aging on interface toughness has been quantified.

It has been demonstrated that a fracture mechanical approach applies to the damage phenomena of TBCs. The results of finite element modelling of the TBC delamination process are compatible with experimental data. It seems reasonable to extend both modelling and experiments.

ACKNOWLEDGEMENT

The authors thank the Deutsche Forschungsgemeinschaft (DFG) for support within the priority program "Gradientenwerkstoffe" and the Deutsches Zentrum für Luft- und Raumfahrt (DLR) Köln for supplying sample material.

REFERENCES

[1] D.M. Nissley, "Thermal barrier coating life modelling in aircraft gas turbine engines", Proc. of Thermal Barrier Coating Workshop, NASA Conference Publication, **3312** 265-281 (1995).

[2] C.Y. Jian, T. Hashida, H. Takahashi, M. Saito, "Thermal shock and fatigue resistance evaluation of functionally graded coating for gas turbine blades by laser heating method", *Composites Engineering,* **5** [7] 879-889 (1995).

[3] I. Hofinger, H.-A. Bahr, H. Balke, G. Kirchhoff, C. Häusler, H.-J. Weiss, "Fracture mechanical modelling and damage characterization of functionally graded thermal barrier coatings by means laser irradiation", *Materials Science Forum,* **308-311** 450-456 (1999).

[4] H. Balke, G. Kirchhoff, I. Hofinger, H.-A. Bahr, H.-J. Weiss, "Damage due to laser irradiation of thermal barrier coatings - fracture mechanical modelling", Mechanical Properties of Films, Coatings and Interfacial Materials, Castelvecchio Pascolie, Italy, June 27 - July 2, 1999, Abstracts, United Engineering Foundation, Inc., New York (1999).

[5] T. Krell, U.Schulz, M.Peters, W.A.Kaysser, "Graded EP-PVD aluminia-zirconia thermal barrier coatings - an experimental approach", *Material Science Forum,* **308-311** 396-401 (1999).

[6] MARC Volume A: Ver. K7.3. MARC Analysis Research Corporation. (1999).

[7] P.P.L. Matos, R.M. McMeeking, P.G. Charalambides, M.D. Drory, "A method for calculating stress intensities in bimaterial fracture", *Int. Journal of Fracture,* **40** 235-254 (1989).

[8] J.R. Rice, "Elastic fracture mechanics concepts for interfacial cracks", *Journal of Applied Mechanics,* **55** 98-103 (1988).

[9] P.G. Charalambides, J. Lund, A.G. Evans, R.M. McMeeking, "A test specimen for determining the fracture resistance of bimaterial interfaces", *J. Appl. Mechanics,* **56** 77-82 (1989).

[10] I. Hofinger, M. Oechsner, H.-A. Bahr, M.V. Swain, "Modified four-point bending specimen for determining the interface fracture energy for thin, brittle layers", *Int. Journal of Fracture,* **92** 213-220 (1998).

THERMAL SHOCK OF FUNCTIONALLY GRADED
THERMAL BARRIER COATINGS

Klod Kokini, Jeffrey DeJonge and Sudarshan Rangaraj   Brad Beardsley
Purdue University   Caterpillar, Inc.
School of Mechanical Engineering   Peoria, IL 61656-1875
West Lafayette, IN 47907-1288

ABSTRACT
An experimental study was conducted to develop an understanding of the thermal fracture behavior of plasma sprayed graded thermal barrier coatings when subjected to a thermal shock loading. For this purpose, two coating architectures *with similar thermal resistances* were studied. The thermal loading was applied using a continuous CO_2 laser. The heating period for each beam shaped specimen was four seconds, followed by cooling in ambient air. The two architectures considered were a two-layer system and a four-layer system. The results showed that the grading increased the temperature at which interface cracks initiated. Also, the length of interface cracks for a given surface temperature was shorter for the graded specimen. It was concluded that the grading of thermal barrier coatings with *similar thermal resistance* causes an increased resistance to surface and interface cracking under thermal shock loading.

INTRODUCTION
Thermal barrier coatings (TBC's) are used and being developed to increase the operating temperature of systems such as a diesel engine in order to achieve an increase in the efficiency, performance and durability of the engine. Also, with increased surface temperatures, the exhaust gasses can be used for turbo-charging.

However, the ceramic (usually yttria stabilized zirconia) that makes up the TBC and the substrate material (such as steel) has different thermal expansions. The difference in thermal expansions and the applied temperature gradients usually result in thermal stresses on the coating surface and at the interface between the coating and the substrate. These stresses can be large enough to initiate surface and interface cracks. Cyclic application of thermal loads can cause these cracks to propagate resulting in spallation of the coating and loss of thermal protection to the substrate.

One method of reducing the thermal fracture in TBC's is to produce a compositionally graded structure, which provides a more gradual change in properties compared to the sudden change of properties in a monolithic TBC system. There have been many studies, which considered different aspects of functionally graded materials (FGM's). Many of these studies have been presented in the international conferences on FGM's initiated in 1990.[1]

The present authors have also performed several studies related to the mechanisms of crack initiation and propagation of functionally graded TBC's.[2-8] However, the crack initiation and propagation mechanisms of graded TBC systems, *with equal thermal resistance, but different architectures*, have not been considered.

In this paper, the effect of a thermal shock on graded thermal barrier coatings with two different coating architectures were studied. The thermal resistances of the two architectures were similar. The surface cracks and horizontal cracks resulting from thermal shock loading were determined as a function of increasing surface temperature.

EXPERIMENTS

All the specimens used in this study were manufactured by Caterpillar Inc. An Inconel steel plate was plasma sprayed with graded zirconia/bond coat coatings. These plates were then cut into beam shaped specimens using a CNC controlled water jet. These beam shaped specimen were 31 mm. wide, 32 mm. high and 3 mm. thick.

Experimental Procedure

Each specimen was then polished on both sides in order to observe and measure cracks before, during and after the thermal shock tests. The coating surface was painted with a high temperature paint (Pyromark Series 2500 silicon based paint) to ensure correct temperature readings by the pyrometer during testing. It was determined that the emissivity of the paint was 0.99 at the operating wavelength of the pyrometer (4.8 – 5.2 μm) and the temperature range (500°C - 1300°C) considered.[9] Before the specimens were subjected to the laser heating, they were observed under the microscope to ensure no cracks developed during the curing process.

The thermal shock experiments were performed by applying a concentrated laser heat flux at the center of the top surface of the coating for 4 seconds, followed by ambient cooling (**Figure** 1). A 1.5 kW CO_2 laser (10.6 μm), manufactured by Covergent Energy, was used to heat the surface of the coating. At this wavelength, the absorptivity of the zirconia coating is one and the laser beam is absorbed at the surface of the coating.[10] The heat flux distribution provided by the laser was found to be of Gaussian shape described as

$$q(x) = q_{max} \exp \{ -2(x/w_1)^2 \} \tag{1}$$

In Equation (1), q is the heat flux and w_1 is a length parameter, which was determined by changing it in the analytical model until a close approximation for the surface temperature was achieved.[9]

Experimental Set Up

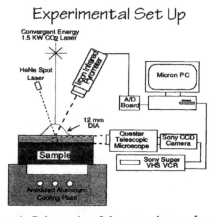

Figure 1: Schematic of the experimental setup.

The surface temperature of the specimen was recorded using an Ircon infrared pyrometer (Modeline 7000). The data from the pyrometer was recorded in a personal computer using the Labview software.

Following the thermal shock procedure, each of the specimens was observed under an optical microscope at 200X magnification. All of the cracks produced during testing were measured and recorded.

Description of Coating Architectures

The material for the substrates in all the specimens was Inconel steel. However, for the TBC, two different layer configurations, which included a two-layer system and a four-layer system were considered. In the two-layer specimen, a 100% bond coat layer was deposited on the steel substrate. A 100% zirconia layer was then deposited over this bond coat layer. The thickness of each layer is shown in **Table I.**

The architecture of the four-layer specimen consisted of a 100% bond coat layer deposited over the substrate. A layer of 50% zirconia / 50% bond coat was deposited over the pure bond coat layer, a 75% zirconia / 25% bond coat layer was deposited next. Finally the top-layer was comprised of 100% zirconia. The thickness of each layer is shown in **Table I.**

Table I: Architecture of two and four-layer TBC specimens.

Layer composition	Layer thickness	
	Two-layer specimen	**Four-layer specimen**
100% Zirconia	0.6	0.25
75% Zirconia / 25% Bond Coat		0.29
50% Zirconia / 50% Bond Coat		0.31
100% Bond Coat	0.2	0.14
Total coating thickness	**0.8**	**0.99**

RESULTS AND DISCUSSION

Thermal Resistance

The thermal resistance of a material is related to its thermal conductivity, surface area, and thickness. When there are multiple layers of different thermal resistances, the total thermal resistance of the layers is the sum of each individual resistance. Thus, the total thermal resistance for the graded thermal barrier coatings was calculated as

$$R_{total} = \sum_{i=1}^{n} \frac{L_i}{A_i k_i} \qquad (2)$$

In Equation (2), R_{total} is the total thermal resistance of the TBC comprised of n layers. L_i, k_i and A_i are the thickness, thermal conductivity and surface area respectively of each layer.

While calculating the total thermal resistance for the different coating architectures the surface area for each layer was considered to be constant. Also, the thickness of each layer was assumed to be equal and was calculated as L/n, where, L represented the total thickness and n was the total number of layers in the coating. Since the bond coat in each model had the same thickness and its thermal resistance was relatively very low, its thermal resistance was neglected. **Figure** 2 shows the calculated thermal resistances of the two coating architectures used in the experiments. The thermal conductivity data (for each layer) used in the calculations, was provided by the manufacturer of the coatings (Caterpillar).

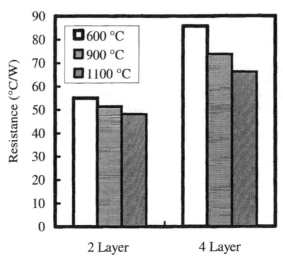

Figure 2: Total thermal resistance for the specimens.

The temperature variation as a function of time was measured at the center of the top surface of the specimen. The spatial temperature distribution throughout the specimen was calculated. Laser heat flux with a Gaussian profile was applied to the surface of the different models. This distribution was modeled based on the CO_2 laser used for experimentation of coated specimens.

Since the laser heat flux applied to the surface had a Gaussian shape the maximum temperature occurred at the center of the top surface of the specimen (x=0). The intensity of the laser heat flux applied to the surface was varied to obtain a constant surface temperature. When comparing different values of total thermal resistance, a model with a high value of resistance required a smaller heat flux to obtain a surface temperature comparable to a model with lower thermal resistance.

The transient maximum temperature which occurs at the center of the top surface, for three different models, with thermal resistances of 50°C/W and 150°C/W is shown in **Figures** 3 and 4 respectively, for different coating architectures. These resistances cover the range of resistances for the experimental samples. It can be noted that the temperature variation in all cases are quite similar.

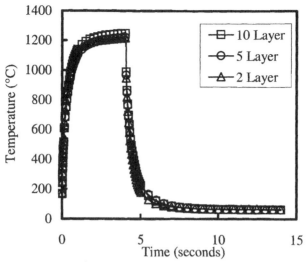

Figure 3: Temperature vs. time at the center on the coating surface for R=50 °C/W

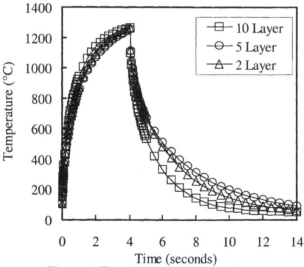

Figure 4: Temperature vs. time at the center on the coating surface for R=150 °C/W

Experimental Results

The coating configurations selected were designed to have similar thermal resistances. However, the variations obtained in the final configurations, shown in **Figure** 2, were due to processing variability, material property variations, etc.

The cracks resulting from the thermal shock experiments were measured for each specimen and presented as the "crack ratio" which represents the ratio of the final crack length to the total coating thickness (**Table** I). In order to study the effects of increasing heat flux, the power of the laser was gradually increased for different specimens. However, each specimen was subjected to only one laser exposure.

For the 2-layer architecture, a photograph of a surface and interface crack is shown in **Figure** 5. The crack ratio in each specimen obtained after thermal shock is presented as a function of the maximum temperature experienced by the center of the surface of the coating in **Figure** 6.

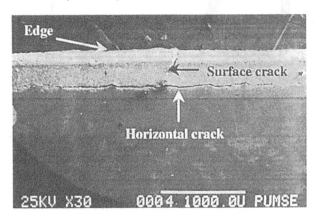

Figure 5: Two-layer specimen after thermal shock.

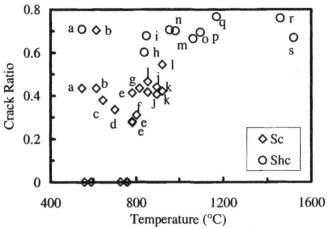

Figure 6: Crack ratio for two-layer specimens.

In **Figure** 6 above, Shc denotes surface cracking along with horizontal cracking, whereas, Sc denotes surface cracking only. It can be noted from **Figure** 6 that the cracks initiate near a surface temperature of about 500°C for the 2-layer specimens and gradually increase with the severity of the test. In most experiments, it was noted that the surface cracks formed as a single crack or as multiple cracks (2–4 cracks). In order to identify this cracking behavior, each specimen was labeled with a letter as shown in **Figure** 6. Thus, for the 2-layer architecture, the majority of surface cracks occurred as single cracks across the entire temperature spectrum.

The cracks, which form in the 4-layer specimen after the thermal shock, are shown in **Figure** 7. The crack ratio vs. maximum surface temperature for this case is presented in **Figure** 8.

Figure 7: Four-layer specimen after thermal shock.

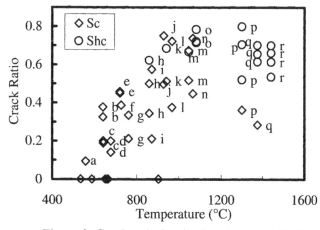

Figure 8: Crack ratio for the four-layer architecture.

In **Figure** 8, Shc represents surface cracking along with horizontal cracking, whereas, Sc represents surface cracking only. As seen in **Figure** 8, for the 4-layer specimens, surface cracks start forming near 550°C and gradually become longer as the surface temperature increases.

Several of the 4-layer specimens were found to have two surface cracks at maximum temperatures below 1000°C. However, above this temperature, the number of surface cracks increased to three or four for all the specimens.

The formation of multiple cracks at the higher temperatures indicates that as the grading of the coating is introduced, and the top zirconia layer becomes thinner, this top layer starts behaving similar to a thin single-layered zirconia coating. Such a behavior was previously observed by Choules[9] for single-layered zirconia coatings. Similarly, as the number of surface cracks increases, the propensity for the initiation and propagation of horizontal cracks decreases. Again, a similar behavior was shown to occur in single-layered zirconia coatings subjected to high heat fluxes.[9] The comparison of horizontal crack lengths for the two coating architectures is shown in **Figure** 9.

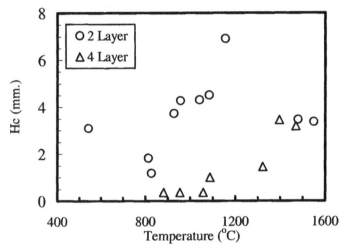

Figure 9: Horizontal crack length (Hc) vs.
maximum surface temperature

It can be noted from **Figure** 9 that the first horizontal crack in the 2-layer coating occurs near 500 °C, compared to around 850 °C for the 4-layer coating. It is also clear that, *for the same surface temperature*, the horizontal cracks in the 4-layer coating are significantly shorter.

Grading of the coating at constant thermal resistance reduces the thickness of the top ceramic layer. Thus the 4-layer specimens had a thinner top ceramic layer compared to the 2-layer specimen (Table I). Multiple surface cracks formed in the 4-layer specimens while most of the 2-layer specimens showed single surface cracks. The formation of multiple surface cracks with compositional gradation is related to the distribution of thermal stresses on the coating surface. These stresses are mainly influenced by the thermally activated time-dependent (viscoplastic) deformation behavior of the top ceramic layer and well as the subsequent lower layers.[11-12] These effects, in the 4-layer specimen may be quite different compared to the 2-layer specimen. Efforts to better understand and model this phenomenon are currently underway.

The transition in thermo-mechanical properties from the top ceramic layer to the substrate is more gradual in a 4-layer specimen as compared to a 2-layer specimen (Table I). Moreover, the 4-layer specimens had multiple surface cracks. A combination of these effects is mainly responsible for the increased resistance of the 4-layer specimens to horizontal cracking compared to the 2-layer specimens. Development of analytical and numerical models to understand these effects better are much needed and are planned for the future.

CONCLUSIONS

The thermal shock experiments performed with two different TBC configurations of similar thermal resistances, but increasingly graded architectures, show that the grading affects the initiation of surface cracks and interface cracks. In particular, the length and morphology of the surface cracks are affected by the grading. Similarly, the surface temperature at which such cracks initiate increases with the grading. The horizontal cracks formed at the end of the thermal shock process also exhibit significant differences. The increased grading increases the temperature at which horizontal cracks initiate. The final lengths of such horizontal cracks, at similar surface temperatures, are shorter for the more graded TBC architecture.

An analytical approach to study more carefully the mechanisms that yield such results and to enable designers of such coatings to predict their life, under the application of repeated thermal cycles is under way. It is clear nevertheless, that the functional grading of TBC's reduces horizontal crack initiation and propagation, thus resulting in a longer life.

ACKNOWLEDMENTS: We thank Caterpillar for support of this project.

REFERENCES

[1]M. Yamanouchi, M. Koizumi, T. Hirai and T. Shiota, *Proceedings of the First Symposium on functionally Gradient Materials*, FGM Forum, Japan, 1990.

[2]Y.R. Takeuchi and K. Kokini, "Thermal Fracture of Multilayer Ceramic Thermal Barrier Coatings", *ASME Transactions, Journal of Engineering for Gas Turbines and Power*, **116**, pp. 266-271, January 1994.

[3]K. Kokini and Y.R. Takeuchi, "Initiation of Surface Cracks in Multilayer Ceramic Thermal Barrier Coatings Under Thermal Loads", *Journal of Materials Science and Engineering A*, **189**, pp. 301-309, 1994.

[4]K. Kokini and B.D. Choules, "Surface Thermal Fracture of Functionally Graded Ceramic Coatings: Effect of Architecture and Materials", *Composites Engineering*, **5[7]**, pp. 865-877, 1995.

[5]K. Kokini and Y.R. Takeuchi, Interface Cracks in Thermally Loaded Multilayer Ceramic Coatings", *Fracture Mechanics: 25th Volume*, pp. 177-190, 1995.

[6]B.D. Choules and K. Kokini, "Architecture of Functionally Graded Ceramic Coatings Against Surface Thermal Fracture", *ASME Transactions, Journal of Engineering Materials and Technology*, **118(4)**, pp. 522-528, 1996.

[7]K. Kokini and M. Case, "Initiation of Surface and Interface Edge Cracks in Functionally Graded Ceramic Thermal Barrier Coatings", *ASME Transactions, Journal of Engineering Materials and Technology*, **119**, pp. 148-152, April 1997.

[8]Y.R. Takeuchi and K. Kokini, "Multiple Surface Thermal Fracture of Graded Ceramic Coatings", *Journal of Thermal Stresses*, **21[7]**, pp. 715-726, October 1998.

[9]B.D. Choules, K. Kokini and T.A. Taylor, "Thermal Fracture of Thermal Barrier Coatings in a High Heat Flux Environment", *Surface and Coatings Technology*, **106**, pp. 23-29, 1998.

[10]T. Makino and T. Kunitomo, "Thermal Radiation Properties of Ceramic Materials, *Transactions of the Japan Society of Mechanical Engineers*, **50**, No. 452, pp. 1045-1052, 1984.

[11] Kokini, K., Choules, B. D., and Takeuchi, Y. R., 'Thermal fracture mechanisms in ceramic thermal barrier coatings', *Journal of Thermal Spray Technology*, **6[1]**, 43-49, 1997.

[12]Kokini, K., Takeuchi, Y. R., and Choules, B. D., 'Surface thermal cracking of thermal barrier coatings owing to stress relaxation: zirconia vs. mullite', *Surface and Coatings Technology*, **82**, pp. 77-82, 1996.

THERMO-MECHANICAL MODELLING OF FUNCTIONALLY GRADED THERMAL BARRIER COATINGS

N. Nomura and M. Gasik
Helsinki University of Technology
P.O.Box 6200, FIN-02150 HUT
Espoo, Finland

A. Kawasaki and R. Watanabe
Tohoku University
2 Aoba, Aramaki, Aoba-ku
Sendai, 980-8579, Japan

Abstract

The introduction of oxygen diffusion barrier such as alumina is one of the promising solutions to use Ni-based superalloys with TBC for longer periods. Three types of TBC systems containing alumina are proposed in this study. Residual thermal stresses at room temperature were calculated for the proposed systems using a linear plate model. In the case of a layered composite system (stabilized zirconia / alumina / bond coat), tensile stresses in the bond coat are shown to be much higher than its yield limit. If a pure bond coat layer was replaced by an alumina / bond coat graded layer was under zirconia layer, the stress distribution significantly depends on the profile of concentration. TBC systems containing metal-rich graded layer were found to have substantially reduced thermal stresses. The system of stepwise FGM is more promising from the standpoint of thermal stress relaxation.

1. Introduction

In order to prevent global warming, it is necessary to decrease CO_2 emissions. This problem should be solved for power generation by improving the efficiency of the turbine systems, e.g. increasing the operating temperature. Ni-base superalloys have been widely used as high temperature structural materials for turbines, but it is difficult to use them for the higher temperature condition. In this case, thermal barrier coating (TBC) is required to protect effectively the hot parts from the combustion gas. This coating system usually consists of yttria-stabilized ZrO_2 (YSZ) for thermal insulation and MCrAlY (M = Ni and/or Co) alloy for corrosion resistance, which is called bond coat (BC). This coating system has been used in power plants and jet engines. The problem of the TBC system is the fracture and delamination of the coating under thermal stress and oxidation. Recently, the concept of functionally graded materials (FGM) was introduced to TBC[1]. Material properties such as coefficient of thermal expansion (CTE) and elastic modulus are expected to change gradually in order to decrease the thermal stress. Although the FGM-TBC has shown longer lifetime than non-FGM TBC under thermal cycling test[2], the weight gain problem caused by oxidation has not been solved yet[3].

From the point of oxidation protection of TBC, it is necessary to introduce the diffusion barrier (DB) for oxygen in TBC. DB should be located between YSZ and BC. Al_2O_3 is one of the candidates as DB, since it has lower oxygen diffusivity[4] (10^{-21} m$^2 \cdot$s^{-1}) than ZrO$_2$ (10^{-11} m$^2 \cdot$s^{-1}) and has relatively close CTE[5] (8 x 10^{-6} K^{-1}) to ZrO$_2$ (10 x 10^{-6} K^{-1}). The basic idea has been proposed by some authors[6], nevertheless the appropriate system of DB in TBC has not been implemented yet.

In this paper, some different FGM systems, which have DB for oxygen, are proposed. Thermal stress profiles are calculated in linear elastic state. Those systems are discussed from the point of the thermal stress distribution.

2. System design and description

Three kinds of systems for TBC are proposed as following.
(i) Layered YSZ (300 μm) / Al$_2$O$_3$ (200 μm) / BC (100 μm) / IN738
(ii) YSZ (300 μm) / Al$_2$O$_3$-BC FGM (200 μm) / IN738
(iii) YSZ-Al$_2$O$_3$-BC stepwise FGM (700 or 800 μm) / IN738

Schematic drawings of these systems are shown in Fig. 1. The thickness of IN738 is fixed at 2 mm for each system. Relatively thick Al$_2$O$_3$ layer is introduced in the case (i), since stress concentration can occur at thin oxide layer on the BC when Al$_2$O$_3$ exists as thermally grown oxide[7]. In the case of (ii), the BC layer on IN738 is removed in order to avoid the plastic deformation of BC because of its low strength. Taking the fracture toughness into account, introducing the pure monolithic Al$_2$O$_3$ in TBC is not favorable because of its brittleness (K$_{Ic}$ < 3 MPa$\cdot$m$^{1/2}$)[8]. It has been reported that the fracture toughness of Al$_2$O$_3$ based composites is increased by the addition of ZrO$_2$[9], so that ZrO$_2$ toughened Al$_2$O$_3$ composite layer (ZTA) is expected to have superior properties of both oxidation protection and fracture toughness. From this point of view, two kinds of stepwise FGM are proposed: 100 vol% Al$_2$O$_3$ layer is introduced in the system of (iii)-1, which has 800 μm in thickness at the stepwise region (8 layers). Another case (iii)-(2) does not have this layer, i.e., the thickness is 700 μm (7 layers).

In order to calculate the thermal stress of these proposed systems, flat and infinite layered

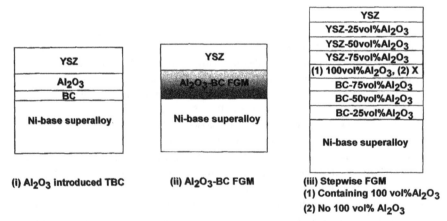

(i) Al$_2$O$_3$ introduced TBC   (ii) Al$_2$O$_3$-BC FGM   (iii) Stepwise FGM
(1) Containing 100 vol%Al$_2$O$_3$
(2) No 100 vol% Al$_2$O$_3$

Fig. 1 Schematic drawings of TBC for Ni-base superalloy.

specimens, which are perfectly bonded with sharp interfaces and compositional gradients, are assumed in this study. Edge effect is neglected. Linear elastic material behavior is considered. A biaxial stress model is used to calculate the thermal stress ($\sigma(z)$), which is given by[10]

$$\sigma(z) = \frac{E(z)}{1-\nu(z)}\left[\varepsilon(z) - \alpha(z)\Delta T(z)\right] = \frac{E(z)}{1-\nu(z)}\left[\varepsilon_0 + \kappa z - \alpha(z)\Delta T(z)\right] \tag{1}$$

where z is the height of the coated specimen, $E(z)$ is the elastic modulus, α is the thermal expansion coefficient, ΔT is the change in temperature from the initial stress free state, ε_0 is the normal strain at $z=0$ and κ is a curvature of the plate in its plane. ε_0 and κ are calculated from force balance and moment balance of specimen, assuming the applied axial force and the applied bending moment are equal to zero for pure thermal loading. Thus, thermal stress profiles of TBC systems are calculated after cooling to room temperature.

Materials properties, which have temperature and porosity dependence, are taken from elsewhere[5, 8, 11-15]. ZrO_2 and Al_2O_3 are assumed to contain 11.9 % and 18.8 % porosity, respectively. Properties of graded layer are evaluated by using the local representative volume elements (LRVE) model[16].

All coated samples are assumed to have stress-free state at 1118 K, which corresponds to the annealing temperature of IN738[17]. The temperature of IN738 at the "cold" side is fixed at 1073 K for $Q = 0.8$ MW•m^{-2}.

3. Results and discussion

Fig. 2 shows temperature profiles of several TBC systems at $Q = 0.8$ MW•m^{-2}. The surface temperature of the TBC depends on the thermal resistivity of the coatings.

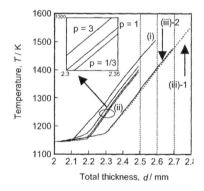

Fig. 2 Temperature profiles of several TBC systems during heating at $Q = 0.8$ MW/m^2.

Fig. 3 shows calculated stress distribution of system (i). Thermal stress about 600 MPa in tensile would generate at bond coat after cooling to room temperature. Plastic deformation may occur since this value is above the yield stress for BC[18]. Ductile / brittle transition temperature of NiCrAlY is around 900 K[19], so that crack may occur in the bond coat. At the Al_2O_3 layer, relatively high compressive stress is generated, which is caused by CTE difference between YSZ, Al_2O_3 and BC. On the other hand, tensile stress is generated at YSZ layer. Vertical crack possibly occurs by the stress, since the fracture strength of porous ZrO_2 is around 50 MPa[14]. Therefore, the BC layer and YSZ layer in this system are preferable place to fracture after cooling to room temperature.

Fig. 4 shows stress distribution of system (ii) at room temperature for various "p" value (1/3, 1 and 3). In this case, "p" is determined as follows.

$$V_{Al2O3} = 1 - (X / X_{FGM})^P \qquad (2)$$

Here V_{Al2O3} is volume fraction of Al_2O_3, X_{FGM} is total thickness of FGM layer, and X is position of FGM from the hot side of IN738. By removing the pure bond coat layer, tensile stress is suppressed compared to that of system (i) at the interface. The tensile stress is decreased below the yield strength of BC. It should be noted that compressive stresses generated in FGM layer are remarkably decreased with the decrease of "p" value. Moreover, compressive stress is generated at the YSZ layer at every p value. Therefore, crack initiation and propagation would be suppressed at the YSZ and alumina-rich layer. The residual stress of TBC depends on the p value; lower p value (BC-rich FGM) is better in this case. There is, however, high stress

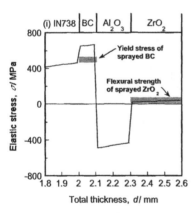

Fig. 3 Elastic stress profiles of (i) DB introduced TBC coating after cooling to room temperature.

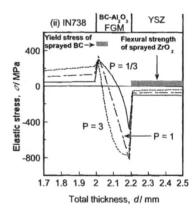

Fig. 4 Elastic stress profiles of (ii) FGM-TBC coatings after cooling to room temperature.

difference at the interface between FGM layers and YSZ, which may cause spalling of the coating at this place.

Fig. 5 shows the stress profiles of system (iii). Stress jump is decreased by introducing stepwise Al_2O_3-ZrO_2 FGM. Tensile stress at the BC-rich layer is also decreased below the yield stress of BC as well. Compressive stresses in region of stepwise FGM for (iii)-2 are lower than the system (ii) of any p value, irrespective of higher temperature difference in the coating, as shown in Fig 2. Comparing these two systems, there are two clear differences. One is that the absolute values of the stress generated for (iii)-(1) are higher than that of (iii)-2 at the BC-rich and Al_2O_3-rich layer. Another one is that tensile stress is generated at YSZ, exceeding the flexural strength of ZrO_2. Layer of 100 vol% Al_2O_3 makes higher stress for BC and YSZ, although this layer surely works more effective as DB. It is considered that the system (iii)-(2) can be more promising from the point of thermal stress relaxation, in addition to the fracture toughness. It should be still clarified which is dominant reason(s) for the fracture of DB introduced TBC, oxidation of BC or thermal fatigue at DB from experimental results such as cyclic thermal shock testing.

Fig. 6 shows the microstructure of Al_2O_3 introduced stepwise FGM-TBC, manufactured by plasma spraying. The coating composition was successfully changed from yttria-stabilized ZrO_2, Al_2O_3 to NiCrAlY. This demonstrates a successful prototype of system (iii), although further modification for the processing in addition to the TBC evaluation is necessary.

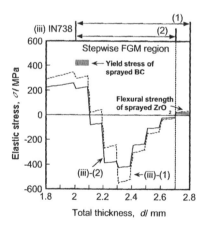

Fig. 5 Elastic stress profiles of (iii) stepwise FGM-TBC coatings after cooling to room temperature.

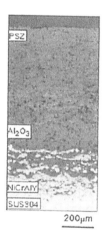

Fig. 6 Microstructure of graded PSZ / Al_2O_3 / NiCrAlY TBC.

4. Conclusions

Several systems of DB introduced TBC are discussed in this work. From the point of thermal stress calculation, based on the Al_2O_3 layer effect for thermal stress distribution in TBC coatings, the conclusions can be made as follows:

1. High tensile stress is generated at BC layer when simply thick Al_2O_3 layer is introduced, but substituting BC and Al_2O_3 layers for BC-Al_2O_3 FGM layer, tensile stress is decreased significantly.
2. Thermal stress in FGM layer is decreased with the decrease of "p" value.
3. Stepwise FGMs can be used to avoid the stress jump compared to that of system (ii).
4. The system (iii)-2 can be more promising from the point of the residual thermal stress.

Acknowledgements

The authors would like to thank National Technology Agency (Tekes) and Academy of Finland for financial support. This work was also supported by Japan Society for the Promotion of Science (JSPS) and Helsinki University of Technology.

Reference

[1] A. Kawasaki and R. Watanabe, "Finite Element Analysis of Thermal Stress of The Metal/Ceramic Multilayer Composites With Controlled Compositional Gradients," *J. Japan Inst. Metals*, **51** [6] 525-29 (1987).

[2] A. Kawasaki and R. Watanabe, "Cyclic Thermal Fracture and Spallation Life of PSZ/NiCrAlY Functionally Graded Thermal Barrier Coatings," Proc. 5th Intern. Symp. on FGM, Ed. W. A. Kaysser, TransTech publ., Lausanne, Schweiz, *Materials Science Forum Vols. 308-311*, 402-09 (1999).

[3] M. W. Porada and R. Rorchert, "An Oxidation Resistant Metal-Ceramic Functional Gradient Material: Material Concept and Processing Requirements," Proc. 5th Intern. Symp. on FGM, Ed. W. A. Kaysser, TransTech publ., Lausanne, Schweiz, *Materials Science Forum Vols. 308-311*, 422-27 (1999).

[4] Research Report in Kyusyu National Industrial Research Institute, "Microstructure and Strength of C-C composites," [3] pp.35 (1993).

[5] R. G. Munro, "Evaluated Material Properties for a Sintered α-Alumina," *J. Am Ceram. Soc.*, **80** [8] 1919-28 (1997).

[6] W. Y. Lee, D. P. Stinton, C. C. Berndt, F. Erdogan, Y. D. Lee and Z. Mutasim, "Concept of Functionally Graded Materials for Advanced Thermal Barrier Coating Applications," *J. Am Ceram. Soc.*, **79** [12] 3003-12 (1996).

[7] V. Sergo and D. R. Clarke, "Observation of Subcritical Spall Propagation of a Thermal Barrier Coating," *J. Am Ceram. Soc.*, **81** [12] 3237-42 (1998).

[8] P. Auerkari, "Mechanical and physical properties of engineering alumina ceramics," *VTT RESEARCH NOTES*, Espoo, Finland, **1792** 8-22 (1996).

[9] J. Wang and R. Stevens, "Zirconia-toughned alumina (ZTA) ceramics," *J. Mater. Sci.*, **24** [10] 3421-40 (1989).

[10] S. Suresh and A. Mortensen, "Thermoelastic Deformation of Graded Multilayers"; pp.

99-104 in *Fundamentals of Functionally Graded Materials*, IOM communications Ltd., London, (1998).

[11]T. Ostrowski, A. Ziegler, R. K. Bordia and J. Rodel, "Evolution of Young's Modulus, Strength, and Microstructure during Liquid-Phase Sintering," *J. Am. Ceram. Soc.*, **81** [7] 1852-60 (1998).

[12]S. Q. Nuisier and G. M. Newaz, "Analysis of Interfacial Cracks in a TBC / Superalloy System Under Thermomechanical Loading," *ASME J. Eng. Gas Turbine and Power*, **120** [4] 813-819 (1998).

[13]K. S. Ravichandran, K. An, R. E. Dutton and S. L. Semiatin, "Thermal Conductivity of Plasma-Sprayed Monolithic and Multilayer Coatings of Alumina and Yttria-Stabilized Zirconia," *J. Am. Ceram. Soc.*, **82** [3] 623-82 (1999).

[14]ASM SPECIALITY HANDBOOK, *Heat Resistant Materials*, pp. 244, ASM International, Materials Park, OH, 1997.

[15]K. Takagi, "The evaluation for the fracture of plasma sprayed FGM-TBC during thermal cyclic tests," Master's thesis in Tohoku University, Sendai, Japan (1997).

[16]M. Gasik, "Micromechanical modelling of functionally graded materials, " *Computational Materials Science*, **13** 42-55 (1998).

[17]D.A. DeAntonio, D. Duhl, T. Howston and M. F. Rothman, Heat Treating of Superalloys, *Heat Treating*, Vol. 4, 793-814, ASM Handbook, ASM International, Materials Park, OH, (1991).

[18]Y. Itoh, M. Saito and M. Tamura, "Characteristics of MCrAlY Coatings Sprayed by High Velocity Oxygen-Fuel Spraying System," *J. Eng. Gas Turbines and Power*, **122** [1] 43-49 (2000).

[19]K. Shimotori and T. Aisaka, "The Trend of MCrAlX Alloys for High-Temperature Protective Coatings -On the Effects of Alloy Compositions," *Tetsu-to-Hagane*, **69** [10] 1229-41 (1983).

Powder Processing

PECULIARITIES OF FUNCTIONALLY GRADED TARGETS FORMATION IN THE SHS-WAVE FOR PVD PROCESSES WITH AN OPERATING LAYER IN THE SYSTEMS Ti-Si-B, Ti-Si-C

E.A. Levashov, B.R. Senatulin, H.E. Grigoryan
and A.S. Rogachev
SHS-Center of Moscow Steel and Alloys Inst.
Leninsky prospect, 4, Moscow 117936

J.J. Moore
Colorado School of Mines
Golden, Colorado 80401 -
1887

ABSTRACT

The results of investigation of SHS – process macrokinetics characteristics in Ti-B-Si system are presented. It's shown that combustion rate is increased with rise of specific Ti+2B part in the charge materials. Bi-layer FGM – targets of different composition were fabricated by SHS. Peculiarities of structure and phase formation of the end products in the combustion wave of Ti-Si-B, Ti-Si-C systems were carried out. The paper gives the experimental data on the composition, structure, and properties of three-layer FGM - targets with working layer in Ti-Si-C system.

INTRODUCTION

The development and application of promising coatings take on more and more important significance in the technologies of surface engineering. The last works and reviews [1, 2] showed a necessity of developing new materials, satisfying by their properties the various technical and practical demands. For example, protective coatings to be applied in the tribology should simultaneously meet the requirements of high wear resistance and hardness, high strength, low friction coefficient along with corrosion resistance, high-temperature strength, stability, and heat resistance. Super-hard materials can be subdivided into groups, such as diamond (HV = 70÷100 GPa), cubic boron nitride (c-BN, $HV \approx$ 48 GPa) and some ternary compounds in the B–N–C system. Investigations of synthesis were recently carried out by reactive magnetron sputtering of thin-film compositions with hardness above 50 GPa of the following kinds: Me_n N/a-Si_3N_4 (where Me = Ti, W, V, and other nitrides of transition metals, a-Si_3N_4 – amorphous silicon nitride): TiN/BN; TiN/TiB$_2$; Ti(BN)$_x$/SiN$_x$; TiN/SiN$_x$ [2–8]. Despite such a variety of promising nano-crystalline super-hard film compositions, the question of optimum composition of both films themselves and composite targets for magnetron sputtering is left open.

In the present work the authors made an attempt to execute the targets synthesis by the self-propagating high-temperature synthesis (SHS) of different compositions in the Ti–B–Si , Ti-C-Si systems with studying their structure and properties. The most interesting compositions of the targets will be used in the process of reactive magnetron sputtering for multi-component coatings in the Ti–B–Si–N, Ti-C-Si-N systems.

1. EXPERIMENTAL PROCEDURE

The following powder components were used: PTM grade titanium (less than 50 μm) of SPA "Tulachermet" production; P804T grade carbon black with specific surface of 15 m^2/g, amorphous brown boron; monocrystalline silicon ground down to a size fraction of less than 50

impregnation and migration takes place in the working target layer of composition 1 (85 % TiB$_2$ + 14 % Si + 1 % Ti$_5$Si$_3$).

In a two-layer specimen with working layer 5 (22 % TiB$_2$ + 78 % Ti$_5$Si$_3$), a very fair quantity of silicon and titanium melt can participate in mass transfer along the boundary surface. It is related to the fact that correspondingly less quantity of refractory substructure TiB$_2$ (22 % only) is formed in charging 6.2 % instead of 24.8 % boron. The rest products being as liquid during a certain time period (till crystallization of Ti$_5$Si$_3$ silicide from saturated melt) can participate in impregnation and migration processes into the TiC–TiB$_2$ framework using a transition zone.

The Ti, Si, and C concentration profile lines over the section of two-layer specimens with working layer of compositions 1 and 5 are plotted as an example in Fig. 3. Methodically these profile lines were plotted point by point in the following manner. Initially, titanium, silicon, and carbon contents were determined in several points of the section of two-layer specimens with a spacing of 10^{-3} m. Concentration of boron as a light element is failed to be determined with using this procedure. It is apparent that silicon impregnated depth was enough large, since its traces were found even in the center of the sublayer. It can be seen from Fig. 3 that there was observed no distinction in kind in element distribution profile lines in two-layer specimens. So, the transition zone is wider in the case of specimen of composition 5. At the same time formation of such wide transition zones in the combustion wave is undoubted advantage of the SHS-technology used in production of the targets as compared to the traditional processes of powder metallurgy. The nonporous middle layer, i.e. the transition zone, fast binds the working layer of high heat conductivity and strong ceramics TiC–TiB$_2$ securing the gradient junction from layer to layer as of thermal-expansion coefficient.

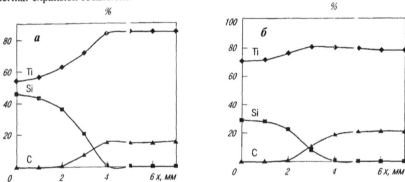

Fig.3. Distribution of Ti, Si and C through the thickness of bilayer targets with working layer 1 (a) and 5 (b)

Let us consider the microstructures of fracture in the two-layer SHS-targets photographed in different places of the specimens with working layer of composition 1 and composition 5 . The cause is formation of the bulk of liquid phase in the combustion wave. The working layer of the target that has composition 1 represents cut but equiaxed titanium diboride grains of 2–3 μm in size located in a silicon binder. In the transition zone the cut TiB$_2$ grains run into 6–8 microns in size. Finally, the sublayer consists of compact cut TiC and TiB$_2$ crystals of 3–5 microns in average size. The structure of the fracture in the two-layer target of composition 5 has distinctions of principle. In the structure of fracture of the layer composed of 22 % TiB$_2$ + 78 % Ti$_5$Si$_3$, a brittle component prevails that contains large titanium silicide crystals against the background of very small TiB$_2$ grains. Their size can run to 7–8 microns and that of TiB$_2$ is less than 10^{-6} m. A difference in the size of Ti$_5$Si$_3$ and TiB$_2$ grains are equalized along the boundary between the

μm. Upon drying at $t = 90$ °C for 6 hr the powders were mixed in ball mills of 3 liters in capacity with a charge weight-to-ball weight ratio of 1:6. The charges of different compositions in the Ti-Si-B system shown in Table 1 were prepared in such a manner. Briquettes 125 mm in diameter were compacted from every charge up to relative density of 56–60 %, and then compacted products of chemical synthesis were produced according to the power SHS-compaction [9] in special "sand" molds in the hydraulic presses of 160 ton-force in loading with simultaneous registration of burning rate using a photodiode. This press is also fitted with a time synchronization unit that makes it possible to automate quick-running processes of burning initiation, premolding, molding delay, major molding, holding time under pressure, and pressure relief. Cooling of the half-finished products was carried out in a muffle furnace from 800 °C in the air to relax internal stresses in the products.

Table 1. Compositional effect of reaction mixture in working layer of target on composition of synthesis products in the Ti–B–Si system

No.	Composition of charge, %			Percentage of phases, %			Combustion rate U,
	Ti	Si	B	TiB_2	Ti_5Si_3	Si	10^{-2} m/s
1	55.2	20.0	24.8	85.0	1.0	14.0	3.7
2	71.5	13.0	15.5	63.0	37.0	0.0	4.2
3	72.5	18.2	9.3	27.5	72.5	0.0	3.7
4	72.0	15.2	12.4	49.3	50.7	0.0	3.9
5	73.0	20.8	6.2	22.0	78.0	0.0	3.7
6	48.3	30.0	21.7	81.1	0.7	18.2	3.0
7	41.4	40.0	18.6	73.2	2.0	24.8	3.4

The quantitative composition of end products was determined by the X-ray phase analysis in DRON-3 diffractometer and following calculation using the ASTM database. The metallographic analysis of cross sections and fractures was carried out in Neophot-32 light microscope with magnification up to 2000 as well as in scanning electron microscopes of JSM-35 "Jeol" and JSM-U3 with magnification up to 4000. Microhardness was measured according to the standard procedure in the PMT-3 durometer. The spectra of element distribution in different sites across thickness of two-layer targets were photographed from the JSM-U3 microscope fitted with a detachable analyzer having semiconductor silicon-lithium detector EUMEX.

2. RESULTS AND DISCUSSION ON THE Ti-Si-B SYSTEM
2.1. Combustion process macrokinetics. Structure and phase formation of synthesis products

Table 1 shows that the combustion rate and phase composition of synthesis products are dependent of the composition of an initial charge. An experimental composition of the mixture was chosen, on the one hand, by variation of silicon concentration «y» in chemical equation $(100 - y)(Ti + 2B) + y$Si calculated for formation of titanium and silicon diborides. In this case, in accordance with the thermodynamics, silicon acts as inert filler, inasmuch as chemical affinity of titanium to boron is stronger than to silicon [10]. On the other hand, the composition of the mixture was chosen by variation of «x» concentration in chemical equation $x(Ti + 2B) + (100 - x)(5Ti + 3Si)$ calculated for formation of titanium diboride TiB_2 and titanium silicide Ti_5Si_3.

Table 1 and Figures 1–2 represent data on the effect of stoichiometric coefficients «x» and «y» on the composition of synthesis products and U. Enough evident behavior is observed: the more relative portion of $(Ti + 2B)$ in the mixture, the higher value U. At the same time, the titanium diboride concentration rises in the finished products. Taking the conceptual approach for the structure formation of synthesis products in the process of SHS-compaction [11] into account, one can propose the following. The titanium diboride grain size should decrease as the filler (silicon in our case) concentration increases. Indeed, an average grain size of TiB_2 decreases from 3.5–4.0 microns in alloy 1 with initial silicon concentration of 20 % down to 1.5–2.0 microns in alloy 7 with initial concentration 40 % Si in the charge mixture.

The formation of Ti_5Si_3 phase simultaneously with titanium diboride in the combustion wave due to proceeding of parallel and successive chemical reactions interferes with TiB_2 grain growth too. This can be explained by the fact that very small TiB_2 and Ti_5Si_3 grains can nucleate in the wave practically simultaneously and interfere with each other later on. When their concentrations are much the same, the grains have the smallest size, and when relative content of one phase increases, the size of this phase grows. Thus, coincidence of time of two reactions $Ti + 2B \rightarrow TiB_2$ and $5Ti + 3Si \rightarrow Ti_5Si_3$, in each of which a solid-phase product is formed, can serve as an effective tool to control the structure of SHS-composite materials.

It can be seen from the microstructures that Ti_5Si_3 titanium silicide phase plays the prevailing role in formation in alloy 5 (see Table 1). This phase is characterized by round grains with average size of 5 microns, and titanium diboride is formed along the grain boundaries as small substructure of 1–1.5 microns in size.

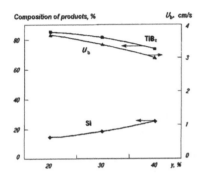

Fig.1. Dependences of combustion rate and composition of synthesis products in the Ti-B-Si system on coefficient"y" in expression $(100 - y)(Ti + 2B) + ySi$

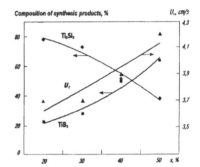

Fig. 2. Dependences of combustion rate and composition of synthesis products in the Ti-B-Si system on coefficient "x" in expression $x(Ti + 2B) + (100 - x)(5Ti + 3Si)$

2.2. Features of composition and structure formation in compact two-layer SHS-targets with working layer of Ti–Si–B composition

The following stage of the work is production of two-layer targets with a working layer made of synthesis products in the Ti–Si–B system according to the SHS-compaction process. An alloy of 60 % TiC + 40 % TiB_2 composition that is formed as a result of proceeding SHS reaction in the mixture of titanium, carbon black, and boron was used as an sublayer. The two-layer synthesis product is formed during one production operation, and the combustion wave runs simultaneously over both layers of two-layer charge mixture. It should be noted that the main purpose of using the ceramic TiC–TiB_2 sublayer is in improvement of impact toughness, thermal resistance, and thermal conductivity of the composite material. These properties allow us to repeatedly use the target in high-power magnetron sputtering units.

Strong cohesion between the layers is secured due to a definite transition zone that is a result of high combustion temperature and fair quantity of liquid phase in the wave along the boundary surface. The melts of products and semiproducts impregnate porous frames formed by more refractory reaction products. For example, mass transfer of silicon melt at the expense of

working layer and the transition zone. Together with Ti_5Si_3, cut elongated crystallites of TiB_2 hexagonal phase, which grow to 8–10 microns as approached to the boundary surface between the transition zone and the sublayer, appear in the transition zone. Sublayer consists of smaller TiC and TiB_2 grains. Their average size is less than 2–3 microns.

The dynamics of TiB_2 grain size change can be also derived from the microstructures photographed from the etched section surface of the two-layer specimen with the working layer of composition 1. Being predominantly equiaxed, cut titanium diboride grains with average size of 2–3 microns essentially vary their sizes and orientation in the transition zone running to 6–8 microns. Crystals are predominantly oriented in the direction perpendicular to the combustion direction. The similar tendency in the dynamics of structure formation of synthesis products can be connected to mutual direction of heat removing and crystallization vectors. Crystallization vector is directed top-down, since the maximum combustion temperature in Ti–C–B mixture of the sublayer is sufficiently higher that the real temperature in the mixture of the working Ti–B–Si layer.

2.3. Microhardness of two-layer targets

The distribution of microhardness across the thickness of a gradient target formed during SHS-compaction of two-layer specimen undoubtedly gives an idea of running inflow processes, migration, formation of diffusive transition layer and makes it possible to carry out a qualitative estimate of internal stresses. Average microhardness value of all the studied two-layer specimens are given in Table 2 and. It is evident that absolute microhardness values and a ratio of values in the working layer, transition zone, and sublayer are different. The lower hardness of $TiC–TiB_2$ as compared to the working layer and the transition zone can be explained by residual porosity at the level of 5 %. Based on the given data, one can design different variants of targets. So, the target with the working layer of composition 1 has the greatest hardness and targets of compositions 4, 5, and 7 have the least internal stresses. The best variant of targets can be recommended for the process of reactive magnetron sputtering of nano-crystalline ultra-hard coating of the Ti–B–Si–N composition.

Table 2. Microhardness of FGM – targets (Ti-B-Si)/(Ti-B-C)

No.	Composition, %			Microhardness, GPa		
	Ti	B	Si	Ti-B-Si	Transition zone	Ti-B-C
1	55,2	24,8	20,0	22990+4520	24750+7100	16702+4570
2	71,5	15,5	13,0	22990+8220	22990+8220	16700+4570
3	72,5	9,3	18,2	17882+6438	17882+6438	12680+4874
4	72,0	12,4	15,6	15332+6987	13544+3670	12255+3167
5	73,0	6,2	20,8	12663+2305	12663+2305	11316+2098
6	48,3	21,7	30,0	21715+7604	15605+3362	12456+4851
7	41,4	18,6	40,0	11495+2388	11866+6722	12663+3946

3. FGM – TARGETS WITH Ti-Si-C WORKING LAYER

3.1. Features of composition and structure formation for compact multilayer SHS-targets

Macrokinetics characteristics of combustion process in the Ti-Si-C system and structure and phase formation during SHS of uniform products were studied recently [12].

Three compositions of the charge (in %): 72.6Ti + 14.1Si + 13.1C (A); 71.3Ti + 13.9Si + 14.8C (B); 42.1Ti + 36.8Si + 21.1C (C) were chosen in order to produce three-layer targets with the working layer from the synthesis products in the Ti–Si–C system. A mixture of Ti + 0.5C was taken to form the second layer on the basis of synthesis of non-stoichiometric titanium carbide $TiC_{0.5}$. The third (last) layer was a mixture of titanium and boron of SHIM-4 composition. All cited compositions of the targets were synthesized from lamellar charge briquettes during a process cycle. The main purpose of two additional layers $TiC_{0.5}$ and SHIM-4 is to raise impact toughness and heat resistance of a composite target as a whole, making it possible to use them many times in the high-power sputtering plants. Such an approach to the creation of similar gradient targets is original and new. The fracture of a three-layer SHS-target of composition C obtained by artificial destruction was investigated. It is not difficult to see quite satisfactory joint of all three layers to one another. Porousless and relatively thin middle layer of titanium carbide combine fast the working layer to highly heat-conducting and strong alloy SHIM-4, thereby providing a smooth transition in thermal-expansion coefficient from layer to layer.

Quantitative X-ray phase analysis of synthesis products allows the following phases in the working layer of the target to be identified:

Composition A: TiC (36%) + Ti_3SiC_2 (58.5%) + $TiSi_2$ (5.5%),

Composition B: TiC (46.5%) + Ti_3SiC_2 (50%) + SiC (3.5%),

Composition C: TiC (24%) + Ti_3SiC_2 (24%) + SiC (48%) + $TiSi_2$ (4%),

The working layer of composition C having a large content of silicon carbide is characterized by relatively high residual porosity (up to 15 %). This fails to appreciably affect the sputtering rate and properties of thin-film coatings produced.

The concentration profile lines of the element distribution of (across thickness of target) near the boundary surface between the working layer and sublayer of non-stoichiometric titanium carbide were plotted to validate causes of such a strong joining all the three ceramic layers of the gradient target one to another. Comparative investigations were carried out in the scanning electron microscope "Camebax" . Photographing was performed under the following conditions: accelerating voltage 15 kV, beam current 100 nA, analysis zone 10^{-6} m, accumulation period 1 sec. Figure 4 shows the spectra of titanium and silicon distribution for all three compositions of the targets with the working layers A, B, C. It is evident that in the case of the working layers B and C there arises a relatively wide (approximately 3.10^{-4} m) diffusion zone between them and the titanium carbide layer with the smooth distribution of main elements.

3.2. Microhardness of three-layer targets

Similar to the concentration profile lines, the microhardness distribution across the thickness of the gradient targets gives a representation about existence of diffusion intermediate layers and allows comparative qualitative estimation of internal stresses near the boundary surface between the layers.

Average microhardness values of all three-layer specimens are given in Table 3. Averaging of microhardness was carried out for 50 measurements in each layer. It is evident that the target of composition A has the microhardness values closest between the layers. y. The smooth growth of microhardness from the working layer (7950 MPa) to the layer of non-stoichiometric titanium carbide (9820 MPa) and then to the layer SHIM-4 on the basis of orthorhombic phase of titanium boride (10940 MPa) is typical for the target of composition B.

Table 3. Microhardness in layers of three-layer specimens

Specimen No.	Charge mixture, %			H_μ, MPa		
	Ti	Si	C	Ti-Si-C	$TiC_{0.5}$	SHIM-4
1	42.1	36.8	21.1	11850	13900	11850
2	72.6	14.1	13.3	12680	14650	12440
3	71.3	13.9	14.8	7950	9820	10940

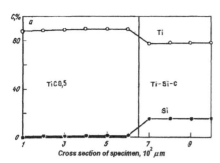

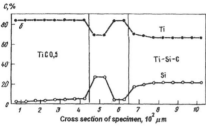

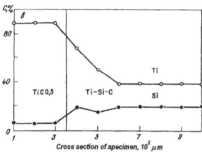

Cross section of specimen, 10^2 μm

Fig. 4. Ti - and Si - concentration profile lines across the two-layer specimen TiC$_{0.5}$: (72.6 % Ti, 14.1 % Si, 13.3 % C) (a), (71.3 % Ti, 13.9 % Si, 14.8 % C) (b), and (42.1 % Ti, 36.8 % Si, 21.1 C) (c) 1 – cross section of specimen, 10^{-4} m

The characteristics of the layers of each target are considered below.

Target with working layer of composition A (72.6Ti + 14.1Si + 13.3C)

The microhardness in the layer Ti–Si–C is within the range of 10640–14270 MPa with its average value of 12680 MPa. Microhardness in the layer of titanium carbide TiC$_{0.5}$ varies from 14270 MPa up to 15050 MPa, and the average value being 1465 MPa goes beyond the microhardness of carbide layer in the target with the working layer of composition C. Lower porosity (4 % instead of 6 %) and lesser grain size are a probable cause of the case. Average value of microhardness in this layer amounts to 13900 MPa. Microhardness in layer SHIM-4 amounts to 12140–12880 MPa.

Target with working layer of composition B (71.3Ti + 13.9Si + 14.8C)
Microhardness in the layer Ti–Si–C is within the range of 7330–8240 MPa with its average value of 7,950 MPa. Upon passing to layer TIC$_{0.5}$, the print diameter decreases, microhardness varies from 8580 MPa up to 10640 MPa, and average value is 9820 MPa in this layer. Finally, microhardness in the layer SHIM-4 is in the range of 10640–11140 MPa with average value 10940 MPa.

Target with working layer of composition C (42.1Ti + 36.8Si + 21.1C)
Microhardness of the working layer is in the range 11400–12500 MPa with its average value 11850 MPa. Accumulation of pores is observed along the surface boundary of the working layer of composition C with titanium carbide and microhardness in layer TiC$_{0.5}$ increases from 13540 MPa up to 15900 MPa.

CONCLUSIONS

1. The macrokinetics of the SHS-processes has been studied in the Ti–B–Si system. It was shown that the combustion rate increases with the growth of specific part of (Ti + 2B) in the charge mixture.

2. The distinctions were found for phase and structure formation of FGM - end products with the working layer made of the Ti–B–Si and Ti-Si-C systems.

3. Functionally graded targets of different compositions were produced according to the force SHS-pressing process. It was shown that relatively wide, up to $(1–2)x10^{-3}$ m, transition zones are formed between the working layers Ti-Si-B, Ti-Si-C and sublayers irrespective of the composition of the working layers of the target. The latter fact made it possible to produce high-quality targets for magnetron sputtering.

4. Microhardness of different composition targets were studied which made it possible to design and forecast the properties of articles.

REFERENCES

[1] M. Stüber, H. Leiste, S. Ulrich, A. Skokan, "Development of Tailored Coating Concepts for CVD and PVD Deposition of Multifunctional Coatings-A Review", *Zeitschrift Metallkunde*, [10] 774–779 (1999).

[2] S. Veprek, "The Search for Novel, Superhard Materials", *Journal of Vac. Sci. and Technol. A.* V. **17** [5] 2401–2420 (1999).

[3] E. A. Levashov, D. V. Shtansky, A. N. Sheveiko, J. J. Moore, "SHS of Composite Sputtering Targets and Structure and Properties of PVD Thin Films", *Plansee Proceedings 14 th Int. Plansee Seminar*, Edited by G. Kreminger, P. Rondhammer, P. Nilhartitz, **3** 276–289 (1997).

[4] D. V. Shtansky, E. A. Levashov, A. N. Sheveiko, J. J. Moore, "The Structure and Properties of Ti-B-N, Ti-Si-B-N, Ti-Si-C-N, and Ti-Al-C-N Coatings Deposited by Magnetron Sputtering Using Composite Targets Produced by Self-Propagating High-Temperature Synthesis (SHS)", *Journal of Materials Synthesis and Processing*, **6** [1] 61–72 (1998).

[5] D. V. Shtansky, E. A. Levashov, A. N. Sheveiko, A.H. Grigiryan, J.J. Moore, "Comparative Investigation of Multicomponent Films Deposited Using SHS Targets", *International Journal SHS*, **7** [2] 249–262 (1998).

[6] D. V.Shtansky, E. A. Levashov, A.N. Sheveiko, J. J. Moore, "Optimization of PVD Parameters for the Deposition of Ultra Hard Ti-Si-B-N Coatings", *Journal of Materials Synthesis and Processing*, **7** [3] 187-193 (1999).

[7] D. V.Shtansky, E. A. Levashov, A. N. Sheveiko, J. J. Moore, "Synthesis and Characterization of Ti-Si-C-N Films", *Metallurgical and Materials Transaction A*, **30A** 2439-2447 (1999).

[8] D. V. Shtansky, E. A. Levashov, A. N. Sheveiko, "Mit einem SHS-Legierungs- Target abgeschiedene Mehrkomponentenschichten Ti-B-N, Ti-Si-B-N, Ti-Si-C-N und Ti-Al-C-N fur unterschiedliche technologische Anwendungen", *Galvanotechnik*, [2] 3368-3378 (1997).

[9] I. P. Borovinskaya, V. I. Ratnikov, G. A. Vishnyakova, "Some Chemical Aspects of the Force SHS - Pressing ", *Inzhenerno-Fizichesky Zhurnal (Physical-Engineering Journal)*, **63** [5] 517–524(1992).

[10] G. V. Samsonov, I. M. Vinnitsky, *"Refractory Compounds: Handbook"*, Metallurgia, Moscow, 1976

[11] E. A. Levashov, Yu. V. Bogatov, A.S. Rogachev, et al. "Regularities of Synthetic Hard Instrumental Materials Structure Formation during the SHS-pressing process", *Inzhenerno-Fizichesky Zhurnal (Physical-Engineering Journal)* **63** [5] 558–576 (1992).

[12] H. E. Grigoryan, A. S. Rogachev, V. I. Ponomarev, E. A. Levashov, "Product Structure Formation at Gasless Combustion in Ti-Si-C System", *International Journal of SHS*, **7** [4] 507–517 (1998)

SYNTHESIS AND DENSIFICATION OF CERAMIC FGMS IN ONE STEP

Ellen M. Carrillo-Heian, Jeffery C. Gibeling, and Zuhair A. Munir[*]
University of California, Davis
Davis, California 95616

Glaucio H. Paulino
University of Illinois, Urbana-Champaign
Urbana, Illinois

ABSTRACT
Dense layered FGMs have been made from elemental powders via simultaneous application of pressure and pulsed direct current to a layered compact in a graphite die. The technique is known as field-activated, pressure-assisted combustion synthesis. Reaction times are on the order of 20 minutes. Five-layer samples have been produced in two sizes: disks 44.5 mm in diameter by 10 mm tall and oblongs 12.7 mm in width, 63.5 mm in length, and 15 mm in height. In the $MoSi_2$-SiC system, densities obtained ranged from 95 to 99% of the theoretical density, and interfaces are very strong, with no cracking evident.

INTRODUCTION
Functionally Graded Materials
    Functionally graded materials (FGMs) have always existed in nature. Two examples are bamboo and bone. Humans have also made FGMs for centuries, by hardening wooden spears in the fire, glazing pottery, and case-hardening steel, for example. Systematic research on manufactured FGMs has been conducted since the mid-1980's, when Japan funded a national research program to develop FGMs for the U.S.-Japan Space Plane [1].
    It is a technical challenge to create a gradual variation in material properties from one surface of an object to another. The field-activated, pressure-assisted combustion synthesis technique discussed in this paper begins with elemental powders mixed in proportion and stacked in layers. The powders are then caused to react, forming a solid ceramic product with varying composition. The technique may be used to create stepped profiles that follow many different curves.

$MoSi_2$-SiC
    The chemical system explored by this paper is the molybdenum disilicide – silicon carbide system. Molybdenum disilicide has been used commercially as heating elements in high-temperature oxidizing-atmosphere furnaces since Kanthal received a patent in 1956 [2]. This is possible because $MoSi_2$ forms a protective SiO_2 coating that does not begin to soften until about 1700°C [3]. Various efforts to improve the room-temperature fracture toughness and high-temperature strength and creep resistance of $MoSi_2$ have been initiated. One of the most heavily researched is the addition of a brittle phase in the form of fibers or particles to reinforce the $MoSi_2$ matrix. The mechanism for fracture toughness improvement due to brittle reinforcement may be due to crack deflection-crack branching processes or to residual stress at the particle-matrix

---

[*] Corresponding author

interfaces [2]. SiC has been found to be particularly effective as a reinforcement material, due to the robust character of interfaces formed between it and MoSi$_2$.

Different techniques to add SiC to MoSi$_2$ have been used by several researchers [4], including chemical vapor infiltration/ deposition, reactive vapor infiltration [5], XD™ [6], and plasma spray [7] as well as hot-pressing, HIPing, sintering [8], and mechanical alloying [9]. A technique related to field-activated, pressure-assisted combustion synthesis is reactive powder sintering (co-synthesis) starting from related compounds, such as Mo$_2$C and Si [10].

The direct synthesis of MoSi$_2$-SiC composites from the elements has been previously investigated by combustion synthesis, with mixed results [12-14]. In recent work, conventional self-propagating high-temperature combustion synthesis (SHS) as well as simultaneous combustion (thermal explosion) on mixtures of molybdenum, silicon, and carbon extinguished when the anticipated fraction of SiC was larger than 33 mol% [13]. Mechanical alloying was used to activate elemental powders before hot pressing, resulting in complete combustion in samples with 20 and 40 vol% SiC (33 and 56 mol%) [12]. In contrast, electric field-activated SHS has produced porous composites with compositions ranging from 100 mol% MoSi$_2$ to 100 mol% SiC, with almost no intermediate phases [14]. A complete discussion of the need for activation via electric field, mechanical alloying, or other means is presented elsewhere [15].

The advantage of field-activated pressure-assisted combustion synthesis is in the rapid heating rate afforded by the direct application of electric current to the sample. Heat is generated in the graphite die immediately surrounding the sample in the case of low-conductivity starting powders, or in the starting powders themselves, if the conductivity is high enough. This rapid heating rate reduces solid state diffusion and the occurrence of other phases such as Mo$_5$Si$_3$.

PROCEDURE

Samples were produced in two sizes: disk-shaped samples 44.5 mm in diameter (height 10 mm) and oblong samples measuring 12.7 mm in width, 63.5 mm in length, and 15 mm in height. These geometries were used to produce beams for mechanical testing in 3- or 4-point bending. The sample geometries are shown in Figure 1. Five-layer samples were produced in two combinations with 10 mol% increments between adjacent layers within the following limits: 100% MoSi$_2$ - 60% MoSi$_2$ and 90% MoSi$_2$ - 50% MoSi$_2$. Single-composition samples were also made in order to measure material properties of each mixture [16].

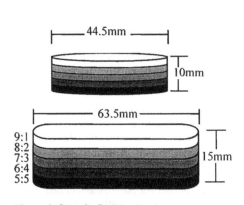

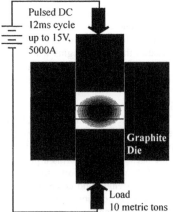

Figure 1. Sample Geometries   **Figure 2. SPS Schematic**

The apparatus utilized in this research is the Spark Plasma Sinterer (SPS) manufactured by Sumitomo Coal Mining Company, Inc. of Japan [17]. This machine is a uniaxial 100kN press combined with a 15V, 5000A pulsed DC power supply, to simultaneously provide current and pressure to a conductive die and sample. A schematic of the sample and die assembly is shown in Figure 2. The pulse cycle utilized in this work was 12 ms on and 2 ms off. Joule heating of the die and the sample results in the combustion of the sample, and the applied pressure densifies the sample at the same time. The SPS may be controlled via a temperature controller or manually by current. Temperature is measured on the surface of the die by means of a single-color pyrometer with a range of 600 to 3000°C. The SPS also has computer data collection of the voltage, current, load applied, temperature, displacement (shrinkage) and displacement rate.

Elemental powders of Mo, Si, and graphite were mixed in stoichiometric proportion to form products of the composition:

$$xMo + (1+x)Si + (1-x)C \rightarrow xMoSi_2 + (1-x)SiC \qquad (1)$$

where x is varied in 10% increments from 100% to 50%. The properties of the starting powders are listed in Table I. The powders were mixed in either a turbula mixer for 1 hour or a rolling mill for 24 hours, in both cases using 7mm diameter zirconia-stabilized tetragonal alumina balls with no solvent. The powders were then sieved to remove the balls. To form a sample, the first powder mixture was poured by hand into a graphite foil-lined double-action graphite die, tapped to level, and then pressed by hand. The next layer was then poured in on top, pressed by hand, and so forth until 5 layers were in place. The entire assembly was then cold pressed to 13.34N (3000 lbs.) before being placed in the SPS.

Table I: Properties of Starting Powders

Material	Source	Particle Size/ Classification	Purity, %
Molybdenum	Alfa Aesar (Ward Hill, MA)	3-7 μm	99.95
Silicon	Alfa Aesar	-325 mesh	99.5
Graphite Carbon	Asbury Graphite Mills (Asbury, NJ)	0.6 μm (average)	99

After packing, the samples were placed in the SPS under a uniaxial pressure of 25 MPa. Samples were heated to an external temperature of between 1200 and 1300°C in 20 to 25 minutes then furnace cooled. Due to the large thermal mass of the dies, the control was not very stable, resulting in some temperature oscillations.

After the reaction, samples were examined via X-ray diffraction (XRD) and electron probe microanalysis (EPMA). Density was determined by geometric and Archimedes' (submersion in methanol) methods.

RESULTS AND DISCUSSION

The reaction of Mo and Si to form $MoSi_2$ in this work was always marked by an abrupt compaction event, usually audible and always accompanied by an abrupt decrease in resistance. This is believed to be the consequence of a combustion reaction between the powders. This event happens at an exterior temperature of between 1150 and 1300°C in the SPS. Other researchers have reported this compaction event as well [18]. In our work, the magnitude of the compaction decreased with an increase in the amount of SiC in the sample. Samples that were quenched immediately after the compaction event contained the desired phases plus some unreacted silicon.

X-ray diffraction reveals that furnace-cooled reactions were always complete, with another phase appearing as the amount of SiC increased. The third phase is $Mo_{4.8}Si_3C_{0.6}$, a so-called Nowotny phase that has been reported by several researchers [19]. It is hexagonal with space group $P6_3/mcm$ and is stable down to room temperature, as reported by Parthé, et al. [20]. Figure 3 shows x-ray diffraction patterns for three composite samples in which x is equal to 0.3, 0.5, and 0.8. The figure shows the relative abundance of the third phase as a function of SiC content. When x=0.8, there is little evidence of the presence of this carbosilicide, as seen from the small peaks in the 2θ range of 42-44°. These peaks increase slightly as x decreases to 0.5 and become significant when x=0.3. This phase is not necessarily detrimental to the mechanical properties of the composite. The presence of Mo_5Si_3 at the grain boundaries of pure $MoSi_2$ contributes to high-temperature ductility, and similar behavior could occur in the $MoSi_2$-SiC composite.

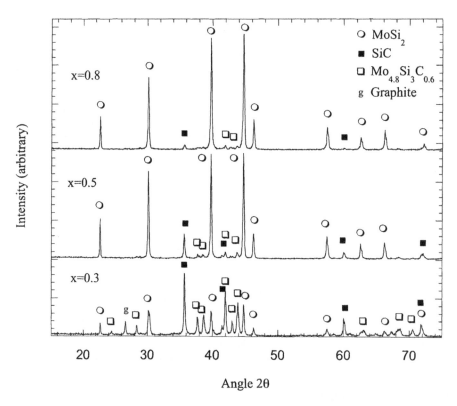

Figure 3. XRD patterns for three compositions: 80%, 50%, and 30% $MoSi_2$

Figure 4 (a) through (f) are backscatter electron micrographs and X-ray dot maps of the two layers in a bilayer sample: (a), (b), and (c) are of a 70% $MoSi_2$ layer, and (d) through (f) are of a 30% $MoSi_2$ layer. (b) and (e) are Mo-l_α dot maps and (c) and (f) are Si-k_α dot maps. By comparing images (d), (e) and (f) it may be seen that there are pores (black) in (d) as well as gray $MoSi_2$, dark gray SiC and white $Mo_{4.8}Si_3C_{0.6}$. In contrast, no $Mo_{4.8}Si_3C_{0.6}$ and no pores are visible in (a) through (c), the $MoSi_2$-rich layer.

The density of single-composition samples varied from a maximum of 98.6% of the theoretical composite density for an 80 mol% $MoSi_2$ sample down to 76% of the theoretical density for a 20 mol% $MoSi_2$ sample. The density dropped dramatically below 50 mol% $MoSi_2$, and therefore this was the minimum $MoSi_2$ content used to make the FGMs. The average density of the five-layer FGMs was between 95 and 97%. Figure 5 is an optical micrograph of a five-layer FGM. The leftmost layer is 90 mol% $MoSi_2$ and the rightmost layer is 50 mol% $MoSi_2$. The rightmost layer exhibits a variation in porosity that was often visible when the SiC content was high. The layer was made thicker than the others so that this porous region could be cut off before analysis.

The interfaces in Figure 5 show no evidence of cracks or delamination. This was found to be true when the composition difference between two layers was no more than 10 mol%. The coefficient of thermal expansion (CTE) for SiC is just over half that of $MoSi_2$, which can cause a large thermal mismatch. However, taking a linear rule of mixtures to calculate the CTE of each layer, we find that when the composition varies by 10 mol%, the CTE varies by just about 3%, small enough to prevent separation at the interfaces between layers and radial cracks in the $MoSi_2$-rich layers.

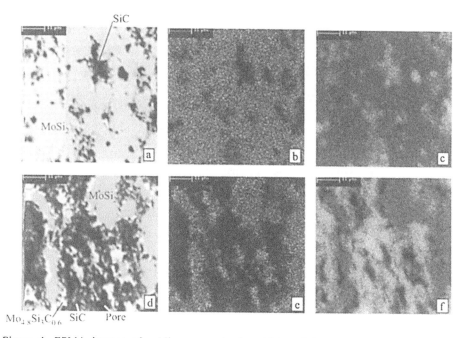

Figure 4. EPMA images of a bilayer sample for which x=0.7 (a-c) and x=0.3 (d-f). (a) backscattered electron image (b) Mo-l_α X-ray dot map. (c) Si-k_α X-ray dot map. (d) backscattered electron image (e) Mo-l_α X-ray dot map. (f)) Si-k_α X-ray dot map.

EPMA was used to measure the concentration of Mo and Si across a five-layer sample to determine the uniformity of the layers and whether any mixing took place at the interfaces. Figure 6 shows the concentration profile of Si across a sample with layers that varied from 100% $MoSi_2$ to 60 mol% $MoSi_2$. The concentration changes at the interfaces look like they are more gradual than they were before the reaction, but closer examination of more samples reveals sharp interfaces, indicating that no mixing took place. Also included in the figure are backscatter images of each layer, showing the microstructure. The lighter regions are $MoSi_2$ and the darker are SiC. A few very bright regions are the Nowotny phase $Mo_{4.8}Si_3C_{0.6}$. The interfaces between grains of SiC and $MoSi_2$ are always quite cohesive.

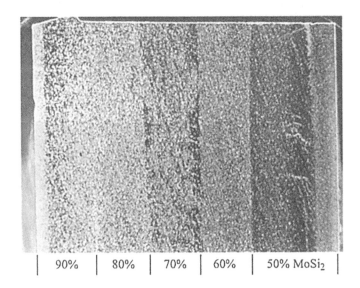

Figure 5. Optical micrograph of a 5-layer $MoSi_2$-SiC FGM.

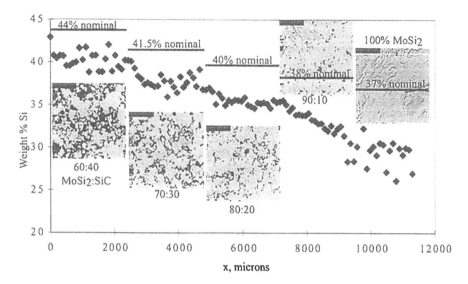

Figure 6. Si concentration profile and BSE micrographs of a MoSi$_2$-SiC FGM.

CONCLUSIONS

Layered FGMs composed of MoSi$_2$ and SiC may be formed with MoSi$_2$ concentrations down to 50 mol% without significant decrease in density by the field-activated pressure-assisted combustion synthesis technique using pulsed direct current. A third phase, Mo$_{4.8}$Si$_3$C$_{0.6}$, forms at higher SiC concentrations. Interfaces between layers are cohesive, as are those between SiC and MoSi$_2$ grains. No mixing at the interfaces between layers is evident.

REFERENCES
1. A. Mortensen, and S. Suresh, "Functionally Graded Metals and Metal-Ceramic Composites: Part I Processing," *International Materials Reviews*, **40** [6] 239-265 (1995).
2. A.K. Vasudevan, and J.J. Petrovic, "A Comparative Overview of Molybdenum Disilicide Composites," *Materials Science and Engineering A*, **A155** 1-17 (1992).
3. G. Sauthoff, *Intermetallics*. 1995, Weinheim, Germany: VCH Verlagsgesellschaft mbH.
4. N.S Stoloff, "An Overview of Powder Processing of Silicides and Their Composites," *Materials Science and Engineering A*, **A261** 169-180 (1999).
5. N. Patibandla, *et al.*, "In-situ Processing of MoSi$_2$-Based Composites," C.L. Briant, J.J. Petrovic, and B.P. Bewlay, Editors. 1994.
6. C.R. Feng and D.J. Michel, "Microstructures of XD(tm) MoSi$_2$+SiC$_P$ Composites, in High-Temperature Ordered Intermetallic Alloys," V Symposium, I. Baker, *et al.*, Editors. 1993, Materials Research Society: Boston, MA, USA. p. 1051-6.
7. Y-L Jeng, E. J. Lavernia, J. Wolfenstine, D. E. Bailey, A. Sickinger, "Creep Behavior of Plasma-Sprayed SiC-Reinforced MoSi$_2$," Scripta Metallurgica et Materialia, **29**(1) 107-111 (1993).
8. J.M .Ting, "Sintering of Silicon Carbide/Molybdenum Disilicide Composites Using Boron Oxide as an Additive," *Journal of the American Ceramic Society*, 77 [10] 2751-2752 (1994).

9. S.E. Riddle, S. Jayashankar, and M.J. Kaufman, "Microstructural Evolution in Compositionally Tailored $MoSi_2$/SiC Composites"; pp. 291-6 in *High Temperature Silicides and Refractory Alloys Symposium*, edited by C.L. Briant, Materials Research Society, Boston, MA, USA, 1994.

10. C.H.Henager, Jr., J.L. Brimhall, and J.P. Hirth, "Synthesis of a $MoSi_2$/SiC Composite in Situ Using a Solid State Displacement Reaction between Mo_2C and Si," *Materials Science and Engineering A*, **A155** 109-114 (1992).

12. S. Jayashankar, S.E. Riddle, and M.J. Kaufman, "Synthesis and Properties of In-Situ $MoSi_2$/SiC Composites"; pp. 33-40 in *High Temperature Silicides and Refractory Alloys Symposium*, edited by C.L. Briant, *et al.*, Materials Research Society, 1994.

13. K Monroe, S. Govindarajan, J. J. Moore, B. Mishra, D. L. Olson, J. Disam, "Combustion Synthesis of $MoSi_2$ and $MoSi_2$ Composites"; pp. 113-118 in *High Temperature Silicides and Refractory Alloys Symposium*, edited by C.L. Briant, *et al.*, Materials Research Society, 1994.

14. S. Gedevanishvili, and Z.A. Munir, "An Investigation on the Combustion Synthesis of $MoSi_2$-βSiC Composites through Electric Field Activation," *Materials Science and Engineering A*, **A242** 1-6 (1998).

15. E. M. Carrillo-Heian, Dissertation, University of California, Davis 2000.

16. E.M. Carrillo-Heian, R. D. Carpenter, G. H. Paulino, J. C. Gibeling, Z. A. Munir, "Dense Layered $MoSi_2$/SiC Functionally Graded Composites Formed by Field-Activated Synthesis," Journal of the American Ceramic Society, 84 [5] 962–968 (2001).

17. M. Tokita, "Development of Large-Size Ceramic/Metal Bulk FGM Fabricated by Spark Plasma Sintering," *Materials Science Forum*, **308-311** 83-88 (1999).

18. T.Y. Um, Y. H. Park, H. Hashimoto, S. Sumi, T. Abe, R. Watanabe, "Fabrication of Mo-Si System Intermetallic Compounds by Pulse Discharge Pressure-Combustion Synthesis," *Powder and Powder Technology*, **44**(6) 530-534 (1997).

19. P. Villars, A. Prince, and H. Okamoto, *Handbook of Ternary Alloy Phase Diagrams*. 1995, Materials Park, OH: ASM International.

20. E. Parthe, W. Jeitschko, and V. Sadagopan, "A Neutron Diffraction Study of the Nowotny Phase Mo_5Si_3C," *Acta Crystallographica*, **19** 1031-1037 (1965).

FUNCTIONALLY GRADED HARDMETALS AND CERMETS

W. Lengauer, J. Garcia and V.Ucakar
Vienna University of Technology
Getreidemarkt 9/161
A-1060 Vienna, Austria
www.tuwien.ac.at/physmet

K. Dreyer, D.Kassel and H.-W.Daub
WIDIA GmbH
Münchener Straße 125
D-45145 Essen, Germany.

ABSTRACT

The diffusion-controlled preparation of functionally graded hardmetals (FGHMs) by in-situ modification of the sintering atmosphere is described. Four different types of near-surface microstructures were found and three intermediate types. The formation of the different graded zones depends on the starting formulation and is governed by diffusion of nitrogen, carbon, titanium and tungsten in the liquid phase above 1320°C in turn depending on the nitrogen equilibrium pressure of the compacts. FGHMs can show a superior performance in service and - from an economically point of view - they are a promising alternative to conventionally-coated hardmetals because of the lower productions costs.

INTRODUCTION

Hardmetals and cermets used for cutting and milling operations are high performance heavy-loaded materials. Hence, their response to thermal and mechanical load is very important for their service lifetime. Therefore, much effort has been made in hardmetal research to find novel techniques and material formulations to increase the performance of such materials.

The most successful step for increasing the lifetime of hardmetal tools has been made with the invention of hard-layer coatings which is already about 30 years old. Layers of TiC, TiN, Ti(C,N), (Ti,Al)N,... were deposited by means of chemical and physical vapour deposition techniques (CVD, PVD) which increased the lifetime substantially, yielded a better surface of the finished products, reduced the number of finishing cycles and the necessary machining power and sometimes made cutting fluids unnecessary avoiding wastes. Modern coatings are composed of many individual compound layers [1] (multilayer coatings) because some hard-phase materials show a better adherence to the hardmetal, others show a lower friction coefficient and diffusion welding towards the workpiece.

Although some layers were prepared with a gradual change of composition, e.g. a changing C/N ratio in a Ti(C,N) layer, all these (multilayer)coatings contain two-dimensional interfaces. At such interfaces the physical properties such as thermal expansion, hardness and elastic moduli, which are most important in performance, change discontinuously. This discontinuity can lead to material failure upon thermal and/or mechanical loading because of the pile-up of stress at these interfaces.

In order to circumvent such interfaces materials with a gradual change in phase composition as a function of distance could have superior performance and hence research on gradient materials [2] has reached recently also the field of hardmetals and cermets used for cutting operations. The gradual change promotes a dissipation of the stress at the grain boundaries over a three-dimensional volume fraction of the tool. Hardmetals and cermets consist of a multiphase microstructure (binder phase with sometimes different hard-phase particles, Fig.1) so a graded materials means a changing phase composition as a function of depth (microstructurally graded). The stress situation of such a *microstructurally (phase-compositionally) graded* material can then be compared to a *compositionally-graded* material (e.g.: Yang and Munz [3]) in which the pile up of stress at the interfaces of two materials bonded via an interlayer can be avoided by use of a graded interlayer, in which the composition changes smoothly from one side into the other, instead of a homogeneous interlayer.

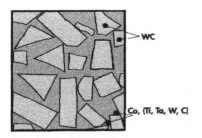

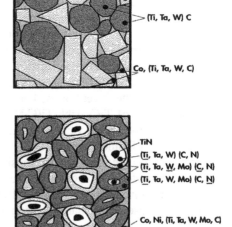

Figure 1
Microstructures of hardmetals and cermets.
Top: typical hardmetal structure,
top right: hardmetal with fcc carbides
right: cermet containing fcc carbonitrides

EXPERIMENTAL

Hard material powders such as TiC, TiN, (Ti,W)C, Ti(C,N), (Ta,Nb)C were mixed with 6-17wt% Co as a binder metal by state-of-the-art industrial techniques. Some formulations were located in the area of hardmetals, i.e. at WC-rich composition, others in the area of so-called cermets, i.e. at Ti(C,N)-rich compositions and some in between. Although these designations are used in industry it should be mentioned that no clear boundary between hardmetals and cermets exist and some intermediate compositions show microstructural features of both type materials, i.e. facetted WC crystallites such as in hardmetals together with spherical Ti(C,N) particles such as in cermets.

Green compacts of the powders mixtures were then subjected to a sintering cycle in a laboratory furnace in which the gas phase composed of N_2 and CO could be modified with respect to pressure (up to 1000mbar) and composition. Most experiments were performed by using nitrogen atmosphere after occurrence of the liquid phase at about 1320°C and sintering cycles close to the time-temperature profile used in conventional hardmetal sintering.

The materials were subjected to metallography, XRD, glow-discharge optical emission spectroscopy GDOS (depth profiling of composition), EPMA and surface roughness measurements.

RESULTS

As a first important result it turned out that the near-surface modification of the microstructure to yield a graded near-surface zone can be performed within the sintering cycle so as to yield dense-sintered graded hardmetals materials [4], some of which had a very low residual porosity which made post treatment by hipping unnecessary. The surface of some of these inserts (Fig.2) turns to violet or bronze which stems from the formation of Ti(C,N) which is yellow at the TiN boundary and turns over bronze to violet and grey if the carbon content increases [1]. The inserts with metallic lustre contained a WC+Co outer rim.

Figure 2
FGHMs (indexable inserts) sintered in reactive atmosphere. Colour from left to right: violet (matt), pink (bright), bronze (bright) and metallic grey. Size: about 12x12x5mm

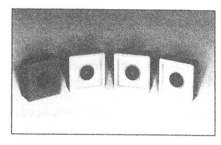

*Colour versions of this image together with the microstructures in the following figures can be inspected at the collection at www.tuwien.ac.at/physmet/welcome/

From about 50 different hardmetal formulations, subjected to more than 30 different sintering cycles, the near-surface microstructures could be divided into four principal types with three intermediate types. These are shown in Fig.3 and described in Table I and II.

Table I. Principal types of FGHMs (compare Fig.3).

type	principal types, microstructural features
1	a graded zone in which a microstructure composed of globular (Ti,W)(C,N) particles, typical for cermets, changes smoothly into a hardmetal-type structure (angular WC grains) or a mixture of angular WC and globular (Ti,W)(C,N); the bulk structure is smoothly attained
2	a microstructure which contains a more or less well differentiated WC-Co surface zone or layer with a following graded microstructure, where the TiN particles are enriched just below the WC+Co zone
3	a microstructure with a Ti(C,N) top layer followed by a WC-Co layer, and then a graded microstructure like to type 2
4	a Ti(C,N) top layer, which formed onto a graded microstructure as described for type 1

Table II. Intermediate types of FGHMs (compare Fig.3).

type	intermediate types, microstructural features
1→ 4	increasing number of nitrogen-rich Ti(C,N) particles at the surface which can finally from a layer and thus change type 1 into type 4
2→ 3	increasing amount of nitrogen-rich Ti(C,N) grains on the WC+Co rim which finally from a Ti(C,N) layer ending up with type 3
3→ 4	decreasing contiguity of the WC grains in the WC+Co layer/zone so that its two-phase layer character fades smoothly away

The diffusion layers of types 3 and 4 show an outer hard-phase layer based on Ti(C,N) interlocked with the interior of the hardmetal yielding a three-dimensional interface. Thus the character is different from CVD and PVD layers where two-dimensional interfaces exist. It is therefore expected that the adherence of the diffusion layer is better and dissipation of heat and stress is increased. A further difference to CVD coatings is the sometimes fine-grained microstructure of the TiN-rich layers of type 3 and 4.

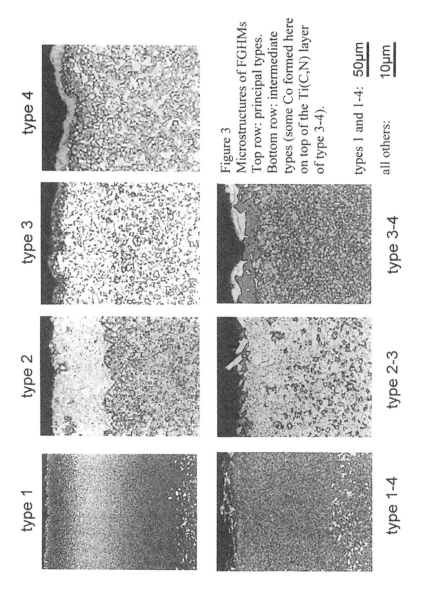

Figure 3
Microstructures of FGHMs
Top row: principal types.
Bottom row: intermediate
types (some Co formed here
on top of the Ti(C,N) layer
of type 3-4).

types 1 and 1-4: $\underline{50\mu m}$

all others: $\underline{10\mu m}$

A plot of the composition vs. type given in Fig.4 shows a clear interdependency. In the hardmetal region of the hard-phase composition triangle layered/graded types are formed whereas in the cermet region only graded type 1 forms. Interestingly, in intermediate cermet/hardmetal regions of the composition triangle, also intermediate types occur.

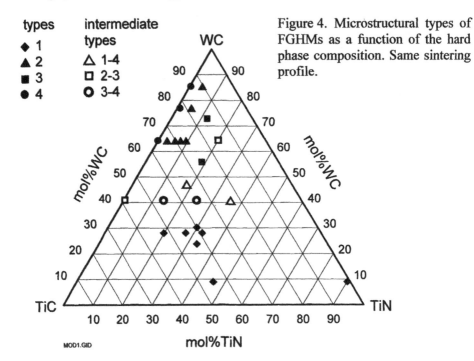

Figure 4. Microstructural types of FGHMs as a function of the hard phase composition. Same sintering profile.

DISCUSSION

The formation of the various types of graded near-surface regions is diffusion-controlled. The in-diffusion of nitrogen changes the equilibria locally so that microstructural features are different from vacuum-sintered materials.

Type 1 FGHMs (compare Fig.3) show the lowest nitrogen equilibrium pressure so that from the TiC or Ti(C,N) phase present in the starting formulation a nitrogen-rich Ti(C,N) phase is already formed at the surface and the nitrogen profile then decreases as a function of depth. Because of the high and uniform activity of Ti there is no Ti diffusion rather than a W diffusion inside. The overall "(Ti,W)(C,N) frame" is rigid and changes in composition only locally by the action of nitrogen which changes the phase equilibria as a function of depth. Upon in-diffusion, carbon is formed from Ti(C,N) by the reaction Ti(C,N) + $N_2 \rightarrow$ TiN

Functionally Graded Materials 2000

+ C and WC is formed from the according tungsten-containing compound [5] (Ti,W)(C,N) + N_2 → Ti(C,N) + WC. As a result of the two reactions W and C dissolve in the binder and WC precipitates inside the material, often some 10-100 µm away from the surface where the nitrogen activity is small. Some features of this type 1 have already been described by Tsuda et al.[6] and are commercially sold under the name CN 8000 by Sumitomo [7]. However, these FGHMs contain a 10-15µm layer of a Ni/Co alloy on the surface because of out-diffusion of the binder phase during sintering. Such a surface layer reduces the performance because of diffusional interaction of the Ni/Co alloy with the workpiece. As we have shown recently, such a surface segregation can be avoided by optimisation of the reactive gas supply upon sintering [8].

Type 2 is closest to vacuum-sintered materials and it occurs at a very high WC contents. Because of the high WC content the nitrogen equilibrium pressure is very high and for the formation of the outer WC+Co rim the difference between a nitrogen-modified sintering cycle and a vacuum sintering cycle is – at a first glance – very small. A WC+Co rim has been described already in literature [9,10] and was yielded by ordinary sintering routines and also several patents exist on this "fcc phase-free" surface zones. However, a closer look to the microstructure of our type 2 FGHMs shows sometimes small inclusions of nitrogen-rich Ti(C,N) within the WC+Co layer, depending of the overall amount of WC phase in the starting formulation and, in addition, a graded layer below the WC+Co. This is due to the reaction of nitrogen with the part of the sample below the WC+Co rim depleted in WC. This material has a lower nitrogen equilibrium pressure and hence response the nitrogen treatment is visible. Hence, nitrogen must in-diffusion through the WC+Co rim.

Schwarzkopf et al. [9], who also described the formation of similar WC+Co rims, used nitrogen only in order to modify the composition of the hard phase at the beginning of the sintering cycles far from the liquidus temperature (the composition could be modified by adding TiN, too). Nitrogen was then pumped off upon further heating so that neither nitrogen-rich grains within the WC-Co zone nor a graded zone below the WC-Co rim occurred. If nitrogen would remain in the furnace a dense-sintered product cannot be obtained.

Type 3 and type 4 functionally graded hardmetals, which appear as completely new types of FGHMs as a whole, show an intermediate nitrogen equilibrium pressure as compared to types 1 and 2. The low concentration of Ti in the starting formulation causes Ti to diffuse out of the material. The affinity of Ti to N causes the out-diffusion of Ti towards the higher nitrogen activity and combining with nitrogen to form almost pure TiN. It depends on the amount whether a layer of just precipitates (intermediate type) of TiN are formed on the surface. The formation of a distinct WC+Co layer in type 3 FGHMs is due to the higher WC content as compared to type 4 FGHMs. The TiN surface layer is of uniform composition because of the formation from the liquid by combination of Ti + N_2 (in solid-state nitrided compacts the C/N ratio of the Ti(C,N) layer changes yielding a compositionally graded layer). It is, however, not yet clear why sometimes a monolithic TiN layer (Fig3, types 4 and 3-4.) and sometimes a small-grained TiN layer (Fig.3, type 3) forms. The formation of nitrogen-rich Ti(C,N) particles below the TiN layer seems to occur from the reaction of TiCN with nitrogen because individual grains form at these positions.

CONCLUSION

Dense functionally graded hardmetals can be prepared in situ by modification of the sintering atmosphere at a special point of the sintering cycle. The formation of graded microstructures of which four different principal types have been obtained occurs mainly in the liquidus state of the compact above ca. 1320°C, depending on the overall composition.

This in situ preparation is an inexpensive route to obtain materials of outstanding properties in which the stress occurring within service is dissipated over a three-dimensional area rather than occurring as a sharp pile-up at two-dimensional interfaces present in coated materials.

REFERENCES

[1] W.Lengauer, Transition metal Carbides, nitrides and carbonitrides, in: *"Handbook of Ceramic Hard Materials"*, Vol.I, p.202-252, ed. R.Riedel, Wiley-VCH, Weinheim (2000)

[2] A.Neubrand and J.Rödel, Gradient materials: an overview of a novel concept, *Z.Metallkunde* **88**, 5 (1997)

[3] Y.Yang and D.Munz, Reduction of the stresses in a joint of dissimilar Materials using graded Materials as interlayer, *Fracture Mechanics* **26**, *ASTM STP*, 572 (1995)

[4] W.Lengauer, J.Garcia, V.Ucakar, L.Chen, K.Dreyer, D.Kassel, H.-W.Daub, Diffusion-controlled surface modification for fabrication of functional-gradient cemented carbonitrides, *Proc.PM²tec 99*, Vancouver, Vol.3, part 10, 85-96 (1999)

[5] A.Doi, T.Nomura, M.Tobioka, K.Takahashi and A.Hara, Thermodynamic evaluation of the equilibrium nitrogen pressure and WC separation in the Ti-W-C-N system carbonitride, *Proc. 11th Plansee Seminar* **1**, 85 (1985)

[6] T.Nomura, H.Moriguchi, K.Tsuda, K.Isobe, A.Ikegaya and K.Moriyama, Material design method for the functionally graded cemented carbide tool, *Int.J. Refract.&Hard Mater.* **17**, 397 (1999)

[7] K.Tsuda, A.Ikegaya, T.Nomura, K.Isobe, N.Kitagawa, M.Chudou, H.Arimoto, Development of a functionally graded hard materials, *Sumitomo Techn.Rev.***41**, 47 (1996)

[8] W.Lengauer, J.Garcia, K.Dreyer, I.Smid, D.Kassel, H.-W.Daub, G.Korb and L.Chen, Diffusion-controlled fabrication of functionally-graded cermets and hardmetals, *Proc. EURO PM 99*, Turin, 475-482 (1999)

[9] M.Schwarzkopf, H.E.Exner, W.Schintlmeister and H.F.Fischmeister, Kinetics of compositional modification of (W,Ti)C-WC-Co alloy surfaces, *Mater. Sci. Eingineer.* **A105/106**, 225 (1988)

[10] P.Gustafson and A.Östlund, Binder-phase enrichment by dissolution of cubic carbides, *Int.J.Refr.Met.&Hard Mater.***12**, 192 (1993-4)

GRADED MATERIALS OF DIAMOND DISPERSED CEMENTED CARBIDE FABRICATED BY INDUCTION FIELD-ACTIVATED COMBUSTION SYNTHESIS

Manshi Ohyanagi, Isao Shimazoe,
Takayuki Hiwatashi, Tetsuya Tsujikami,
and Mitsue Koizumi
Ryukoku University
Seta 1-5, Ohtsu, 520-2194, Japan

Zuhair A. Munir
University of California
Davis, CA 95616, USA

Evgeny Levashov
Moscow Steel and Alloys Institute,
Leninsky pr. 4, Moscow 117936, Russia

ABSTRACT

Graded dispersion of diamond was introduced in WC-10wt%Co cermets by static pseudo isostatic compaction (SPIC) with induction field-activated SHS (IFASHS) to fabricate graded composites consisting of 4 layers of No.1 (50 vol% of diamond) / No.2 (30 vol% of diamond) / No.3 (10 vol% of diamond) / No.4 (no diamond) in which the ratio in vol% was 1/1/1/3. The SPIC was performed using commercial casting sand as the pressure-transmitting medium for the densification of the cermets. The process enabled to simultaneously synthesize and densify the cermet matrix within a few minutes. Diamond mixed with the reactant was fixed in the matrix produced during the SHS. The maximum combustion temperatures were controlled to be approximately 1800 K to prevent transformation from diamond to graphite. An addition of 0.5wt% of copper into the cemented carbide also drastically suppressed the transformation. The diamond particles were found by SEM to be strongly fixed in the matrix even after lapping with a diamond abrasive.

INTRODUCTION

Metal alloys and ceramic materials containing diamond have been expected as advanced materials with extremely hard, highly heat-conductive properties and so on because diamond itself is also attractive as an industrial materials of next century. The bonding is one approach to fabricate diamond composite materials[1]. The other approach in fabrication of the composite materials is to highly disperse diamond particles into the material matrices through combustion synthesis method.[2-3]

Graded dispersion of diamond in TiB_2/Si cermet (90vol% diamond layer / diamond graded dispersed layers / matrix) was achieved by dynamic pseudo isostatic compaction (DPIC) just after self-propagating high temperature synthesis (SHS) or combustion synthesis in our previous report [4]. Diamond particles were strongly fixed in the matrices through grain boundary of SiC formed between diamond and TiB_2/Si. However, these diamond composites were very brittle and it was difficult to reduce completely thermal stress in the layered composites without any cracking.

Good candidate as a matrix to disperse diamond is, of course, cemented carbide with high fracture strength and toughness because poly-crystalline sintered diamond (PCD) is often fabricated with WC-Co cemented carbide under about 6.0GPa at 1500°C. But, in combustion synthesis, both ignition and self-propagation do not occur in reactants with high diamond concentration without an external energy support because the diamond works as reaction diluent. The reactant layer without diamond adhered to the diamond-dispersed matrix of reactant is required as an additional energy source to initiate and self-sustain the reaction through the whole sample. This is the reason why TiB_2 system of highly exothermic reaction was used in the first step on fabrication of diamond composites.

Reactions which are weakly exothermic can not be initiated or sustained, making the SHS process unsuitable for the synthesis of such materials as SiC, B_4C, WC, and others. The unsuitability of the SHS process is also demonstrated in the synthesis of composites. For example, while the synthesis of $MoSi_2$ by SHS is feasible, that of composites of $MoSi_2$-SiC is not. Similar considerations apply to the synthesis of metal-matrix composites, if the metal content is relatively large. These materials systems can be synthesized by other methods, e.g., volume combustion. In this method, the reactants are furnace-heated until the reaction takes place over the entire sample. Another alternative is the preheating of the reactants and the subsequent ignition to initiate SHS reactions. The increase in the ambient temperature in the latter case results in a higher adiabatic temperature and hence makes possible the establishment of the SHS wave. In both of these alternative approaches (the volume combustion and the preheating), their success depends on the rate of heating which influences the formation of pre-combustion phases through solid-state diffusion.

A new alternative to the above methods of activation has been developed recently [5]. The activation in the new method is derived from the application of a voltage across the reactants at the same time as an ignition source is activated. The method, referred to as the field-activated combustion synthesis, FACS, was shown to make possible the establishment of SHS waves in weakly exothermic reaction, e.g., in the synthesis of SiC and others [6]. Other advantages of the FACS method have been also demonstrated [7]. These include the effect of the field on the composition of the product phases and on the elemental distribution in solid solutions. It was shown that in the presence of a field, dense reactant and reactants with relatively large particle sizes can be used. Experimental results and modeling studies have led to the conclusion that the effect of the field is to provide Joule heating. The benefit of this method depends on whether the electric energy is confined to the narrow combustion wave (reaction zone). In reactions where the product is relatively non-conducting, the electrical energy is localized.

In this paper, we describe the use of another field-activated process, that arising from the use of induction [8]. Here a sample is covered with a conductive sheet such as carbon foil and is placed inside an induction coil. The induction current is generated in the foil and in the sample, if it is conducting. Or the foil can raise the sample temperature through conduction to a temperature where it becomes conducting. Since the induction current is localized in a surface layer of the sample, ignition takes place in this region and is expected to propagate in an inward radial direction.

A formation enthalpy of WC is –40.5kJ/mol and W-C stoichiometric reactant mixture with 10wt% of Co does not react without a form of activation. In the reaction of W-C, the adiabatic combustion temperature is less than 1500K. When these reactant compacts covered by carbon sheets were placed inside induction coil in which a high frequency current passed, they could be ignited at the each edge surface by the sheet heated due to its induction current. The resulting combustion wave propagated toward the center. The activation of SHS consists of two factors, that is, preheating of reactant compacts based on heat generated in the carbon sheet by its induction current and the other Joule heat by the induction current passed through the combustion zone with molten species in the sample. In this process, the combustion reaction takes place under static pseudo isostatic compaction through commercial casting sand as a pressure transmitting media. As the result, simultaneously combustion synthesis and densification can be successfully accomplished. It should be also noted in the process that an additional thermal treatment could be performed even after the combustion reaction.

The objective of this research is to fabricate graded materials of diamond dispersed WC-Co composites by the both of combination techniques of SHS for short time processing and following static compaction for densification.

EXPERIMENTALS

Evaluation of residual stresses using FEM: The finite element method (FEM) has been used to evaluate thermal residual stresses at interface of diamond/WC/Co composites. Axisymetric cylindrical specimens were used, allowing two dimensional models to be employed. A model system composed three layers, diamond, diamond in WC/Co and WC/Co. The middle layer was devided into graded layers. The finite element analysis was performed using the original developed software SACOM for composite materials[9]. In this simulation, thermal residual stresses, considering only elastic behavior were calculated, and the diamond/WC/Co composites was cooled from the assumed high temperature service (1630K) to room temperature (298K). Time and temperature dependent properties were neglected. Table 1 shows physical and mechanical properties of diamond, WC and Co relevant to the calculation. Table 2 shows the condition for FEM analysis. The specimen's dimensions were 10 mm long and 40 mm in diameter. Constitutive properties for the composite material interlayer were computed using a rule-of-mixtures.

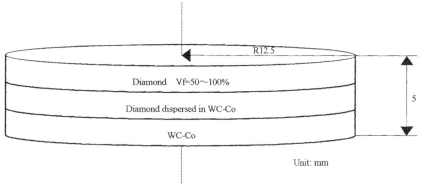

Unit: mm

Scheme 1. Dimension and layer's structure of diamond dispersed cemented carbide

Table 1.Physical and mechanical properties of Diamond, WC and Co.

	Elastic modulus E [GPa]	Coef. of thermal expansion [K^{-1}]	Poisson's ratio ν
Diamond	1050	1.00×10^{-6}	0.3
WC	720	5.70×10^{-6}	0.3
Co	179	1.25×10^{-5}	0.3

Table 2. Condition for FEM analysis

Difference of temperature ΔT	1630-298K
Equation to decide diamond concentration in graded region	$(x/t)^p$ x : position, t : width of graded layer, p : the power law exponent
Element mesh type	Triangle linear element
Number of element and node	11520 elements 5929 nodes
Others	using rule-of-mixtures for calculation of the properties

Materials and Procedure: The materials used in this study included 99.9 % pure graphite powders (with an average particle size of about 7.0 µm, Tokai Carbon Inc., Japan), 99.9% pure powders of tungsten (with an average particle size of about 8.0 µm, Kojundo Chemical Lab. Co. Inc, Japan), and 99.9% pure powders of cobalt (with an average particle size of 2.0µm, Kojundo Chemical Lab. Co. Inc, Japan) and C (diamond: artificial, an average particle size: approximately 30 and 125 µm, >99.9%, Tomei Diamond Co. Ltd.). Tungsten, carbon, and cobalt powders were homogeneously mixed in ethanol for 24 hours using silicon nitride milling media in proportion that gave cemented carbides with a molar ratio of C/W=1.0 and with 10wt% cobalt and were subsequently dried in vacuum at 220°C for 24 hours. The powder mixture was dry-mixed with diamond in given ratios (10 to 50 vol%) by V-type mixing equipment again for 20 min. Disk-like compacts (approximately 30 mm in diameter and 15 mm in height) were prepared in a stainless steel die so that diamond powder was dispersed in the each 3/5 layer of the compact, that is, 3 (matrix) / 1 (50 vol% of diamond) / 1 (30 vol% of diamond) / 1 (10 vol% of diamond) / 3 (matrix) in vol%. In the final product, WC-Co matrix near the highest diamond concentration layer can be eliminated (Approximate dimension of final product: 25 mm in diameter and 5mm in height). The powders were pressed uniaxially at an approximate pressure of 70MPa for 30 s.   The compacts were then cold-isostatically pressed (CIP) at 200MPa for 120 s. The twice-pressed compacts were then outgassed in vacuum for 12 hr at 220°C to eliminate highly volatile species such as adsorbed water. The surface of each compact was covered with two types of carbon foil. The top and bottom surfaces were covered with a foil with a thickness of 0.75 mm and the side surface was covered with a foil with a thickness of 0.4 mm.

The foil-covered compact was then inserted into a silicon nitride die with a 70 mm inside diameter, 140 mm outside diameter, and a 70 mm height. The inside cavity of this die was filled with commercial casting sand, as shown in Figure 1. By pressing uniaxially with a non-conducting piston, a pseudo-isostatic pressure can be loaded on the compact. While under pressure, the reactants can be ignited by an induction current through the carbon sheet on the surface of compact. We refer to this method as "SHS / Static Pseudo Isostatic Compaction, or *SHS/SPIC*", on contrast to the "SHS / Dynamic Pseudo Isostatic Compaction, *SHS/DPIC*" reported previously [4]. But in a general sense, this method can be classified as *Induction Field-Activated SHS* similarly to the *Electric Field Activated SHS* as reported by Munir *et al.*[6] The passage of a high frequency current through the induction heating coil (170A, 70 kHz, 85V), a current is induced through both the carbon foil, and the outer layer of the sample, if it is conducting. This process is done while a pressure of 25 MPa is applied under this pseudo isostatic condition. As an induction current is basically effective on the surface of sample, the

Functionally Graded Materials 2000

ignition took place on the surface closest to the coil. Therefore, after ignition at the outer diameter of the sample, a ring-like combustion wave propagated toward the center. Initially, the induction current can also pass through the combustion zone, providing Joule heating. This arrangement makes it feasible to introduce preheating and annealing stages in the overall process.

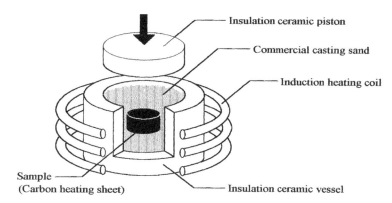

Figure 1. Scheme of an induction field-activated SHS/SPIC equipment

Temperature profiles were measured by a W- 5%Re /W- 26%Re thermocouple. The voltage output from the thermocouple was monitored using a data acquisition recorder. This recorder made it possible to measure and store the data during the SHS/SPIC experiments. The products were analyzed by X-ray powder diffraction (XRD), using CuKα radiation (RINT-2500, Rigaku Co. Ltd.). The product surfaces were lapped using a diamond abrasive and were examined by scanning electron microscopy (SEM) using a JSM-5410, JEOL microscope. The density of products was measured by the Archimedes method after removing the carbon foil and polishing the surface. The disk-like product samples were cut into rectangular bars of 4 mm in width, 3 mm in thickness and 40 mm in length, and their surfaces were mirror-polished using 1 μm diamond paste in preparation for textures observation, bending fracture strength and hardness measurements. The fracture strength was measured at a cross-head speed of 0.2 mm/min using a three-point bending equipment with a span of 30 mm for the WC-Co without diamond. The Young's Modulus was also simultaneously calculated from the stress-strain curve obtained from the applied stress and the resulting strain. The latter was measured by a strain gauge cemented on the sample surface. The hardness was measured by indenting the mirror-finish polished sample surface with a load of 100 N, using a semi-Vickers hardness tester for the WC-Co without diamond.

RESUTLS AND DISCUSSION

Evaluation of residual stresses using FEM: Constitutive properties for the composite material interlayer were computed using the rule-of-mixtures. In two layers system, when the diamond volume fraction of top layer are 100 vol% as shown in Figure 2, the maximum residual tensile stress in WC-Co bottom layer was 5.0GPa, which suggests possibility of fracture in the cemented carbide layer without diamond. So, graded dispersion of diamond in cemented carbide layers is required to fabricate this kind of diamond composite.

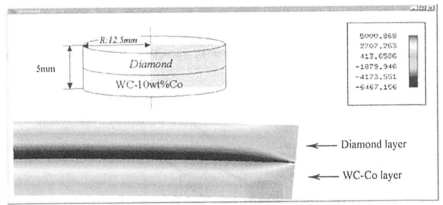

Figure 2. Distribution of radial stress in the two layers system

Next, we calculated the residual stress in the 4 layers system of WC-10wt%Co/ 10vol%diamond/ 30vol%diamond/ 50vol%diamond in which the ratio of each layer's volume was 3/1/1/1, respectively as shown in Figure 3. The larger residual tensile stress could be calculated in the layers containing 10vol% and 0vol% of diamond. However, the maximum tensile stress was 1.2GPa and approximately a half of bending fracture strength of WC-10wt%Co, which suggests possibility of successful fabrication. We attempted the graded dispersion of diamond in this system as a first step on fabrication of diamond dispersed FGM

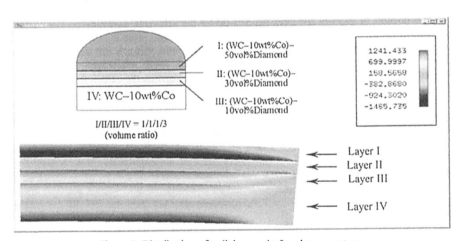

Figure 3. Distribution of radial stress in four layers system

Induction field-activated SHS/SPIC of cemented carbide and the diamond-dispersed FGM: The density of WC-10wt%Co fabricated by IFASHS/SPIC method increased with an increase in the induction time and after 360 s, the product was nearly 100% dense. From X-ray diffraction patterns of the products, in passage times of induction current of 240 and 300 s, the products contain small peaks belonging to the (η) phase Co_3W_3C. This phase is well known to

Functionally Graded Materials 2000

have a deleterious effect on the mechanical properties of the composite. However, with longer induction time of 360 and 420 s, there was no evidence for the presence of this phase. The bending fracture strength, Young's modulus, and hardness of the WC-10wt%Co is 2480MPa, 570GPa and 1500MPa, respectively. These values are similar to those of WC-10wt%Co fabricated by conventional electric furnace or hot press, which indicates an IFASHS/SPIC method of quick processing is very effective to fabricate the matrix of WC-Co as a host of diamond insertion.

Figure 4 shows a temperature profile of diamond-dispersed cemented carbide (diamond particle size: 30~40μm, 50vol%) prepared by IFASHS/SPIC for 360sec of current passage time. The temperature was measured in the center of diamond containing layer. After preheating of the reactant up to about 1250K, the ignition occurred and the temperature increased slightly with an increase in an induction current passage time. Herein, the volume ratio of each layer was 1/1/1. The maximum temperature was controlled not to be higher than 1800K to suppress transformation of diamond to graphite. The least induction current passage time to achieve the complete reaction was also chosen for diamond not to transform.

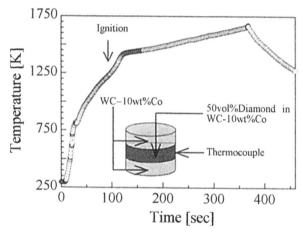

Figure 4.   Temperature profile of diamond dispersed cemented carbide (diamond particle size: 30~40μm) prepared by induction field activated SHS/SPIC: 360sec of current passage time.

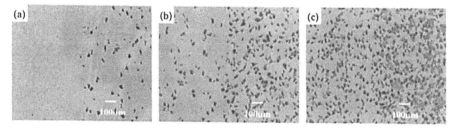

Figure 5. SEM photographs of cross section in the 4 layers composite prepared by IFASHS/SPIC, WC-10wt%Co/10vol%diamond/30vol%diamond/50vol%diamond: 360sec of current passage time; Interfaces between (a) WC-10wt%Co and 10vol%diamond layers, (b) 10vol%diamond and 30 vol % diamond layers, (c) 30vol%diamond and 50vol% diamond layers.

Figure 5 shows SEM photographs of cross section in the five layers composite consisting of WC-10wt%Co/10vol%diamond/ 30vol%diamond/ 50vol%diamond prepared by IFASHS/SPIC method. The fraction of each layer is 3/1/1/1 in vol%. Generally cobalt works as a catalyst of diamond production from the other carbon source just like a graphite under about 6.0GPa, at 1500°C. Of course, herein, cobalt attacked diamond and accelerated the transformation to graphite. However, an addition of small amount of copper to cobalt drastically suppressed the transformation. Figure 6 shows the differences in their textures between with 5wt% of copper in cobalt and without copper. In case without copper, most of the diamond surface was edged by cobalt. On the other hand, in case with copper, the surface was very smooth and was not edged.

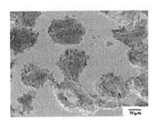

Figure 6. SEM photographs of diamond-dispersed cemented carbide prepared by IFASHS/SPIC, passage time of induction current: 360sec. (a) 50vol% of diamond in WC-10wt%Co, (b) 50vol% of diamond in WC-10wt%Co-0.5wt%Cu (diamond particle size: 100-125μm)

CONCLUSION

Graded materials designed by the evaluation of residual stresses using FEM, of diamond-dispersed WC-10wt%Co were fabricated by an induction field-activated SHS / SPIC method. The 4 layers FGM consisting of 1 (50 vol% of diamond) / 1 (30 vol% of diamond) / 1 (10 vol% of diamond) / 3 (WC-Co) in vol% was successfully fabricated without cracking. An addition of copper into cobalt suppressed drastically transformation from diamond to graphite.

ACKNOWLEDGMENT

We are grateful for the support given to one of us (M.O.) by the Japanese Ministry of Education and the High Tech Research Center, and by the US Army Research Office (to Z.A.M.).

REFERENCES

1. M. Ohyanagi, M. Koizumi et al., *Am. Cer. Soc. Bull.*, **72**, 86 (1993)
2.. M.Ohyanagi, E.A. Levashov et al., *Trans. Mat. Res. Soc. Jpn.*, **14A**, 685 (1994)
3. M.Ohyanagi, M.Koizumi, and E.A.Levashov et al., *Intern. J. SHS*, 4, 387 (1995)
4. M.Ohyanagi, T.Tsujikami, S.Sugahara, E. A. Levashov, I. P. Borovinskaya, *Materials Science Forum*, **Vols. 308-311**, 145(1999)
5. Z.A.Munir, W.Lai, and K.Ewald, U. S. Patent No. 5,380,409, January 10, (1995).
6. A.Feng, and Z.A.Munir, *J. Appl. Phys.*, **76**, 1927 (1994).
7. A.Feng and Z.A.Munir, *Metall. Mater. Trans.*, **26B**, 581 (1995).
8. M.Ohyanagi, T.Takayuki, M.Koizuni and Z.A.Munir, Proceedings of 1st Russia-Japan Workshop on SHS, Karlovy Vary, Czech Republic, Oct.30-Nov.3, 65-69 (1998)
9. M. Zako, T. Ttujikami, Development of Personal Computer Program of Stress Analysis for Composite Materials, Journal of the society of material science, **38**, No.438, (1990)

Processing and Characterization of Functionally Graded Ti-B Based Composites For Armor Applications

M. Cirakoglu, S. Bhaduri, and S.B. Bhaduri
University of Idaho
Department of Materials and Metallurgical Engineering
McClure Hall
Moscow-Idaho 83844-3024

Abstract

In this study, we have attempted to produce FGMs in Ti-B binary system for potential armor applications. Elemental titanium and crystalline boron powders were used as raw materials. Compositionally graded pellets were prepared by uniaxial pressing and cold isostatic pressing. Samples were ignited by using a tungsten coil as an external heat source. A pressurized-air activated metal punch was used to compact the reacted product for further densification. Samples were characterized by X-ray diffraction and hardness of each layer was determined by Vickers indentation. The microstructure of interfaces and graded layers were examined by using a scanning electron microscope and an optical microscope.

Introduction

Functionally graded materials (FGMs) are a class of composite materials, with a variation in the constituents such that their composition and microstructure change in one direction. The graded composition eliminates many problems such as poor mechanical integrity, and poor interfacial adhesion associated with the presence of abrupt interfaces in conventional composites. This distinctive feature of FGMs provides properties that are not offered by monolithic materials [1-3].

From a ballistic standpoint; FGMs offer significant advantages to armor designers. The FGM armor scheme typically consists of a hard frontal surface and softer backing. The hard frontal materials are usually ceramics. The purpose of the hard surface is to blunt and to induce a destructive shock wave on to the projectile upon impact. The softer backing materials act as a "catcher" for residual broken fragments in preventing target penetration. In this type of armor scheme, the hardest frontal material will typically provide the best level of ballistic protection [4,5].

In this work, processing of FGMs has been attempted utilizing a reaction process such as the combustion synthesis method. Combustion synthesis (CS) has various advantages, including its simplicity of the process, higher purity of the products, and relatively low energy requirements. CS process utilizes the energy of chemical reactions between the constituents. In a typical combustion synthesis process, a powdered mixture (which is capable of undergoing an exothermic reaction) is cold pressed into a cylinder then placed into an evacuated chamber to prevent oxidation. Surface of the sample is exposed to an external heat source (e.g. a heated coil, electric match or laser beam) for a short time. Once the mixture is ignited, a strong exothermic reaction liberates enough heat to the adjacent layer of reactants and the reaction becomes self-sustaining and propagates in the form of a combustion wave [6-8]. CS products frequently exhibit a considerable amount of porosity primarily because of volume changes between reactants and products, volatile gases released during the exothermic excursions and short residence times at elevated temperatures. The porosity can be reduced if the reacted CS product is rapidly consolidated while it is still hot. The heat retained in the sample at the time of the compaction promotes sintering. This concept was previously applied for producing dense carbides and borides. Kecskes [9], successfully produced TiC and TiB_2

ceramics in excess of 98% of theoretical density. LaSalvia et al. [10], showed that TiC could be densified to 96% of theoretical density by high velocity forging.

The main objective of this study is to fabricate ceramic/metal FGMs by using Ti and B elemental powders. As processing method we utilized both conventional combustion synthesis and combustion synthesis/dynamic compaction (CS/DC) techniques and compared the results in terms of microstructure and mechanical properties.

Experimental Procedure

In this research, elemental Ti (99% purity, -325 mesh, Johnson Mathey, Ward Hill, MI) and crystalline B (> 99% purity, Cerac, Milwaukee, WI) were used as starting materials. Different compositions were prepared by dry milling for 3 hours in separate polyethylene bottles. The graded layers were prepared by simply stacking up the different compositions in a steel die in a discrete homogeneous layer fashion as shown in Figure 1. The mixture was first uniaxially pressed into cylindrical compacts under a pressure of 525 MPa using a stainless steel die with double acting rams. The resulting pellets had typical dimensions of 12.7 mm and about 14 mm long. The compacts were further densified using a cold isostatic press (ISI Inc., Columbus, OH) at a pressure of 100 MPa. To remove the adsorbed moisture and gases, the as-prepared samples were de-gassed at 600°C for 3 hours under flowing argon using a vacuum furnace (Centorr Assoc. Inc., Suncook, NH) prior to combustion experiments. In the absence of such a treatment, the samples exploded during the reaction because of gas evolution.

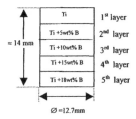

Figure 1. Schematic drawing of a 5 layered FGM.

A schematic of the experimental set-up used in combustion experiments is shown in Figure 2. The set-up consists of a die cavity, a vacuum pump, Ar gas tanks, view port, pressurized air activated punch, and a tungsten coil. The view port facilitates the observation of the combustion wave and recording the combustion process. Before each experiment, the chamber is flushed with argon a few times and subsequently evacuated by using a mechanical pump attached to the system. The green compact is placed into the die cavity, which is made out of steel with a diameter of 1 inch. There is a 5 mm air gap between the sample and the walls of the die. The main function of this cavity is to hold the sample in place during experiments. A tungsten coil with a wire diameter of 2mm (R.D. Mathis Co., Long Beach, CA) placed at 2-3 mm from the top of the sample is used as an external heat source for ignition by passing an electric current through it as shown in Figure 3.

The compaction is performed after the green compact is fully converted into the final product. This is possible when the combustion wave reaches the bottom of the sample as verified by visual observation. A pressurized air activated steel punch is used for this purpose, which is located on the upper section of the reactor assembly. The potential energy stored in pressurized air is transformed into the kinetic energy of the punch and collapses the porous microstructure. The tungsten coil moves out of the way right before the compaction. The applied pressure can be adjusted to the desired level by simply increasing the air pressure in our preliminary experiments. In our experiments we used different pressures (137 and 220 kPa). Since the diameter of the punch is slightly smaller than that of the die, the wall of the die is protected.

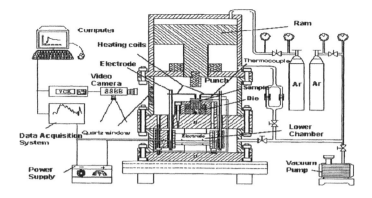

Figure 2. Schematic drawing of the experimental system.

→ W-coil
→ Sample
→ Die cavity

Figure 3. Die assembly used in CS/DC experiments

Characterization

The reacted samples were sectioned and mounted. Samples then were ground using SiC papers from grade size 180 to 4000 and polished with alumina suspension from 5 μm to 0.01 μm and finally lightly etched in a solution of 10 % HF, 40 % HNO_3 and 60 % H_2O so as to improve the visibility of any interfacial reaction products. The phases of each layer were determined by a X-ray diffraction method (Siemens D5000, CuKα radiation). The room temperature microhardness of each layer was determined by Vickers test machine and Vickers hardness numbers (VHN) were obtained. A load of 1 kgf was applied for 30 seconds. The microstructure and graded layers were examined by using an optical microscopy (Leco, Olympus PMG3, St. Joseph, MI) and a scanning electron microscope (AMRAY 1830, Amray Inc., Bedford, MA).

Results and Discussions

X-ray diffraction

The as-fabricated samples are cut and polished. Phase analysis was carried out by X-ray diffraction method on each layer of FGMs produced by conventional combustion synthesis as shown in Figure 4 a-d. It was found that unreacted Ti is present as a matrix phase in all layers. Apparently as the boron content increases the formation of TiB and TiB$_2$ are promoted.

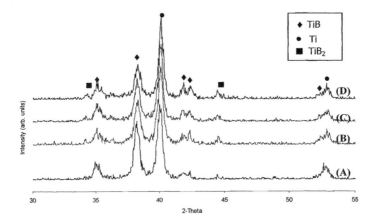

Figure 4. X-ray diffraction patterns **(a)** Ti-5 wt%B **(b)** 10 wt%B **(c)** 15 wt%B **(d)** 18 wt%B

Microstructure

Optical Microscopy Studies

In Figure 5a, optical micrograph of the interface of 1st and 2nd layers of the combustion-synthesized samples is shown. In Figure 5b, the same interface of the CS/DC produced sample is shown. Comparison of Figure 5a with 5b provides an interesting feature of the two types of experiment i.e. conventional combustion synthesis and combustion synthesis followed by dynamic compaction. A noticeable feature of the microstructure in the cross-section is that the particles are irregular in shape and elongated perpendicular to the direction of the application of pressure during processing.

Functionally Graded Materials 2000

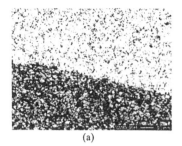 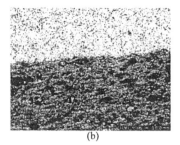

(a)   (b)

Figure 5.   Optical micrograph of the interfaces produced by
    (a) Conventional combustion synthesis
    (b) Combustion synthesis/dynamic compaction

Scanning Electron Microscopy (SEM) Studies

Figure 6 (a) and (b) presents the SEM micrographs of interfaces of samples produced by conventional combustion synthesis. Figure 6a, depicts the interface between the first and the second layers and Figure 6b shows the interface between the second and the third layers. As can be seen from the figures, the cross section consists of somewhat dense layers with a network of small pores and locally sintered areas. The cross sections show that there were no microcracks and the transition between the layers is good as shown in Figures 5-7.

 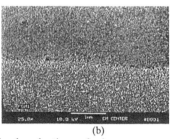

(a)   (b)

Figure 6. SEM of polished sample produced by conventional combustion synthesis
    (a) interface of the 1st and 2nd layers
    (b) interface of 2nd and 3rd layers

In Figure 7, the SEM images of the samples produced by combustion synthesis followed by dynamic compaction are shown. Figure 7a shows the interface region of 1st and 2nd layers. Figure 7b shows the interface between 2nd and 3rd layers.

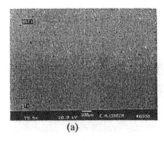
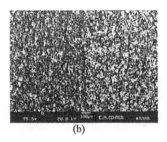

(a)   (b)

Figure 7. SEM of polished sample produced by combustion synthesis/dynamic compaction
(a) interface of the 1st and 2nd layers
(b) interface of 2nd and 3rd layers

The apparent porosity in these figures (Figure 6 and 7) is due to the particle pullout during metallographic preparation. Such phenomena were also observed in all layers. However, it was found that macropores (black regions) also existed in the reacted products. These macropores may be due to the remnant porosity from the green compact or caused by outgassing during the combustion reaction. In the combustion synthesized samples (Figure 6), the microstructure is showing irregular shaped pores in the range of 16-27 µm. On the other hand samples produced by CS/DC, the pore sizes are in the range of 7-18 µm. This indicates that the compaction energy was not sufficient enough to collapse these voids. Also due to the heat losses, the reaction temperature was lower when the combustion front reached these bottom layers in turn resulted in poor conversion and a large number of voids.

Mechanical Properties

In order to compare the effect of compaction on the mechanical properties, hardness tests were carried out on samples produced by both CS and CS/DC. Figures 8 shows the hardness data of the sample produced by conventional combustion synthesis. As the boron content increased, the hardness value is also slightly increased. A maximum hardness of 126 HVN was measured on the 5th layer.

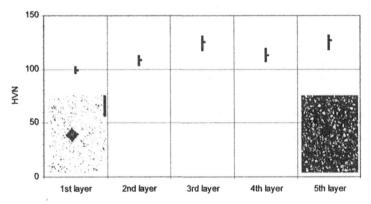

Figure 8. Microhardness values of combustion synthesized samples.

It was noticed that microhardness values were relatively low. These low values may be attributed to the presence of porosity. As was mentioned in the previous sections, due the nature of the CS, the product is usually porous. Therefore, a pressure application is necessary to form more dense products. Figure 9 shows the hardness profile of the FGM layers produced by CS/DC under a compaction pressure of 137 kPa. Similar to CS produced samples, the hardness increases with increasing boron content due to the formation of hard boride phases. It was evident that microhardness of the FGM layers is increased by the application of pressure after the CS indicating a more dense microstructure. Figure 10 shows the hardness data after the application of a pressure of 220 kPa. Increasing the pressure resulted in higher hardness values and a denser product. However, the fluctuations in the error bars quite big especially in the higher boron content layers. This might be indicating microstructural inhomogenities due to the remained porosity. The improvement in density and hardness even with compaction loads of 220 kPa is a promising result for the potential production of tough, dense, near net shaped FGMs.

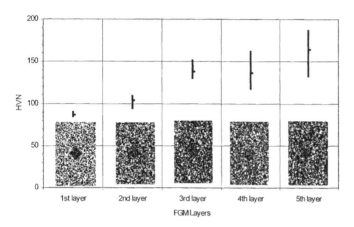

Figure 9. Microhardness values after the application of pressure of 137 kPa.

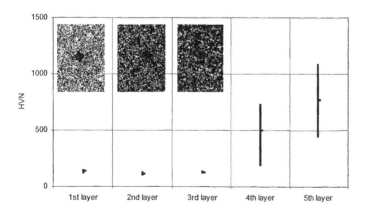

Figure 10. Microhardness values after the application of pressure of 220 kPa.

Conclusions

Our experiments showed that it is feasible to fabricate Ti/TiB/TiB$_2$ FGMs. Both conventional combustion synthesis and combustion synthesis/dynamic compaction technique was utilized as processing technique. Microstructural examination of the FGMs revealed crack-free interfaces. Application of pressure after combustion experiments has a considerable effect on improving the hardness. Further optimization of processing parameters is ongoing.

Acknowledgement

This work is funded by a grant from the Army Research Office, DAAG-55-9810281 with Dr. W.M. Mullins as the monitor.

References:

1. P.R. Marur and H.V. Tippur, *J. of Testing and Evalution*, 26, 6, (1998), 539-545.
2. A.J. Markworth, K.S. Ramesh and W.P. Parks Jr., *J. Mater. Sci.*, 30, (1995), 2183-2193.
3. D.P. Miller, J.J. Lannutti and R.D. Noebe, *J. Mater. Res.*, vol: 8, no: 8, (1993), 2004-2013.
4. W.A. Gooch, B.H.C. Chen, M.S. Burkins, R. Palicka, J. Rubin and R. Ravichandran, *Mater. Sci. Forum*, vols: 308-311, (1999), pp: 277-282, Trans. Tech. Publications, Switzerland, 614-621.
5. E.S.C. Chin, *Mater. Sci.and Eng.* A259, (1999), 155-161.
6. S.C. Deevi, *J. of Mater. Sci.* 26, (1991), 3343-3353.
7. J.J. Moore and H.J. Feng, *Progress in Mater. Sci.* vol: 39, (1995), 243-273 Elsevier Sci. Ltd.
8. H.C. Yi and J.J. Moore, *J. of Mater. Sci.*, 25, (1990), 1159-1168.
9. L.J. Kecskes, T. Kottke and A. Niiler, *J.Am.Ceram.Soc.*,73,5, (1990), 1274-1282.
10. J.C. LaSalvia, L.W.Meyer and M.A. Meyers, *J.Am.Ceram.Soc.*, 75, [3], (1992), 592-602.

Microwave Assisted Processing of Functionally Graded Composites in Ti-B Binary System

M. Cirakoglu, S. Bhaduri, and S.B. Bhaduri
University of Idaho
Department of Materials and Metallurgical Engineering
McClure Hall
Moscow, Idaho 83844-3024

ABSTRACT

Compositionally graded composites have been produced in the Ti-B binary system by using a microwave furnace. Elemental titanium and crystalline boron powders were used as raw materials. Several compositions in this binary system were chosen. Compositionally graded samples were prepared by stacking different compositions in discrete layers followed by uniaxial pressing and cold isostatic pressing. Samples were ignited in a microwave furnace at a power of 1.5 kW. A two-color pyrometer was used for measuring temperature. The phase composition of each layer was characterized by X-ray diffraction. Scanning electron microscope and optical microscope were used to examine the interfaces and graded layers. Hardness was determined by Vickers indentation technique.

Introduction

In recent years, there has been an increasing interest in applying microwave energy for processing of ceramics, glasses and composites. Compared to conventional heating, microwave processing requires less energy and has the potential to heat materials more uniformly [1-4]. In spite of that the microwave processing of ceramics is limited by the fact that many ceramics do not absorb microwaves efficiently. On the other hand, certain materials are known to be strong microwave couplers (e.g. SiC) which are used as "susceptors". By using susceptors, the samples, which initially do not couple efficiently by microwaves, can be heated. In this type of heating (hybrid heating) susceptor is used to conduct heat to the surface of the sample [5].

One of the important characteristics of hybrid microwave heating that it provides rapid and uniform heating. This is especially important for dissimilar materials such as ceramics and metals. A major problem with joining two dissimilar materials is the generation of thermal residual stresses around the interfaces. In order to overcome this problem, the concept of functionally graded materials (FGMs) was developed. In these materials, composition, microstructure and properties change gradually and as a result of this the residual and thermal stresses can be reduced and the bonding strength can be enhanced [6-8]. In the literature several studies on microwave processing of FGMs were reported. Borchert et al [7] achieved FGMs by using microwaves in the systems such as Al_2O_3/Mo and Al_2O_3/steel. In another study Porada et al [9] reported that a ZrO_2/metal FGM produced by infiltration by MW hybrid heating using SiC as a susceptor.

Microwave energy has also been used to initiate internal ignition in mixtures of exothermic compacts. In conventional combustion synthesis, one surface of the sample is exposed to an external heat source. On the other hand, microwave heating provides rapid and uniform heating. The energy is absorbed within the material and after the ignition, microwave energy and combustion reactions assist each other in sustaining the reaction. Since the heating is generated internally, the heating process is not heat transfer limited. As Clark et al. [3] reported, by using microwaves the dependence of combustion reaction on the thermal conductivity and density of the compact is greatly reduced. One other advantage of microwaves over the conventional synthesis is that further densification can be achieved simply by leaving the microwave power on after the reaction. The main objective of this present work is to explore the feasibility of microwave hybrid heating to fabricate compositionally graded ceramic/metal FGMs.

Experimental Procedure

Processing of FGMs

Elemental Ti (-325 mesh, purity: 99%, Johnson Mathey, Ward Hill, MI) and crystalline B (-325 mesh, purity > 99%, Cerac Inc. Milwaukee, WI) powders were used as raw materials. Powder premixes of different mixing ratios were prepared by ball milling for 3 hours in polyethylene bottles with alumina balls. The powder premix was stacked layer by layer in a steel die having a cavity diameter of 12.75 mm as shown in Figure 1a. The gradient consisted usually of 5 layers of approximate the same thickness with 0, 5, 10,15 and 18-wt % boron. Cylindrical samples with 12.75 mm diameter and 16.30-16.50 mm height were obtained after uniaxial pressing under a load of 15,000 lbs. The green compacts were placed in a tube furnace at 550°C for 3 hours under flowing argon gas for degassing.

Microwave experiments were conducted in an industrial microwave furnace (Microwave Materials Technology (MMT), Knoxville, TN). A strip of dense β-SiC was used as a susceptor. SiC, as described earlier, being a microwave absorbing material is used to thermally heat the poorly absorbing material.

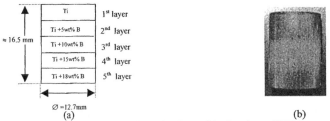

Figure 1. Schematic drawing (a) and a picture (b) of a 5-layered FGM green compact.

The experimental set-up is shown in Figure 2. The system has variable power output magnetron source capable of operating from 0 to 3 KW at 2.45 GHz. The samples were placed in a pyrex glass flask, which was first evacuated and then filled with Ar gas. The gap between the sample and glass flask was filled with an alumina fiber blanket. Some experiments were carried out in air. For the experiments carried out in air, sample was placed into a box made out of low density alumina fiber board and alumina blanket was used for thermal insulation. In this study the maximum power used was 1.5 kW and the power was shut down once the ignition started and combustion wave started to propagate. A two-color pyrometer (Mikron L77, Mikron Inc. Oakland, NJ) was used for temperature measurements (with a temperature range of 900-3000°C).

Characterization

The as-prepared samples were sectioned and mounted in thermosetting resin. Samples were initially polished using SiC papers from grade size 180 to 4000 and finally with alumina suspension from 5 μm to 0.01 μm. A solution of 10 % HF, 40 % HNO₃ and 60 % H₂O was used to etch the surface of the samples. The composition of each layer was characterized by X-ray diffraction technique (Siemens D5000, CuKα radiation). The room temperature hardness of each layer was determined by a Vickers testing machine. A load of 1 kg$_f$ was applied for 30 seconds. The microstructures of graded layers were examined by using an optical microscope (Leco,Olympus PMG3, St. Joseph, MI) and a scanning electron microscope (AMRAY 1830, Amray Inc., Bedford MA). Elemental analysis was performed using an energy dispersive spectroscopy (EDS) attached to the SEM.

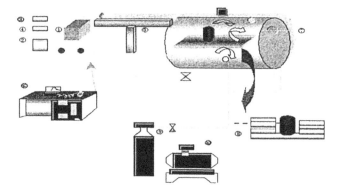

1) Magnetron 2) Main Controller 3) Forwarded Power Indicator 4) Reflected Power Indicator 5) Wave guide 6) Computer Control 7) Optical Pyrometer 8) Sample Insulation 9) Argon Tank 10) Vacuum Pump

Figure 2. A schematic of the experimental set-up.

Results and Discussion
Combustion Characteristics:

The investigation was divided into four categories. The results are summarized below.

(i) *Samples placed without a SiC susceptor :* In this case, samples were placed inside an alumina fiber board enclosure and surrounded with alumina blanket. Alumina fiberboard was chosen for its transparency to microwaves and its stability to withstand relatively high temperatures. The enclosure served for the purpose of containing the heat during the process. The front of the enclosure was covered with and alumina blanket. This assemblage was heated in air. In order to measure the temperature, a small hole was opened on the blanket. The sample temperature never exceeded 900°C (lower limit of the pyrometer used in our experiments).

(ii) *Samples placed on a SiC susceptor with thermal insulation:* Samples were placed on a SiC block, which acts as a susceptor material. Ti side of the FGMs was in contact with the SiC block and once the ignition temperature reached, the samples were ignited. By visual examination we observed that, the ignition started from bottom (Ti side) and propagated towards the top (high ceramic content side). Approximately, 4-5 minutes were required for samples of mass 6-7 grams to reach the ignition temperature in air. A maximum surface temperature of 1900°C was recorded by pyrometer. Upon reaching the top, the sample exploded due to the high exothermicity of the formation of reactions as shown in Figure 3 a.

(iii) *Samples placed on a SiC susceptor with reduced thermal insulation:* In an attempt to control the combustion experiment, we reduced the thermal insulation around the sample. In this arrangement, the thermal insulation around the sample was removed. SiC susceptor was placed on alumina blanket and the front opening in the enclosure was covered with a thin layer of alumina blanket with an enlarged hole to measure the temperature. A maximum temperature of 1180°C was measured and the samples retained their original shapes. The cross section of such sample is shown in Figure 3 b. The heat loss from the reaction front to the environment reduces the heat transfer to the adjacent layers. The maximum temperature recorded was much lower without proper insulation. Therefore, the results showed that the presence of thermal insulation substantially improves the resulting temperature.

(a)   (b)

Figure 3. (a) Photograph of the sample exploded during MW processing (b) By reducing the heat insulation, sample retained its original shape. The cross section of the sample after MW processing.

(iv)   *Samples placed into an Ar filled flask with SiC susceptor and thermal insulation:* To determine the effect of atmosphere on combustion, we have carried out several experiments in argon filled flask. Similar to the experiments carried out in air, a strip of SiC block was used as susceptor and alumina fiber was used for thermal insulation. In these experiments, the main problem was arcing during processing. As is well known, the processing atmosphere (Ar in this case) interacts with the microwave field. Therefore, if a non-oxidizing atmosphere is required, in order to prevent problems associated with arcing nitrogen is often used instead of argon [10]. But in our case, in order to prevent TiN formation, we tried to avoid using nitrogen gas. During our experiments, we observed intense arcing and tiny sparks appeared on the sample surface, which can be attributed to local electric field concentration around surface defects and rough spots. Some of the earlier works [6,11], showed that arcing and the breakdown of gas disturbs the electric field and degrades coupling and as a result of this the sample temperature drops. Under these conditions approximately 11-12 minutes were required for samples of mass 6-7 grams to reach the ignition temperature. The temperature was measured through a small hole on the alumina blanket and a maximum temperature of 1160°C was recorded which was 740°C lower compared to air.

It must be noted here that, these temperatures correspond to the outside temperatures of the samples. As it was pointed out by Clark et al [1] unlike conventional heating microwave heating provides an inverted temperature profile with the highest temperature in the center of the sample. Therefore, due to the heat losses to the surrounding, the measured temperatures do not represent bulk internal temperatures.

Phase Analysis

X-ray diffraction studies were carried out on the individual FGM layers. The patterns of the samples ignited in argon atmosphere and in air are shown in Figure 4 and 5, respectively.

The first (Ti), second (95Ti-5B) and third (90Ti-10B) layers contained mainly Ti. No boride phase was detected by X ray diffraction. Fourth (85Ti-15B) and fifth (82Ti-18B) layers consisted of TiB and TiB$_2$ along with unreacted Ti. The fifth layer contained an increased amount of TiB and TiB$_2$ as expected. The presence of unreacted titanium indicates that despite the combustion reaction, the conversion to titanium boride was not fully completed. This suggests that the temperature did not reach adiabatic temperature due to heat losses. The adiabatic temperature for the formation of TiB$_2$ phase is 3190K. However, the temperature of the specimen would never rise to such a high valve and therefore the amount of TiB$_2$ phase forming was very small.

The entire outer surface of the samples, which were ignited in air, was covered by a layer (about 0.03 mm). As expected the layer was identified as TiO$_2$ by X-ray diffraction. On the other hand, the cross section shows a considerable amount of TiN as a reaction product along with titanium borides. This might be due to the reaction between titanium and air trapped in the pores of the sample. The formation of TiO$_2$ coating on the surface prevented further oxidation.

278   Functionally Graded Materials 2000

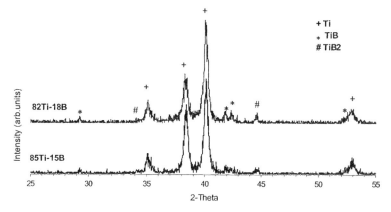

Figure 4. X-ray diffraction patterns of two layers of FGMs ignited in argon.

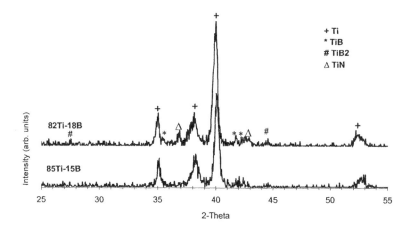

Figure 5. X-ray diffraction patterns of two layers of FGMs ignited in air.

Microstructural Analysis

Both SEM and optical microscopy analysis were carried out on the polished and etched samples. The position of the interfaces can be roughly recognized especially for the fourth and fifth interfaces. Figure 6 is an optical microscope picture showing the interface between the first and the second layers. Over the entire cross-section, the interfaces were continuous and crack-free.

In Figure 7, surface structures of Ti layer of the FGMs ignited in air (a) and in argon (b) are shown. As the figures indicate, the surface consists of dense, locally sintered areas with areas of small pores.

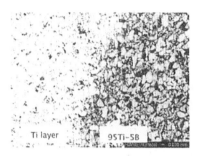

Figure 6. Optical micrograph showing the interface between Ti layer and Ti-5wt% B layer (magnification: x200)

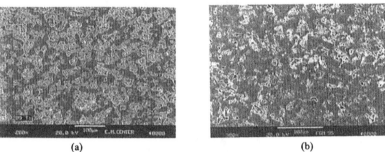

(a)   (b)

Figure 7. SEM image of Ti layer of FGM ignited in air (a) and in argon (b)

The apparent porosity evident in these figures is due to the particle pull-out during metallographic preparation. A similar phenomenon was also observed for the other layers. The microstructure of the FGM produced in argon 7b, indicates a considerable contact between Ti particles.

In Figure 8, the microstructures of fifth layer of the FGMs ignited in air (a) and in argon (b) are shown. A typical EDS spectra obtained from the marked particles is given in Figure 8c. The layer exhibits a microstructure consisting essentially of titanium boride particles surrounded by Ti matrix for both cases. However, in argon atmosphere, the layer contains a higher volume fraction of titanium boride phase.

The EDS analysis has identified the gray particles as Ti and darker particles (indicated with a cross) as Ti and B, as shown in Figure 8c. This indicates the formation of titanium boride. Quantitative identification by EDS has been attempted but was not successful because B gives low levels of X-rays. Therefore, we could not determine the exact chemical formula of this boride phase. As minor elements O, Al and Si were also detected. These are mainly from the polishing stage since we used alumina suspension and SiC abrasive papers.

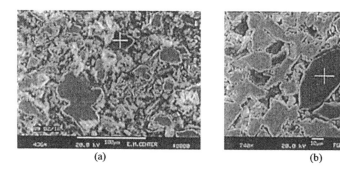

(a)   (b)

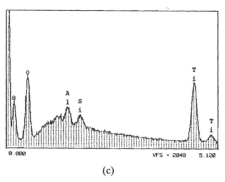

(c)

Figure 8. SEM images of the 5th layers of the FGMs (a) ignited in air (b) ignited in argon (c) EDS spectra

Mechanical Property Determination

A number of hardness measurements were conducted on the individual layers. In Figure 9 and 10, the hardness test results were shown for the FGM samples ignited in argon atmosphere and in air. Increasing the content of boron in the mixture slightly increases hardness. The increased hardness resulted from the formation of borides as revealed by the X-ray diffraction analysis. However, the influence of the boride phases on the hardness values was small because of the small volume fraction of these phases. The samples ignited in air have slightly higher hardness values.

Conclusions

Microwave hybrid heating was used in order to initiate combustion reactions in compositionally graded composites. Combustion experiments carried out under Ar atmosphere resulted in TiB and TiB_2 formation on the other hand under air TiN formation was observed along with borides. Microstructural examination revealed that over the entire cross section, the interfaces were continuous and crack free. The hardness tests showed no considerable variations in hardness values between air and argon atmosphere. Further works is in progress for characterization of the interfaces and improve the microstructure and mechanical properties.

Acknowledgement
This work is funded by a grant from the Army Research Office, DAAG-55-9810281 with Dr. W.M. Mullins as the monitor.

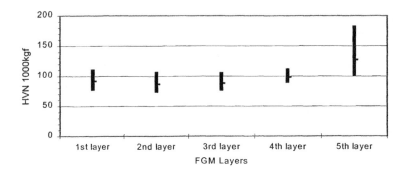

Figure 9. Vickers hardness data of the individual layers in FGM ignited in argon.

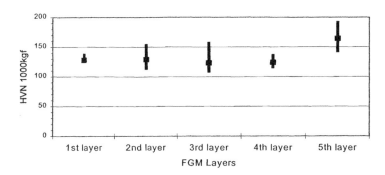

Figure 10. Vickers hardness data of the individual layers in FGM ignited in air.

References

1. K.S. Leiser and D.E. Clark, *Adv. Ceram., Mater., and Structures: A, Ceramic Science and Science Proc.* 23rd Ann. Conf. On Composites, vol: 20, Issue: 3 (1999), 103-109.
2. D. Atong and D.E. Clark, Advanced *Ceramics, Materials, and Structures:A,Ceramic Science and Science Proc.* 23rd Annual Conf. On Composites, vol:20, Issue:3 (1999), 111-118.
3. D.E. Clark, I. Ahmad, and R.C. Dalton, *Mat. Sci and Eng.*A144 (1991) 91-97.
4. B. Vaidhyanathan, D.K. Agrawal, and R. Roy, *J.Mater. Res.*, 15, 4, 2000, 974-981.
5. S. Komarneni, E. Breval, and R. Roy, *Mat. Res. Soc. Symp. Proc*, 124, 1988, 235-238.
6. M. Samandi, and M. Doroudian, *Mat. Res. Soc. Symp. Proc*, 347, 1994, 605-615.
7. R. Borchert, M. Willert-Porada Microwave Theory and Applications in Material Processing IV, Ceram. Trans. Vol: 80, (1997), 491-498.
8. I. Ahmad, R. Silberglitt, W.M. Black, H.Sa'adaldin, and J.D. Katz *Mat. Res. Soc. Symp. Proc,* 269, 1992, 271-276.
9. M.Willert-Porada, T. Gerdes, S. Vodegel *Mat. Res. Soc. Symp. Proc,* 269, 1992, 205-2210.
10. M.A. Janney, H.D. Kimrey, and J.O. Kiggans, *Mat. Res. Soc. Symp. Proc.*,269, 1992, 173-185.
11. Y. Tian, M.E. Brodwin, H.S. Dewan, and D.L. Johnson, *Mat. Res. Soc. Symp. Proc.*, 124, 1998, 213-218.

DEVELOPMENT OF AUTOMATIC FGM MANUFACTURING SYSTEMS BY THE SPARK PLASMA SINTERING (SPS) METHOD

Masao Tokita
Sumitomo Coal Mining Company, Ltd.
208 East KSP, Kanagawa Science Park
3-2-1 Sakato, Takatsu-ku, Kawasaki
Kanagawa, 213-0012 Japan

KEYWORDS: Automatic FGM manufacturing systems, Spark Plasma Sintering (SPS), zirconia/stainless steel FGM/LB, heat resistance

ABSTRACT
The goal of the research is to develop Large-size Ceramic/Metal Bulk Functionally Graded Materials (FGM/LB), featuring thermal stress relaxation, high heat-resistance, high wear-resistance, and high mechanical strength for high efficiency engine component applications and production type Spark Plasma Sintering (SPS) Systems for FGMs. The approach focused on the use of the SPS method, one of the pressurized pulsed current sintering technologies, for FGM/LB fabrication utilizing a Large Experimental SPS Apparatus, Automatic Powder Stacking Equipment and a Fully Automated FGM/LB Manufacturing System.

1. INTRODUCTION
Spark plasma sintering (SPS) is a pressure assisted pulsed electric current sintering process that has recently been attracting attention [1, 2]. It provides excellent features for use in creating new materials such as composite ceramics and FGMs. Recently disk-shaped sintered compacts with diameters of 100 and 150 mm, and thicknesses of approximately 15 and 17 mm $ZrO_2(3Y)$/stainless steel FGM were homogenously consolidated in a shorter sintering time, while maintaining high quality and repeatability within one hour.

The National Institute of Standards and Technology (NIST) in the United States and Mechanical Engineering Laboratory (MEL), Ministry of International Trade and Industry (MITI) in Japan have been carrying out a 4-year joint research project, from fiscal year 1996 through March 2000, to synthesize and evaluate large-size ceramic/metal bulk functionally graded materials (FGM/LB) using a spark plasma sintering technique. In Japan, Sumitomo Coal Mining Co., Ltd. (Izumi Technology Co., Ltd.) and Yanmar Diesel Engine Co. Ltd. have joined the project [3]. Potential applications for this FGM/LB could provide new wear-resistant materials that can withstand high temperature environments and which are suitable for high efficiency engine materials, ceramic turbine components and mold & die materials for industrial use [4, 5]. Research goals include: (1) Development of Heat-resistant FGM/LB Processing Technology, (2) development of FGM/LB Manufacturing Systems, (3) establishment of FGM/LB Evaluation Technology.

This paper introduces SPS process principals, the development of fully automated FGM/LB manufacturing systems and FGM/LB processing technology.

2. MECHANISM OF SPS PROCESSING

Conventional electrical hot press sintering processes use DC or commercial AC power, and the main factors promoting sintering in these processes are the Joule heat generated by the power supply (I^2R) and the plastic flow of materials due to the application of pressure. However, the SPS process is an electrical sintering technique which directly applies an ON-OFF DC pulse voltage and current from a special pulse generator to a powder of particles. It is regarded as a rapid sintering method, using the self-heating action from inside the powder, similar to self-propagating high temperature synthesis (SHS) and microwave sintering. Fig. 1 shows a typical ON/OFF DC pulsed current path using conductive powder materials and a sintering die and punch made of graphite. Fig. 2 illustrates how pulse current flows through the powder particles inside the SPS sintering die. In addition to the factors promoting sintering described above, also the effective discharge between particles of powder occurring at the initial stage of the pulse energizing. When electrical discharges occur, a high temperature sputtering phenomenon generated by spark plasma and spark impact pressure eliminate adsorptive gas and impurities existing on the surface of the powder particles. The action of the electrical field causes high-speed diffusion due to the high-speed migration of ions.

These effects were summarized as follows: The ON-OFF DC pulse energizing method generates (1) spark plasma, (2) spark impact pressure, (3) Joule heating, and (4) an electrical field diffusion effect. In the SPS process, the powder particle surfaces are more easily purified and activated than in conventional electrical sintering processes and material transfers at both the micro and macro levels are promoted, so a high-quality sintered compact is obtained at a lower temperature and in a shorter time than with conventional processes.

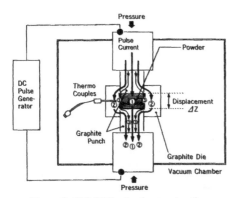
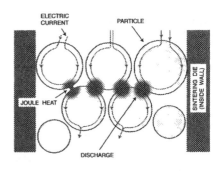

Figure 1: ON-OFF pulsed current path

Figure 2: Pulsed current flow through powder particles

3. AUTOMATIC FGM MANUFACTURING SYSTEMS

The world's first tunnel type automatic FGM/LB manufacturing system was developed in 1999 targeting the optimization and practical implementation of the sintering process for large FGMs with diameters of 100 and 150 mm. Although fabrication of FGMs usually takes many hours, with the development of this new automated SPS system it is now possible to process such materials in a significantly shorter period of time—within 1 hour or less by making use of the features that are inherent in SPS rapid sintering technology.

3.1 System Configuration

The fully automated FGM/LB manufacturing system is composed of a DC pulse power supply, FGM/LB manufacturing system and their hydraulic units and control panels. The FGM/LB manufacturing unit is composed of an automatic FGM powder stacking stage, a tunnel type SPS system atmospheric preheating stage, SPS sintering and FGM cooling stages, an automatic die-releasing stage, an automatic material conveyer system and an automated control system as shown in Fig. 3. All are modular on a per-stage unit basis. Fig. 4 shows an outside view of the fully automated FGM/LB manufacturing system.

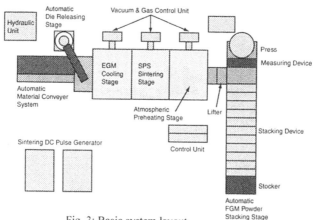

Fig. 3: Basic system layout

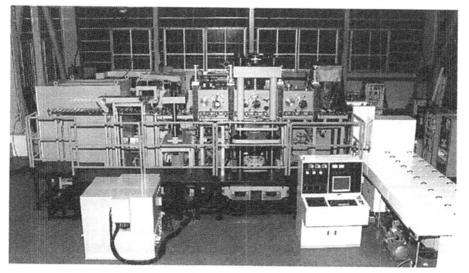

Fig. 4: Outside view of a Fully Automated FGM/LB Manufacturing System

3.2 Automatic FGM Powder Stacking Equipment

Two types of automatic materials handling robot systems and FGM powder stacking systems for the spark plasma sintering (SPS) process were developed for manufacturing of large-size ceramic/metal bulk FGMs. To obtain a uniform stacked powder thickness over a large area, FGM powder stacking and lamination are performed by the volume method. As shown in Fig. 5, this method levels the stacked powder surface in the graphite die cavity with the bottom of the container. Figs. 6 and 7 show stand alone type Automatic Powder Stacking Equipment with a rotary table for FGM/LB and a large-size Spark Plasma Sintering System for experimental use. The sintering graphite die and punch are controlled by a robot mounted on the left side of the Sintering Machine. The modular configuration uses material powder containers of different volume fractions arranged in a line for fully automated FGM manufacturing systems.

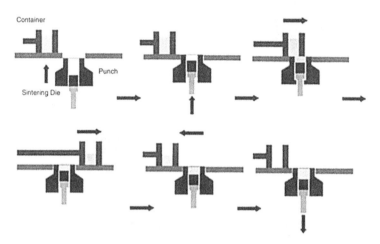

Fig. 5: Principles of automatic FGM powder stacking

| Fig. 6: Automatic Powder Stacking Equipment with rotary table | Fig. 7: 3,000 KN large-size Spark Plasma Sintering System |

3.3 Fully Automated FGM/LB Manufacturing Systems

The tunnel type SPS system consists of automatic FGM powder stacking, preheating, SPS sintering, cooling and die releasing stages. Each of the chambers of the atmosphere control, SPS sintering and FGM cooling stages has an independent vacuum exhaust system and gas introduction system, so that these stages can be used independently in separate inactive gas atmospheres. A gate valve is installed between the chambers.

3.3.1 Atmospheric Preheating Stage

The atmospheric preheating stage has two roles; the first is for the atmospheric preheating adjustment of the sintering chamber that regulates the pressures of the SPS sintering and atmospheric chambers equally, opens the gate valve between the chambers and feeds the sintering die and punch on the carrying tray. The second role is accepting the carrying trays from the FGM powder stacking stage. The atmospheric preheating chamber can also be used as the preheating stage for the sintering die and punch.

3.3.2 SPS Sintering Stage

The SPS sintering stage processes the powder stacked in the FGM sintering die and punch by means of pressurization at a maximum of 2MN and a pulse power supply at a maximum of 20,000A. This mechanism corresponds to the sintering machine in a conventional SPS system. The extremities of the electrodes are integrated with graphite spacers. The window for visual observation of the SPS has a 150 mm, large-diameter wiper/shutter mechanism and is attached to the main body using a removable flange system to improve ease of maintenance. This visual observation window design is also used with the atmospheric chamber and FGM cooling chamber.

3.3.3 FGM Cooling Stage

The FGM cooling stage cools the sintering die, punch and FGM sintered compact after SPS energizing. The temperature of the graphite die and punch material and FGM compact are decreased to a temperature at which the material can be taken out into the atmosphere. The sintering die, punch and sintered compact can be cooled either by releasing argon gas or by heat transfer of a high thermal conductivity material such as copper or aluminum.

3.3.4 Automatic Die Releasing Stage

At the position where the sintering die, punch and FGM sintered compact are separated, the carrying tray is stopped at the home position stopper, then a push rod driven by a hydraulic cylinder is raised from below to separate the sintering die and punch and the FGM sintered compact. The FGM compact pushed up from the sintering die and punch is clamped by a hand driven air cylinder and placed on the slat conveyer.

4. FGM/LB FABRICATION AND EVALUATION

Generally, dimensional size and shape effects of SPS tend to fluctuate in hardness around the outer portion. The degradation increases as the size of the specimen becomes larger. Utilizing advanced SPS Systems, the ϕ100 mm and ϕ150 mm ZrO_2(3Y)/SUS410L stainless steel FGM/LB were fabricated homogeneously, free of cracks, delaminations or distortion.

4.1 Homogenization of $\phi 50$ mm ZrO₂(3Y)/Stainless Steel FGM/LB

This investigation was carried out as a preliminary test to obtain the optimum sintering conditions. As shown in Fig.8, a 3 mol% yttrium partially stabilized zirconia (PSZ) powder, SUS410L stainless steel (SUS) powder and their mixed powders as intermediate layers were stacked on a graphite temperature gradient die of 50 mm internal diameter. Table 1 shows the layer composition and thickness of each layer after sintering. The SUS powder has an average particle size of 9μm while the PSZ powder is of granulated particles with an average particle size of 50μm (its crystalline size is $350\,\text{Å}$). Sintering pressures used were from 20 to 40 MPa, SPS temperature of 1243 or 1293K, with a temperature rise rate of 50K/min. The temperature near the SUS layer expressed the sintering temperature. Sintering was performed in a vacuum of about 10 Pa. Fig. 9 shows the relationships between the sintering pressures at SPS temperatures of 1293K and hardness distribution in the PSZ layer. The FGM specimens sintered with a sintering pressure of 20 MPa presented degradation in hardness in the lower parts of the PSZ layers, the drop in hardness being more severe in the areas near the perimeter. In contrast, the FGM specimens sintered at pressures of 30 MPa or more presented hardly any difference in hardness.

Table 1: Composition and thickness
of FGM layers

Layer No.	Layer Composition(vol %)		Thickness After Sintering (mm)
	SUS410L	PSZ	
1	0	100	3
2	10	90	1
3	20	80	1
4	30	70	1
5	40	60	1
6	50	50	1
7	60	40	1
8	70	30	1
9	80	20	1
10	90	10	1
11	100	0	3

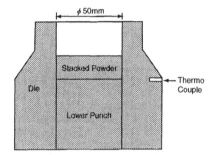

Fig. 8: Graphite temperature gradient
die and punch

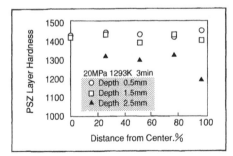

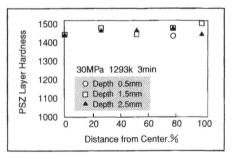

Fig. 9: Relationships between the sintering pressure and PSZ layer hardness distribution at 1293K

4.2 φ150 mm FGM/LB

ZrO_2(3Y) submicron-size partially stabilized zirconia powder containing 3 mol% Y_2O_3 and SUS410L stainless steel powder with an average particle size of 9 μm and 60 μm were used as starting materials in the study. The composition of the stainless steel was 0.03 wt% C, 0.89 wt% Si, 0.13 wt% Mn, 0.011 wt% P, 0.012 wt% S, 0.09 wt% Ni and 13.06 wt% Cr. By stacking nine kinds of mixed-composition powders with ZrO_2(3Y) stainless steel ratios of 90/10, 80/20, 70/30, 60/40, 50/50, 40/60, 30/70, 20/80 and 10/90 vol. % as interlayers between the 100% front and back layers, a total of 11 gradient layers were laminated. Fig. 10 shows the laminating conditions of φ 150 mm FGM/LB by Automatic FGM Powder Stacking Equipment. As the result of the optimization of SPS conditions, it is now possible with the SPS method to fabricate in 54 minutes, including heating up time, holding time and cooling time, a fine ZrO_2(3Y)/stainless steel large-size FGM/LB with a diameter of 150 mm and thickness of 15 mm having 9 interlayers that are free of cracks, distortion and delamination as shown in Fig.11.

Fig. 10: Laminating conditions of
Automatic FGM Powder
Stacking Equipment

Fig. 11: φ150 mm ZrO_2(3Y)/stainless
steel (SUS410L) FGM/LB
sintered compact

5. PROSPECTS FOR SPS SINTERED FGMS

If SPS fabricated FGMs are to become practical, it is of vital importance to develop the technology to form a 3D Near-Net-Shape as shown in Fig. 12. Such technology is indispensable in significantly reducing the processing steps and overall cost of the manufacture of materials for engine components.

Moreover, large-size bulk FGM/LB manufacturing technology has wide-ranging applications in fields other than automobile materials. For example, the FGM/LB has shown a strong potential for stamping die and cutting tool markets. As shown in Fig. 13, WC/Co compositionally graded cemented carbides make possible new FGM hard alloys having both high hardness and high toughness. Along with the scaling-up of size, such SPS compacts are suitable for the manufacture of sputtering target materials, large-size high-quality fine ceramics, and nano-phase bulk materials for related industries and it also makes possible the expansion of applications for the electric vehicle and environmental recycling sectors. Therefore, in the near future, the manufacturing technology of FGMs using the SPS method will open up new markets for advanced materials.

Figure 12: Example of 3 dimensional
Near-Net-Shape forming by SPS (right)

Figure 13: Examples of FGM hard alloys
fabricated by SPS

6. SUMMARY

Development of large-size ceramic/metal bulk FGM (FGM/LB) using the Spark Plasma Sintering (SPS) method have been briefly described. And, the multi-layer composition, optimum sintering conditions and the manufacturing machine systems were investigated. The study resulted in: (1) Development of the world's first fully automated FGM/LB manufacturing system targeting the optimization and practical implementation of the sintering process for FGM/LB with diameters of 100 and 150 mm, (2) the ϕ 150 mm ZrO2(3Y) stainless steel (SUS410L) FGM/LB was sintered with a fully automated FGM system without any cracks or distortion. Use of this automated SPS system made possible the automatic fabrication of a 150 mm diameter, 15 mm thick ZrO2(3Y)/stainless steel FGM/LB having 11 gradient layers every 54 minutes (50 minutes heating-up time and 4 minutes holding time).

To meet industrial needs such as complex shape and properties including 3-dimensional gradient FGM components, however, further development of automatic SPS manufacturing systems will be required.

ACKNOWLEDGEMENT

This work was performed as a part of the "Development of Functionally Graded Materials" Project sponsored by NEDO.

REFERENCE

[1] Tokita M., (1993). Journal of the Society of Powder Technology Japan, 30, 11 pp. 790-804.

[2] Tokita M., (1997). Microwave, Plasma and Thermochemical Processing of Advanced Materials, Proceedings of the International Symposium pp. 69-76.

[3] Enomoto Y., Ichikawa K., Tokita M., Dapkunas S. J. and Smith D. T., (1998) Proc. "Cooperating Internationally: The U.S.-Japan Civil Industrial Tech. Arrangement" Mich. pp. 8-1.

[4] Tokita, M., (1998) Development of Large-Size Ceramic/Metal Bulk FGM Fabricated by Spark Plasma Sintering, Proceedings of the 5th International Symposium on Functionally Graded Materials FGM 98, Dresden, Germany. pp. 83-88.

[5] Tokita, M., (1999) Trends in Advanced SPS Systems and FGM Technology, Proceedings of NEDO International Symposium on Functionally Graded Materials, Tokyo, Japan. pp. 23-33.

DESIGN AND FABRICATION OF SiC/C FGM

Wenbin Cao, Anhua Wu, Jiangtao Li and Changchun Ge
Laboratory of Special Ceramics & P/M, University of Science and Technology Beijing,
Beijing 100083, P. R. China

ABSTRACT

Silicon carbide(SiC)/graphite(C) functionally graded materials (FGM) with dimension of ϕ40mm×6mm were fabricated by hot pressing (HP). Boron and carbide additions were used to densify the SiC and SiC+x·C powders. Residual thermal stress caused by thermal mismatch of graphite and SiC was calculated by finite element method (FEM). Thermal stress distribution was optimized through changing the compositional distribution between SiC and C. Two kinds of SiC/C FGMs were fabricated successfully based upon the theoretical optimization of compositional distribution. Microstructures of fabricated SiC/C FGMs were observed by scanning electron microscopy (SEM). Thermal shock resistance was tested by the water quenching method.

1. INTRODUCTION

Silicon carbide (SiC) is a highly covalently bonded compound. Therefore, it has good properties of high melting point, high mechanical strength and chemical erosion resistance. But SiC ceramic has relatively low thermal conductivity, while graphite, with layered structure, has a higher thermal conductivity but lower mechanical strength, which limits it's structural applications. A new composite, SiC/C FGM, will be composed with good thermal conductivity and better mechanical strength if these two materials could be combined together through FGM concept [1].

In this study, the design and fabrication of SiC/C FGMs were investigated. The residual thermal stress caused by thermal mismatch of graphite and SiC was analyzed by using finite element method (FEM). The residual thermal stress was optimized through changing the compositional distribution between SiC and graphite. Boron and carbide additions were used to densify SiC and SiC+x·C powders. SiC/C functionally graded materials (FGM) with dimension of ϕ40mm×6mm were successfully fabricated by hot pressing (HP). The microstructures of fabricated SiC/C FGMs were observed by SEM. Thermal shock resistance was tested by the water quenching method.

2. ANALYSIS OF RESIDUAL THERMAL STRESS IN SiC/C FGM [2]

Geometric model of SiC/C FGM is schematically illustrated in Fig. 1. The model is fully graphite(C) at the bottom surface and changes to fully silicon carbide (SiC) at the top surface. A commercial finite element code (Algor) was used to analyze the macro thermal residual stresses in the graded structures [3]. In order to simulate the thermal stress by FEM, it was assumed that only elastic deformation would be generated in SiC/C FGMs. Half of the cross section was used for FEM analysis. The modeled configurations of the Non-FGM and FGM specimen are shown in Figure 2. Near the side surface and the interface between layers the meshes were refined to obtain better accuracy [4]. An axial symmetric boundary condition was applied on the boundaries at $r=0$. Axial symmetrical elements were employed in the computation assuming a reference temperature of 1950℃ and a stress calculation temperature of 25℃. FEM was carried out by using the following values of Young's modulus E, Poisson's ratio ν, thermal expansion coefficient α, compressive strength σ_c and tensile strength σ_t, $E_{SiC}=430GPa$, $\nu_{SiC}=0.19$, $\alpha_{SiC}=4.7\times10^{-6}$, $\sigma_{c, SiC}$ =1300MPa, $\sigma_{t, SiC}=400MPa$; $E_C=11.7GPa$, $\nu_C=0.15$, $\alpha_C=3.0\times10^{-6}$, $\sigma_{c, C}=50MPa$, $\sigma_{t, C}=12.5MPa$, where the above physical properties were mean values from different literatures [5]. Young's moduli, Poisson's ratios, thermal expansion coefficients and strength of intermediate layers, SiC+x·C, were obtained by the composite mixture rule.

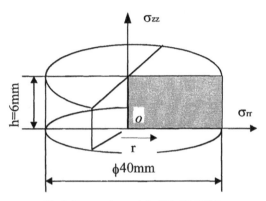

Fig.1 Geometric model of SiC/C FGM

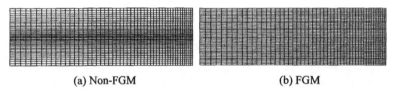

(a) Non-FGM (b) FGM

Fig. 2 Mesh geometry for finite element analysis: (a) Non FGM; (b) FGM

2.1 Effects of the number of FGM layers on the residual thermal stress

Residual thermal stresses of stepwise SiC/C FGMs were calculated for different layers ranging from 2 to 12 layers. Fig. 3(a) shows the relationship between maximum residual thermal stress and the number of FGMs layers. The maximum thermal stress tensors were decreased with increasing the number of FGM layers, except the case of direct bonded SiC-C non-FGM. It is very

Functionally Graded Materials 2000

interesting that maximum thermal stress in SiC-C non-FGM is less than that in 3-layered SiC/C FGM. This is because that Young's modulus of SiC is one order higher than that of graphite. This large difference of Young's modulus results in quite different thermal stress distribution from the normal cases. Generally speaking, FGMs, which are composed of more layers, the less thermal stress will be generated due to thermal mismatch between different layers.

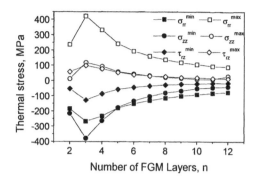

(a) maximum thermal stress

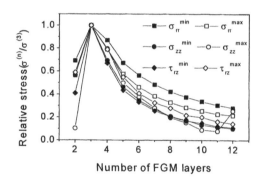

(b) Relative thermal stress

Fig. 3 Relationship between maximum residual thermal stress and the number of FGMs layers

Fig.3 (b) shows the relationship between relative stress and number of FGM layers. Thermal stress of 3-layered FGMs is set as the reference. The longitudinal value, $\sigma^{(n)}/\sigma^{(3)}$, is the ratio of each maximum thermal stress tensor in FGMs to that of 3-layered FGM. The maximum thermal stress in FGMs was greatly relaxed with increasing the number of FGM layers. When number of FGM layers is increased to 7, the maximum stress is about 50% percent of that in 3-layered SiC/C FGMs. The number of FGM layers is one of the main factors that will affect the thermal stress distribution.

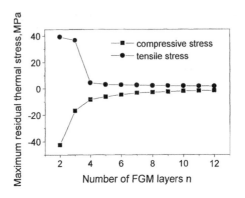

(a) maximum thermal stress

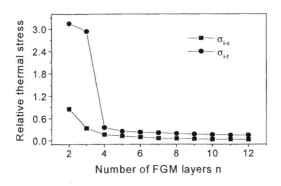

(b) stress- strength ratio σ_{i-c} and σ_{i-c}

Fig. 4 Relationship between thermal stress in graphite layer and number of FGM layers

As graphite is the weak phase in SiC/C composite with lower mechanical strength, it is important that the value of maximum stress must be lower than that of graphite strength. Otherwise, the graphite will be broken due to thermal stress.

Fig. 4 (a) shows the maximum thermal stress distributed in graphite of SiC/C FGMs with different layers. The value of maximum tensile stress exceeds the tensile strength of graphite at the case of $n=2$ and $n=3$. Fig. 4(b) shows the relationship between the value of maximum thermal stress-strength ratio σ_{i-t}, σ_{i-c} in graphite layers and number of FGM layers, respectively. σ_{i-t} is defined as the ratio of maximum tensile stress in graphite layer to tensile strength of graphite. σ_{i-c} is defined as the ratio of maximum absolute value of compressive stress to compressive strength of graphite. Both σ_{i-t} and σ_{i-c} show the resistance of graphite to the tensile stress and compressive stress. If $\sigma_{i-t} > 1$ or $\sigma_{i-c} > 1$, it is regarded that the stress exceed the maximum resistance of graphite to the thermal stress. At this case, at least graphite layer of SiC/C FGM will be broken due to residual thermal stress and can't be fabricated.

Functionally Graded Materials 2000

It is clear that thermal stress in graphite layer is greatly decreased with the increasing of the number of FGM layers in Fig. 4. The value of tensile stress in graphite layer is decreased from about 3.2 times to almost half of the value of the tensile strength for non-FGM SiC-C to 4-layered SiC/C FGMs. At the same time the absolute value of compressive stress in graphite layer is always less than compressive strength of graphite. From non-FGM to 4-layered SiC/C FGMs, σ_{i-c} is decreased from about 0.90 to about 0.2. So, if the number of FGM layers is over 4, the strength of graphite is high enough to resist the thermal stress.

It can be determined from the above calculation that the number of FGM layers should be $n \geq 7$ when fabricating this kind of FGM.

2.2 Effect of the thickness of surface layers on the residual thermal stress distribution in FGMs

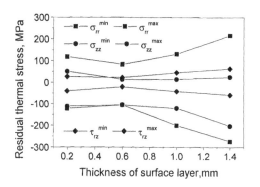

(a) maximum thermal stress

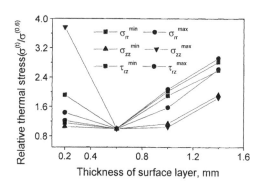

(b) relative thermal stress
Fig.5 Relationship between stress and thickness of surface layer t

Thermal stress in FGMs with different surface thickness was calculated assuming both thickness of top and bottom surfaces are the same. Fig.5 shows effects of the surface thickness on thermal stress in SiC/C FGMs. The maximum thermal stress in the SiC/C FGMs with surface thickness $t_{surface}$=0.6mm are set as the reference. The longitudinal value is the ratio of maximum thermal stress in those FGMs with different surface thickness to those of the reference. It is clear that the maximum thermal stress in FGM with surface thickness $t_{surface}$=0.6 is less than that in other FGMs. So, the thickness of surface was tailored as $t_{surface}$=0.6mm.

2.3 Effect of compositional distribution factor p on the thermal stress distribution in FGMs

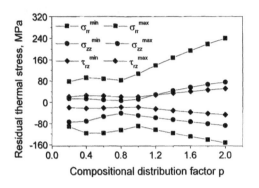

(a) maximum thermal stress

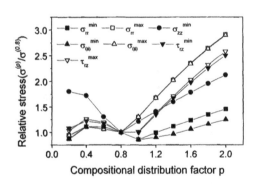

(b) relative thermal stress
Fig.6 Relationship between stress and compositional distribution factor p

Thermal stress in FGMs with different compositional distribution factor p was calculated. Fig. 6(a) shows that the maximum thermal stress varies with the variation of compositional distribution factor p. Fig. 6(b) shows relationship between relative stress $\sigma^{(p)}/\sigma^{(8)}$ and p. The maximum thermal stress tensors of FGM with p=0.8 are set as the reference. The longitudinal value is the ratio of maximum thermal stress in FGM with different factor p to that of FGM with p=0.8. It is found that most of the stress tensors are large than that of FGM with p=0.8, and only σ_{rr}^{min} and $\sigma_{\theta\theta}^{min}$ reaches the minimum value when p=1.0. But the difference of σ_{rr}^{min} and $\sigma_{\theta\theta}^{min}$ at p=0.8 and p=1.0 are very little as indicated in Fig. 6(a). So, p=0.8 is the optimum compositional distribution factor.

In general, the optimum parameters for fabricating SiC/C FGMs are: $n{\geq}7$, $t_{surface}$=0.6mm and p=0.8.

3. FABRICATION OF SiC/C FGMS

3.1 Experimental
The starting materials used were silicon carbide, amorphous carbon and boron powders with an average particle size of 0.05μm, 2.5μm and 1μm, respectively. The powders were wet mixed in predetermined compositions by ball milling for 6 hours, and then dried in air.

Two specimens, T4 and T7 were made according to the above analysis to check the simulation of thermal stress distribution in FGMs. Both T4, which is 4-layered FGM and T7, which is 7-layered FGM, have same compositional distribution factor p=0.8 and surface thickness t=0.6mm. The composition for T4 and T7 are:

T4: 100%C+(60%C+40%SiC)+(20%C+80%SiC)+100%SiC
T7: 100%C+(60%C+40%SiC)+(40%C+60%SiC)+(20%C+80%SiC)+
(10%C+90%SiC)+(5%C+95%SiC)+100SiC

The green compacts was put into a graphite die and sintered at a condition of 1950℃, 25MPa and holding for 1 hour.
Scan360 scanning electronic microscopy was used to observe the microstructures of fabricated T4 and T7 FGMs.

Thermal shock resistance was tested by repeatedly quenching samples from 500℃ to 25℃ water.

3.2 Results and discussion.
Fig.7 shows the microstructures of T4 and T7. From Fig. 7, T4 and T7 were fabricated successfully. No macro cracks were found in T4 and T7 specimens. Both T4 and T7 were fully densified with boron and carbide additions under a relative lower sintering temperature. This also certifies that the design and fabrication of SiC/C FGM are successful.

T4 and T7 were continuously heated to 500℃ in a furnace and then were quenched into room temperature water. T4 cracked after 33 quenching cycles. No cracks were found in T7 specimen after 52 cycles of quenching. It is proved that T7 has better thermal shock resistance than that of T4 due to optimized residual thermal stress distribution. Also, it is proven that FGMs with better thermal stress distribution have better thermal shock resistance.

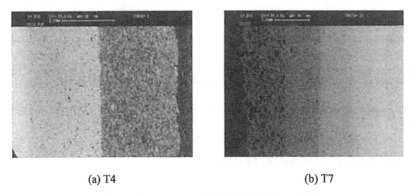

(a) T4   (b) T7

Fig. 7 Microstructure of SiC/C FGM T4 and T7

4. CONCLUSIONS

The residual thermal stress can be reduced by increasing the number of FGM layers. But the tendency of thermal stress relaxation will slow down with the number of FGM layers were continuously increased over 7. The residual thermal stress will vary with the variation of the thickness of surface layer, and will reach a minimum when the thickness of surface layer $t_{surface}$=0.6mm. Thermal stress will be reduced to a minimum when compositional distribution factor p=0.8. That is, the optimized parameters for fabrication SiC/C FGMs are number of FGM layers n≥7, surface thickness t=0.6mm and compositional distribution factor p=0.8. Two kinds of SiC/C FGMs were fabricated with optimized fabricating parameters by hot pressing. The microstructures were observed by SEM. The thermal shock resistance of T4 and T7 were tested by continuously quenching samples at 500℃ to water at 25℃. It was found that T4 was cracked after 33 quenching cycles and no crack was found in T7 after 52 cycle of repeated quenching. It was proven that T7 has better thermal shock resistance than T4 due to optimized residual thermal stress distribution. This also certifies that the design and fabrication of SiC/C FGM are successful.

ACKNOWLEDGEMENT

This project is supported by China High Technology New Materials Committee, the grant number is 863-715-011-0230.

REFERENCES

[1] Niino, M., Hirai T., and Watanabe, R., J. Japan Soc. of Comp. Mater. Vol. 13, 1987, pp257-261
[2] Cao Wenbin, Post doctor research report, 2000, pp19-38
[3] Algor users' manual
[4] Xie Yiquan, Elastic mechanics, Zhejiang University Press, Hangzhou, 1988
[5] Zhang Jiayu, Basic of Carbon Materials, Metallurgy Press, Beijing, 1992

FABRICATION OF PSZ-Al₂O₃ FUNCTIONALLY GRADED DISKS AND PLATES

H.Kobayashi
College of Industrial Technology
1–27–1, Nisikoya
Amagasaki 661–0047, Japan

ABSTRACT

The fabrication of PSZ-Al₂O₃ functionally graded disks and plates has been investigated by powder stacking and pressureless sintering process. PSZ, mixed PSZ and Al₂O₃, and Al₂O₃ powders with a binder were laminated into a mold, and then were compacted by the vibration pressing method under a pressure of 50 MPa, and after that pressed under a pressure of 100 MPa without vibration. Green specimens were sintered pressurelessly for 7200 s at 1723K or 1773 K in air. Functionally graded disks and plates fabricated from 2 layers with PSZ and Al₂O₃ powders had warps without cracks or they had cracks in those bonding interface. However, some FGMs fabricated from 3 layers or multilayer with the composition of PSZ, mixture of PSZ and Al₂O₃, and Al₂O₃ were free from cracks and warps. The structure of FGMs was microscopically examined, and the graded distribution was examined by a X-ray line analysis of Zr and Al. The bonding interfaces were completely bonded by entwining each other with each grain.

INTRODUCTION

The processing of functionally graded materials (FGMs) has been proposed by several methods.[1-5] The powder metallurgical processing with powder stacking and sintering is superior to the other processing for a large scale shape.[6,7] PSZ-Al₂O₃ FGMs with small diameter were successfully fabricated by explosive powder consolidation technique.[8] However, PSZ-Al₂O₃ FGMs for a large scale shape are difficult to fabricate because the difference in those sintering shrinkage causes warps or cracks of FGMs.

It was reported in a previous paper that the wet vibration pressing method with wet powders was effective for the stress relaxation in PSZ-SUS 304 compacts.[9] This method can decrease the applied pressure for compacting in contrast to a conventional die pressing method with dry powders, because the binder content was much more sufficient to allow the freedom of the flow of powders in the state of vibration. In this forming method, green compacts can be fabricated by laminating and compacting in a mold in one process.   The bonding interfaces in green FGM compacts have three dimensional bond, not two dimensional bond.   Therefore, in this work, the fabrication of PSZ–Al$_2$O$_3$ disks and plates without cracks and warps has been investigated by the wet vibration pressing method.

EXPERIMENTAL

(1) STARTING MATERIALS

Many PSZ and Al$_2$O$_3$ powders having different particle size and sintering characteristics were examined in this investigation, because the large difference of their sintering shrinkage causes the warps and cracks of FGMs.   The satisfactory PSZ powder (HSY-3.0B, Daiichi Kigenso Kagaku Kogyo Co., Ltd) for fabrication of flat disks and plates was composed of partially stabilized zirconia containing 3 mol % of Y$_2$O$_3$ with a mean particle size of 0.64 μm.   Al$_2$O$_3$ powder (AKP-50, Sumitomo Kagaku Kogyo Co., Ltd) was 0.2 μm in diameter.   These chemical compositions and the firing shrinkage of sintered compact for 7200 s at 1723 K are shown in Tables I and II.

Wet powder materials containing PSZ, Al$_2$O$_3$, and mixtures of PSZ and Al$_2$O$_3$ were prepared by blending the powders and a binder in a mortar.   The binder was an aqueous type binder with 20 mass% effective ingredient (Seruna WD-830, Chukyo Yushi Co., Ltd) and its added content was15 mass% for powders.

Table I .   Characterization of PSZ powder.

Chemical composition (mass %)					Mean particle	Firing shrinkage of
ZrO$_2$	Y$_2$O$_3$	Na$_2$O	CaO	L.O.I.	size (μm)	sintered compact (%)
93.24	5.27	0.02	0.01	0.16	0.64	20.1

Table II .   Characterization of Al$_2$O$_3$ powder.

Chemical composition(ppm)						Mean particle	Firing shrinkage of
Al$_2$O$_3$	Si	Na	Mg	Cu	Fe	size (μm)	sintered compact (%)
99.99%	15	2	2	1	7	0.2	17.0

(2) EXPERIMENTAL PROCEDURE

The vibration pressing method with wet powders was used for the forming of green specimens in this investigation.   This is a new plastic pressing method with vibration, and it could decrease the applied pressures for compacting in contrast to a conventional die pressing method with dry powders, because the binder content was sufficient to allow the freedom of the flow of powders.   Cyclic powder compression has been reported before.[9,10]

The vibration pressing machine for the forming of green specimens is shown in Figure 1.   This machine consists of a vibration top plate with an oil cylinder and a bottom plate, both of which are fixed with 4 poles, and 4 top and bottom springs supported with 2 plates.   The mechanical vibrator with a vibrating force of 300 kg is attached to the bottom of a vibration plate, and it consists essentially of 2 contra-rotating shafts with an out-of-balance weight at each end.   The die set in bottom vibration plate, and then the vibration and pressure were applied simultaneously by the top and bottom punches.

PSZ, PSZ and Al$_2$O$_3$ mixture, and Al$_2$O$_3$ wet materials were laminated into a die, and then were compacted by a vibration pressing method.   120 sec is taken to reach a fixed pressure after the die was set in the vibration plates.   The forming vibration time after a fixed pressure was 60 s, amplitude was 0.9 mm, frequency of vibration was 60 Hz, and applied pressure was 50 MPa.   After vibration pressing, the specimens were pressed in a die under a pressure of 100 MPa without vibration.

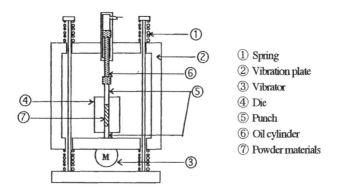

① Spring
② Vibration plate
③ Vibrator
④ Die
⑤ Punch
⑥ Oil cylinder
⑦ Powder materials

Fig.1.   Schematic layout of vibration pressing machine.

The obtained PSZ–Al$_2$O$_3$ graded green disks were 30 to 50 mm in diameter and the green plates were 20×50 mm and approximately 4 to 20 mm in height.   The green specimens were dried at 373 K, and then sintered pressurelessly for 7200 s at 1723K or 1773 K with a ramp rate of 5 K per min in air.   The relative density, apparent porosity and linear shrinkage of monolithic sintered specimens were measured.   The microstructure of sintered specimens was microscopically examined.   The graded distribution was examined by an X-ray line analysis of Zr and Al.   Furthermore, the resistance to thermal shock of FGMs was investigated.

RESULTS AND DISCUSSION

Firstly, disks and plates with 2 layers of PSZ and Al$_2$O$_3$ were fabricated.   They were free from cracks with large warps, or they had cracks in the bonding interface. It is thought that in case of directly bonding two different materials, a tensile stress occurs at the side of PSZ layer while a compressive stress does at the Al$_2$O$_3$, because the linear shrinkage after 1723 K sintering of PSZ is larger than that of Al$_2$O$_3$ as shown in Tables I and II. A tensile or compressive stress caused the warps of sintered compacts. These stresses exceeding its own bonding strength limit leads to the horizontal cracks in boundaries, However, FGMs fabricated from 3 layers were free from cracks, for example, FGMs fabricated from 3 layers with the composition of PSZ, 60 mass% PSZ-40 mass% Al_2O_3 or 50 mass% PSZ-50 mass% Al_2O_3 and Al_2O_3 had smaller warps without cracks than that of 2 layers FGMs. FGMs fabricated from PSZ-Al_2O_3 multilayer with more than 4 layers had slight warps without cracks. It is thought that a more continuously graded FGMs possesses the stronger bonding strength between each layer than that of FGMs with 2 steps in composition, and their FGMs possesses slight warps without cracks. These FGMs were expected to possess smaller warps than that with large graded distribution for the small difference of the linear shrinkage, thermal expansion, thermal conductivity of monolithic composites. However, FGMs without warps couldn't be fabricated, although fabrication of 3 layers or multilayer FGMs without warps has been investigated by the changing of applied forming pressure.

Secondly, the side of enriched Al_2O_3 of green specimens formed were cut, and specimens were sintered after that. These sintered FGMs possessed more slight warps than that without cutting, but FGMs without warps couldn't be fabricated. The appearances of sintered PSZ-Al_2O_3 functionally graded plates fabricated from 3 layers without cutting and with cutting are shown in Figure 2. Each layer was bonded smoothly by appearance. FGMs with cutting had more slight warps than that without cutting. Figure 3 shows SEM micrographs of each layer in FGMs fabricated from 3 layers. In the PSZ layer, fine grains of PSZ are shown (a). In the 50 mass % PSZ-50 mass % Al_2O_3 layer, the grain growth of Al_2O_3 was inhibited by pinning of PSZ (b). In the Al_2O_3 layer, Al_2O_3 grains grow to ca. 1~2 μm (c).

Figure 4 shows each bonding interface of FGMs. The SEM micrograph of interface with PSZ and 50 mass% PSZ- 50 mass% Al_2O_3 layer shows PSZ grains in the left part, and shows monolithic composite in the right part (a). The interface (b) shows the structure of monolithic composite in the left part, and Al_2O_3 in the right part. Each interface is completely bonded by entwining each other with each grain.

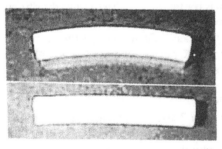

without cutting

with cutting

Fig. 2. Appearances of FGMs fabricated from 3 layers without and with cutting at the side of enriched Al_2O_3.

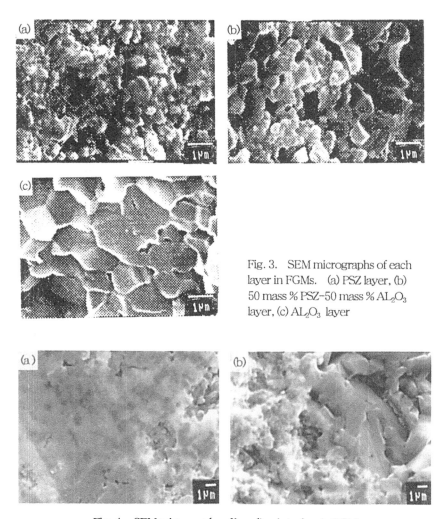

Fig. 3. SEM micrographs of each layer in FGMs. (a) PSZ layer, (b) 50 mass % PSZ-50 mass % AL$_2$O$_3$ layer, (c) AL$_2$O$_3$ layer

Fig. 4. SEM micrographs of bonding interface in FGMs.

These FGMs had the sharp graded distribution of Zr and Al in the each interface from the X-ray line analysis profile obtained by EPMA on the cross-section of that.

Next, PSZ material was changed from HSY-3.0B to the mixture of 50 mass% HSY-3.0B and 50 mass% KYZ-3, and also Al$_2$O$_3$ material was changed from AKP-50 to the mixture of that and its calcined Al$_2$O$_3$ in order to give the small difference of the linear shrinkage between PSZ and Al$_2$O$_3$. KYZ-3 is nearly equal to HSY-3.0B in chemical composition, but its mean particle size is 1.26 μm in diameter. In case of use of these materials, green specimens were sintered at 1773 K, because specimens were unsatisfactorily sintered at 1723 K.   Table 3 shows the linear sintering shrinkage of sintered compacts with each starting material and monolithic sintered compacts after 1723 K or 1773 K sintering.

Table III.   Linear sintering shrinkage of sintered compacts with each starting material.

	1723 K sintering (%)	1773 K sintering (%)
PSZ compacts		
(A) HSY–3.0B	20.1	21.5
(B) KYZ–3	13.0	14.7
(C) 50 mass % (A)– 50 mass % (B)	18.1	19.1
Al$_2$O$_3$ compacts		
(D) AKP–50	17.0	17.2
(E) AKP–50 calcined at 1073 K	19.1	19.5
(F) 50 mass % (D)– 50 mass % (E)	17.4	18.5
monolithic compacts		
(G) 50 mass % (A)– 50 mass % (D)	16.2	17.7
(H) 50 mass % (C)– 50 mass % (D)	15.9	16.8
(I) 50 mass % (C)– 50 mass % (E)	17.2	18.4

Consequently, some disks and plates fabricated from 2 layers with these PSZ (C) and Al$_2$O$_3$ materials (E) or (F) in Table III were free from cracks and they had slight warps.   Some FGMs fabricated from 3 layers or multilayer after 1773 K sintering were free from cracks and warps.   The value of apparent porosity of FGMs fabricated from 3 layers was 3 ~5 %.   These graded plates with 3 layers had smaller cracks than that of PSZ alone after thermal shock test for 600 s at 1073 K for and water quenching, therefore they had better resistance to thermal shock than that of PSZ plate alone.

CONCLUSIONS

The fabrication of PSZ–Al$_2$O$_3$ functionally graded disks and plates was investigated by powder stacking and pressureless sintering process.   The laminated green compacts were formed by the vibration pressing method for the stress relaxation in compacts.   Some functionally graded disks and plates fabricated from 3 layers or multilayer with the compositions of PSZ, mixture of PSZ and Al$_2$O$_3$, and Al$_2$O$_3$ were free from cracks and warps.   These bonding interfaces were completely bonded by entwining each other with each grain.   These functionally graded plates with 3 layers had better resistance to thermal shock than that of PSZ plate alone.

Functionally Graded Materials 2000

REFERENCES

[1] M.Sasaki, Y.Wang, T.Hirano and T.Hirai, J.Ceram. Soc.Jpn.,97,539-43 (1989).

[2] M.Sato, S.Tanaka, I.Hashimoto, E.Setoyama, T.Gejo and T.Sato, Hitach Hyoron, Vol.68, 821 (1986).

[3] S.Kitaguchi, H.Hamatani, N.Shimada, Y.Ichiyama and T.Saito, Keisha Kino Zairyo Shinpojiumu FGM 94, (1991) pp.149-52.

[4] A.Takahashi, K.Tanihara, Y.Miyamoto, M.Oyanagi, M.Koizumi and O.Yamada, Funtai oyobi Funmatsuyakin, 37.263-66 (1990).

[5] R.Watanabe, Proc. Shinzairo Sosei Toronkai, No.6, Jpn. Inst. Metals (1988) pp.9-13.

[6] H.Takebe, T.Teshima, M.Nakashima and K.Morinaga, J. Ceram. Soc. Jpn., 100[4]387-91(1992).

[7] M.Sasaki and T.Hirai, J.Ceram.Soc.Jpn.,99[10]1002-13 (1991).

[8] M.Omori, H.Sakai, A.Okubo, M.Kawahara and T.Hirai, Keisha Kino Zairyo Ronbunshu FGM 94 (1994) pp.99-104.

[9] H.Kobayashi, Functionally Graded Materials 1996, (1997) pp.209-14.

[10] H.Kobayashi, Trans. Materials Research Soc. Jpn., Vol.5, 109-18 (1992).

FGM Research in LSCPM of China

C.C.Ge, W.P.Shen, J.T.Li, W.B.Cao, Z.J.Zhou, G.Y.Xu, Y.H.Ling, A.H.Wu

Laboratory of Special Ceramics & Powder Metallurgy (LSCPM), University of Science and Technology Beijing (USTB), Beijing, 100083, China

ABSTRACT

Beginning from 1993, 6 national projects and 2 international cooperation projects on FGM in LSCPM have been approved and carried out. The work included design and prediction for FGM, fabrication of various systems of FGM with evaluation of properties. 8 processing technologies have been employed and developed for different systems or several technologies for a same system.

1. INTRODUCTION

LSCPM, USTB was established in 1985. It devotes to the research and development of new advanced ceramics and powder metallurgy (P/M). The FGM research in LSCPM was begun in 1993. C.C.Ge had applied for a project titled "Ceramic-Metal Composites and FGM with gas Pressure Sintering (GPS)" approved by the China National Committee of High Technology New Materials (CNCHTNM). Other project titled "composite design and structure control of TiB_2-Cu FGM" was approved by State Education Commission of China. Considering the inter-disciplinary feature of FGM research, C.C.GE established a cooperation between our group and experts working on mathematics and computer-aided design(CAD).

On the basis of the first project, three projects sequentially were approved by National Natural Science Foundation of China (NNSFC), they are: (1) Structure formation mechanism of SHS FGM (1996), (2) Investigation on CAD and fabrication of symmetrically compositional FGM (1997-1999), (3) CAD and fabrication of thermo-electric FGM (1998-2000).

On 1996, C.C.Ge had made a proposal to the government for establishing a project on FGM research for the first wall material of Chinese Fusion Experimental Facility, which was approved by CNCHTNM in 1997.

Since 1997, a project on SHS FGM between LSCPM and ISMAN (Institute of Structural

Macrokinetics and Materials Science, Russian Academy of Sciences) has been approved by National Natural Science Foundation of China (NNSFC) and Russian Foundation of Basic Research (RFBR)(1997-1999) and is being carried out. Some aspects of above-mentioned projects are reported in this paper.

2. DESIGN AND PREDICTION OF PROPERTIES FOR FGM

By means of the analysis model used for calculation of thermal stresses in FGM based on thermo-elastic mechanics and the classical lamination theory, the distribution of temperatures and stresses was defined. The effects of the volume distribution factor (p), the intermediary layer thickness (t), surface layer thickness (t$_1$, t$_2$) and the gradient layer numbers (n) on thermal stress distribution were investigated. According to the principle of minimum thermal stresses, the optimum parameters of p, t, t$_1$, t$_2$ and n were determined with CAD system on Microsoft Windows Platform, which was first developed for TiB$_2$-Cu FGM[1]. Analytical procedure for empty cylinder of FGM with axial symmetry has also been investigated. Temperature distribution and thermal stresses of FGM in the ring section have been calculated [2].

As the FGM is a kind of heterogeneous composites, its properties are affected by various factors, such as composition and its distribution, layer numbers and thickness, process parameters, etc. The underlying relationships among these factors are subtle, complicated, and very difficult to find and describe as formulas or rules. To solve these problems which lack existing solutions, we have applied the Artificial Neural Network (ANN) technology into property estimation of the FGM design. At first ANN was used for estimation of density of synthesized FGM product through self-learning the experimental data with sufficient accuracy. It is a novel, approach for FGM property prediction on the basis of adequate experimental data [3].

In contrary to conventional thermal-relaxed FGM, Symmetrically Compositional FGM (SCFGM) has symmetrically graded structures, which is designed to increase residual compressive stress on both SCFGM surfaces. The analytical model for SCFGM was established based on elasticity mechanics, mechanics of composite materials and computing mathematics. Triangle series were employed to describe distributions of residual thermal stresses on both surfaces of FGM caused by the difference of thermal expansion coefficients of surface layers and central layers. Based on the established analytical SCFGM model, SCFGM CAD system on Microsoft Windows System was first reported [4].

As a practical, effective method of calculation and analysis, the finite element method (FEM) has been paid more and more attention as an important method of structural analysis of FGM. On the basis of FEM, LSCPM established a SiC/C FGM designed model and tailored the compositions distribution. The optimal parameters of bulk SiC/C FGM were determined[5].

3. FABRICATION OF VARIOUS SYSTEM OF FGM

Various fabrication processes were used for different material systems, or for a same material system. Following are some typical examples:

3.1. SHS/HIP for TiB$_2$-Cu FGM and SiC/Cu FGM [6]

It is well known that TiB$_2$-Cu cannot be fabricated with conventional sintering or hot pressing. It is not only due to great difference of sintering temperature of TiB$_2$ and Cu, but also due to TiB$_2$ unwettable with liquid Cu. SHS/HIP has been proved as an effective technology for fabricating TiB$_2$-Cu FGM in our experiments. It enables simultaneous, rapid synthesis and densification of FGM sample sealed in a capsule and embedded in the chemical oven in a pressurized gas atmosphere.

Thermodynamic calculation was carried out before experiments with the aim to adjust the combustion temperatures of different layers for reducing the thermal stress during processing and preventing the macro-defects of products. For adjusting the combustion temperatures of different layers in FGM, we use the reaction product TiB$_2$ as diluent. Then:

$$Ti + 2B + a TiB_2 + bCu = (1+a) TiB_2 + bCu$$

Where a is mole content of TiB$_2$ diluent, b is mole content of Cu.

Specimens with 26 in diameter ×10mm thick were processed with SHS/HIP on the first China-made (2000°C, 200MPa) HIP apparatus R120.

3.2 Hot-pressing for SiC/C [7] [8]

SiC/C FGMs generally are being fabricated with CVD or CVI process, but the processing cycle is long (in some cases 100h is required). Cost is expensive, and it is difficult to obtain thick graded layers. Many systems of FGM have been made with powder-stacking-hot pressing process, but hot-pressed C/SiC FGM is scarcely found in literatures.

At first, specimens of monolithic SiC and C and C/SiC composites with 100% SiC, 75%SiC+25%C, 50%SiC+50%C, 25%SiC+75%C, 100%C, and 2-layered C/SiC were hot pressed. SiC, Al-B-C was used as sintering aids of SiC. Table.1 listed the density and relative density of SiC/C composites with different compositions for sintering schedule: 1900°C, 20MPa, 1h. Based on these experiments, design according to the finite-element calculation. Both 4-layered (T4) and 7- layered (T7) specimens were fabricated. C/SiC specimens with 40mm in diameter ×6mm thick and optimum parameters: compositional distribution exponent p=0.8-1.0, layer number n=7, graded layers thickness t=0.8mm, surface layer thickness =0.6mm, were fabricated. The bonding between various layers of a C/SiC FGM specimen is good, and no macro-defects are found in or between the layers.

Thermal shock resistance was measured on C/SiC FGM specimens, which were heated to

500°C, and quenched in water at room temperature repeatedly. For specimen T4, cracks and peeling occurred after 33 cycles, while no crack or peeling appeared in T7 after 52 repetitions. The result is consistent with the FGM design prediction.

Table2 Density and relative density of SiC/C composites with different compositions

	75%SiC+25%C	50%SiC+50%C	25%SiC+75%C	100%C
ρ	2.82	2.50	2.19	1.87
ρ / ρ_{th}	98.8	98.6	98.8	98.9

3.3 Plasma Spraying for B₄C/Cu Coating FGM [9]

Commercial B_4C powder with particle size of 20μm and Cu powder with particle size of -200 mesh were used as coating materials, which were sprayed on Cu substrates with dimensions of 10mm×10mm×2mm , 10mm×20mm×2 mm and 15mm in diameter×2mm thick. Before coating with B_4C, the Ni-Al alloy was applied to the substrates. The thickness of B_4C coating layer varied from 160μm to 260μm. B₄C/Cu non-FGM coating (BC-1), and B₄C/Cu coating FGM with different composition distribution exponents (P=0.2; 1.0; 2.0 for BC-2, BC3, BC-4) were fabricated with this process. Thermal shock resistance was measured through repeatedly heating and quenched in cooled water with ΔT=500K. BC-2 and BC-3 were completely peeled after 2 cycles, while no peeling occurred for BC-3 after 20 cycles

Hot impact experiments were conducted on non-FGM and FGM specimens in an electron-beam apparatus with the experimental parameters: Number of pulses 30, Pulse width 2ms, Electron current 400mA, Energy of electron beam 5000eV. Cracks occurred in non-FGM specimen after 30 pulses, while basically no damage occurred in FGM coating specimens with p=1 after 250 pulses. For coating FGM specimen with p=1 and coating thickness 300-400μm, no cracks and only small damages appeared after 1000 cycles with $6.4MW/m^2$ and 1000ms pulse in the hot impact experiment.

3.4 Porous Skeleton with Liquid Phase Cu Infiltration-welding for W/Cu FGM [10]

W powders with particle sizes of 3μm, 7μm and 15 μm and Cu powder of –200 mesh were used. W powders with different particle sizes were mixed separately with pore-making agent in agate mill. A green compact of 40mm in diameter x 3mm thick was pressed in a steel die, then put into a graphite crucible embedded with Al_2O_3 powder and sintered at 1300 ℃ for 1 h under H_2 atmosphere. A porous graded W skeleton was formed. Cu powder was then placed on the top of the W skeleton and sintered at 1300 ℃ for 30 min. Liquid Cu flowed into the pores of W skeleton, and excess Cu remained on the top surface of the specimen. Then, a W plate was welded on the other surface and a W/Cu FGM with 100%Cu -70%Cu-30%Cu-5%Cu-100%W was formed. The specimen was heated in a $MoSi_2$ furnace to 500 ℃, kept for 5 min and quenched in cooled water, no cracks appeared after 30 cycles. It is interesting to notice that thermal expansion coefficient is $13.13\times10^{-6}/°C$, which is close to that of Cu $(17\times10^{-6}/°C)$, and far from that of W $(4.5\times10^{-6}/°C)$.

3.5 Spark Plasma Sintering (SPS) for Fabrication of Mo/Cu FGM [10]

The SPS process is a new material processing technology recently developed by Sumitomo Coal Mining Co. Ltd., Japan. In addition to Joule heat and external pressure applied in the specimen during conventional hot-pressing, a direct-current pulse voltages from a special electric source is overlapped and spark plasma charge occurred between particles producing local high temperature and causing melt and evaporation of particle surfaces and particle welding. Mo, Cu and Ni powders with -200 mesh were used, and 4-layered green compact of 10mm in diameter with 2.5mm thick for each layer was made with powder stacking process. The compositions of each layer are as follows: (wt%): Mo 99% + Ni 1%; Mo 69.5% + Ni 1% + Cu 29.5%; Mo 49.5% + Ni 1% + Cu 49.5%; Mo 29.5% + Ni 1% + Cu 69.5%.

The SPS parameters were: 850°C, 5min, 30MPa. It is noticed that the interfaces between layer are clear, indicating no evident migration of Cu occurred due to absence of liquid copper and short time sintering.

3.6 SHS/pseudo-HIP (PHIP) for Symmetrically Compositional SHS-cermet Systems [11]

Symmetrically Compositional FGM (SCFGM) consists of outer layers with lower thermal expansion coefficients (CTE), and inner layers with higher CTE in order that compressive stress is created in the outer layers, resulting in enhancement of the strength and fracture toughness of the FGM. One step SHS/PHIP technology is very attractive in making SCFGM as it concentrated the advantage of SHS and PHIP process in one step with short duration which is favored in keeping the designed compositional distribution. Fig.1 is SEM micrograph of TiB_2 +TiN+Ni SCFGM which is formed through SHS and PHIP in one step.

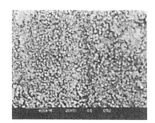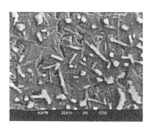

Fig.1 Ti+BN+NiAl from 5%(surface, left) to 30%(center, right)

3.7 UHPC Combined Graded Sintering for SiC/Cu[12] and W/Cu FGM

FGMs, which are most difficult to be sinter was those with constituents of high melting point ceramics and relatively low melting point metals. These FGMs are impossible to be simultaneously sintered with conventional sintering or hot pressing, such as SiC/Cu, or W/Cu etc..

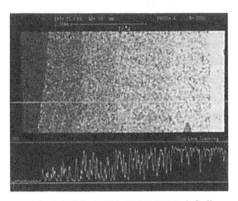

Fig. 2 SEM of W/Cu FGM (BSD, left: line scanning of W; right: tungsten layer×2000)

A novel technology, UHPC combined graded sintering, is proposed by authors in this paper for sintering of these FGMs by using a UHPC system. The UHPC system consisted of a mechanical press and a tungsten-carbide multi-anvil system capable of 5 GPa, and associated electronic and hydraulic systems. Pyrophyllite containing the green compact was placed in the multi-anvil system. Pressure was applied on the sample assembly and electric current with input of 13 kW passed through and heated compacts. The FGM sample consists of layers with different electrical resistance, which constitutes the prerequisite condition for gradient sintering. The metal rich layers had lower resistance, which was just required for a lower sintering temperature. A very short duration of 40 seconds was used for sintering.

Two kinds of FGM, i.e. W/Cu, SiC/Cu were performed with this technology, Fig. 2 shows SEM of W/Cu FGM.

3.8 Sintering-hot pressing processes for thermoelectric FGM Sb-Te system and Pb-Te system[13]

For useful thermoelectric elements, the figure of merit should be high over a wide temperature range. In order to achieve thermoelectric performance, FGM approach is attracting many material scientists. One of the most important issues on FGM thermoelectric technology is concerned with the stability of FGM performance for a long time, for which a basic way is to make a diffusion barrier for each layer of multi-stage graded FGM. Here is a typical example of research on thermoelectric FGM in LSCPM. P-type $(Bi_{0.15}Sb_{0.85})_2Te_3$ and $(Bi_{0.10}Sb_{0.90})_2Te_3$ two segments compositionally thermoelectric FGM with different metals as Fe, Co, Cu, and Al as barriers between two segments and were fabricated with hot pressing. Stoichiometric composition of elementary Bi, Sb and Te powder of purity higher than 99.999% were used as raw materials of <250μm and mixed in mortar. X-ray shows that metal mixture has been fully converted into $(Bi_{0.10}Sb_{0.90})_2Te_3$, as there is no of other phases. The relative density is >92.4% of theoretical. It is indicated by electron microprobe analysis that Fe and Al are better to be used as barriers of Be_2Te_3-based FGM due to their low diffusion with the layers.

4.CONCLUSIIONS

Beginning from 1993, 6 national projects and 2 international cooperation projects on FGM in LSCPM have been approved. Among them, 4 national projects and 1 international cooperation projects have been completed, and others are being studied. They involved thermal-relaxd FGM,(TiB_2-Cu etc.),symmetrically compositional FGM (SCFGM, TiB_2-TiN-Ni etc.), FGM for Fusion facility ($B_4C/C(Cu)$, SiC/C(Cu), W/Cu etc.), and thermal-electric FGM (PbTe, SbTe etc.). By means of analysis-modal, and CAD system on Microsoft Windows platform for calculating thermal stress of thermal relaxed FGM (TiB_2–Cu) and SCFGM have been established, while FEM design model has been established for SiC/C FGM. 8 various processing technologies have been employed and developed for different material systems or several technologies for a same material system. These works creates foundation for selecting optimal processing technologies and parameter for different systems.

5.Acknowlodgents

This work was supported by NNSFC,CCNCHHTM and State Education Committee.

REFERENCES
1. C. C. Ge, Z. X. Wang, Z. C. Mu, Design for SHS TiB_2-Cu FGM, Int. J. Self-Propagating High-Temperature Synthesis, V.6, No.3, 1997
2. X . D. Zhang, T. Q. Liu, C. C. Ge., Mathematical Model for Axial-Symmetical FGM, Functionally Graded Materials 1996, edited by I. Shitota, and M. Y. Miyamoto, 1997 Elsevier Science B. V., pp301-306
3. Z. C. MU, C. C. Ge, W. B. Cao, Artificial Neural Network Used for FGM Design, Functionally Graded Materials 1996, edited by I. Shitota, and M. Y. Miyamoto, 1997 Elsevier Science B. V., pp65-68
4. W. B. Cao, Doctoral dissertation of USTB, 1998
5. W. B. Cao, Postdoctoral Report of USTB, 2000
6. C. C. Ge, Z. X. WANG, W. B. Cao, Thermodynamic Calculation and Processing of TiB_2-Cu FGM, Functionally Graded Materials 1996, edited by I. Shitota, and M. Y. Miyamoto, 1997 Elsevier Science B. V., pp301-306
7. C. C. Ge, Y. L. Tang, M. X. Wang, N. M. Zhang, Hot Pressing C/SiC FGM for Fusion

Technology, Proc. The First China Int. Conference High-Performance Ceramics (October, 1998, Beijing), Edited by D. S. Yan and Z. D. Guan, Tsinghua Univ. Press, Beijing, 1999. 12, pp262-265

8. W. B. Cao, A. H. Wu, J. T. Li and C. C. Ge, Design and Fabrication of SiC/C FGM, accepted by FGM 2000.

9. C. C. Ge, Y. K. Li, Fabrication and Plasma Relevent Characteristics of B_4C/Cu Coating FGM, The 5th China-Japan Symp. Materials for Advanced Energy System and Fission and Fusion engineering, Nov. 2-6, 1998, Xi'an

10. C.C.Ge, Z.J.Zhou, J.T.Li, X.Liu, Z.Y.Xu, Fabrication of W/Cu and Mo/Cu FGMs Plasma-facing Materials, Japan-China Joint Seminar-Designing, Fabrication and Application of FGM, Oct, 28-29, 1999, Sendai

11. Wen-Bin Cao, Wei-Ping Shen, Chang-Chun Ge, E. H. Grigoryan, A. E. Sytschev, A. S. Rogachev, Combustion Wave Propagation During SHS of Bi-Layered System, Int. J. Self-propagating High Temperature Synthesis, Vol. 9, No. 1, 2000

12. Y. H. Ling, C. C. Ge, J. T. Li and C. Huo, Fabrication of SiC/Cu FGM by Graded Sintering, accepted by FGM 2000.

13. G. Y. Xu and C. C. Ge, Thermoelectric Properties of N-type Bi_2Te_3-PbTe Graded Thermoelectric Materials with Different Barriers, accept by FGM 2000

FABRICATION OF Ti-Sc SYSTEM FGM WITH DENSITY GRADIENT

C. J. Deng   H. Tao   L. M. Zhang   R. Z. Yuan
State Key Laboratory of Advanced Technology for Materials Synthesis and Processing,
Wuhan University of Technology, Wuhan, Hubei, P. R. China, 430070

ABSTRACT

Ti-Sc alloys with differently designed ratios are hot -press sintered at 1473K under a pressure of 30Mpa for two hours in a vacuum furnace. Under the same experimental condition, Ti-Sc system FGM is successfully fabricated. Its density range changes quasi-continuously from 3.0g/cm^3 to 4.5g/cm^3 in the thickness direction. 1.5% Mg additive can greatly increase the density of Ti-Sc alloys.

1.INTRODUCTION

In dynamic technology, we need a flier-plate with impedance gradient to create a quasi-isentropic compression.[1-2] According to the density dependence of impedance, a new kind of functionally graded material (FGM) with density gradient can be used as a flier-plate in dynamic technology to obtain super high pressures and velocities.[3-5]

Previous works show us that the density range of the flier-plate changes gradually from 19.3 g/cm^3 to 10.2 g/cm^3 [6] and from 10.2 g/cm^3 to 4.5 g/cm^3 [7]. A fully dense W-Mo-Ti system flier-plate with graded impedance in its thickness direction was successfully fabricated by the method of powder metallurgy. The result of the impact experimer.. on a light gas gun showed that dynamic quasi-isentropic compression had been created [8]. In order to obtain more super high pressure, Xiong et al have researched Ti-Al system FGM with density gradient[9]. However, the brittle TiAl intermetallic compounds are formed in the Ti-Al system FGM during sintering process. The strength and stiffness of the flier-plate of moderate impedance decrease greatly, which is still an unsolved problem. Since the density of Sc is 3.0g/cm^3, it is eligible as a component of the Ti-Sc FGM. According to Sc-Ti binary phase diagram [10], there is no intermetallic compound between Ti and Sc. Besides, continuous Ti-Sc solid solutions are formed at W-Mo-Ti system FGM experimental condition (1473K). Therefore, a one-step sintering of W-Mo-Ti-Sc FGM is possible at 1473K.

The purpose of this research is to fabricate a Ti-Sc system FGM whose density changes gradually from 4.5g/cm³ to 3.0 g/cm³ in the thickness direction.

2.EXPERIMENT

(1)Raw materials

Fresh powders of Ti, Sc and Mg are used in the experiment. The particle size and purity are shown in Table 1.

Table 1 Particle size and Purity of the raw materials

	Particle size/μm	Purity /mass%
Ti	30	>99.90
Sc	≤74	>99.90
Mg	≤100	>99.00

(2)Experimental method

We blended the fresh powders of Ti and Sc mechanically in a designed ratio for half an hour to make them homogeneous. In order to improve the density, we added 0.5-3.0% Mg. The mixture was then put into a graphite mould and hot pressed in a vacuum furnace. According to W-Mo-Ti system FGM with density gradient experimental condition, the sintering condition of Ti-Sc system FGM was 1473K, 30Mpa and 2hrs. A pressure of 30MPa was exerted on the sample. The temperature was increased to 1473K at 20K/min and maintained for 2h. We obtained the samples of A1(Ti), A2(87.5mol%Ti+12.5 mol %Sc), A3(75 mol %Ti+25 mol % Sc), A4(50 mol %Ti +50 mol %Sc), A5(25 mol %Ti +75 mol %Sc) and A6(Sc). The Ti-Sc system alloys were 16mm in diameter and 2-3mm in thickness. Under the same experimental condition, the Ti-Sc FGM was also produced, which was 32mm in diameter and 2mm in thickness.

(3)Testing

We measured the density of all samples using the water-immersion technique. The phase composition of the sample of Ti-Sc alloys was investigated by means of XRD (D/MAX –RB, by Rigaku Co. Japan). The microstructure and element distribution of Ti-Sc FGM were researched by means of EPMA (JXA-8800R, Hitachi Co. Japan) +EDS (Link-ISIS, Oxford University, England).

3.RESULTS AND DISCUSSION

On the experimental conditions stated above, fully dense samples of A1, A2, A3, A4, A5 and A6 were obtained. Under the same condition, a Ti-Sc FGM was produced as well.

(1) XRD analysis

Figure 1 shows that the Ti-Sc alloys are the mechanical mixtures of α-Ti and α-Sc at room temperature.

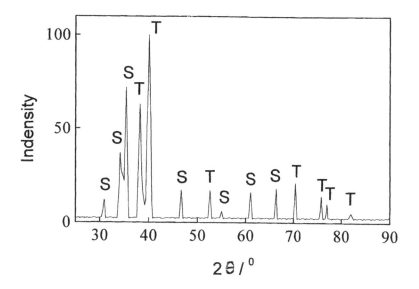

Fig.1 The pattern result of XRD of Ti-Sc alloyθ

(T--α-Ti and S--α-Sc)

(2)Densification

Figure 2 shows the density of A1 sample with different Mg content. The density of Ti alloys was improved with the increase of the content of Mg addition. Ti alloy was densified when Mg added was 1.0-2.0%. When 3% Mg additive was used, the relative density of Ti was over 100%. This reason can be that Ti-Mg solid solution was formed at the sintering process, and part of Mg was vaporized.

Figure 3 and Figure 4 show the density of Ti-Sc alloys with the increase of the content of Mg addition. When no Mg was added, we got relatively low density values of the Ti-Sc alloys. With the increase in amount of Mg, the density of Ti-Sc alloys was improved greatly. When 1.5mol% Mg was added, the relative density of the synthesized Ti-Sc alloys reached or even exceeds 100%. The reason may be that Mg particles melted first and then Ti-Sc-Mg solid solutions were formed during the sintering process.

According to the experimental results, when 1.5% Mg was added , the relative density of sintered Sc was 99.35% at 1473K. Consequently, when 1.5% Mg was the addition as the sintering aid, every layer of Ti-Sc FGM can be densified at 1473K (shown in Fig.5).

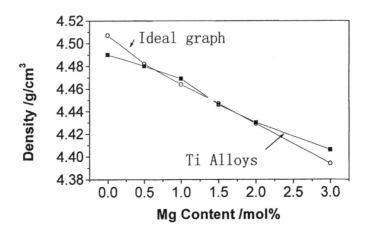

Fig.2   The Density Change of A1 with Different Mg Content

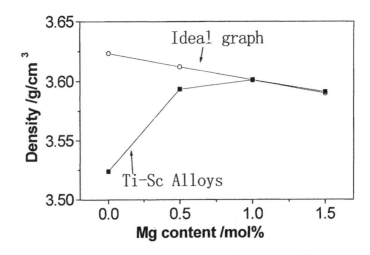

Fig.3   The Density Change of A3 with Different Mg Content

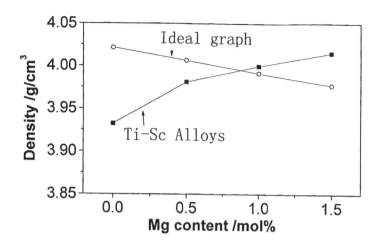

Fig.4   The Density Change of A4 with Different Mg Content

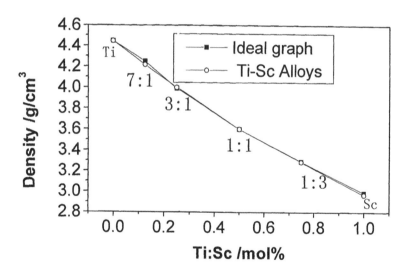

Fig.5   Density of Ti-Sc system alloys Vs different Ti:Sc (mol%)

The real density value of the Ti-Sc FGM was 3.755g/cm³. According to the result of calculation, the relative density of the Ti-Sc FGM was 99.88%. The density change of the fully dense Ti-Sc FGM can be seen in Fig.6. With the increase in Sc content and the decrease in Ti content, the density values of the Ti-Sc FGM changed gradually from 4.5 g/cm³ to 3.0 g/cm³.

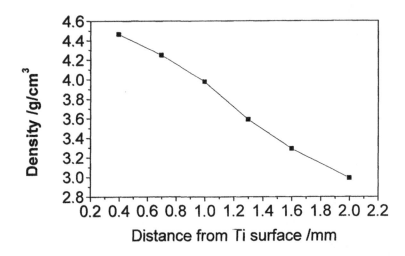

Fig.6 The change of density of the Ti-Sc system FGM with Distance from Ti surface

(3) Results of EPMA+EDS

The secondary electron image of a Ti-Sc FGM obtained at 1473K was shown in Fig.7 (a). The area distribution images of element Ti and Sc (Fig.7 (d) and (c)) indicated that homogeneous Ti and Sc mixtures were present at room temperature. This was because that liquid Mg promoted the diffusion of Ti and Sc to each other and the formation of Ti-Sc solid solution at 1473K. But when the sample was cooled to room temperature, Ti-Sc solid solutions turn into Ti and Sc mixtures.

The back-scattered electron image of the Ti-Sc FGM was shown in Fig.8 (a) together with the element line distribution images of Sc and Ti (Fig. 8 (b)) in the thickness direction. It can be seen that the content of Sc increased quasi-continuously in the thickness direction with the gradual decrease of Ti content. The content change of Sc and Ti was stepwise and the interface of every lay of FGM was not clear. This may be that Ti and Sc have plastic flowed within 30MPa pressure in sintering process.

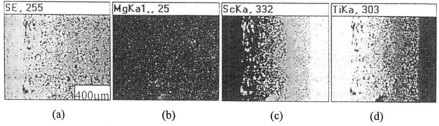

(a)   (b)   (c)   (d)

Fig.7   Secondary electron image of a Ti-Sc FGM(a)
and element area distribution images of Mg (b), Sc (c), and Ti(d)

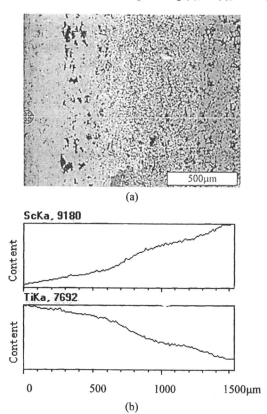

(a)

(b)

Fig.8 The back-scattered electron image of Ti-Sc FGM (a)
and element line distribution images of Sc and Ti in the thickness direction(b)

4 CONCLUSIONS

(1) Ti-Sc system alloys with high relative density can be fabricated by hot -press sintering at 1473K under 30MPa for two hours.

(2) 1.5% Mg additive can greatly increase the density of sintered Ti-Sc system alloys.

(3) A fully dense Ti-Sc FGM can be successfully fabricated. Its density value changes gradually from $4.5g/cm^3$ to $3.0 g/cm^3$ in the thickness direction.

(4) In Ti-Sc alloys and the Ti-Sc system FGM with density gradient, there is no Ti-Sc intermetallic compound.

ACKNOWLEDGEMENT

This research was sponsored by National Natural Science Foundation of P. R. China (No, 59771028) and Foundation of State Key Laboratory of Advanced Technology for Materials Synthesis and Processing, Wuhan University of Technology.

REFERENCES

[1]Barker L. M. and Scott D. D, "Development of a light –pressure quasi- isentropic plane wave generating capability," SAND 84-0432.

[2]Asay J. R and Parker L. M, "shear strength of tungsten under shock and quasi- isentropic loading to 250GPa," SAND 88-0306,UC-704.

[3]Chhabildas L. C, in "Bulletin of the 1995 APS topic conference" on *shock compress of condensed mater*, Scatter, Washington, 1995.

[4]Chhabildas L. C, Barker L. M, "Dynamic quasi- isentropic compression of tungsten in Schmid," S C, Holmes N C, Eds, *Shock waves in condensed Matter 1987*, Elservier Science Publishers B. V. pp111-114 (1988).

[5]Chhabildas L. C, Kmetyk L. N, Reihart W. D, "Enhanced hypervelocity launcher- capabilities to 16km/s," *International Journal of Impact Engineering*, 17pp183-194(1995).

[6]Huaping XIONG, Lianmeng ZHANG, Junguo Li, "Densification of the tungstem heavy alloys at 1473K and Fabrication of W-Mo system FGM with density gradient," *J. Mater. Sci. Technol.* 15[3] 229-232(1999).

[7]Xiong Huaping, Zhang Lianmeng, Shen Qiang, "Farication of Mo-Ti functionally graded materials," *Trans. Nonferrous Met. Soc. China.* 9[3] 582-585(1999).

[8] Shen Qiang, Zhang Lianmeng, Xiong Huaping, "Fabrication of W-Mo-Ti system flier- plate with graded impedance for generating quasi- isentropic compression," *Chinese Science Bulletin*, 45[8], 878-881(2000).

[9] H. P. Xiong, L. M. Zhang, R. Z. Yuan, "Connection of TiAl/Al and Fabrication of Ti/TiAl/Al system FGM," *Acta Metallurgica Sinica*, 35[10] 1053-1056(1999).

[10]Hugh Baker Editor, "ASM Handbook, Volume 3:Alloy Phase Diagrams," The Materials Information Society, 1992.

PROCESSING OF SILICON NITRIDE-TUNGSTEN PROTOTYPES

Guoping He and Deidre A. Hirschfeld
Department of Materials and Metallurgical Engineering
New Mexico Institute of Mining and Technology
Socorro, NM 87801

J. Cesarano III and John N. Stuecker
Advanced Materials Lab
Sandia National Laboratories
Albuquerque, NM 87106

ABSTRACT

In selected advanced engine applications, silicon nitride ceramic components must be joined to metallic alloys. Current technology utilizes a stiff tungsten layer brazed between the ceramic and metal. An alternative approach is to create a functionally graded material (FGM) where the silicon nitride component is manufactured with the tungsten incorporated in the appropriate places using solid freeform fabrication techniques. This study describes the design of an intermediate layer with graded composition and the effect of processing parameters on the interfacial microstructure, properties, and shrinkage of silicon nitride-tungsten FGM made using robocasting, a novel solid freeform fabrication technique.

INTRODUCTION

During the past several decades, the gas turbine community has made considerable progress in using ceramic components in high temperature environments.[1] The need for enhanced performance of ceramic components for advanced gas turbine engines is ever increasing. Silicon nitride is considered to be a suitable ceramic for structural and advanced heat engine applications due to its excellent thermomechanical properties, but in some applications it must be joined to metallic alloys. Current technology utilizes a stiff tungsten layer brazed between the ceramic and metal. It is possible to achieve the metal to ceramic transition by grading. Functionally graded materials (FGM) are produced by gradually modifying the composition of two or more different materials. They have attracted much interest for their ability to exhibit not only the combined functionality of their component materials, but also additional functionality derived from the graded composition.[2]

Recently, silicon nitride[3] and tungsten[4] parts have been fabricated by robocasting,[5-6] a novel solid freeform fabrication technique developed by Sandia National Laboratories, and evaluated. However, there has been no report of Si_3N_4-W graded materials in the literature. In this study, graded compositions were designed and cofired. Robocasting has the ability to mix any composition of two materials, creating parts in a layerwise manner. Samples 10x10x10 mm were created in minutes and cofired by the next day. Microstructure and chemical composition profiles of intermediate region between Si_3N_4 and W were observed and identified.

EXPERIMENTAL PROCEDURE

The Si_3N_4 powder used in this study was GS-44 powder (AlliedSignal Inc., Torrance, CA). The GS-44 is a mixture containing proprietary sintering additives. The average particle diameter was 0.77 μm. A commercial available dispersant, Darvan 821A (R.T. Vanderbilt Company, Norwalk, CT), was used to disperse the GS-44 Si_3N_4 powder in aqueous solution at 52 vol% solids. The tungsten powder used was 99.95 wt% W content and 0.6 ~ 0.7 μm average particle size (Strem Chemicals Inc., Newburyport, MA). Standard reagent grade nickel nitrate hexahydrate [$Ni(NO_3)_2$ $6H_2O$] (Strem Chemicals Inc., Newburyport, MA) was used to introduced nickel from aqueous solution as a sintering agent for W. Polyethyleneimine (PEI) (Polysciences, Inc., Warrington, PA), with an average molecular weight of 10,000 and chemical structure as [$-CH_2-CH_2-NH-$]$_n$, was used as a cationic dispersant for W aqueous slurries at 33 vol% solids. Ammonium alginate, Collatex ARE (ISP Alginates Inc., San Diego, CA), was used as a deflocculant-binder for W. The pH of all slurries was adjusted with analytical grade nitric acid (1.0 N) and ammonium hydroxide solution (40%). Deionized water was used throughout this study.

Robocasting graded materials requires two appropriate compositions in highly concentrated slurry form. The materials are then fed into a small sealed rotary mixing chamber and deposited through a 1.36mm tip. FGM Si_3N_4-W samples were prepared from slurries of Si_3N_4 and W by robocasting at the Advanced Materials Lab, Sandia National Laboratories, Albuquerque, NM. Square samples were fabricated in a layerwise fashion, adjusting the composition in each vertical layer. Green samples of graded Si_3N_4 –W materials were placed in a graphite crucible embedded in a mixed powder bed, consisting of 50 wt% Si_3N_4 and 50 wt% BN. The crucible was heated to 1720°C and soaked for 60 min in a high purity N_2 atmosphere at 12 psi pressure and 2.8 l/m flow rate. Linear shrinkage, weight loss, and sintered density were measured. Cofiring conditions were selected based on the sintering of Si_3N_4 ceramics.

The microstructure of the interface between the layers and the atomic element distributions in the cross-sections were evaluated by scanning electron microscopy (SEM) and electron microprobe microanalysis (EMPA) (CAMECA SX-100, Trumbull, CT). X-ray diffraction (XRD) (SIEMENS D500, Germany) analysis was used to identify the new phases present in the intermediate layer.

RESULTS AND DISCUSSION
Robocasting of Si_3N_4-W with Graded Composition Design

Graded compositions of 5, 6, 7, 9, and 11 layers were selected for fabrication with the 5 layer part analyzed for this particular discussion. Figure 1 shows the 5-layer design. The top layer is pure tungsten and the bottom layer is pure silicon nitride.

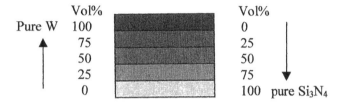

Figure 1. Design of graded composition of Si_3N_4 and W with 5 layers.

Table 1. Slurry characteristics of W and Si_3N_4 for robocasting graded materials

	W	Si_3N_4
Powders	99.95% Purity	GS-44
Particle size d_{50} (µm)	0.6-0.7	0.77
Solids loading (vol%)	33	52
Additives (wt% by dry weight basis)		
	PEI: 3	Darvan 821A: 1
	ARE: 0.15	Al-nitrate: 0.4
	Ni-nitrate: 0.2 Ni	
pH	~ 5.5	~ 8.5

Slurry batch compositions for robocasting graded W-Si_3N_4 materials are shown in Table 1. It was found that the slurry batch compositions of silicon nitride and tungsten for robocasting single material structures had to be changed when robocasting graded composition structures. At such high solids loading, slight changes in batch composition can drastically affect slurry build characteristics. The solids loading and pH values are most critical. For the Si_3N_4 slurries, viscosity increases with decreased pH and increased solids loading.[3] For the W slurries, the viscosity increases with increased pH and increased solids loading.[4] It is therefore appropriate to use slurries that are comparable, otherwise, as they are introduced to each other during the mixing procedure, viscosity may increase and stop material flow through the dispensing tip.

Cofiring of Graded Si_3N_4-W in N_2 by Pressureless Sintering

Green and cofired robocast 5-layer Si_3N_4-W samples are shown in Figure 2. Each layer was distinguishable due to the different colors created by different compositions. No cracking occurred between two neighboring layers during curing of the green parts or cofiring. Average green density of robocast graded Si_3N_4-W is about 4.25 g/cm^3 which is about 40% of theoretical. Table 2 shows the cofiring results of graded materials with different numbers of layers. Average linear shrinkage of the middle layer of cofired Si_3N_4-W was 23%. The average % theoretical density was approximately 70%. The W layer exhibited a slightly larger shrinkage than that of Si_3N_4 layer. It may be possible to match shrinkage values by adjusting the slurry solids loading of either material. The weight loss was due to either decomposition of Si_3N_4 or oxidation of W during pressureless cofiring.

 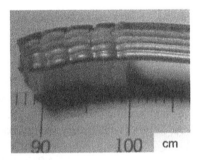

Figure 2. Green and cofired robocast FGM Si_3N_4-W. Note that pure W is bottom fired layer.

Microstructure and Elemental Distribution on the Cross-sections

SEM micrographs of the cross section fracture surfaces of graded Si_3N_4-W materials are shown in Figure 3. The microstructure of the pure Si_3N_4 layer

(Figure 3a) shows elongated β-Si$_3$N$_4$ grains, which are characteristics of typical dense Si$_3$N$_4$ ceramics.

Table 2. Cofiring results of graded Si$_3$N$_4$–W materials at 1720°C/1h in N$_2$

Total layers	Linear shrinkage of Intermediate layer (%)	Weight loss (%)	Sintered density (g/cm^3)
5	22	4.5	7.69
7	24	6	7.55
9	24	6	7.60

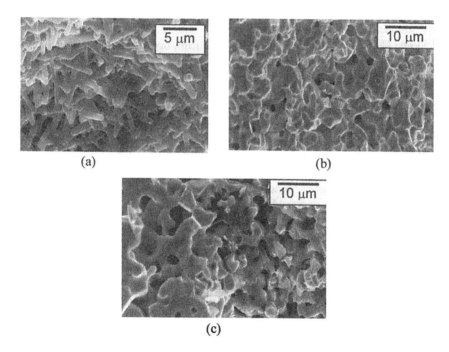

(a) (b)

(c)

Figure 3. SEM images of the fractured surface of the FGM Si$_3$N$_4$-W; (a) pure Si$_3$N$_4$ layer, (b) pure W layer, (c) intermediate layer.

The microstructure of the pure W layer (Figure 3b) shows equiaxed W grains which are agglomerated or connected with some liquid phase. Brophy et. al.[7] studied the nickel-activated sintering of tungsten and believed that W particles are coated with a uniform Ni layer and sintering involved the movement of W on or through the Ni carrier phase. Tungsten dissolves preferentially into the Ni layer at points of particle contact and diffuse outward in the interface between the carrier phase layer and the particle itself. The result is a decrease in distance between adjacent particle centers and overall shrinkage of the compacts.

The microstructure of the middle layer containing 50 vol% Si_3N_4 and 50 vol% W (Figure 3c) shows a similar structure as the W layer but with larger grains and larger pores. However, this layer exhibits large connected grains and more pores is rough and uneven, showing an interlocking structure. This structure probably is due formations of intermetallic phases.

Figure 4 shows the backscattered electron (BSE) image and X-ray images of the polished cross-section of the graded Si_3N_4-W materials. Tungsten exhibits high BSE intensity because of its high atomic number (Figure 4a) and thus shows up brighter than Si_3N_4. Figure 5 shows an electron microprobe line scan of Si and W on the cross-section along line AA in Figure 4. It is clear that the intensities of Si and W, Si $K\alpha$ and W $L\alpha$, are different from one surface to another. No W was found in the Si_3N_4 layer, however, some Si was found near the surface of the supposedly pure W layer. It is possible Si penetrated the W layer during the cofiring because the part was embedded in a Si_3N_4 powder bed, necessary for firing the Si_3N_4 phase but may have reacted with the W phase. Both Si_3N_4 and W sinter through a liquid-phase, but the liquid phase of the W has a lower viscosity than that of Si_3N_4. It is possible that during cofiring, Si diffuses through the W phase interfaces between layers where are low interfacial free energies, explaining the buildup of Si seen in Figures 4 and 5. It is also possible that segregation occurred during slurry deposition due to the large density mismatch, or that nonuniform mixing occurred.

Figure 6 shows an XRD pattern of the middle in the 5 layer Si_3N_4-W FGM. The peaks of W, β-Si_3N_4, and tungsten silicides, such as W_5Si_3 and WSi_2, are observed. The identification of tungsten silicides suggests that direct reactions observed. The identification of tungsten silicides suggests that direct reactions between W and Si_3N_4 occur during cofiring. Suggestions for the intermetallic chemical reactions are expressed with the following equations:

$$5W + Si_3N_4 = W_5Si_3 + 2N_2\uparrow \qquad (1)$$
$$3W + 2Si_3N_4 = 3WSi_2 + 4N_2\uparrow \qquad (2)$$

CONCLUSION

Graded Si_3N_4-W materials have been successfully robocast using highly concentrated slurries and cofired using pressureless sintering at 1720°C in N_2. Parts with graded composition of 5 to 11 layers were fabricated with each layer yielding a different composition and a correspondingly different microstructure. The pure Si_3N_4 layer showed typical elongated β-Si_3N_4 grains and the W layer showed more equiaxed grains. Intermediate layers showed interlocking structures with larger grain and larger pores. X-ray images and EMPA line scans indicate that Si and W elements were distributed gradually in the cross section with segregation of some elements found in the interfaces. Tungsten silicides were found in the intermediate layers due to reactions between Si_3N_4 and W.

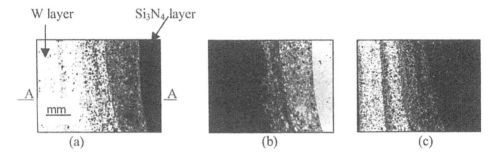

Figure 4. Backscattered electron image (a) and X-ray images of Si (b) and W (c) of a 5 layer Si_3N_4-W FGM.

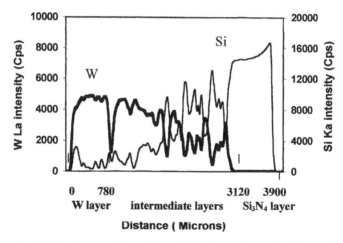

Figure 5. EMPA line profile of Si and W along line as in Figure 4(a).

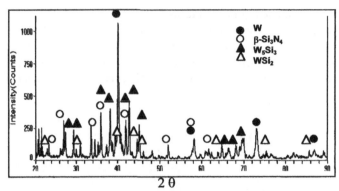

Figure 6. XRD pattern of intermediate layers of Si_3N_4-W FGM.

ACKNOWLEDGMENTS

This work was supported by Sandia National Laboratories, Albuquerque, New Mexico. Sandia is a multiprogram laboratory operated by Sandia Corporation, a Lockheed Martin Company, for the United States Department of Energy under contract DE-AC04-94AL85000. The authors would like to thank Dr. Ronald Loehman and Dr. Nelia Dunbar for their assistance and advice.

REFERENCES

[1]V.K. Pujari, D.M. Tracey, M.R. Foley, N.I. Paille, P.J. Pelletier, L.C. Sales, C.A. Wilkens and R.S. Yeckley, "Reliable Ceramic for Advanced Heat Engines," *Am. Ceram. Soc. Bull.*, **74** [4] 86-90 (1995).

[2]K. Tsuda, A. Ikegaya, K. Isobe,N. Kitagawa, and T. Nomura, "Development of Functionally Graded Sintered Hard Materials," *Powder Metallurgy*, **39** [4] 296-300 (1996).

[3]G. He. D.A. Hirschfeld, and J. Cesarano III, "Processing and Mechanical Properties of Si_3N_4 Formed by Robocasting Aqueous Slurries," in proceedings of *24th International Conference & Exposition on Engineering Ceramics and Structures*, paper No. C-071-00, Cocoa Beach, FL, Jan. 27, 2000.

[4]G. He, "A New Freeform Fabrication of Silicon Nitride Ceramics and Tungsten Metals," *M.S. Thesis*, New Mexico Institute of Mining and Technology, Socorro, New Mexico, May, 2000.

[5]J. Cesarano III, R. Segalman and P. Calvert, "Robocasting Provides Moldless Fabrication from Slurry Deposition," *Ceramic Industry*, **148** [4] 94-102 (1998).

[6]J. Cesarano III and P. Calvert, "Freeforming Objects with Low-Binder Slurry," US *Patent No.* 6,027,326, Feb. 22, 2000.

[7]J.H. Brophy, L.A. Shepard, and J. Wulff, "The Nickel-Activated Sintering of Tungsten," pp. 113-135 in *Powder Metallurgy*, Edited by W. Leszynski, Interscience Publishers, New York, 1961.

FABRICATION OF SiC/Cu FUNCTIONALLY GRADIENT MATERIAL BY GRADED SINTERING

Yunhan Ling   Changchun Ge   Jiangtao Li   Chao Huo

Laboratory of Special Ceramics & P/M,
University of Science & Technology, Beijing 100083, P.R.China

ABSTRACT

SiC/Cu functionally gradient material (FGM) is a new composite that has excellent mechanical properties since it can efficiently mitigate interlayer thermal stress induced by their mismatch thermal expansion coefficients during high heat flux. It integrates the advantages of silicon carbide such as high melting point and strength, as well as copper's high heat conductivity and plasticity. Thus it exhibits satisfactory heat corrosion and thermal shock resistance and will be a prospective candidate as first-wall materials facing plasma in thermo-nuclear reactors. Due to the dramatic difference of melting point between silicon carbide and copper, conventional processes meet great difficulties in fabricating this kind of FGM. A novel approach termed graded sintering is proposed and for the first time a near dense SiC/Cu FGM that contains 0-100% compositional spectrums has been successfully fabricated. In this paper, the temperature distribution in FGM during self-resistance sintering was deduced by Fourier's heat conduction law; various means were used to characterize the sintering effects of SiC/Cu FGM, with special emphasis on relative density and microstructure analysis.

INTRODUCTION

Fusion energy is considered an ideal alternative source that can secure increasing needs in the future and magnetic-confinement of a deuterium-tritium (D-T) plasma fusion reactor such as Tokamak will be a promising device to achieve this goal [1]. Materials, especially those directly exposed to high heat flux such as the first wall and divertor, play key role among all factors concerned. However, in order to realize sustainable operation, they are firmly confined to those that have low atomic number (Z), low activation, high melting point, physical and chemical sputtering resistance, structural stability and insensibility to magnetic field [2].

As a low-Z material, SiC has a series of advantages for use in fusion reactors, such as excellent high temperature properties, corrosion resistance, low density, and especially low induced activation after irradiation [3-6]. It is well known that copper has very good thermal conductivity and plasticity. A material combining both advantages of SiC and Cu will exhibit satisfactory heat corrosion and thermal shock resistance, and will be a prospective candidate for future use in thermo-nuclear fusion devices. However, to join SiC and Cu as heat sinks meets great difficulties due to their dramatic difference of properties as listed in Table I. The first one is the mismatch thermal expansion coefficients which will cause great thermal stress that always leads to cracks or peeling on SiC-Cu interface during fabrication or service. Besides, the great melting point discrepancy and no overlap of sintering temperature ranges makes it difficult to fabricate a full compositional spectrum (0-100%) SiC/Cu FGM even by delicate means [7-8], and SiC/metal graded composite was thought to be eliminated from consideration through simultaneous sintering [9].

In this paper, the idea of FGM was adopted and the possibility of fabricating SiC/Cu FGM, utilizing their distinct electrical resistance, by gradient sintering under ultrahigh pressure (GSUHP) is explored. The temperature distribution was analyzed by Fourier heat conduction law. Various means were used to characterize the sintering effects of SiC/Cu FGM, with special emphasis on the relative density and microstructure analysis.

Table I Main physical properties of SiC and Cu (at room temperature)

Materials	Density (g/cm^3)	Melting point (K)	Thermal expansion coefficient $(10^{-6}/K)$	Thermal conductivity, $(W/m \cdot K)$	Elastic modulus (GPa)	Electric resistivity $(x10^{-8}$ $\Omega \cdot m)$	Tensile strength (MPa)	Poisson ratio
SiC	3.2	2873*	4.7	50-75	430	10^6	400	0.19
Cu	8.9	1356	17	400	85	1.78	314	0.33

* decompose

PRINCIPLES OF GRADED SINTERING

In view of the large resistivity difference between SiC and Cu, it can be expected that a gradually ascending resistance distribution and thus an ever-increasing temperature zone will establish with strong electric current passing through FGM green compact along its compositional changing orientation (from Cu to SiC). Fig.1 shows the illustrative model of SiC/Cu FGM. It is reasonable to assume that no macroscopic liquid migration within FGM sample for very short duration. To simply analyze the sintering process, we suppose:

(1) FGM green compact was circumferentially surrounded with heat insulator, so we consider an adiabatic process in radial direction, and temperature is only a function of height (Z)

(2) Radiation neglected, only thermal conduction exists along Z coordinate and conforms to Fourier's heat conduction law

(3) Temperature influence omitted, and resistivity (ρ) increases linearly along Z and take the form $\rho = k' \cdot Z$, where k' is assumed constant

(4) Heat conduction is in steady state, heat conductive coefficient is not affected by temperature but is also a function of Z and briefly defined in the following way: $k = k'' \cdot Z$, k'' is assumed constant.

Choose a tiny thickness of $\triangle Z$ along Z coordinate , then

(a) The heat input by conduction at Z is $\pi R^2 q_z|_z$, while the output at $Z+\triangle Z$ is $\pi R^2 q_z|_{z+\triangle Z}$, where q_z is heat flux (units, W/m^2)

(b) Exothermic velocity of Joule heat of unit $\triangle Z$ is $\pi R^2 \cdot \triangle Z \cdot J^2 \cdot \rho$, in which J is current density (units: A/m^2). When heat conduction is in equilibrium, we obtain:

$$\pi R^2 q_z|_{z+\triangle Z} - \pi R^2 q_z|_z + \pi R^2 \cdot \triangle Z \cdot J^2 \rho = 0 \qquad (1)$$

or

$$\frac{q_z|_{z+\triangle z} - q_z|_z}{\triangle Z} = -k' J^2 \cdot Z . \qquad (2)$$

When $\triangle Z$ approaches zero, Eq.2 can be expressed, after taking the limit form, as:

$$\frac{dq_z}{dZ} = -k' J^2 \cdot Z \qquad (3)$$

According to the hypothesis (2) and (4), Eq.3 gives:

Functionally Graded Materials 2000

$$q_z = -k'' \cdot Z \cdot \frac{dT}{dZ} \quad \text{, accordingly}$$

$$\frac{dq_z}{dZ} = -k'' \cdot \frac{dT}{dZ} - k'' \cdot Z \cdot \frac{d^2T}{dZ^2} \qquad (4)$$

Whence Eq.5 follows by replacing Eq.3 with Eq.4

$$\frac{d^2T}{dZ^2} + \frac{1}{Z} \cdot \frac{dT}{dZ} - \frac{k'}{k''} \cdot J^2 = 0 \qquad (5)$$

By integrating above differential equation two times, it follows that:

$$\frac{dT}{dZ} = \frac{k'}{2k''} J^2 \cdot Z + \frac{C_1}{Z} \qquad (6)$$

$$T = \frac{k' \cdot J^2}{4k''} \cdot Z^2 + C_1 \ln Z + C_2 \qquad (7)$$

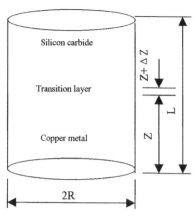

Fig.1 Sketch of SiC/Cu FGM model

Where, C_1 and C_2 are integrating constants, respectively.

When $Z=0$, $\frac{dT}{dZ}$ cannot be infinite; by setting $T=T_{Cu}$ then gives $C_1=0$ and $C_2 = T_{Cu}$. Consequently:

$$T - T_{Cu} = \frac{k' \cdot J^2}{4k''} Z^2 \qquad (0<Z<L) \qquad (8)$$

Where, $T_{Cu,}$ and L are temperature of copper side in steady state and thickness of FGM respectively.

It is clear from Eq.8 that an escalating temperature profile will be constructed when electric current passes through FGM green sample. In other word, by controlling resistance distribution, graded sintering of FGM may be feasible.

Strictly speaking, the de facto temperature distribution pattern, which will be mentioned in a later section, is by no means so simple. Even so, the gradient trend of temperature profile in SiC/Cu FGM is beyond doubt.

EXPERIMENTAL PROCEDURES

To satisfy main demands of the above hypotheses, an experimental setup was designed to fabricate SiC/Cu FGM, which is shown in Fig.2. This device consisted of a mechanical press, pressure vessel and associated electrical and hydraulic system. FGM assembly, which contained SiC/Cu FGM green compact, was placed in the pressure vessel. The pressure was applied by raising the bottom anvil with force provided by lower hydraulic ram. The sample assembly was encapsulated with pyrophyllite sleeve, which was used as heat/electric insulator in graded sintering. Graphite and steel platelets were used as sealing and pressure reinforced components, respectively, as well as electric conductors. The alternating current (A.C.) passed through the sample assembly and sintered the FGM by Joule heat.

Ultra-fine SiC powder with average particle size of 150nm, purity of more than 99% and Cu powder with particle size of -44μm, purity of more than 99.9% were used. B$_4$C powder (15wt%)with mean particle size of 800 nm and purity of about 95% was employed as

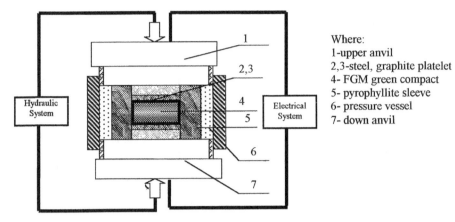

Where:
1-upper anvil
2,3-steel, graphite platelet
4- FGM green compact
5- pyrophyllite sleeve
6- pressure vessel
7- down anvil

Fig.2 Schematic illustration of the experimental setup

sintering additive to activate densification of SiC and improve its toughness. Chromium is considered good wettable agent [10], so a modicum of Cr (1.5wt%) was added to copper.

Silicon carbide, boron carbide and copper powder were mixed and milled, and powders with different compositions were stacked layer by layer in a steel mould preinstalled with cylindrical graphite foil of thickness 0.2mm to form green compact of $\phi 20 \times 10$ mm^3.

The volume ratio of ceramics in the graded layer was determined according to the design of compositional distribution: $C = (x/d)^p$, where C is the volume fraction, x is the relative distance from the surface, d is the thickness of the FGM layer, and p is the compositional distribution factor. In the present work, 6-layered FGM with compositional spectrum ranging from 0-100% of different p, i.e., p=0.6, 1.0, 1.4,1.8, were prepared.

Graded sintering was performed under pressures of 3000-5000MPa, electric power input of about 12kW(7.5V, 1600A) and sintering time 30-60 seconds. Ultrahigh pressure and high power input are prerequisite for shortening sintering time to hinder diffusion in SiC/Cu FGM. The density of specimens of SiC/Cu FGM was measured by Archimedes' method, and scanning electron microscopy (SEM) was used to examine the microstructure.

RESULTS AND DISCUSSION
Densification Effect

When strong current passed through the SiC/Cu FGM green compact, a gradient temperature profile quickly established, the highest temperature was located in ceramic side without question. It can be inferred that part of Joule heat will conduct to Cu-rich side, due to their great melting point difference (about 1500K), and the densification of SiC/Cu FGM should be expected to be increased from SiC side to Cu side. But it is not always that case, as seen from Fig.3. The relative densities drop from 2nd to 4th layer (20-60 vol% Cu) when p=0.6 and p=1.0, that may be caused by high temperature compared to Cu melting point that results in Cu partially evaporating or liquid Cu dispelling from SiC matrix, another reason for the phenomenon may be attributed mainly to the loss of heat absorbed by phase transition of copper, which reduce the general temperature of that compositional part. Nevertheless, the densification of FGM increased from ceramic side to copper side when p>1.0, this is due to lower Cu content in overall FGM compared to that of p<1.

From Fig.3, it is also pronounced that the densification of ceramic-rich side dropped when p increased. Heating style is causative of this change, as we know, the resistivity of silicon carbide is far higher than that of graphite (about $10^{-5}\Omega \cdot m$), accordingly, the electric current applied to SiC/Cu FGM will bypass graphite foil at that thickness Z of larger SiC resistance compared to that of graphite and thus the ceramic rich layers are generally sintered by energy from outside heater. In this way, the temperature distribution can also be deduced by Fourier's heat conduction law of non-steady state (to be discussed in a separate paper), and was associated with time dependent term, thermal conductivity, density and heat capacity of ceramics. The sintering efficiency of indirect heating styles is always lower than that of direct one, as a result, we may conclude that transient sintering is not conducive to the densification of SiC/Cu FGM of a higher p value. By way of parenthesis, because the resistance of graphite is much higher than that of copper (in our present work, R_C/R_{Cu} is about 7000, while R_{SiC}/R_C is about 40), the highest temperature still drops on the ceramic side; thus the general gradient temperature profile in SiC/Cu FGM remains unchanged.

The sintering effect of the whole SiC/Cu FGM was shown in Fig.4. It is clear that pressure has important influence on the FGM densification. When p increases, the densification of SiC/Cu FGM declines by the same token above mentioned.

Heat treatments of sintered SiC/Cu FGM were conducted and the results were shown in Fig.5. We can learn that a near dense SiC/Cu FGM was achieved when secondary power of 60-80% of initial input and the extension of 30 seconds were applied. Heat treatment can, by analogy, improve ceramic sintering and promote liquid copper relocating or infiltrating into the ceramic matrix, that means the further densification of sintered SiC/Cu FGM proceeds. Even in this case the gradient temperature distribution of SiC/Cu FGM still remains, for the macroscopic composition of SiC/Cu FGM retained under ultrahigh pressure and very short heating duration.

Microstructure Analysis

Fig.6 to Fig.12 reveal the development of SEM morphology of SiC/Cu FGM. Fig.6 depicts the overall 6-layered FGM backscattering image and a good graded compositional transition is found. This reflects no macro elemental migration during short time sintering. Fig.7 demonstrates good sintering with fine and homogeneous particle distribution of monolithic ceramic layer. From a higher magnification image of Fig.8, it is obvious that ultra-fine SiC has been sintered and bonded well together, while B_4C and pores were dispersed uniformly among them. Fig.9-Fig.12 present images of 20vol%Cu to 80vol%Cu content, comparison of these figures shows that graded composition is appreciable and residual pores are mostly concentrated on copper phase or its boundary, that might be attributed to the residual gas which could not be removed under high pressure and short duration.

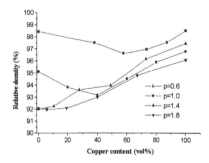

Fig.3 Effects of copper content on relative density
(7.5V, 1600A; 5000MPa; 40s)

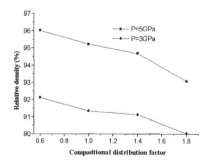

Fig.4 Relative density as a function of different p value
(7.5V, 1600A; 40s)

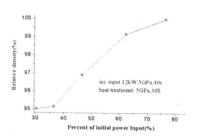

Fig.5 Effects of power of heat treatment on densification of p=1.0 SiC/Cu FGM

Fig.6 Backscattering image (BSI) of overall 6-layered SiC/Cu FGM

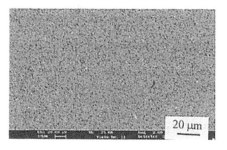

Fig.7 BSI of monolithic ceramics layer

Fig.8 Magnification of ceramic layer (BSI)

Fig.9 BSI of 20vol%Cu layer

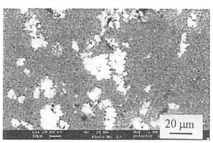

Fig.10 BSI of 40vol%Cu layer

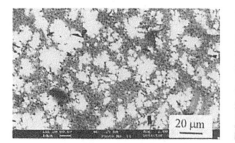

Fig.11 BSI of 60vol%Cu layer

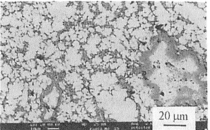

Fig.12 BSI of 80vol%Cu layer

Functionally Graded Materials 2000

Interface Investigation

As it is known that SiC and Cu are immiscible, their bonds may be poor in service. The improvement should be focused on introducing phase binder. It should be pointed out that compounds produced by reaction of binder and major phases in most cases are harmful. In the present work, Cr was employed to be wettable agent of SiC and Cu. XRD patterns (Figure not shown) of each layer of SiC/Cu FGM demonstrate no interface reaction. Fig.13 and Fig.14 delineate the SEM of 20vol%Cu layer and elemental distribution. The matrix of fine SiC and B_4C particles was impregnated by liquid copper, which revealed a good permeating ability of copper under ultrahigh pressure despite its poor wettability under atmosphere. Fig.15 and Fig.16 exhibit the relations of SiC, Cr and Cu. it can be seen that Cr slightly diffused into copper and reflects good wettability between SiC and Cu. With reference to Fig.12 and Fig.15, an interesting chromium ring is found, the explanation for that awaits further analysis.

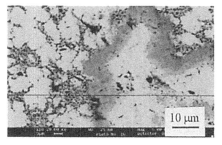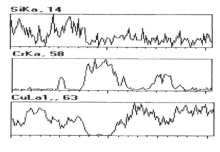

Fig.13 SEM of 20vol%Cu layer of SiC/Cu FGM
(Dots from central particle number position 1 to 5)

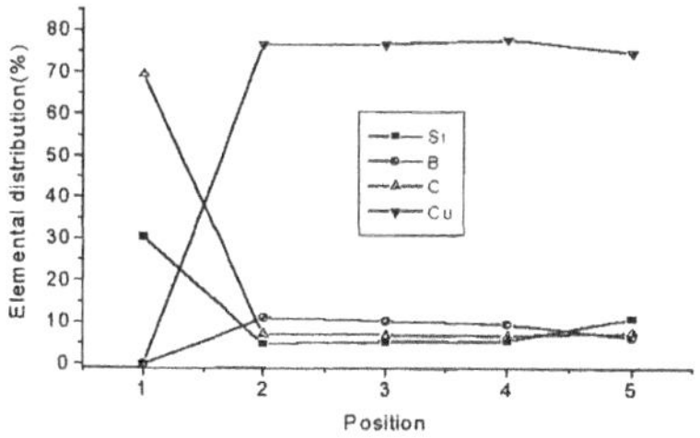

Fig14. EDS analysis of dots of Fig.13

Fig.15 BSI of 80vol%Cu layer

Fig.16 Elements linear scanning of Fig.15

CONCLUSIONS
(I) Temperature distribution in SiC/Cu FGM during self-resistance sintering was simply deduced and it is proportional to the square of thickness and current density applied.
(II) Based on the gradient temperature profile, a novel approach termed graded sintering is proposed, with which nearly dense SiC/Cu FGM that contained compositional spectrum ranging 0-100% has been successfully fabricated.
(III) Sintering effect of SiC/Cu FGM varies with different compositional distribution factors (p), increase of pressure and the power input of heat treatment will promote densification.
(IV) Microstructure analysis shows that a good sintered and graded composition of SiC/Cu FGM has been obtained.

ACKNOWLEGEMENT
The authors would like to express their thanks for the financial support of China National Committee of High Technology New Materials under grant No.863-715-011-0230.

REFERENCES
[1] M. Meade Dale, "Tokamak Fusion Test Reactor D-T Results", *Fusion Engineering & Design*, **30**,13(1995)

[2] Z.Xu and H.W.Ni, "Ceramics for Nuclear Energy"; pp.200 in *Modern Functional Ceramics*, National defense Industry Press, Beijing, 1998.

[3] A.Donato and R.Andreani, "Material requirements and perspectives for future thermonuclear fusion reactor", *Fusion Technol.*, **26**, 58-72(1996)

[4] L. L.Snead, R. H. Jones, A.Kohyama, P.Fenici, "Status of silicon carbide composites for fusion", *J. Nucl. Mater,* **233-237**, 26-36(1996)

[5] P.Fenici, Rebelo A.J. Frias, R.H.Jones, A.Kohyama, L.L.Snead, " Current status of SiC/SiC composites R&D", *ibid*, **258-263**, 215-225(1998).

[6] Everett E Bloom, " The challenge of developing structural materials for fusion power system", *ibid*, **258-263**, 257-264 (1998).

[7] P.Czubarow and D.Seyferth, "Application of poly (methylsilane) and Nicalon polycarbosilane precursor as binder for metal/ceramic powders in preparation of functionally graded materials", *J. Mater. Sci.*, **32**[8] 2121-2130(1997).

[8] Y.F.Lee, S.L.Lee, C.L.Chuang, J.C.Lin, "Effects of SiCp reinforcement by electroless copper plating on properties of Cu/SiCp composites", *Powder Metallurgy*, **46**[10] 3491-3499 (1998).

[9] A.Mortensen and S.Suresh, " Functionally graded metals and metal-ceramic composites: part I Processing", *Int. Mater. Rev.*, **40**[6] 239-265(1995).

[10] P.Xiao and B.Derby, "Wetting of silicon carbide by chromium containing alloys", *Acta Mater.*, **46**[10] 3491-3499 (1998).

Functionally Graded Materials 2000

SINTERING OF FGM HARDMETALS IN DIFFERENT CONDITIONS: SIMULATION AND EXPERIMENTAL RESULTS

Michael Gasik and Baosheng Zhang
Helsinki University of Technology,
P. O. Box 6200,
FIN-02015 HUT, Finland

ABSTRACT

A model for describing deformation behaviour and stress evolution of graded powder compacts during the entire sintering cycle was proposed. The constitutive equation, based on thermally coupled elastic-visco-plastic approach, was implemented into ABAQUS code. A WC-Co hardmetal plate was simulated with various parameters, such as mean particle size, green density distribution and cobalt gradation parameter. The results show that the deformation behaviours and stress history of graded powder compacts during heating, sintering and cooling could be predicted for optimisation of sintering process.

INTRODUCTION

The powder metallurgy is a very important processing method for fabrication of Functionally Graded Materials (FGM).[1-3] Here special attention should be given to sintering behaviour of FGM, because of higher probability of appearance of bending or distortion of the sintered bodies, or cracks, due to the difference in shrinkage between the phases.[1-5] In these works, the stress state and cracks in heterogeneous powder compacts induced by non-uniform shrinkage have been analysed.[1-5] Finite element analysis for sintering process has been widely used to predict deformation behaviour during isothermal sintering of graded compacts, including WC-Co hardmetals.[3-7] Graded compacts may exhibit different behaviour (elastic to visco-plastic and then back to elastic) at different stages of sintering.

The deformation and stress during entire cycle (heating, isothermal holding and cooling) are thus necessary to know for optimisation of the sintering process and prevention of defects and undesired distortions. In this work, the finite element analysis based on the thermal elasto-visco-plastic sintering model proposed by the authors[8,9] was carried out for sintering of functionally graded hardmetals. A gradated hardmetal plate sintering was simulated with various parameters, like particle size, green density distribution and cobalt gradation parameter.

2. MODEL FOR SINTERING OF FGM

The behaviour of sintering graded bodies is considered to be adequately represented by elasto-visco-plastic model.[8,9] The basic feature of the full equation is that the deformation is divided into four different parts: elastic, viscous, sintering and thermal deformation[9], that is a substantial extension comparing to other models.[5-7] This model was applied for calculation of the distribution of temperature, density, strains/strain rates and displacements, as well as normal and shear stresses:

$$\{d\sigma\} = [D^*](\{d\varepsilon\} - [\eta]^{-1}\{\sigma\}dt - \{d\varepsilon^s\} - d\{\alpha T\})$$ (1)

where $\{d\sigma\}$, $\{d\varepsilon\}$ are stress and strain increment matrixes, $[\eta]$ is viscoplastic matrix, $d\{\alpha T\}$ is thermal strain increment, $\{d\varepsilon^s\}$ is volumetric sintering strain increment, and $[D^*]$ is a modified elastic compliance matrix.[6,8,9]

The constitutive equation (1) was also coupled with the heat transfer, which is based on the energy balance equation and material parameters like CTE (coefficient of thermal expansion) α, thermal conductivity (λ), specific heat (c) and density (ρ). Here the CTE and other parameters as a function of cobalt volume fraction and porosity were calculated using an original micromechanical model.[10]

For the solid state sintering, the dominant mechanism of mass transport is assumed to be grain boundary diffusion. The grain growth during sintering was assumed to be described as

$$d^3 = d_0^3 + \int_0^\tau C_1 T(\tau)\exp(\frac{C_2}{T(\tau)})d\tau \quad ,$$ (2)

where d_0 is the initial grain size, $T(\tau)$ is the current temperature, and the constants C_1, C_2, are obtained from experimental data.[5,6,8,9]

3. SINTERING OF GRADED WC-Co PLATE

For the stress and deformation analysis, a graded WC-Co plate was considered (Fig. 1). This plate of $25 \times 31 \times 6$ mm is being sintered horizontally with 6 wt. % Co at the top and 10 wt. % Co at the bottom. The cobalt concentration change across the thickness is governed by a power law with the gradation parameter p:

$$\%Co(z) = \%Co(0) + (\%Co(H) - \%Co(0)) \cdot \left(\frac{z}{H}\right)^p$$ (3)

where H is plate thickness, z - vertical coordinate and $p = 0.5... 2.5$ (i.e. $p = 1$ means the linear function).

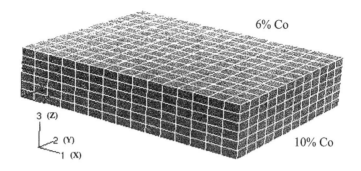

6% Co

3 (Z)

2 (Y)

1 (X)

10% Co

Fig. 1. FEM mesh for simulation of sintering of FGM plate specimen

In these simulations, the program code was implemented into ABAQUS (Standard and CAE), where quadric elements have been chosen for the meshing for fully coupled thermal displacement analysis. Heating and cooling rates have been chosen of 10 K/min and sintering was performed at 1573 K for 30 min. The same parameters (except cooling rate) have been also used in dilatometric experiments.[8,9]

4. EXPERIMENTAL

Sintering was performed in an optical dilatometer[9] in argon flow. The measurements have been done with a CCD-camera and a video frame grabber on a Pentium PC. The resolution of the screen was 1280 x 1024 pixels, which corresponds to the raw image resolution of 1.3±0.05 μm/mm of the specimen length. The reproducibility of length measurements in recent experiments was about 0.15%.

Homogeneous specimens of 10 mm diameter and 10 mm height were made by CIP of Nanocarb[TM] and Mycrocarb[TM] WC-Co powders. Specimens with 10% Co have higher expansion during heating than of 6% Co but very close shrinkage to each other during isothermal soaking at 1300°C and 1350°C (Fig. 2). Axial shrinkage is also starting between 800 and 900°C with the highest rate between 1100 and 1250°C. In all these cases, shrinking continues during cooling, which is a combined effect of both sintering and thermal contractions. These data have been used to evaluate strain differences induced in FGM specimens in the simulated sintering conditions.

5. RESULTS OF SIMULATION

Fig. 3 shows the maximal tensile and compressive stresses level in the gradated plate of Fig. 1. When p value decreases, plane stresses are significantly increasing, approaching their maximal values for bi-layer (non-graded) material (p = 0). Fig. 4 presents combined (von Mises) stresses contours (not to scale) in the ¼ of the plate and its shape after heating (a) and after sintering and cooling (b). One may see that the plate is bent toward higher cobalt content (10% Co) after processing.

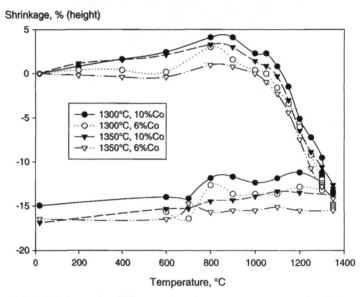

Fig. 2. Axial shrinkage for different sintering temperatures and cobalt content.

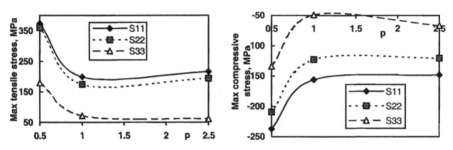

Fig. 3. Maximal tensile and compressive stresses in the plate (Fig. 1) vs. gradation parameter ($S11 = \sigma_x$, $S22 = \sigma_y$, $S33 = \sigma_z$).

In order to prevent bending and decrease residual stress level, further optimisation of the component parameters is necessary. This could be made in several ways[4-9] like green shape adjustment, green density and/or particle size distribution, position of the plate in the furnace, proper substrate selection, etc. Here, variation of the average particle size was considered as an example.

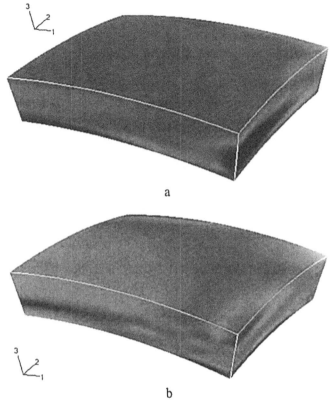

Fig. 4. Gradated plate shape after heating up (a) and cooling (b).

The new plate design (Fig. 1) additionally included particle size gradients. The mean particle size was chosen to vary from 0.3 μm (top; 6% Co) to 0.8 μm (bottom; 10% Co). The idea was to hinder shrinkage of the cobalt-rich part to obtain smaller warpage of the plate. Fig. 5 shows the final shape of this plate design after cooling.

In this case, the plate warped in the opposite direction, to cobalt-poor side, on the contrary of the previous case (Fig. 4, b). One may conclude that the particle size is also a very important parameter to control sintering and to obtain a compliant shrinkage in FGMs. Similar methods have been used[3,4] for stainless steel/zirconia FGM to control sintering rates and to get uniform shrinkage.

However, recent simulation results also exhibit higher stress levels, associated with the particle size gradient. Generally, the larger is the difference of particle size between the two sides, the higher the stress will be in FGM components. When Fig. 5 and Fig. 4,b, are compared, it could be possible to find the optimal combination of cobalt concentration and WC-Co particle size gradients, which may result in the flat shape of the plate after sintering.

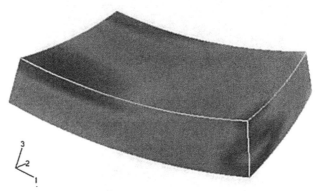

Fig. 5. The distortion of the FGM plate (Fig. 1) due to the particle size gradient.

6. CONCLUSIONS

The behaviour of the graded compacts during heating, sintering and cooling can be simulated with FEM technique using developed thermal elasto-visco-plastic model. If a compositional gradient (e.g. cobalt for the WC-Co FGM hard-metals) is only used, the final distortion and residual stresses of the sintered component may be significant.

For the WC-Co plate with both graded cobalt concentration and particle size it was demonstrated that an optimal combination of these gradients exists, which may retain the flat shape of the plate after sintering.

The optimisation of FGM design should consider also additional gradients of such parameters like distribution of green density, particle size and possibly FGM green shape to minimise residual stresses incurred during entire sintering cycle.

ACKNOWLEDGEMENTS

This work is part of the project BE97-4176, supported by the European Commission under the contract BRPR-CT97-0505. The dilatometric experiments have been carried out by Mrs. P. Kervinen and Mr. J. Ylikerälä.

REFERENCES

[1] *Functionally Graded Materials: Design, Processing and Applications*, Eds. Y. Miyamoto, W.A. Kaysser, B.H. Rabin, A. Kawasaki, and R.G. Ford. Kluwer Academic Publishers, Bosten/Dordrecht/London, 1999, 320 pp.

[2] S. Suresh and A. Mortensen, *Fundamentals of Functionally Graded Materials*, IOM Communications Ltd., London, 1998, 166 pp.

[3] R. Watanabe, "Powder processing of functionally gradient materials", MRS Bull., **20** [1] 32-34 (1995).

[4] R. Watanabe and A. Kawasaki, "Design, fabrication and evaluation of functionally gradient material for high temperature use"; pp. 285-299, in *Mechanics and Mechanisms of Damage in Composites and Multimaterials*, Ed. D. Baptiste. Mechanical Engineering Publications Ltd., London, 1991.

[5] H. Riedel and T. Kraft, "Distortions and Cracking of Graded Components During Sintering", *Mater. Sci. Forum*, **308-311** 1035-1040 (1999).

[6] K. Shinagawa and Y. Hirashima, "A Constitutive Model for Sintering of Mixed Powder Compacts", *Mater. Sci. Forum*, **308-311** 1041-1046 (1999).

[7] N. Favrot, J. Besson, C. Colin and F. Delannay, "Cold Compaction and Solid-State Sintering of WC-Co-Based Structures: Experiments and Modelling", *J. Amer. Ceram. Soc.*, **82** [5] 1153-61 (1999).

[8] M. M. Gasik and B. Zhang, "Modelling and Sintering of Functionally Graded WC-Co Materials", *Proc. of European Conf. on Advances in Hard Materials Production: EURO PM 99*, Turin, Italy. European Powder Metallurgy Association, 1999, 449-454.

[9] M. M. Gasik and B. Zhang, "A Constitutive Model and FEM Simulation for the Sintering Process of Powder Compacts", *Computat. Mater. Sci.* **18** 93-101 (2000).

[10] M. M. Gasik, "Principles of Functional Gradient Materials and Their Processing by Powder Metallurgy", *Acta Polytechn. Scand.*, **Ch** 226 1-73 (1995).

FABRICATION OF Al$_2$O$_3$/TiC/Ni GRADED MATERIALS BY PULSED-ELECTRIC CURRENT SINTERING

Youping Ren, Junshan Lin
and Yoshinari Miyamoto
Joining & Welding Research Institute
Osaka University
Ibaraki, Osaka 567-0047
Japan

Guanjun Qiao and Zhihao Jin
State Key Laboratory for
Mechanical Behavior of Materials
Xi'an Jiaotong University
Xi'an 710049
China

ABSTRACT

The sintering condition for symmetric FGMs of the Al$_2$O$_3$/TiC/Ni and (Al$_2$O$_3$-WC/Co) /TiC/Ni systems by pulsed-electric current sintering method was optimized and dense FGMs were fabricated at 1300°C, 10 minutes, and 30 MPa with a heating rate of 100 °C/min. The grain growth was effectively prevented and the fine mirostructure obtained by this rapid sintering. The residual stress produced in the outer Al$_2$O$_3$ and Al$_2$O$_3$-WC/Co layers, which was induced by the thermal expansion mismatch between the inner TiC/Ni and outer layers, was in the range of −180 MPa to −300 MPa. The compressive stress and the dispersion of WC/Co particles in the Al$_2$O$_3$ matrix enhanced the toughness of the outer ceramic layers and developed steep R-curve behavior.

INTRODUCTION

High resistances against heat, wear, oxidation, and corrosion as well as high specific strength are important attributes of engineering or structural ceramics. However, the brittle nature limits applications. Much effort has been devoted over the past couple decades to improve the toughness of ceramics[1]. It has been manifested that the compressive stress existing in the surface of ceramics has many benefits including increased fracture stress and weibull modulus [2,3], thermal-shock resistance [3], toughness [4,5], rolling-contact fatigue life and wear resistance [6]. The compressive stress can be produced by some methods including thermal treatments such as quenching [7], application of low thermal expansion coatings or adopting layered structures which employ the transformation of unstabilized tetragonal ZrO$_2$ to monoclinic phase in the outer layers upon cooling or the thermal expansion mismatch between the outer and inner layers of a symmetric or sandwiched structure.

New symmetric FGMs such as the TiC/Ni/TiC and Cr$_3$C$_2$/Ni/Cr$_3$C$_2$ were fabricated by Pityulin and Merzhannov in Russia [8], and the Al$_2$O$_3$/TiC/Ni by Miyamoto and J.S. Lin in Japan [9,10]. These symmetric FGMs were fabricated by SHS compaction and SHS/HIP, respectively. Strong residual compressive stress is preferable for restraining the initiation or growth of microcracks. This concept of the stress-enhanced toughening has been applied to manufacturing of cutting tools, resulting in the extension of tool life [11].

In the present study, the symmetric FGMs of the Al$_2$O$_3$/TiC/Ni system were fabricated by pulsed-electric current sintering and their mechanical properties were evaluated and discussed. Pulsed-electric current sintering is a new process providing a means by which ceramic powders can be easily sintered at lower temperature and shorter time comparing with conventional sintering methods [12].

EXPERIMENTAL

FGMs Fabrication

The starting materials used are Al_2O_3, TiC, Ni, Mo_2C, WC, and Co powders with an average particle size of 0.4 μm, 1.4 μm, 1.0 μm, 1.0 μm, 1.5μm and 1.0μm, respectively. These powders were wet-mixed in pre-determined compositions for over 48 hours by ball milling, and then dried in a vacuum oven. The sample was designed to have a symmetric five-layers structure. The compositions of every layer are shown in Fig. 1. In order to increase the compressive residual stress of the outer layer, the Al_2O_3-WC/Co composite with a lower thermal expansion coefficient than Al_2O_3 was also used. The mixed powders were carefully placed into a graphite die with 30 mm diameter and subjected to pulsed-electric current sintering. The sintering condition optimized was 1300 °C with a heating rate of 100 °C /min for 10 minutes under pressure of 30MPa.

According to the structure design of FGM, much of nickel exists in the central layer (about 20vol.%). Nickel formed a eutectic liquid with TiC at 1150 °C – 1200 °C in this sintering process though the melting point of nickel is as high as 1400 °C, which caused squeezing out of the molten nickel phase from a cylindrical sample and damage to a carbon mold. In order to prevent the squeezing out of molten nickel, the mixture of equi-weight BN + ZrO_2 powders was successfully used to wrap the sample. Figure 2 is a schematic diagram of this sample arrangement for sintering.

Testing

The sintered sample was a disk with 30mm in diameter and 6 mm in thickness, which was polished to a diamond surface finish of 3μm. The compressive stress at the surface of FGMs was determined by the $\sin^2\psi$-2θ method using an X-ray diffraction peak (416) of Al_2O_3. Hardness and indentation toughness were measured by using a Vickers hardness-testing machine with an indentation load of 98N for 10 seconds. The indentation-induced crack length, 2c, was measured using an optical microscope and the indentation toughness, K_c, was calculated using the following equation [13]:

$$(K_c\phi/H_v a^{1/2})/(H_v/E\phi)^{2/5}=0.129(c/a)^{-3/2} \tag{1}$$

where ϕ is a material-independent constant. H_v, E, and a are Vickers hardness, Young's modulus and half-diagonal length of the indentation, respectively.

The sintered sample was cut into rectangular bar specimens with the size of 2x6x25 mm. The surface of a specimen was ground and the edges were slightly chamfered and the tensile face was polished to a 1μm finish. Flexural strength was measured by means of three-point bending test with a span length of 18 mm and crosshead speed of 0.5 mm·min^{-1} using Instron 1185 machine. The indentation-strength method was used to evaluate the R-curve behavior of FGMs [14,15]. The indentations with loads of 9.8-196N were applied at the center of a sample beam, and subsequently fractured by three-point bending. The indentation strength, crack length and load in bending experiments were used to calculated R-curves: $K_R=k(\Delta c)^m$, where K_R is the fracture resistance, Δc is crack extension, and k and m are material constants. The fractured surface was observed by scanning electron microscopy.

RESULTS AND DISCUSSION

Optimization of Sintering Condition

Figure 3.1 shows the curves of relative densities for Al_2O_3 and Al_2O_3-WC/Co ceramics as functions of temperature at a fixed holding time of 10 minutes and a pressure of 30 MPa. The densities of two kind ceramics increase with the sintering temperature. In addition, the density of Al_2O_3-WC/Co ceramic was higher than Al_2O_3 at the same condition, that is probably due to the

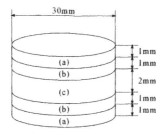

(a) Al_2O_3, $Al_2O_3+7.62vol.\%WC+2.38vol.\%Co$
(b) $Al_2O_3+46.8vol.\%TiC+2.5vol.\%Ni$
(c) $TiC+19.8vol.\%Ni+6.5vol.\%Mo_2C$

Fig. 1.   Structure and composition of a sintered FGM body.

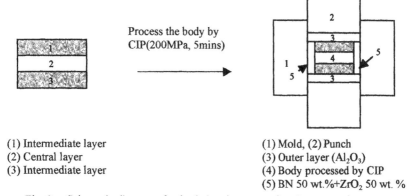

Process the body by
CIP(200MPa, 5mins)

(1) Intermediate layer
(2) Central layer
(3) Intermediate layer

(1) Mold, (2) Punch
(3) Outer layer (Al_2O_3)
(4) Body processed by CIP
(5) BN 50 wt.%+ZrO_2 50 wt. %

Fig. 2.   Schematic diagram of pulsed-electric current sintering processing.

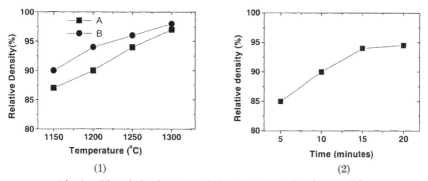

(1)   (2)

Fig. 3.   The relation between relative density and sintering conditions.
(1) Relative density versus temperature (30 MPa, 10 minutes)
A: Al_2O_3, B: $Al_2O_3+7.62vol.\%WC+2.38vol.\%Co$
(2) The relative density of Al_2O_3 versus holding time

binder action of Co metal for sintering. The plot of the density of Al_2O_3 versus holding time at 1200 C° and 30MPa is shown in Fig. 3.2. Holding time is an important parameter for densification, however, the density showed no apparent change after 15 minutes. Based on these sintering behaviors of the monolithic Al_2O_3 and the Al_2O_3-WC/Co composite, we have decided an optimal sintering condition for our FGMs as 1300°C, 30 MPa, 10 minutes and a heating rate of 100 °C/min.

Microstructure Observation

Figure 4 shows the fracture surfaces of two different FGMs of the Al_2O_3/TiC/Ni and the Al_2O_3-WC/Co/TiC/Ni systems. The photos (a) and (b) show the surface layers of the Al_2O_3 and the Al_2O_3/WC/Co, respectively. Photo (c), (d), (e) and (f) show the interfacial regions of the first/second layer, second layer, interfacial region of the second/central layers and central layer of Al_2O_3/TiC/Ni system, respectively. The grain growth could be effectively prevented and fine microstructures obtained, which resulted from the densification condition of lower temperature and shorter time by pulsed-electric current sintering. No microcrack was observed at the layer boundaries.

Mechanical Properties

The mechanical properties of FGMs are listed with reference to the monolithic Al_2O_3 in Table 1. Because of the compressive stress, the local fracture toughness of every outer layer was effectively enhanced. The value of toughness increased with increasing compressive stress in the outer layers. Especially for the Al_2O_3-WC/Co FGM, the bridging and deflection effects due to WC/Co particles to the crack propagation are considered to act besides the stress effect as seen in Fig.5. The WC particles coagulated with Co metal and dispersed in the Al_2O_3 matrix. As a crack reached the WC/Co particles or the Al_2O_3/WC interface, the ductile cobalt metal around WC particles lead to crack blunting or deviation along the interfaces.

Residual Stress

The residual stress can be determined according to the indentation fracture mechanics for a half–penny flaw [16]. The indentation toughness, K_c, and crack length can be related by the following equation:

$$K_C = K_C^0 - 2\sigma_R(C/\pi)^{1/2} \qquad (2)$$

where K_c^0 is the toughness of a ceramic without stress, σ_R is the residual stress in a ceramic and C is the indentation crack length. K_c^0 and σ_R can be obtained by a linear fitting associated with K_c and $C^{1/2}$. Figure 6 shows the plots of K_c versus $C^{1/2}$ of the two FGMs. The indentation loads are 49N, 98N, and 196N. For the Al_2O_3/TiC/Ni and the Al_2O_3-WC/Co/TiC/Co FGMs, the toughness rose with the increase of crack length. The calculated and measured K_C^0 and σ_R are listed in Table 2. Both values showed good coincidences.

R-Curve Behavior

R-curve behavior arises because the additional energy is consumed in the process zone of a crack besides the fracture energy dissipated at the crack tip. The shape of an R-curve reflects the ability of ceramic to tolerate the crack extension and thus the strength reliability. It is very important, therefore, to characterize and understand the R-curve behavior of ceramics as well as the development of materials having an appropriate R-curve behavior. It has been manifested that the surface compressive stress can introduce a steep R-curve in a ceramic by a three-dimensional finite element analysis [17]. In this study, the compressive stress was induced by the thermal

Fig. 4.   SEM images of fractured surface.
(a) Outer Al$_2$O$_3$/WC/Co layer, (b) Outer Al$_2$O$_3$ layer, (c) Interfacial region of Al$_2$O$_3$/ Al$_2$O$_3$-TiC layer, (d) Al$_2$O$_3$-Ti layer, (e) Interfacial region of Al$_2$O$_3$-TiC/TiC-Ni layer, (f) Center TiC-Ni layer

Table I.   Mechanical properties of symmetric FGMs and an Al$_2$O$_3$ ceramic.

Properties \ Sample	Al$_2$O$_3$-WC/Co FGM	Al$_2$O$_3$ FGM	Al$_2$O$_3$
K$_s$(MPa·m$^{1/2}$) by IF(98N)	8.4	6.3	4.5
Hardness(GPa) (98N load)	19.8	19.5	17.5
Flexural Strength(MPa)	670	670	500
Residual stress(MPa) (measured by X-ray)	-300	-180	0

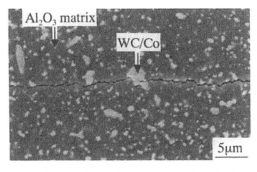

Fig. 5.   A SEM image showing the crack deflection and bridging in the outer layer of Al$_2$O$_3$-WC/Co/TiC/Ni FGMs.

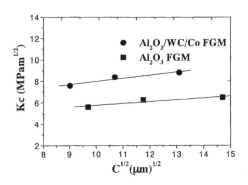

Fig. 6. Indentation toughness vs crack length.

Table II. Comparison of calculated and measured K_o^0 and σ_R.

Materials	$K_o^0(MPam^{1/2})$ calculated	$K_o^0(MPam^{1/2})$ measured	$\sigma_R(MPa)$ calculated	$\sigma_R(MPa)$ measured (by x-ray)
Al_2O_3 FGM	4.0	4.5	-150	-180
Al_2O_3-WC/Co FGM	5.0	—	-260	-300

Table III. Parameters defining the R-curve for two different FGMs and a monolithic Al_2O_3.

Material	k	m
Monolithic Al_2O_3	11.9	0.112
Al_2O_3FGM	31.9	0.188
Al_2O_3-WC/CoFGM	80	0.278

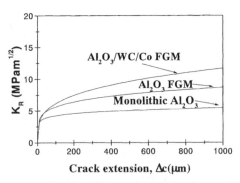

Fig. 7. R-curves as a function of the crack extension for two different FGMs and a monolithic Al_2O_3.

expansion mismatch between the outer and inner layers. Due to the addition of about 20 vol.% nickel whose thermal expansion coefficient is as high as $16 \times 10^{-6}/°C$, an apparent difference in thermal expansion coefficient of $1 \sim 2 \times 10^{-6}/ °C$ exists between the central and outer layers of the $Al_2O_3/TiC/Ni$ FGM, that causes a strong compressive stress of $-180 \sim -300$ MPa in the outer layers. The central layer of TiC/Ni has a tensile stress of $300 \sim 400$ MPa [6], which it can withstand because of the high strength of about 1000MPa. The fracture resistance, K_R, and the crack extension length, Δc, satisfies the following relationship[10]:

$$K_R = k(\Delta c)^m \qquad (3)$$

where k is a constant and m is a characteristic exponent that describes the sensitivity of R-curve behavior. When m is zero, K_R is invariant with the crack extension. The exponent m can be obtained from a slope β of a log-log plot of post-indentation strength, σ, versus indentation load, P.

$$m = (1-3\beta)/(2+2\beta) \qquad (4)$$

$$k = Y\alpha(\beta\gamma)^{-\beta}(1+\beta)^{(1+\beta)} \qquad (5)$$

(α is obtained from the intercept of the log-log plot; $\gamma = P/a_I^{2/(1+\beta)}$ where a_I is the initial crack length) The ratio of the crack length, a_C, and the initial crack length, a_I, can be obtained by the initial crack length, a_I, from the following equation:

$$a_C/a_I = [4/(1-2m)]^{2/(3+2m)} \qquad (6)$$

Table 3 gives the values of k and m of the two FGMs and the monolithic Al_2O_3. Figure 7 shows their R-curves. The FGMs exhibited steep R-curves with the residual compressive stress in the outer layers, that can lead to higher crack growth resistance and damage tolerance.

CONCLUSIONS

Dense symmetric FGMs of the $Al_2O_3/TiC/Ni$ and Al_2O_3-WC/Co/TiC/Ni systems were successfully fabricated at 1300 °C, 10 minutes and 30MPa by pulsed-electric current sintering. A severe problem of squeezing out of the molten nickel from the sample which occurs frequently in densification of ceramic/metal composites by pulsed-electric current sintering was successfully prevented by wrapping it with the ceramic powders of $BN+ZrO_2$. This rapid sintering resulted in effectively preventing the grain growth. Because of the residual compressive stress tailored in the outer ceramic layers, which is induced by the thermal expansion mismatch of the inner and outer layers of FGMs, the outer ceramic layers were strongly toughened. The compressive residual stress in the surface can develop a steep R-curve behavior of FGMs and significantly enhance the crack growth resistance and damage tolerance.

REFERENCES:
[1]A. G. Evans, "Perspective on the Development of High-Toughness Ceramics", *Journal of the American Ceramic Society*, **73**, 187-206 (1990).
[2]R. A. Culter, J. D. Bright A. V. Virkar and D. K. Shetty, "Strength Improvement in Transformation Toughened Alumina by Selective Phase Transformation", *Journal of the American Ceramic Society*, **70**[10] 714-18 (1987).
[3]R. A. Culter, C. B. Brinkpeter, A. V. Virkar and D. K. Shetty, "Fabrication and

Characterization of Slip-Cast Layered Al_2O_3-ZrO_2 Composites", pp. 397-408 in the Proceeding of the 4th Symposium on Ceramic Materials and Components for Engines (Goteborg. Sweden) Edited by R.Carlsson Elsevier. London, U. K. (1992).

[4]F.F Lange, " Compressive Surface Stresses Developed in Ceramics by an Oxidation-Induced Phase Change", *Journal of the American Ceramic Society*, **63** [1-2] 38-40 (1980).

[5]Y. Miyamoto, Z.Li, Y.Kang and K.Tanihata, "Fabrication and Properties of Alumina-Based Hyperfunctional Ceramics with Symmetrically Compositional Gradient Structures", *Journal of The Japan Society of Powder & Powder Metallurgy*, **42** [2] 933-938 (1995). (in Japanese)

[6]L. Y Chao, R. Lakshminarayanan, D K Shetty, and R. A. Culter, "Rolling Contact Fatigue and Wear of CVD-SiC with Residual Surface Compression", *Journal of the American Ceramic Society*, **87**[9] 2307-13 (1995).

[7]H. P. Kirchner, Strengthening of Ceramics, Treatments, Test, and Design Applications. Marcel Dekker, New York, 1979.

[8]A.G. Merzhanov, "Advanced SHS Ceramics: Today and Tomorrow Morning", pp.395 in *Proceedings of the Centennial International Symposium on Ceramics:Toward The 21st Century*, The Ceramic Society of Japan, (1991).

[9]Y.Miyamoto, K.Tanihata, T.Kawai and K.Nishida, "Gas-Pressure Combustion Sintering (SHS/HIP) Using Silicon Fuel", pp.275 in *Proceedings of International. Conference on Hot Isostatic Pressing' 93*, Elsevier, (1994).

[10]J.S.Lin and Y. Miyamoto, " Notch Effect of Surface Compression and Toughening of Graded Al_2O_3/TiC/Ni Materials", *Acta Materialia*, **48**, 767-775(2000).

[11]H.Moriguchi, A.Ikegaya, T.Nomura, Y.Miyamoto, Z.Li and K.Tanihata, "Cutting Performance of Hyperfunctional Ceramics", *Journal of The Japan Society of Powder & Powder Metallurgy* , **42** [12] 1389-1393 (1995) (in Japanese).

[12]M.Tokita, "Development of Large-Size Ceramic/Metal Bulk FGM Fabricated by Spark Plasma Sintering", pp.83-88 in *Proceedings of the 5th International Symposium on Functionally Graded Materials* (1998).

[13]K. Niihara, R. Morena and D. P. H. Hasselman, "Comment on Elastic/Plastic Indentation Damage in Ceramics: The Median/Radial Crack System", *Journal of the American Ceramic Society*, **65**, C-116 (1982).

[14]R.F. Krause, "Rising Fracture Toughness from the Bending Strength of Indented Alumina Beams", *Journal of the American Ceramic Society*, **71**,338-343(1988).

[15]Y.W.Kim and M.Mitomo, "R-curve Behaviour of Sintered Silicon Nitride", *Journal of Materials Science*, **30**, 4043-4045(1995).

[16]D.B.Marshall and B.R.Lawn, "An Indentation Technique for Measuring Stress in Tempered Glass Surface", *Journal of the American Ceramic Society*, **60**, 86-87(1977).

[17]R.Lakshminarayanan, D.K.Hetty and R.A.Cutler, "Toughening of Layered Ceramic Composites with Residual Surface Compression", *Journal of the American Ceramic Society* ,**79,** 29-87(1996).

356

FGM FABRICATION BY SURFACE THERMAL TREATMENTS OF TiC-Ni₃Al COMPOSITES

T. N. Tiegs, M. L. Santella, C. A. Blue, and P. A. Menchhofer
Oak Ridge National Laboratory
Oak Ridge, TN 37831-6087

F. Goranson
Southern Illinois University
Carbondale, IL 62901

ABSTRACT

TiC-Ni₃Al composites containing 30-50 vol. % of Ni₃Al alloy as the binder phase were subjected to both laser and infrared surface treatments in order to assess the effect on microstructure and properties. The laser treatment produced significant thermal transients which resulted in dendritic growth of the TiC grains. Because the laser was operated in a pulsed mode with over-lapping spots, some areas were melted and remelted as the laser was scanned across the surface. The interface between the melt-remelt areas revealed differences in the TiC grain size depending on the local heating and cooling conditions experienced.

INTRODUCTION

Previous studies have shown that TiC-Ni₃Al composites have an excellent combination of strength, fracture toughness, hardness and corrosion resistance [1-9]. As a result, there is interest in using these types of materials for wear applications in diesel engines. Materials of interest contain 30-50 vol. % of Ni₃Al alloy as a binder phase because these levels have thermal expansion characteristics similar to the steel components in the engines. Preliminary wear testing indicates that improved wear resistance could be achieved by decreasing the grain size of the TiC. Achieving fine grain size with the high binder contents is difficult because of the large inter-grain distances and uninhibited grain growth during high temperature processing. In addition, it was thought that changing the TiC grain shape from a highly faceted one to a more rounded equiaxed grain would reduce localized stress at sharp corners. This, in turn, would improve abrasion resistance from any wear debris.

Various methods have been used with conventional and experimental TiC-based cermets to reduce grain size in the dense materials [9-11]. These include reduction of the initial TiC particle size; use of additives to change the interface behavior of the growing TiC grains; employing additives to physically inhibit grain growth; and rapid sintering to minimize high temperature exposure. As an alternative, earlier results showed post-densification thermal treatments were capable of altering the microstructure of the near-surface region [11]. These preliminary experiments were done with a pulsed laser source that was scanned across the surface. Laser surface treatments are characterized by extremely high heating and cooling rates on the order of 1000°C/s. Localized melting and resolidification were found to take place at the surface. Consequently, additional testing was performed to further characterize the microstructure and determine the applicability of this technique to change the near-surface properties.

EXPERIMENTAL PROCEDURE

The samples were fabricated by milling fine TiC* powder, with prealloyed $Ni_3Al^†$ powder, drying, pressing into discs and sintering at 1450°C to densities of >98% T. D. Details of the fabrication can be found in previous references [7-9].

Samples of dense TiC-Ni$_3$Al were surface treated by a ruby laser operated at 18.5 kW in a pulsed mode (7 ms duration at 40 pulses/s) which was scanned across the surface at 0.84 cm/s. Several defocus conditions were used to vary the beam intensity impinging on the surface. The surfaces had been machined prior to the laser treatment. In conjunction with the laser treatments, limited infrared heating tests were also done, but the intensities were too low to significantly alter the near-surface region microstructure. Thus, further characterization work of these samples was not done. Current infrared heating devices are now capable of higher intensities and consequently some additional study may be done. Hardness testing was done with a Vickers diamond indenter at a load of 20 kg. Scanning electron microscopy (SEM) was done on both polished sections and etched bulk samples using back-scattered electron (BSE) imaging and with energy dispersive x-ray analysis (EDAX). Etching of the bulk samples was done with concentrated nitric acid to dissolve away the Ni$_3$Al binder phase.

RESULTS AND DISCUSSION

Macroscopic Observations – The macroscopic effect of the laser treatment on the composite surface is shown in Fig. 1. Significant changes are obvious in the center ~1 mm of the treated bar with some observable changes extending out another ~100 µm. At more intense laser settings, cracks within the laser affected zone (LAZ) are observed as shown in Fig. 2. A cross-section of the LAZ is shown in Fig. 3. Surface distortion is observed due to the localized melting and solidification that takes place. While there is no available data for the Ni$_3$Al-TiC system, in the Ni-TiC system, a eutectic is predicted at ~1280°C and for the compositions used in the present study, liquidus temperatures range from 2100-2500°C [11]. Thus, for melting to occur, the peak temperatures attained in the LAZ were in excess of 1280°C and based on the results of the microstructure, the temperatures were probably closer to the liquidus temperatures for this system. It is also observed from Fig. 3 that the LAZ has a conical shape extending into the bulk from the surface to a depth on the order of 1200 µm. Typically the depth of the affected zone ranged from 700-1400 µm. The conical shape is the same as for a classic heat-affected zone observed in welding operations.

Microstructure – The typical microstructure of TiC-Ni$_3$Al composites consists of equiaxed TiC grains surrounded by the Ni$_3$Al binder phase (Fig. 4). The TiC grains have a core-rim structure due to solution-reprecipitation that occurs during liquid phase sintering. The microstructure within the LAZ is shown in Fig. 5. As shown, there is a dramatic change from the base material compared to the microstructure in the LAZ. Complete melting of the composite occurs locally within the LAZ and during cooling

---

* Kennametal, Grade 2000, average particle size 1.2-1.3 µm (Latrobe, PA)

† Homogeneous Metals, Cohoes, NY, Alloy IC-50 (Ni-11.3 w/o Al-0.6 w/o Zr-0.02 w/o B), ≤75 µm

dendritic growth of the TiC occurs as it rapidly precipitates from solution. These dendrites are more easily observed in an etched sample where the Ni_3Al binder phase has been removed (Figs. 6 and 7). The formation of the dendrites indicates that the temperatures attained in the LAZ were at or above the liquidus temperatures for these systems which are all >2000°C.

Because the laser was operated in a pulsed mode with over lapping spots, some areas were melted and remelted as the laser was scanned across the surface. The overlapping spots can be distinguished in Fig. 8. The interface between the melt and unmelted areas reveals interesting differences in the TiC grain size depending on the local heating and cooling conditions experienced (Fig. 9). In the area adjacent to an unmelted area, the TiC dendrites are very fine because of the high cooling rate. Whereas, the grains in the area not melted grow in size because of the heat-treatment received from being near the laser spot. Such a gradient in dendrite size can also be observed across the microstructure in Fig. 5.

Mechanical Properties – Indent hardness measurements were performed on composites containing 30 and 50 vol. % Ni_3Al in the LAZ (indent sizes were ~200 μm). The results showed the hardness increased slightly for both compositions: for the 30 vol.% sample from 8.5 GPa for the baseline to 9.0-11.8 GPa in the LAZ, and for the 50 vol. % sample from 7.1 GPa for the baseline to 7.3-7.9 GPa in the LAZ. These preliminary tests indicate that laser surface treatments may be applicable to refining the effective grain size in $TiC-Ni_3Al$ composites and improve hardness. However, the indents also produced increased crack propagation than normally observed for these types of composites. Accurate crack lengths were not possible, but the increases suggest a decrease in the fracture toughness.

CONCLUSIONS

Laser surface treatments produced significant thermal transients in $TiC-Ni_3Al$ composites and localized melting and resolidification. Cracking was associated with high intensities and low defocus conditions. Appropriate conditions would be subject to individual equipment and process parameters for various laser systems. The high temperatures attained in the LAZ resulted in dendritic growth of the TiC grains. Because the laser was operated in a pulsed mode with over lapping spots, some areas were melted and remelted as the laser was scanned across the surface. The interface between the melt-remelt areas revealed differences in the TiC grain size depending on the local heating and cooling conditions experienced. Increased hardness within the LAZ was observed.

ACKNOWLEDGMENTS

Research sponsored by both the Propulsion System Materials Program, DOE Office of Transportation Technologies under contract DE-AC05-00OR22725 with UT-Battelle. The research also used the ORNL SHaRE User Facility supported by the Division of Materials Sciences, U.S. Department of Energy.

REFERENCES

1. T. N. Tiegs, K. B. Alexander, K. P. Plucknett, P. A. Menchhofer, P. F. Becher and S. B. Waters, "Ceramic Composites With a Ductile Ni_3Al Binder Phase," Mater. Sci. Eng., Vol. A209, No. 1-2, 243-47 (1996).

2. K. P. Plucknett, T. N. Tiegs, P. A. Menchhofer, P. F. Becher and S. B. Waters, "Ductile Intermetallic Toughened Carbide Matrix Composites," Ceram. Eng. Sci. Proc., 17[3]314-321 (1996).

3. T. N. Tiegs, K. P. Plucknett, P. A. Menchhofer and P. F. Becher, "Development of Ni_3Al-Bonded WC and TiC Cermets", pp. 339-357 in Internat. Symp. Nickel and Iron Aluminides, S. C. Deevi, V. K. Sikka, P. J. Maziasz, and R. W. Cahn (eds.), ASM International, Metals Park, OH (1997).

4. R. Subramanian and J. H. Schneibel, "Processing Iron-Aluminide Composites Containing Carbides or Borides," J. Metals, 49[8] 50-54 (1997).

5. T. N. Tiegs, P. A. Menchhofer, K. P. Plucknett, P. F. Becher, C. B. Thomas and P. K. Liaw, "Comparison of Sintering Behavior and Properties of Aluminide-Bonded Ceramics." Ceram. Eng. Sci. Proc.,19[3] 447-455 (1999).

6. K. P. Plucknett, P. F. Becher, and R. Subramanian, "Melt-Infiltration Processing of TiC/Ni_3Al Composites," J. Mater. Res., 12 [10] 2515-2517 (1997).

7. T. N. Tiegs, F. Montgomery, F. Goranson, P. A. Menchhofer, D. L. Barker, and D. E. Wittmer, "Microstructure and Properties of TiC-Ni_3Al Composites With Alternate Binder Compositions," to be published in Ceram. Eng. Sci. Proc., Am. Ceram. Soc., Westerville, OH (2000).

8. T. N. Tiegs, P. A. Menchhofer, C. B. Thomas and P. K. Liaw, "Effect of Alloying Additives on Fabrication and Properties of Ni_3Al and FeAl-Bonded Hardmetals,", Adv. Powd. Met. Partic. Mater.-1999, Metal Powder Industries Fed., Princeton, NJ (1999).

9. D. E. Wittmer, F. Goransson, T. N. Tiegs, and J. L. Schroeder, "Comparison of Batch and Continuous Sintering of Aluminide-Bonded Titanium Carbide," pp. 3-237-3-248 in Advances in Powder Metallurgy and Particulate Materials-1999, Metal Powd. Indus. Fed., Princeton, NJ (1999).

10. P. Chantikul, G. R. Anstis, B. R. Lawn and D. B. Marshall, "A Critical Evaluation of Indentation Techniques for Measuring Fracture Toughness: II, Strength Method," J. Am. Ceram. Soc., 64 [9] 539-543 (1981).

11. Phase Equilibrium Diagrams, Vol. 10,A. E. McHale (ed.), Fig. 9017, pp. 345, Am. Ceram. Soc.,Westerville, OH (1994).

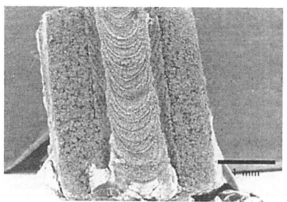

Fig. 1. TiC-40 vol. % Ni₃Al composite showing effect of laser treatment on surface. Sample was etched in nitric acid after treatment.

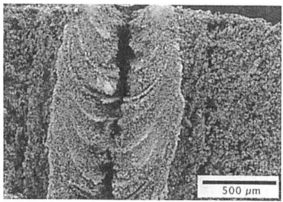

Fig. 2. TiC-40 vol. % Ni₃Al composite showing effect of laser treatment on surface and cracking in LAZ. Sample was etched in nitric acid after treatment.

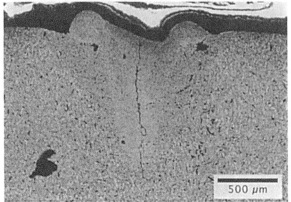

Fig. 3. Polished cross-section of a TiC-40 vol. % Ni₃Al composite showing effect of laser treatment on surface.

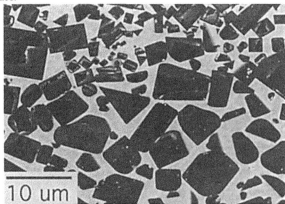

Fig. 4. Polished section of a TiC-50 vol. % Ni₃Al composite showing typical microstructure.

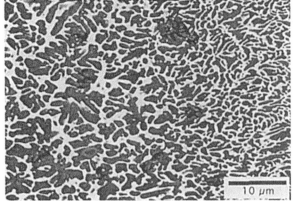

Fig. 5. Polished section of a TiC-50 vol. % Ni₃Al composite showing microstructure in LAZ.

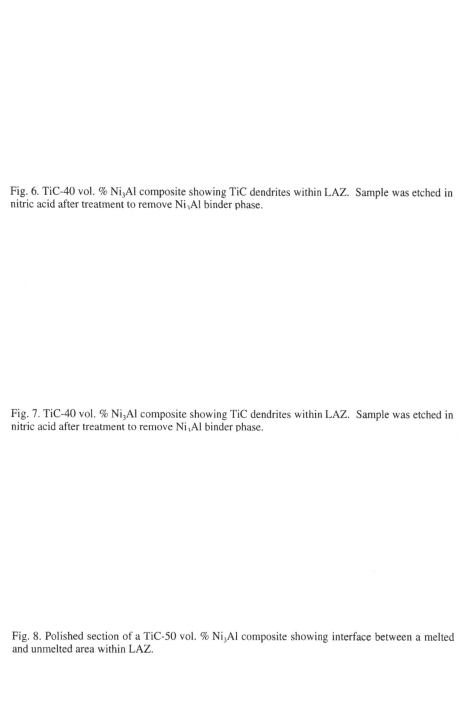

Fig. 6. TiC-40 vol. % Ni₃Al composite showing TiC dendrites within LAZ. Sample was etched in nitric acid after treatment to remove Ni₃Al binder phase.

Fig. 7. TiC-40 vol. % Ni₃Al composite showing TiC dendrites within LAZ. Sample was etched in nitric acid after treatment to remove Ni₃Al binder phase.

Fig. 8. Polished section of a TiC-50 vol. % Ni₃Al composite showing interface between a melted and unmelted area within LAZ.

FORMATION AND CONTROL OF TI-MO FGM WITH CONTINUOUS TRANSITIONAL COMPOSITION

Z.M. Yang[1]   L.M. Zhang[1]   F. Tian[1]

[1] State key Lab of Advanced Technology for Materials Synthesis and Processing, Wuhan University of Technology, Wuhan 430070, P. R. China

L.D. Chen[2]   Toshio Hirai[2]
[2] Tohoku University, Sendai 980-8579, Japan

ABSTRACT

Ti-Mo functionally graded material with continuous transitional composition was fabricated by the sedimentation of Ti and Mo particles with various sizes. The main factors that influence the formation of gradient structure were discussed. On the basis of theoretical analysis, the compositional distribution of Ti-Mo FGM was calculated. The results show that the theoretical calculation values matched the experiments well.

INTRODUCTION

Ti-Mo FGM can be applied in dynamic high-pressure technology [1]. Presently, a popular fabrication method is powder stacking [2], which results in the formation of interfaces in the body of Ti-Mo FGM, thus limiting its application. Although other methods such as physical vapor deposition, plasma spraying, can be used to produce FGM with continuous transitional composition, problems such as high cost of fabrication and high porosity of products restrict their applications[3, 4]. In the present paper, the continuous arrays of Ti and Mo particles were realized by sedimentation method [5, 6, 7]. After sintering, Ti-Mo FGM with compositionally-graded structure was then obtained.

EXPERIMENTAL

Pure ethanol was used as the dispersant in this study. Ti and Mo powders (Table 1 and Figure 1) were carefully weighed and placed in a large beaker containing pure ethanol to make a suspension with a volume concentration of

0.6% [1]. To break up powders into individual particles, the suspension was ultrasonically treated. Subsequently it was poured into a sedimentation tube with a diameter of 45 mm and a height of 1000mm. Due to the gravity Ti and Mo particles began to settle.

After particles completely settled out, the liquid was drawn off and the deposit was thoroughly air dried until no odor of ethanol remained. The central 45mm diameter of deposit was installed in a pressure chamber and compacted. Then the compact body was transferred into graphite die and hot pressed for an hour at 1673K under a pressure of 20MPa in an argon atmosphere.

The compositional distributions of final sample were measured by electron probe and the results are showed in Fig. 2.

Table 1. Experimental description

	Ti powder		Mo powder		
Sample	Particle size distribution	$D_{50mas.\%}$ (μm)	Particle size distribution	$D_{50mas.\%}$ (μm)	$Vol._{Ti}/Vol._{Mo}$
$1^{\#}$	Fig.1(a)	8.63	Fig.1(b)	3.10	1:1
$2^{\#}$	Fig.1(a)	8.63	Fig.1(b)	3.10	2:1
$3^{\#}$	Fig.1(a)	8.13	Fig.1(b)	5.03	2:1

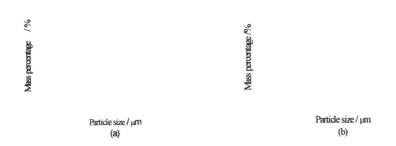

Fig. 1 Particle size distribution
（a） Ti （b） Mo

SENDIMENTATION GRADIENT SIMULATION

The motion of particles in a dilute suspension can be described by Stokes' Law [8].

$$V_D = D^2 g \left(\rho_{particle} - \rho_{liquid} \right) / \left(18\eta \right) \tag{1}$$

where V_D is the particle settling velocity, D is the diameter of particle; ρ particle, ρ liquid are the particle and suspension density, respectively, and η is the viscosity of suspension.

From equation (1), we can clearly see that the settling velocity of a particle depends not only on its density, but also on its size.

During settling process, the sedimentation mass M of particles is a function of time and is given by

$$M(t) = m \times \int_{Dt}^{D\,max} f(D)dD + m \times \int_{D\,min}^{Dt} \alpha f(D)dD \qquad (2)$$

Where m and f(D) are the mass and mass frequency distribution of one powder, respectively, α is the deposit fraction of particles of a certain size, which is related to the particle diameter and the settling height; D_{max} and D_{min} are the maximum and minimum diameters of a powder, respectively, D_t is a critical diameter, which is the size of the particles which have settled completely at a time t. The first part at the right hand of equation (2) is the mass of the particles with diameters from D_t to D_{max}, the second part is the sedimentation mass of the particles with diameters ranging from D_{min} to D_t.

For two kinds of powders A and B, the content of powder A in the sedimentation layer within a given unit time can be calculated by assuming no interactions between particles.

$$C_v(t) = \frac{\dfrac{1}{\rho_A}\dfrac{dM_A}{dt}}{\left(\dfrac{1}{\rho_A}\dfrac{dM_A}{dt} + \dfrac{1}{\rho_B}\dfrac{dM_B}{dt}\right)} \qquad (3)$$

where $C_v(t)$ is the content of powder A at a time t. ρ_A and ρ_B are the densities of powder A and B, separately.

It is assumed that the final sample is dense after sintering. So the deposit thickness at an arbitrary time t is given by

$$h(t) = \frac{\left[\dfrac{M_A(t)}{\rho_A} + \dfrac{M_B(t)}{\rho_B}\right]}{S} \qquad (4)$$

where S is the cross-section area of sedimentation tube.

Substituting equations (1) and (2) into equations (3) and (4), we can calculate the content $C_v(t)$ of component and corresponding place h(t) in the deposit when particle size distributions of raw materials powders and settling height are known. Based on these equations, a simulation program that has been developed was used to describe settling process and predict the graded composition of final products.

RESULTS AND DISCUSSION

Fig. 2 shows the graded composition of Ti-Mo profiles along their cross-sections. As expected, the continuously graded structures with elements Ti and Mo are formed. It is very obvious that Fig. 2(c) is different from Fig. 2(a) and Fig. 2(b). The difference is caused by the particle size distributions of Ti and Mo powders. When the settling height is fixed, the settling time of particles is controlled by their sizes. Consequently the array orders of Ti and Mo particles on the bottom of sedimentation tube are also controlled by their sizes. However, the deposit mass in a certain time depends only on the content of every size of particles. If the size distributions of Ti and Mo powders are changed, not only the particle sizes but also the contents of particles of each size are changed. Such changes result in the compositional variation of final samples.

Fig. 3 shows the calculation results for samples 1[#] and 2[#] by the equations outlined above. The calculation results are in approximate agreement with experiment ones. Furthermore, Comparing the experimental results between Fig. 3(a) and Fig. 3(b) we can see that they are not same. We can also draw the same conclusion by observing microscopic structures showed in Fig. 2 (a) and Fig. 2 (b). So different mass ratios of powder Ti to Mo also generate different gradient structures when their particle size distributions are fixed. This is because the variation of mass ratio between Ti and Mo powders can directly change the contents of Ti and Mo in the deposit.

All these analyses mentioned above demonstrate that the purpose to adjust the composition of gradient layer was to be achieved by proper control of particle size distributions and mass ratios.

From Fig. 3 we also find that the calculation values are higher than the experimental values on the Ti-rich side, while on the Mo-rich side they are reversed. This phenomenon can be explained as follows: The average size of Ti powder is larger than that of Mo, which is showed in Table 1 and Fig. 1. During settling process, the large particles of Ti deposit on the bottom of sedimentation tube first and form a loose structure. Part of the following Mo particles filled in the voids of the loose structure, which result in a decrease of Mo content in the upper part of the deposit and an increase in the lower part. Consequently, the difference in the particle sizes between Ti and Mo powders should be kept small.

CONCLUSIONS

(1) Ti-Mo FGM with smoothly varying composition was fabricated by sedimentation method. The composition of gradient layers can be adjusted by proper control of the size distributions of Ti and Mo and their mass ratios.

(2) The calculated results are in approximate agreement with those measured by electron probe.

REFERENCES

[1] Barker LM and Scott D D, *SAND 84-0432*.

[2] H.P. Xiong, L.M. Zhang, Q. Shen and R.Z. Yuan, "Fabrication of Mo-Ti Functionally Graded Material". *Trans. Nonferrous Met. Soc. China.* 9[3] 582~585 (1999). (in Chinese)

[3] L.M. Zhang, Q. Shen . "Research Development and New Trend of Functionally Graded Materials", *Advanced ceramics*, 2[1] 23~26 (1995). (in Chinese)

[4] X.T. Huang and M. Yan, "Functionally Graded Materials: Review and Perspective", *Materials Science and Engineering*, [15] 35~38 (1997). (in Chinese)

[5] L.M. Zhang, Z.M. Yang, Q. Shen and F. Tian. "Analysis of the Possibility of Eliminating the Interfaces of Functionally Graded Materials by Sedimentation", *Bulletin of the Chinese Ceramic Society*, [6] 60~62 (1999). (in Chinese)

[6] T. Jüngling, J. Bronback and B. Kieback, "Ceramic-Metal Functional Gradient Materials by Sedimentation"; pp. 161~166 in *3rd International Symposium on Structural and Functional Gradient Materials*. edited by B.Ilschner and N.Cherradi. Polytechniques et universitarires romandes press, Lausanne, 1994.

[7] J.H. Huang, J. Zhao, J.F. Li, Kawasaki Akira and Watanabe Ryuzo. "Fabrication of Metal/Ceramic Functionally Graded Materials by Sedimentation", *Chinese Science Bulletin*, 43[5] 550~553 (1998).

[8] M.J. Wang and L. Gao (Translated), pp.60~62 in *Separation of Solid-Liquid*, Atomic power press, Beijing, 1982. (in Chinese)

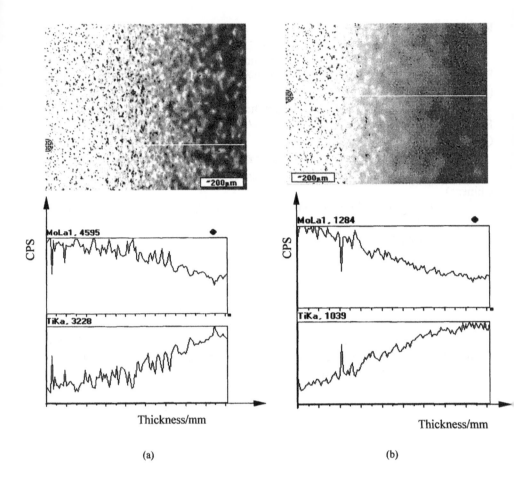

(a)

(b)

Functionally Graded Materials 2000

200μm

(c)

Fig. 2 Back scattered electron images of the cross-section for sample
(a) 1$^{\#}$ (b) 2$^{\#}$ (c) 3$^{\#}$

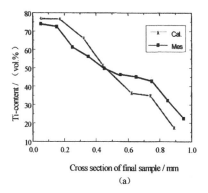

(a)

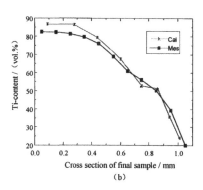

(b)

Fig. 3 Comparison of measured and calculated results
(a) 1$^{\#}$ (b) 2$^{\#}$

PREPARATION OF W-MO-TI GRADED DENSITY FLIER-PLATE MATERIALS

Q.Shen
State Key Lab. of Materials Synthesis and Processing, Wuhan University of Technology, Wuhan 430070, P.R.China;
Institute for Materials Research, Tohoku University, Sendai 980-77, Japan

L.M.Zhang and H.P.Xiong
State Key Lab. of Materials Synthesis and Processing, Wuhan University of Technology, Wuhan 430070, P.R.China;

L.D.Chen and T.Hirai
Institute for Materials Research, Tohoku University, Sendai 980-77, Japan

ABSTRACT
A new kind of flier-plate material with graded density has come to show great potential for the application in dynamic high-pressure technology. In the present paper, the W-Mo-Ti system was chosen to prepare such FGMs. The experimental results showed that the Fe-Al additives were suitable for the hot-pressing of W-Mo alloys and Mo-Ti alloys at a low temperature of 1473 K. The mechanism of the densification could be expressed as following, part of the addition of Fe dissolved into the liquid of Al and displayed a more active feature during the sintering, thus promoted the formation of bonding-phases in W-Mo alloy and solid solutions in Mo-Ti alloy, and evidently improved their relative densities. Finally, a wholly dense W-Mo-Ti graded density flier-plate material was successfully fabricated. Its density changed gradually from 4.5×10^3 kg/m^3 up to 17.0×10^3 kg/m^3.

INTRODUCTION
Recently, a new kind of flier-plate material with density gradients has come to show great potential for the application in dynamic high-pressure technology. By using such flier-plate, quasi-isentropic loading to target materials can be realized, and extreme conditions of high pressures or velocities can be provided at a much cooler temperature in contrast to that of shock wave loading techniques[1-3]. In fact,

the design idea of density gradient in the flier-plate was as the same as the concept of FGM, although the appearance of graded density flier-plate was earlier than the promotion of FGMs concept. In 1984, L.M.Barker et al[1] reported the preparation of the graded density flier-plates via a sedimentation technique of Ta, Cu, Al and TPX-plastic, green bodies were then densified under a pressure of 400 MPa and a temperature just above the melting point of TPX powder, but the processing was too complicate to be manipulated and the flier-plate had a low mechanical strength. After that, specific studies on this type of FGMs are rather few until 1995. Z.H.Yong et al[4] fabricated Fe-based graded density materials by constructing a pore gradient, resulting in the corresponding change in the density of the material, while the density range was too narrow, only from 6.03×10^3 kg/m^3 to 6.65×10^3 kg/m^3. Al-Cu system graded density materials[5] were reported in 1996, but Al and Cu tend to form alloy compounds, leading to brittleness of the FGMs. In addition, these FGMs with low and middle density values or narrow density ranges should be improved to meet the demand in impact experiments. In the present paper, the graded density flier-plate materials with higher density values and a wider range of density were investigated, and the densification mechanisms of the different parts of the flier-plate were discussed.

BASIC CONSIDERATION

The density of pure W is as high as 19.30×10^3 kg/m^3. According to binary phase diagram, there is no trend to form brittle compounds between W and Mo, which has a density of 10.22×10^3 kg/m^3. Ti is a light metal with a density of 4.51×10^3 kg/m^3. Moreover, Ti and Mo form no compounds either. Taking these factors into account, we chose W-Mo-Ti system to investigate the graded density flier-plates.

The liquid-phase sintering method usually applied to densify W alloys with additives such as Ni-Fe, Ni-Cu, Fe-Cu, etc. The corresponding sintering temperatures are above 1673 K. The sintering temperatures of Mo and Ti alloys are slightly lower than those of W alloys[6-9]. In order to make all selected elements densified at the same sintering conditions, Fe-Al were determined as the sintering additives. Fe, which plays an activating role, is a suitable additive for sintering W alloys and Mo alloys. In the meanwhile, the sintering densities of Ti alloys remain high when the weight content of Fe is below 4%[10]. The purpose of introducing element Al, which has a low melting point, is to obtain fully dense bodies by forming liquid phases.

EXPERIMENTS AND MEASUREMENTS

The fresh powders of W, Mo, Ti, Fe and Al with average particle size of 1.7, 2.4, 30, 30 and 20 µm respectively were used. The powders were mechanically mixed according to designed ratios and were hot-pressed in a flowing argon atmosphere. The sintering parameters were 1473 K-30 MPa-60 min. The size of the sintered sample is 20 mm in diameter and about 2 mm in height. Ten transient layers with certain weight ratios were inserted between W side and Ti side when the W-Mo-Ti system flier-plate was fabricated. The same additives, 3%Fe-1.5%Al were used for both sides and all transient layers, and the thickness of each transient layer was fixed at 0.15 mm.

The densities of sintered samples were measured by the water-immersion technique. The microstructures of W-Fe-Al, Mo-Fe-Al and Ti-Fe-Al alloys were observed by scanning electron microscope. The secondary electron images of W-Mo alloy and Mo-Ti alloy as well as back-scattered electron image of W-Mo-Ti system flier-plate were investigated by electron probe microscope. The area and line distributions of the essential elements were determined by X-ray wave-dispersion spectrometer (XRWDS).

RESULTS AND DISCUSSIONS

The relative densities of W alloys and Mo alloys with only 3%Fe exhibited low values of 91.3% and 89.7%, respectively, as shown in figure 1. When the weight content of Fe was fixed at 3%, the densities of W alloys and Mo alloys increased rapidly with the addition of Al. With 3%Fe-1.5%Al additives, the peak relative densities of W alloys and Mo alloys reached 99.5% and 97.8% respectively. It was also found that Fe-Al additives are of aid to the sintering of Ti alloy, and its relative density comes up to 99.2%. It is considered that, during the sintering process, Al powders melted first when heated to its melting point 933 K, and, part of the Fe adjacent to Al was dissolved into the liquid phase due to the further increasing of temperatures. Consequently, the flow of the liquid made it easier for Fe to contact with W, Mo or Ti particles. The sintering activator Fe in liquid form displayed more active feature than in solid form, which is helpful to increase the relative densities of the samples. Figure 2 gives the microstructures of W95.5Fe3Al1.5, Mo95.5Fe3Al1.5 and Ti95.5Fe3Al1.5 alloys. It can be seen obviously that fully dense W alloy, Mo alloy and Ti alloy were obtained at the same hot-pressing conditions.

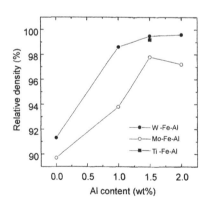

Figure 1 The changes of the relative densities (Fe is fixed at 3%)

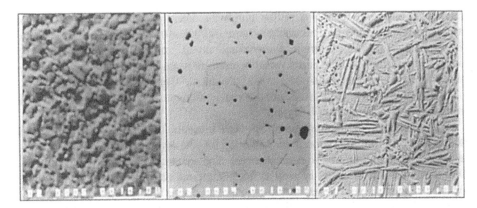

W95.5Fe3Al1.5   Mo95.5Fe3Al1.5   Ti95.5Fe3Al1.5
Figure 2 Microstructures of W95.5Fe3Al1.5, Mo95.5Fe3Al1.5 and
Ti95.5Fe3Al1.5 alloy (polished and etched)

Because the designed Fe-Al additives are suitable for the densification of W, Mo and Ti alloys, high relative densities of various ratios of W-Mo alloys and Mo-Ti alloys are also obtained with the same additives. The secondary electron images and area distributions of elements of W-Mo alloy and Mo-Ti alloy are shown in figure 3 and figure 4, respectively. It can be found from figure 3 that at the low sintering temperature of 1473K, W and Mo formed no solid solution but mechanical intermixture. Fe-Al sintering additives undoubtedly activate the

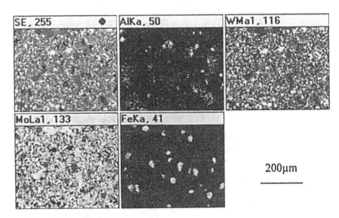

Figure 3 The secondary electron image of W38.2Mo57.3Fe3Al1.5 alloy
and the area distributions of W, Mo, Fe and Al elements

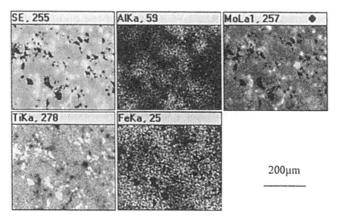

200μm

Figure 4 The secondary electron image of Mo38.2Ti57.3Fe3Al1.5 alloy
and the area distributions of Mo, Ti, Fe and Al elements

diffusions of W and Mo, and small amounts of bonding-phases are produced at
the boundaries between W and Mo phases, thus good coherence is achieved. In
Mo-Ti alloy, Fe-Al additives exist more in the matrix phases of Mo and Ti than at
the boundaries. This may due to the strong reaction between the additives and the
matrix components. Al is a fundamental element in Ti alloys, and Fe-Al additives
may interact concurrently with Mo and Ti. So, some solid solution of (Mo,Ti) is
formed. From figure 4, it can also be seen that part of Mo and Ti were dispersed
in the solid solution of (Mo,Ti), which verifies that at the low temperature of
1473K, the solid solution reaction is inadequate. The formation of bonding-phases
in W-Mo alloy and solid solutions in Mo-Ti alloy evidently promoted their
densification.

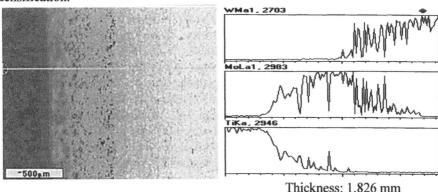

Thickness: 1.826 mm
Figure 5 The back-scattered electron image of W-Mo-Ti system flier-plate and the
line distributions of W, Mo and Ti elements in the thickness direction

A cross-section of the fully dense W-Mo-Ti flier-plate and the line distribution of the essential elements are presented in figure 5. It can be seen that, with the alternate content change of W, Mo and Ti, density of the flier-plate increases gradually from 4.5×10^3 kg/m^3 to 17.0×10^3 kg/m^3 in the 1.826 mm thickness range. On this basis, flier-plates with arbitrary density distribution profiles can be fabricated by well adjusting the ratios of raw powders, which is of great importance to generate quasi-isentropic compressive waves in impact experiments.

CONCLUSIONS

1) By adding suitable amount of Al, part of the additive of Fe could dissolve and display a more active feature during the liquid-phase sintering.

2) Fe-Al additives promoted the formation of bonding-phases in W-Mo alloys and solid solutions in Mo-Ti alloys, thus densification was achieved at a low temperature of 1473 K.

3) A fully dense W-Mo-Ti graded density flier-plate was successfully fabricated. Its density changed gradually from 4.5×10^3 kg/m^3 up to 17.0×10^3 kg/m^3.

ACKNOWLEDGEMENT

This work was supported by the National Natural Science Foundation of China(Grant No.59771028) and the Doctor Foundation of the Education Ministry of China(Grant No.1999049702).

REFERENCES

[1] L.M.Barker and D.D.Scott, "Development of a high-pressure quasi-isentropic plane wave generating capability," *SAND* 84-0432 (1984).

[2] L.C.Chhabildas, J.R.Asay and L.M.Barker, "Shear strength of tungsten under shock- and quasi-isentropic loading to 250GPa, "*SAND* 88-0306 (1988).

[3] Q.Shen, L.M.Zhang, H.P.Xiong, J.S.Hua and H.Tan, "Fabrication of W-Mo-Ti system flier-plate with graded impedance for generating quasi-isentropic compression," *Chinese Science Bulletin*, 45 [15] 1421-4 (2000).

[4] Z.H.Yong and X.J.Lu, "Preparation of functionally graded density materials(in Chinese), " *Materials Engineering*, 9 [7] 14-5 (1995).

[5] R.Tu, Q.Shen, J.S.Hua, L.M.Zhang and R.Z.Yuan, "Fabrication of Al-Cu system with functionally graded density profiles"; pp.307-311 in *Proceedings of the 4th International Symposium on Functionally Graded Materials*, edited by Ichiro Shiota & Yoshinari Miyamoto, ELSEVIER, 1997.

[6] J.K.Park, S.L.Kang, K.Y.Eun and D.N.Yoon, "Microstructural change during liquid phase sintering of W-Ni-Fe alloys," *Metallurgical Transactions*, 20A 837-

45 (1989).

[7]Q.Shen, C.B.Wang and L.M.Zhang, "Liquid phase sintering and densification of W-Ni-Fe-Cu alloys at a low temperature(in Chinese)," *Journal of Wuhan University of Technology*, **22** [1] 1-3 (2000).

[8]T.Sakamoto, "Effect of boron and carbon addition on densification of sintered Mo-Ni powder compacts(in Japanese)," *Journal of the Japan Society of Powder and Powder Metallurgy*, **38** 839-43 (1991).

[9]S.Mizunuma, "Sintering and forging characteristics of sintered Ti-6Al-4V alloy produced by blended elemental method(in Japanese)," *Journal of the Japan Society of Powder and Powder Metallurgy*, **43** 918-23 (1996).

[10]K.Majima, "A study on sintered tianium alloy produced by blended elemental process(in Japanese)," *Journal of the Japan Society of Powder and Powder Metallurgy*, **36** 917-925 (1989).

SINTERING BEHAVIOR OF WET CHEMICALLY DERIVED ZINC OXIDE VARISTOR

Katsuyasu Sugawara and Takuo Sugawara
Faculty of Engineering and Resource Science,
Akita University
1-1 Tegata Gakuen-cho, Akita 010-8502 Japan

Tadashi Ogasawara
Electronic Device Division,
TDK Corp.,
Hirasawa, Nikaho-machi, Yuri-gun, Akita 018-0402 Japan

ABSTRCT
Effects of dopants on sintering behavior and nonlinear V-I characteristics were investigated for ZnO varistor prepared by coprecipitation. Changes of grain size and dopants distribution were followed with sintering. Thermomechanical analysis elucidated not only the effective temperature region of each dopant but also the interaction of dopants in sintering of ZnO. Nonlinearlity coefficient, 20~30 were obtained for the ternary dopant systems.

INTRODUCTION
Variable resistor ceramics, based on zinc oxide and silicon carbide compositions, have been widely used as a protection device for electrical circuits because of their excellent surge-energy absorbing capabilities. It is well known that chemical and physical inhomogenities of crystalline cause extraordinarily dropping in electrical performance of varistor. Wet chemical methods have some advantages for preparation of high-grade and homogeneous materials. However, only a few papers have been reported on the preparation and characteristics of chemically derived oxide varistor with dopants.

The objective of the present study is to clarify the effects of praseodymium, cobalt, chromium and molybdenum as dopants on sintering behavior of zinc oxide precursor prepared by coprecipitation. Changes of grain size of zinc oxide and shirinkage of green pellet were followed with the contents of dopants and temperature in sintering.

EXPERIMENTAL

Preparation procedure of zinc oxide varistor is shown in Fig.1. Zinc acetate, cobalt acetate and chromium acetate, praseodymium oxide and molybdenum oxide are used as starting materials. Triethylamine was added to the mixed aqueous solution of zinc acetate and cobalt acetate in order to coprecipitate homogeneously. The precipitant was filtered and dried at 110°C, and then calcined at 500°C for 2 h in air. Praseodymium oxide and molybdenum oxide are mixed with the calcined powder.

The calcined powder, by adding PVA, was pelletized by a press of 0.5 ton/cm^2 in a mold. The obtained green pellet was heated up to 1300°C with a heating rate of 5°C/min and holding time for 2 h. Average grain size was measured by an image analysis with FESEM. Themomechanical analysis was applied to observing shirinking behavior of the pellet with sintering.

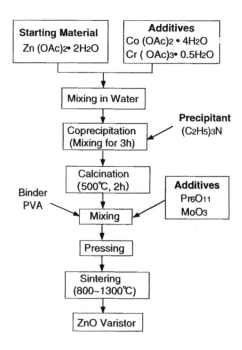

Fig.1 Preparation procedure of ZnO varistor

RESULTS and DISCUSSION

Figure 2 shows examples of change in grain size with sintering temperature for single dopant systems of ZnO-Cr and ZnO-Mo. Chromium and molybdenum contents were in the range from zero to 0.02 mol %. No appreciable difference was observed for the grain growth with and without chromium. For the ZnO-Mo system, molybdenum accerelerates the grain growth in the sintering temperature from 1000℃. While praseodymium also enhanced the grain growth of zinc oxide, effective temperature region was higher than that of ZnO-Mo system. The grain growth rate became large with the addition of molybdenum.

The grain growth behavior with sintering temperature is shown in Fig.3 for ternary additive systems, ZnO-Co-Pr-Cr and ZnO-Co-Pr-Mo.

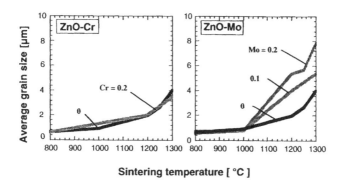

Fig.2　Grain growth with sintering temperature for single additive systems

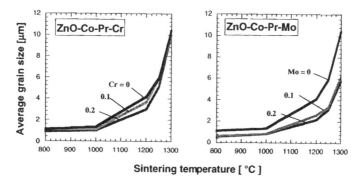

Fig.3　Grain growth with sintering temperature for ternary additive systems

Chromium inhibits the grain growth of zinc oxide in the ternary additive system. This inhibition effect becomes lower above 1250°C. In contrast to the grain growth behavior of ZnO-Mo system, distinct inhibition was observed for the grain growth of ZnO-Co-Pr-Mo system.

Figure 4 shows the contraction rate of single dopant systems obtained by thermomechanical analysis. The contraction rate curve of dopant-free zinc oxide indicates that shrinking starts at 900°C and the rate attains the maximum at 1000°C. Cobalt addition lowers the starting temperature of contraction. On the other hand, the contraction curve moves to higher temperature region with the addition of praceodymium. Very sharp curve of contraction was observed around 1300°C in ZnO-Mo system.

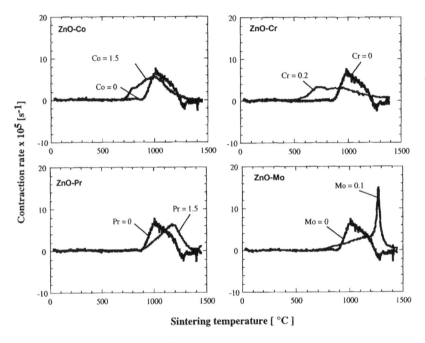

Fig. 4 Contraction rate of single additive systems

Figure 5 represents change in contraction rate of ternary dopant systems. Chromium and molybdenum addition causes the smooth contraction in the wide temperature range. In the contrast to the single dopant system, for example ZnO-Mo, the sharp peak was not observed in the ternary dopant system, ZnO-Co-Pr-Mo. These results indicate the exsistance of interaction between the dopants.

Changes in nonlinearity coefficient and varistor voltage with sintering temperature were shown in Fig.6. Both ternary dopant samples shows the nonlinearity coefficient of 20 to 30. Varistor voltage shows a decreasing tendency with sintering temperature which is related to the grain growth of ZnO.

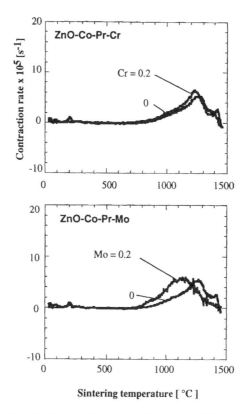

Fig.5 Contraction rate of ternary additive systems

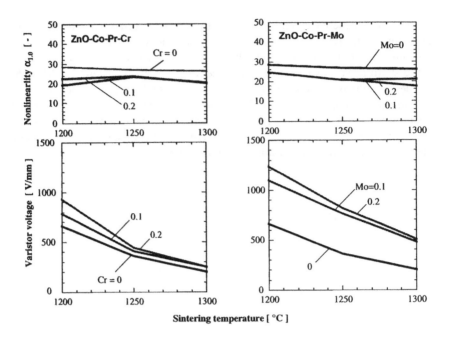

Fig.6 Changes in nonlinearity coefficient and varistor voltage with sintering temperature

ACKNOWLEDGMENT

The authors are grateful to the Electronic Device division of TDK Corp. for financial support, and to T. Kato and T. Haraya for their assistance.

Infiltration Processing

PROCESSING OF POROSITY GRADED SILICON CARBIDE EVAPORATOR TUBES BY PRESSURE FILTRATION

Rainer H. Oberacker, Michael Dröschel and Michael J. Hoffmann
Universität Karlsruhe, Institut für Keramik im Maschinenbau
Haid und Neustr. 7
D-76706 Karlsruhe (Germany)

ABSTRACT

Porous silicon carbide (SiC) ceramics are promising materials for liquid fuel evaporator tubes in gas turbine combustors. Due to thermal stresses, a porosity gradient is favorable. For tubes with porosity gradients, a new processing route based on pressure filtration was developed. The formation of one or more dimensional wax concentration gradients in the filter cake (corresponding to porosity and pore size gradients after sintering) is adjusted by controlling the composition of a mixture of SiC and SiC-wax slips (slurries) and the filtration pressure. The design concept of a laboratory casting apparatus which uses this new approach and an experimental/ numerical method for the derivation of the process parameters are described in this paper.

INTRODUCTION

Fig. 1 shows a concept for a gas turbine combustor with premix burner. In contrast to a diffusion burner system, the evaporation of the liquid fuel occurs outside of the burning zone.[1] The fuel is sprayed on the outer surface of an evaporator tube. The air stream from the compressor, forms a thin film of fuel on the outer surface of the evaporator, which vaporizes completely. The homogeneous mix of air and fuel vapor enters the reaction zone inside the tube where the combustion takes place. It is known, that a rough or porous evaporator surface allows much higher evaporation rates compared to a smooth surface. This was studied quantitatively for sintered SiC with a broad variation of porosity and pore sizes.[2] The application of porous SiC could thus result in advantages with respect to the design of the combustion system, if its reliability is maintained despite of the porosity. As the evaporator tube has to be gas tight on the inner side, either a two layer joint or a porosity gradient has to be introduced.

In previous investigations,[3] FEM calculations were carried out in order to find the optimum porosity gradient for a hypothetical tube-shaped burner component with an inner diameter of 45 mm. The inner surface consists of a dense SiC layer with 2.0 mm thickness which is followed by a porous SiC-layer with either constant porosity or a radial porosity gradient. The material data (density, Young's modulus, Poisson's ratio, strength, thermal expansion coefficient, thermal conductivity and specific heat) used in finite element calculations are based on data from homogeneous porous SiC specimens, which were experimentally determined in a former study.[4] For this component, thermal stresses occur due to the inhomogeneous temperature distribution, which appear both in steady state and transient mode. An initial component temperature of 150°C is assumed, in accordance with the temperature of the incoming compressor air. After ignition of the flame in the combustor the inner surface of the evaporator tube is heated. The steady state is reached after 2.2 to 13.0 seconds, depending on the porosity gradient, at which the temperature of the inner surface is 1500°C and the temperature of the outer surface is 550°C. The maximum stress occurs always in the transient mode.

nozzle

compressed air
liquid fuel
evaporator tube

Fig. 1: Concept of a gas turbine combustor with premix burner[1]

It could be clearly demonstrated by the calculations, that a porosity gradient is superior compared to the two-layer concept. Table 1 shows the relative failure probabilities of the investigated design variations (see Eq. 2 for gradient functions). Compared to the two layer concept, the failure probability of the graded materials is reduced by 2 to 6 orders of magnitude. It is interesting to mention, that the reduced failure probability of the graded tubes results not from a decrease in local stress, but from an increased local strength at the radial coordinate of maximum tensile stresses.[3]

Table 1: Normalized failure probability for the two layer and the graded tubes

Gradient function	two layer (n=0)	declining (n=0.5)	linear (n=1)	progressive (n=2)
rel. failure probability	1	$4*10^{-2}$	$8*10^{-5}$	$1*10^{-6}$

The use of a tailored transition function for porosity (declining radial porosity gradient) should therefore be advantageous for the investigated component. However, manufacturing of evaporator tubes with tailored porosity gradients requires a new processing route such as pressure filtration.

MANUFACTURING OF DEFINED POROSITY GRADIENTS BY PRESSURE FILTRATION

Several papers concerning the production of different functionally graded materials by using slip casting technique have been published.[5-13] One dimensional gradients were achieved by changing the slurry composition sequentially or continuously during the casting process. The filtration pressure resulted from capillary suction of the porous molds[5-9] or by vacuum pumps.[10-12] With none of these methods could a higher dimensional concentration gradient (parallel and perpendicular to the cake forming direction) be adjusted.

For evaporator tubes with defined porosity gradients in the radial and axial direction a new processing route has been developed. This process uses continuous pressure filtration of aqueous slurries containing SiC and wax particles. The latter act as pore formers which are burned out of the cake prior to sintering. By controlling and adjusting the local filtration pressure and the wax concentration in the slurry, defined gradients of wax content in the filter cake are realized in and perpendicular to the cake forming direction. The local porosity and pore size in the sintered material correspond to the wax concentration and wax particle size in the filter cake. This forming method can also be used for manufacturing other types of two- or multiphase materials with concentration gradients. It requires a new type of pressure casting apparatus and a numerical process control.

Laboratory Apparatus for Pressure Filtration of Two-Dimensionally Graded Components

An apparatus was designed and constructed to realize axial-symmetrical porosity or pore size gradients in tubes with an outer diameter of 65 mm, a length of 55 mm and a wall thickness of about 15 mm (after forming). Its function is depicted in Fig. 2.

The molding chamber at the bottom of the apparatus can be replaced for the fabrication of parts with other geometry, e.g. plates. A maximum gas pressure of 10 MPa can be applied. The SiC and SiC-wax slurries are stored and mixed inside the pressure chamber. The wax concentration of the SiC-wax slurry at the cake-suspension interface can be continuously changed. To manipulate the local cake forming rate, the molding filter elements are segmented in the axial direction. The filtrate flow of each segment, is measured by flow meters and can be controlled by throttle valves.

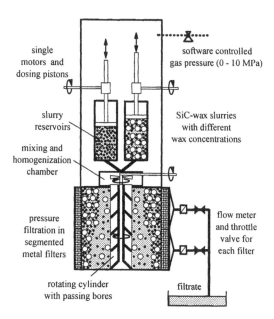

Fig. 2:   Apparatus for forming of tubes with two dimensional concentration gradients

Forming of tube shaped components with compositional gradients in the radial and axial directions requires adjusting the chamber pressure, the wax concentration at the filtration front and the local flow resistance in accordance to the desired local wax concentration in the tube at any given time of the process. As there is no general solution for the correlation between the process variables, a numerical simulation model had to be developed for process control.

Calculation of the Process Parameters for One-Dimensional Gradients

The growth of a filter cake during pressure filtration is schematically shown in Fig. 3. The cake forming rate (dh/dt) during pressure filtration is described by the basic equation of cake filtration.[13]

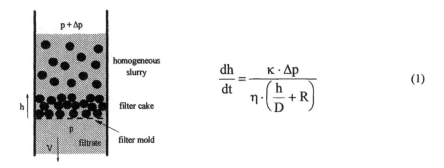

$$\frac{dh}{dt} = \frac{\kappa \cdot \Delta p}{\eta \cdot \left(\dfrac{h}{D} + R\right)} \qquad (1)$$

Fig. 3:   Schematic diagram of the cake filtration process

Cake growth can be controlled through the process parameters time t and filtration pressure Δp. The solids loading c and the filtrate viscosity η are known process parameters, which are determined by the slurry. R is the flow resistance of the filter mold, which can be measured easily in a calibration experiment. The porosity ε and the permeability D of the filter cake depend on the cake microstructure. κ is a constant determined by the solids loading c and the porosity ε ($\kappa = c/(1-\varepsilon-c)$). ε describes the fraction of space between the SiC and wax particles, which is different from the porosity P of the sintered component. The cake structure parameters D and ε can be derived from experiments with homogeneous suspensions.[14] Knowing these parameters, Eq. 1 can be integrated numerically.

Aqueous SiC and SiC-wax slurries, based on Norton SiC FCP 15 and Hoechst wax TP P300, with a constant solids loading of c = 32.5 vol% were used for the experimental determination of D and ε. The slurries are electrostatically stabilized, their viscosity is about 10 mPa·s at a pH of 7.5. The SiC particles have a mean diameter of $d_{50} = 0.8$ μm, the spherical wax particles have a d_{50} of 150 μm. These slurries were consolidated under constant filtration pressures of 0.5 to 4 MPa in a laboratory system, which allows to monitor continuously the filtrate flow. D and ε can be derived from the filtration kinetics[14], which is described in more detail in ref. 3.

In the investigated range of filtration pressures (0.5 to 4 MPa), the cake porosity ε and permeability D were practically pressure independent. This is an indication for incompressible cake structures. The SiC/wax ratio, however, has significant influence for these parameters as shown in Fig. 4. The cake porosity ε decreases up to a wax concentration of 40 vol%. This can be explained by the increased packing efficiency of the bimodal particle mixtures. The permeability D increases by a factor of 4 (from $1 \cdot 10^{-18}$ m² to $4 \cdot 10^{-18}$ m²) when the wax volume fraction of the solids reaches 60 vol%, despite of the reduction in porosity. This is an indication for increased pore channel diameters of the cakes with bimodal particles. The strong influence of the wax content on the cake structure parameters ε and D can not be neglected in the solution procedure for Eq. 1 and makes numerical integration necessary.

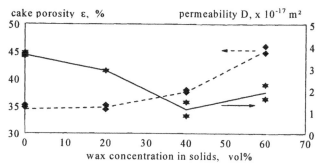

Fig. 4: Dependence of cake structure parameters on the SiC/wax ratio

The calculations were carried out for some of the porosity gradients of the sintered components shown in Fig. 5. The transition function of the porosity P inside the graded zone is described by a power law, according to Eq. 2.

$$P(r) = P(r_a) \cdot \left(\frac{r - r_p}{r_a - r_p} \right)^n \tag{2}$$

with: $r_p \leq r \leq r_a$, $r_p = 24.5$ mm, $r_a = 38.8$ mm

$P(r_p) = 0\,\%$, $P(r_a) = 50\,\%$

r is the tube radius, r_p stands for the radius, where the graded porous zone begins and r_a is the outer radius of the tube. The given dimensions relate to the sintered component. Therefore, drying and sintering shrinkage has to be taken into account. The wet cake has a wall thickness of $h = 20.9$ mm. Eq. 1 can then be brought to an incremental form (Eq. 3) and solved for short time increments Δt_i.

$$\frac{\Delta h_i}{\Delta t_i} = \frac{\kappa_i \cdot \Delta p_i}{\eta \cdot \left(\frac{\Delta h_i}{D_i} + R_{tot,i} \right)} \tag{3}$$

The permeability D and the filter cake porosity ε (and thus the concentration constant κ) depend on the wax content at the filtration front. During filtration, the total flow resistance $R_{tot,\,i}$ which contains the flow resistance of the metal filter and that of the previously formed filter cake increment Δh_i increases according to Eq. 4.

$$R_{tot,i} = R + \sum_{j=0}^{i-1} \frac{\Delta h_j}{D_j} \tag{4}$$

By integration of Eq. 3 and Eq. 4 the filtration kinetics for any desired one-dimensional wax concentration gradient in the cake can be simulated. The solution provides either the course of the filtration pressure for a given function of the cake forming rate, or the course of the cake forming rate for a given function of the filtration pressure. With respect to a low level of casting defects, casting at constant cake forming rates seems to be the better way. A constant cake forming rate of dh/dt = 0.1 mm/s was therefore chosen for the subsequent calculations. The course of the wax concentration in the slurry shows a linear dependency on the desired gradient function. This finding is valid for the special case of constant cake forming rate.

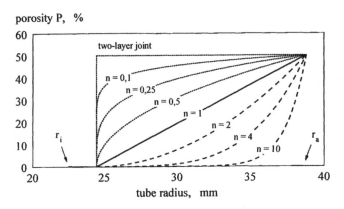

Fig. 5:   Porosity gradients chosen for simulation of pressure filtration

Rather complicated functions are calculated for the course of the filtration pressure (Fig. 6 and 7). From cake thickness and constant cake forming rate the total filtration time is 209 seconds in all cases. For homogeneous suspensions and cake structures, a linear pressure - time dependency would be expected in the case of a constant cake forming rate.

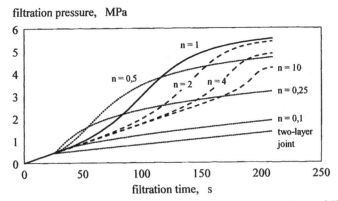

Fig. 6:   Simulation of the filtration process for the porosity gradients of Fig. 5

With increasing exponent n > 0 the value of the maximum filtration pressure is raised up to a maximum of $\Delta p_{max} = 5.57$ MPa for n = 1.25. For exponents n > 1.25 the maximum filtration pressure is reduced again and reaches a value of $\Delta p_{max} = 3.61$ MPa for n → ∞. The kinetics of pressure filtration and the sequence shown in Fig. 7 cannot be understood only on the basis of permeability D, as it depends strongly on the concentration parameter κ.

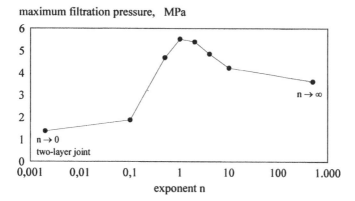

Fig. 7:   Influence of the exponent n on the maximum filtration pressure (model simulation)

The process simulation can be carried out for any other slurry mixture if the structure parameters D and κ of the resulting cakes are known. By using SiC-wax slurry mixtures with different wax concentrations and/or waxes with different particle sizes, defined gradients in the pore structure (porosity and/or pore size) of the sintered SiC evaporator tube can be adjusted. This is under investigation in present pressure casting experiments, in which plates with a thickness of 15 mm and a linear porosity gradient could be produced.

SUMMARY
Silicon carbide evaporator tubes with porosity gradients are promising materials for liquid fuel in gas turbine combustors. Finite element method calculations show that for these devices a tailored porosity gradient is necessary to meet the local stress/strength requirements. To realize such components a new processing route based on the continuous pressure filtration has been developed. With this process it is possible to produce defined one- and two-dimensional concentration gradients of a second particle type in a filter cake. The filtration process parameters have to be derived by a numerical integration of the basic filtration equation. The required cake structure parameters can easily be measured in constant pressure experiments with homogeneous filter cakes. This simulation concept can also be used for process control in fabrication of two-dimensional gradients.

ACKNOWLEDGEMENT
The financial support of the Deutsche Forschungsgemeinschaft (Ob 104/6) is gratefully acknowledged.

REFERENCES

[1]M. Brandauer, A. Schulz, S. Wittig, "Optimization of fuel prevaporization on porous ceramic surfaces for low NO_x-combustion", Combustion technology for a clean environment, Lisboa, Portugal, (1995)

[2]M. Dröschel, R. Oberacker, M.J. Hoffmann, "Eignung poröser SiC-Keramiken als Werkstoffe für die Filmverdampfungs-Brennkammerwand", pp. 37-42 in *Werkstoffwoche 98, Vol. III.* Edited by A. Kranzmann, and U. Gramberg. Wiley - VCH, Weinheim, 1999.

[3]M. Dröschel, M.J. Hoffmann, R. Oberacker, W. Schaller, Y.Y. Yang and D. Munz, "SiC-Ceramics with Taylored Porosity Gradients for Combustion Chambers", pp. 149-162 in *Engineering Ceramics: Multifunctional Properties.* Edited by P. Sajgalík and Z. Lencés. Trans Tech Publications, Switzerland, 2000.

[4]M. Dröschel, "Grundlegende Untersuchungen zur Eignung poröser Keramiken als Verdampferbauteile", doctoral thesis, University of Karlsruhe, IKM 022, ISSN 1436-3488, (1998) http://www.ubka.uni-karlsruhe.de/cgi-bin/psview?document=1998/maschinenbau/1

[5]J. Requena, R. Moreno, J. S. Moya, "Alumina and Alumina/Zirconia Multilayer Composites Obtained by Slip Casting", *Journal of the American Ceramic Society,* 72 [8] 1511-1513 (1989)

[6]J. S. Moya, A. J. Sánchez-Herencia, J. Requena, R. Moreno, "Functionally gradient ceramics by sequential slip casting", *Materials Letter,* 14 [5, 6], 333-335 (1992)

[7]J. Chu, H. Ishibashi, K. Hayashi, H. Takebe, K. Morinaga, "Slip Casting of Continuous Functionally Gradient Material", *Journal of the Japanese Ceramic Society,* 101 [7] 841-844 (1993)

[8]B. R. Marple, J. Boulanger, "Graded Casting of Materials with Continuous Gradients", *Journal of the American Ceramic Society,* 77 [10] 2747-2750 (1994)

[9]B. R. Marple, J. Boulanger, "Slip Casting Process and Apparatus for Producing Graded Materials", United States Patent, No. 5498383, (1994)

[10]H. Mori, Y. Sakurai, M. Nakamura, S. Toyama, "Formation of Gradient Composites Using the Filtration Mechanism of Binary Particulate Mixtures", *Proceedings of 3rd International Symposium on Functionally Graded Materials, FGM-3*, Lausanne, Switzerland, 1994, ISBN 2-88074-290-0, pp. 173-178, (1995)

[11]K. Taka, Y. Murakami, T. Ishikura, N. Hayashi, S. Watanabe, Y. Uchida, S. Higa, T. Imura, D. Dykes, "Development of stainless Steel / PSZ functionally graded materials by means of an expression operation", *Proceedings of 4th International Symposium on Functionally Graded Materials, FGM-4, Tsukuba, Japan, 1996*, ISBN 0-444-82548-7, pp. 343-348, (1997)

[12]S. Watanabe, T. Ishikura, A. Tokumura, Y. Kim, N. Hayashi, Y. Uchida, S. Higa, D. Dykes, G. Touchard, "The Use of a Functionally Graded Material in the Manufacture of a Graded Permittivity Element", *Proceedings of 4th International Symposium on Functionally Graded Materials, FGM-4, Tsukuba, Japan, 1996*, ISBN 0-444-82548-7, pp. 373-378, (1997)

[13]H. Gasper, *Handbuch der industriellen Fest/ Flüssig-Filtration,* ISBN 3-7785-1784-8, (1990)

[14]Verein Deutscher Ingenieure, "Filtering properties of suspensions: The determination of filter cake resistance", *VDI-Richtlinie* 2762, (1997)

ROLE OF ELABORATION PARAMETERS ON THE POROSITY FRACTION OF ALUMINA GRADED PREFORMS.

G. Kapelski and A. Varloteaux
Institut National Polytechnique de Grenoble / Université Joseph Fourier
Génie Physique et Mécanique des Matériaux (GPM2), UMR CNRS 5010
ENSPG, BP 46, 38402 Saint Martin d'Hères, France.

ABSTRACT

Porosity graded alumina preforms were prepared by co-sedimentation of a dispersed mixture of alumina and graphite powders with subsequent heat treatments to burn graphite and to sinter alumina. Molten metal can then be pressure infiltrated in the open graded porosity. This investigation studies the influence of two elaboration parameters, the graphite volume fraction and the alumina grain size distribution, on the porosity gradient . The gradient slope and the maximum porosity value both increase with the graphite volume fraction. A narrower alumina grain size distribution modifies the porosity gradient curvature. Experimental results are compared with simulations of the porosity gradient in the alumina preforms.

INTRODUCTION

Gravity sedimentation of a dispersed mixture of ceramic and metal powders can be used to produce, at low cost, concentration graded deposits. However, due to sintering behavior differences between the metal and the ceramic, problems such as warping or cracking [1] often occur during the densification step. Furthermore, the same differences limit considerably the choice of the metal [2] as well as the different gradient profiles obtainable. An alternative way consists in preparing a porosity graded preform by sedimentation of ceramic and pyrolyzable graphite powders used as a pore-forming agent during sintering. This graded preform can then be pressure infiltrated by a molten metal, leading to a concentration graded cermet with interpenetrating networks.[3] Moreover, the choice of the metal and the infiltration step do not affect the resulting gradient. In the elaboration of an alumina (Al_2O_3) preform by sedimentation, the influence of the graphite (Cg) proportion and that of the alumina grain size distribution on the graded porosity are investigated .

EXPERIMENTAL PROCEDURE

Powders.

Two grain size distributions of alumina powder have been tested: that as-provided by ALTECH (Aluminium Pechiney) which lies between 10 and 0.1 µm (Fig. 1, curve a) and that obtained by sedimentation separation of the same powder which narrowed the distribution to a range between 2 and 0.1 µm (Fig. 1, curve b). Synthetic graphite (TIMCAL), with a roughly equiaxed grain morphology, was used as a porosity inducing element. The graphite grain size distribution lies between 8 and 1 µm (Fig. 1 curve c).

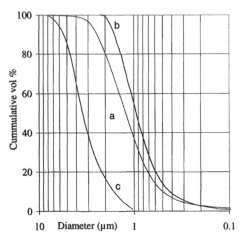

Figure 1: Grains size distribution of: (a) as-provided alumina, (b) narrow distribution alumina, and (c) graphite, as-measured by laser diffraction technique.

Elaboration.

As-provided grain size distribution. In a solution of 100 g of distilled water and 0.1 g of citric acid [4], 6 g of alumina powders were dispersed. Different graphite volume fractions, defined as the ratio between graphite and total solid volumes, were obtained by varying the graphite content. Three fractions were tested: 0%, 5% and 15%. The graphite powder was dispersed in 50 g of distilled water with polyvinyl alcohol, respectively 0.05 g and 0.15 g for 5%Cg and 15%Cg. An ultrasonic bath was used to break the agglomerates of each slurry. The two suspensions were mixed and introduced in a plastic tube closed at the bottom end by an alumina crucible. The sedimentation height is 16 cm. After 20 h, 80% of the powder is deposited on the crucible (Fig. 2). The slurry with the remaining 20% consisting of the smallest alumina grains is removed.

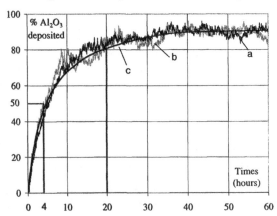

Figure 2: (a) and (b) Experimental sedimentation curves of alumina, (c) simulation (See Modeling).

Narrow alumina grain size distribution. This size distribution was obtained by removing the largest grains by sedimentation of 12 g of alumina powders for 4 hours. This procedure deposits 6 g (50%) of the largest alumina grains (Fig. 2). The slurry containing the remaining 6 g of the alumina will then have a narrower grain size distribution and this was recovered for use in the same elaboration procedure.

After drying for 4 hours at 45°C, the deposits follow the heat treatment described in figure 3. The dispersing agents and graphite were first removed at controlled low heating rates and temperature holdings to prevent cracking. The alumina preform was then sintered in air, at 1600°C for 30 minutes.

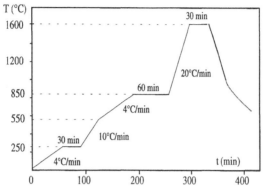

Figure 3: Thermal cycle.

The final sintered alumina preforms are lightly warped disks, 35 mm in diameter and 3 mm thick. A considerable open porosity is observed at the bottom surface while on the top, the alumina is very dense . The porosity gradients were studied by Scanning Electron Microscopy (SEM) on polished surfaces along the thickness of the samples. To facilitate the sample preparation and observations, the open porosity of the preform is pressure infiltrated with liquid tin (Sn), following the method described in [3].

RESULTS

Porosity fraction analysis.

SEM observations of an Al_2O_3 / Sn gradient are shown in figure 4, for a sample elaborated with 15% graphite. The metal appears in white, the ceramic in grey. In the upper dark zone, the alumina does not contain tin. Some non-infiltrated pores are observed (Fig. 4b) but their concentration is low (about 1%). However, near the upper alumina rich zone, their concentration increases to about 8%, From SEM BSE micrographs at magnifications of 1000 or 2000, the porosity fraction was determined by image analysis which takes into account both tin and non-infiltrated pores. For the sample shown in figure 4a, the metal concentration decreases from 40% to 0% from the bottom through a height of 1000 µm, while the mean infiltrated pore diameter decreases from about 7 to 0.7 µm.

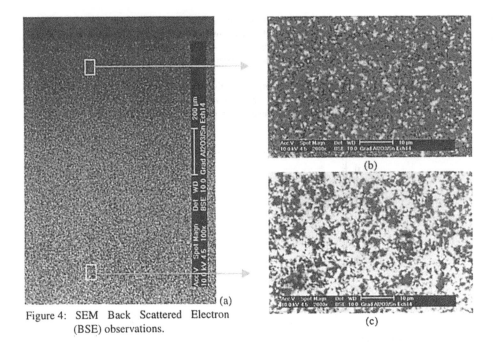

Figure 4: SEM Back Scattered Electron (BSE) observations.

Graphite fraction influence.

The influence of the graphite fraction on the porosity gradient can be seen in figure 5. Higher pore fractions were observed at the bottom of the sample, i.e., 30%, 42% and 55% respectively for 0%, 5% and 15% of graphite. All the curves are convex with a mean slope increasing in absolute value with increasing graphite fraction. The boundary between the alumina zone and the alumina/tin zone lies at about 70% of the theoretical deposit height obtained if the alumina is considered to be entirely in the deposit.

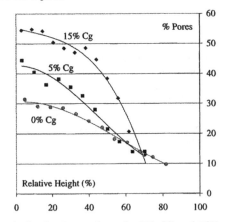

Figure 5: Porosity gradient in the alumina preforms for 0%, 5% and 15% of graphite.

The 0% Cg curve corresponds to sedimentation of alumina alone. For this sample, the observed porosity gradient is due to the grain size gradient; the mean grain diameter decreases from the bottom to the upper surface.[5] This grain size gradient results in different shrinkage kinetics leading to relative densities ranging from 70% and 96%.

Alumina grain size influence.
The narrow alumina grain size distribution gives rise to the porosity gradients represented in figure 6. Two graphite volume fractions, 0% and 5%, were used. Several differences with the results obtained for the as-provided grain size distribution can be noted:
(i)   the curves are concave instead of convex,
(ii)   the boundary between alumina and alumina/tin zones lies at a shorter relative height of about 60%,
(iii) for the 5% Cg sample, the maximum porosity concentration reaches higher values, i.e. 50% instead of 42%.

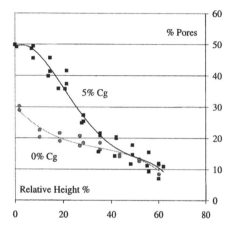

Figure 6: Porosity gradient in the alumina preforms elaborated with the narrow alumina grain size distribution and for 0% and 5% graphite.

MODELING
For a spherical particle of diameter d, the sedimentation time t for a height h can be determined from Stokes equation:

$$t = \frac{h \, 18 \, \eta}{d^2 \, \Delta\rho \, g} \qquad (1)$$

where $\Delta\rho$ is the relative density of the particle in the liquid of viscosity η, and g = 9.81 m²/s.
One can consider two sets of alumina and graphite particles, randomly dispersed along the sedimentation height, in accordance with a given graphite fraction. The diameter of each particle is selected such as to give the required powder grain size distribution. The time needed by each particle to settle can be calculated from equation (1) and this determines its position in the deposit. One can then roughly calculate:

- the concentration gradient of graphite in alumina as a function of the height of the deposit for different graphite volume fractions (curves C_5, C_{15}, C_{m5}, Fig. 7, 8a and 8b),
- the grain size gradient,
- the cumulative sedimentation versus time (Fig. 2 curve c).

It is assumed that there is no interaction between particles during sedimentation. The concentration gradient of graphite in alumina in the deposit does not correspond to the porosity gradient in the alumina preform because of the complex shrinkage kinetics during sintering (compare curve C_5 and experimental data for 5% of graphite in Fig. 7). However, an approximation of the latter gradient can be calculated from the experimental measurement of the graded porosity in the sample obtained by sedimentation of as-provided alumina without graphite (Fig. 5 curve 0%Cg, now referred to as P(x)), by the relation:

$$S_5(x) = P(x) + C_5(x) \, (P(x)+d_g)/100 \qquad (2)$$

with d_g the relative green density and x the relative height.
The term $(P(x) + d_g)/100$ takes into account shrinkage during sintering.

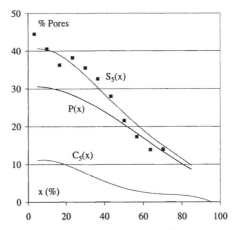

Figure 7: $C_5(x)$: modeling of graphite concentration gradient in the deposit containing 5% graphite,
P(x): experimental porosity gradient in the sample obtained by sedimentation of as-provided alumina alone (id Fig. 5, curve 0%Cg),
$S_5(x)$: modeling of the porosity gradient in the sintered preform for 5% graphite (equation(2)),
■ corresponding experimental data.

The same calculation was applied to as-provided alumina for 15% Cg (Fig. 8a) and to the narrow alumina grain size distribution for 5% Cg (Fig. 8b). For the latter, $P_m(x)$ corresponds to the experimental porosity gradient in alumina elaborated without graphite (Fig. 6, curve 0%Cg).

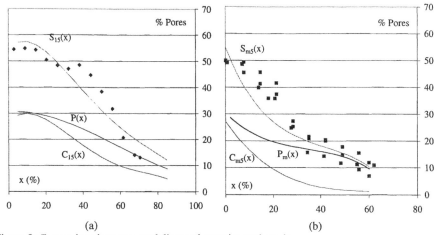

(a)   (b)

Figure 8: Comparison between modeling and experimental results:
(a) for the as-provided alumina grain size distribution – 15%Cg,
(b) for the narrow alumina grain size distribution - 5% Cg.

The different modeling curves are in quite good agreement with the corresponding experimental results. Some differences can however be related to the assumption of non interaction between the particles. Despite the low mass fraction of powder in water (less than 5%) and the dispersing agents, agglomerates can be formed which sediment faster than elementary particles. The dispersion of the graphite powder is particularly difficult. The choice and the ratio of the dispersing agents can be better optimized.

CONCLUSION - PERSPECTIVES
Various porosity graded preforms can be obtained by an elaboration process based on sedimentation of alumina and graphite powders, followed by pyrolyzing the graphite and sintering of the alumina.
The porosity gradient is studied as a function of the graphite fraction and the alumina grain size distribution. The maximum porosity fractions lie between 30% and 55% obtained respectively for 0% and 15% of graphite. The gradient curvature is reversed by a narrower alumina grain size distribution.
The role of other elaboration parameters can be studied, among them, the graphite grain size distribution, the sedimentation parameters like viscosity, height, free height, or the sintering parameters. Modeling can help in the choice of these parameters. Such an investigation should lead to the elaboration of controlled gradient profiles.

REFERENCES
[1] R. Bernhardt, F. Meyer-Olbersleben and B. Kieback, "The Influence of Hydrodynamic Effects on the Adjustment of Gradient Patterns through Gravity Sedimentation of Polydisperse Particle Systems in Newtonian and Viscoelastic Fluids"; Materials Science Forum vols, 308-311 (1999) pp 31-35, © Trans Tech Publications, Switzerland.

[2] S. Suresh, A. Mortensen, "Powder densification process", pp 16-33 in *Fundamentals of functionally graded Materials – Processing and thermomechanical Behaviour of Graded Metals and Metal-Ceramics Composites,* pub. by IOM Communications Ltd, 1998.

[3] G. Kapelski, A. Varloteaux, "A Technique of Elaboration of Functionally Graded Cermets", Materials Science Forum vols, 308-311 (1999) pp 175-180, © Trans Tech Publications, Switzerland.

[4] P.C. Hidber, T.J. Graule, L.J. Gauckler, "Citric Acid - A Dispersant for Aqueous Alumina Suspensions"; *J. Am. Ceram. Soc.,* **79** [7] 1857-67 (1996).

[5] K. Darcovich and C.R. Cloutier, "Processing of Functionnally Gradient Ceramic Membrane for Enhanced Porosity"; *J. Am. Ceram. Soc.,* **82** [8] 2073-79 (1999).

Functionally Graded Materials 2000

TWO-PHASE TiC+TiB$_2$ GRADED CERAMIC PREFORMS

Yehuda Seidman, Natalie Frumin, Naum Frage and Moshe P. Dariel*
Department of Materials Engineering, Ben-Gurion University of the Negev,
Beer-Sheva, Israel

ABSTRACT
A varying rate of sintering within a green compact can be taken advantage of in
order to generate a graded porosity in ceramic preforms. The rate of sintering in
ceramic mixtures often depends on the relative ratio of the components. TiC and
TiB$_2$ do not interact up to 2500°C and are well wetted by liquid Al. Oxygen-free
TiC may be fully densified at 1600°C, whereas TiB$_2$ undergoes sintering only at
T>1800°C. The sintering behavior of TiC+TiB$_2$ mixtures was established and a
synergistic densification effect observed. Graded preforms were built by
stacking layers that had a different ratio of the two components. The sintering
rate of each layer was a function of the relative content of its components. The
sintered preforms were infiltrated with molten Al and hardness profiles were
determined in the resulting graded ceramic-metal composites.

INTRODUCTION
Functionally Graded Materials (FGMs) have been the subject of significant
scientific interest during the last years. The presence of a pre-determined
gradient in properties sets these materials apart from homogeneous materials and
results in an improved overall performance that is achieved by optimizing the
properties as a function of the spatial coordinates. One approach for improving
the toughness of a brittle ceramic material is achieved by combining it with a
metal phase to form a ceramic-metal composite (Cermet). Since many
mechanical properties of cermets are strongly dependent on the ceramic-to-metal
ratio, graded hardness may be achieved by varying the ratio between these
phases. One approach for producing graded cermets, known as the "infiltration"
approach, is based on creating a ceramic preform with graded porosity and
infiltrating it with a liquid metal. The graded porosity may be achieved by
superposing layers of powder mixtures that have different ratios of component
phases and, consequently, sinter at a different rate. Sintering at an intermediate
temperature at which only one of the two phases undergoes shrinkage results in a
final porosity that is governed mainly by the composition of the powder mixture.

The TiB$_2$-TiC-Al system was chosen for this study. Both TiB$_2$ and TiC
show very high hardness of above 2500 HV and a density of 4.5-4.9 g/cm^3, good
wetting by several liquid metals and are, therefore, used in many ceramic/metal
systems, such as TiC-Ni, TiB$_2$-Fe, to produce high performance cermets.

TiB$_2$/TiC composites were also studied by Holleck [1] and showed excellent wear resistance and improved fracture toughness, presumably due to crystallographic matching at the interface between the two phases. In addition, these phases differ considerably in their sinterability and coexist up to 2620°C. Oxygen free TiC can be sintered at 1500°C, while TiB$_2$ undergoes sintering only at T>1800°C. Therefore, it was expected that sintering mixtures of these phases in the 1400-1600°C temperature range, in which only TiC undergoes considerable mass transport, allows to determine the final porosity in the preforms by controlling the composition of the powder mixtures. Both phases are also well wetted by liquid Al at temperatures above 1000°C under vacuum. These features make the TiB$_2$-TiC-Al system suitable for the production of graded cermets via pressureless sintering of a ceramic preform with a pre-designed porosity profile, followed by free infiltration with liquid Al in order to obtain graded (TiB$_2$-TiC)-Al composites.

EXPERIMENTAL

Powders supplied by Alfa Aesar®, TiB$_2$ 99.5% pure 325 mesh (CAS # 12045-63-5), with an average particle size of about 15 μm, and TiC 99.5% pure (CAS # 12070-08-5), with a typical particle size of about 2 μm were used for preparing the preforms.

Two sets of blended powder mixtures of TiB$_2$ and TiC with compositions in the 0-100%TiC range were prepared. The first set was ball milled in a stainless steel cup with steel balls for 4 hours in alcohol and then dried at 100°C. The second set was blended in a plastic bottle for 24 hours at low speed in order to avoid any additions to the powder and particle size reduction.

Monolithic preforms weighing about 3 grams were uniaxially pressed at 35 MPa and sintered the 1400-1600°C-temperature range, for durations from 30 min to 120 min, in a vacuum furnace at about 10^{-2} Pa. Infiltration with aluminum was done at the same vacuum at 1050°C for 30 min.

In addition, several graded preforms were prepared. Layers of different mixtures of powders were stacked in the die, with compositions of 0, 20, 40, 60, 80 and 100%TiC and pressed at 35 MPa. The layers, before compaction were approximately 1.5 mm thick and were uniform within 10%. These preforms were sintered in the above-mentioned temperature and time ranges.

RESULTS

Porosity

After sintering, the porosity of the preforms was measured using the liquid displacement method in distilled water. The results are shown in Fig.1. It is apparent that the final porosity depends strongly on the TiB$_2$/TiC ratio, and that the larger the TiC fraction in the powder mixture, the lower the final porosity.

Functionally Graded Materials 2000

For samples that were prepared from milled powder, the final density was higher, and the final porosity considerably reduced, from about 35% for TiB₂ to about 5% for TiC.

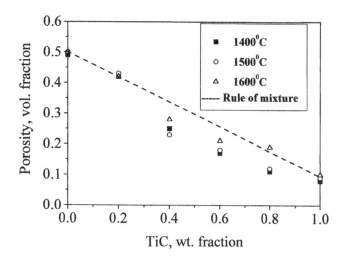

Figure 1. The porosity of preforms sintered at various temperatures for 60 min.

Synergistic Effect During Sintering

The presence of an 'inert' component (TiB₂) is expected to affect negatively the density not only by the reduced volume of the densifying material (TiC), but also by the retarding sintering stress that the inert component applies [2]. Actually, the observed porosity, as shown in Fig.1, is lower than expected if the porosity followed the two-phase "Rule of Mixture". Improved sinterability of the two-phase powder mixtures was observed by Holleck [1] and is confirmed in the present study.

The maximum deviation from the law of mixtures occurs around equal fraction of the two phases. Holleck [1] suggested that this behavior was due to the higher homologous temperature of the compositions around the eutectic point at 57mol%TiC (60wt%TiC). However, the temperature range at which sintering took place was by about 1000°C lower than the eutectic temperature (2620°C). Other mechanisms seemed responsible for the synergistic sintering effect. It is noteworthy that the surface of contact between TiB₂-TiC particles is a maximum around the equi-molar composition.

Auger depth profiling and thermodynamic analysis: Electron spectroscopy surface analysis was carried out in order to gain further insight into the mechanism that led to the improved sinterability that was observed in the powder mixtures. The results of this analysis are shown in Fig.2. The TiC powder initially had a high oxygen concentration on the surface that was significantly reduced after a 1600°C heat treatment in vacuum of 10^{-2} Pa. The TiB$_2$ particles also had a high surface concentration of oxygen but it was not reduced under the same treatment, as shown in Fig. 2b. However, when a mixture of TiB$_2$ and TiC was sintered at 1600°C in vacuum, the oxygen concentration on the surface of the particles was greatly reduced, as can be seen in Fig. 2c. It appears as if sintering TiB$_2$ in the presence of TiC, led to a reduction of the surface oxides, possibly by forming CO with carbon that originated in TiC. The result of this reaction is a cleaner TiB$_2$ surface and possibly some reduction in the near-surface stoichiometry of TiC. Sub-stoichiometric TiC is well known to have a higher rate of sintering then stoichiometric TiC [3]. The combination of these effects is responsible for the overall improved sinterability of the two-phase mixtures.

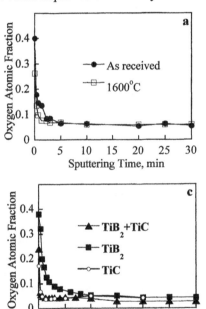

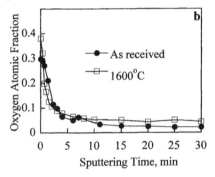

Figure 2. Oxygen content as determined by AES analysis of sample surface before and after a heat treatment at 1600°C at 10^{-2} Pa vacuum. **a.** TiC powder. **b.** TiB$_2$ powder. **c.** TiB$_2$+TiC powder mixture.

A thermodynamic analysis was carried out in order to verify the feasibility of reduction of the boron or titanium surface oxides by carbon originating in TiC. Reduction of the oxides takes place according to reactions:

$$B_2O_3 + \frac{3+x}{1-x}TiC = TiB_2 + 3CO\uparrow + \frac{2}{1-x}TiC \qquad (1)$$

$$TiO_2 + \frac{2+x}{1-x}TiC = \frac{3}{1-x}TiC_x + 2CO\uparrow \qquad (2)$$

Assuming that TiC can be considered as a solid solution between C and Ti, eqs. (1) and (2) can be written as:

$$B_2O_3 + 3[C] + [Ti] = TiB_2 + 3CO\uparrow \qquad k(3) = \frac{P_{CO}^3}{a_{Ti} \cdot a_C^3} \qquad (3)$$

$$TiO_2 + 2[C] = [Ti] + 2CO\uparrow \qquad k(4) = \frac{P_{CO}^2 \cdot a_{Ti}}{a_C^2} \qquad (4)$$

where a_i is the activity of component I and the 'k's are the equilibrium constants of reactions 3 and 4, respectively.

The activities depend on the composition of the titanium carbide phase (x) and were reported previously [4]. The values of the free energy changes (ΔG_0) and of the equilibrium constants were calculated on the basis of the elemental reactions and using the thermodynamic data [5].

The partial pressure of CO for both reactions was calculated for the temperature range of processing up to 1800 K. The results have shown that the CO pressure of the reactions exceeds even the overall pressure in the furnace chamber during sintering for most of the homogeneity range of the TiC phase. It is, therefore, expected that these reactions will occur under the sintering conditions. Combining these results with the one obtained with the AES depth profiles suggests that surface oxide reduction is a possible mechanism for enhanced sintering of the mixed powders.

MECHANICAL PROPERTIES

The hardness and bending strength are shown in Figs. 3 and 4. The hardness, determined at a 2 kg load, shows an increase from about 140 HV for a composite of TiB₂/aluminum, to about 2100 HV for a composite of TiC/aluminum. The increase in hardness is attributed to the lower porosity in preforms that sintered at a higher rate due to their large TiC content. The bending strength, determined by 3 pt bending on 4x12x1.5 mm size samples, ranges between 350 MPa for the TiB₂/aluminum sample to about 600 MPa for the TiB₂-40%TiC sample, decreasing again to about 400 MPa for the

TiC/aluminum sample. This is a typical behavior for a metal-ceramic composite as a function of the metal content. Five samples were tested for each composition.

The fracture toughness was estimated according to the method of Evans and Charles [6] and is shown in Fig.5. The fracture toughness increases unexpectedly from about 2.5 MPa·m$^{0.5}$ for the TiB$_2$/Al composites that contains about 50 vol.% Al to about 9 MPa·m$^{0.5}$ for the (TiB$_2$-80%TiC)-Al composites that contains only about 8 vol.% Al. In a two component ceramic-metal composite, the fracture toughness usually decreases as the metal content of the composite decreases. However, when a three-component composite is produced, the resulting toughness should be considered as the sum of contributions from several mechanisms operating simultaneously. The presence of two or more phases in the ceramic matrix with different properties (e.g., the thermal expansion and elastic moduli), may affect the residual thermal stresses and alter the toughness behavior of the composite. The residual stresses in the ceramic matrix were estimated according to Selsing's equation [7]:

$$\sigma_R = \frac{E_m(\alpha_p - \alpha_m)\cdot \Delta T}{\dfrac{E_m}{E_p}(1 - 2\nu_p) + \dfrac{1}{2}(1 + \nu_m)} \tag{3}$$

where E is the Young's modulus, ν is Poisson's ratio, α is the thermal expansion coefficient and T is the temperature, the subscripts m and p stand for the matrix and the particles, respectively. Using this expression, the residual stresses that form upon cooling from the sintering temperature, in the vicinity of the interface between TiB$_2$ and TiC, were calculated to be of the order of 400-450 MPa. These values are of the order of the strength of the ceramic components and may, in principle, lead to the formation of microcracks.

GRADED COMPOSITES

Using the database that was established for the sintering behavior of TiB$_2$+TiC powder mixtures, graded composites were fabricated. Several layers of TiB$_2$+TiC powder mixtures, with ratio of components that was determined according to the database and a pre-determined design, were stacked and cold-pressed in order to form green samples of about 20 mm in diameter. The samples were sintered and subsequently infiltrated with Al in order to form graded composites. Hardness profiles of two such samples are shown in Fig. 6. Specific profiles can be designed and implemented by stacking layers with appropriate composition and thickness.

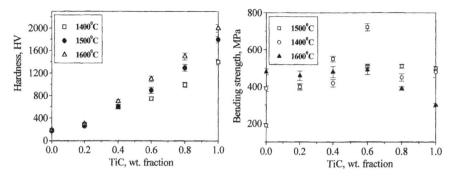

Figure 3. The hardness of the (TiB$_2$-TiC)-Al composites.

Figure 4. The bending strength of (TiB$_2$-TiC)-Al composites.

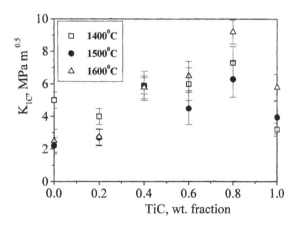

Figure 5. The fracture toughness of (TiB$_2$-TiC)-Al composites.

The composites showed structural integrity with no apparent cracks. Larger plate-like 50x50 mm composites, graded along their thickness, were also fabricated but deformed severely due to the non-uniform shrinkage of the superposed layers.

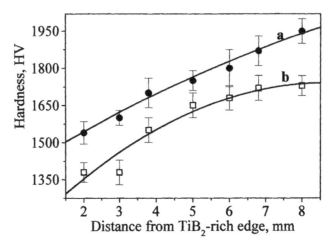

Figure 6. Hardness profiles of the graded composites. **a.** preform constructed from 4 layers in the 40-100%TiC composition range. **b.** preform constructed from 4 layers in the 20-80%TiC range, with a double thickness of the 80%TiC layer.

SUMMARY AND CONCLUSIONS

A database was established for the sintering behavior of TiB_2+TiC powder mixtures in the 1400-1600°C temperature range. The porosity that was obtained after sintering decreased continuously with the relative fraction of TiC, from 50 vol.% for TiB_2 preforms to about 7 vol.% for TiC preforms.

A synergistic effect of densification was observed during the sintering of TiB_2+TiC powder mixtures. This effect was confirmed by AES depth profiling and was attributed to the reduction of the TiB_2 surface oxides by carbon originating in the TiC phase. The improved sinterability of the mixtures is the result of the reduced stoichiometry of TiC_{1-x}, the cleaner surface of the particles and the *in situ* formation of TiB_2, as predicted by thermodynamic analysis.

A database for the mechanical properties of $(TiB_2\text{-}TiC)\text{-}Al$ composites was also established. The hardness of the homogeneous composites ranged from 130 HV for TiB_2-Al samples to 2200 HV for the TiC-Al samples, and correlates with the relative fraction of the metal phase in the composites. The observed bending strength was in the 300-600 MPa range. The fracture toughness was estimated to vary in the 2 to 9 $MPa\cdot m^{0.5}$ range.

Based on these databases, graded composites of 20 mm diameter and also 50x50 mm square plaques were designed and prepared. Hardness values in these composites ranged from 1200 to 2000 HV. These graded composites showed structural integrity and did not crack, thus proving the feasibility of the approach for producing functionally graded ceramic/metal composites.

REFERENCES

[1] H.Holleck, H.Lesite and W.Schneider, "Significance of Phase Boundaries in Wear Resistant TiC/TiB$_2$ Materials". *International Journal of Refractory & Hard Materials,* **6** [3] p.149-154 (1987).

[2] R.M.German, *"Sintering Theory and Practice",* John Wiley, New York, pp.192-194, 1996.

[3] M.P.Dariel, S.Sabatello, L.Levin and N.Frage, "Functionally Graded TiC-based Cermets"; pp. 493-499 in *Functionally Graded Materials 1998,* Edited by W.A.Kaysser, Trans Tech Publications LTD, Uetikon-Zuerich, 1999.

[4] N. Frage, N.Frumin, L.Levin, M.Polak and M.P. Dariel, "High Temperature Phase Equilibria in the Al-rich Corner of the Al-Ti-C System", *Metallurgical and Materials Transactions,* **29A**, 1341-1345 (1998).

[5] Y.K.Rao, *"Stoichiometry and Thermodynamics of Metallurgical Processes",* Cambridge University Press, Cambridge, pp. 882-889, 1985.

[6] A.G.Evans and E.A.Charles, "Fracture Toughness Determination by Indentation", *Journal of the American Ceramic Society,* **59** [7-8] 371-372 (1976).

L.S.Sigl, "Microcrack Toughening in Brittle Materials Containing Weak and Strong Interfaces", *Acta Mater.,* **44** [9] p.3599-3609 (1996).

GRADED BORON CARBIDE-ALUMINUM CERMETS

N. Frage, L.Levin and M.P.Dariel

Department of Materials Engineering, Ben-Gurion University of the Negev,
Beer-Sheva, 84105, Israel

ABSTRACT

One possible approach for the fabrication of graded boron carbide-aluminum cermets is based on the infiltration of molten aluminum into a ceramic preform that displays graded porosity. Unidirectional graded porosity of the ceramic perform was achieved by sintering stacked layers of B_4C + TiO_2 powder mixtures. Reactive sintering takes place between the two components at a rate that strongly depends on the relative amount of TiO_2. Upon infiltration of the porous preform with the molten metal, a graded ceramic-to-metal ratio can be constructed. The ability to vary the ceramic-to-metal ratio, according to required specifications, allows generating desired property profiles and paves the way for a variety of specifications. Hardness profiles of the final composite that can be built depend on the nature of the ceramic preform, on the processing parameters and on the characteristics (thickness, number) of the initially stacked ceramic powder layers.

INTRODUCTION

Extensive efforts have been made towards implementing the FGM (Functionally Graded Materials) concept in various ceramic-metal combinations[1]. The gradual changes that take place in a FGM can, in principle, eliminate the discontinuity at the interface of dissimilar materials and relax the thermal stress induced by the temperature variation both in the course of fabrication and service. One approach for the generation of a graded ceramic-to-metal composite is based on the fabrication of a ceramic preform with a graded porosity and its subsequent infiltration with a molten metal. The resulting ceramic-metal composite displays a varying ceramic-to-metal (C/M) ratio and properties that change according to their dependence on that ratio. The objective of the present communication is to describe the experimental methods that were used to generate graded B_4C-based preforms. The approach relies on layer-by-layer stacking of appropriate powder mixtures. The powder layer stack is compacted and at elevated temperature a reactive sintering reaction takes place between the constituents of each layer. The sintering rate depends on the relative amount of the constituents, ultimately giving rise to a cohesive body that displays a unidirectional variation of porosity. The porosity ratio is determined by several parameters such as the thickness of each layer, the ratio of the constituents in each layer and their rate of sintering. In the present investigation, the porosity gradient was generated by taking advantage of the composition-dependent rate of reactive sintering of B_4C-TiO_2 powder mixtures.

THEORETICAL BACKGROUND

Boron carbide is an extremely promising material for a variety of applications that require elevated hardness values, good wear and corrosion resistance. Widespread use of the material is limited, however, by the very high sintering temperatures that are required for its densification. Actually, nearly fully dense boron carbide can be prepared only by hot pressing above 2300 °C. Boron carbide exists over a wide composition domain that extends from B_4C (20 at%C) to a sub-stoichiometric compound with 9 at%C. Recently we have carried out a study aimed at finding out whether, in the absence of an extraneous carbon addition, titanium oxide could be reduced by carbon originating from boron carbide, thereby altering the composition of the latter[2]. The specific objective of the study was to determine the contribution of the *in situ* developed deviations from stoichiometry on the sinterability of B_4C. In order to provide an appropriate frame of reference, two sets of samples, namely boron carbide with, (a) additions of Ti, for *in situ* generation of TiB_2; (b) additions of TiO_2, leading to the formation of TiB_2 simultaneously with depletion of the carbon content in the carbide phase, were studied. The study included an extended thermodynamic analysis of the interaction of stoichiometric boron carbide with titanium and of the deviations from stoichiometry due to the reduction of TiO_2. The experimental results supported the theoretical predictions, namely, that TiO_2 is reduced by carbon originating in the boron carbide phase, thus depleting the carbon content of the latter according to the reaction:

$$B_4C + TiO_2 \rightarrow B_4C_{1-x} + TiB_2 + (CO \text{ or/and } CO_2)\uparrow \qquad (1)$$

The composition range of the boron carbide phase after the reaction is determined by the relative amount of the additive and may be predicted on the basis of the simple stoichiometric relations. According to Fig.1, up to 32.5 wt.% TiO_2, if CO gas is formed and 52.0 wt.% TiO_2, if CO_2 is formed, can be added without the generation of any free boron.

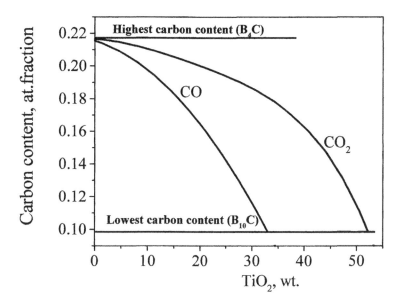

Fig. 1. The dependence of the carbon content in the boron carbide phase on the amount of TiO_2 that was added to the initial powder mixture.

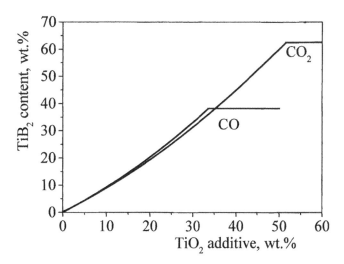

Fig. 2. The TiB_2 content in the sintered preforms as a function of the TiO_2 additive, when CO or CO_2 gases are formed.

The calculated TiB_2 content in the sintered preforms as a function of the amount of TiO_2 additive is shown in Fig.2.

The reactive sintering leads to formation of a two-phase ceramic material consisting of nonstoichiometric boron carbide and TiB_2. Composition-dependent reactive sintering of the B_4C-TiO_2 powder mixtures enables production of preforms with graded porosity.

EXPERIMENTAL PROCEDURES

Two sources of boron carbide powder were used. The first powder was supplied by Cerac, (99+%, 5-7 μm particle size) and the second by Starck (99%, 3.5 μm mean particle size, HS grade).The TiO_2 powder (Merck, 99.8%, 2 μm and less particle size) was used to prepare ceramic mixtures containing 5 to 40 wt.% TiO_2. Sedimentation of the coarse B_4C (Cerac) powder from aqueous solutions with varying pH values allowed preparing a powder lot with a narrow particle size distribution around 12 m^2/g specific area, as measured by the BET technique. The fine B_4C and TiO_2 powders were dry-blended and uniaxially compacted with no binder in steel dies at 80 MPa. The standard sample preparation consisted of a two-step treatment: (a) heating to 1500°C for two hours in a vacuum furnace to ensure the completion of the reaction between the B_4C and TiO_2, (b) sintering for 1h, performed in an "ASTRO" furnace with graphite heat elements under an argon atmosphere in the 2100 to 2190°C temperature range. Graded preforms were prepared by stacking B_4C+TiO_2 mixtures with different ratio of the component powders (0, 10, 20,30, 40 and 50 wt. % TiO_2). Two-phase ceramic bodies, sintered to different average levels of graded porosity were used as preforms and infiltrated with molten aluminum. Infiltration was carried out in a graphite furnace under vacuum 10^{-2} Pa at 1100 °C.

RESULTS AND DISCUSSION

The dependence of the relative density of the sintered powder compacts made from B_4C powder after sedimentation, as a function of the amount TiO_2 added for three different sintering temperatures, is shown in Fig.3. Noteworthy are the elevated values of density attained by pressureless sintering below 2200°C.

The SEM image of the fracture surface of the sample with 30%TiO_2 additive (82-85% relative density is shown in Fig.4a. The very efficient densification of a sample with 40% TiO_2 additive, sintered at 2160°C with 92-94% relative density, is illustrated in the SEM image of a fracture surface in Fig.4.b, showing the presence of a few closed pores and some larger ones, part of the open channel network. The microstructure and the mechanical properties of the homogeneous B_4C-TiB_2 preforms infiltrated with Al were discussed previously [3]. The microstructure of four different areas of a graded composite is shown in Fig.5. The two ceramic component phases can be clearly distinguished in the polished samples after infiltration with molten aluminum. The average density of the pre-infiltration preform was about 80%. The size of the B_4C particles is of the order of 10 μm and that of the TiB_2 grains less than 5 μm. The formation of TiB_2 particles seems to inhibit the growth of the B_4C matrix grains. The interparticle channels underwent full infiltration by the molten metal, in agreement with the well-established wetting behavior of both ceramic phases B_4C and TiB_2 by liquid Al. The hardness profile within the graded part is shown in Fig.6.

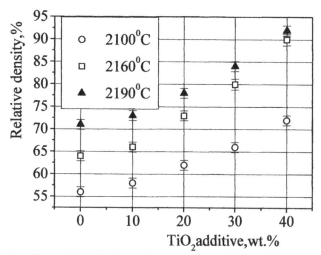

Fig.3. Relative density of the B_4C preforms with a TiO_2 additive, sintered at various temperatures. The specific surface area of the B_4C powder was about 12 m^2/g.

Fig. 4. SEM image of B_4C+30%TiO_2 (a) and B_4C+40%TiO_2 (b) preforms sintered at 2160 °C for 1h.

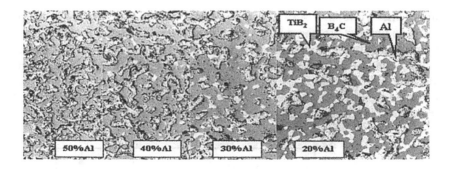

Fig. 5. The microstructure of the infiltrated two-phase ceramic preform at different locations along the graded composite (x 400)

Noteworthy is the sintering behavior of the B_4C powder supplied by Starck. A relative density of 95-96 % was observed in B_4C samples that underwent pressureless sintering without additives at 2190°C. This finding led us to determine the composition and particle size distribution of the powder (Table 1 and Fig.7).

TABLE 1. Composition (wt.%) of the B_4C powder supplied by Starck (HS grade)

Element or compounds	B_2O_3	Si	N	O	Compound C	Free C	Total B	Total C
%	0.46	0.09	0.426	2.04	20.69	1.28	74.96	21.97

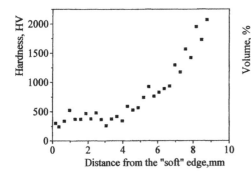

Fig. 6. The hardness profile in a graded $(B_4C+TiB_2)/Al$ composite.

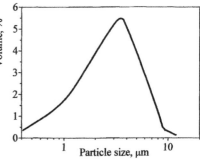

Fig.7. Particle size distribution for B_4C (Starck HS grade).

The outstanding sintering behavior of the Starck powder may be attributed to its free carbon and nitrogen contents and/or to the presence of a fraction of small size particles. The addition of TiO_2 powder led to almost fully dense ceramic bodies sintered at about 2160°C. Compacted green stacks were sintered at 2100°C to give performs with graded porosity. The microstructures of graded infiltrated composites, consisting of three layers with different composition (30, 40 and 50% TiO_2 additive), are shown in Fig.8 and the hardness profile in Fig.9.

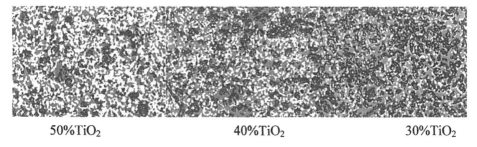

50%TiO_2 40%TiO_2 30%TiO_2

Fig.8. The microstructure of the infiltrated two-phase ceramic preform based on B_4C powder (Starck HS grade) at different locations along the graded composite (x400)

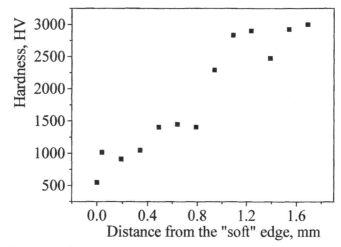

Fig.9. The hardness profile in a graded two-phase (B_4C+TiB_2)/Al composite based on B_4C powder (Starck HS grade).

A roughly linear hardness profile was obtained in this case, in contrast to the "concave" profile shown in Fig.6. Extremely high hardness (more than 3000HV) was achieved by using the mixtures based on HS grade Starck B_4C powder. Obviously, the hardness profile can be pre-designed by varying the thickness of each layer and its composition.

SUMMARY AND CONCLUSIONS

The reactive sintering approach induces variations in the rate of sintering as a function of the amount of TiO_2 added to B_4C powder. Moreover, at a relatively large fraction of the additive, (40% TiO_2), a very significant decrease of the sintering temperature with no concomitant grain growth was observed. By infiltrating the porous preforms with molten metals, a graded ceramic-to-metal ratio can be constructed. The ability to vary the ceramic-to-metal ratio according to required specifications allows generating desired property profiles and paves the way for a variety of specifications. The intrinsic drawback of this approach is the deformation that accompanies a densification process that depends on the spatial coordinates. The deformation can be corrected, to some extent, by applying compensating hot deformation procedures. Within these limitations, graded ceramic preforms can be designed and constructed by using a powder layer stacking approach. The hardness profiles of the final composite that can be built depend on the nature of the ceramic preform, on its processing parameters and on the characteristics (thickness, number) of the initially stacked ceramic powder layers.

REFERENCES

[1] Y.Miyamoto, W.A.Kaysser, B.H.Rabin, A.Kawasaki, and R.G.Ford, *"Functionally Graded Materials: Design, Processing and Applications"*, Kluwer Academic Publ., Dordrecht, 1999.

[2] L. Levin, N. Frage and M.P.Dariel, "The Effect of Ti and TiO_2 Additions on the Pressureless Sintering of B_4C" *Met. Mat. Trans.* **30A**, 3201-3210 (1999).

[3] L. Levin, N. Frage and M.P.Dariel, "A Novel Approach for the Preparation of B_4C-based Cermets", *International Journal of Refractory Metals & Hard Materials,* **18**, 131-135 (2000).

GRADED MULTILAYER BORON CARBIDE-ALUMINUM COMPOSITES

F. Zhang, K. P. Trumble and K. J. Bowman
School of Materials Engineering
Purdue University
West Lafayette, IN 47907

ABSTRACT

Fully dense boron carbide-aluminum composites with graded multilayers were prepared by liquid aluminum infiltration of boron carbide preforms produced by centrifugal casting. The amounts of secondary phases were effectively suppressed by rapid spontaneous infiltration. Young's modulus, microhardness, SEVNB fracture toughness and flexural strength were measured and correlated to microstructure and fractography. The results are compared with models for composite properties.

INTRODUCTION

Boron carbide is well known for its high hardness and low density. When it is combined with lightweight metal, such as aluminum, to produce a metal ceramic composite, it is a very promising material in the situations where high stiffness, high hardness, wear resistance and light weight are demanded. It is also of special interests in armor applications.[1]

Some research and patents have been reported to develop boron carbide–aluminum composites.[1,2,3] Liquid aluminum infiltration methods are most commonly used in the preparation. Chemical reactivity between aluminum and boron carbide, and the wetting angle variation with temperature and processing time have also been investigated.[4,5] It is reported that Al and B_4C are highly reactive at elevated temperatures and that the contact angle at short time drops dramatically with increasing temperature under vacuum and inert gas atmospheres. Special structured composites could be obtained by infiltrating ceramic performs with tailored structures. Petrovic and coworkers[1] have made graded boron carbide-aluminum composites by varying the amount of sintering aids and starting powder particle size to form gradient ceramic preforms.

In the present work, graded multilayer boron carbide-aluminum composites were developed and their mechanical properties assessed. Multilayer boron carbide performs were prepared by a centrifugal casting technique, partially densified, and infiltrated with pure aluminum by rapid spontaneous infiltration, which ensure a minimum amount of reaction products.

EXPERIMENTAL PROCEDURE

Centrifugal casting was used to prepare the ceramic performs. Boron carbide powder (BO-301, 1500 grit, Atlantic Equipment Engrs., NJ, USA) was mixed with de-ionized water by stirring to form a slurry of 20 vol.% solid content, followed by ultrasonic treatment to achieve homogeneity and the pH was adjusted to about 7.6 to 8.0 using NH_4OH. Slurry was then added into a centrifuge casting bucket with a plaster bottom plate (Figure 1), which enhances the uniform drying of the cast preforms. Boron carbide green bodies were cast layer-by-layer. After casting each layer, the top supernatant was removed and new slurry for next layer was transferred into the bucket. A centrifugal acceleration force of ~2000 times standard gravity was used in casting. Drying was carried out at room temperature for 24 hours before the casting was removed from the bucket and finally dried in an oven at 60°C until completely dry. A typical casting is about 14 mm thick and 38 mm in diameter, with each layer 600 to 700 μm in thickness.

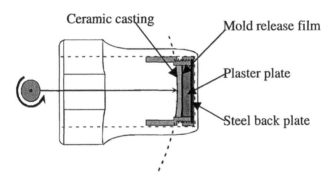

Figure 1 Configuration of centrifuge casting bucket.

The dried bodies were then hot-pressed at 1800-1900°C/10MPa to partially densify and obtain an interconnected boron carbide network with layered structure. The preforms were then infiltrated with molten aluminum (99.99% pure shot, Alcoa Specialty Metals, Banton, NC) under vacuum or flowing argon in a temperature range of 1200 to 1300°C. The temperature control profile was chosen so that a fast ramp rate, ~30°C/min, and short hold processing time was used to minimize the reactions between boron carbide and aluminum.

After infiltration, the composite samples were cut for microstructural and mechanical evaluation. Samples were polished using SiC, diamond and colloidal silica abrasives. Grain size was determined with quantitative image analysis software, NIH image. Ultrasonic technique was employed to measure the Young's modulus. Strength was measured by 4-point bending using 3 x 4 x 24 mm beams, where the tensile surface was polished through 1 μm diamond paste. Fracture toughness was measured by 4-point bending on the same size beams using the single edge V-notch beam (SEVNB) method. Figure 2 shows the test configuration schematically along with micrographs of a typical V-notch. The V-notch preparation followed the procedure used by Kübler,[6] wherein a pre-notch was made with a diamond blade and then a razor blade machining technique was used to produce the sharp V-notch. Reproducible V-notch tip radii of 15 to 20 μm can be achieved.

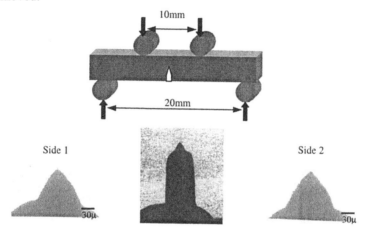

Figure 2 Configuration of SEVNB fracture toughness test and micrographs showing side views of a typical V-notch.

RESULTS AND DISCUSSION

The effects of processing time, temperature, and atmosphere on the formation of the composite were investigated. As reported by Halverson,[5] aluminum spontaneously infiltrates boron carbide when the temperature is high enough to promote the drop of contact angle. In the present work, at ~1200°C under rough vacuum (~13 Pa (100 mTorr)), aluminum on the top of a porous boron carbide preform was observed to spread over and spontaneously infiltrate the perform in a few minutes. It is worth noting that the oxide skin on the aluminum may play a role in keeping it from direct contact with the boron carbide. Spontaneous infiltration was not observed under flowing argon atmosphere, even at higher temperatures. This result is consistent with Halverson's measurement of contact angle change in flowing argon, where even after 100 minutes at 1200°C the contact angle was still greater than 70°.[5] From calculation for close-packed spheres, Hilden[7] and Trumble[8] found that contact angle of ~50° or less is necessary for spontaneous (pressureless) infiltration.

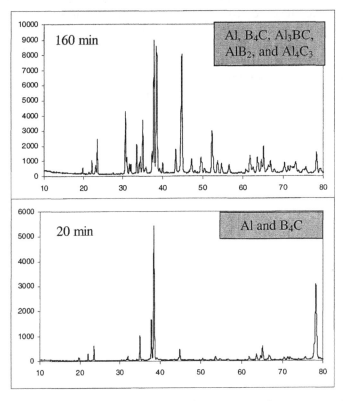

Figure 3 XRD pattern for boron carbide-aluminum composites processed at 1200°C for different times.

High temperature and long process time help drop the contact angle and promote spontaneous infiltration, but they also lead to more reaction products, which can deteriorate the composite mechanical properties. Figure 3 shows X-ray diffraction patterns of composites produced at 1200°C for 160 minutes and 20 minutes. After 160 minutes, many different kinds of reaction products were found whereas after 20 minutes no phase other than boron carbide and aluminum was detected.

Optical micrographs taken from a cross section of the boron carbide-aluminum composite are shown in Figure 4. Two layers of the composites are shown at lower magnification in the background. The ceramic and metal phase network can be observed. The multilayered structure and the variation of the particle size within each layer are clearly shown from the four higher magnification micrographs of the selected areas along a trans-layer direction. The gradient in particle size is a characteristic of the centrifugal casting processing. During centrifugal casting, the large particles settle faster than small particles, preferentially depositing on the bottom of the each layer. The particle size variation as measured by quantitative stereology is shown in Figure 5. In contrast to the gradient in boron carbide particle size across the layers, the volume fraction of boron carbide is almost constant through the composite, suggesting that its packing is uniform and independent of the particle size in this process.

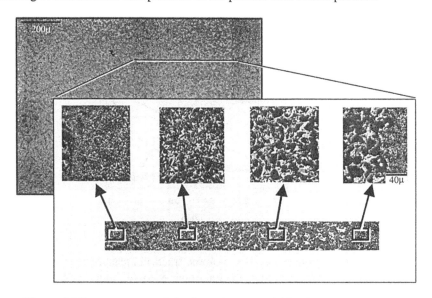

Figure 4 Microstructure of the layer boron carbide-aluminum composite.

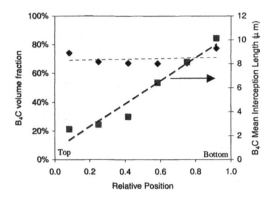

Figure 5 Boron carbide particle size distribution and volume fraction across a single layer of composite.

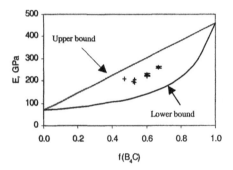

Figure 6 Young's modulus vs. B$_4$C volume fraction measured for layered boron carbide aluminum composite and upper & lower bound analyses.

Mechanical properties were measured for the layered boron carbide–aluminum composite. Young's modulus measured by ultrasonic technique is shown on Figure 6. All the data points from different samples fall in the middle of the upper bound and lower bound calculated from Reuss and Voigt models. For 53 vol% B$_4$C-Al composite processed at 1200°C for 20 min Young's modulus in the transverse direction was measured to be 197 GPa, assuming isotropy. Despite the layered structure, measurements in directions parallel to the layers differed by less than two percent of those in the transverse direction.

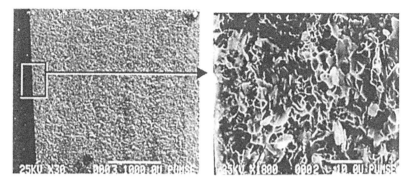

Figure 7 Fracture surface of a SEVNB toughness test sample. Fracture was initiated from the V-notch tip.

Flexural strength and fracture toughness were measured for 65 vol% B_4C-Al composites infiltrated at 1200°C for 20 min. The flexural strength was 498 MPa ± 31 MPa (average of 5 samples). Fracture toughness was 7.2 ±0.2 $MPa.m^{1/2}$ (average of 7 samples), measured by SEVNB. Pizik and co-workers[2,9] reported flexural strength ranging from 250 to 620 MPa for homogenous (non-layered) B_4C-Al composites with widely varying, but unfortunately poorly described microstructures. Those most similar to the present samples (70 vol%, connected B_4C, minimal reaction phases[2]) had reported flexural strength of 550 MPa, which agrees well with the present results. The fracture toughness reported for these materials measured by SENB compares very well with that of the layered materials in the present study, although the toughness values reported by Pyzik[2] are probably exaggerated due to the use of a much large radius notch (~250 µm saw cut) in the SENB test compared to the SEVNB test. Figure 7 shows photos of the fracture surface after the fracture toughness test, where the fracture initiated from the tip of the V-notch. The layer structure can be observed in the low magnification picture. Broken boron carbide network and dimples of aluminum can be observed on the fracture surface. The high fracture toughness could be attributed to the plastic deformation of the aluminum, and it is also benefited from contributions by the graded layer ceramic structure.[10] Further studies are ongoing to quantify the effects of layering and layer gradient microstructure on mechanical properties.

SUMMARY

Graded multilayer boron carbide composites were successfully prepared and their mechanical properties were evaluated. Centrifugal casting process formed the particle size gradient within each layer, while the volume fraction was uniform through the layers. Aluminum infiltration of partially densified boron carbide preforms can be fully achieved spontaneously at 1200°C in a few minutes under rough vacuum. Longer processing time can lead to more reaction products, which may be deleterious to the mechanical property of the composites.

For 65 vol% B4C-Al layered composites infiltrated at 1200°C for 20 min. K_{IC} of 7.2 ± 0.2 MPa·m$^{1/2}$, measured by SEVNB method, and flexural strength of 500 ± 30 MPa were obtained. The increased fracture toughness may be attributed to the plastic deformation of the aluminum in addition to the benefits contributed by the gradient layer ceramic structure.

ACKNOWLEDGEMENT

This research is supported by the United States Army Research Office MURI grant No. DAAH04-96-1-0331.

REFERENCES

[1] J. J. Petrovic, K. J. McClellan, C. D. Kise, R. C. Hoover, and Scarborough, "Functionally Graded Boron Carbide," MRS 1997 meeting, Boston, MA, 1-5 Dec., 1997

[2] A. J. Pyzik. and D. R. Beaman, "Al-B-C Phase Development and Effects on Mechanical Properties of B4C/Al-Derived Composites", J. Am. Ceram. Soc., 78 [2] 305-12, 1995.

[3] A. J. Pyzik, U. V. Deshmukh, S. D. Dunmead, J. J.Ott, T. L. Allen, and H. E. Rossow, US patent 5521016.

[4] J. C. Viala, J. Bouix, G. Gonzalez, and C. Esnouf, "Chemical Reactivity of Aluminum with Boron Carbide," J. Mater. Sci. 32 (1997) 4559-4573.

[5] D. C. Halverson, A.J. Pyzik, and I. A. Askay, "Processing and Microstructure Characterizaton of B4C-Al Cermets," *Ceram. Eng. & Sci. Proc.*, 6, [7-8], 736-741, 1985

[6] J. Kübler, "Fracture Toughness Using The SEVNB Method: Preliminary Results," *Ceram. Eng. & Sci. Proc.*, 18 [4] 155-162 (1997).

[7] J. Hilden, and K. P. Trumble, "Spontaneous Infiltration of Non-cylindrical Porosity: large Pores," Materials Science Forum, Vols 308-311 (1999) p127-162.

[8] K. P. Trumble, "Spontaneous Infiltration of Non-cylindrical Porosity: Close-Packed Spheres," Acta Mater. 46, [7], 2363-2367, 1998.

[9] A. J. Pyzik, I. A. Aksay and M. Sarikays, "Microdesigning of Ceramic-Metal Composites," Ceramic Mircostructure'86, Role of interfaces, Plenum press, 45-54, 1987.

[10] R. Moon, Ph. D Thesis, Purdue University, 2000

TRANSFORMATION STABILITY AND MECHANICAL PROPERTIES OF INFILTRATED FUNCTIONALLY GRADED MULLITE/YTTRIA-STABILIZED TETRAGONAL ZIRCONIA POLYCRYSTAL/ALUMINA COMPOSITES

Zhenbo Zhao, Cheng Liu and Derek O. Northwood*
Mechanical, Automotive & Materials Engineering
University of Windsor
Windsor, Ontario, Canada, N9B 3P4

*Also Faculty of Engineering & Applied Science
Ryerson Polytechnic University,
Toronto, Ontario, Canada M5B 2K3

ABSTRACT

A novel infiltration method for processing mullite/yttria-stabilized tetragonal zirconia polycrystal/alumina (mullite/3Y-TZP/Al$_2$O$_3$) composite with graded transformation stability is described. This process involves infiltrating porous 3Y-TZP/Al$_2$O$_3$ composite preforms with a solution of ethyl silicate, followed by sintering at 1600°C and 1700°C for 2.5 hours. The resultant material has a homogeneous 3Y-TZP/Al$_2$O$_3$ core encased with a graded and heterogeneous layer of mullite/3Y-TZP/Al$_2$O$_3$. Analysis by X-ray diffraction and energy-dispersive spectrometry has revealed the existence of a concentration gradient of mullite, with the concentration decreasing with increasing depth into the sample. The graded composite displays a gradual change in thermal expansion values due to the presence of mullite. The partial loss of the Y-TZP/Al$_2$O$_3$ composite's mechanical strength and fracture toughness due to the introduction of Al$_2$O$_3$, can be prevented.

INTRODUCTION

The low temperature environmental degradation phenomenon occuring in yttria-stabilized tetragonal zirconia polycrystal (Y-TZP) has been extensively investigated for the last two decades[1-9]. Much attention has been focused on retaining the commercial viability of Y-TZP by elimination of the low temperature environmental degradation phenomenon. Generally, these preventive methods can be grouped as either "bulk" or "surface" methods[3, 9]. The bulk methods are those which render the bulk Y-TZP stable against degradation. However, although the degradation of the Y-TZP is inhibited by using these methods, a partial loss of its mechanical properties usually results because of the reduced transformability. The other group of methods involve surface modification of the Y-TZP by forming a surface layer that is resistant to the degradation, since the degradation of Y-TZP initiates at the surface. These surface methods can both prevent degradation and maintain the mechanical properties of the bulk. However, many of these surface modification methods yield bodies in which there is an abrupt transition in materials composition and properties across the boundary between the modified region and the bulk. This often results in sharp local concentrations of stress which can disrupt the integrity of the bond between the two zones or lead to failure within the zones due to the existence of a mismatch in the properties (such as thermal expansion coefficient).

The liquid precursor infiltration method for surface modification of ceramics has been shown to be an efficient surface modification method in fabricating a variety of unique microstructures with graded thermomechanical properties[9-10]. This new approach deliberately seeks to produce strong bonding between the surface modified zone and the bulk with a gradual change in thermal expansion coefficient and forms surface compressive stresses since the difference in thermal expansion coefficients is greater for mullite and Y-TZP/Al$_2$O$_3$ composite than for Al$_2$O$_3$ and zirconia. Thus, surface modification of Y-TZP/Alumina composites by incorporation of mullite as a second phase was attempted by fabricating an infiltrated functionally graded mullite/yttria-stabilized tetragonal zirconia polycrystal/alumina composite.

EXPERIMENTAL METHODS

80wt% co-precipited zirconia powders containing 3mol% Y_2O_3 (3Y-TZP) and 20wt% alumina were milled and uniaxially pressed at 120 MPa to form standard rectangular specimens. It was found that it was necessary to partially sinter the specimens at 1200°C before infiltration to impart enough strength so that they could withstand the subsequent processing steps[10]. The procedures followed in fabricating surface modified mullite/Al_2O_3/3Y-TZP specimens are the same as that was used by the authors in ref.[9]. The porous Al_2O_3/3Y-TZP composites were infiltrated by immersing them in an ethyl silicate solution (PS912, the percentage of SiO_2 is 28vol%.) made by Petrarch Systems, Inc., Bristol, PA, USA. The depth of liquid penetration can be controlled by the length of time that the specimen is left in the solution (10 seconds to 24 hours were applied.). The specimens were then sintered at 1600°C and 1700°C for 2.5 hours in air to decompose the infiltrant, bring about the mullite formation reaction, and densify the specimens. The methods for specimen fabrication, measurement and calculation for the bending strength and fracture toughness, are same as that were used by authors in ref.[9]. Selected samples sintered at the different temperatures were annealed in a hydrothermal corrosion environment (in an autoclave with water) at 180°C and 10 bar for 144 hours. Surface phase analysis was carried out by X-ray diffraction techniques for surface mullite-modified 3Y-TZP/Alumina composites after hydrothermal treatment. The fraction of monoclinic phase on the surface was determined as follows [11]:

$$X_m = [I_m(111) + I_m(11\bar{1})] \quad / \quad [I_m(111) + I_m(11\bar{1}) + I_c(111)] \tag{1}$$

where, $I_m(111)$ and $I_m(11\bar{1})$ are the intensity of the monoclinic (111) and $(11\bar{1})$ line, $I_c(111)$ is the intensity of cubic (111) line in the X-ray diffraction pattern. Energy-dispersive spectrometry was used to identify the concentration distribution of mullite with infiltration depth in combination with electron microprobe analysis. The bulk and relative density of each sample was measured using Archimedes method. The average grain size for 3Y-TZP and the aspect ratio for mullite at the surface were determined by scanning electron microscopy and the line intercept method [12] since the typical shape of grains for mullite is elongated when sintering temperatures are high enough (at 1600-1700°C).

RESULTS AND DISCUSSION

The average grain size for 3Y-TZP, the aspect ratio for mullite at the surface and relative density for surface mullite-modified 3Y-TZP/Alumina composites sintered at different temperatures are given in Table 1. The relative density of the surface mullite-modified 3Y-TZP/Alumina composite increases slightly as the sintering temperature increases. The effect of the incorporation of mullite into 3Y-TZP/Alumina composites on its relative density was overcome when relatively high sintering temperatures (such as 1700°C) were applied (as shown in Table 1, 99.1% relative density was obtained.). Also, relatively larger average grain sizes for 3Y-TZP and aspect ratios for mullite were observed at the higher sintering temperatures. An increase in grain size with increasing sintering temperature is not surprising, and is typically observed since grain growth depends mainly on time and temperature[13]. The concentration distribution of mullite for different infiltration times is shown in Fig. 1.

Table 1 Average grain size for 3Y-TZP and aspect ratio for mullite at surface and relative density for samples of surface mullite-modified 3Y-TZP/Alumina composites sintered at 1600°C or 1700°C.

Method of treatment		Sintered 1600°C	Sintered 1700°C
Relative density (%)		98.5	99.1
Grain size (μm)	(for 3Y-TZP only)	1.5	2.5
Aspect ratio	(for mullite only)	1.7:1	2.5:1

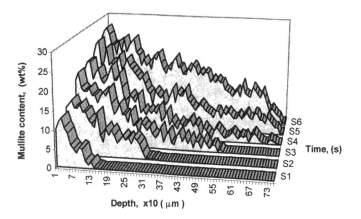

S1 -- 10 s; S2 -- 60 s (1min); S3 -- 1800 s (30min); S4 -- 14400 s (4hs); S5 -- 43200 s (12hs); S6 -- 86400 s (24hs)

Fig.1. Concentration distribution of mullite content as a function of infiltration time

The concentration gradients of mullite show a decreasing concentration with increasing depth into the sample for all infiltration times. Three possible reasons for this result have been proposed[14], namely: (1) filtering of the sol as it infiltrates the body, (2) incomplete filling of the pores initially, with backfilling behind the infiltration front, and (3) liquid redistribution within the preform at some point in the processing after the infiltration step. It is believed that contributions from any, or all, of these could give rise to gradients in composition.

The effects of the exposure time in a hydrothermal environment on the monoclinic phase content (tetragonal phase stability) for different infiltration times are shown in Fig.2. It can be seen from Fig. 2 that only when the infiltration time is long enough (>12hrs), (i.e. the depth of infiltration layer reachs certain level (>500μm) and the average volume fraction of infiltrated mullite within the infiltration zone is more than 10wt%) the t→m phase transformation can be effectively inhibited (see Figs. 2 and 3). Further comparisons of the m-phase content for 3Y-TZP[15] and 3Y-TZP/Mullite-Alumina[7] with surface mullite-modified 3Y-TZP/Alumina composites after hydrothermal treatment at 180°C and 10 bar are shown in Fig. 3. In contrast to 3Y-TZP, which shows a poor resistance to hydrothermal corrosion (more m-phase is formed) and bulk mullite-doped 3Y-TZP/Alumina composites, both surface mullite-modified 3Y-TZP/Alumina composites show excellent resistance (almost no m-phase was detected) to hydrothermal corrosion.

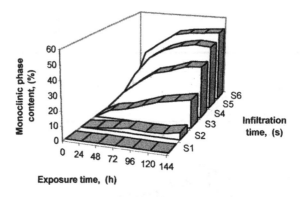

Exposure time, (h)

S1 --- 86400s;   S2 --- 43200 s;   S3 --- 14400 s;   S4 ---1800s;   S5 --- 60 s;   S6 --- 10 s

Fig. 2. Comparison of m-phase content for surface mullite-modified 3Y-TZP/Alumina composites for different infiltration times after hydrothermal treatment at 180°C and 10 bar.

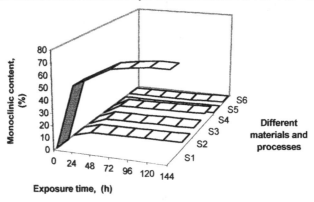

Exposure time, (h)

S1 -- 3Y-TZP**;   S2 -- Composite sintered at 1500°C*;   S3 -- Composite sintered at 1600°C*
S4 -- Composite sintered at 1700°C*;   S5 -- Surface modification at 1600°C;   S6 -- Surface modification at 1700°C

Fig. 3 Comparison of m-phase content for 3Y-TZP (**Data from Ref. 15) and 3Y-TZP/Mullite-Alumina (* Data from Ref. 7) and surface mullite-modified 3Y-TZP/Alumina composites after hydrothermal treatment at 180°C and 10 bar.

The infiltration method for introducing mullite (with a very low coefficient of thermal expansion in comparison to that of 3Y-TZP and alumina) into a 3Y-TZP/Alumina composite was chosen for this work because it was felt that the formation of a composite layer in the surface region (with a lower coefficient of thermal expansion) could lead to the formation of residual compressive surface stresses during heat treatment and, hence, strengthening. To precisely measure geometric change of an functionally graded materials (FGM), or the distribution of local thermal expansion, two- or three-dimensional very complex measurement techniques must be used, such as laser interferometry or digital image correlation[16]. In order to simplify the calculation, the most simple model, the Voight estimate was selected since no model exists that is completely accurate for calculating thermal expansion coefficient (TEC) for the full range of compositions [16].

$$\alpha_C = V_A \bullet \alpha_A + V_B \bullet \alpha_B \qquad (2)$$

where α_C is the TEC of a composite, V_A and V_B are the volume fraction of A and B, α_A and α_B are the TEC of A and B, respectively. The average value of TEC for mullite, 3Y-TZP and alumina (i.e. 5.7×10^{-6}/K, 9.75×10^{-6}/K and 7.9×10^{-6}/K) were used as a reference[17]. The calculated TEC values for specimens infiltrated for 24hrs (corresponding to the volume fraction of mullite for S6 in Fig. 1) are shown in Fig. 4. The graded composite displays a gradual change in TEC due to the presence of mullite. The lower TEC in the surface region will lead to a decrease in thermal stress.

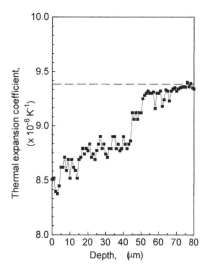

Fig. 4 The distribution of the thermal expansion coefficient with the depth of infiltrated mullite

The bending strengths and fracture toughness of surface mullite-modified 3Y-TZP/Alumina composites before and after hydrothermal corrosion are compared with 3Y-TZP and bulk mullite-doped 3Y-TZP/Alumina composites in Figs.5 and 6. Due to the higher Young's modulus and the dispersion effects of Al_2O_3, the bending strength for both surface mullite-modified 3Y-TZP/Alumina composites and bulk mullite-doped 3Y-TZP/Alumina composites, were slightly increased compared to 3Y-TZP even before annealing in hot water. Although the initial strength of 3Y-TZP is high enough for commercial application, the low strength of 3Y-TZP after annealing precludes its potential application in hydrothermal environments. There is little, or no, further degradation on hydrothermal corrosion for the surface mullite-modified 3Y-TZP/Alumina composites, thus demonstrating their high resistance to degradation.

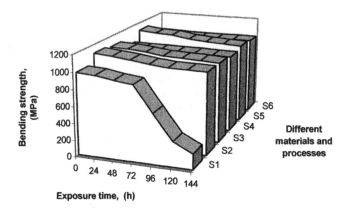

Fig. 5 Bending strengths for 3Y-TZP (**Data from Ref. 15) and 3Y-TZP/Mullite-Alumina (* Data from Ref. 7) and surface mullite-modified 3Y-TZP/Alumina composites before and after hydrothermal treatment at 180°C and 10 bar for different times.

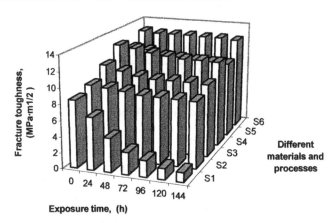

Fig. 6 Fracture toughness for 3Y-TZP (**Data from Ref. 15) and 3Y-TZP/Mullite-Alumina (* Data from Ref. 7) and surface mullite-modified 3Y-TZP/Alumina composites before and after hydrothermal treatment at 180°C and 10 bar for different times.

Although the fracture toughness of surface mullite-modified 3Y-TZP/Alumina composites decreases somewhat after annealing, they remain as tough, or tougher, than non-annealed 3Y-TZP. It is shown that the introduction of mullite as a second phase in the surface of composite is effective in inhibiting the low temperature degradation of Y-TZP/Al_2O_3 composite. The partial loss of Y-TZP/Al_2O_3 composite's mechanical strength and fracture toughness due to the introduction of

Al_2O_3[18], is prevented (see Fig.6). The difference in the thermal expansion coefficients is greater for mullite and Y-TZP/Al_2O_3 as discussed previously. It is believed that this increase is very important in generating surface compressive stresses. Also, since the Al_2O_3 necessary for forming mullite (as shown in reaction (3) [10]) is already present in the composite, pores need only be used to introduce SiO_2 into the system. Therefore, the Al_2O_3 already in the composite will tend to maximize the amount of the second phase that can be formed and makes this method quite simple and effective.

$$3Al_2O_3 + 2SiO_2 \rightarrow 3Al_2O_3 \bullet 2SiO_2 \qquad (3)$$

The better resistance of surface mullite-modified 3Y-TZP/Alumina composites to hydrothermal corrosion can be attributed to the contributions of Al_2O_3. Due to its higher Young's modulus (about 390 GPa[19]), the free energy change for the strain free energy term will be increased with the addition of Al_2O_3: for a detailed analysis see Ref. 7. Also, SEM-EDS analysis of the grain-boundary fracture surfaces of 3Y-TZP/Mullite-Alumina composites by Zhao and Northwood [7] show that Al is present. However, Al has seldom been detected on polished surfaces of 3Y-TZP/Mullite-Alumina composites. Furthermore, an enrichment of Y_2O_3 at the fractured grain boundaries was also observed[7]. Ruhle et al.[20] suggested that a certain compound or phase of ZrO_2-Al_2O_3 and/or ZrO_2-Al_2O_3-SiO_2 (the starting mullite powder contains SiO_2) may form at grain boundaries since the amorphous grain boundary phase often contains Y_2O_3, Al_2O_3, and SiO_2 in these composites. The formation of a phase containing Y_2O_3, Al_2O_3, and SiO_2 at the grain boundaries may induce a much stronger cohesion between grains than that between grains in 3Y-TZP[20]. Thus, the increased cohesion between grains in 3Y-TZP containing Al_2O_3 will give rise to a constraining effect from the surrounding matrix which prevents the formation of the monoclinic phase except at the specimen surface. Therefore, it is concluded that Al_2O_3 segregation at grain boundaries is one possible reason that degradation is inhibited [7].

CONCLUSIONS

A mullite/yttria-stabilized tetragonal zirconia polycrystal/alumina (mullite/3Y-TZP/Al_2O_3) composite with graded transformation stability was fabricated by an infiltration method. It is believed that the graded transformation stability can be attributed to the existence of a concentration gradient of mullite, with the concentration decreasing with increasing depth into the sample. The graded composite displays a gradual change in thermal expansion values due to the presence of mullite. The partial loss of the Y-TZP/Al_2O_3 composite's mechanical strength and fracture toughness due to the introduction of Al_2O_3, can be prevented. It was found that the introduction of mullite as a second phase in the surface of the composite is very effective in inhibiting the low temperature degradation of the Y-TZP/Al_2O_3 composite when the average volume fraction of infiltrated mullite within the infiltration zone is more than 10wt% and the depth of infiltrated mullite is more than 500 μm.

REFERENCES

[1]O. T. Masaki, H. Kuwashima, and K. Kobayashi, "Phase Transformation and Change of Mechanical Strength of ZrO_2-Y_2O_3 by Aging," in Abstracts of the Annual Meeting of the Japanese Ceramic Society, A-3, Japanese Ceramic Society, Tokyo, 1981.

[2]F. F. Lang, G. L. Dunlop, and B. I. Davis, "Degradation During Aging of Transformation-Toughened ZrO_2-Y_2O_3 Materials at 250°C," J. Am. Ceram. Soc., 69 [3] 237-40 (1986).

[3]T. Chung, H. Song, G. Kim, and D. Kim, "Microstructure and Phase Stability of Ytttria-doped Tetragonal Zirconia Polycrystals Heat Treated in Nitrogen Atmosphere," J. Am. Ceram. Soc., 80 [10] 2607-12 (1997).

[4]Z. Zhao, and D. O. Northwood, "Mechanism of Low Temperature Environmental Effects on Transformation-Toughened Zirconia Ceramics," Ceramic Engineering and Science Proceedings, **20**[3-4] 95-103 (1999).

[5]T. Sato, and M. Shimada, "Crystalline Phase Change in Yttria-Partially-Stabilized Zirconia by Low-Temperature Annealing," J. Am. Ceram. Soc., **67** [10] C-212-C-213 (1984).

[6]T. Sato, and M. Shimada, "Transformation of Yttria-Doped Tetragonal ZrO_2 Polycrystals by Annealing in Water," J. Am. Ceram. Soc., **68** [6] 356-59 (1985).

[7]Z. Zhao, and D. O. Northwood, "Combined Effects of Mullite and Alumina on the Stability and Mechanical Properties of a Y-TZP Ceramic in a Hydrothermal Corrosion Environment," Advances in Ceramic-Matrix Composites V, Ceramic Transactions, **103**, 503-514 (2000).

[8]Z. Zhao, and D. O. Northwood, "The Effect of Surface GeO_2-Doping and CeO_2-Doping on the Degradation of 2Y-TZP Ceramic on Annealing in Water at 200°C," Materials & Design, **20**, [6] 297-301 (1999).

[9]Z. Zhao, C. Liu and D. O. Northwood, "Surface Modification of Yttria-Stabilized Tetragonal Zirconia Polycrystal/Alumina Composites by Incorporation of Mullite as a Second Phase," Ceramic Engineering and Science Proceedings, **21**[3] 619-626 (2000).

[10]B. R. Marple, and D. J. Green, "Incorpration of Mullite as a Second Phase into Alumina by an Infiltration Technique," J. Am. Ceram. Soc., **71** [11] C-471-C-473 (1988).

[11]R. C. Garvie and P. S. Nicholson, "Phase Analysis in Zirconia Systems," J. Am. Ceram. Soc., **55**[6] 303-305 (1972).

[12]R. L. Fullman, "Measurement of Particle Size in Opaque Bodies," J. Metal. Trans., AIME, **197**[3] 447-52 (1953).

[13]W. D. Callister, Jr., "Dislocation and Strengthening Mechanisms (Chapter 7)," pp. 183, Materials Science and Engineering--An Introduction, Second Edition, John Wiley & Sons Inc., Toronto, 1991.

[14] B. R. Marple, and D. J. Green, "Graded Compositions and Micorstructures by Infiltration Processing," J. Mater. Sci., **28**, 4637-43 (1993).

[15]C. Liu, Z. Zhao, and S. Yu, "Corrosion Behavior of Fine-Grained (Mg,Y)-PSZ Ceramics in Hot Steam Environment," J. Function Mater., **18**, 653-655 (1997).

[16]Y. Miyamoto, W.A. Kaysser, B.H. Rabin, A. Kawasaki and R.G. Ford, "Modeling and Design, The Characterization of Properties (Chapters 4 and 5)," pp.68-99, Functionally Graded Materials: Design, Processing and Application, Kluwer Academic Publishers, London, 1999.

[17]"Guide to Engineered Materials," Advanced Materials & Processes, **156** [6] 143 (1999).

[18]M. Hirano, "Inhibition of Low Temperature Degradation of Tetragonal Zirconia Ceramics- A Review," Br. Ceram. Trans. J. , **91**, 139 (1992).

[19]J-. F. Li, and R. Watanabe, "Fracture Toughness of Al_2O_3-Particle-Dispersed Al_2O_3-Partially Stabilized Zirconia," J. Am. Ceram. Soc., **78**[4] 1079-1082 (1995).

[20]M. Ruhle, N. Claussen, and A. H. Heuer, "Microstructural Studies of Y_2O_3-Containing Tetragonal ZrO_2 Polycrystals (Y-TZP)," pp. 352-370 in Advances in Ceramics, Vol. 12, Science and Technology of Zirconia II. Edited by N. Claussen, M. Ruhle, and A. H. Heuer. American Ceramic Society, Columbus, OH, 1984.

Deposition and Casting

CONFORMAL ENCAPSULATION OF FINE BORON NITRIDE PARTICLES WITH OXIDE NANOLAYERS

Jeffrey R. Wank and Alan W. Weimer
University of Colorado
Department of Chemical Engineering
Boulder, CO 80309

John D. Ferguson and Steven M. George
University of Colorado
Department of Chemistry
Boulder, CO 80309

Abstract

Alumina (Al_2O_3) was deposited on boron nitride (BN) particles with atomic layer control using alternating exposures of $Al(CH_3)_3$ and H_2O. The sequential surface chemistry was monitored in vacuum using transmission Fourier transform infrared (FTIR) spectroscopy studies on sub-micron sized BN particles. Transmission electron microscopy (TEM) studies revealed extremely uniform and conformal nanometer thick Al_2O_3 coatings on the BN particles. X-ray photoelectron spectroscopy (XPS) analysis was consistent with conformal Al_2O_3 coatings. These results illustrate the potential of sequential surface reactions to deposit ultra-thin Al_2O_3 coatings on BN particles.

Process scale-up is currently being studied using a vibration-assisted, low pressure (<50 torr) fluidized bed apparatus. One consideration is the uniform coating of primary particles rather than aggregates. The fine individual BN particles form lightly bonded larger aggregates that remain in the bed during fluidization. A pulsed laser system has been used to investigate the mechanism of aggregate formation. Agglomerate size is dependent on vibrational force applied and superficial velocity, but not on pressure.

INTRODUCTION

Atomic Layer Deposition

Coating BN particles with ultrathin films allows the surface of the particles to be altered while maintaining bulk properties. Boron nitride has a high thermal conductivity, and is thus desirable as a filler in epoxy to make high thermal conductivity composite materials for electronic packaging [1, 2]. One difficulty with BN particles is the relative inertness of their surfaces [3, 4], which limits the loading of BN particles in the polymer composite. Higher loadings are needed as the microelectronics industry develops faster and denser integrated circuits that produce more heat, and may be achieved by coating the BN particles with a more chemically reactive film. This deposited film should interact strongly with the coupling agents in the epoxy, and should be ultrathin to maintain the bulk properties of the BN particles. Since chemical coupling agents have been designed to interact strongly with oxide surfaces [5, 6], ultrathin Al_2O_3 films on BN particles should increase BN particle loading in the polymer composite. Ultrathin Al_2O_3 films can be deposited on BN particles using Atomic Layer Deposition (ALD) [7, 8].

ALD techniques have been developed for Al_2O_3 [9-12] using sequential surface reactions. Al_2O_3 can be deposited using the following ABAB... reaction cycles with $Al(CH_3)_3$ and H_2O [9-12]

A) $AlOH^* + Al(CH_3)_3 \rightarrow AlOAl(CH_3)_2^* + CH_4$ (1)

B) $AlCH_3^* + H_2O \rightarrow AlOH^* + CH_4$ (2)

where * indicates a surface species. Each of the surface reactions is self-limiting. Sequential exposure to $Al(CH_3)_3$ and H_2O can deposit ~1.1 Å per AB cycle at 450 K [13]. The resulting films are observed to be smooth and conformal with a roughness similar to the roughness of the initial substrate [10]. The ALD of Al_2O_3 on BN particles is contingent on the initiation of the sequential surface reactions on the BN surface. Previous FTIR studies have shown that the BN surface contains BOH^* and BNH_2^* species [14, 15]. The BOH^* and BNH_2^* species can be utilized to start the sequential surface reactions for Al_2O_3 and SiO_2 ALD in the following manner:

$BOH^* + Al(CH_3)_3 \rightarrow BOAl(CH_3)_2^* + CH_4$ (5)

$BNH^* + Al(CH_3)_3 \rightarrow BNAl(CH_3)_2^* + CH_4$ (6)

Following these initiation reactions, the BABA... reaction cycles for Al_2O_3 ALD can be employed to coat the BN particles.

The initiation of the sequential surface reactions on the BN particles and the subsequent Al_2O_3 growth were examined using transmission Fourier transform infrared (FTIR) spectroscopy. Additional transmission electron microscopy (TEM) and x-ray photoelectron spectroscopy (XPS) investigations evaluated the uniformity and conformality of the Al_2O_3 coatings on the BN particles. These FTIR, TEM and XPS studies reveal the ability of sequential surface reactions to deposit ultrathin and conformal coatings on BN particles.

Fluidization Properties of Fine BN Powders

Fine powders self-agglomerate into larger particles when exposed to gas flows in a fluidized bed apparatus, even under vacuum conditions [16-18]. These light agglomerates are formed by Van Der Waals (cohesive) forces [19], and the size of the agglomerates may depend on several factors such as individual particle type, size, shape, and density, as well as gas flow, vacuum, and vibration applied to the bed. The agglomerates are the actual entities that fluidize in the bed, not the individual particles. Thus, the size, shape, and density of the agglomerates under different conditions are important to determine. Some work has been done in this regard [18, 20-22] but there has been no attempt to actually measure the size of agglomerates in a fluidized bed while the bed is being fluidized under conditions of vacuum and vibration.

The size of agglomerates in a fluidized bed of fine BN powder was determined using a technique called Particle/Droplet Image Analysis or PDIA to measure the size distribution of the agglomerates. In this technique, a laser is used to illuminate a region near the upper surface of the fluidized bed from behind and shadow images of the agglomerates are taken with a specially calibrated digital camera. We vary vacuum, gas flow rate, and level of vibration to determine the effects of these parameters on agglomerate size.

EXPERIMENTAL

A vacuum apparatus designed for *in situ* transmission FTIR studies was employed to study the surface chemistry and coat the BN particles with Al_2O_3 and SiO_2 films [9]. A schematic of this apparatus is shown in Figure 1. Hexagonal BN particle agglomerates from Advanced Ceramics Corporation were used for these FTIR investigations. These agglomerates are ~10 μm in diameter

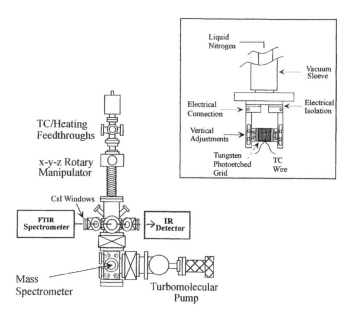

Figure 1. Apparatus used to perform ALD studies.

and composed primarily of ~0.1 - 0.5 μm BN crystals. These BN particles have a total surface area of ~40 m²/g. The vibrational spectroscopic studies were performed with a Nicolet 740 or a Nicolet Magna 560 Fourier Transform Infrared (FTIR) spectrometer and an MCT-B infrared detector. The transmission electron microscopy (TEM) and x-ray photoelectron spectroscopy (XPS) analyses were performed by Dr. Paolina Atanassova at the Center for Micro-Engineered Materials at the University of New Mexico. The TEM results were obtained with a HRTEM JEOL 2010 high resolution transmission electron microscope in combination with electron dispersive spectroscopy and a GATAN digital micrograph with a slow scan CCD camera. The x-ray photoelectron spectra (XPS) were recorded on an AXIS HSi Kratos Analytical XPS spectrometer.

An illustration of the experimental fluidization setup is shown in Figure 2. The fluidized bed is made of borosilicate glass 40 mm in diameter and 1 m tall. The distributor is constructed of coarse porous glass 0.3 m from the bottom of the tube. Vacuum in the fluidized bed is controlled by varying the pump inlet diameter with a throttle valve. The fluidizing gas used is obtained from a high purity liquid nitrogen source. Gas flow rate is controlled and measured with an MKS® 1179

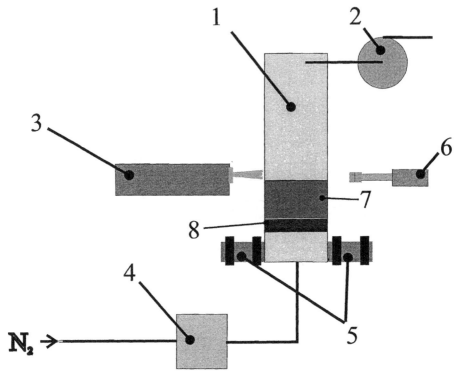

Figure 2. Schematic of the experimental fluidization apparatus. (1) Reactor, (2) Vacuum pump, (3) IR laser, (4) Mass flow controller, (5) Vibro-motors, (6) CCD camera, (7) fluidized bed, (8) porous frit.

series mass flow controller. Vibration frequency is controlled using an ACS-140 speed controller from ABB Drive and Power Products.

The powder used in the fluidized bed tests is hexagonal 7-11 μm diameter BN particles obtained from Advanced Ceramics Corporation. This powder is comprised of individual BN particles that take the shape of hexagonal flat plates, 5-11 μm in diameter, with a surface area of 15-20 m^2/g. The fluidization experiments are all carried out utilizing the decreasing velocity method. The PDIA system used was the Oxford Lasers VisiSizer™ system. An external pulse from the CCD camera strobe output triggers the laser pulse; hence the image capture is synchronized by and with the camera. The captured image is thresholded such that the software can distinguish between the particles and the illumination background. One of these images is shown in Figure 3. The pixel area of the particle(s) is then measured and prior calibration of the system allows the agglomerate or particle diameter distribution to be reported.

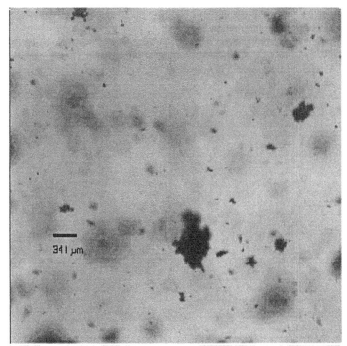

Figure 3. An image taken with the VisiSizer™ system. The black bar near the bottom right denotes 341 μm length.

RESULTS

Negative absorbance in the AlO-H stretching vibration region corresponds to the loss of AlOH* surface species after Al(CH₃)₃ exposure. Positive absorbance in the AlC-H₃ stretching vibration region is also evident and consistent with the gain of AlCH₃* species [9]. Also, the lack of the BO-H stretching vibration at 3680 cm⁻¹ and the BN-H₂ stretching vibrations at 3430 and 3575 cm⁻¹ indicates a lack of those species on the surface (data not shown). Following the 9th H₂O exposure, the FTIR difference spectrum reveals that the AlO-H stretching vibrations reappear as positive absorbance. This behavior is consistent with the addition of AlOH* species. Negative absorbance in the AlC-H₃ stretching region corresponds with the loss of AlCH₃* species after the H₂O exposure [9]. This switching between AlCH₃* and AlOH* species with alternating Al(CH₃)₃ and H₂O exposures is expected from the reactions given by Eqns. 1 and 2. These spectra were omitted for brevity. See [7] for a complete description.

Transmission electron microscopy (TEM) was used to evaluate the conformality of the Al₂O₃ deposition on the BN particle. Figure 4 shows a TEM image of an Al₂O₃-coated BN particle. This Al₂O₃ film was deposited by 50 AB cycles at 450 K. The TEM image reveals an ~90 Å Al₂O₃ coating that is extremely uniform and conformal to the surface of the BN particle. The Al₂O₃ is deposited equally well to the basal and edge planes of the BN particle. The TEM image also displays the crystalline graphitic planes in the BN particle. In contrast, the Al_2O_3 coating on the BN particle is amorphous.

X-ray photoelectron spectroscopy (XPS) measurements were also conducted on the same Al_2O_3-coated BN particles. The XPS analysis observed the expected photoelectrons at 190 eV (B, 1s) and 397 eV (N, 1s) for uncoated BN particles. The B and N photoelectrons were almost

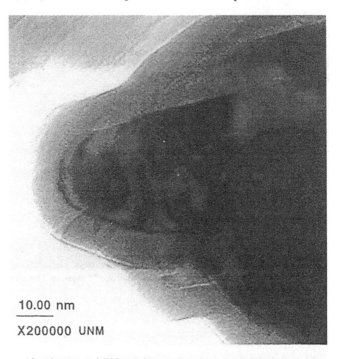

10.00 nm

X200000 UNM

Alumina coated BN powder

Figure 4. TEM image of BN powder coated with Alumina. The dark region is crystalline BN; the lighter region on the outside is amorphous Al_2O_3.

totally attenuated in the photoelectron yield from the Al_2O_3-coated BN particles. This behavior is expected if the Al_2O_3 coating completely covers the BN particle [23, 24]. The prominent photoelectrons observed on the Al_2O_3-coated BN sample were at 119 eV (Al, 2s), and 531 eV (O, 1s). In agreement with the TEM images, depth-profiling experiments indicated that the B and N XPS signals did not reappear until ~100 Å of material was removed by Ar^+ sputtering.

The size of BN agglomerates is directly related to the cohesion forces between the primary particles and the vibrational force applied. As the forces that counter the cohesion forces increase, then the agglomerate size should drop; alternatively, less vibration should mean larger agglomerates. We varied the vibrational force by varying the frequency applied to the vibro-motors between 30-60 Hz. Varying the vibrational force applied creates an apparently linear change in the agglomerate size, as shown in Figure 5.

CONCLUSIONS

Ultrathin Al_2O_3 films were deposited with atomic layer control on BN particles using sequential surface reactions. Al_2O_3 was deposited with alternating $Al(CH_3)_3/H_2O$ exposures. Fourier transform infrared (FTIR) vibrational studies observed the progressive switching between AlOH* and AlCH3* species during Al_2O_3 deposition. The growth of Al_2O_3 bulk vibrational modes was also monitored in the FTIR spectra versus number of alternating reactant exposures.

Transmission electron microscopy (TEM) was utilized to evaluate the deposited Al_2O_3 films. The TEM images showed very uniform Al_2O_3 films deposited conformally on the BN particles. X-ray photoelectron spectroscopy (XPS) analysis was in agreement with highly

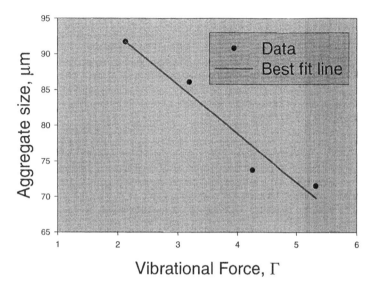

Figure 5. Agglomerate size of PT180 at 30 Torr, 6.7 cm/s, vs. non-dimensional vibrational force, Γ where Γ is the ratio of vibrational force applied to force of gravity.

conformal Al2O3 coatings. The atomic layer deposition of ultrathin Al_2O_3 films on BN particles may be important for improving polymer composites for thermal management applications. Al2O3 film thickness can be controlled to within ~1 Å on BN particles using atomic layer deposition techniques. These ultrathin coatings should enhance the chemical reactivity and increase the loading of the BN particles in polymer composites without degrading the thermal conductivity of the BN particles.

Fine BN particles form larger light agglomerates under conditions of fluidization. These light agglomerates stay in the bed during fluidization; the amount of vibration applied and the fluidizing gas velocity affect their size. As vibrational force is increased or fluidizing gas velocity is increased, the agglomerate size decreases.

REFERENCES

1. Hatta, H., *et al.*, Thermal Diffusivities of Composites with Various Types of Filler, J. Comp. Mater., 1992. **26**(5): p. 612-625.
2. Tsutsumi, N., N. Takeuchi, and T. Kiyotsukuri, Measurement of Thermal diffusivity of Filler-Polyimide Composites by Flash Radiometry, J. Polymer Sci. B, 1991. **29**: p. 1085-1093.
3. Arya, S.P.S. and A. D'Amico, Preparation, Properties and Applications of Boron Nitride Thin Films, Thin Solid Films, 1988. **157**: p. 267-282.
4. Shetty, R. and W.R. Wilcox, Boron nitride coating on fused silica ampoules for semiconductor crystal growth, J. Cryst. Growth, 1995. **153**: p. 97-102.
5. Abboud, M., *et al.*, PMMA-based composite materials with reactive ceramic fillers Part 1.--Chemical modification and characterisation of ceramic particles, J. Mater. Chem., 1997. **7**(8): p. 1527-1532.
6. Gun'ko, V.M., *et al.*, Modification of some oxides by organic and organosilicon compounds, J. Adhesion Sci. Technol., 1997. **11**(5): p. 627-653.
7. Ferguson, J., A. Weimer, and S. George, Atomic Layer Deposition of Ultrathin and Conformal Al2O3 films on BN Particles, Thin Solid Films, 2000. **371**: p. 95-104.
8. Ferguson, J., A. Weimer, and S. George, Atomic Layer Deposition of Al2O3 and SiO2 on BN Particles Using Sequential Surface Reactions, Applied Surface Science, 2000. **162/3**: p. 280-292.
9. Dillon, A.C., *et al.*, Surface-Chemistry of Al2O3 Deposition Using Al(CH3)(3) and H2O in a Binary Reaction Sequence, Surface Science, 1995. **322**(1-3): p. 230-242.
10. Ott, A.W., *et al.*, Atomic layer controlled deposition of Al2O3 films using binary reaction sequence chemistry, Applied Surface Science, 1996. **107**: p. 128-136.
11. Ott, A.W., *et al.*, Al3O3 thin film growth on Si(100) using binary reaction sequence chemistry, Thin Solid Films, 1997. **292**(1-2): p. 135-144.
12. Higashi, G.S. and C.G. Fleming, Sequential surface chemical reaction limited growth of high quality Al2O3 dielectrics, Appl. Phys. Lett., 1989. **55**(19): p. 1963-1965.
13. Ott, A.W., *et al.*, Al2O3 thin film growth on Si(100) using binary reaction sequency chemistry, Thin Solid Films, 1997. **292**: p. 135-144.
14. Baraton, M.I., *et al.*, Surface Activity of a Boron Nitride Powder: A Vibrational Study, Langmuir, 1993. **9**: p. 1486-1491.
15. Baraton, M.I., *et al.*, Nanometric Boron Nitride Powders: Laser Synthesis, Characterization and FT-IR Surface Study, J. Eur. Ceram. Soc., 1994. **13**: p. 371-378.
16. Morooka, S., *et al.*, Fluidization State of Ultrafine Powders, Journal of Chemical Engineering of Japan, 1988. **21**(1): p. 41-46.
17. Morooka, S., T. Okubo, and K. Kusakabe, Recent Work On Fluidized-Bed Processing of Fine Particles As Advanced Materials, Powder Technology, 1990. **63**(2): p. 105-112.
18. Noda, K., Y. Mawatari, and S. Uchida, Flow patterns of fine particles in a vibrated fluidized bed under atmospheric or reduced pressure, Powder Technology, 1998. **99**(1): p. 11-14.
19. Visser, J., Vanderwaals and Other Cohesive Forces Affecting Powder Fluidization, Powder Technology, 1989. **58**(1): p. 1-10.
20. Castellanos, A., *et al.*, Flow regimes in fine cohesive powders, Physical Review Letters, 1999. **82**(6): p. 1156-1159.
21. Venkatesh, R.D., M. Grmela, and J. Chaouki, Simulations of vibrated fine powders, Powder Technology, 1998. **100**(2-3): p. 211-222.

22.   Zhou, T. and H.Z. Li, Estimation of agglomerate size for cohesive particles during fluidization, Powder Technology, 1999. **101**(1): p. 57-62.

23.   Cimino, A., D. Gazzoli, and M. Valigi, XPS quantitative evaluation of the overlayer/support intensity ratio in particulate systems, J. Elec. Spec. Rel. Phen., 1994. **67**: p. 429-438.

24.   Cross, Y.M. and J. Dewing, Thickness Measurements on Layered Materials in Powder Form by Means of XPS and Ion Sputtering, Surf. Interface Anal., 1979. **1**(1): p. 26-31.

CONTINUOUSLY GRADED METAL - CERAMIC GEOMETRIES USING
ELECTROPHORETIC DEPOSITION (EPD)

W.E. Windes and A.W. Erickson   Jeramy Zimmerman
INEEL   Colorado School of Mines
2351 N. Boulevard   1537 Secrest Ct
Idaho Falls, ID 83415-2218   Golden, CO 80401

ABSTRACT
    A graded Ni-Al$_2$O$_3$ structure was deposited on a nickel substrate using Electrophoretic Deposition (EPD) methods. Simultaneous deposition of metallic and ceramic particles was achieved by adjusting the solid particle density, electrolytic bath, and electrical potential allowing similar electrophoretic behavior between metallic and ceramic powders in the slurry. Deposition parameters for alumina-nickel cermet compositions were determined allowing a certain degree of control over the layer geometry and compositional gradient. Layered and continuously graded FGM structures transitioning from pure nickel to pure alumina were fabricated. After deposition "green" deposits were dried, consolidated in a hot press at 1350 °C, and characterized for compositional consistency and morphology in the microstructure of the layers.

INTRODUCTION
    The Idaho National Engineering and Environmental Laboratory (INEEL) has been investigating the mechanical response and residual stresses in Functionally Graded Materials (FGMs) for a number of years. Currently, the FGMs are fabricated by sequential stacking of dry powder cermet compositions (nickel and Al$_2$O$_3$) to form layered structures. These techniques have yielded samples with large overall dimensions and uniform layers through the gradient. However, the process is not conducive to fabricating structures with complex shapes or continuous gradients.
    A new fabrication process was therefore required to produce these complex components with gradient structures. Studies have demonstrated Electrophoretic Deposition (EPD) is capable layered or continuous coatings as well as depositing uniform layers on electrodes with complex geometries.[1,2,3,4] EPD was investigated to determine the viability of producing samples adequate for macroscopic mechanical testing and characterization. Desiring to use the same type of particles as in the dry powder processing method, the electrophoretic and deposition parameters for these powders were investigated to determine suitable EPD characteristics.
    EPD is a colloidal process where green bodies are directly formed by deposition of a relatively dilute suspension of powder particles from a stable slurry.[5] Electrophoretic potential forces are employed to move the powder particles through the suspension while the deposition phase of the process keeps the green body structure stable during post-deposition handling. In practice, EPD uses an induced electric field on a slurry suspension to drive particles to one or another of the electrodes where deposition takes place as shown in Figure 1.
    The electrophoretic behavior of particles in a slurry is derived from a complex relationship between the liquid media, applied electric field, and the particle characteristics. In theory, given the appropriate surface charges on a particle and a strong enough electric potential, the particles should be deposited on one of the electrodes in the suspension. However, additional parameters

such as the slurry pH, conductivity, particle size, solid solubility, and viscosity are important factors as well. These parameters must be considered before adequate depositions can be achieved.

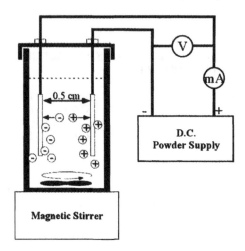

Figure 1. Electrophoretic deposition apparatus with charged particles depositing onto electrodes.

The basic premise behind EPD is that the surface charges on a particle will attract oppositely charged ions (i.e. counterions) from the surrounding suspension media. This atmosphere of counterions surrounding the particle is described as a "double-layer" or lyosphere and dictates the electrophoretic behavior of the particles in the slurry. [2,5,6] A large lyosphere surrounding a particle corresponds to a larger driving force on the particle in an applied electric field.

As expected, the solid particle lyosphere is strongly dependent upon the liquid media and any added impurities. Aqueous solutions have been used for ceramic particles in the past since the water provides adequate electrical conductivity and charged ions that attach to the surfaces of the particles providing additional charging. However, voltages need to be kept fairly low due to hydrogen bubbles being formed from the electrolysis of the water in the deposits. [7] Organic solutions are much more stable at higher electrical potentials and are more desirable for fully dense coatings. An ethanol-based solution was used for this work.

Ceramic powders usually have high surface areas and relatively low solubility in a liquid suspension, while the surface chemistry tends to control the charging behavior. The surface of the particles become charged through 1) desorption of ions at the surface of the solid, 2) chemical reactions between the surface and the liquid media which tends to alter the surface composition, and 3) preferential adsorption of specific additive or impurity ions from the solution. [8] These mechanisms should also hold true for any oxide films adherent to metallic particles allowing stable suspension and adequate electrophoretic properties for both powders. The difficulty lies in manipulating the slurry suspension so both ceramic and metallic particles have similar surface charge and are deposited simultaneously.

EXPERIMENTAL
ALCOA SG-16 alumina and INCO-123 Nickel powders were used for these initial studies. The powders were similar to those used previously in dry powder stacking processes at the INEEL. Particle size distribution was measured using a Coulter LS130 Particle Diffraction

Analysis System. To break apart any soft agglomerates formed in the powder a dilute suspension of powder and ethanol were mixed in an ultrasonic bath for 10 minutes before analysis.

Dilute slurries of 0.5 - 1 vol% nickel, alumina, or cermet powder in 99.999% pure ethanol with 2 mM acetic acid were used for depositing all layers. Since the nickel particles tended to settle during deposition the mixture was mechanically stirred to keep the nickel particles suspended in the slurry as illustrated in Figure 1. Square nickel electrodes, 2 cm x 2 cm, were fabricated from 0.0254 mm thick foil and positioned 0.5 cm apart in the powder-ethanol slurry.

A series of voltages ranging from 10 - 45 V were applied across the electrodes from a variable powder supply while the current was held constant at ~ 10 mA for all tests. The 0.5-cm distance separating the electrodes resulted in applied electric field strengths of 20 - 90 V/cm.

After deposition, the green bodies were sintered either in a horizontal tube furnace or hot pressed at 1350 °C for 20 minutes. A pressure of 41 MPa was applied to those samples consolidated in the hot press to reduce cracking and deformation in the deposited layers during sintering. Sintered green bodies were cross-sectioned and the composition in the individual layers was determined using image analysis techniques.

RESULTS AND DISCUSSION

Deposition Parameters

Both powder types were characterized to determine size, distribution, and morphology of the individual powders. Particle analysis determined the alumina powder to have a bimodal distribution (0.7 μm and 70 μm peaks) while the nickel powder was measured at ~ 0.7 μm. In addition, the morphology of both the alumina and nickel powder was a regular shaped particle with moderate surface area (the nickel displayed a spiked or dentritic surface). The size, distribution, and morphology indicated a good possibility for high packing densities and thus dense coatings for these powders.

In order to achieve dense depositions the optimal parameters to move and deposit both the ceramic and metallic particles on the electrodes were determined. An extensive series of tests were conducted to determine the optimal solid concentrations, voltages, and current levels for deposition of both the ceramic and metallic powders. The apparent coating thickness for each test was determined by visual observation.

It was observed that while deposition did occur in the ethanol/acetic acid solution the amount of deposited powders was inadequate to provide noticeably thick layers. To enhance the deposition rate electrolytic additions in the form of metal salts were added to the slurry. The optimal EPD parameters required to deposit thick layers are summarized in Table I for a deposition time of 5 minutes.

Table I. Optimal EPD parameters for deposition of Ni and Al_2O_3 particles.

Layer Composition	Particle Conc. (vol%)	Electric Field (V/cm)	Current (mA)	Max. Deposition (g/cm^2)
100% Ni	0.5 - 1	80 - 90	5 - 10	0.006
50% Ni	0.75 - 1.1	80 - 90	7 - 10	0.05
0 % Ni	0.5 - 1.1	70 -80	7 - 10	0.04

Adding even dilute amounts of metal salts (1 - 3 mM) to the ethanol solution significantly enhanced the deposition of the nickel/alumina powders. Various salts including $AlCl_3$, $MgCl_2$, and ZnCl were possible candidates for use in the slurry. The magnesium chloride salt was selected due to a slightly better deposition rate of the powders once in solution[9]. Additionally, adding 1- 2 vol% of water to the slurry also seemed to help stabilize the deposit characteristics. These additions consistently yielded smooth and relatively heavy depositions on the nickel electrode.

It is assumed the metal salts increased the ion concentration and overall conductivity of the slurry.[6] This creates larger double-layer charges surrounding the particles allowing greater driving forces to move the particles toward the cathode. In addition, the added water tends to adsorb to ceramic particle surfaces in lower alcohol solutions yielding increased surface charges.[5] Oxide films on the metallic particles are also likely surfaces for the water to attach. This combination of increased surface charges and added ionic concentration tends to increase the deposition of the particles on the electrodes. Table II lists the final parameters arrived at for simultaneous deposition of both the alumina and nickel powder.

Table II. Final Deposition Parameters for Depositing Ni-Al$_2$O$_3$ Powders

Parameter	Value
Particles	Submicron
	- 0.7 μm and 70μm (Al$_2$O$_3$)
	- 0.7 μm (Nickel)
Solid conc.	0.5 - 1.0 vol%
Electric field	80 V/cm
Current	10 mA
Liquid solution	Ethanol (99.999% pure)
	2 mM distilled water
	1 mM acetic acid
	1 mM MgCl$_2$

Composition of Deposits

After establishing the deposition parameters for the metallic and ceramic particles the composition and microstructure of the coatings was determined. The rate of deposition differs for alumina and nickel powders producing layer concentrations different from the slurry mixture. This required compositional characterization of each layer in order to control the gradient profile in the FGM structure. Image analysis was used to determine the deposited layer compositions from a series of slurry solid concentrations. Results from these tests are summarized in Figure 2.

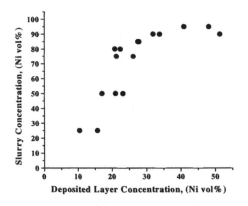

Figure 2. Compositions of deposited coatings from different slurry concentrations.

These tests illustrate the difficulties associated with depositing the denser nickel particles on the electrode surfaces. The nickel particles tended to rapidly settle out of the slurry making the

Functionally Graded Materials 2000

nickel unavailable for deposition. Thus, the resulting slurry composition was too dilute to produce noticeably thick nickel-rich layers. The only way to produce the nickel-rich layers for these "static" conditions was to manually add very small quantities of alumina to a pure nickel suspension. This provided a crude way of depositing the nickel-rich layers until the slurry concentration was high enough to permit uniform deposition of both powders (at approximately 60% nickel vol% in the slurry). This proved unsatisfactory and lead to the design of a continuously circulating slurry apparatus that is described in a later section.

Continuous Gradient

This study was primarily intended to explore the challenges of fabricating a continuous FGM structure. Figure 3 is a cross-sectional view of a continuously graded structure deposited as a result of using the deposition and compositional parameters established previously. The electrode was suspended in a sequence of slurries that gradually increased in alumina content (and subsequently decreased in nickel). The slurries were composed of 80%, 60%, 40%, 20%, and 0% by volume nickel. Deposition times of 90 seconds for each slurry bath resulted in a total deposition time of approximately 7 minutes to produce a 300-µm thick coating. The samples were consolidated in a hot press at 1350 °C and 41 MPa for 20 minutes to produce fully dense composites.

As noted previously, the nickel-rich region was not adequately reproducible by manual additions of alumina. However, a smooth transition from about 60% nickel was achievable for the rest of the sample. The samples exhibited some cracking and delamination in the alumina layer adjacent to the gradient structure but generally were completely dense and exhibited good sintered strength.

To mitigate the problem with the nickel settling from the slurry either a chemical or mechanical means of suspending the particles needed to be implemented. Organic binders or deflocculants would assist with the suspension but would also add organic contaminants to the final deposited layer. This would increase undesirable porosity in the final sintered product. Another approach was to create a dynamic suspension of nickel/alumina by continuously circulating the entire slurry through a pump. The mechanical mixing of the particles will allow the dense nickel particles to stay suspended in the slurry while eliminating organic contaminants. A concept to achieve a continuously circulated slurry is illustrated in Figure 4. In addition, as shown in Figure 4, by discharging the pumped slurry into the top of the bath the settling action of the nickel particles will enhance the availability of the particles for deposition. This apparatus has been fabricated and will be used in future studies to assist in producing a smooth gradient structure.

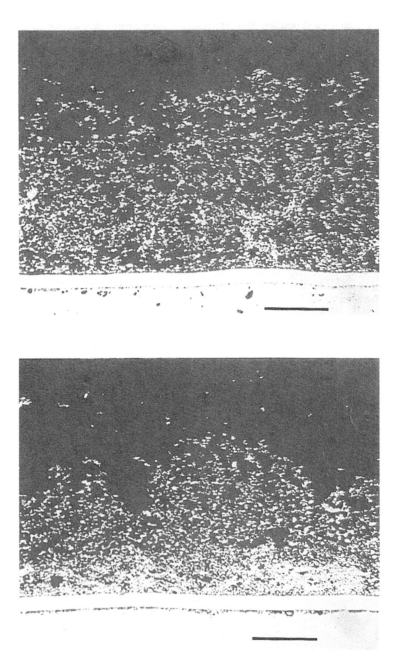

Figure 3. Graded nickel-alumina structure formed a) without manual additions of nickel and b) with manual additions of nickel powder. Nickel is seen as the white phase (ref. mark = 100 μm).

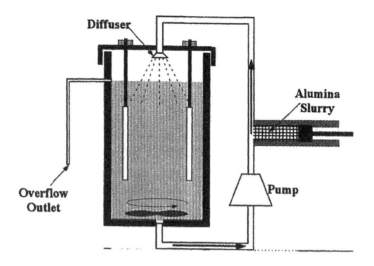

Diffuser

Alumina Slurry

Overflow Outlet

Pump

Figure 4. Schematic of continuously circulated slurry to increase deposition for the nickel-rich layers.

CONCLUSIONS

Electrophoretic deposition parameters to create a continuously graded Ni-Al$_2$O$_3$ structure were established for an organic, ethanol-based slurry suspension. Both metallic and ceramic particles were capable of being deposited on the cathode at these deposition conditions. The thickness for all deposited cermet layers was enhanced with the minute addition of metallic salts to the slurry. The thicker layers allowed the samples to be sintered and characterized for microstructure composition. Compositional analysis of deposited monocomposition layers found that nickel-rich layers were very difficult to deposit no matter what the solid concentration was in the slurry. Settling of the nickel particles out of the slurry contributed to this nickel-poor deposition.

A continuously graded structure was deposited, consolidated, and analyzed. All deposited samples were completely dense with good microstructural development. However, the nickel-rich regions were similar to the monocomposition tests in that they did not have enough nickel to form a completely interconnected microstructure. This necessitated the development of a new method to suspend the nickel particles. An apparatus to continuously circulate the slurry was fabricated and will be used in future work to assist in the deposition of the nickel-rich layers.

ACKNOWLEDGEMENTS

This work was supported by the U.S. Department of Energy, under the DOE Idaho Operations Contract DE-AC07-99ID13727.

REFERENCES

[1] P. Lessing, A. Erickson, D. Kunerth, "Electrophoretic Deposition (EPD): Applied to Reaction Joining of Silicon Carbide and Silicon Nitride Ceramics", Journal of Materials Science, 35, pp 2913 – 2925, 2000

[2] P. Sarkar, S. Datta, and P. Nicholson, "Functionally graded ceramic/ceramic and metal/ceramic composites by Electrophoretic Deposition", Composites Part B-Engineering 28: (1-2) 49-56 1997.

[3] B. Ferrari, A.J. Sánchez -Herencia, and R. Moreno, "Electrophoretic Forming of Al_2O_3/Y-TZP Layered Ceramics from Aqueous Suspensions", Materials Research Bulletin, Vol. 33, No. 3, pp 487-499, 1998.

[4] B. Ferrari, A.J. Sánchez-Herencia, and R. Moreno, " Aqueous Electrophoretic Deposition of Al_2O_3/ZrO_2 layered ceramics", Materials Letters 35, pp 370 - 374, 1998

[5] P. Sarkar and P. Nicholson, "Electrophoretic Deposition (EPD): Mechanisms, Kinetics, and Application to Ceramics", J. Am. Ceram. Soc., 79 [8], pp1987-2002, 1996.

[6] J. Widegren and L. Bergström, "The effect of acids and bases on the dispersion and stabilization of ceramic particles in ethanol", Journal of the European Ceramic Society, 20 pp 659-665, 2000

[7] G. Wang, P. Sarkar, and P. Nicholson, "Influence of Acidity on the Electrophoretic Stability of Alumina Suspensions in Ethanol" J. Am. Ceram. Soc, 80 [4], pp 965 – 972, 1997

[8] J. Reed, Principles of Ceramic Processing, 2nd Edition, John Wiley & Sons, Inc., New York, 1995.

[9] D. Brown and F. Salt, "The Mechanism of Electrophoretic Deposition", J. appl. Chem., 15 pp 40 - 48, January,1965.

ELECTROPHORETIC DEPOSITION OF FUNCTIONALLY GRADED HARDMETALS

S. Put, J. Vleugels and O. Van der Biest
Department of Metallurgy and Materials Engineering
Katholieke Universiteit Leuven
Kasteelpark Arenberg 44
BE-3001 Leuven, Belgium

KEYWORDS : Functionally Graded Materials, Electrophoretic Deposition, Hardmetals

ABSTRACT
The possibility to manufacture functionally graded hardmetals by electrophoretic deposition (EPD) is investigated. WC-Co graded materials with a gradient in cobalt content were processed by EPD, cold isostatic pressing and sintering. Most attention is focused on obtaining a fully dense material without losing the cobalt gradient during sintering. The resulting graded material showed a continuous variation in composition, microstructure and mechanical properties. The cobalt content increases from 6wt% on the hard side to 17wt% on the soft side. The Vickers hardness is found to increase continuously from 9 to 21GPa.

1. INTRODUCTION

In today's highly demanding technology environment, one of the main challenges in materials design and processing is combining irreconcilable properties of materials in the same component. Therefore functionally graded materials (FGM) have been developed to combine desirable material properties. The initial idea of a functional graded material was to combine the incompatible properties of heat resistance and strength with low internal stress, by producing a compositional gradient structure of distinct ceramic and metal phases [1]. A wide range of processes, such as PVD, CVD, plasma spraying, common powder metallurgy and colloidal processing have been used for FGM production.

Among these processes, electrophoretic deposition (EPD) is a low cost, fairly rapid process capable of producing continuously graded materials of a complex shape [2]. Electrophoretic deposition consists of two processes: the movement of charged powder particles suspended in a liquid (electrophoresis) followed by the deposition of these particles on one of the electrodes. The deposit, naturally, consists of a powder compact, which needs to be sintered afterwards in order to obtain a dense material. Gradient materials can be obtained since the composition of the next powder layer that deposits is determined by the composition of the suspension at that time. Numerous applications of EPD have been developed for fabrication of ceramics, including the production of coatings [3-4], laminated SiC/C materials [5-6] and functionally graded materials, as for example Al_2O_3/ZrO_2 [7].

Recently, there has been an increasing tendency in the cutting tool industry to use finer grain sized materials in order to increase the properties of WC-Co, which is a hard, tough and wear resistant material. These properties are determined by the composition, microstructure, porosity, etc. A reduction of the tungsten carbide grain size generally gives a marked increase in hardness, wear resistance and even transverse rupture strength. In this paper, the possibility to EPD nanocrystalline WC-Co composites with a functional gradient of mechanical properties such as hardness and toughness is investigated.

2. EXPERIMENTAL

The WC-Co materials used are Nanocarb grade WC-6Co (nominally 6wt% Co) and Nanocarb grade WC-10Co (nominally 10wt% Co), both composite powders (Nanodyne). A WC-25Co was prepared by mixing Nanocarb WC-10Co with Co powder (Union-Minière grade EF with a particle size of 1.3µm) in a multidirectional mixer for 48 hours in n-propanol in a polyethylene bottle. To break the agglomerates in the starting powder, hardmetal milling balls were added to the container. After mixing, the n-propanol was removed by means of a rotating evaporator.

For the preparation of flat functionally graded materials with a cobalt gradient from 6wt% to 17wt%, an electrophoretic deposition set-up as illustrated in Figure 1 was used. The deposition cell is made of polytetrafluorethylene (PTFE), containing 2 stainless steel electrodes. Both electrodes have a surface area of 9cm² and the separation distance between the electrodes is 3.5cm. A starting suspension of 200ml acetone containing 100g/l Nanocarb WC-6Co is pumped in a circulation system through the deposition cell by pump 2. A second acetone suspension with 400g/l WC-25Co powder is added by pump 1 to the starting suspension, thus changing its cobalt concentration with respect to WC continuously from 6wt% Co towards approximately 17wt% Co at the end. The second suspension was added by using two different pump rates: for the first FGM (FGM1) the second suspension was pumped with an addition rate of 1.68ml/min, creating a shallow gradient in the deposit. The second FGM (FGM2) with a steeper gradient was made by pumping at a rate of 6.54ml/min. Since the composition of the deposit at a certain time is given by the concentration of the suspension, this EPD technique makes it possible to produce a functionally graded hardmetal with a continuous gradient.

Before starting EPD experiments a graphite coating was applied on the surface of the deposition electrode to facilitate the removal of the deposit from the electrode and avoid cracking of the deposit during drying. Electrophoretic deposition experiments were performed by applying a constant DC voltage of 800V. After depositing for 1 minute, the second suspension was added creating a gradient in the deposit. The total deposition time was 15 minutes for FGM1 and 11 minutes for FGM2. During electrophoretic deposition, the mixed suspension circulates along the direction indicated by the arrows shown in Figure 1.

The deposits were removed from the electrode and dried in an acetone-containing atmosphere for 1 day. After drying, the samples were Cold Isostatically Pressed (CIP) at 300MPa for 3 minutes. Finally the green bodies were pressureless sintered in vacuum at different temperatures (1290°C, 1340°C and 1400°C) for 1 hour.

The gradient in composition, hardness and indentation fracture toughness of the FGMs was characterised. Microstructural investigation was performed by scanning electron microscopy (SEM, XL30 FEG, Philips, Eindhoven, The Netherlands) and electron probe microanalysis (EPMA, Model Superprobe 733, Jeol, Tokyo, Japan). The density of the specimens was measured in ethanol, according to the Archimedes method (BP210S balance, Sartorius AG, Germany). The Vickers hardness, $H_{V0.5}$ and H_{V10}, was measured on a Zwick hardness tester (model 3202, Zwick, Ulm, Germany) with a load of respectively 0.5 and 10kg.

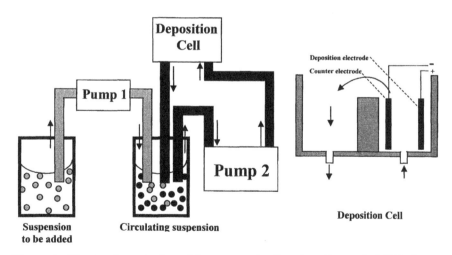

Figure 1: Electrophoretic deposition set-up for functionally graded WC-Co hardmetals.

3 RESULTS AND DISCUSSIONS

3.1 EPD characteristics of the WC-Co powders

Before performing the electrophoretic deposition experiments the most important properties of both Nanocarb powders were measured. As summarized in Table 1, the WC grain size equals 40nm, whereas the average particle size (d_{50}) is about 1µm. The particle size is important for EPD because sedimentation can occur during deposition.

Another very important powder property for EPD is the point of zero charge (pzc). The point of zero charge is the pH of the suspension where the powder carries no net charge and can be determined by a potentiometric titration [8]. The pzc is crucial in order to prepare a stable suspension, because a powder will only charge sufficiently when the pH of the suspension medium is controlled to differ enough from the point of zero charge. The results of the potentiometric titration are presented in Figure 2a, showing a pzc of 9.44 for WC-6Co and 8.85 for WC-10Co.

The natural pH of a powder is the pH at which the suspension equilibrates when a powder is suspended in water. Since the natural pH is the pH before addition of any acid or base, impurities make the natural pH diverge from the value of the pure powder surface. Acid impurities for example decrease the pH of the suspension with respect to the natural pH of a pure powder surface. Hence beside the pzc, the natural pH determines the choice of an acid or alkaline suspension medium. The powder is negatively charged when the natural pH > pzc and positively charged when natural pH < pzc. In the first case, an alkaline suspension medium will be needed whereas in the latter case, an acid medium will be used for EPD. For the Nanocarb WC-Co powders, the natural pH > pzc (Figure 2b) indicating the need for an alkaline suspension. Different deposition experiments showed indeed that EPD is only successful from an acetone suspension, whereas there is no deposition from ethanol (see Table 1). Figure 2b illustrates that the natural pH as well as the pzc are very powder specific and can even vary from batch to batch.

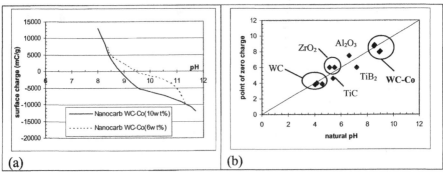

Figure 2: Surface charge on Nanocarb WC-6Co and WC-10Co powders as a function of pH of the suspension (a) and charging map of different powders illustrating pzc as a function of natural pH (b).

Table I: Properties of the WC-Co starting powders from Nanodyne.

	Nanocarb WC-6Co	Nanocarb WC-10Co
Co content (wt%)	6	10
WC grain size (nm)	40	40
d_{50} (μm)	0.83	1.20
d_{90} (μm)	1.89	3.36
BET (m^2/g)	2.90	1.75
Pzc (pH)	9.44	8.85
Natural conductivity (μS/cm)*	111	77
EPD in ethanol	no	no
EPD in ethanol + acetic acid	no	no
EPD in acetone	yes	yes

*measured for a concentration of 100g powder/1000ml demineralised H_2O.

3.2 EPD of WC-6Co/WC-25Co FGMS

Two WC-Co graded FGM plates (FGM1 and FGM2, as defined above) of 35x35mm with a thickness of about 2mm were produced containing a gradient layer of about 1.5mm. The results of density measurements are given in Table 2. Residual porosity was observed after sintering at 1290°C and 1340°C. The influence of the sintering temperature on the hardness gradient was investigated by sintering the FGMs at 1290°C, 1340°C and 1400°C for 1 hour. Whereas the gradient totally disappeared after sintering at 1400°C, it remained after sintering at 1290°C. Backscattered electron micrographs showing the cobalt gradient of FGM1 are given in Figure 3. The cobalt gradient after sintering at 1290°C varied from 17wt% Co on the soft side to about 3.5wt% Co on the hard side. However, one would expect a gradient from about 6wt% to 17wt%. The reason for the lower cobalt content is probably due to a not optimal carbon content in the sample, causing migration of cobalt during sintering. Optimization of the carbon content by addition of carbon black to the suspension is ongoing. The difference in composition profile between FGM1 and FGM2 is clearly illustrated in Figure 4. The $HV_{0.5}$ hardness profile for FGM1 is shown in figure 5, revealing hardness values from 21GPa on the low cobalt side down to 9GPa on the high cobalt side. HV_{10} measurements on the low and high cobalt sides revealed values of 19.5GPa and 8.8GPa respectively (see Figure 5).

Table II: Density measurements of FGM1 and FGM2 after sintering at 1290°C, 1340°C and 1340°C for 1hour (in g/cm³).

	1290°C	1340°C	1400°C
FGM1	14.20	14.40	14.45
FGM2	13.90	14.05	14.15

Functionally Graded Materials 2000

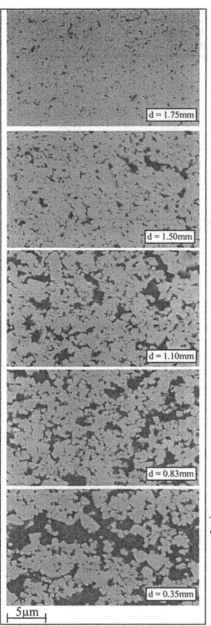

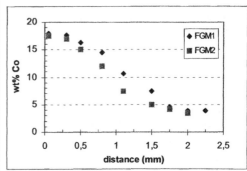

Figure 4: Compositional profile along the thickness of the WC-6Co/WC-25Co FGMs.

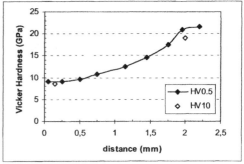

Figure 5: Vickers hardness $HV_{0.5}$ and HV_{10} of FGM1 after sintering at 1290°C for 1hour.

Figure 3: Backscattered SEM observation of WC-6Co/WC-25Co FGM1. (d = distance from high cobalt side)

Functionally Graded Materials 2000

During sintering at 1290°C and 1340°C the FGMs warped as shown in figure 6, caused by a different sintering rate and final shrinkage between the high and low cobalt side. These results are in accordance with the results obtained by Colin et al [9] for laminated composite plates. The FGMs sintered at 1400°C did not warp, due to the homogenization of the cobalt content during liquid phase sintering.

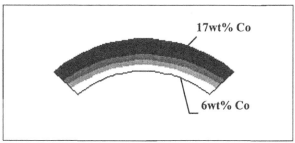

Figure 6: Schematic view of deformation (exaggerated) of the FGMs after sintering at 1290°C and 1340°C.

4 FINAL REMARKS AND CONCLUSIONS

Electrophoretic deposition (EPD) is found to be a possible processing technique for continuously graded hardmetals with a cobalt gradient. The slope, length and microstructural composition of the gradient can be engineered by changing the EPD parameters and concentration of the suspension. The production of crack-free FGM plates of 35x35mm is improved when a graphite coating is applied on the surface of the deposition electrode and drying of the green deposits is performed in a controlled atmosphere.

The green EPD bodies were CIPed and sintered in order to obtain a near-fully dense material. WC-25Co/WC-6Co FGMs, with HV_{10} hardness values from 8.8 up to 19.5GPa were produced. It was shown that the FGM green plates could be sintered to closed porosity with a density of >98%, opening the prospect for complete densification using a sinter-hip cycle. Whereas the FGMs did not deform after liquid phase sintering at 1400°C, they warped after sintering at 1290°C and 1340°C.

Further research will be focused on the remaining problems concerning deformation of the FGM, carbon control and control of the cobalt migration during liquid phase sintering.

ACKNOWLEDGEMENT

This work was supported by the Flemish Institute for the Promotion of Scientific and Technological Research in Industry (IWT) and by the Brite-

Euram programme of the Commission of the European Communities under project contract No. BRPR-CT97-0505.

REFERENCES

1. W.A. Kaysser and B. Ilschner, "FGM Research Activities in Europe", *MRS Bulletin*, **20** [1], 22-26(1995).
2. O. Van Der Biest and L. Vandeperre, *Ann. Rev. Mater. Sci.*, **29**, 327 (1999).
3. I. Zhitomirsky, "Electrophoretic and Electrolytic Deposition of Ceramic Coatings on Carbon Fibers", *J. Eur. Ceram. Soc.*, **18**, 849-56 (1998).
4. Z. Wang, J. Shemilt and P. Xiao, "Novel Fabrication Technique for the Production of Ceramic/Ceramic and Metal/Ceramic Composite Coatings", *Scripta Mat.*, **42**, 653-59 (2000).
5. L. Vandeperre and O. Van Der Biest, *Key Eng. Mat.*, **127-131**, 567 (1997).
6. L. Vandeperre and O. Van Der Biest, "Electrophoretic forming of Silicon Carbide Laminates with Graphite Interfaces", *Fourth Euro Ceramics*, **1**, 359-66 (1995).
7. C. Zhao, J. Vleugels, L. Vandeperre, B. Basu and O. Van Der Biest, "Graded Tribological Materials formed by Electrophoretic Deposition", *Mat. Sci. Forum*, **308-311**, 95-100 (1999).
8. I.E. Dzjaloshinsky, E.M. Lifshits and L.P. Pitaevsky, *Zhurn. Eksp. Teor. Fiz.*, **37**, 229-40 (1959).
9. C. Colin, L. Durant, N. Favrot, J. Besson, G. Barbier, F. Delannay, "Processing of Composition Gradient WC-Co Cermets", *Proc. of the 13th Int. Plansee Sem.*, **2**, 522-35 (1993).

INFLUENCE OF CENTRIFUGAL CASTING PARAMETERS ON THE STRUCTURE AND PROPERTIES OF Al-Si/SiC$_p$ FGMs

L.A. Rocha, A.E. Dias, D. Soares
Department of Mechanical Engineering, University of Minho
Campus de Azurém, 4800-058 Guimarães, Portugal

C. M. Sá
Center of Materials, University of Porto
Rua Campo Alegre, 823, 4100 Porto - Portugal

A. C. Ferro
Department of Materials Engineering, Instituto Superior Técnico
Av. Rovisco Pais, 1096 Lisboa Codex, Portugal

ABSTRACT

Al-Si metal matrix composites selectively reinforced at the surface (SiC) may be considered as advanced materials, aimed to be used in the automotive industry in components requiring high wear resistance, high bulk toughness or even a thermal barrier at the surface.

In this work centrifugal casting was used to produce Al-Si/SiC$_p$ FGMs. The main aim of the work was to study the influence of the centrifugal casting parameters on the structure and properties of the FGM. Centrifugal acceleration, pouring temperature and mould temperature were the studied variables. Both the individual contribution of each parameter, and of their combination on the evolution of the structure and hardness were investigated. Results suggests that for a given alloy/particle mixture, the characteristics of the material depends essentially on the pouring temperature and centrifugal acceleration. By the correct control of the above variables it is possible to produce FGMs with reproducible properties.

INTRODUCTION

The automotive industry is particularly interested in the development of aluminum alloy-based MMCs for use in cast engine and other automobile components such as cylinder liners, pistons and valves. This interest is mainly motivated by the potential that these materials present for weight savings over conventional iron-based components, together with the attractive tribological properties of the MMCs surface. In addition, for some applications it is desirable to have high wear resistance coupled with high bulk toughness, in order to allow the component to absorb impact loads [1]. Al-Si FGMs selectively reinforced at the surface (SiC) are strong candidates to fulfil those requirements.

Casting under a centrifugal force is one of the most effective methods for processing SiC reinforced Al-based FGMs [2]. However, accurate control of the distribution of the particles and of the mechanisms leading to their distribution is not well understood [2]. As a consequence, manufacturing feasibility and/or reliability have partially impaired the industrial application of these materials.

For a given alloy/particle mixture, several processing parameters influences the final characteristics of FGMs obtained by centrifugal casting, namely temperature of the melt, temperature of the mould, atmosphere in the neighborhood of the melt, pouring rate, centrifugal force, and solidification rate [2-4]. These parameters will have influence on several phenomena that will occur when the melt containing solid particles solidifies in the mould: interaction and/or chemical reactions between the solidification front and the moving particles (phenomena occurring in opposite directions), the rate of variation of the viscosity of the liquid during solidification, possible interactions between particles and the liquid, and the initial position of the particles in the mould before starting their movement in the liquid are the most important [2,5]. All these phenomena will dictate the final spatial distribution of the particles, the overall microstructure, and the mechanical properties of the FGM.

In this work, preliminary results concerning the influence of centrifugal casting parameters, namely centrifugal acceleration, pouring temperature and temperature of the mould, on the structure and properties of Al-Si/SiCp FGMs are presented and discussed. Results suggests that the pouring temperature, and centrifugal acceleration are the most important parameters to be considered. By the correct control of those variables, it is possible to produce FGMs with reproducible properties.

EXPERIMENTAL

Materials

All experiments were carried out using samples retrieved from an ingot of F3S-20S aluminum matrix composite *Duralcan™* from Alcan Aluminum Ltd, USA. This composite is designed for general propose gravity casting. In Table I the most relevant characteristics of this composite are presented.

Table I. Characteristics of the F3S-20S *Duralcan™* composite [6].

Nominal chemical composition (wt%)	8.5-9.5% Si; 0.45-0.65% Mg; 0.2% Ti; 0.2% Fe (max); 0.2% Cu (max)
Density (g/cm^3)	2.77
Ultimate strength (MPa)	221-359
Yield Strength (MPa)	165 - 338
Elongation (%)	2.8 – 0.4
Elastic Modulus (GPa)	98.6
Hardness (HRB)	73
Solidus/Liquidus Temperatures (°C)	568/609

Centrifugal Casting Apparatus

A high frequency induction centrifugal casting furnace (*Titancast 700 μP Vac*, from *Linn High Therm*, Germany), equipped with a vacuum system, was used for melting and casting the samples. The furnace was instrumented in order to monitor and control the temperature of the melt, measure the rotational speed of the casting arm, and control the temperature of the mould. A schematic representation of the furnace layout is presented in Fig. 1.

468

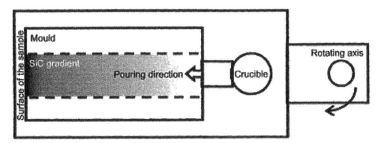

Fig. 1 – Schematic representation of the casting arm of the furnace. The region of segregated SiC particles is indicated in the scheme. This region is referred in the text as "surface".

Unlike traditional systems in which the liquid melt is poured into a rotating mould, in the equipment used in this work both the melt (crucible) and the mould are positioned in a centrifugal arm which rotates around a central axis. This system allows the production of cylindrical samples (40 mm diameter and 80 mm long), in which the gradient of particles is obtained at the surface of the cylinder, at the opposite side of the pouring hole. A charge of ca. 200 g of material was used in each experiment, which was always performed under vacuum (P < 0.2 Pa). After melting, the material was poured into the mould as soon as the desired pouring temperature was achieved. The temperature of the melt was monitored and controlled by a chromel-alumel thermocouple (1.5 mm diameter) placed inside of an alumina tube. The thermocouple was connected to a *Shimaden FP21* controller. The material was poured into a graphite mould, which was placed inside a small furnace allowing the control of the temperature of the mould.

Sample Analysis
All samples were longitudinally cut in two halves, and polished for hardness measurements and metallographic examination. A good correlation was found between the hardness of the material and the SiC particles area fraction determined by image analysis.
Quantitative analysis of the microstructure of the material was carried out in order to obtain information on particles area fraction and particle size distribution in depth. Image analysis was performed on gray scale digitized images (8 bit, 512x512 pixels), using an image analysis system developed at CEMUP (*PAQI*). Bright field images (x200) were first acquired by a TV camera attached to a *ZEISS Axioplan* reflection optical microscope. Images were captured at 2 mm intervals, both in the transverse and longitudinal directions of the samples. Image processing and segmentation prior to analysis involved a sequence of operations including: delineation (for contrast and improved phase segmentation), thresholding (for particle segmentation into a binary image), morphological open-close filtering (for noise suppression), fill-holes filtering (for incomplete damaged carbide shape recovery) and finally particle de-agglomeration (for carbide particle separation, using a watershed based algorithm). In the final segmented (binary) image two main parameters were measured: particle surface fraction and mean particle dimensions.

RESULTS AND DISCUSSION
Fig. 2 a) shows the evolution of the angular rotational velocity (ω) as a function of the position of the knob of the casting arm controller. Casting time was kept constant (90 sec.). During casting, the furnace allows a maximum angular rotational speed of 44 s^{-1}. The main difference between each position is related to the time taken by the motor to achieve the maximum speed, i.e. the force associated to the torque (except for position 1 in which a maximum angular speed of 29 s^{-1} is

achieved). The dependence of the acceleration with casting time is shown in Fig. 2 b). For position 1, acceleration varies continuously during casting, achieving a maximum of 11.5 g (112.6 m/s²). A continuous evolution of the acceleration with casting time was observed for the other casting conditions, the difference being the achievement of a maximum acceleration of 24.5 g (240.4 m/s²) some time after the rotation of the casting arm starts.

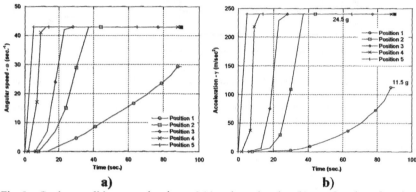

a)   b)

Fig. 2 – Casting conditions: rotational speed (a) and acceleration (b) as a function of casting time.

In Fig. 3 the influence of acceleration (Fig. 3 a), pouring temperature (Fig. 3 b) and temperature of the mould (Fig. 3 c) on the longitudinal hardness profile of the samples is presented.

Considering the influence of acceleration (Fig. 3 a) for samples poured at 750° C in the mould at ambient temperature, the maximum hardness is always obtained at some distance from the surface (15 – 20 mm). Results obtained for position 4 are those presenting a higher hardness near the surface, and a less pronounced increase in hardness to the interior of the sample.

As described by *Lajoye and Suéry* [5], particle segregation to the surface of the casting during centrifugal casting occurs owing to the difference in densities of the SiC (3.2 g/cm³) and the melt (2.3-2.7 g/cm³). Under a constant acceleration, the velocity (υ) of a spherical particle of radius R_p may be estimated using Stokes' law [7]:

$$\upsilon = \frac{2 R_p^2 (\rho_p - \rho_l) \gamma}{9 \eta} \tag{1}$$

where: ρ_p and ρ_l = densities of the particle and of the liquid

γ = acceleration

η = viscosity of the liquid

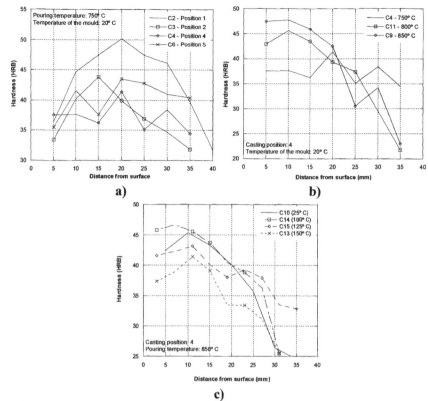

c)

Fig. 3 – Influence of the casting conditions (a), pouring temperature (b) and temperature of the mould (c) on the longitudinal hardness profile of the cast.

If the liquid contains a certain amount of particles, its viscosity will be affected and, according to Eq. 1, the velocity will be reduced. The apparent viscosity (η_{app}) as a function of the particles volume fraction is given by [7]:

$$\eta_{app} = \eta_l \, (1 + 2.5 \, V_p + 7.6 \, V_p^2) \qquad (2)$$

where: η_l = viscosity of the liquid
          V_p = volume fraction of particles

Considering that all experiments were carried out at the same pouring temperature and using the same kind of material, according to Eq. (1), the only variable influencing the speed of the particle is acceleration. This model is not taking into account the effect of temperature changes in the process. In fact, each particle is subjected to two main forces acting on opposite directions: the centrifugal force, and the viscous drag force due to the increase in viscosity of the melt during solidification [2,5,7]. If the solidification rate of the melt upon contact with the mould walls is high, the solidification front may impel the SiC particles to the interior of the sample. Results presented in Fig. 3 a) suggests that this is the phenomenon occurring for the experimental conditions used in this work.

Considering the graph presented in Fig. 3 b), it can be seen that the particle gradient seems to be strongly affected by the pouring temperature: as pouring temperature increases a higher surface hardness is obtained, and a relatively smooth gradient to the interior of the sample is obtained. The increase in the pouring temperature affects the solidification delay, thus the viscosity of the liquid. Together with acceleration (position 4; Fig. 3 a), the increase in the temperature of the melt seems to modify the balance of forces acting on the particles, in order to assist segregation of the particles towards the surface of the samples.

Raising the mould temperature seems to have little effect on the hardness profile of the samples. In fact, observing Fig. 3 c), it can be seen that some improvement in the hardness profile near the surface (10 mm) can be obtained by heating the mould to 100° C, but from that distance inward the hardness profiles are similar. However, a further increase in mould temperature may have a deleterious effect on the hardness gradient, as it is shown in the graph for the profiles obtained at 125 and 150°C. Raising the mould temperature should have an effect on particle segregation behavior similar to that obtained by increasing the pouring temperature. However, it seems that the simultaneous increase of both parameters may be deleterious for a smooth gradient to be obtained. Further investigation will focus on this aspect: the mould is being instrumented in order to allow the cooling profile curves to be obtained along the samples. In fact, this behavior suggests that the time the melt takes to start solidification in the mould may influence the particle distribution along the sample (in the experimental setup used in this work, the casting arm rotates no more than 90 seconds).

The macroscopic approach using hardness profiles may be useful for rapidly screening the best casting conditions. However, this method does not give information on the gradient at the surface of the sample. In Fig. 4, two micrographs captured at different depths (surface and 12 mm from surface) in a sample poured at 850° C into the mould at room temperature, with a centrifugal acceleration corresponding to position 4, are presented. The first remark concerns the absence of porosity, which was a characteristic for all the tested conditions. Also, the presence of undesirable interfacial reaction products resulting from the reaction of Al with SiC particles, such as Al_4C_3, was not detected by SEM observation in all tested conditions. It is known that a high silicon activity in the melt, which is characteristic of Al-Si alloys, hindered the formation of deleterious interfacial compounds [8].

In Fig. 5, the particle area fraction is plotted as a function of the distance from surface. Each point in the graph corresponds to the average obtained in 20 images. Measurements were performed in depth at 2 mm steps. The graph presented in Fig. 5 corresponds to measurements performed in sample C9, from which hardness results are plotted in Fig. 3 b). The first hardness point plotted in Fig. 3 b) was obtained at 5mm depth (47 HRB) corresponding to a particle area fraction of ca. 32%. At 10 mm a hardness of 48 HRB was obtained which corresponds to 32.8% particles. However, at the surface, the particle area fraction is of about 26% corresponding to a hardness of 38 HRB (value determined at the top of the sample). Considering wear resistance these differences are important, as demonstrated by the work of *Gomes et al.* [9].

Quantitative image analysis was also used in order to obtain information of the size distribution of particles as a function of the depth below surface (Fig. 5). The analysis of this graph suggests that larger particles are preferentially segregated to the surface of the samples, while those having smaller dimensions tend to be kept at the interior. Considering Eq. 1, it can be seen that velocity of a particle increases with the square of the radius of the particle, this dependence being originated by the influence of the mass of the particle under acceleration. This particle

472

distribution may have some influence on the mechanical behavior of the samples. Furthermore, these results indicates that it might be possible to obtain a wider range of gradients by acting on the mean diameter of ceramic particles introduced in the MMC, as suggested by other authors [2,5].

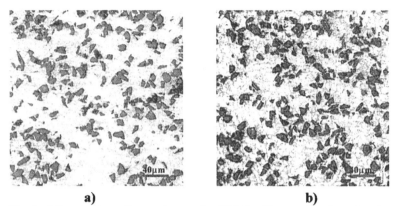

a)   b)

Fig. 4 – Microstructure of a sample poured at 850° C, with a centrifugal acceleration corresponding to position 4, into a mould at room temperature. Surface of the sample (a) and 12 mm from surface (b)

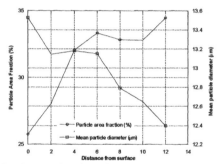

Fig. 5 – Particle area fraction as a function of the depth for a sample poured at 850° C, using centrifugal casting position 4, in a mould at room temperature.

CONCLUSIONS

In this work the influence of some centrifugal casting parameters (centrifugal acceleration, pouring temperature and mould temperature) on the structure and properties Al-Si/SiC$_p$ FGMs was studied. Results allow the following conclusions to be drawn:

-   Using the described experimental setup it is possible to produce samples without visible pores by centrifugal casting.
-   Centrifugal acceleration (maximum 24.5 g), produced by the centrifugal arm is unable, by itself, to produce the desirable gradient: density of particles decreasing from the surface of the sample.

- Pouring temperature together with centrifugal acceleration allow the gradient of particles to be modified.
- Raising both the pouring temperature and the temperature of the mould has a deleterious effect on the particle distribution.
- A preferential segregation of the larger particles towards the surface of the samples was observed.

REFERENCES

[1] M.R. Jolly, "Opportunities for aluminum based fibre reinforced metal matrix composites (FRMMCs) in automotive castings", *The Foundryman*, Nov. 1990, 509-513.

[2] Y. Watanabe, N. Yamanaka and Y. Fukui, "Control of composition gradient in a metal-ceramic functionally graded material manufactured by the centrifugal method", *Composites Part A*, **29A**, 595-601 (1998).

[3] M. Mamoru, K.-I. Abe and T. Inoue, "Processing of metal matrix composite by centrifugal casting: technique and evaluation of the elastic properties", *J. of the Soc. of Materials Science, Japan*, **46** [8] 946-951 (1997).

[4] C. G. Kang and P. K. Rothatgi, "Transient thermal analysis of solidification in a centrifugal casting for composite materials containing particle segregation", *Metallurgical and Materials Transactions B*, 27 [2], 277-285, (1996).

[5] Lajoye, L; Suéry, M., "Modelling of particle segregation during centrifugal casting of Al-matrix composites", in *Cast Reinforced Metal Composites*, Proceedings of the International Symposium on Advances in Cast Reinforced Metal Composites, (S. G. Fishman and A. K. Dhingra (Edts.) ASM International, pp. 15-20, (1988).

[6] "Composite Casting Guidelines", Duralcan, USA, (1990)

[7] S. Suresh and A. Mortensen, "Fundamentals of Functionally Graded Materials: Processing and Thermomechanical Behaviour of Graded Metals and Metal-Ceramic Composites", IOM Communications Ltd., 1998.

[8] A.C. Ferro and B. Derby, "Wetting behaviour in the Al-Si/SiC system: Interface reactions and solubility effects", *Acta Metall. Materialia*, 43, 3061-3073 (1995).

[9] J.R. Gomes, A.S. Miranda, D. Soares, A.E. Dias, L.A. Rocha, S.J. Crnkovic and R.F. Silva, "Tribological characterization of Al-Si/SiC$_p$ composites: MMCs vs. FGMs", 6th International Symposium on Functionally Graded Materials, Estes Park, Colorado (2000).

ACKNOWLEDGEMENTS

This work was sponsored by Fundação para a Ciência e Tecnologia (FCT - Portugal) under the program PRAXIS XXI (contract PRAXIS P/CTM/12301/1998). The financial support for traveling costs from Fundação Luso-Americana para o Desenvolvimento (FLAD – Portugal) obtained by L.A. Rocha is gratefully acknowledge.

MICROSTRUCTURE IN NiAl/STEEL JOINT
PRODUCED BY REACTIVE CASTING METHOD

Kiyotaka Matsuura and Masayuki Kudoh
Division of Materials Science and Engineering
Hokkaido University
Sapporo, Hokkaido 060-8628, Japan.

Hiroshi Kinoshita and Heishichoro Takahashi
Center for Advanced Research of Energy
Hokkaido University
Sapporo, Hokkaido 060-8628, Japan

ABSTRACT

A NiAl block has been produced by pouring aluminum and nickel liquids onto a steel block placed in the bottom of a crucible. Due to an exothermic reaction between the elemental liquids, the temperature of the liquid mixture rises very quickly and exceeds 2800 K, which is extremely high compared to the pouring temperatures of 1023 K for aluminum and 1773 K for nickel. The liquid mixture melts the surface of the steel block and soon solidification of NiAl containing small amount of iron starts at the steel surface, and finally the NiAl and steel are strongly joined after solidification. The microstructure at the joint interface has a very fine dual-phase structure of rod-like beta-NiAl and gamma-ferrite. The morphology of the beta-NiAl gradually changes from rod to round grain, as the distance from the interface increases. This gradual change in microstructure brings about a high joint strength exceeding 220 MPa.

INTRODUCTION

An intermetallic compound of nickel monoaluminide, NiAl, has attracted considerable interest as a potential high-service-temperature material for applications such as in hot-end components of the turbine blades of aeronautic engines, because it has the alluring combinations of low density (5.86 g/cm^3), high melting temperature (1911 K), high strength, good corrosion and oxidation resistance, high thermal conductivity and low cost. [1-5] However, NiAl has very low ductility at temperatures below 600 K, which is one of the most significant disadvantages of NiAl and impedes the practical use of NiAl as a structural material. [6,7] Although the ductility of NiAl can be improved by reducing the grain size or alloying some elements, the degree of the improvement is not very remarkable. [8,9]

For an application of NiAl to new corrosion and oxidation resistant structural materials, it will be useful to join NiAl to ductile structural materials such as steels and superalloys. From this viewpoint, protective coating using NiAl has been studied. A variety of processing techniques have been proposed regarding the NiAl coating, such as pack cementation, chemical vapor deposition, slurry cementation, hot dipping, and so on. [10] However, when a thick NiAl coating is required, those coating techniques based on the diffusion mechanism are not suitable, because they all demand a high temperature and an extremely long period of time.

The present authors have studied a thick NiAl coating by a reactive sintering technique, which is based on sintering of a mixture of elemental powders of aluminum and nickel on the surface of the base material under a pressure. [11-13] Because a remarkably exothermic reaction between the aluminum and nickel powders leads to the synthesis of NiAl and the simultaneous joining between the NiAl and the base material, the coating process starts at a low temperature and ends in a very short time. Moreover, the present authors have studied the NiAl coating by a reactive casting technique, which is based on pouring of aluminum and nickel liquids onto the surface of the base material. [14] The reactive casting method has some advantages over the reactive sintering method: (1) heat generated from the exothermic synthesis reaction is larger, which brings about deeper melting of the surface of the base material and therefore stronger joint between the synthesized NiAl and the base material, (2) premixing and compacting of the elemental powders and pressure application during coating are not necessary, which brings about more simple coating process, and (3) the use of elemental liquids instead of elemental powders reduces the cost of the raw material and increases the strength of the synthesized NiAl because of low cost and oxygen level of the raw materials.

In this work, the microstructure of the NiAl/steel joint produced by the reactive casting method is investigated and the effect of the microstructure on the joint strength is discussed.

PRODECURE

Figure 1 shows a schematic drawing of the reactive casting method, which is based on the pouring of the elemental liquids onto the base material. Nickel pellets of 99.97 wt% and aluminum ingots of 99.99 wt% in purity were used as the raw materials for the liquids, while an ultra-low carbon steel block containing 0.001 wt% of carbon, 0.008 wt% of silicon and 0.01 wt% of manganese was used as the base material. The molar ratio of the elemental liquids was fixed at Ni:Al=1:1 to synthesize the stoichiometric NiAl. The steel block was placed in the bottom of a porous alumina crucible of 30 mm in inner diameter and 95 mm in depth.

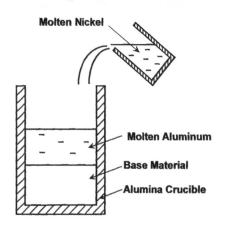

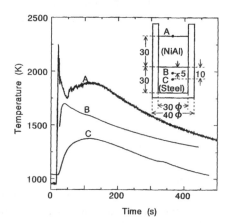

Fig. 1 Schematic drawing of the procedure of the synthesis and simultaneous joining based on pouring of the elemental liquids onto the base material.

Fig. 2 Change in temperature of the sample. Preheated temperature of the base material: 1023 K. Thickness of the synthesized NiAl: 30 mm.

Aluminum liquid was first poured onto the steel, and then nickel liquid was soon poured into the aluminum liquid. The pouring temperatures of the aluminum and nickel liquids were 1023 K and 1773 K, superheated by approximately 90 K and 50 K, respectively. The initial temperature of the steel was varied from room temperature to 1023 K. The estimated thickness of the synthesized NiAl was varied from 10 to 50 mm, while the thickness of the steel block was fixed at 30 mm. Melting of aluminum and nickel was carried out in an argon atmosphere, while pouring in air.

Chemical analysis using an electron probe microanalyzer (EPMA) and microstructure observation using an optical microscope (OM), a scanning electron microscope (SEM) and a transmission electron microscope (TEM) were carried out. A microanalysis using an energy dispersive spectroscopy (EDS) was performed to identify the phase appearing in the TEM observation field, and an X-ray diffraction (XRD) analysis for a bulk sample was also performed to identify the reaction product. A four-point bending test was carried out at room temperature to evaluate the joint strength at the interface. The tensile faces of the bend specimens were polished using #1200 emery paper before testing. Polish direction was parallel to the longitudinal direction. The conditions of the bending test were: 3×4 mm^2 in rectangular cross-sectional area of the specimen, 15 mm in distance between the supporting points, 4 mm in distance between the loading points and 0.5 mm/min in cross-head speed.

RESULTS AND DISCUSSION
Heat Generation and Heat Transfer
Figure 2 shows the change in temperature of the sample during and after pouring of the elemental liquids. Temperature was measured at three different points: point A at the surface of the NiAl, point B in the steel 5 mm away from the interface and point C in the steel 10 mm away from the interface, by using a spot thermometer TR-630, MINOLTA, for point A and type B thermocouples for points B and C.

The temperature of the liquid mixture suddenly rises and exceeds 2250 K, which is much higher than the pouring temperatures of the elemental liquids, i.e. 1023 K for aluminum and 1773 K for nickel. This extreme increase in temperature is due to a large formation enthalpy of -1380 J/g for NiAl.[15] The calculated maximum temperature based on the adiabatic condition using the thermal properties listed in Table I was 2876 K. The difference between the measured and calculated maximum temperature can be explained from the fact that the heat transfer from the liquid to the base material, crucible and the atmosphere is not considered in the calculation. When the temperature of the liquid decreases after reaching the maximum, a recalescence of temperature is observed due to the latent heat of solidification (Figure 2, curve A).

Table I   Thermodynamic data used in the calculation [15]

Property	Value	
Specific Heat of Al, C_{Al}	0.899	J/(g·K)
Specific Heat of Ni, C_{Ni}	0.439	J/(g·K)
Specific Heat of NiAl, C_{NiAl}	0.53+0.00023T	J/(g·K)
Heat of Fusion of Al, ΔH_{Al}	711	J/g
Heat of Fusion of Ni, ΔH_{Ni}	309	J/g
Heat of Fusion of NiAl, ΔH_{NiAl}	200	J/g
Heat of Formation, ΔH_{p298}	-1380	J/g

Functionally Graded Materials 2000

Melting of the Steel Surface

As shown in Figure 2, the heat from the exothermic reaction is transferred to the steel, and the temperature of the steel is elevated from 1023 K to approximately 1700 K at a point of 5 mm from the initial interface, for example. This heat is sufficient to melt the surface of the steel. The depth of the melted steel increases with the thickness of the synthesized NiAl, as shown in Figure 3. The depth of the melted steel increases also with the preheating temperature of the steel block, as shown in Figure 4.

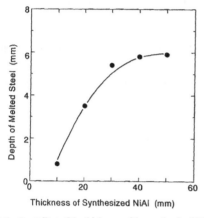 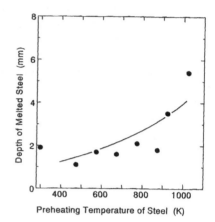

Fig. 3 Effect of the thickness of the synthesized NiAl on the depth of the melted steel. Preheated temperature of the steel: 1023 K.

Fig. 4 Relation between the preheating temperature of the steel block and the depth of the melted steel. Thickness of the synthesized NiAl: 30 mm.

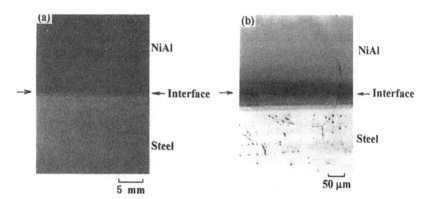

Fig. 5 Microstructure on a longitudinal section of a sample photographed (a) at a low magnification and (b) at a high magnification. Initial temperature of the steel: room temperature. Thickness of the synthesized NiAl: 30 mm.

Functionally Graded Materials 2000

The depth of the melted steel may affect the chemical and mechanical properties of the synthesized NiAl, because the increase in depth brings about an increase in iron content of the NiAl. A controlled depth may be achieved by choosing the thickness of the synthesized NiAl and the preheating temperature of the steel. The pouring temperatures of the elemental liquids will also affect the depth, although it was not investigated in this study.

Microstructure and Microanalysis

Figure 5 shows microstructures from a longitudinal section of a sample produced under a condition that the thickness of the synthesized NiAl is 30 mm and the preheating temperature of the steel is 1023 K. There are no casting defects such as blowholes in the synthesized NiAl and no cracks at the joint interface, which implies that both the synthesis and joining were successful. It is likely that the sound casting and joining are brought about by low thermal stress due to close thermal expansion coefficients between NiAl and steel.[16,17] Change in microstructure at the joint interface does not look sharp, but gradual, even at a higher magnification (Figure 5(b)).

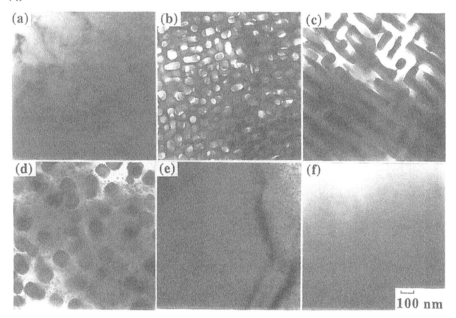

Fig. 6   Microstructures photographed using a TEM showing gradual change in microstructure across the joint interface. The positions of observation are (a) 900 μm, (b) 300 μm, (c) 100 μm away from the interface into the NiAl, and (d) 20 μm, (e) 40 μm, (f) 100 μm into the steel. The sample is same as that shown in Figure 5.

More detailed microstructures near the joint interface were observed using a TEM. The TEM disk was prepared from the sample shown in Figure 5, by slicing, grinding and electro-chemical polishing. Photographs taken using the TEM are displayed in Figure 6, showing the change in microstructure across the joint interface.

The NiAl side away from the joint interface consists of a single-phase structure (Figure 6 (a)). The chemical composition of this position measured by the EDS was 41.8 Al, 12.2 Fe and 46.0 Ni in at.pct, which corresponds to a solid solution of β-NiAl, according to an aluminum-iron-nickel ternary phase diagram[18]. In the NiAl near the interface (Figure 6 (b)), light gray ellipsoidal particles of 50 to 100 nm in diameter are dispersed in the dark matrix. At the interface (Figure 6 (c)), a mixture of dark and light gray phases with a rod-like shape is observed. The results of the EDS analysis carried out for this position were 35.9 Al, 27.6 Fe and 36.5 Ni in at.pct for the dark phase and 20.6 Al, 56.3 Fe and 23.1 Ni in at.pct for the light gray phase. Those concentration values involve an experimental error, because each phase is very small. However, it is undoubted that the dark phase corresponds to β-NiAl and the light gray phase to γ-iron, according to the aluminum-iron-nickel ternary phase diagram. On the steel side of the interface (Figure 6 (d)), dark square particles are dispersed in the light gray matrix. As the distance from the joint interface increases, the particles size of the dark phase decreases (Figure 6 (e)), and finally they disappear completely (Figure 6 (f)). The very fine dark particles appearing in Figure 6 (e) are also observed in Figure 6 (d) together with the square dark phases. The ternary phase diagram suggests that the dual structure shown in Figures 6 (b) through 6 (d) was formed by a eutectic reaction that Liquid → β-NiAl +γ-iron, and that the very fine dark particles observed in Figures 6 (d) and 6 (e) were formed by precipitation from the matrix phase of γ-iron during cooling after the eutectic reaction.

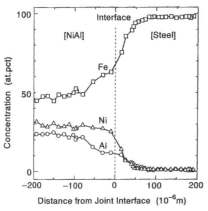

Fig. 7 Concentration profiles across the joint interface. Preheating temperature of the steel: 1023 K. Thickness of the synthesized NiAl: 30 mm.

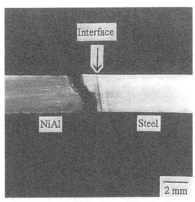

Fig. 8 Fractured specimen of the bending test. Preheating temperature of the steel: 1023 K. Thickness of the synthesized NiAl: 30 mm.

Figure 7 shows concentration profiles across the joint interface obtained using an EDS for the same sample as shown in Figure 6. Due to melting of the steel, iron dissolves in the NiAl at approximately 50 at.pct near the joint interface. The concentrations of aluminum and nickel in the NiAl are approximately 20 at.pct and 30 at.pct, respectively. The chemical composition of the NiAl is very close to a eutectic line in the ternary phase diagram. On the steel side, on the other hand, aluminum and nickel defuse into the steel, resulting in concentrations of 10 at.pct or less and a diffusion depth of 50 to 70 μm. A composition of 10 Al, 80 Fe and 10 Ni in at.pct is within a solid solution of γ-iron at 1323 K. At 1023 K, however, it is within a three-phase region of

Functionally Graded Materials 2000

γ-iron, α-iron and β-NiAl, which indicates that the γ to α transformation in iron phase and the precipitation of β-NiAl phase occurs during cooling.

Joint Strength

Bend specimens machined from the sample shown in Figure 6 were bent until fracture, and the fracture strength was measured. In all bending tests, fracture occurred in the NiAl near the joint interface, as shown in Figure 8. The fracture strength was approximately 220 MPa on average, which is close to the inherent strength of polycrystalline NiAl of 260 MPa at 300 K[7].

The joint strength must be higher than the fracture strength, because the joint interface was intact when the specimen was fractured. In the bending tests carried out in our previous study for a NiAl/steel joint produced by reactive sintering of the elemental powders, fracture occurred also in the NiAl near the joint interface, but the average fracture strength was as low as approximately 60 MPa[12]. The difference in fracture strength between our present and previous studies indicates that the reactive casting can bring about a much sounder NiAl than the reactive sintering. It is considered that when powders are used as the starting materials, impurities such as oxygen reduce the strength of the reaction product.

The joint strength is usually affected by the microstructure at the joint interface. The microstructure does not change drastically at the joint interface but gradually, as shown in Figures 5 and 6. The change in chemical composition across the joint interface is similarly gradual, as shown in Figure 7. Moreover, the microstructure at the joint interface consists of a very fine dual-phase structure. It is suggested that the gradual changes in microstructure and chemical composition across the interface and the interpenetrating fine dual-phase structure at the interface brought about the high joint strength.

CONCLUSIONS

By pouring aluminum and nickel liquids (Ni:Al=1:1) onto a surface of a steel block, the synthesis of nickel monoaluminide, NiAl, and joining of it to the steel block have been carried out and the changes in microstructure and chemical composition across the joint interface were investigated. The results are summarized as follows.

The aluminum and nickel liquids exothermically react and produce an extremely superheated NiAl liquid. The steel near the surface is melted by the heat generated by the reaction, and iron from the melted steel dissolves in the NiAl liquid and, consequently, an iron-containing NiAl-base intermetallic compound, (Ni, Fe)Al or β-phase, is produced. The depth of the melted steel increases with the increase in both preheating temperature of the steel and thickness of the NiAl on the steel.

The (Ni, Fe)Al is strongly joined to the steel after solidification, showing a joint strength exceeding 220 MPa. The high joint strength is due to the gradual change in microstructure and chemical composition across the joint interface and the interpenetrating fine dual-phase structure at the joint interface.

ACMOWLEDGEMENT

The authors gratefully acknowledge the financial support of this work by the Izumi Science and Technology Foundation, Osaka, Japan. Appreciation is extended to Mr. H. Jinmon for his assistance in performing many examinations.

REFERENCES

1   W. F. Gale and S. V. Orel, "Microstructural Development in NiAl/Ni-Si-B/Ni Transient Liquid Phase Bonds" *Metall. Trans. A*, **27A** [7] 1925-1931 (1996).

2   R. Darolia, "NiAl Alloys for High-Temperature Structural Applications," *J. Met.* **43** [3] 44-49 (1991).

3   R. D. Noebe, A. Misra and R. Gibala, "Plastic Flow and Fracture of B2 NiAl-based Intermetallic Alloys Containing a Ductile Second Phase," *ISIJ International*, **31** [10] 1172-1185 (1991).

4   M. F. Singleton, J. L. Murray and P. Nash, "Aluminum-Nickel"; pp. 181-184 in *Binary Alloy Phase Diagrams*, vol. 1, ed. by T. B. Massalski, Am. Soc. Met., Materials Park, Ohio, 1990.

5   R. L. McCarron, N. R. Lindblad and D. Chatterji, "Environmental Resistance of Pure and Alloyed γ'Ni$_3$Al and β-NiAl," *Corrosion*, **32** [12] 476-481 (1976).

6   T. Takasugi, S. Watanabe and S. Hanada, "The Temperature and Orientation Dependence of Tensile Deformation and Fracture in NiAl Single Crystals," *Mater. Sic. Eng.*, **A149** 183-193 (1992).

7   R. D. Noebe and M. K. Behbehani, "The Effect of Microalloying Additions on the Tensile Properties of Polycrystalline NiAl," *Scr. Metall. Mater.*, **27** [12] 1975-1800 (1992).

8   E. M. Schulson and D. R. Barker, "A Brittle to Dictile Transition in NiAl," *Scripta Metall.*, **17** [4] 519-522 (1983).

9   R. Darolia, D. Lahrman and R. Field, "The Effect of Iron, Gallium and Molybdenum on the Room Temperature Tensile Ductility of NiAl," *Scripta Metall. Mater.*, **26** [7] 1007-1012 (1992).

10 P. C. Patnaik, "Recent Developments in Aluminide Coatings for Superalloys"; pp. 169-204 in *Advances in High Temperature Structural Materials and Protective Coatings*, National Research Council of Canada, Montreal, 1994.

11 K. Matsuura, K. Ohsasa, N. Sueoka and M. Kudoh, "In-Situ Joining of Nickel Monoaluminide to Iron by Reactive Sintering," *ISIJ International*, **38** [3] 310-315 (1998).

12 K. Matsuura, K. Ohsasa, N. Sueoka and M. Kudoh, "Reactive Sintering of NiAl and Simultaneous Joining to Steel"; pp. 2419-2424 in *Proc., 3rd Pacific Rim Int. Conf. Adv. Mater. Processing*, Vol. 2, ed. by M. A. Iman, R. DeMale, S. Hanada, Z. Zhong and D. N. Lee, TMS, 1998.

13 K. Matsuura, K. Ohsasa, N. Sueoka and M. Kudoh, "Nickel Monoaluminide Coating on Ultra-Low Carbon Steel by Reactive Sintering," *Metall. Mater. Trans. A*, **30A** [6] 1605-1612 (1999).

14 K. Matsuura, H. Jinmon and M. Kudoh, "Fabrication of NiAl/Steel Cladding by Reactive Casting," *ISIJ International*, **40** [2] 167-171 (2000).

15 T. A. Nielsen, Ph.D. Thesis, Wayne State University, Detroit, MI, 1995.

16 M. I. Sandakova, V. M. sandakov, G. I. Kalishevich and P. V. Gel'd, "Thermal Expansion of NiAl"; pp. 8.3-8.4 in *Properties of Intermetallic Alloys*, Vol. 1, ed. By J. Payne and P. D. Desai, Metals Information Analysis Center, West Lafayette, Ind., 1994.

17 M. Yoshida, "Thermal Properties of Steels"; pp. 299-306 in Tekkou Binran (Handbook of Steels) 3rd ed., Iron & Steel Inst. Jpn., Maruzen, Tokyo, 1981.

18 P. Budberg and A. Prince, "Aluminum-Iron-Nickel"; pp. 309-322 in *Ternary Alloys*, Vol. 5, ed. G. Petzow and G. Effenberg, VCH Verlagsgesellschaft, Weinheim, 1993.

FABRICATION OF ZIRCONIA-NICKEL FUNCTIONALLY GRADED MATERIAL BY DIP-COATING

Jingchuan Zhu, Mingwei Li and Zhongda Yin
School of Materials Science and Engineering, Harbin Institute of Technology,
Harbin 150001, China

Jae-Ho Jeon
Department of Materials Engineering, Korea Institute of Machinery & Materials,
Changwon 641-010, Korea

ABSTRACT
    A slurry dip-coating technique was developed for fabrication of zirconia-nickel functionally graded material (FGM). The rheological behavior of ZrO_2-Ni-ethanol slurry was characterized by viscosity test. The amount of polyvinyl butyral (PVB) additives, which served as the dispersant and binder in ZrO_2-Ni-ethanol slurry, was optimized. The stainless steel substrate was coated several times by dipping in the slurries, and followed by drying in air every dipping. After debinding in Ar, the coated FGM plate was finally hot pressed at 1300°C for 1 hour under 5MPa. Microstructural observations of the sintered FGM specimens reveal that the graded layers were formed on the substrate, in which no defects such as small cracks and residual pores were observed.

INTRODUCTION
    Thermal expansion coefficient mismatch between different materials, which may lead to excessive stresses at the interface, is the main problem in the metal-ceramic joining. This is also a common problem for ceramic coatings applied on metals, such as thermal barrier coating (TBC). Functionally graded material (FGM) has been suggested to eliminate this problematic interface [1].
    Since material properties, such as thermal, electrical and mechanical performance, are strongly dependent on composition and microstructure, the constitutional profile in a FGM should be properly controlled. Various techniques have been employed to fabricate FGMs [2]. Powder metallurgy (PM) is one of the most useful method in FGM manufacturing, for its versatility allows composition and microstructure in the FGM to be well controlled even from the earliest stage

in the process. For example, slurry dip-coating that form coatings by a slip casting mechanism [3] is a potential method in FGM preforming because of its simplicity and low cost. The keys to successful fabrication of FGMs are (1) well-dispersed ceramic and metal particles in the slip; (2) layer constituents control in the pre-coatings; and (3) avoidance of fracture of FGM layers during the following drying and sintering process. Therefore, the present paper aims to investigate the dip-coating process in ZrO_2-Ni FGM, and explore the main factors to control the microstructure.

EXPERIMENTAL PROCEDURE
Powders of partially stabilized (PSZ, doped with 3mol% yttria) and nickel were chosen as the raw materials (Table I). The suspending liquid was pure ethanol, and polyvinyl butyral (PVB) was used as both dispersant and binder in the powder suspension.

Table I. Raw powders used in the experiment

Powder	Purity (%)	Particle size (μm)	Apparent density (g/cm⁻³)	Specific surface area (m²/g)
PSZ	99.8	<0.5	1.3	10 – 15
Ni	99.6	2.5 - 4	0.8 - 1.0	—

The slurries with different compositions were prepared as follows: (1) The different raw powders and the solvent were milled with varying amounts of PVB dispersant in a polyethylene jar for 12 h; (2) then the binder (also PVB) was added to the slurries and continued to mill for 24 h.

The viscosity of slurries was measured using a rotary viscometer (Model NDJ-1). At the same rotational speed, the particles will disperse more homogeneously with a lower suspension viscosity.

The stainless steel plate (36×36×1.5mm) was adopted as the substrate for dip-coating. The dip-coated FGM specimens were dried at room temperature in air, and then heated in Ar at 400℃ to burn out the organic additives. Finally, the FGM specimens were sintered at 1300°C for 1h under a pressure of 5MPa.

Microstructures of dip-coated and sintered samples were observed by optical microscope and Hitachi S-570 scanning electron microscope (SEM).

RESULTS AND DISCUSSION
Behavior of Suspensions
Particle dispersibility: The effectiveness of PVB as a dispersing agent for ZrO_2-Ni-ethanol suspensions was evaluated by studying the rheological property of the suspension under various conditions. Fig.1 shows slurry viscosity versus

the amount of dispersant for single zirconia or nickel particle slips at 50wt% solids loading. Both of the curves exhibit a distinct viscosity minimum, corresponding to the point of optimal dispersity. Thus, the optimal dispersant content for zirconia and nickel slips can be determined to be 0.5 and 0.3 wt% of powders, respectively.

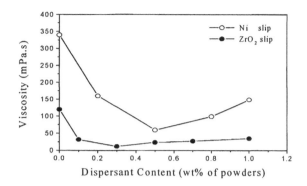

Fig.1 Effect of dispersant content on viscosity of nickel and zirconia slurries at 30 rpm rotary rate of viscometer.

It is known that the adsorption process of dispersant on particle surfaces is irreversible. So, from above result of rheological behavior of zirconia and nickel single particle slips, the zirconia and nickel mixture slips can be prepared by blending the two well-dispersed single particle slips.

Binder concentration: In dip-coating process, the coats were formed by a slip casting mechanism. But the optimal slurry viscosity for tape casting ranges from 500 mPa·s to 25 000 mPa·s [4]. Therefore, it is necessary to adjust binder amount so that the slurry viscosity can be maintained at suitable value.

Experiments indicate that the suitable PVB binder concentration is 4 – 6 wt% for nickel slips and 3 – 5 wt% for zirconia slips. When binder content was less than the lower limit value, the viscosity of both zirconia and nickel slips was below 500 mPa·s and the dip-coatings exhibited some cracks. On the other hand, when binder concentration exceeded the upper limit value, there were more air bubbles in the slurries, which will decrease the density of dip-coatings. Furthermore, the coat layer after each dipping thickened with increasing viscosity, which would raise the coat stress and lead to cracks on coated layers. At the same time, it is more difficult to prevent air bubbles formed in slurries at higher binder content. As a result, the amount of PVB binder added to the slips is fixed at 4wt% of powders in this study.

Characteristics of Dip-coatings
SEM micrographs of nickel and zirconia dip-coatings' surface (both with 4.5wt% PVB additive) are shown in Fig.2. It reveals that the dip-coatings are quite smooth and uniform. Some flocculated particles existed in nickel coating but the particles dispersed very well in zirconia coating.

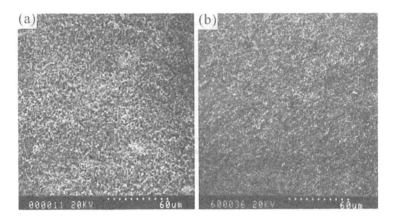

Fig.2 SEM micrographs of the dip-coatings' surfaces:
(a) nickel coating and (b) zirconia coating (both with 4.5wt% PVB).

Fig.3 shows relative density of ZrO_2-Ni single-layer dip-coatings with different nickel content. When nickel content rises from 0 to 20vol%, the relative density decreases from 52 to 44.2%. But the density changes little when nickel content increases over 20vol%. It may be in favor of the FGM dip-coating due to the little shrinkage difference among FGM layers.

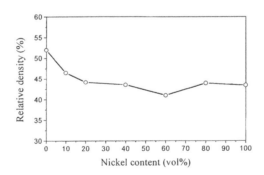

Fig.3 Variation of relative density versus nickel content in
ZrO_2-Ni single-layer dip-coatings

Binder Burnout

The TGA plot shows that PVB decomposes in three stages when heating at 5 $^{\circ}C\,min^{-1}$ in argon(Fig.4). Moreover, the coexistence of powders will decrease the initial decomposition temperature of PVB [5]. Based on these results, the binder removal procedure for ZrO_2-Ni dip-coatings was heating in argon with three stages: below 220 $^{\circ}C$, 220 - 500 $^{\circ}C$ and 500 $^{\circ}C$.

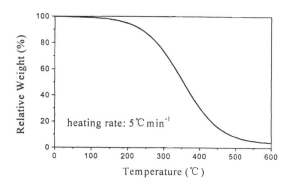

Fig. 4 TGA plot of PVB decomposed in argon

Table II gives the effect of binder burnout heating rate on the surface cracking of ZrO_2-Ni FGM green dip-coatings. According to Table II, the FGM dip-coatings can be free from surface cracking during binder removing process when heating at 5 $^{\circ}C\,min^{-1}$ below 220 $^{\circ}C$, then at 1 $^{\circ}C\,min^{-1}$ between 220 - 500 $^{\circ}C$ and finally heating at 500 $^{\circ}C$ for 24 hours.

Table II. The effect of heating rate on the surface status of dip-coatings

Heating temperature ($^{\circ}C$)	Heating Rate ($^{\circ}C\,min^{-1}$)	Surface status of the FGM dip-coatings
	1	No cracks
	3	No cracks
< 220	5	No cracks
	7	Less cracks
	10	Many cracks
	0.5	No cracks
220-500	1	No cracks
	3	Less cracks
	5	Many cracks

Sintering

The dip-coated specimens were hot pressed at $1300\,^{\circ}C$ under a pressure of 5MPa for 1 hour in nitrogen atmosphere. The microstructure of the sintered ZrO_2-Ni FGM coating is shown in Fig.5, in which the light colored phase is nickel and the dark one is zirconia. It is observed that the microstructure gradually changes from nickel to zirconia with varying composition. The microstructure of each layer in the FGM are quite fine and homogeneous. There are no defects such as small cracks and residual pores were observed in the FGM layers preformed by dip-coating. Both nickel and zirconia components are continuous even on the FGM layer interfaces, which eliminate the macroscopic interface just like that in traditional ceramic/metal joint. Furthermore, the FGM exhibits a microstructural transition, which is typically characteristic of powder metallurgical materials.

Ni $\longrightarrow$ ZrO_2

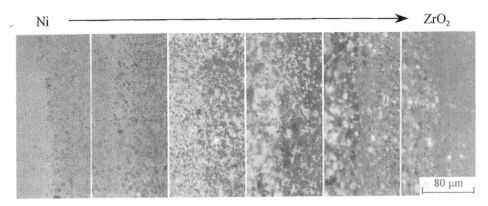

Fig.5 Cross-sectional microstructure of ZrO_2-Ni FGM
by dip-coating and hot-pressing

CONCLUSIONS

(1) In ZrO_2-Ni-ethanol slurry system, a little polyvinyl butyral (PVB) was added as dispersant and binder. The optimal PVB dispersant content is 0.5 and 0.3wt% of powders for zirconia and nickel, respectively; the suitable PVB binder concentration is 4wt% of powders in all slurries.

(2) The green dip-coatings are quite smooth and uniform, in which the PVB binder can be burned out by three stage slow heating in Ar and free from surface cracking.

(3) Microstructural observations of the hot-pressed ZrO_2-Ni-stainless steel FGM compacts, which preformed by dip-coating, reveal that the graded layers were formed on the substrate, in which no defects such as small cracks and residual pores were observed.

REFERENCES

[1] C. R. C. Linma and R. E. Trevisan, "Graded Plasma Spraying of Premixed Metal-Ceramic Powders on Metallic Substrates," *Journal of Thermal Spray Technology,* **6** [2] 199-204 (1997).

[2] J. C. Zhu, Z. D. Yin and Z. H. Lai, "Fabrication and Microstructure of ZrO_2-Ni Functional Gradient Material by Powder Metallurgy," *Journal of Materials Science,* **31** 5829-5834 (1996).

[3] H. Takebe and K. Morinaga, "Fabrication of Zirconia-Nickel Functionally Gradient Materials by Slip Casting and Pressureless-Sintering," *Materials and Manufacturing Processes,* **9** [4] 721-733 (1994).

[4] H. Hellebrand, "Slip Casting"; pp.242 in *Processing of Ceramics Part I,* Edited by J. B. Brook. Weinheim. VCH Press, Weinheim. 1996.

[5] M. Steven, D. C. Paul, E. R. Wendell and H. K. Bowen, "Effect of Oxides on Binder Burnout During Ceramics Processing," *Journal of Materials Science,* **24** 1907-1912 (1989).

PARTICLE SIZE DISTRIBUTIONS IN *in-situ* Al-Al₃Ni FGMs FABRICATED BY CENTRIFUGAL *in-situ* METHOD

PARTICLE SIZE DISTRIBUTIONS IN *in-situ* Al-Al$_3$Ni FGMs FABRICATED BY CENTRIFUGAL *in-situ* METHOD

Koichi MATSUDA and Yoshimi WATANABE
Department of Functional Machinery and Mechanics, Shinshu University
3-15-1 Tokida, Ueda 386-8567, Japan

Yasuyoshi FUKUI
Department of Mechanical Engineering, Kagoshima University
1-21-40 Korimoto, Kagoshima 890-0065, Japan

ABSTRACT

In-situ Al-Al$_3$Ni FGMs were fabricated by the centrifugal *in-situ* method. The microstructures were observed and the composition gradients were studied. A detailed evaluation of particle size distributions within FGMs was also conducted because the particle size distribution in composite materials plays an important role in controlling the mechanical properties. It is revealed that both volume fraction and particle size had graded distributions within the *in-situ* FGMs. Smaller particle size was obtained in case of the larger applied G number and the smaller initial Ni content in an Al-Ni master alloy. It is concluded that the particle size distribution within the FGMs arises from a difference in cooling rate.

INTRODUCTION

Many processing methods have been used to fabricate functionally graded materials (FGMs) [1, 2]. The authors have proposed a centrifugal method as an FGM fabrication method [3-5]. It has been shown that intermetallic compound-dispersed FGMs can be successfully fabricated by the centrifugal method [6-11]. The fabrication of the intermetallic compound-dispersed FGMs by the centrifugal method can be classified into two categories by taking into account the processing temperature [5, 9, 10]. If the process is done at a temperature within liquid-solid co-existing condition, the intermetallic compound remains solid in a liquid alloy. This situation is similar to ceramic-dispersed FGMs, and this method is referred to as a centrifugal solid-particle method [5, 8]. On the other hand, if the process is done at full resolved condition, centrifugal force can be applied during the solidification to both the intermetallic compound and the matrix. This solidification is similar to the production of *in-situ* composites using the crystallization phenomena,

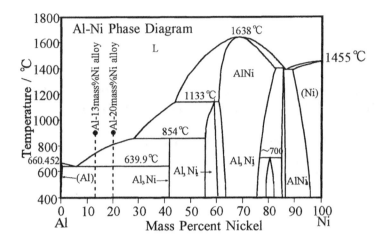

Fig. 1 Al-Ni phase diagram after Massalski [14].

and this method is, therefore, named as a centrifugal *in-situ* method [5-7, 11].

It is well known that the particle size distribution in particle-reinforced or dispersion-strengthened composite materials plays an important factor in achieving superior mechanical properties. Therefore, a detailed knowledge of particle size distributions in the FGMs is required to predict the mechanical properties. Particle size distributions in ceramic-dispersed FGMs made by the centrifugal solid-particle method have been studied [12]. The particle size gradually changed and the average particle size at the outer regions were greater than that at the inner regions. Moreover, the particle size gradient in the FGMs showed a steeper change both the applied G number (ratio of centrifugal force to gravitation) was increased and the mean volume fraction of particles was decreased. These experimental results were well explained by Stokes' law [12]. Similar results have also been reported for an Al-Al₃Ti FGM which was fabricated by the centrifugal solid-particle method [13].

The centrifugal force is applied to a completely liquid phase and the intermetallic compound particles crystallize from the liquid phase directly during casting in case of the *in-situ* FGMs. Since the sizes of the crystallized particles are influenced by the solidification process, a variation of the particle size in the *in-situ* FGMs fabricated by the centrifugal *in-situ* method should be different from those in the FGMs fabricated by the centrifugal solid-particle method. Thus, it must be useful to obtain information on particle size distributions in the *in-situ* FGM both from the viewpoint of a fundamental understanding and the perspective of technological application.

In the present study, *in-situ* Al-Al₃Ni FGMs were fabricated by the centrifugal *in-situ* method using two kinds of master Al-Ni alloys. The microstructures of the *in-situ* Al-Al₃Ni FGMs were observed with an optical microscope, and the composition gradients within the *in-situ* Al-Al₃Ni FGMs were studied. Moreover, particle size distributions were determined by measuring the size of each particle in two-dimensional micrographs. The origin of particle size distributions in the *in-situ* FGMs is discussed considering the experimental results.

EXPERIMENTAL METHODS

The *in-situ* Al-Al₃Ni FGMs were fabricated by the centrifugal *in-situ* method [3-11]. Two master ingots of Al-13 mass% Ni and 20 mass% Ni alloys were used (see Figure 1) and the applied G numbers were 30, 50 and 80 in the experiments. The other casting conditions applied in the present study and notations of the specimens are listed in **Table 1**. Since the melting point of the Al₃Ni intermetallic compound is lower than the processing temperature of 900 °C, the centrifugal force was applied during the solidification of the Al₃Ni intermetallic compound. The products of cast thick-walled rings were dimensions of 90 mm in outer diameter, roughly 25 mm in wall thickness, and 30 mm in length. The details of the *in-situ* Al-Al₃Ni FGMs fabrication process can be found elsewhere [6, 7].

Table 1   Casting conditions used in the centrifugal *in-situ* method.

Initial master ingots	Al-13 mass% Ni (13mass% FGM) and Al-20 mass% Ni (20mass% FGM) alloys
Pouring temperature	900 °C
Preheated mold temperature	600 °C
Weight of molten Al alloys poured	400 g
G number	30, 50 and 80
Cooling	Mold preheating furnace is moved after pouring (Cooling in air until complete solidification)
Cooling rate	0.27 °C/s (G=30), 0.29 °C/s (G=50) and 0.37 °C/s (G=80)

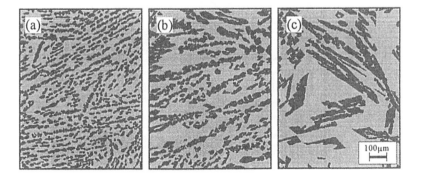

Fig. 2 Typical microstructures of the *in-situ* Al-Al₃Ni FGM (20mass% FGM) fabricated by the centrifugal *in-situ* method. Initial master ingot was Al-20 mass% Ni, and applied G number was 50. (a), (b) and (c) are taken at the outer, the interior, and the inner regions of the ring, respectively.

The microstructures of the fabricated *in-situ* Al-Al₃Ni FGMs were observed using an optical microscope along the plane perpendicular to the rotation axis. The width along thickness direction (centrifugal force direction) was divided into ten regions of equal width, and the volume fractions and particle size distributions of Al_3Ni primary crystal particles in each region were estimated directly from the micrograph.

RESULTS AND DISCUSSION

Microstructures of *in-situ* Al-Al₃Ni FGMs

Typical microstructures of the *in-situ* Al-Al₃Ni FGM (20mass% FGM) fabricated by the centrifugal *in-situ* method are shown in **Fig. 2**. Figures 2 (a), (b) and (c) were taken at the outer, the interior, and the inner regions of the ring, respectively. The black particles are stoichiometric Al_3Ni intermetallic compound and white region is Al matrix. From Fig. 2, the volume fraction of the Al_3Ni primary crystal particles changes depending on the radial position within the casting.

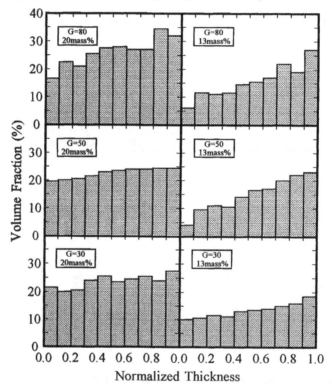

Fig. 3 The distributions of Al_3Ni primary crystal particles in the *in-situ* FGMs. The abscissa represents the position in the radial direction of the ring, normalized by the wall thickness, *i.e.*, 0.0 and 1.0 correspond to the inner and outer peripheries, respectively.

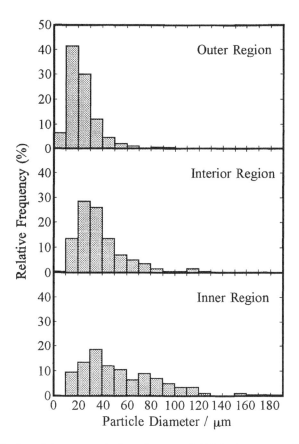

Fig. 4 The distribution of the area-equivalent diameter of Al_3Ni primary crystal particles at each region. Initial master ingot was Al-20mass%Ni alloy, and the FGM was fabricated under G=50.

Figure 3 shows the variations of Al_3Ni primary crystal particles in the *in-situ* FGMs. Here, the abscissa represents the position in the radial direction of the ring, normalized by the wall thickness, *i.e.*, 0.0 and 1.0 correspond to the inner and outer peripheries, respectively. It reveals a tendency that the volume fraction of the Al_3Ni primary crystal particles increases towards the ring's outer region. Moreover, the steeper distribution profiles of the Al_3Ni primary crystal particles are formed in the larger G number specimens. These results are in good agreement with a previous study [7].

Another important feature found in Fig. 2 is that the size of Al_3Ni primary crystal particles varies depending on the radial position. The average area-equivalent diameter of Al_3Ni primary crystal particles at the ring's outer region is greater than that at the inner region as shown in **Fig. 4**.

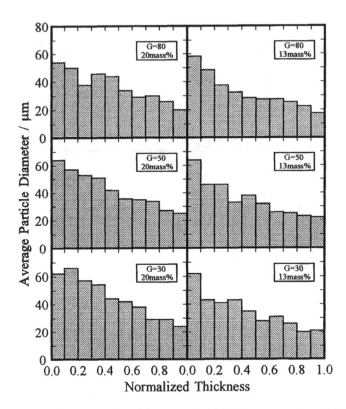

Fig. 5 The average three-dimensional diameter of Al$_3$Ni primary crystal particles in the *in-situ* Al-Al$_3$Ni FGMs as a function of ring position.

In the case of spherical shaped particle, it is known that multiplication of $4/\pi$ by the average two-dimensional diameter gives the average three-dimensional diameter [15]. **Figure 5** shows the average three-dimensional diameter of Al$_3$Ni primary crystal particles as a function of ring position. It must be noted here that the average three-dimensional diameter is distributed in these specimens in a graded manner. Moreover, the larger G number and/or the smaller initial Ni content in the master alloys give a smaller average particle size, especially at the outer regions of the ring. Thus, it is confirmed that the present *in-situ* Al-Al$_3$Ni FGMs contains a gradation of the particle-size, as well as the volume fraction.

Origin of Particle Size Gradient

In this study, it was found that the average three-dimensional diameter was distributed in these specimens in a gradually graded manner, and the average particle size at outer regions was smaller than that at inner regions. In the following, the origin of particle size distributions within the *in-situ* FGMs is discussed.

As previously mentioned, the motion of solid particles, i.e., ceramic particles, in a viscous liquid under centrifugal force obeys Stokes' law, and the average particle size varies gradually depending on the radial position in the FGMs fabricated by the centrifugal solid-particle method. Because larger particles migrate faster than smaller particles and the average particle size at outer region of the ring is larger than that at inner region in that case. Such results conflict with the present results that the average particle size at outer region is smaller than that at inner region. Thus, the origin must be different in case of the FGMs fabricated by the centrifugal solid-particle method, which can be explained according to the Stokes' law.

Another possibility is that the particle size may be influenced by a local chemical composition. A process of the graded composition formation in the *in-situ* FGM, e.g., Al-Cu system, under the centrifugal force has been discussed recently [11]. It is concluded that a chemical composition gradient is formed before the crystallization of the primary crystal due to the density difference between co-existing atoms, e.g., Al and Cu, in the liquid state. The primary crystal in the matrix appears depending on local chemical composition, and the primary crystals migrate according to density difference, and a further compositional gradient is formed. We will apply the above discussion for Al-Ni system. Since the partial separation of Al and Ni in the liquid state occurs during the early stage of the centrifugal casting, the chemical composition of molten alloy at the ring outer becomes Ni-rich before the crystallization of the Al_3Ni primary crystal. Then, the crystallization of primary crystal should occur at higher temperature at ring outer region, because the initial master alloys are hyper-eutectic compositions, as shown in Fig. 1. Consequently, the larger particles should be formed at the outer part of the ring and *vice versa*. This conclusion contradicts with the present experiment. However, the above discussion is partially supported by the experimental facts that the particle sizes in the 13mass%Ni FGMs are smaller than those in the 20mass%Ni FGMs, as shown in Fig. 5. At any rate, it is impossible to explain why the smaller particle is formed at ring outer region from a theory of the gradation of the chemical composition.

The third possibility is based on the well-known concept that the particle size of the crystallized particles is influenced by the solidification process. The temperature at the outer region of ring is lower than that at the inner region in case of centrifugal casting [16]. Therefore, the graded distribution of Al_3Ni primary crystal particles should be caused by the difference in cooling rate within the ring. In this study, cooling rate was found to increase with G number, as shown in Table 1. As a consequence, the particle size at the ring outer region becomes smaller for the specimens fabricated under a larger G number. This means that the difference in the particle size distribution within the *in-situ* FGM should be caused by a difference in cooling rate.

CONCLUSIONS

(1) The average three-dimensional diameter of Al_3Ni primary crystal particles within the *in-situ* Al-Al_3Ni FGMs varies gradually. Both volume fraction and particle size have graded distributions, which can be manufactured by the centrifugal *in-situ* method.

(2) The larger G number and the smaller initial Ni content in Al-Ni master alloy give a small particle size.

(3) The particle size distribution within the *in-situ* FGM depended on the difference in cooling rate.

Acknowledgment
This work was supported by Grant-in-Aid for COE Research (10CE2003) by the Ministry of Education, Science, Sports and Culture of Japan.

REFERENCES

[1] S. Suresh and A. Mortensen, "Fundamentals of Functionally Graded Materials, Processing and Thermomechanical Behaviour of Graded Metals and Metal-Ceramic Composites," IOM Communications Ltd, London, 1998.

[2] Y. Miyamoto, W. A. Kaysser, B. H. Rabin, A. Kawasaki, and F. G. Ford, (Editors), "Functionally Graded Materials: Design, Processing and Applications," Kluwer Academic Publishers, Boston, 1999.

[3] Y. Fukui, "Fundamental Investigation of Functionally Gradient Material Manufacturing System using Centrifugal Force," *JSME Int. J. Series III*, 34 [1] 144-48 (1991).

[4] Y. Watanabe, N. Yamanaka and Y. Fukui, "Control of Composition Gradient in a Metal-Ceramic Functionally Graded Material Manufactured by the Centrifugal Method," *Composites Part A*, **29A** [5-6] 595-601 (1998).

[5] Y. Watanabe and Y. Fukui, "Fabrication of Functionally-Graded Aluminum Materials by the Centrifugal Method," *Aluminum Trans.*, 2 [2] 195-209 (2000).

[6] Y. Fukui, K. Takashima and C. B. Ponton, "Measurement of Young's Modulus and Internal Friction of an *in situ* Al-Al$_3$Ni Functionally Gradient Material," *J. Mater. Sci.*, 29 [9] 2281-88 (1994).

[7] Y. Fukui, N. Yamanaka, Y. Watanabe and Y. Oya-Seimiya, "Fabrication of in-situ Al-Al$_3$Ni Functionally Graded Material by Centrifugal Method," *J. Jpn Light Met.*, 44 [11] 622-27 (1994).

[8] Y. Watanabe, N. Yamanaka and Y. Fukui, "Orientation of Al$_3$Ti Platelets in Al-Al$_3$Ti Functionally Graded Material Manufactured by Centrifugal Method," *Z. Metallkd.*, 88 [9] 717-21 (1997).

[9] Y. Fukui, H. Okada, N. Kumazawa and Y. Watanabe, "Near Net Shape Forming of Al-Al$_3$Ni FGM over Eutectic Melting Temperature," *Metall. Mater. Trans. A*, **31A** [10] 2627-2636 (2000).

[10] Y. Watanabe and T. Nakamura, "Microstructures and Wear Resistances of Hybrid Al- (Al$_3$Ti+ Al$_3$Ni) FGMs by Centrifugal Method," *Intermetallics*, 9 [1] 33-43 (2001).

[11] S. Oike and Y. Watanabe, "Development of *in-situ* Al-Al$_2$Cu Functionally Graded Materials by a Centrifugal Method," *Inter. J. Mater. Prod. Tech.*, 16 [1-3] 40-49 (2001).

[12] Y. Watanabe, A. Kawamoto and K. Matsuda, "Particle Size Distributions of Functionally Graded Materials Fabricated by Centrifugal Method," to be submitted.

[13] K. Yamashita, I. Fujimoto, S. Kumai and A. Sato, "Plastic Deformation of D0$_{22}$ Ordered Al$_3$Ti in a Centrifugally Cast Al-Al$_3$Ti Composite," *Mater. Trans. JIM*, 39 [8] 824-33 (1998).

[14] T. B. Massalski, (Editor) "Binary Alloy Phase Diagrams," ASM International, Materials Park, 1988.

[15] R. T. DeHoff, "Measurement of Number and Average Size in Volume"; pp. 128-48 in *Quantitative Microscopy*, Edited by R. T. DeHoff, F. N. Rhines. McGraw-Hill Book Company, New York, 1968.

[16] C. G. Kang and P. K. Rohatgi, "Transient Thermal Analysis of Solidification in a Centrifugal Casting for Composite Materials Containing Particle Segregation," *Metall. Mater. Trans. B*, **27B** [4] 277-85 (1996).

Properties Modeling

BOUNDARY INTEGRAL ANALYSIS FOR FUNCTIONALLY GRADED MATERIALS

L. J. Gray, T. Kaplan and J. D. Richardson
Computer Science and Mathematics Division
Oak Ridge National Laboratory
Oak Ridge, TN 37831-6367

Glaucio H. Paulino
Department of Civil and Environmental Engineering
University of Illinois
Urbana-Champaign, IL 61801-2352

ABSTRACT

Free space Green's functions are derived for functionally graded materials (FGM's) in which the thermal conductivity varies exponentially in one coordinate. Closed form expressions are obtained for the time independent (steady state) equation in three dimensions. The corresponding boundary integral equation formulation for this problem is derived and has been implemented numerically using a Galerkin approximation. The results of test calculations are in excellent agreement with exact solutions and finite element simulations.

INTRODUCTION

Boundary integral analysis could potentially provide an effective computational method for designing specific FGM systems, and for understanding FGM properties. However, the formulation in terms of integral equations relies upon having, as either a closed form or a computable expression, a fundamental solution (Green's function) of the partial differential equation. Application of the boundary integral technique has therefore been limited, almost exclusively, to homogeneous (or piece-wise homogeneous) media.

The fundamental solutions traditionally employed in boundary integral analysis are 'free space' Green's functions, satisfying the appropriate differential equation everywhere in space except at the site where a point load driving force is applied. Derivations for some of the basic Green's functions can be found in [1, 2]. In this paper we derive the free space fundamental solution for the FGM Laplace equation, assuming that the thermal conductivity varies exponentially, $Ae^{\alpha z}$. The corresponding boundary integral equation, which turns out to be somewhat different from the homogeneous media case, is also obtained.

STEADY STATE ANALYSIS

Time independent isotropic diffusion analysis is governed by the equation

$$\nabla \cdot (k\nabla\phi) = 0 \ . \tag{1}$$

Here $\phi = \phi(x, y, z)$ is the temperature function, and we assume the functionally graded material is defined by the thermal conductivity

$$\kappa(x, y, z) = \kappa(z) = \kappa_0 e^{-2i\alpha z} \ , \tag{2}$$

where α is real. This assumption of a purely imaginary exponent is apparently necessary for the derivation that follows. Nevertheless, once the solution is obtained it is readily seen to be valid for any complex α and a real solution is retrieved by setting $\alpha = i\beta$. Substituting into Eq. (1), the temperature satisfies

$$\nabla^2\phi - 2i\alpha\phi_z = 0 \ , \tag{3}$$

where ϕ_z denotes the derivative with respect to z.

The corresponding Green's function equation can be derived by constructing the boundary integral formulation of Eq. (3). Thus, following the standard procedure Eq. (3) is multiplied by an arbitrary function $f(x, y, z) = f(Q)$ and integrated over a bounded volume V. Integrating by parts, and denoting the boundary of V by Σ,

$$
\begin{aligned}
0 &= \int_V f(Q) \left(\nabla^2\phi(Q) - 2i\alpha\phi_z(Q)\right) dV_Q \\
&= \int_\Sigma \left\{ f(Q)\frac{\partial}{\partial n}\phi(Q) - \phi(Q)\frac{\partial}{\partial n}f(Q) - 2i\alpha n_z(Q)\phi(Q)f(Q) \right\} dQ \\
&\quad + \int_V \phi(Q) \left(\nabla^2 f(Q) + 2i\alpha f_z(Q)\right) dV_Q \ ,
\end{aligned}
\tag{4}
$$

where $n(Q) = (n_x, n_y, n_z)$ is the unit outward normal for Σ. Thus, by requiring that $f(Q) = G(P, Q)$ satisfy the Green's function equation

$$\nabla^2 G(P, Q) + 2i\alpha G_z(P, Q) = -\delta(Q - P) \ , \tag{5}$$

where δ is the Dirac delta function, the remaining volume integral becomes simply $-\phi(P)$ and we obtain the boundary integral equation

$$\phi(P) \; + \; \int_{\Sigma} \phi(Q) \left(\frac{\partial}{\partial n} G(P,Q) + 2i\alpha n_z G(P,Q) \right) dQ =$$
$$\int_{\Sigma} G(P,Q) \frac{\partial}{\partial n} \phi(Q) \, dQ \; . \tag{6}$$

Note that this differs in form from the usual integral statements by the presence of the additional term multiplying $\phi(Q)$.

Let $\hat{f}(\omega)$ denote the Fourier transform of a function $\mathcal{F}(Q)$,

$$\hat{f}(\omega) = \int_{\mathcal{R}^3} f(Q) e^{-i\omega \cdot Q} \, dQ \tag{7}$$

where $\omega = (\omega_x, \omega_y, \omega_z)$ is the transform variable. Transforming Eq. (5) and solving for $\hat{G}(\omega)$ (the transform of G with respect to Q), yields

$$\hat{G}(\omega) = \frac{e^{-i\omega \cdot P}}{\omega^2 + 2\alpha \omega_z} \; , \tag{8}$$

where $\omega^2 = \omega \cdot \omega$, and the dot represents the inner product. Applying the inverse transform,

$$G(P,Q) = \frac{1}{(2\pi)^3} \int_{\mathcal{R}^3} \frac{e^{i\omega \cdot (Q-P)}}{\omega^2 + 2\alpha \omega_z} \, d\omega \; . \tag{9}$$

where $d\omega$ is shorthand for $d\omega_x d\omega_y d\omega_z$. Changing variables $\omega_z \to \omega_z - \alpha$ and setting $R = Q - P$, $R_z = Q_z - P_z$, we obtain

$$G(P,Q) = \frac{1}{(2\pi)^3} e^{-i\alpha R_z} \int_{\mathcal{R}^3} \frac{e^{i\omega \cdot R}}{\omega^2 - \alpha^2} \, d\omega \; , \tag{10}$$

which can be conveniently split into two terms,

$$G(P,Q) = \frac{e^{-i\alpha R_z}}{(2\pi)^3} \left[\int_{\mathcal{R}^3} \frac{e^{i\omega \cdot R}}{\omega^2} \, d\omega + \alpha^2 \int_{\mathcal{R}^3} \frac{e^{i\omega \cdot R}}{\omega^2 (\omega^2 - \alpha^2)} \, d\omega \right] \; . \tag{11}$$

The first integral is Eq. (9) with $\alpha = 0$, and is therefore recognized as the Green's function for the constant κ Laplace equation, the point source potential:

$$\frac{e^{-i\alpha R_z}}{(2\pi)^3} \int_{\mathcal{R}^3} \frac{e^{i\omega \cdot R}}{\omega^2} \, d\omega = \frac{e^{-i\alpha R_z}}{4\pi r} \; , \tag{12}$$

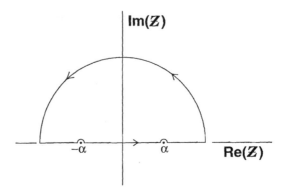

Figure 1: Contour path in the complex plane.

where $r = \|R\|$ is the distance between Q and P.

To evaluate the second term in Eq. (11), it is convenient to employ spherical coordinates (ρ, θ, ψ), with the axis defining the pole $\psi = 0$ taken as the direction R/r instead of the z-axis. The limits of integration are $-\infty < \rho < \infty$, $0 \leq \psi \leq \pi/2$, and $0 \leq \theta \leq 2\pi$. Noting that $\omega \cdot R = \rho r \cos(\psi)$ and that the integrand is independent of θ, we therefore obtain

$$\frac{\alpha^2 e^{-i\alpha R_z}}{(2\pi)^2} \int_0^{\pi/2} \sin(\psi)\, d\psi \int_{-\infty}^{\infty} \frac{e^{i\rho r \cos(\psi)}}{\rho^2 - \alpha^2}\, d\rho \ . \tag{13}$$

Using the contour shown in Fig. 1, the ρ integration is a straightforward exercise in residue calculus, yielding

$$\int_{-\infty}^{\infty} \frac{e^{i\rho r \cos(\psi)}}{\rho^2 - \alpha^2}\, d\rho = -\frac{\pi}{\alpha} \sin\left(\alpha r \cos(\psi)\right) \ . \tag{14}$$

The final integration,

$$-\frac{\pi}{\alpha} \int_0^{\pi/2} \sin(\psi) \sin\left(\alpha r \cos(\psi)\right)\, d\psi \tag{15}$$

follows from a simple change of variables, and thus the second term is seen to be

$$\frac{e^{-i\alpha R_z} \cos(\alpha r)}{4\pi r} - \frac{e^{-i\alpha R_z}}{4\pi r} \ . \tag{16}$$

Including Eq. (12), we find the simple result

$$G(P, Q) = \frac{e^{-i\alpha R_z} \cos(\alpha r)}{4\pi r} \ . \tag{17}$$

Although this result was derived assuming that α is real, it is a simple matter to now check, by direct calculation, that Eq. (17) satisfies Eq. (5) for any complex α. It is useful to observe that

$$G(P,Q) = e^{-i\alpha R_z} \frac{e^{-i\alpha r}}{4\pi r} \tag{18}$$

is an equally valid solution of Eq. (5) for α real. Moreover, the added $\sin(\alpha r)/r$ term is regular as $r \to 0$, and thus does not alter the delta function at $Q = P$. Replacing α by $i\beta_0$, β_0 real, we obtain

$$G(P,Q) = \frac{e^{\beta_0(r+R_z)}}{4\pi r} \tag{19}$$

as the Green's function for $\kappa(z) = e^{2\beta_0 z}$.

1   Numerical Examples

The three-dimensional steady state fundamental solution has been incorporated into a boundary element (BEM) algorithm. The integral equation Eq. (6) is numerically approximated via the Galerkin method [1], together with standard six-node iso-parametric quadratic triangular elements to interpolate the boundary and boundary functions. For the numerical examples, the conservation equation given in Eq. (1) will be taken as energy conservation in a functionally graded media under the condition of steady state heat conduction without volumetric generation.

1.1   Unit Cube: Constant ϕ on Two Planes

For the first example problem, the geometry is a unit cube with the origin of a Cartesian system fixed at one corner. The thermal conductivity in this example is taken to be

$$\kappa(z) = \kappa_o e^{2\beta z} = 5e^{3z}. \tag{20}$$

The top face of the cube $[z = 1]$ is maintained at a temperature of $T_1 = 100°$ while the bottom face $[z = 0]$ is maintained at $T_0 = 0°$. The remaining four faces are insulated (zero normal flux).

With these imposed boundary conditions, the problem is one-dimensional in nature (independent of x and y) and has a simple exact solution. The temperature field and heat flux vectors are given by

$$\phi = T_1 \frac{1 - e^{-2\beta z}}{1 - e^{-2\beta}}$$

$$\mathbf{q} = -\kappa_o T_1 \frac{2\beta}{1 - e^{-2\beta}} \widehat{\mathbf{k}} \tag{21}$$

where $\widehat{\mathbf{k}}$ is a unit vector in the $z-$direction.

Numerical solutions for the temperature profile for this problem are shown in Fig. 2. The plot also includes the results obtained from an FEM simulation

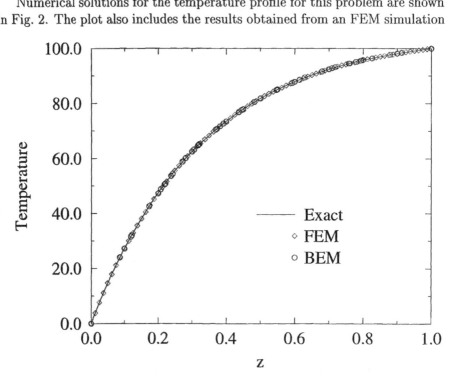

Figure 2: Temperature distribution in the FGM unit cube.

using a commercial package. In the FEM simulation, forty homogeneous layers were used to approximate the continuous grading; the conductivity of each layer was computed from Eq. 5 where z was taken as the $z-$coordinate of the layer's centroid. The FEM elements which were used were 20-node quadratic brick elements and each of the forty layers contained 400 brick elements. In the boundary element solution, a uniform grid consisting of isoceles right triangles, each leg having length 0.1, was employed. This results in a total of 1200 elements.

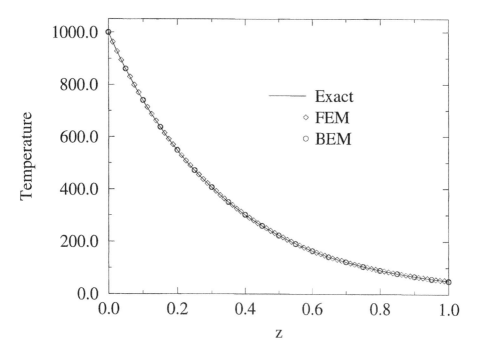

Figure 3: Temperature distribution in the FGM unit cube along the edge $[x = 1, \ y = 1]$.

1.2 Unit Cube: Linear Heat Flux

The second problem once again employs the unit cube previously described with the thermal conductivity again given by Eq. 5. However, the cube is now insulated on the faces $[y = 0]$ and $[y = 1]$, while uniform heat fluxes of $5000 \left[\frac{\text{POWER}}{\text{AREA}}\right]$ are added and removed, respectively, at the $[x = 1]$ and $[x = 0]$ faces. In addition, the $[z = 0]$ face is specified to have an x–dependent temperature distribution $T = 1000 \ x^\circ$ and at $[z = 1]$ a normal heat flux of $q = 15000 \ x$ is removed. The analytic solution for this problem is

$$
\begin{aligned}
T &= 1000 \ xe^{-3z} \\
\mathbf{q} &= -5000\widehat{\mathbf{i}} + 15000 \ x\widehat{\mathbf{k}}
\end{aligned}
\tag{22}
$$

where $\widehat{\mathbf{i}}$ is a unit vector in the x–direction.

The results of the numerical simulations for the temperature distributions along an edge are shown in Fig. 3. The FEM and BEM discretizations for this problem are identical to those used in the previous example.

1.3 FGM Rotor

The final numerical example will be a rotor with eight mounting holes. Due to the anticipated axisymmetric nature of the problem only one-eighth of the rotor will be modeled as drawn Fig. 4. The grading direction for the rotor

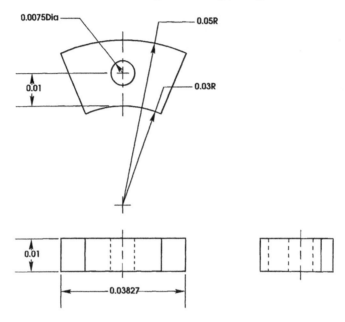

Figure 4: Geometry of the functionally graded rotor.

is parallel to its line of axisymmetry which will be taken as the z-axis. The thermal conductivity for the rotor will vary according to

$$k(z) = 20e^{330z} \ \frac{\text{W}}{\text{m} \cdot {}^\circ\text{K}}. \tag{23}$$

The BEM solution will be compared with an FEM solution obtained from the same package used in the previous examples using ten-node tetrahedral elements to handle the geometric complexity of the rotor. Due to resource limitations, the FEM model was limited to 12 layers which resulted in the rather crude conductivity profile shown in Fig. 5.

A schematic for the thermal boundary conditions is shown in Fig. 6. The

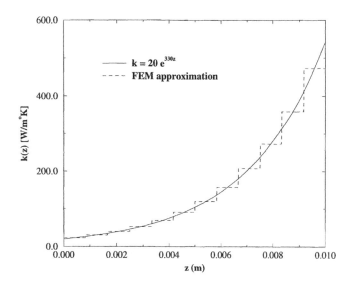

Figure 5: Thermal conductivity profiles for the computational models.

temperature is specified along the inner and outer radii and a uniform heat flux of 5×10^5 W/m^2 is added on the bottom surface where $z = 0$. All other surfaces are insulated as shown.

The temperature distribution around the hole is shown in Fig. 7. The angle θ is measured from a line passing through the line of axisymmetry for the problem and the center line of the hole as shown in Fig. 7. Though the surface nodal positions in the two models were not coincident in general, the plot shows a strong agreement in the two solutions.

The radial heat flux along the line shown as the interior corner in Fig. 6 is plotted in Fig. 8. The negative sign indicates that the flow of heat is toward the interior of the rotor. A limitation on the use of piece-wise constant conductivities in FEM models may be evident in the plot where the FEM nodal value at $z = 0.01$ seems to fall out of line with the other values on the curve. The behavior should be fully expected, however, given the local error associated with the piece-wise constant approximation seen near $z = 0.01$ in Fig. 5. As should also be expected, the nodal flux values from the BEM solution seem to fall onto a single curve even in the region of the steepest conductivity gradient.

ACKNOWLEDGMENTS

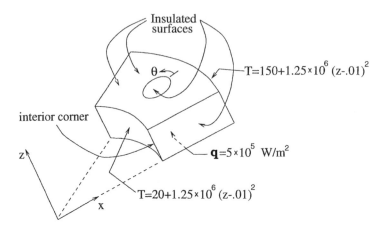

Figure 6: Thermal boundary conditions on the rotor.

This research was supported in part by the Applied Mathematical Sciences Research Program of the Office of Mathematical, Information, and Computational Sciences, U.S. Department of Energy under contract DE-AC05-00OR22725 with UT-Battelle, LLC. Additional support was provided by the Laboratory Directed Research and Development Program of the Oak Ridge National Laboratory. G. H. Paulino acknowledges support from the National Science Foundation under grant CMS-9713008 (Mechanics and Materials Program).

REFERENCES

[1] M. Bonnet. *Boundary Integral Equation Methods for Solids and Fluids.* Wiley and Sons, England, 1995.

[2] G. Barton. *Elements of Green's Functions and Propagation.* Oxford University Press, Oxford, 1999.

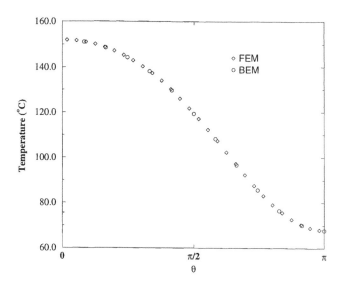

Figure 7: Temperature distribution around the hole on the $z = 0.01$ surface.

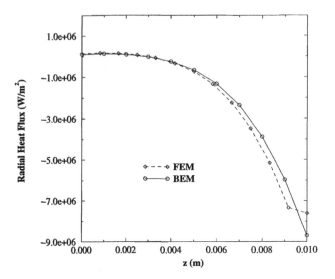

Figure 8: Radial heat flux along the inside corner.

COMPUTATIONAL MICROMECHANICS OF FUNCTIONALLY GRADED MATERIALS

Peng-Cheng Zhai, Qing-Jie Zhang and Run-Zhang Yuan
State Key Lab. of Advanced Technology for Materials Synthesis and Processing
Wuhan University of Technology, Wuhan, 430070, China

Shin-Ichi Moriya
National Aerospace Lab.
Kakuda Research Center, Miyagi, 981-15, Japan

ABSTRACT
In this paper, the computational micromechanical method, which is combined with the Digital Image Processing Technique, the Mesh Auto Generation Technique and the Finite Element Analysis Method, is developed for prediction of the effective properties of the interlayers of functionally graded materials and for the local response analysis of functionally graded materials under special loads. Emphasis is put on the local response analysis. As a numerical example, the microscopic responses of a ZrO2/Ni FGM coating under thermal shock load are analyzed by the computational micromechanical method.

INTRODUCTION
Micromechanics is a foundation of the design of Functionally Graded Materials (FGMs). The purposes of micromechanics are: (1) determination of the effective properties according to the microstructure and micro properties of the materials, and (2) determination of the response of materials under special loading in micro scale. In this paper, the Computational Micromechanical Method (CMM) is used to calculate the effective stress-strain curves of the interlayers of ZrO_2/Ni FGM coatings and the microscopic analysis of the FGM coating under thermal shock loading.

The CMM is combined with the Digital Image Processing Technique, the Mesh Auto Generation Technique and the Finite Element Analysis Method to deal with the effective properties' problems and the micro response problems. The fulfilling procedures of the CMM are briefly restated as follows:

Step 1: Digital Image Processing

In the CMM, the analysis is based on the real microstructure of materials. For this subject, SEM pictures of each interlayers of the FGM are required. After converting the SEM pictures into a digital format, such as JPEG, GIF or BMP format, image processing software is used to identify the interfaces or boundaries between the component materials. The geometry of each interlayer is obtained in this step.

Step 2: Mesh Generation

In this step, results of Step 1 are introduced to a pre-processing Finite Element Analysis software. According to the characteristics of the geometry of each interlayer, a triangle or quad mesh is auto generated. In order to analysis the micro response of the FGMs under special loading, these meshes of interlayers will be connected carefully and will construct a macroscopic model for finite element analysis.

Step 3: Finite Element Analysis

According to the purpose of the analysis, proper boundary conditions and initial conditions are applied on the model obtained in Step 2. For the purpose of predicting the effective properties, the boundary conditions and initial conditions must simulate the real experimental process. For the purpose of determining of the micro response, the conditions must simulate the real thermal and mechanical loading of the materials.

In the second section of this paper, the CMM is described in detail especially focussing on the digital image processing procedure. In the third section, the effective stress-strain curves of the interlayers of a ZrO2/Ni FGM coating are calculated. The microscopic responses of the FGM coating under thermal shock loading are analyzed in the fourth section. In that section, the macroscopic responses are also analyzed by a typical analysis method. The differences of the responses between the microscopic and macroscopic analysis are discussed and some suggestions for the materials optimum design have been made.

METHOD DESCRIPTION

For the purpose of exact determination the effective properties and the microscopic responses of materials, the key of the analysis by CMM is the generation of the microscopic model of the materials. There are two methods to obtain a rational microscopic model of the material:

1) If the material has been fabricated, a photo of the microstructure such as a SEM picture will serve as the microscopic model base;

2) If the material has not been fabricated, an imagined configuration constructed by a random mathematical algorithm and an automatic geometry formation technique will serve as the microscopic model base.

In this imagined configuration, the distributions, orientations and shapes of the inclusions are constructed according to experience.

In Ref.[1] and [2], the present authors have given a clear description of the second way. In the present paper, all microscopic models are based on SEM photos.

Fig.1 shows the SEM images of the interlayers of a ZrO_2/Ni FGM coating fabricated by Low Pressure Plasma Spray(LPPS) processing. These images were made by the National Aerospace Lab. of Japan Ref.[3].

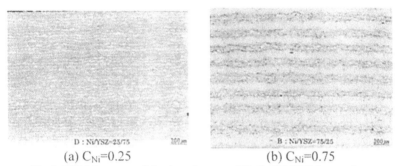

(a) C_{Ni}=0.25   (b) C_{Ni}=0.75

Fig.1 SEM images of the interlayers of ZrO_2/Ni FGMs coating

These SEM images are stored as gray-level pictures in BMP format. The modeling procedures based on the BMP images are as follow:

Step 1: Definition of the Threshold

In order to identify the interfaces of the two component materials, an exact threshold value of the gray-level must be defined. In the present study, the gray-levels of each pixel are counted except the boundary pixels of the image. According to the volume fractions of Ni and ZrO2, a suitable threshold of gray-level can be defined.

Step 2: Selection of the Representative Area

A representative area is selected to represent the material. In the present paper, the area is 200×200μm. The gray-level values of the pixels in the representative area will be counted again according to the threshold value in order to ensure the correct volume fraction.

Step 3: Identification of the Interface

The constant threshold method is used to identify the pixels on the interfaces between the Ni and the ZrO2 materials according to the threshold value defined in Step 1. Then these interfaces are simplified to lines and arcs by a curve fit algorithm. These lines and arcs are stored in a formatted file and can be imported to the preprocess Finite Element Analysis software. In the present paper, AutoCAD DXF file are used. The geometry of the microscopic model is obtained after this step.

After import the geometry of the microscopic model, the mesh can be generated by the preprocess software. Fig.2 shows the meshes of the microscopic model corresponding to the materials shown in Fig.1.

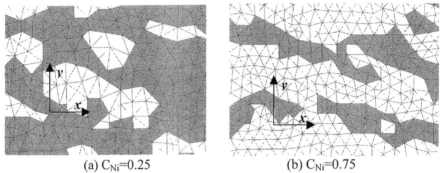

(a) C_{Ni}=0.25   (b) C_{Ni}=0.75

Fig.2 The mesh for the microstructure in Fig.1 in a local exploded form

EFFECTIVE STRESS-STRAIN CURVES

The effective elastoplastic stress-strain curves of the interlayers of a ZrO2/Ni FGM coating are calculated by CMM. In the present study, the metal Ni is assumed to be a linear strain hardening metal, with an initial yield stress of 150 MPa and hardening parameter of 667 MPa. The Young's modulus and Poisson's ratio are 207 GPa and 0.32, respectively. The ceramic is assumed to remain elastic, with a Young's modulus of 186 GPa and Poisson's ratio of 0.337. The computational results are plotted in Fig.3. In this paper, two effective elastoplastic stress-strain curves are calculated for each interlayer, these two curves correspond to the properties along the X and Y axes that are displayed in Fig.2. In the present study, the influence of the temperature on the materials' properties is neglected to simplify the problem.

From Fig.3, the effective stress-strain relationship of an interlayer is in a curve form when the metal is considered as a elastoplastic material. At different strain values, the secant modulus is quite different. For example, for the interlayer with 75% Ni, the secant modulus in the X direction is 164.4 GPa when the strain is 500μ. When the strain is 1000μ, the respective value is 139.7GPa. The secant modulus will decrease rapidly as the strain increases. On the other hand, the effective stress-strain relationships are also different in different directions. It is obvious that the stiffness along the thickness is smaller. For the interlayer with 75% Ni, the difference of secant modulus between two directions is about 4% for almost each strain value.

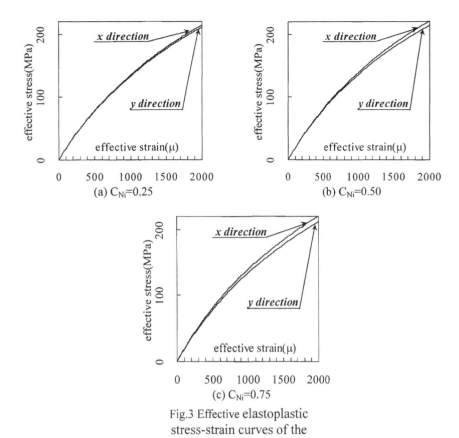

Fig.3 Effective elastoplastic
stress-strain curves of the

MICROSCOPIC ANALYSIS FOR THERMAL SHOCK LOADING

In this section, the thermal and thermoelastic responses of the ZrO2/Ni FGM coating under thermal shock loading are analyzed by the CMM and the local stress is obtained.

Consider an infinite FGM coating, the thickness of all interlayers and the pure ceramic layer are 0.2mm. The heated ceramic surface is simulated by a uniformly distributed heat flux input. The heat flux magnitude q_0 is $5MW \cdot m^{-2}$ and the heated time is 1.0 seconds. The metal surface is simulated as an ambient boundary, the ambient temperature and thermal transfer coefficient are 298K and $10^4 Wm^{-2}K^{-1}$, respectively. The initial temperature is assumed to be 298K and the reference temperature (stress-free temperature) is also assumed to be 298K. Due to the characteristics in the thermal load and geometry, a planar strain analysis for the response is employed. In the analysis, the microscopic model is constructed by the models shown in Fig.2. The thermal properties of Ni and ZrO2 are listed in Table 1. In this section, the materials are assumed to remain elastic.

Table 1: Thermal properties of ZrO2 and Ni

	λ(W/mK)	C_P(J/kgK)	$\rho(10^3 \text{kg/m}^3)$	$\alpha(10^{-6} \text{K}^{-1})$
ZrO$_2$	3.1	455.5	5.754	9.93
Ni	90.5	439.0	8.900	13.70

A macroscopic analysis is also made in this section by a layered model. In the macroscopic analysis, each interlayer is simplified to an isotropic material and the properties of each interlayer are obtained from Ref.[3]. The results by the layered model are macroscopic and global ones.

The temperature response is indicated in Figure 4. In Fig.4 and following figures, interface 1 means the interface between the pure ceramic and the first interlayer with 25% Ni. Comparing the results in Fig.4 (a) and (b), the characteristics of the temperature response by the two different models are very similar. There exists differences between the two results on the temperature value. The maximum temperature of the macroscopic analysis is about 1550K and the microscopic analysis is about 1350K. The difference is about 200K. This disagreement is because the material porosity was neglect in the microscopic model.

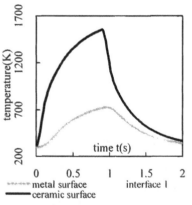
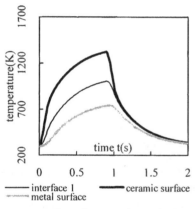

(a) macroscopic analysis   (b) microscopic analysis
Fig.4 The temperature response of the materials

The equivalent von Mises stress at the ceramic surface, metal surface and first interface are plotted in Fig.5 as a function of time. The global stress by macroscopic analysis is higher than the local one by the microscopic analysis, especially for the ceramic surface and the first interface. At the metal surface, the results by the two models are almost the same.

To describe the stress distribution in detail, the equivalent von Mises stress distribution of the interlayer with 25% Ni is given in Fig.6. From Fig.6, the local stress distribution is very complex and quite different to the global one. In the macroscopic analysis, the maximum stress appears at the interface between the pure ceramic and the first interlayer. According to the microscopic analysis, the maximum local stress doesn't appear at the interface but at the micro-interface between the ceramic and the metal.

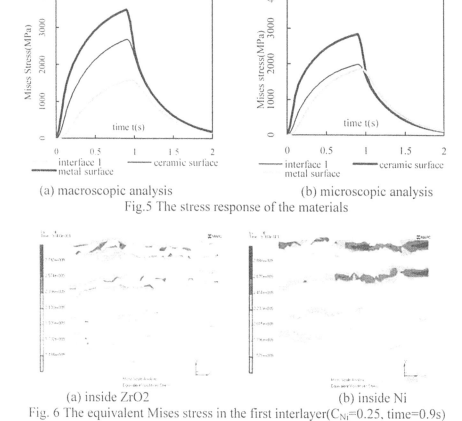

(a) macroscopic analysis   (b) microscopic analysis

Fig.5 The stress response of the materials

(a) inside ZrO2   (b) inside Ni

Fig. 6 The equivalent Mises stress in the first interlayer(C_{Ni}=0.25, time=0.9s)

According to the macroscopic analysis, the normal stress σ_y which is in the direction of the thickness is smaller than 50MPa and can be ignored in the design of the materials. Fig.7 gives the corresponding results calculated by microscopic analysis. From Fig.7, it's obviously that distribution of the normal stress σ_y is also very complex. At point on the micro interface between the ceramic and the metal, the stress σ_y is tensile   and the value is about 300MPa. Since the tensile strength of the interface is lower than the strength of the ceramic and metal, this tensile stress maybe cause the interface to crack. According to the results by CMM, the delamination of FGMs under thermal shock loading could be explained more reasonably.

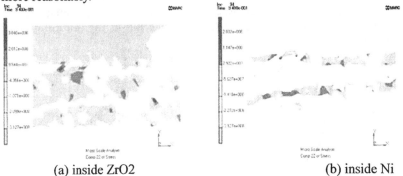

(a) inside ZrO2   (b) inside Ni

Fig. 7 The stress σ_y in the first interlayer(C_{Ni}=0.25, time=0.94s)

CONCLUSIONS

The computational micromechanical method is used to investigate the effective property of the interlayers of FGMs and the microscopic response of FGM coating under thermal shock loading. The CMM is based on the real microstructure of the materials and the finite element method. With the help of the Digital Image Processing Technique, the CMM can simulated the microstructure more exactly. The analysis results show that the CMM is a powerful method for the calculating of effective properties and local stress analysis.

ACKNOWLEDGMENT

This work was supported by the National Natural Science Foundation of China and the ministry of Education of China.

REFERENCES

[1] P.C.Zhai, Q.J.Zhang and R.Z.Yuan, "A Random Microstructure Finite Element Method For Effective Properties Of Functionally Graded Materials", Materials Science Forum, Vol.308-311, 995-1000, 1999

[2] P.C.Zhai, Q.J.Zhang and R.Z.Yuan, "Random microstructure finite element method and its verification for effective properties of composite materials", Journal of Wuhan University of Technology (Mater. Sci. Ed.), 15(1), 6-12, 2000.

[3] Nippon Steel Technology Research(NSTR), "Experimental report for the properties of ZrO2/Ni FGM coating fabricated by LPPS ", National Aerospace Lab., 1997(in Japanese)

DETERMINATION OF FGM PROPERTIES BY INVERSE ANALYSIS

Toshio Nakamura[1] and Sanjay Sampath[2]
[1]Department of Mechanical Engineering
[2]Department of Materials Science and Engineering
State University of New York at Stony Brook, NY 11794

ABSTRACT

As the use of FGMs increases, an effective method to characterize their properties must be sought. In general, when FGMs are manufactured, their composition profile and effective material properties need to be verified. However, due to spatial variation of their properties, it is often difficult to make direct measurements of parameters which define FGM properties. In order to alleviate the difficulties, a new procedure based on the inverse analysis, which solely relies on micro-indentation records, is introduced. Specifically, the Kalman filter technique is utilized to estimate the FGM composition profile through-thickness and the stress-strain transfer parameter that defines the effective mechanical properties of FGM. Our simulation study shows promising results when records from two differently sized indenters are employed.

INTRODUCTION

The material gradients induced by spatial variations of properties make FGMs behave differently from common homogeneous materials and traditional composites. Due to this behavior, quantification of FGM properties is more difficult. Recently, several indentation methods were introduced to measure some FGM properties. Suresh and coworkers[1,2] have developed analytical and experimental tools to estimate the linear elastic properties of graded materials using instrumented micro-indentation. The characterization of *nonlinear* or *elastic-plastic* FGMs requires additional material parameters. Unlike elastic FGMs, not only the elastic modulus but also the yield stress and the strain hardening which vary through location must be determined. In order to overcome these difficulties, we propose a new procedure based on the inverse analysis. It makes an effective use of micro-indentation data and uses the Kalman filter theory to estimate the FGM composition profile and effective properties of elastic-plastic FGMs.

One of the techniques that offer flexible and economic means for producing FGMs is the plasma spray[3]. This method has been used to apply layered and graded deposits to enhance the survivability of thick ceramic thermal barrier coatings. By controlling feeding rates of different materials, the thermal spray can produce linear, parabolic or stepped profile grading to suit needs of various applications[4]. In the case of thermal barrier coatings, it is desirable to maximize the ceramic content in the FGM to enhance the thermal properties and a parabolic profile would be suitable.

FGM PARAMETERS

In our effort to establish a robust but simple procedure to determine properties of FGM, we modeled a thin FGM (PSZ- NiCrAlY composite) placed on top of a thick steel substrate as shown in Fig. 1. Although there are various methods to measure mechanical responses of FGMs, we consider instrumented micro-indentation whose requirements for specimen preparation are minimal. Our proposed Kalman filter process requires only the measured load-displacement records. The aim is to develop an effective procedure with a relatively simple experimental effort.

In our finite element analysis PSZ is assumed to be linear elastic ($E = 50\text{GPa}$) while NiCrAlY is modeled as bi-linear elastic-plastic model ($E = 30\text{GPa}$, $\sigma_y = 50\text{MPa}$, $H = 5\text{GPa}$). The Poisson's ratios of both materials are assumed to be the same at $\nu = 0.25$ and constant throughout the FGM layer. At any location within the FGM where two phases coexist, material response is elastic-plastic. The spherical indenter is assumed to be made of a hard material (e.g., tungstein carbide). Two different radius sizes of 100μm and 500μm are considered.

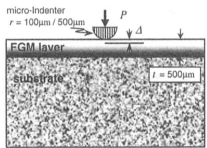

Figure 1. Schematic of micro-indentation of FGM layer on steel substrate.

Compositional Variation through Thickness

One of the important FGM parameters to be determined is the composition of multiple phases through-thickness. It is assumed that the compositional variation of ceramic can be idealized by a power-law expression as, $V_c = (z / t)^n$. Here V_c is the volume fraction of ceramic phase, z is the location measured from the substrate, t is the thickness of FGM layer, and n is the power exponent. A possible range of n is assumed to be 1/3 to 3. In order to estimate this variable in the Kalman filter, this power exponent is redefined as $N = \log n / 2\log 3$ ($-0.5 \leq N \leq 0.5$). The profiles of the compositional variation through-thickness for various N are shown in Fig. 2(a).

Effective Mechanical Properties of FGM

Another important variable in FGMs is a parameter which defines the effective property of ceramic-metal composite. The mechanical response of two-phase material is dependent on the shape, contiguity and spatial distribution of each phase. In the current analysis, so-called 'modified rule-of-mixture' described by Suresh and Mortensen[5] is adopted. The effective properties of FGM with given V_c and V_m ($= 1 - V_c$) can be determined by

$$E_f = \frac{V_c E_c + V_m E_m R}{V_c + V_m R}, \quad \sigma_{yf} = \sigma_{ym} \cdot \left(V_m + \frac{E_c}{E_m} \cdot \frac{V_c}{R} \right) \quad \text{where} \quad R = \frac{q+1}{q + E_m / E_c} \quad (1)$$

Here E_f, E_c and E_m, are the Young's moduli for FGM, ceramic and metal phases, respectively. It is clear that the stress-strain transfer parameter q plays a key role in defining the properties. It depends on many factors including composition and microstructural arrangement and its value may range from zero to infinity. In the Kalman filter, it is suitably represented as $Q = (1 - e^{-q})^{1/2}$ ($0 \leq Q \leq 1.0$). The modified rule-of-mixture can be extended to elastic-plastic composites as shown for the effective yield stress σ_{yf} expression. Here σ_{ym} is the yield stress of pure metal phase. The effective strain-hardening coefficient of the composite H_f can be obtained similarly. If the stress-strain transfer parameter and the volume fractions are known, the above formulations can be used to describe the complete effective mechanical properties of the composite. Although there are other formulations that are more complex, this approach requires only the single parameter to be determined. The stress-strain relations for different Q are shown in Fig. 2(b).

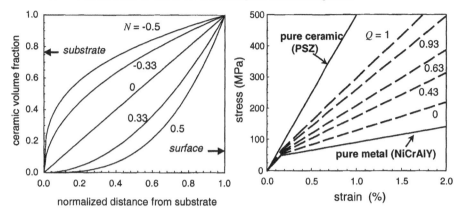

Figure 2. Two unknown FGM parameters determined in the inverse analysis. (a) N: variation of ceramic volume fraction through-thickness, (b) Q: stress-strain transfer parameter. Stress-strain curves are pure ceramic and metal, and equal mixture composite ($V_c = V_m = 0.5$) with various Q.

KALMAN FILTER

The Kalman filter was developed as an optimal recursive signal processing algorithm[6], and it has been used in a wide range of engineering applications. It provides an efficient computational solution based on the least-square theory. To estimate unknown parameters, state and covariance variables are updated at each increment with indirect measurement data (e.g., indented displacements). This method is very effective when measurements contain random error or noise since it is designed to filter them out. This method was proven to be effective in seeking unknown parameters of other materials[7, 8].

In the current analysis, the two unknown parameters range $0 \leq Q \leq 1.0$ and $-0.5 \leq N \leq 0.5$, respectively. In the Kalman filter, $x_t = (Q, N)^T$ is set as the state vector containing these unknown components. The subscript 't' denotes the load increment (P_t) when the measured indented displacement vector Δ_t^{meas} is obtained. The measured data is assumed to contain error as $\Delta_t^{meas} = \Delta_t + \Delta_t^{err}$. The dimensions of vectors and matrices depend on the number of different P-Δ curves (from different indenters) used in the Kalman filter. Initially a record from a single indenter is considered. In this case, $\Delta_t = (\Delta_t)$. In a separate analysis, we consider two separate P-Δ records from two indenters, $\Delta_t = (\Delta_t^A, \Delta_t^B)^T$. Here the measurements from two indenters are denoted as A and B, respectively. In the Kalman filter, the state vector is updated as

$$x_t = x_{t-1} + k_t [\Delta_t^{meas} - \Delta_t(x_{t-1})] \qquad (2)$$

and the Kalman gain matrix k_t is calculated as

$$k_t = P_t \delta_t R_t^{-1} \quad \text{where} \quad P_t = P_{t-1} - P_{t-1} \delta_t^T (\delta_t P_{t-1} \delta_t^T + R_t)^{-1} \delta_t P_{t-1} \qquad (3)$$

Here P_t and R_t are the *matrices of covariance*. The matrix δ_t contains the gradients of Δ_t with respect to x_t. Both Δ_t and δ_t must be available for the Kalman filtering. In the initial matrix of measurement covariance P_0 and the error covariance matrix R_t, the components are set as,

$$P_0 = \begin{pmatrix} (\Delta Q)^2 & 0 \\ 0 & (\Delta N)^2 \end{pmatrix} \quad \text{and} \quad R_t = \begin{cases} R^2 & \text{for single indenter} \\ \begin{pmatrix} R^2 & 0 \\ 0 & R^2 \end{pmatrix} & \text{for two indenters} \end{cases} \tag{4}$$

Here, $\Delta Q = Q_{max} - Q_{min}$ and $\Delta N = N_{max} - N_{min}$ (i.e., $\Delta Q = \Delta N = 1$ in the present case). When the error bound or measurement tolerance (white noise) is assumed to be constant throughout the loading, R can be set constant and independent of t as in the present analysis. The increments are carried out for $t = 1, 2, \ldots, t_{max}$. The state vector x_t at t_{max} contains the final estimates of Q and N from the Kalman filter.

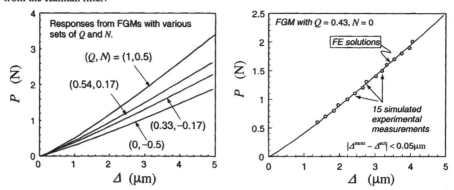

Figure 3. (a) Indented load-displacement curves of FGM with various Q and N. (b) Simulated displacement measurements are created by adding random errors to the finite element solutions. The maximum measurement error/noise is set at $0.05\mu m$.

Since the Kalman filter compares the measured data with known solutions, the P-Δ relation for given q and n, or Q and N, must be available. This relation can be obtained either analytically or numerically. In our proposed procedure, this information or the reference data source is generated by finite element calculations since no closed-form solution exists for the elastic-plastic FGMs. By varying the values of Q and N, stiff to compliant responses can be obtained. Figure 3(a) shows P-Δ relations of FGMs with four different sets of Q and N. As shown in (2), the Kalman filter requires Δ and its derivatives with respect to Q and N at load P. These values can be determined from finite element calculations of various Q and N. In order to minimize such effort, we adopt cubic interpolation functions to calculate displacement and its gradients.

IDENTIFICATION OF FGM PARAMETERS
Single Indenter Procedure
        To verify and quantify the accuracy of this procedure, simulated experimental data is generated from the finite element solutions. At first, an arbitrarily set of Q and N are chosen for the FGM model and the indentation process is simulated. Then suitable load increments for the displacement measurements are selected. After several trials, 15 load increments ranging from $P = 0.6N$ to $2.0N$ with interval of $0.1N$ are chosen. In order to replicate experimental white noise, *random* errors within a bound are added to these 15 displacement records. The error bound or

tolerance of measurement is assumed to be $|\Delta_t^{\text{err}}|_{\text{max}} = 0.05\mu m$, which is about 1~4% of the indented displacements. Figure 3(b) shows the simulated 15 displacement record for the FGM with the properties $Q = 0.43$ and $N = 0$.

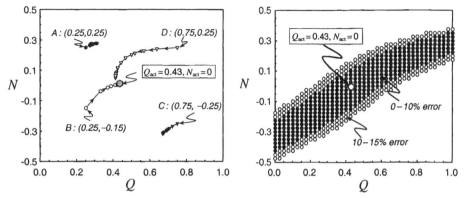

Figure 4. (a) Converging trends of different initial estimates on the Q-N domain. The diverged cases (A and C) stray away from the actual values ($Q = 0.43$ and $N = 0$). (b) Shaded circles correspond to initial estimates that achieved convergence within 10% of the actual values while open circles came within 15%. They make up domain of convergence for the Kalman filter.

Using this data as input, the Kalman filter process is carried out. The component of the error covariance matrix R_t is set to be $R^2 = (0.05\mu m)^2$, which equals the square of measurement tolerance. The Kalman filter also requires initial estimates to be assigned to the unknown parameters. Figure 4(a) shows the converging trends of four sets of arbitrary chosen initial values, denoted as A, B, C, D on the Q-N domain. Each symbol indicates estimated values of Q and N at an increment. This figure shows two initial estimates, B and D, continuously move toward the actual values while the others, A and C, never approach toward the actual solutions. Unlike the finite element simulation, the actual solutions are not known *a priori* in real FGMs. This means, in order for the inverse procedure to be useful, many initial estimates should lead to the exact solutions. We investigate this aspect by assigning many different initial estimates through the Kalman filter. The initial values of Q and N are varied every 0.025 within the Q-N domain (i.e., 41 × 41 = 1681 times in $0 \leq Q \leq 1.0$, $-0.5 \leq N \leq 0.5$). With every set of initial estimates, the deviation or error of its final estimates Q_{est} and N_{est} are computed and measured by circles in Fig. 4(b). Here the shaded circles represent initial estimates whose final values converged within 10% of the actual values while the open circles are those came within 15%. These circles essentially make up a 'domain of convergence' for the Kalman filter. It can be observed that these circles form a strip along a diagonal direction and the converged domain occupies less than 1/3 of the total Q-N area. This means that the majority of initial estimates never yield solutions close to the actual values. Such a procedure is not effective, and an improvement is necessary. One way to enhance the accuracy is to supply more measurement information to the Kalman filter as discussed next.

Double Indenter Procedure
Our aim is to establish a method which requires a minimal experimental effort. Because of this reason, additional measurements from strain gages and optical methods are not considered although they should certainly improve the results. Instead, measurements from another indenter with a different radius are considered. It is assumed different P-Δ characteristics can be obtained

from such an indenter and is sufficient to improve the accuracy. The choice of a new indenter radius is made carefully. Its size must be substantially different from the first one ($r = 100\mu m$) while it can not be too large compared to the layer thickness ($t = 500\mu m$). After several trials, the new radius is chosen to be $r = 500\mu m$. The load increments for this large indenter are set between 5~12N with 0.5N intervals. At first, as in the smaller indenter, the reference data is created by carrying out separate finite element calculations with 16 sets of different Q and N. Again the loading to the FGM layer is applied through increasing the displacement of nodes on the top of the larger indenter. The same measurement tolerance is used. Prior to running the Kalman filter using *both* small and large indenter data as shown in Fig. 5(a), the convergence behavior of the large indenter alone was inspected. The resulting domain of convergence was very similar that of the small indenter shown in Fig. 4(b).

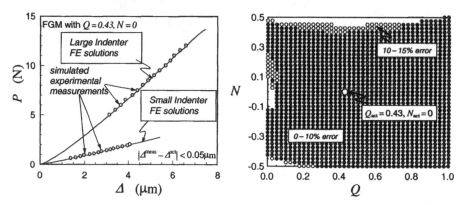

Figure 5. (a) Simulated displacement measurements from two indenters with different radii ($r = 100$ and $500\mu m$). Random errors up are added to the solutions. (b) Domain of convergence for combined small and large indenter case. The size is significantly enlarged as compared with the single indenter case.

The combined indenter procedure is carried out with the Kalman filter equations for the two indenter case. In the Kalman filter program, the reference P-Δ data for both small and large indenters are supplied initially. After initial estimates of Q and N are input, the program processes the two displacements of the first increment from the small and large indenters simultaneously (i.e, Δ^{meas} at 0.6N for the small indenter and Δ^{meas} at 5N for the large indenter). These values are used to update the estimates of Q and N. This process is repeated for 15 increments to obtain the best estimates in the current procedure. As in the single indenter case, its accuracy is evaluated by generating a domain of convergence. The results are shown in Fig. 5(b), where a dramatic improvement can be observed. Almost all of the initial estimates lead to the actual solutions during the Kalman filter procedure. Alternatively, one can assign any initial values for Q and N in the Kalman filter to obtain very accurate estimates of Q and N. These results support the usefulness of the proposed method. The feasibility of this method is also examined by choosing different Q_{act} and N_{act}. Their convergence behaviors are similar to the one in Fig. 5(b). When combined P-Δ data from the small and large indenters are used, more than 80% of initial estimates have achieved the convergence to within 10% error of the actual solutions.

One of the major strengths of the Kalman filter is its ability to process data containing substantial measurement noises or errors. We examined this feature by re-creating simulated measurements with much greater random errors as shown in Fig. 6(a). Here the measurement tolerance or the error bound is set at $|\Delta_r^{err}|_{max} = 0.2\mu m$ in both small and larger indenter results. This magnitude is four times as large as the previous cases and the error can be 3~15% of the displacements. Using these data, the Kalman filter is carried out and the resulting domain of convergence is shown in Fig. 6(b). Obviously the accuracy is reduced due to the worsened data quality. However, the domain including the open circles has maintained nearly its size. More than 80% of the total Q and N domain is still covered by the filled or open circles. These results prove the strength of this method even when relatively large error/noise are present in the measurements.

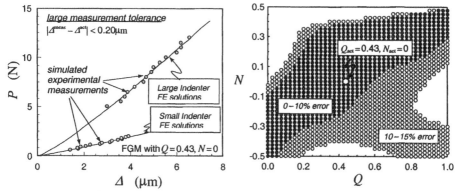

Figure 6. (a) Simulated displacement measurements from two indenters with large random errors (tolerance) of up to 0.20μm. (b) Domain of convergence obtained from the combined indenter case with large errors. Although the accuracy declines, the size is still large.

DISCUSSIONS

A new procedure based on the Kalman filter and instrumented indentation is introduced. This method allows more material information to be extracted from measured load-displacement curves. For the verification and optimization of the procedure, detailed finite element analyses are carried out to determine unknown parameters of elastic-plastic FGMs. Using the idealized model, the parameters which define the compositional variations and effective mechanical properties through-thickness are estimated by the inverse analysis. The former parameter is for the power law compositional distribution and the latter parameter sets the stress-strain transfer parameter of modified rule-of-mixture properties.

As in many inverse analyses, we found that the model is initially 'ill-posed' or 'ill-conditioned' (i.e., not being able to achieve good convergence characteristics) with the single indenter measurements. This problem is overcome by the use of an additional indenter with a larger radius. Furthermore, it is shown that the Kalman filter is well suited for the present nonlinear or elastic-plastic problem since it processes over *multiple* load increments to best estimate the solutions. Required steps in the proposed procedure are summarized below.

- Determine the material constants of homogeneous phases. The ceramic and metal properties can be obtained by separate indentation procedures[9].

- Perform preliminary finite element calculations to determine suitable ranges of load increments for two differently sized indenters.

- Establish reference data from numerically generated $P-\Delta$ records with 16 different Q and N. Use cubic Lagrangian functions to interpolate the displacements and their gradients.

- Perform instrumented indentation on the FGM layer using two spherical indenters with different diameters.

- Assign initial estimates of Q and N. Set values for P_o (with ΔQ, ΔN) and R_t (with $R = 10|\Delta^{err}|_{max}$ where $|\Delta^{err}|_{max}$ is the measurement error/noise bound).

- Carry out the Kalman filter process for 15 increments to obtain the final estimates of Q and N.

One important question in the inverse analysis is how to identify the accuracy of the final estimates in real tests where actual solutions are unknown. Although the uniqueness cannot be proven, we can propose the following procedure. First, carry out the Kalman filter with some initial estimates. Instead of terminating at this point, carry out another Kalman filter procedure using the final estimates of the first attempt as the initial estimates of the second attempt. This procedure can be repeated for a few times. If the final estimates are similar in all processes, then it is likely that they are close to the actual solutions.

Implementation of the procedure in the real FGM specimen is being prepared currently. Although the proposed procedure is described for the elastic-plastic FGMs, a similar procedure can be used to determine other parameters, including thickness, yield stress, anisotropic properties of layers and coatings. Furthermore, the number of unknown parameters in the inverse analysis can be also increased.

ACKNOWLEDGMENTS

The authors acknowledge the National Science Foundation under award CMS-9800301 and the Army Research Office under grant DAAD19-99-1-0318 for the support of this work. The computations are carried out on HP7000/C180 and C360 workstations using the finite element code ABAQUS which was made available under academic license from Hibbitt, Karlsson and Sorensen, Inc.

REFERNCES

[1]S. Suresh, A.E. Giannakopoulos and J. Alcala. "Spherical indentation of compositionally graded materials: theory and experiments," *Acta Mater.*, **45** 1307-1321, 1997.

[2]E. Giannakopoulos and S. Suresh. "Indentation of solids with gradients in elastic properties: part II. axisymmetric indenters," *Int. J. Solids Struct.*, **33** 2393-2428. 1997.

[3]Sampath, W. C. Smith, T. J. Jewett and H. Kim. "Synthesis and characterization of grading profiles in plasma sprayed NiCrAlY-zirconia FGMs," *Mat. Sci. Forum,* **308-11** 383-388, 1999.

[4]Sampath, H. Herman, N. Shimoda, and T. Saito. "Thermal Spray Processing of FGMs," *MRS Bulletin* January, pp.27-31, 1995.

[5]S. Suresh and A. Mortensen. *"Fundamentals of Functionally Graded Materials,"* London: IOC Communications Ltd, 1998.

[6]R. E. Kalman. "A new approach to linear filtering and prediction problems," *Trans. of ASME—J. of Basic Engin.* 35-45, 1960.

[7]S. Aoki, K. Amaya, M. Sahashi , T. Nakamura. "Identification of Gurson's material constants by using Kalman filter," *Comp. Mech.,* **19** 501-506, 1997.

[8]S. Aoki, K. Amaya and F. Terui. "A new method for identifying elastic/visco-plastic material constants," in *Proceeding of 4th WCCN. Computational Mechanics: New Trends and Application.* Ed. Onate and Idelsohn.; pp. 82/1-82/5, 1998

[9]A. E. Giannakopoulos and S. Suresh. "Determination of elastoplastic properties by instrumented sharp indentation," *Scripta mater.*, **40** 1191-1198, 1999

USING ADJOINT EQUATIONS TO OPTIMIZE COMPOSITIONALLY GRADED INTERLAYERS

Dj. Boussaa
LMA, CNRS
31, chemin Joseph Aiguier
F-13402 Marseille Cedex 20
France

H. D. Bui
LMS
École Polytechnique
F-91128 Palaiseau Cedex
France

ABSTRACT

We consider an assemblage consisting of 2 homogeneous parts made of dissimilar materials and a functionally graded interlayer inserted between the two parts for thermal stress relief. A method is proposed to find the interlayer composition profile that minimizes stresses in the assemblage when its temperature is varied from an initial, uniform temperature to another uniform temperature. The optimization problem is (1) set as an optimal control problem whose control is the volume fraction of one of the two materials, (2) discretized using the finite element method, and (3) solved using the adjoint equations. A numerical example is investigated for which it is found that the optimal composition has jumps at the interface between the interlayer and the homogeneous parts.

INTRODUCTION

By their definition, functionally graded materials (FGMs) entails optimization: composition and/or microstructure of the FGM are spatially graded in the component so that its overall performance is optimized [1]. Many optimization approaches have been proposed to optimize FGMs composition profiles, ranging from classical descent techniques as applied to single- and multiobjective programming [2, 3] to more recent heuristic techniques such as genetic algorithms and neural networks [4, 5].

The paper, a contribution of the former class of approaches, proposes a method based on the use of adjoint equations [6, 7, 8] to compute the profile that minimizes stresses in an assemblage in which a functionally graded interlayer is inserted between 2 parts made of dissimilar materials. The problem is formulated in Section 2, where the gradient of the objective function is expressed in terms of the state and adjoint fields. In Section 3, a numerical example is discussed for which it is found the materials properties profile corresponding to the stress minimizer composition is not continuous.

PROBLEM FORMULATION

Consider the assemblage described in Figure 1 that consists of two homogeneous parts Ω_1 and Ω_2 joined by a graded interlayer whose material is a composite made up by combining the materials

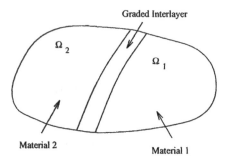

Figure 1: Assemblage with graded interlayer

of parts 1 and 2, and whose composition is allowed to vary spatially. The assemblage, initially at a uniform temperature, has its temperature raised or lowered to another uniform temperature. Our objective is to find the interlayer composition profile that minimizes the stresses induced in the structure by the variation of temperature.

The main assumptions are as follows:

- The structure remains elastic during the temperature change.

- The anisotropy of the FGM is given and the axes of anisotropy of the FGM, if any, are assumed to be fixed. That is, the microstructure and its orientation are fixed, and ruled out as design variables.

- The compliance tensor and the thermal expansion tensor of the FGM are assumed to be determined in terms of ξ, the volume fraction of one of the two materials, say material 2, and the properties of materials 1 and 2 through a given micromechanical model. The body forces in the interlayer, if any, are also assumed to have a known dependence on ξ.

Under these assumptions, the minimization problem can be stated as follows.

$$\text{Find } \xi(\mathbf{x}) \text{ which minimizes } J(\xi)$$

subject to

$$\text{div } \boldsymbol{\sigma} + \mathbf{b}(\xi) = \mathbf{0} \qquad\qquad \text{in } \Omega, \qquad (1)$$

$$\boldsymbol{\varepsilon}(\mathbf{u}) = \mathsf{S}(\xi)\,\boldsymbol{\sigma} + \boldsymbol{\epsilon}^{\text{th}}(\xi) \qquad\qquad \text{in } \Omega, \qquad (2)$$

$$\boldsymbol{\sigma}\mathbf{n} = \mathbf{0} \qquad\qquad \text{on } \partial\Omega, \qquad (3)$$

and

$$0 \le \xi(\mathbf{x}) \le 1 \qquad\qquad \text{in } \Omega, \qquad (4)$$

where $\mathbf{x}$ is the position, J is the objective functional, $\boldsymbol{\sigma}$ is the stress tensor, $\mathbf{b}$ is the body forces, S is the compliance tensor, $\mathbf{u}$ is the displacement vector, $\mathbf{n}$ is the outward normal to the boundary of the assemblage, and $\boldsymbol{\epsilon}^{\text{th}}$ the incompatible thermal strain defined as

$$\boldsymbol{\epsilon}^{\text{th}} = \int_{T_i}^{T_f} \boldsymbol{\alpha}(\xi, \theta) \, d\theta, \qquad (5)$$

allowing for a possible temperature dependence of the materials properties between initial temperature T_i and final temperature T_f.

The minimization problem thus stated is an optimal control problem whose control is the volume fraction ξ, the state fields are displacement $\mathbf{u}$ and stress σ, and the state equations are the balance and the constitutive equations (1) and (2).

The objective functional J, though typically a stress measure, is assumed here to have the general form

$$J(\xi) = \int_\Omega j(\xi, \sigma(\xi), \mathbf{u}(\xi); \mathbf{x})\, d\Omega. \tag{6}$$

We introduce the adjoint stress τ and the adjoint displacement $\mathbf{v}$ and form the augmented criterion

$$J_A(\xi) = \int_\Omega \left[j + \mathbf{v} \cdot (\operatorname{div} \sigma + \mathbf{b}) - \tau \cdot \left(\varepsilon(\mathbf{u}) - \mathsf{S}\,\sigma - \epsilon^{\text{th}} \right) \right] d\Omega, \tag{7}$$

where "$\cdot$" denotes the usual dot product ($\mathbf{a} \cdot \mathbf{b} = a_i b_i$ and $\mathbf{A} \cdot \mathbf{B} = A_{ij} B_{ij}$).

Now, consider the variation δJ_A of J_A due to the (weak) variations in the volume fraction ξ. Using the following relationships between the variations:

$$\operatorname{div} \delta\sigma + \frac{\partial \mathbf{b}}{\partial \xi}\, \delta\xi = \mathbf{0}, \tag{8}$$

$$\left(\frac{\partial \mathsf{S}}{\partial \xi} \sigma + \frac{\partial \epsilon^{\text{th}}}{\partial \xi} \right) \delta\xi + \mathsf{S}\, \delta\sigma = \varepsilon(\delta\mathbf{u}), \tag{9}$$

$$\delta\sigma\, \mathbf{n} = \mathbf{0}, \tag{10}$$

applying Green's theorem, rearranging terms, and denoting $\partial\Omega$ the boundary of Ω, yield

$$\delta J_A = \int_\Omega \left\{ \left[\frac{\partial j}{\partial \xi} + \mathbf{v} \cdot \frac{\partial \mathbf{b}}{\partial \xi} + \tau \cdot \left(\frac{\partial \mathsf{S}}{\partial \xi} \sigma + \frac{\partial \epsilon^{\text{th}}}{\partial \xi} \right) \right] \delta\xi + \right.$$
$$\left. \left(\frac{\partial j}{\partial \sigma} + \varepsilon(\mathbf{v}) - \mathsf{S}\tau \right) \cdot \delta\sigma + \left(\frac{\partial j}{\partial \mathbf{u}} + \operatorname{div} \tau \right) \delta\mathbf{u} \right\} d\Omega - \int_{\partial\Omega} \tau\mathbf{n} \cdot \delta\mathbf{u}\, d\partial\Omega. \tag{11}$$

In this equation, the variation of the compliance tensor and the thermal strain can readily be computed from the equations of the micromechanical model. The same comment applies to body forces $\mathbf{b}$. In contrast, the dependence of $\delta\sigma$ and $\delta\mathbf{u}$ on $\delta\xi$, which is nonlocal, in the sense that a variation of ξ in the graded interlayer induces a variation of σ and $\mathbf{u}$ everywhere in the structure, is more complicated to determine.

To circumvent this difficulty, choose $\mathbf{v}$ and τ such that

$$\operatorname{div} \tau + \frac{\partial j}{\partial \mathbf{u}} = \mathbf{0} \tag{12}$$

$$\varepsilon(\mathbf{v}) = \mathsf{S}\tau + \frac{\partial j}{\partial \sigma} \tag{13}$$

$$\tau\mathbf{n} = \mathbf{0} \tag{14}$$

so that

$$\delta J = \int_{\Omega_{\text{FGM}}} \left[\frac{\partial j}{\partial \xi} + \mathbf{v} \cdot \frac{\partial \mathbf{b}}{\partial \xi} + \tau \cdot \left(\frac{\partial \mathbf{S}}{\partial \xi} \sigma + \frac{\partial \epsilon^{\text{th}}}{\partial \xi} \right) \right] \delta \xi \, d\Omega. \tag{15}$$

where the subscript A has been dropped since J and J_A coincide when equilibrium and constitutive equations are satisfied by the state fields. In (15), the term between square brackets can be identified as the gradient of J in the integral inner product sense.

The integral in (15) is taken over the volume of the graded interlayer because the volume fraction is constant in the homogeneous parts of the assemblage ($\xi = 0$ in part 1 and 1 in part 2).

To proceed numerically with the solution of the minimization problem, we assume that ξ has been discretized in some way; that is, that ξ has been made to depend on a finite number of parameters $\mathbf{a} = (a_1, a_2, \ldots, a_n)$, with n being finite. Then

$$\delta \xi = \frac{\partial \xi}{\partial \mathbf{a}} \cdot \delta \mathbf{a}, \tag{16}$$

and

$$\delta J(\xi) = \left(\int_{\Omega_{\text{FGM}}} \left[\frac{\partial j}{\partial \xi} + \mathbf{v} \cdot \frac{\partial \mathbf{b}}{\partial \xi} + \tau \cdot \left(\frac{\partial \mathbf{S}}{\partial \xi} \sigma + \frac{\partial \epsilon^{\text{th}}}{\partial \xi} \right) \right] \frac{\partial \xi}{\partial \mathbf{a}} \, d\Omega \right) \cdot \delta \mathbf{a}, \tag{17}$$

and the gradient of the objective function, in the sense of the usual euclidean inner product, can be written as

$$G(\xi) = \int_{\Omega_{\text{FGM}}} \left[\frac{\partial j}{\partial \xi} + \mathbf{v} \cdot \frac{\partial \mathbf{b}}{\partial \xi} + \tau \cdot \left(\frac{\partial \mathbf{S}}{\partial \xi} \sigma + \frac{\partial \epsilon^{\text{th}}}{\partial \xi} \right) \right] \frac{\partial \xi}{\partial \mathbf{a}} \, d\Omega. \tag{18}$$

We now specialize this result to the case of a unidirectional gradient and consider some simple interpolation schemes. If

$$\xi(\eta) = \sum_{i=1}^{n} a_i \varphi_i(\eta) \tag{19}$$

where η is a coordinate along the direction of the gradient, and the φ_i's, for $i = 1, \ldots, n$ are interpolating functions. Then the ith gradient component reads

$$G_i(\xi) = \int_{\Omega_{\text{FGM}}} \left[\frac{\partial j}{\partial \xi} + \mathbf{v} \cdot \frac{\partial \mathbf{b}}{\partial \xi} + \tau \cdot \left(\frac{\partial \mathbf{S}}{\partial \xi} \sigma + \frac{\partial \epsilon^{\text{th}}}{\partial \xi} \right) \right] \varphi_i \, d\Omega. \tag{20}$$

In the case of a piecewise constant interpolation (the interlayer is subdivided into n sublayer over each of which the volume fraction $\xi = a_i$ is constant), the ith component of the gradient is

$$G_i(\xi) = \int_{\Omega_{\text{FGM}}^i} \left[\frac{\partial j}{\partial \xi} + \mathbf{v} \cdot \frac{\partial \mathbf{b}}{\partial \xi} + \tau \cdot \left(\frac{\partial \mathbf{S}}{\partial \xi} \sigma + \frac{\partial \epsilon^{\text{th}}}{\partial \xi} \right) \right] d\Omega, \tag{21}$$

where Ω_{FGM}^i denotes the domain occupied by the ith sublayer.

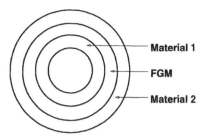

Material 1

FGM

Material 2

Figure 2: The three-layer assemblage.

A NUMERICAL EXAMPLE

We now apply the results of the previous section to optimizing the composition profile of an interlayer inserted between two concentric cylinders made of dissimilar materials. The resultant three-layer assemblage is described in Figure 2. The plane strain assumption was adopted and the finite element method was used to discretize the problem. The other particulars of the problem are described below.

The Objective Functional

The following expression for J was used:

$$J(\xi) = \frac{1}{[\text{meas}(\Omega)]^{(1/q)}} \left(\int_\Omega |f(\xi, \sigma)|^q \, d\Omega \right)^{(1/q)}, \tag{22}$$

where meas(Ω) is the volume of Ω, $f(\sigma)$ is a Drucker-Prager-like effective stress,

$$f = \lambda(\xi) \sqrt{J_2} + \mu(\xi) I_1, \tag{23}$$

with $J_2 = \frac{1}{2} s \cdot s$, s being the deviatoric part of σ, I_1 its trace, λ and μ are two material parameters that can depend on the volume fraction ξ, and q is a large even positive integer.

Although (22) is slightly different from the form (6), the derivation of the gradient in the former case is basically the same as for the latter as can be seen from the expression for the variation of J,

$$\delta J = \frac{\left(\int_\Omega f^q \, d\Omega \right)^{(1/q-1)}}{[\text{meas}(\Omega)]^{(1/q)}} \int_\Omega f^{q-1} \left(\frac{\partial f}{\partial \sigma} \cdot \delta\sigma + \frac{\partial f}{\partial \xi} \delta\xi \right) d\Omega. \tag{24}$$

The objective functional (22) can be viewed as a regularization of the L_∞-norm. The objective functional is thus an approximation to the maximum of $|f|$ over the structure.

Composition Profile Interpolation

The graded interlayer was divided into n concentric sublayers, over each of which the composition is assumed to be constant. In the absence of free edges, the piecewise constant interpolation is legitimate without taking any specific precautions.

The Effective Properties

The Reuss bound was used for computing the effective compliance tensor of the graded interlayer and Levin's formula for computing its effective thermal expansion tensor. Under these assumptions, the micromechanical model amounts to the law of mixture,

$$S(\xi) = (1 - \xi)S_1 + \xi S_2, \tag{25}$$

$$\alpha(\xi) = (1 - \xi)\alpha_1 + \xi \alpha_2. \tag{26}$$

But, of course, any micromechanical model can be implemented in conjunction with the adjoint equations. Body forces were ignored.

Optimization Algorithm

The NAG routine E04UCF, which uses a sequential quadratic programming method, was invoked to solve the finite-dimensional nonlinear programming problem resulting from the discretization. The routine, whose default parameters were adopted, takes as input a user defined subroutine that computes the function to be minimized and its gradient given a set of design parameters. The organization of our "user-defined" subroutine is schematically as follows:

Step 1. Compute the state variables by solving a thermoelastic problem with ϵ^{th} as an initial strain.

Step 2. Compute the adjoint state variables by solving a thermoelastic problem with $-\frac{\partial j}{\partial \sigma}$ as an initial strain. Use the factorized stiffness matrix obtained in Step 1. The main task in this step is the assemblage of the internal load equivalent to the adjoint thermal strain.

Step 3. Compute the gradient using (21) and (24)

Numerical Values

The following nondimensional data were used: inner radius = 1; inner layer: thickness = 0.15, Young's modulus = 5, Poisson ratio = 0.2, thermal expansion coefficient = 0.1; interlayer: thickness = 0.15; outer layer: thickness = 2, Young's modulus = 2., Poisson ratio = 0.3, thermal expansion coefficient = 1; exponent $q = 100$; coefficients in f (assumed to be independent of ξ): $\lambda = 1$ and $\mu = 1/\sqrt{6}$; temperature variation $T_f - T_i = -1$. The interlayer was divided into $n = 30$ sublayers.

Results

The optimal profile is shown in Figure 3. The algorithm converged to this profile from various initial guesses, including $\xi_i = 0$ and $\xi_i = 1$, for $i = 1, \ldots, 30$, with sublayer 1 being the innermost.

Note that the optimal profile is not a continuous transition between Material 1 and Material 2.

The evolution during iterations of the objective function and the norm of the projected gradient from the initial guess $\xi_i = 0$, for $i = 1, \ldots, 30$, is represented in Figure 4. The reduction in the objective function is about 20%.

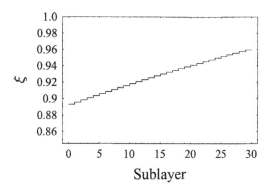

Figure 3: Optimal profile

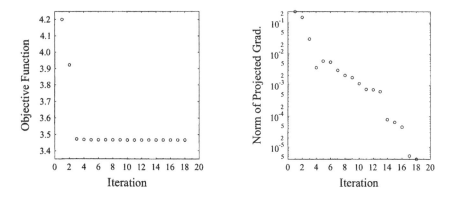

Figure 4: Evolution during iterations of objective function and norm of projected gradient

REFERENCES

[1] S. Suresh and A. Mortensen. *Fundamentals of Functionally Graded Materials, Processing and Thermomechanical Behaviour of Graded Metals and Metal-Ceramic Composites.* IOM Communications Ltd, 1998.

[2] K. Tanaka, Y. Tanaka, K. Enomoto, V.F. Poterasu, and Y. Sugano. Design of thermoelastic materials using direct sensitivity and optimization methods. Reduction of thermal stresses in functionally gradient materials. *Computer Methods in Applied Mechanics and Engineering,* 106:271–284, 1993.

[3] K. Tanaka, H. Watanabe, Y. Sugano, and V. F. Poterasu. A multicriterial material tailoring of a hollow cylinder in functionally gradient materials: Scheme to global reduction of thermoelastic stresses. *Computer Methods in Applied Mechanics and Engineering,* 135:369–380, 1996.

[4] Y. Ootao, R. Kawamura, Y. Tanigawa, and R. Imamura. Optimization of material composition of nonhomogeneous hollow sphere for thermal stress relaxation making use of neural network. *Computer Methods in Applied Mechanics and Engineering,* 180:185–201, 1999.

[5] Y. Ootao, Y. Tanigawa, and O. Ishimaru. Optimization of material composition of functionally graded plate for thermal stress relaxation using a genetic algorithm. *Journal of Thermal Stresses,* 23(3):257–271, 2000.

[6] H. D. Bui. *Inverse Problems in the Mechanics of Materials : An Introduction.* CRC Press, Boca Raton, FL, 1994.

[7] Z. Mróz. Variational methods in sensitivity analysis and optimal design. *European Journal of Mechanics,* 14(4-suppl.):115–147, 1994.

[8] G. I. Marchuk. *Adjoint Equations and Analysis of Complex Systems,* volume 295 of *Mathematics and Its Applications.* Kluwer Academic Publishers, Dordrecht, The Netherlands, 1995.

FEM SIMULATION OF THE BEHAVIOUR OF GRADED MATERIALS WITH MACROSCOPIC COMPOSITION GRADIENT DURING DEFORMATION PROCESSES

Sven Raßbach and Wolfgang Lehnert
Freiberg University of Mining and Technology
Institute of Metal Forming
Bernhard-von-Cotta-Strasse 4
Freiberg, D-09596
Germany

ABSTRACT
The development of the graded materials is motivated by the local adaptation of the material composition to the specific application. Many of the processes used for the manufacture of graded materials have procedure-specific limitations concerning component complexity or gradient design. Therefore, graded materials should further be processed by deformation operations. For this, fundamental investigations for the deformation behaviour of graded materials were carried out. These investigations show promising results. It is expected that technological compressive deformation processes can be applied in order to produce components with a graded material composition.

INTRODUCTION
New developments in mechanical and plant engineering or car technology demand components which must cope with greater and greater loadings. Mechanical, tribological, thermal and chemical loadings that often occur in a combined manner represent increased requirements to material properties. Conventional materials can satisfy these demands only in a limited manner. Graded materials with improved service properties are a solution for this. Thus high-performance materials can be reduced onto the zones of a component which are a heavy strain on. In addition, the possibility of a more independent constructional designing of the component results from this.

Different melt and powder metallurgical procedures are known for the manufacture of graded materials [e.g. 1-4]. In particular the powder metallurgical processes stacking and pressing [5] and wet powder spraying [6] with subsequent consolidation step appear to be suitable in order to fabricate graded semi-finished products of larger dimensions in a purposeful manner. But many of these processes have procedure-specific limitations, e.g. a simple geometry of the components as well as a simple gradient design in spite of extensive process control.

A way to reduce these restrictions is the possibility, though hardly used so far, to further treat graded materials by deformation processes, e.g. die forging or extrusion operations. Such forming certainly enlarges the field of application of graded materials since, for example, components with more complex geometric shapes can be produced. The microstructure with regard to improved properties can be influenced as soon as the position and shape of the gradient can be modified.

To this, fundamental investigations were carried out for hot upsetting of graded specimens. The results are presented at the example of the powder metallurgical material system 316L/430L. These investigations represent an assumption, which concern further considerations at more complex deformation processes.

SOLUTION APPROACH

Graded materials are characterized by an inhomogeneous composition and so different material zones have different deformation behaviour. For the simulation of the deformation behaviour, the graded material is subdivided into a finite number of zones in which homogeneous composition and definite material behaviour are assumed. The yield strength of these zones are determined experimentally and described in a mathematical model. The determined data following the model are integrated into a simulation. The corresponding yield strength can be assigned to every zone. Thus the plastic deformation behaviour of graded materials will be predetermined.

EXPERIMENTAL PROCEDURE

The investigations were carried out among other things at the material system 316L/430L. The specimens were produced powder metallurgically by stacking and pressing from powder mixtures of the steel powders 316L and 430L. The specimens were then sintered (1000 °C, 30 minutes in vacuum, then 1300 °C, 120 minutes in vacuum + 60 minutes at 9 MPa of argon pressure). The sintering technology was optimized so that the measured densities of all specimens have values of $\rho_{theor.} \geq 98.5$ %. Consequently, dense materials can be accepted. This condition ensures a continuum mechanical consideration of the material system. The experimental details are presented in Tab. I.

Table I. Experimental details of the examined 316L/430L graded specimens

	Experimental details
Fractional composition of the layers [vol.-% 316L/430L]	100/0; 80/20; 60/40; 50/50; 40/60; 20/80; 0/100
Gradient	quasi-continuously (7-layered)
Specimen dimensions [mm]	$\varnothing$ 7.55 x 15.4
Deformation temperature [°C]	700, 1000
Deformation rate [s^{-1}]	0.1, 1.0
Forming degree [-]	0.3, 0.6, 0.9

The heating of the specimens to deformation temperature occurred in a muffle furnace. After this they were upset with different deformation parameters (Tab. I) in the servo hydraulic test system SHM 400.

The FEM program MARC/AutoForge was used for the simulation of deformation behaviour. An isothermal hot upsetting with distance control was calculated which corresponds to the characteristic curve of the available servo hydraulic test system. The computation of the model employed single axial symmetry and utilized an elastic-plastic material law. The von Mises flow criterion was used. The model was generated by discretization of the respective composition gradient into layers of homogeneous composition. The corresponding material behaviour was assigned to the individual layers. Values from literature according to [7,8] for Young's modulus and Poisson's ratio and the yield stresses from hot upsetting of homogeneous materials prepared by fuzzy approximation [9] were applied. The workpiece was meshed at high density in order to avoid the necessity of remeshing. This is required, because the current version of the FEM program does not yet support a remeshing of several workpieces (in this case - layers).

In order to draw a comparison with the FEM calculations with regard to the geometrical shape as well as the change of the composition gradient upset graded specimens were examined. On the one hand, the outlines of the top and bottom end faces and of the centre planes (prepared division plane in axial direction) were determined. This was carried out on a 3D-measuring microscope. On the other hand, the element concentrations (Fe, Cr, Ni, and Mo) were measured by electron probe microanalysis (EPMA). These were measured in the prepared centre plane on prescribed lines parallel to the specimen longitudinal axis.

RESULTS AND DISCUSSION

In accordance with the plastic material behaviour determined, the simulation of hot upsetting showed expected results. In contrast to the upsetting of homogeneous materials, in graded materials it was found a heterogeneous material flow across the specimen height. This is due to the different yield strength on account of the composition gradient changing across the specimen height. In material zones with lower yield strength (in all experiments, the specimen bottom was prescribed for this – zone of ferritic steel 430L), a stronger material flow appears than in the zones with higher yield strength (in all experiments, the specimen top – zone of austenitic steel 316L). The top of the specimen is almost undeformed and behaves like a rigid die, compared to other material zones, until the yield stress is also attained in this zone. This is exemplary confirmed in Fig. 1 for an upsetting sequence on a 316L/430L gradient.

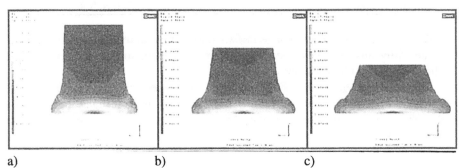

a)   b)   c)

Fig. 1:   FEM simulation of upsetting a 7-layered 316L/430L graded specimen
          [$\vartheta = 1000\ °C$, $\varphi^* = 1.0\ s^{-1}$ and a) $\varphi = 0.3$, b) $\varphi = 0.6$, c) $\varphi = 0.9$].

The FEM results were checked at experimental upsetting tests. First, a comparison of the geometrical dimensions was carried out. The coordinates of the specimen outline at the top and bottom end faces as well as the centre plane were determined.

The heterogeneous material flow across the specimen height calculated in advance by FEM could be also found at the experimental specimens (Fig. 2).

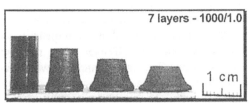

Fig. 2:   Material flow of an upsetting sequence of a 7-layered 316L/430L graded
          specimen ($\vartheta = 1000\ °C$, $\varphi^* = 1.0\ s^{-1}$, from left to right $\varphi = 0.0$, $\varphi = 0.3$,
          $\varphi = 0.6$, $\varphi = 0.9$).

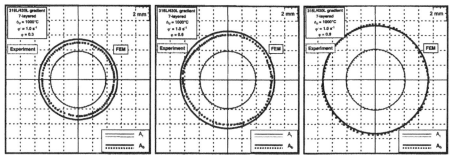

Fig. 3: Comparison of the outlines of specimen top and bottom from simulation and experiment for the upsetting sequence of a 7-layered 316L/430L graded specimen ($\vartheta = 1000$ °C, $\varphi^{\bullet} = 1.0$ s^{-1}).

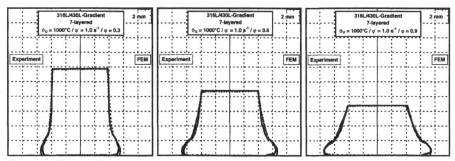

Fig. 4: Comparison of the centre plane outline from simulation and experiment for the upsetting sequence of a 7-layered 316L/430L graded specimen ($\vartheta = 1000$ °C, $\varphi^{\bullet} = 1.0$ s^{-1}).

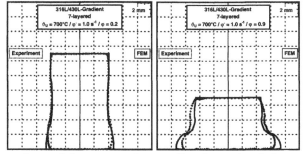

Fig. 5: Comparison of the centre plane outline from simulation and experiment for the upsetting sequence of a 7-layered 316L/430L graded specimen ($\vartheta = 700$ °C, $\varphi^{\bullet} = 1.0$ s^{-1}).

The expected results are confirmed. On account of its higher yield strength, the top of the specimen with the austenitic steel 316L is almost undeformed. Whereas the bottom of the specimen (ferritic steel 430L) flows more strongly.

The experiments show a regular radial material flow in the horizontal cross-section of all specimens during the upsetting process. This is to be seen exemplary in Fig. 3 for the same upsetting sequence as represented in Fig. 2.

Fig. 3 also shows a good agreement of the experimental specimens with the FEM simulation. Some larger deviations are to be found in the area of the specimen bottom (stronger deformations). This is to be traced mainly to measuring mistakes. In the area of the specimen top, the differences are smaller.

Fig. 4 and 5 verify for the vertical sections of the centre plane that the geometry is simulated correctly. For the upsetting sequence represented in Fig. 4, the yield stresses of the individual material zones decrease continuously in vertical direction from the specimen top to the bottom. The material flow increasing in this direction is represented correctly by the conical vertical section.

A test with the same gradient structure, but deviating deformation condition compared to Fig. 4 is represented in Fig. 5. The central layers in vertical direction have higher yield strength as the zones of top and bottom in the case of a forming degree of $\varphi = 0.2$. The top and bottom zones flow more strongly than the central specimen area. In the case of $\varphi = 0.9$, conditions exist in which the yield stresses of the material zones decrease continuously from specimen top to bottom. Therefore, the radial material flow increases in this direction. This is confirmed by the results of FEM and experimental upsetting tests (Fig. 5).

The applied axially symmetric FEM model can reflect the reality during the course of an upsetting sequence for the 316L/430L gradients. The geometric shape changes can be simulated very well for different deformation conditions by means of FEM.

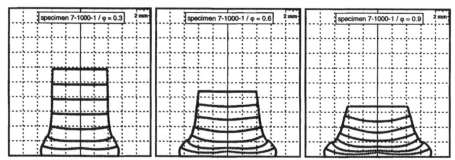

Fig. 6:   Simulated change of the gradient showing as layer boundaries for the deformation conditions $\vartheta = 1000\ °C$ and $\varphi^{\cdot} = 1.0\ s^{-1}$.

In addition, the change of the composition gradient during deformation was examined by FEM. The gradient was initially assumed as axially one-dimensional. A change of the plane layers to a shape similar to vaulted glass bowls was proved. This is to be seen exemplary in Fig. 6.

To the proof from real upset specimens, the element concentrations were determined by means of EPMA at the prepared centre planes. First results of the 316L/430L system show good agreement between data from FEM and experiment. The comparison of the iron content along particular lines (across specimen height) is demonstrated exemplary for a deformed specimen in Fig. 7. Measured lines were compared to the corresponding lines converted from the FEM analysis.

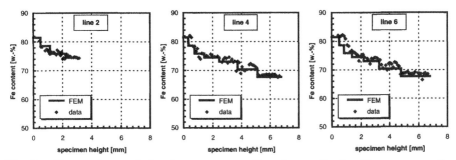

Fig. 7: Comparison of the iron concentration for particular lines across the specimen height of a deformed 7-layered specimen ($\vartheta = 1000$ °C, $\varphi^* = 1.0$ s^{-1}, $\varphi = 0.9$).

It can be estimated on account of the present investigations that changes of the gradient during deformation can be simulated with an appropriate expenditure by means of FEM.

CONCLUSIONS AND OUTLOOK

The investigations of the behaviour of graded materials during deformation have shown, that an inhomogeneous shape change (material flow) occurs in dependence of the composition gradient. In addition a change of the initial gradient is provable. It is possible to simulate these facts by means of FEM. Good agreement exists between simulation and experimental results for the considered deformation parameters. The axially symmetric FEM model reflects the real relations during deformation for 316L/430L graded materials.

The present state of the investigations approves the conclusion, that the behaviour of graded materials during compressive deformation can be simulated, evaluated and quantified by means of FEM.

In a next step, it is aspired to use the present results as a basis in order to simulate the fabrication of more complex components and to check this experimentally. For this purpose, a corresponding constructional element should be deformed, e.g. by die forging or extrusion.

ACKNOWLEDGEMENT

The authors wish to thank the Deutsche Forschungsgemeinschaft (DFG) for funding these investigations under project no. Le 872/13. They also thank Dr. D. Heger, Institute of Physical Metallurgy of the Freiberg University, for the realization of EPMA.

REFERENCES

[1] A. Neubrand, "Gradientenwerkstoffe – ein Zwischenresümee nach 10 Jahren internationaler Forschung"; pp. 291-300 in *Verbundwerkstoffe und Werkstoffverbunde*, Edited by K. Schulte and K.U. Kainer. Wiley-VCH-Verlag GmbH, Weinheim, Germany, 1999.

[2] A. Mortensen and S. Suresh, "Functionally graded metals and metal-ceramic composites: Part 1 Processing", *International Materials Reviews*, 40 [6] 239-265 (1995).

[3] A. Neubrand and J. Rödel, "Gradient materials: An overview of a novel concept", *Zeitschrift für Metallkunde*, 88 [5] 358-371 (1997).

[4] T. Jüngling and B. Kieback, "Pulvertechnologische Konzepte zur Herstellung von Werkstoffen mit gradierter Struktur", *Werkstoffwoche 96, Neue Werkstoffkonzepte*, Vol. 9 139-143 (1997).

[5] G. Hinzmann, "Schichtverbundpressen auf CNC-gesteuerten Maschinen", *Fortschritte bei der Formgebung in Pulvermetallurgie und Keramik – Pulvermetallurgie in Wissenschaft und Praxis*, Vol. 7 94-106, VDI Verlag, Düsseldorf, 1997.

[6] B. Kieback and F. Meyer-Olbersleben, "Bauteile mit konträren Eigenschaften aus Gradientenwerkstoffen herstellbar", *Maschinenmarkt* 105 [1/2] 38-40 (1999).

[7] W. Helling in *Landolt-Börnstein Vol. VI/2c*, p. 53 ff, 6th ed. Edited by H. Borchers and E. Schmidt, Springerverlag Berlin Göttingen, Heidelberg, New York, 1965.

[8] C. Kammer, *Aluminium-Taschenbuch Vol. 1*, 15th ed. Aluminium-Zentrale, Düsseldorf, 1995.

[9] S. Raßbach, "Fuzzy-Approximation von Fließkurven", ch. 5 in *Proc. of MEFORM98*, Freiberg, Germany, 1998.

FINITE ELEMENT ANALYSIS IN DESIGN OF A GRADED COATING SYSTEM FOR GLASS FORMING DIES AND TOOLS

D. Zhong, G.G.W. Mustoe, and J.J. Moore
Colorado School of Mines
1500 Illinois Street
Golden, CO 80401

S. Thiel and J. Disam
Schott Glas
Hattenbergstrasse 10, P.O. Box 2480
D-55014 Mainz, Germany

ABSTRACT
Finite element analysis (FEA) is being used as an integral part of an overall research program that is being conducted to develop a non-sticking, oxidation resistant, and wear resistant coating system for glass molding dies and forming tools. The FEA was performed with the general purpose FE code in the MARC K7.32 system. In this work, a non-linear thermomechanical FE model has been used to analyze the residual stresses generated in the coating system during a simulated glass molding process, and to predict an optimal coating architecture with minimized residual stress and optimized stress distribution. The results suggested that the proposed graded coating architecture could benefit the performance of the coating system for glass forming dies and tools. The methodology described in this paper can be used to explore the effects of die geometry, die materials, coatings, and process parameters, etc. on die life.

INTRODUCTION
It is well known that the integrated stress present in a coating is an important issue since the stress often controls the mechanical behavior and adhesion of thin films, and is frequently held responsible for malfunction or failure of technologically important coating systems[1]. For instance, buckling / warping, stress cracking, and loss of adhesion are commonly observed in thin films. In addition to contact stress due to external load, other stresses in compositionally uniform thin films are: coherency stress due to lattice mismatch between film and substrate; thermal stress due to mismatch in coefficient of thermal expansion (CTE) between film and substrate; and intrinsic growth stresses produced during deposition[2,3]. Stress generation of the types described for compositionally uniform thin films can also occur in multilayered thin films[4]. The contact and thermal stresses in a coating system can be modeled directly in finite element modeling (FEM). The residual internal stresses (coherency stress, thermal stress during deposition, and intrinsic growth stress) built into the coating can be measured (e.g., by glancing incidence x-ray diffraction) and incorporated into the FEM as an initial condition. Therefore, most types of possible stresses present in a coating system can be modeled. Finite element analysis (FEA) can predict the extent of the residual stress, while the distribution of the stress/strain field generated in the FE model during mechanical and/or thermal processes, can be used to propose an optimal coating architecture to accommodate the residual stresses.

Development of a high temperature coating system for glass molding dies and forming tools is required to meet three criteria: non-sticking by molten glass, oxidation resistance, and wear resistance. Inevitably, a functionally graded thin film architecture is needed to meet these requirements. The success of the coating system design depends on the synergy of the properties of each layer to provide an optimum combination of functions. Although the performance and reliability of the system can only be truly verified by field and service tests, FEA is being developed to simulate the service conditions in an effort to substantially reduce the number of possible experimental coating systems that will be investigated. In this work, a non-linear thermomechanical FE model has been used to analyze the residual stresses generated in the coating system during a simulated glass molding process, and to predict an optimal coating architecture with minimized residual stress and optimized stress distribution. Another motivation for developing this analytical procedure is to have a tool to investigate the mechanisms behind the thermal fatigue problem presented in the glass molding process. This methodology can then be used to explore the effects of die geometry, die materials, coatings, and process parameters, etc. on die life. The commercial software MARC K7.32 was employed in this finite element analysis[5].

DESCRIPTION OF THE FINITE ELEMENT ANALYSIS

Figure 1 shows a hypothetical coating architecture proposed intuitively from a consideration of chemical bonding. Three types of film architectures on a 304 stainless steel substrate were analyzed in this FEA: (1) a 2-μm NiAl single layer; (2) a 2-μm NiAl layer with a 50-nm Ti binding layer; and (3) a NiAl/FGM(TiAlN/TiN)/Ti multilayer (as shown in Figure 1). The physical model used in the FEA is a disk shape punch with its surface coated (Figure 2a). Due to its geometric nature, an axisymmetric FE model was used. The boundary conditions are shown in Figure 2b. A press force is evenly distributed at the coating surface. It was assumed that the substrate was resting on a smooth support so that movement in the y-direction was not allowed for the nodes at the bottom of the substrate. The axisymmetric geometrical model automatically prevents displacements in the x-direction for the nodes along the symmetric axis.

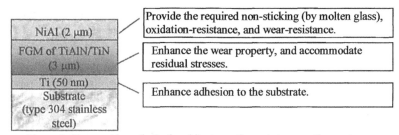

Figure 1. Proposed hypothetical architecture of a prototype coating system for glass molding dies and forming tools.

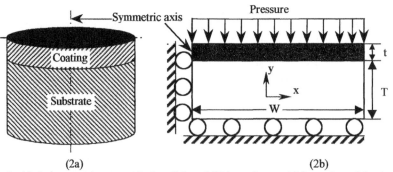

(2a)   (2b)

Figure 2. (a) Axisymmetric geometrical model, and (b) boundary conditions adopted for the FEA.

In the current FEA, the temperature and loading profiles shown in Figure 3 have been used to simulate the glass molding process. The die surface is thermally cycled between 450°C and 1000°C. The load used to press the glass through the die is 4 MPa which is approximately that force used to produce optical glass elements[6].

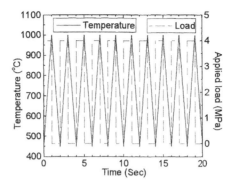

Figure 3. Profiles of temperature cycling and pressing load used in the FEM to simulate the glass molding process. A total of 10 cycles is shown.

The material properties used in the FEA are listed in Table I[7,8]. They are average values that are independent of temperature. Plasticity effects in titanium, type 304 stainless steel, and NiAl have been incorporated into the FEM. No appropriate material property data has been found for TiAlN. Therefore, the data for TiAlN has been calculated as the average of TiN and AlN. This method is only valid for composites, whereas TiAlN is a metastable compound. A FGM TiAlN/TiN intermediate layer was also modeled. The material properties in FGM region are assessed by using the linear rule of mixtures, i.e.,

$$P_{FGM} = V_{TiAlN} \times P_{TiAlN} + V_{TiN} \times P_{TiN} = V_{TiAlN} \times P_{TiAlN} + (1 - V_{TiAlN}) \times P_{TiN} \quad (1)$$

where P represents the appropriate material property and V_{TiAlN} is the volume fraction of TiAlN which is defined by the power law equation:

$$V_{TiAlN} = (x/t)^p \quad (2)$$

where t is the total thickness of the FGM region, x is the distance from the 100% TiN interface, and p is the power law exponent.

Table I. Material properties (20~1000°C) used in the FEA

Materials	Young's modulus (GPa)	Poisson's ratio	CTE α ($\times 10^{-6}$ K^{-1})	Yield strength (MPa)
NiAl	169	0.32	13.2	400
TiAlN	265	0.3	7.48	Not yield
AlN	330	0.3	5.6	Not yield
TiN	200	0.3	9.35	Not yield
Ti	113	0.3	10.1	140
304 stainless steel	130	0.3	20	200

In the FEA, four-node quadrilateral axisymmetric elements were used (as illustrated in Figure 4). The mesh was refined gradually from the substrate to the films. Other assumptions made in the FEA were: (1) perfect adhesion between film and substrate; (2) perfect thermal conductivity in all materials, i.e., the temperature is uniform everywhere in the model; and (3) internal residual stresses built in the coatings during deposition are excluded.

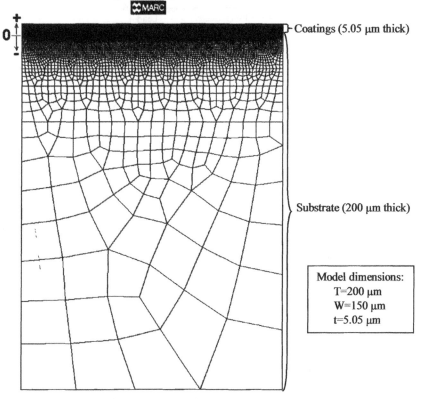

Figure 4. The mesh composed of four-node elements used for the FEA.

RESULTS

Convergence

Using FEA in thin films, there are two very important questions that needed to be answered: (1) how to deal with the huge dimensional difference between coatings and substrate, i.e., how to choose substrate thickness relative to the film stack thickness; and (2) what is the correct boundary condition to approximate a coating system with a large lateral extent? These major dimension ratios can be defined as "thickness ratio" and "geometric aspect ratio":

$$\text{Thickness ratio} = T/t \tag{3}$$

$$\text{Geometric aspect ratio} = W/t \tag{4}$$

Functionally Graded Materials 2000

where t and T are the film stack thickness and substrate thickness, and W is the width, as shown in Figure 2(b). It is well known that FE models are very sensitive to boundary conditions. The convergence procedure was conducted for various thickness ratios and geometric aspect ratios in this FEA. The thickness ratio will affect the stress/strain distribution in the coating and substrate a lot. It was found that the thickness ratio must be at least 20, in order to get the stress distribution in good agreement with the typical stress distribution in thin films discussed by Thornton and Hoffman[1]. The geometric aspect ratio will reflect the free-edge influence on the stress/strain at the center region of the model, i.e., along the symmetric axis. It was found that a geometric aspect ratio of 30 is sufficient to allow the calculation to rapidly converge. Figure 5 illustrates the effects of thickness ratio and geometric aspect ratio on the calculated stresses in coating systems.

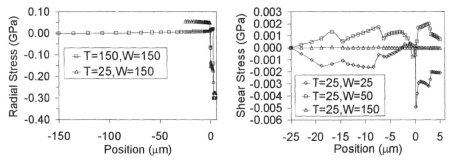

Figure 5. (a) Effect of thickness ratio on the radial stress distribution, and (b) effect of geometric aspect ratio on the shear stress distribution along the symmetric axis in the NiAl/FGM/Ti coating architecture (after 10 cycles of thermomechanical loading).

Note that the mechanical behavior of the film system includes plasticity during thermomechanical cycling and is path dependent in general. Therefore, although the material properties are not temperature dependent in this model, a certain number of loading cycles have to be used in order to get convergent results. For the model with dimensions of t=5.05μm, T=200μm, and W=150μm, the stress and strain results after modeling 9 cycles are different from those after modeling 8 cycles by only 0.1%. Thus, cycling ten times is sufficient to approach convergence. Consequently, models with dimensions of T=200 μm and W=150μm were used to analyze the architecture effects on stress/strain distribution in the coating system.

Comparison of Different Architectures

In this FEA, the contour plots demonstrate that in the center region around the symmetric axis of this axisymmetric model, the residual axial and shear stresses are each approximately zero, and the residual radial stresses are relatively uniform throughout the thickness of each layer. The interface shear stress is seen to be concentrated at the film edge. This result is in good agreement with the typical stress distribution in thin films discussed by Thornton and Hoffman[1].

Figure 6 demonstrates the effect of a layered structure/architecture on the residual stress/strain results after 10 cycles of operation. It is shown that: (1) the introduction of a 50 nm Ti layer changes the compressive radial stress and the total equivalent plastic strain in the outer NiAl layer a little; (2) using an intermediate layer of a linear (p=1) FGM TiAlN/TiN considerably alters the stress and strain distribution, and decreases the level of radial stress and total equivalent plastic strain in the coatings.

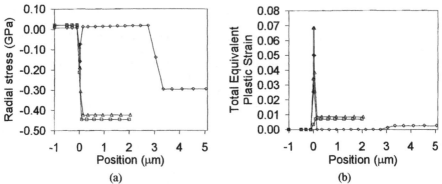

(a)   (b)

Figure 6. Comparison of (a) radial stresses and (b) total equivalent plastic strains along the symmetric axis at the end of the 10th cycle for different film architectures: NiAl monolayer (□), NiAl/Ti bilayer (Δ), and NiAl/FGM($p=1$)/Ti multilayer (◊).

As shown in Figure 7, tensile radial stresses are generated in the coatings at the highest temperature (1000°C) in the 10th cycle of the simulated glass molding process. The tensile radial stresses at the outer NiAl layer are approximately the same for the three coating architectures studied. The tensile stresses at the TiAlN/TiN FGM region is very high (the highest level computed is about 2 GPa). This is high enough to possibly result in failure of the TiAlN/TiN FGM coatings. The tensile strength of these materials should be further tested to make sure if this level of tensile stress is likely to cause any problem.

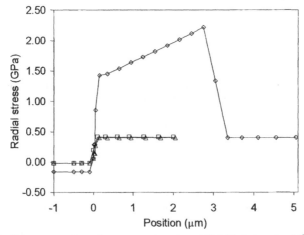

Figure 7. Radial stresses along the symmetric axis at 1000°C during the 10th cycle for different film architectures: NiAl monolayer (□), NiAl/Ti bilayer (Δ), and NiAl/FGM($p=1$)/Ti multilayer (◊).

FEA was also conducted to investigate the effect of different p values in the power law equation (2) for the FGM intermediate layer. Changing the power (p) in the FGM region, the stress levels in the substrate will change little, especially at the highest temperature (1000°C). However, as shown in Figure 8, this will affect the stress distribution profile in the FGM region considerably at highest temperature (1000°C), and alter the stress level in the outer NiAl layer at lowest temperature (450°C). Namely, the stress distribution profile $\sigma(x)$ in the FGM region will reflect the trends of equation (2). For example, $\sigma(x)$ versus position x is a linear relationship when p is equal to 1. This phenomenon resulted from the way the material properties were assessed for the FGM layer. Consequently, it can be used to engineer the stress profile in the transition region between the outmost working layer and the substrate. The case with $p=0.2$ showed the smallest compressive residual stresses in the NiAl layer after 10 cycles.

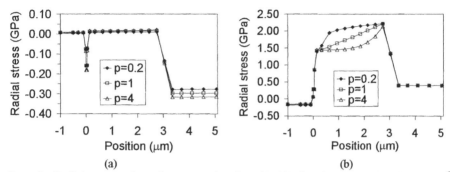

Figure 8. Radial stresses along the symmetric axis at (a) 450°C and (b) 1000°C during the 10th cycle for the graded film architectures, NiAl/FGM/Ti, with varied p values in the power law equation.

CONCLUSIONS

In this preliminary FEA, the following results have been obtained:

(1) A geometric aspect ratio of 30 was needed to eliminate the free-edge effect on the region close to the symmetric axis, and a thickness ratio of 30 was also needed to obtain the correct stress distribution in a coating system.

(2) Along the symmetric axis, i.e., the center region of a disk-geometry model, axial and shear stresses remain approximately zero. The residual radial stresses change from compressive at 450°C to tensile at 1000°C during a simulated glass molding cycle.

(3) Compared to a NiAl single layer and NiAl with a Ti binding layer on an AISI 304 stainless steel substrate, the introduction of a TiAlN/TiN FGM ($p=1$) intermediate layer decreases the level of residual stress and strain in the films, and changes the stress transition profile from the substrate to the outmost working surface layer.

These results suggest that the graded coating architecture will benefit the performance of the coating system, and it is possible to find the optimized residual stress level and stress transition by engineering the compositional gradient of the FGM layer.

Further modifications of this FE model should be: (1) include temperature dependent material properties into the FEM, and Young's moduli of films measured by nano-indentation as well; (2) experimentally measure the residual deposition stresses in the coatings, e.g., by GIXRD, and incorporate these stresses into the FEM; and (3) include thermal conductivity values and produce a thermal-mechanical coupled FEA. As a result, an optimal coating architecture can be proposed, while potential failure mechanisms and thermal fatigue performance can be predicted.

ACKNOWLEDGEMENTS

The authors gratefully acknowledge the financial support provided for this project by Schott Glas, Mainz, Germany, and a Li Foundation Fellowship for D. Zhong.

REFERENCES

[1] J.A. Thornton and D.W. Hoffman, "Stress-Related Effects in Thin Films," *Thin Solid Films*, 171 5-31 (1989).

[2] R. Koch, "The Intrinsic Stress of Polycrystalline and Epitaxial Thin Metal Films," *J. Phys.: Condens. Matter*, 6 9519-9550 (1994).

[3] H. Windischmann, "Intrinsic Stress in Sputtered Thin Films," *J. Vac. Sci. Technol. A*, 9 [4] 2431-2435 (1991).

[4] R.C. Cammarata, J.C. Bilello, A.L. Greer, K. Sieradzki, and S.M. Yalisove, "Stresses in Multilayered Thin Films," *MRS Bulletin*, February 34 (1999).

[5] MARC User Manual (Version K7.32), MARC Analysis Research Corporation, Palo Alto, 1997.

[6] Kashiwagi et al., "Method of Manufacturing a Die for Use in Molding Glass Optical Elements Having a Fine Pattern of Concavities and Convexities," U.S. Patent No. 5 405 652, April 11, 1995.

[7] J.F. Shackelford et al., *CRC Materials Science and Engineering Handbook (2nd Edition)*, CRC Press, 1994.

[8] J.R. Davis et al., *ASM Specialty Handbook*, ASM International, 1994.

MODELLING OF ALUMINA/COPPER FUNCTIONALLY GRADED MATERIAL

M.Gasik and M.Friman, Helsinki
University of Technology,
Vuorimiehentie 2, Espoo, P.O.Box
6200, FIN-02015 HUT-Finland

M.Kambe, Central Research Institute of
Electric Power Industry (CRIEPI) 11-1,
Iwado-Kita, Komae-shi, Tokyo, 201-8511
Japan

ABSTRACT

An alumina-copper functionally graded material (FGM) joint was examined to determine the optimum grading of composition. The aim was to find the best grading profile in regard to minimal residual thermal stress and maxmal thermal conductivity. The modelling of the Al_2O_3-Cu system was done at two application temperatures, 20°C and 700°C. Thermal stress is generated in the joint due to the different thermal expansion coefficients of Al_2O_3 and Cu. The linear plate model supplemented with a micromechanical model was used for thermoelastic behavior of the joint. The properties of the materials were evaluted with the micromechanical model for different grading functions in the FGM layer.

A thin Al_2O_3-rich FGM layer is found to be better from the point of view of "stress-electrical resistance-thermal conductivity" combination. Increasing the amount of copper in the graded layer makes residual stress dominant though thermal conductivity increases.

INTRODUCTION

A thermoelectric module converts thermal energy (heat) to the electric energy. It has no moving parts, which makes it reliable in long-term use. It does not break due to mechanical wear, but material aging may become a problem. Long service periods without maintenance makes thermoelectric module interesting option for electric production for the industry, if the conversion effeciency and energy density are good enough. For space applications, thermoelectric conversion systems have been successfully used.

Thermoelectric module combines electrical conductors, semiconductors and insulators. Semiconductors are connected to copper/alumina pads in series electrically and in parallel thermally. The alumina layer insulates the module electrically from the energy source. The insulator must be capable to operate and provide a good electrically insulating layer at the high temperature and resist as little as possible to heat flow. Thermal energy flows through the thermoelectric module forming a thermal gradient in the thermoelectric modules, thus producing electric current, Figure 1.

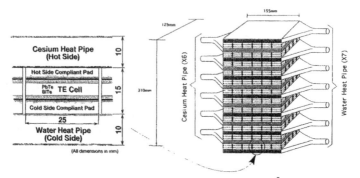

Figure 1: Thermoelectric module structure.[2]

Residual stresses are generated in the joint due to the different thermal expansion coefficients of alumina and copper. Thermal stresses are also caused by a temperature gradient in the thermoelectric element. On the hot side the temperature is usually much higher than on the cold side of the thermoelectric module, e.g. 840°C and 530°C[1,3].

It is not enough that thermal stresses are low at the operating temperature. The component also has to withstand thermal cycling between standby and operating temperatures. The problem is to find a good compromise between thermal conductivity and thermal stress for operation and standby temperatures. It is possible to preserve high thermal conductivity, electrical resistance and fracture toughness by tailoring thermal expansion coefficients in the joint by grading alumina functionally in the copper matrix.

In this work, the plate model supplemented with a micromechanical model was used for thermoelastic modelling of the joint. An alumina-copper system with different gradients and temperatures was examined to determine the optimum grading of compositions for an alumina-copper FGM joint. The aim was to find the best grading in regard to residual elastic thermal stress and thermal conductivity.

CALCULATION

The alumina and copper joint properties were calculated with different gradient functions and thicknesses. The alumina thickness was kept constant at 1 mm. The copper and FGM layers were placed symmetrically on both sides of the alumina layer, and the thickness of the component was calculated by adding up these layers, Figure 2.

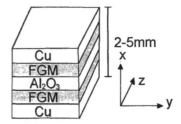

Figure 2: Structure of the calculated FGM component.

The graded layer was composed of copper and alumina. The phase distribution of the components in the layer was defined by:

$$V(x) = \left(\frac{x}{h}\right)^p \tag{1}$$

The p value defines the shape of the material distribution (linear, convex, concave), h is the thickness of the graded layer and $V(x)$ presents the volume fraction of Al_2O_3.

After forming the gradient distribution function, a material property function can be determined. The contribution of the component's properties is not necessarily the same as may be estimated from the volume fraction of the components.

The modulus of elasticity can be calculated for copper- and alumina-rich phases of FGM in the graded layer from[4,5]:

$$E_{Cu}^G(x) = E_{Cu} \cdot \left(1 + \frac{V(x)}{FE_{Cu} - (V(x))^{\frac{1}{3}}}\right), \quad E_{Al_2O_3}^G(x) = E_{Al_2O_3} \cdot \left(1 + \frac{1 - V(x)}{FE_{Al_2O_3} - (1 - V(x))^{\frac{1}{3}}}\right), \tag{2,3}$$

where

$$FE_{Cu} = \frac{1}{1 - \dfrac{E_{Cu}}{E_{Al_2O_3}}}, \quad FE_{Al_2O_3} = \frac{1}{1 - \dfrac{E_{Al_2O_3}}{E_{Cu}}} \tag{4,5}$$

(Copper and alumina are assumed to be homogeneous phases.).

The total modulus of elasticity for the graded layer is calculated from these moduli of elasticity[4,5]:

$$E^G(x) = \left(\frac{V(x)}{E_{Cu}^G(x)} + \frac{1 - V(x)}{E_{Al_2O_3}^G(x)}\right)^{-1} \tag{6}$$

The plate theory analyzes stresses and curvatures which have been developed during mechanical and thermal loading of the layered solid. The component investigated is divided into layers.

The layers change in the direction of the x axis. The thickness of the layer is much smaller than the other dimensions of the layer. The edge effects of the plate are ignored. Strain is considered to be a function of x and only the initial stage is stress-free,

$$\varepsilon_x(x) = \frac{2 \cdot v(x)}{1 - v(x)} \cdot \varepsilon_y(x) + \frac{1 + v(x)}{1 - v(x)} \cdot \alpha(x)\Delta T \tag{7}$$

where $\varepsilon_y = \varepsilon_z$ is the strain in the direction of the y and z axes:

$$\varepsilon_y(x) = C_1 \cdot x + C_2 \tag{8}$$

where C_1 is the curvature of the plate and C_2 is the strain at $x=0$. Stress σ can be calculated from:

$$\sigma(x) = \frac{E(x)}{1 - v(x)}\left(\varepsilon_y(x) - \alpha(x)\Delta T(x)\right) \tag{9}$$

The equilibrium state from stress-strain relations gives:

$$\begin{pmatrix} J_0 \\ J_1 \end{pmatrix} = \begin{pmatrix} I_0 & I_1 \\ I_1 & I_2 \end{pmatrix}\begin{pmatrix} C_2 \\ C_1 \end{pmatrix} \tag{10}$$

These two equilibrium equations result in a linear system in C_1 and C_2, where I_i is calculated from:

$$I_i = \int_{-\frac{h}{2}}^{\frac{h}{2}} \frac{E(x)}{1 - v(x)} \cdot x^i dx, \ (i = 0,1,2) \tag{11}$$

and J_i is calculated from:

$$J_i = \int_{-\frac{h}{2}}^{\frac{h}{2}} \frac{E(x)}{1 - v(x)} \cdot x^i \cdot \alpha(x) \cdot \Delta T dx, \ (i = 0,1) \tag{12}$$

The change of elastic modulus in the layers is notified as a function of the x axis.[6,7]

RESULTS

The thermal stresses analysis was done to get the stresses in one component and to determine whether the FGM layer reduces them significantly. It is possible to estimate whether the local thermal stress exceeds the endurance of the material locally, which will lead to the breaking of the joint.

The thermal resistance indicates the ability of thermal energy (heat) to flow through a component. A low thermal resistance is desired, because the function of the alumina layer is to act as an electrical insulator, not as a thermal insulator.

The calculations at 20°C indicate that residual thermal stress relaxation for the 5 mm component with 0.5 mm FGM layer is almost as good as with 1.5 mm FGM layer. The relaxation compared to the bimaterial system (without FGM) is significant around the linear FGM compositional profile ($p\approx1$), Figure 3. Lowering the thickness of the component decreases the residual thermal stress in the bimaterial system.

The thermal resistance of the 5 mm component varies greatly on the alumina-rich side of the graded layer, and the size of the FGM layer affects thermal conductivity significantly. Variation decreases with increasing copper quantity. On the copper-rich side the thermal conductivity is nearly as good as that of bimaterial systems.

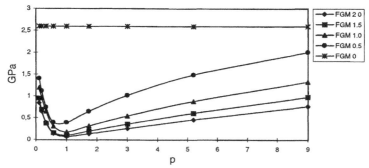

Figure 3: Residual maximum thermal stress differences (GPa) between neighbour layer points for the 5 mm component with different FGM thicknesses (mm) and gradients (p) at 20°C.

The thermal stress differences vary little with a p value of 1 and different gradient layer sizes and component thicknesses, Figure 4. With lower p values the variation remains small. Increasing a p value enlarges the differences in thermal stress differences.

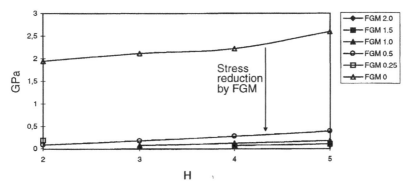

Figure 4: Effect of different components (H) and graded layer thicknesses (mm) on the maximum thermal residual stress differences (GPa) between neighbour layer points at 20°C and $p=1$.

Residual thermal stress differences and thermal resistivity dependences were analysed to get an optimum p range for the gradient function where stress-resistance product is minimal (the artificial figure of merit).

The residual thermal stress data show that the FGM layer thickness directly affects the stress differences. A 2.0 mm FGM layer has a lower stress than a 0.5 mm layer, and the p value changes primarily the magnitude of stress difference between the layers.

Thermal resistance decreases almost linearly in the logarithmic scale with different FGM layer thicknesses when the p value increases. The effect of the amount of copper in the FGM layer is the strongest in the alumina-rich grading range of the FGM layer, Figure 5.

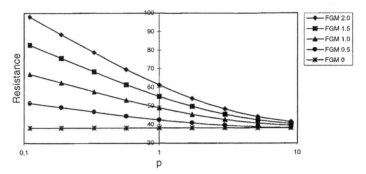

Figure 5: Thermal resistance ($\times 10^{-6}$ m^2K/W) for the 5 mm component with different FGM thicknesses and gradient functions at 20 °C.

The stress and thermal resistivity results show the effect of the FGM layer thickness on the stress-resistance product. A thin FGM layer gives a better result in the alumina-rich area than a thicker FGM layer. When copper amount increases, in thicker FGM layers stress relaxation properties becomes dominant over the importance of resistance which becomes similar regardless of the FGM layer thickness. The effect is clearly seen also with 3 and 4 mm components, but it disappears on the alumina-rich side of the grading with thin FGM layers and 2 mm component. The optimum p range for such joint (Figure 2) is from 0.75 to 0.9 at 20°C and from 0.6 to 0.7 at 700°C.

EXPERIMENTAL

Test samples were made by powder metallugy method and sintered by hot pressing and spark plasma sintering. All specimens were crack-free and show good electrical resistivity (>40 MΩ) through the thickness.Testing of other properties of the samples is planed to do in future of this project, Figure 6.

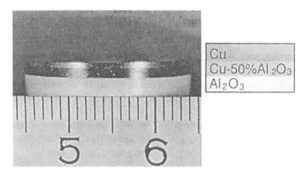

Figure 6: Manufacturing of Cu/Al_2O_3 FGM sample was achieved under the direction of Prof. Kawasaki of Tohoku University.[8]

CONCLUSIONS

The calculations show that the graded layer reduces thermal stresses in the Cu/Al$_2$O$_3$/Cu component, and thermal conductivity is proportional to the amount of copper in the structure. Thermal stress reductions are significant in the FGM case. With good thermal stress relaxation, thermal conductivity may be kept still high in relation to an equivalent bimaterial (non-draded) component. From the point of view of stress, the best p value for the grading function is less than 1.0 at 20°C and at 700°C it is shifted more towards lower p values.

The cold-side (20°C) component can be manufactured with a p range 0.7-1.0 of the grading function. The FGM layer thickness can be equal to 0.5 mm or more and the total thickness of the component can be selected from the point of view of energy conversion module design. The thermal conductivity will be good enough in this case.

The hot-side (700°C) component can be manufactured with a p range 0.5-0.9 of the grading function. The FGM layer should be as thin as possible, but not less than 0.25 mm, because then the residual thermal stresses will increase significantly. High temperatures reduce the thermal conductivity and therefore a thinner component is preferable.

ACKNOWLEDGEMENTS

This work is supported by National Technology Agency of Finland (TEKES), Scandinavia-Japan Sasakawa Foundation, Foundation of Helsinki University of Technology, Japan Science and Technology Corporation as well as Academy of Finland in the Framework of Finnish-Japanese research cooperation.

REFERENCES

[1]M. Kambe, "High Energy Density Thermoelectric Energy Conversion Systems by Using FGM Compliant Pads for Space and Terrestrial Applications," 48th International Astronautical Congress October 6-10, 1997/Turin, Italy.

[2]M. Kambe, "Thermoelectric Direct Energy Conversion System for Fast Reactors," CRIEPI Report (in Japanese), T94032, May 1995, 23 pp.

[3]M. Kambe, "High Energy Density Thermoelectric Energy Conversion Unit by Using FGM Compliant Pads," Proceedings of the 5th International Symposium on Functionally Graded Materials, Dresden, Germany, October 1998, p 1-6.

[4]M. Gasik, "Principles of the Functional Gradient Materials and their Processing by Powder Metallurgy," Acta Polytechnica Scandinavica, Chemical Technology Series No.226, Helsinki University of Technology, Helsinki, 1995, 73 pp.

[5]M. Gasik, "Micromechanical Modelling of Functionally Graded Materials," Computational Material Science **13** 42-55 (1998).

[6]Y. Stavsky and N.J. Hoff , "Mechanics of Composite Structures,"; pp 5-59 in: Composite Engineering (ed.) A.G.H. Dietz, The MIT Press, London, 1969.

[7]B.A. Boley and J.H. Weiner, "Theory of Thermal Stresses," 2nd Ed. R.E. Krieger Publishing Co., Malabar, FL, 1985, 586 pp.

[8]M. Kambe et al., "Conceptual design of a FGM Thermoelectric Energy Conversion System for High Temperature Heat Source (2)", CRIEPI Report (in Japanese), T96041, June 1997, 25 pp.

Properties Characterization

INVESTIGATIONS OF THE RESIDUAL STRESS STATE IN MICROWAVE SINTERED FUNCTIONALLY GRADED MATERIALS

D. Dantz*, Ch. Genzel and W. Reimers   T. Buslaps
Hahn-Meitner-Institut Berlin   European Synchrotron Radiation Facility
Glienicker Str. 100   BP 220
D-14109 Berlin, Germany   F-38043 Grenoble Cedex, France

ABSTRACT

The residual stress distribution in microwave sintered functionally graded materials (FGM) consisting of Ni and 8Y-ZrO_2 was analyzed by non-destructive diffraction experiments. In order to evaluate the residual stress state in the near surface region as well as in the bulk of the material, complementary methods were applied. Using conventional X-ray sources the residual stresses at the surface were investigated by the $\sin^2\psi$-method. For the interior of the material high energy synchrotron radiation was used allowing high spatial resolution. The residual stress state was found to be related to the compositional distribution of the present phases, their volume fraction and their coefficients of thermal expansion as well as to the additive $ZrSiO_4$ in the ceramic rich region of the gradation. Additionally, the line broadening was analyzed by a single-line method with respect to plastic deformation of the metallic phase, in order to characterize the microstructure of the samples.

INTRODUCTION

In order to meet the demand for adjusting defined property gradients within technical parts the material concept of functionally graded materials (FGM) has been proposed [1]. Residual stress concentrations at the interface area or at the free edge of the FGM can be avoided by a continuous variation of the compositional distribution and, therefore, the related properties [2]. But as a result of the different coefficients of thermal expansion (CTE) and the elasto-plastic behavior of the components forming the material strong micro stresses may arise as a consequence of thermal or mechanical loads. The mechanical properties of the material are well-known to depend on these stresses significantly. Therefore the residual stresses must be critically regarded with respect to failure, because they are often superimposed under load conditions [3].

The present paper deals with the experimental determination of micro and macro residual stresses in Ni/8Y-ZrO_2 FGMs by diffraction methods. The investigations at the surface were performed by the $\sin^2\psi$-method using conventional X-ray sources [4]. The experiments concerning the residual stresses in the bulk of the material were carried out

*Now at Wacker Siltronic AG, P.O. 1140, 84479 Burghausen, Germany, dirk.dantz@wacker.com

using high energy X-ray diffraction (HXRD) [5, 6]. The penetrating power of high energy X-rays enables through-thickness strain profiling without any material removal. In addition the diffraction line broadening was estimated by a single-line method for profile analysis concerning the mean squared stresses $\langle e \rangle$ and the size of the crystallites [7].

BASIC PRINCIPLES OF THE APPLIED DIFFRACTION METHODS

X-ray Residual Stress Analysis (XSA)

XSA at the surface of polycrystalline materials by the classical $\sin^2\psi$-method is based on the determination of the lattice spacings $d(hkl)$ for different angle sets (φ,ψ) with respect to the sample system P. According to HOOKE's law from the lattice strains $\varepsilon(hkl) = d(hkl)/d_0(hkl) - 1$ (d_0 – lattice spacing of the stress free state) obtained at various inclination angles ψ the corresponding stresses are calculated by linear regression [4]. In multi-phase materials the average phase specific residual stresses $\langle\sigma\rangle_{\alpha,\beta}$ are experimentally accessible. The average with respect to the volume fraction yields the macro residual stresses $\sigma^I = V_\alpha\langle\sigma\rangle_\alpha + V_\beta\langle\sigma\rangle_\beta$ (V_α - volume fraction of the phase α, β). From the average phase specific and the macro residual stresses the micro residual stresses $\langle\sigma^{II}\rangle_{\alpha,\beta} = \langle\sigma\rangle_{\alpha,\beta} - \sigma^I$ can be derived.

Investigations Using High Energy Synchrotron Radiation

Residual stress analysis in the bulk of the material In energy dispersive X-ray diffraction the correlation between the lattice spacings $d(hkl)$ and the corresponding energy peak $E(hkl)$ is obtained by $d(hkl) = hc/[2\sin\theta \cdot E(hkl)]$, where h is PLANCK's constant and c is the light velocity. The average values of the phase specific stresses can be calculated according to HOOKE's law for the triaxial stress state [4]. Recording the complete energy spectrum available at some fixed detector position 2θ offers the possibility to include a series of reflections in the stress evaluation enhancing the reliability of the results [6].

Line profile analysis The line profiles of diffraction patterns are influenced by instrumental broadening and by the microstructure of the crystals. The line breadth is affected by both the size of the coherent reflecting domains as well as the micro strain in the crystal lattice. Assuming that the diffracted profile is a convolution of the reference and the physical profile, which can be described by a VOIGT-function, the single-line method was applied. Since the CAUCHY-contribution of the corrected integral breadth is mainly affected by size effects, the GAUSS contribution predominantly reflects the line broadening due to the mean squared micro strain [8]. This method is converted to the energy dispersive case. For details see [7] and [9].

EXPERIMENTAL DETAILS

Samples

The samples of the material system Ni/8Y-ZrO$_2$(ZrSiO$_4$) investigated consist of 11 sublayers, in which the ceramic content was successively varied by steps of 10 vol.% from 0 to 100 vol.%. The graded powder specimens were pressureless sintered in a mi-

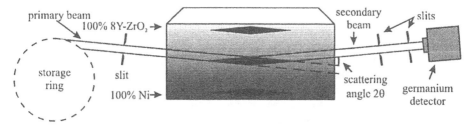

Fig. 1: Experimental setup for residual stress analysis using high energy synchrotron radiation [12]

crowave field and subsequently cooled down to room temperature. Due to the different absorption of the microwave radiation by the ceramic and metal phase, a temperature gradient of 100-200°K from the ceramic to the metal side arises during the sintering process [10]. The parametric description of the phase distribution along the gradient direction is described by the expression $V_\alpha = (x/t)^n$. V_α is the volume fraction of component α, t the thickness of the whole gradation and n denotes the distribution exponent. Specimens consisting of Ni/8Y-ZrO$_2$ were investigated, two of them with a non-linear gradation ($n = 1,3$ and $n = 2$), and one with a linear metal/ceramic gradation ($n = 1$), in which the 8Y-ZrO$_2$ was partially substituted by ZrSiO$_4$ in the sub-layers with more than 50 vol.% ceramic volume fraction [11].

Diffraction Methods

X-ray residual stress analysis The experiments using conventional X-ray sources were performed with CoK$_\alpha$-radiation and a Huber Ψ-diffractometer, which was equipped with a φ-rotation table as well as with a parallel beam unit consisting of a vertical soller-slit followed by (001)-LiF analysator in order to suppress fluorescence as well as the horizontal divergence of the diffracted beam for measurements at high ψ-tilt angles (77°). The measurements were performed at the Ni (222)-reflection ($2\theta \approx 123.3°$), at the 8Y-ZrO$_2$ (511)-reflection ($2\theta \approx 129.6°$) and at the ZrSiO$_4$ (312)-reflection ($2\theta \approx 63.1°$).

Experiments using high energy synchrotron radiation Due to the rather low absorption of high energy synchrotron radiation, penetration depths in the range of centimeters can be achieved and the experiments can be performed in the transmission mode. The investigations in the bulk were carried out at the beamline ID15A at the European Synchrotron Radiation Facility (ESRF). In Fig. 1 the experimental setup is shown. An energy dispersive Ge-detector was used to record the diffraction pattern at a fixed angle of $2\theta = 7.6°$ within an energy range from 30 to 200 keV. Due to the high photon flux of the synchrotron radiation the gauge volume could be limited to 0.11 x 0.12 x 1.6 mm^3 by slits in the primary as well as in the secondary beam. The high spatial resolution allows measurements in each sublayer perpendicular and parallel of the gradation direction as well as in an individual sublayer [13]. In addition the microstructure of the metal phase was investigated at the Ni-(222)-reflection by a single-line method of the profile analysis. For the residual stress analysis in the interior as well as at the surface of the specimens the diffraction elastic constants (DEC) have been calculated by the KRÖNER-model [14].

RESULTS AND DISCUSSION

Prior investigations showed that the residual stress state parallel to the sublayer interfaces is of rotational symmetry. In Fig. 2a the results of the residual stress analysis in the bulk of the sample in the material system Ni/8Y-ZrO$_2$ are presented. The out-of-plane stresses perpendicular to the interfaces have the same amount as the in-plane components of the phase specific residual stresses (Fig. 2a). Therefore, an almost hydrostatic residual stress state may be concluded for the interior of the material. The high compressive stresses in the 8Y-ZrO$_2$ decrease with increasing ceramic volume fraction. They are partially compensated by tensile stresses in the Ni. These tensile residual stresses in the metal phase are nearly constant for a ceramic content of about 30 vol.% up to amounts of

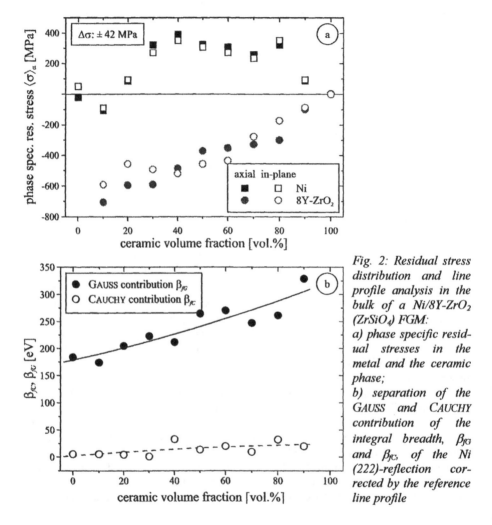

Fig. 2: Residual stress distribution and line profile analysis in the bulk of a Ni/8Y-ZrO$_2$ (ZrSiO$_4$) FGM: a) phase specific residual stresses in the metal and the ceramic phase; b) separation of the GAUSS and CAUCHY contribution of the integral breadth, β_{fG} and β_{fC}, of the Ni (222)-reflection corrected by the reference line profile

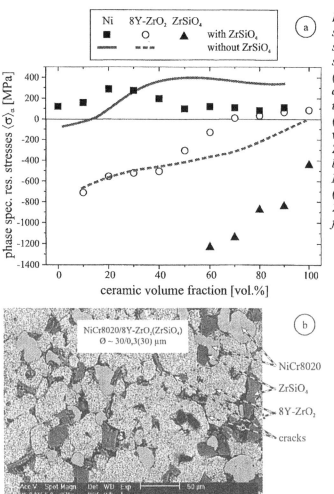

Fig. 3: Residual stress state in the bulk of the sample in the material system Ni/8Y-ZrO₂ (ZrSiO₄):
a) phase specific residual stress distribution, (dashed lines – sample without the addition of ZrSiO₄)
b) crack network in a NiCr8020/8Y-ZrO₂ (ZrSiO₄) FGM, 20 vol% metal volume fraction

80 vol.%. This points to a relaxation of the stresses due to plastic deformation of the metal component.

In order to analyze these plastic effects, the line broadening was investigated by a single-line method, which was converted to the energy dispersive case. The results are shown in Fig. 2b. The CAUCHY contribution of the corrected integral breadth β_{fC}, which is predominantly affected by size effects [8], remains nearly constant over the whole cross section of the graded plate. On the other hand, the GAUSS contribution β_{fG} which mainly reflects the line broadening due to the mean squared micro strain $\langle e \rangle$ caused by lattice defects such as dislocations and stacking faults, shows higher values. With de-

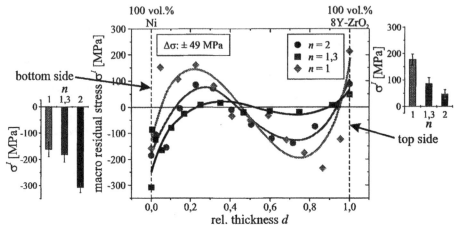

Fig. 4: Macro residual stresses in the bulk (in the large plot) and on the top and bottom side (on the dashed lines of the large plot and in the small plots) of Ni/8Y-ZrO$_2$(ZrSiO$_4$) FGMs with different compositional distribution profiles

creasing metal volume fraction the GAUSSian integral breadth increases, which is an additional indication for a stress relaxation in the metal phase due to plastic deformation.

Fig. 3a shows the phase specific residual stresses in the bulk of the sample with a linear gradation and partially substituted 8Y-ZrO$_2$ by ZrSiO$_4$. Due to the very low coefficient of thermal expansion high compressive stresses are observed in the silicate phase. At low ceramic contents the results of the stresses in the metal and the zirconia are in good agreement to the findings obtained for the FGM without the silicate additive (Fig. 2a). With regard to the phase specific stresses this holds true for both the run as well as the absolute amount of the profiles in both samples.

At the ceramic rich region in the zirconia lower compressive stresses can be found in comparison to the sample without the additive, which are finally changing to tensile values. This indicates a compensation of the high compressive stresses in the silicate by the zirconia matrix. On the other hand, the phase specific stress in the metal phase is decreasing significantly at the presence of ZrSiO$_4$. First of all these findings are surprising, because the average CTE of the ceramic phases is decreasing due to the additional silicate phase, higher or even equal tensile stresses are expected. But, during cooling down from sintering to room temperature high tensile stresses arise in the 8Y-ZrO$_2$ near the interfaces to the silicate grains due to the mismatch of the CTE's. At low temperature these stresses can exceed the strength of the zirconia matrix and initiate cracking. As shown in Fig. 3b these cracks start at the silicate grains and, finally, lead to a network of cracks connecting the ZrSiO$_4$ particles in the 8Y-ZrO$_2$ matrix and, therefore, to relaxation of the tensile phase specific stresses in the Ni.

The residual stress distributions on the macroscopical scale in the bulk of FGMs with different compositional profiles are shown in Fig. 4. Additionally the macro residual stresses on the top side (pure ceramic) and on the bottom side of the sample (pure metal) are plotted on the dashed lines as well as in the small diagrams.

On the top tensile macro residual stresses and on the bottom side compressive stresses can be found irrespective of the compositional distribution. This is in good agreement to the expected macroscopic stress profile due to the curvature resulting from the average values of the coefficients of thermal expansion across the gradation [15]. Due to the temperature gradient during sintering a similar residual stress distribution can be developed. During cooling down an additional temperature difference of 100 K occurs between the pure ceramic and the pure metal side.

In the bulk of all samples compressive stresses can be found in the ceramic rich region which are balanced by tensile stresses in the metal dominated region. In the range of the experimental deviation the stress free layer is in the geometrical middle of the samples. On the one hand this means 50 vol.% ceramic content for a linear gradation, while on the other hand for an exponent of $n = 2$ this corresponds to a ceramic volume fraction of more than 60 vol.%. Both, the compressive stresses in the ceramic rich bulk and the tensile stresses on top of the sample are decreasing with an increasing exponent n. The macro stresses in the ceramic rich region can be spread to the higher cross sections of these layers. Since at the pure metal bottom side of the sample higher compressive stresses can be observed with an increasing exponent, as expected from the smaller thickness of the layers, in the bulk of the metal rich area lower tensile macro residual stresses are developed. This indicates a relaxation of the tensile stresses by plastic deformation of the metal phase, which leads furthermore to the shifting of the tensile peak macro stresses to higher ceramic contents.

CONCLUSIONS

The residual stress state in microwave sintered metal/ceramic graded specimens can be characterized by different diffraction methods. It is shown, that high energy X-ray diffraction with its high spatial resolution is a well suited method for the evaluation of the residual stress state in the bulk of FGMs even with steep compositional gradients.

The results show, that the residual stress state is influenced by a multitude of parameters. These different parameters have both, benefits and disadvantages with respect to the residual stress distribution. An increasing exponent n leads to decreasing macro stresses in the bulk of the graded plates, while the tensile peak stress is shifted to the ceramic rich region. On the one hand the tensile stresses in the Ni decrease due to the admixture of the silicate phase, on the other hand the compressive stresses in the silicate itself have to be compensated by tensile stresses in the brittle zirconia. The long range stresses caused by the different thermal expansion behaviour are superimposed by them originated by the temperature gradient. This leads to decreasing values in the bulk and to increasing macro stresses at the top and bottom side of the samples. The ideal residual stress distribution in FGMs depends on the demands from both, the manufacturing process and the service conditions. By a variation of the individual parameters the maximum as well as the position of the residual stresses can be varied and, therefore, the thermomechanical properties of functionally graded materials can be improved.

ACKNOWLEDGEMENT

The authors are obliged to Prof. Dr. M. Willert-Porada and Dr. R. Borchert for the provision of the specimens. The financial support of the German Research Foundation (DFG) is gratefully acknowledged, reference no. Re 688/23, 1-3.

REFERENCES

[1]   R. L. Williamson, B. H. Rabin, J. T. Drake, "Finite Element Analysis of Thermal Residual Stresses at Graded Ceramic-Metal Interfaces. Part I. Model, Description and Geometrical Effects" *Journal of Applied Physics* **74** 1310-20 (1993)

[2]   S. Suresh, A. Mortensen, "Functionally Graded Metals and Metal-Ceramic Composites: Part 2: Thermomechanical Behaviour" *International Materials Reviews* **42** [3] 85-116 (1997)

[3]   A. E. Giannakopoulos, S. Suresh, M. Finot, M. Olsson, "Elastoplastic Analysis of Thermal Cycling: Layered Materials with Compositional Gradient" *Acta Metallurgica et Materialia* **43** [4] 1335-1354 (1995)

[4]   E. Macherauch, P. Müller, " The sin²ψ-Method of the X-ray Residual Stress Analysis" *Zeitung für angewandte Physik* **13** [7] 305-312 (1961)

[5]   D. Dantz, Ch. Genzel, W. Reimers, "Microstructural Investigations and Residual Stress Analysis of ZrO₂/Ni Functionally Graded Materials" *Microstructural Investigations and Analysis Euromat '99* **4** 157-163 (1999)

[6]   W. Reimers, M. Broda, G. Brusch, D. Dantz, K.-D. Liss, A. Pyzalla, T. Schmackers, T. Tschentscher, "Evaluation of Residual Stresses in the Bulk of Materials by High Energy Synchrotron Diffraction" *Journal of Non-Destructive Evaluation* **17** [3] 129-140 (1998)

[7]   J. I. Langford, "A Rapid Method for Analysing the Breadths of Diffraction and Spectral Lines using the Voigt Function" *Journal of Applied Crystallography* **11** 10-14 (1978)

[8]   J. I. Langford, R. Delhez, T. H. de Keijser; E. J. Mittemeijer, *Australian Journal of Physics* **41** 173-187 (1988)

[9]   J. W. Otto, " On the Peak Profiles in Energy-Dispersive Powder X-ray Diffraction with Synchrotron Radiation " *J. Appl. Cryst.* **30** (1997) 1008-1015

[10]   M. A. Willert-Porada, R. Borchert; "Microwave Sintering of Metal-Ceramic FGM" *Proceedings of the 4ᵗʰ International Symposium on FGM '96* Tsukuba, Japan 349 (1996)

[11]   R. Borchert, M. Willert-Porada, "Pressureless Microwave Sintering of Metall-Ceramic Functional Gradient Materials", *Microwaves: Theory and Application in Materials Processing IV*, Hrsg.: E.D. Clark, D. Lewis, W. Sutton, B. Krieger 491-498 (1997)

[12]   D. Dantz, Ch. Genzel, W. Reimers, "Analysis of Macro and Micro Residual Stresses in Functionally Graded Materials by Diffraction Methods" *Materials Science Forum* **308-311** (Proc. 5ᵗʰ Int. Symp. on FGM '98, Dresden, Germany) 829-836 (1998)

[13]   D. Dantz, "Residual Stresses in Microwave Sintered Ni/8Y-ZrO₂ and NiCr8020/8Y-ZrO₂ Functionally Graded Materials", Dissertation Technical University Berlin August 2000

[14]   E. Kröner, "Calculation of the Elastic Constants of the Polycrystal from the Constants of the Single Crystal" Zeitung für Physik **151** 504 (1958)

[15]   S. Suresh, A. Mortensen, "Fundamentals of Functionally Graded Materials", IOM Communications Ltd. London **698**, University Press Cambridge UK (1998)

SPATIALLY RESOLVED THERMAL DIFFUSIVITY MEASUREMENTS FOR FUNCTIONALLY GRADED MATERIALS

Hans Becker and Theo Tschudi
Darmstadt University of Technology
Institute of Applied Physics
Hochschulstr. 6
D-64289 Darmstadt, Germany

Achim Neubrand
Darmstadt University of Technology
Ceramics Group
Petersenstr. 23
D-64287 Darmstadt, Germany

ABSTRACT

Knowledge of the position dependent thermal diffusivity of FGMs is important in situations where these materials are exposed to thermal shock - however measurements with conventional techniques require uniform samples and have no spatial resolution. A laser-based photothermal beam-deflection scheme is presented that is capable of thermal diffusivity measurements with a spatial resolution down to 50 μm. For homogeneous materials, the thermal diffusivity values retained by this method agree very well with the well-established laser flash technique. The new method was used to determine local thermal diffusivity values on graded Al_2O_3/Al composites and a graded AlCu alloy with a high spatial resolution.

INTRODUCTION

The availability of thermophysical property data of FGMs as a function of position (or composition) is crucial for modelling their behavior of under external loads. Typically, mechanical properties such as Young's modulus or Poisson's ratio can be calculated with a reasonable accuracy using suitable micromechanical models, e.g. Eshelby's theory for inclusion type microstructures or Tuchiinski's method for interpenetrating network microstructures. Thermal conductivity of FGMs can be calculated using the Maxwell-Eucken law or using Linear Representative Volume Element Micromechanics [1].

However, for non-ideal microstructures and large property differences values calculated by micromechanical models may have a considerable degree of uncertainty. Moreover, for graded glasses or solid solutions micromechanical approaches cannot be applied and local property measurements are inevitable. Some properties such as hardness can be easily measured "in situ". Elastic moduli have been determined indirectly from the sound velocity which can be measured with high spatial resolution by acoustic microscopy [2,3]. Thermal expansion coefficients as a function of composition have been successfully measured by position-resolved diffractometry at different temperatures [4]. Position dependent thermal conductivity measurements require accurate temperature measurements with a high spatial resolution. They can be achieved using infrared thermography [5]. This method employs a steady heat flow in a cylindrical sample with a radial slit, and the temperature distribution is observed with an infrared camera. It has been successfully used to measure the thermal conductivity profile in various metal/ceramic FGMs. In the present contribution a complementary method is presented which does not require a special sample geometry: the thermal diffusivity of a material is determined with high spatial resolution by laser beam deflection. This method has a high versatility which makes it applicable to a variety of materials including 2D-FGMs.

PHOTOTHERMAL BEAM DEFLECTION

For years photothermal beam deflection (PBD) has been used to determine the absorption of optical thin films [6]. In a configuration with a probe beam parallel to the surface it has also been used to measure thermal diffusivity of materials [7]. In the present work a configuration where the probe laser is reflected at a high angle of incidence from the specimen surface is used. Fig. 1 shows the principle of the PBD thermal diffusivity measurement. A pump laser beam (Argon-ion-laser operating at 514 nm) is periodically intensity modulated with a frequency f by an acousto optic modulator. The pump laser radiation is absorbed at the sample surface and causes a temperature profile both in the sample and in the air. The temperature distribution in the sample causes a small thermoelastic surface deformation of the probe. The temperature profile in the air causes a profile in the index of refraction that acts similar to a lens.

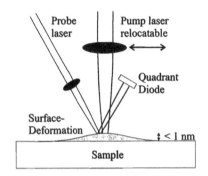

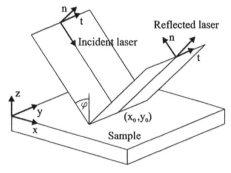

Figure 1: Principle of the PBD measurement.

Figure 2: Parameters describing the geometry of the laser beam.

The probe laser is a HeNe laser operating at 633 nm. Its beam is deflected both in the index of refraction profile in the air and reflected at the surface deformation. These two effects superimpose but can be separated using a special geometry which is shown in Fig. 2. The pump laser is located in the origin of the coordinate system. The probe laser beam falls at an angle φ onto the probe surface and is reflected. A four-quadrant-diode is used to detect the very small deflection of the probe laser beam caused by the aforementioned effects. The deflection of the beam can be separated into two perpendicular components. These components are shown as the normal- (n) and transverse (t) component in Fig. 2. The normal component of the deflection in air varies with $\tan^2 \varphi$ for small angles of incidence φ while the normal component of the surface reflection is independent of the angle of incidence. This allows to neglect effects of the air lens above the sample surface. For small angles of incidence the dominating effect for thermal diffusivity measurements is thus the deflection of the probe laser caused by the surface deformation of the probe.

Figure 3 depicts the complete experimental setup. In order to improve the signal-to-noise ratio, lock-in technology [8] was used to detect the probe beam deflection which was only of the order of several μrad. Stepping motors were used to focus the laser beams

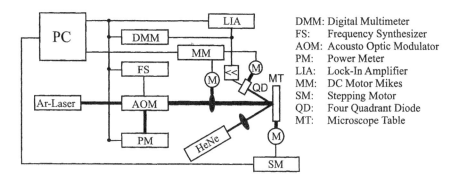

Figure 3: Block diagram of the main components of the experimental setup.

precisely onto the surface of the investigated samples which were mounted on a microscope table.

The surfaces of all samples studied were polished. Rough sample surfaces disturb the light wave fronts and produce speckle patterns on the quadrant diode. Samples with an insufficient optical reflection were coated with a thin Al film. Since the film thickness (< 1 μm) is much smaller than the penetration of heat into the sample, no influence on the determined diffusivity data was observed if the coating adhered well to the sample. This was experimentally verified using samples measurable both with and without a coating.

The correlation between the surface displacement and the thermal diffusivity of the sample is complicated and establishing it requires numerical calculations. Here, these calculations can only be outlined. The surface displacement of the sample can be described by coupling the Fourier equation of heat conduction and the Navier-Stokes displacement equation. Fortunately, in the present case, the coupling between the Fourier equation and the Navier-Stokes equation is so weak that they can be solved independently. The Fourier heat conduction equation was solved assuming two semi infinite half spaces (air and sample). The boundary conditions were continuity of temperature and heat flux at the sample-air interface. The laser heat source in the sample has a Gaussian profile decreasing exponentially with depth. The solution of the Fourier equation is achieved using a spatial Fourier transformation. The inverse transformation back into laboratory coordinates can only be carried out numerically.

This solution is then inserted in the Navier-Stokes equation. Here, the boundary conditions are absence of stresses normal to the sample surface and no displacement of the surface at infinity. For our experiment, only the vertical component u_z of the surface displacement is of interest. The derivative of the surface displacement $m = \partial u_z / \partial r$ can be determined by an inverse transformation shown in (1) and can only be evaluated numerically.

$$m = \frac{-(1 + \nu)\alpha_{th}\alpha P}{\pi \lambda_P} \int_0^\infty \frac{\delta^2 J_1(\delta r)\exp(-\frac{(\delta a)^2}{8})[\lambda_s(\alpha + \delta + \beta_s) + \lambda_a\beta_a]}{(\lambda_s\beta_s + \lambda_a\beta_a)(\beta_s + \delta)(\alpha + \delta)(\beta_s + \alpha)} d\delta \qquad (1)$$

Here ν is the Poisson ratio, α_{th} and α the thermal expansion- and optical absorption coefficient, P and a the heat laser power and spot size, $\lambda_{s,a}$ and $k_{s,a}$ the thermal conductivity and diffusivity of sample or air, ω the angular frequency and $\beta_{s,a}$ is defined as:

$$\beta_{s,a} = \sqrt{\delta^2 + \frac{i\,\omega}{k_{s,a}}} \tag{2}$$

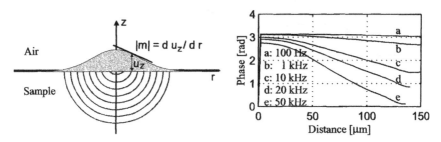

Figure 4: Thermoelastic surface deformation caused by the temperature profile.

Figure 5: Phase of surface deformation calculated for different modulation frequencies.

Figure 4 illustrates the slope m of the surface deformation given by (1). The surface deformation oscillates periodically as the pump laser is modulated with a frequency f. The phase ϕ of this surface deformation, as observed by the deflection of the probe laser beam, contains the information on the thermal diffusivity of the sample. Fig. 5 shows the numerically calculated phase as a function of the distance from the center of the pump laser beam for different modulation frequencies. If the modulation frequency is chosen adequately a region exists where the phase of the signal increases linearly with distance r from the pump laser. The derivative $\partial\phi/\partial r$ is larger for high modulation frequency. Since a numerical fit of all measured data would have required excessive numerical calculations a simple method for evaluation is presented that yields very satisfactory results: The thermal diffusivity data presented in the results section were calculated only from the measured slope of the phase in the linear region of (1).

A simple formula was used to convert the slope of the phase into thermal diffusivity:

$$p = \frac{1}{\mu} = \sqrt{\frac{\pi f}{\gamma k}} \tag{3}$$

Equation (3) connects the measured linear slope p, which is the reciprocal thermal diffusion length μ, in a simple way to the modulation frequency f and the thermal diffusivity k. For $\gamma = 1$ this formula is only valid for the homogeneous heat diffusion equation. The introduction of the correction factor γ allows to apply this formula to the surface deformation problem.

In a thermal diffusivity measurement, the slope p of the phase is determined by a linear fit as shown in the upper half of Fig. 6. The lower half illustrates the calculation of the measurement error from the standard deviation and length of the scan. Typically the error is in the range of one to three percent.

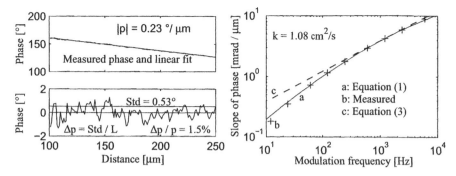

Figure 6: Accuracy of a phase measurement. Upper Half: Linear fit, Lower Half: Deviation of measured phase from Linear Fit.

Figure 7: Slope of the phase p as function of frequency. Theoretical and experimental values.

Figure 7 shows measurements performed on the same copper sample with different modulation frequencies using (3) for evaluation. The measured slope is described very well by (1) for all employed modulation frequencies if a thermal diffusivity value of 1.08 cm^2/s is used. If the factor γ is chosen as 2 the correct value for the thermal diffusivity 1.08 cm^2/s can be also retained by (3) for a wide range of modulation frequencies between 250 Hz and 4 kHz. For lower or higher modulation frequencies the diffusivity retained by (3) would be higher than the actual one. A suitable range of modulation frequency where (3) can be applied was determined in the following way: Both, (1) and (3) depend only on the quotient of modulation frequency and thermal diffusivity. Therefore the same results as in Fig. 7 can be achieved for different materials by adapting the modulation frequency to the thermal diffusivity of the material. For the geometry of the used setup a frequency should be selected where the measured phase slope is 4 mrad/μm, which corresponds to a thermal diffusion length of 250 μm. However, due to the arguments given above, some variation in meeting this condition can still lead to precise thermal diffusivity values.

RESULTS AND DISCUSSION

	Laser-Flash	Beam-Deflection
Copper	0.939 ± 0.047	0.935 ± 0.021
Aluminium	0.567 ± 0.017	0.570 ± 0.012
Brass	0.342 ± 0.008	0.330 ± 0.006
Iron	0.174 ± 0.004	0.164 ± 0.004
Steel	0.039 ± 0.001	0.040 ± 0.001

Table 1: Comparison of Laser-Flash and Beam-Deflection Results. Values are in cm^2/s.

In order to check the precision of the thermal diffusivity values received with the PBD method a set of commercially available Cu, Al and Fe alloys with a wide range of thermal diffusivities between 0.01 cm^2/s and 1 cm^2/s was investigated and the results obtained were compared with results obtained using the well established laser flash method (Table 1).

At room temperature laser-flash measurements show a considerable scatter. Table 1 shows thus mean values of 20 laser flash measurements. The statistical error calculated from the standard deviation of the laser flash measurement is 2% for the lower thermal diffusivity samples rising to 5% for the copper alloy. As can be seen the values determined using the PBD technique are in excellent agreement with the laser flash results if the error of the laser flash measurement is taken into account.

The thermal diffusivity of an Al_2O_3/Al -FGM as a function of position is shown in Fig. 8. The FGM was prepared using a foam replication method described in these proceedings [9]. The maximum aluminum concentration was 31.1% and decreased linearly to 3.9%. It can be seen that the trend of the thermal diffusivity values mimics the gradation profile. The measured thermal diffusivity shows considerable fluctuations even in the region of the composite with a constant composition. These fluctuations are much larger than the error of the measurement. They are thus a real property of the sample - it has to be kept in mind that the information on the material's diffusivity is recorded in a volume of about $(100\mu m)^3$. Indeed the investigated composite has a microstructure with Al ligaments of about 40 μm diameter. At low metal volume fractions below 30% as encountered in this material it is likely that only very few metal ligaments will be found in the volume investigated in a single measurement. Obviously the fluctuations in thermal conductivity encountered are thus due to the poor statistics of the number, size and orientation of the aluminum ligaments. The measured thermal diffusivity values were converted to thermal conductivity in Fig. 9 and plotted against the volume fraction of Al. The observed values fall between the upper and lower Maxwell-Eucken bound. Thus, for the Al_2O_3/Al composite the thermal conductivity could have been determined equally well by using Maxwell bounds.

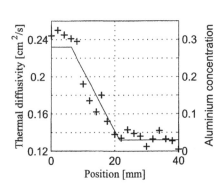

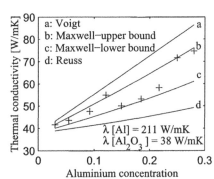

Figure 8: Spatially resolved thermal diffu-sivity measurement of a functionally graded Al_2O_3/Al sample. + Measured Values; - Al concentration profile

Figure 9: Heat conductivity of functionally graded Al_2O_3/Al sample according to differ-ent models. + Experimental values

As a second example the thermal diffusivity of a graded Al-Cu alloy was determined. The alloy was produced by directed solidification of AlCu4 in a vertical Bridgman furnace. Natural thermosolutal convection was used to homogenize the melt. Details of the process are described in Ref. [10]. The resulting copper concentration gradient was determined

Functionally Graded Materials 2000

in a scanning electron microscope equipped with EDX (Fig. 10). The composition on the left corresponds to the original alloy AlCu4 which was not molten. The copper content of the specimen then drops within a few mm to about 0.7% and then slowly increases to about 6% Cu. The measured thermal diffusivity profile reflects these changes in Cu concentraton: The diffusivity of in the AlCu4 region is determined as 0.75 cm^2/s with very little change. As the Cu content is decreased the measured diffusivity increases. This is due to the fact that the AlCu alloy forms a solid solution in this concentration range and the Cu atoms on Al sites effectively reduce the electric and thermal conductivity of the alloy. Indeed the maximum value of thermal diffusivity of 0.88 cm^2/s is observed at the lowest Cu concentration of about 0.7%. This value is already very close to the literature value for pure Al which is 0.89 cm^2/s. During the directed solidification process the Cu concentration increases again and, as a consequence the thermal conductivity decreases to values as low as 0.5 cm^2/s. The irregularity in the thermal diffusivity of the sample near the position of 24 mm is most likely not an artefact of the measurement. Indeed, forced convection may become less stable as the molten region becomes smaller which can lead to local fluctuations in Cu concentration.

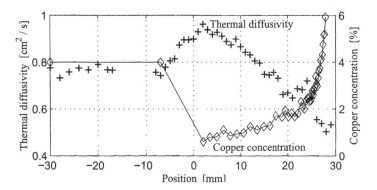

Figure 10: Spatially resolved thermal diffusivity measurement of a functionally graded AlCu alloy. + Thermal diffusivity data. - Copper concentration profile.

CONCLUSIONS

(1) The probe beam deflection method is capable of measuring thermal diffusivity with a precision better than 3% and with a high spatial resolution. For practical reasons the resolution of presented measurements was about 100 μm, but values as low as 10 μm seem to be feasible with this method.

(2) The probe beam deflection method requires a smooth specimen surface that can reflect and absorb laser radiation. The second requirement can be met even for glasses by applying a very thin metal coating, the first requirement precludes investigations on macroporous materials.

(3) The probe beam deflection method can determine thermal diffusivity profiles of FGMs accurately. The resolution of the measurement can give interesting insights in local fluctuations of properties of graded materials.

ACKNOWLEDGEMENTS
Financial support of the Deutsche Forschungsgemeinschaft is gratefully acknowledged. We are indebted to B. Siber and M. Rettenmayr for supplying the graded AlCu alloy.

REFERENCES

[1] M.M. Gasik; K.R. Lilius, "Evaluation of properties of W-Cu functional gradient materials by micromechanical model", Computational Mater. Sci. 3, 41-49 (1994)

[2] T. Mihara; T. Sato; Y. Kitamura; K. Date, "Local elastic constant measurement of functionally gradient materials by line-focus-beam acoustic microscope", Proc. of the IEEE Ultrasonics Symposium 1993, 617-622, IEEE, Baltimore, USA 1993

[3] J. Ndop, T.J.Kim and W. Grill, "Mechanical Characterisation of Graded Materials by Ultrasonic Microscopy with Phase Contrast", in Functionally Graded Materials 1998, Materials Science Forum 308-311, W.A. Kaysser (ed.), p. 873-878

[4] T. Ota; I. Yamai; T. Hayashi, "Nepheline gradient solid solutions", J. Mat. Sci. 30, 2701-2705 (1995)

[5] T. Ishizuka, S. Okada, and S. Wakashima, "In Situ Measurement of Through-the-Thickness Thermal Conductivities in ceramic/Metal Composition-Graded Multilayers by means of Thermography", Proc. of the FGM Syposium 1994, B. Ilschner & N. Cherradi (eds.), Presses Polytechniques et Universitaires Romandes, p.453-458

[6] M. Commandre, L. Bertrand, G. Albrand and E. Pelletier, "Measurement of absorption losses of optical thin film components by photothermal deflection spectroscopy", SPIE 805 Optical Components and Systems, 128-135 (1987)

[7] A. Salazar, A. Sanchez-Lavega and J. Fernandez, "Theory of thermal diffusivity determination by the mirage technique in solids", J. Appl. Phys. 65 (11), 4150-4156 (1989)

[8] EGG Signal Recovery, "The Analog Lock-in Amplifier, Technical Note TN 1002", P.O.Box 2565, Princeton, NJ 08543-2565, USA, (1998)

[9] T.-J. Chung, A. Neubrand, J. Rödel, T. Fett, "Fracture Toughness and R-Curve behavior of Al2O3/Al FGMs", these proceedings

[10] B. Siber, M. Rettenmayr and C. Müller, "Concentration gradients in Aluminium alloys generated by directional solidification and their effects on fatigue crack propagation", in Functionally Graded Materials 1998, Materials Science Forum 308-311, W.A. Kaysser (ed.), p. 211-216

TRIBOLOGICAL CHARACTERIZATION OF Al-Si/SiC$_p$ COMPOSITES: MMC's vs. FGM's

J.R. Gomes, A.S. Miranda, D. Soares, A.E. Dias and L.A. Rocha
Department of Mechanical Engineering, University of Minho
Campus de Azurém, 4800-058 Guimarães, Portugal

S.J. Crnkovic
Department of Materials and Technology, UNESP
Campus de Guaratinguetá–12500 000, São Paulo - Brazil

R.F. Silva
Department of Ceramics and Glass Engineering, UIMC, University of Aveiro
3810-193 Aveiro, Portugal

ABSTRACT

In this work, aluminum based materials were tested against gray cast iron in a pin-on-disk tribometer. Cast SiC particulate reinforced F3S-20S aluminum matrix composite as prepared by *Duralcan*, the same material after centrifugal casting to obtain functionally graded properties and the non reinforced Al-Si (Al12SiMg) alloy, were tested without lubrication, at room temperature, under a normal load of 5 N and constant sliding speed of 0.5 ms^{-1}. Because of the poor mechanical response of non-reinforced Al-Si alloy, catastrophic wear is rapidly attained. For aluminum matrix composites, the wear resistance strongly increases due to the combined effect of reinforcing particles as load bearing elements and the formation of adherent tribolayers. FGM composites exhibited superior wear resistance when compared to homogeneous MMCs. It was also evidenced that the best tribological system tested corresponds to a FGM with ca. 33% of SiC particles at the contact surface. The wear mechanisms were investigated using SEM/EDS and a comparative analysis is established between the functionally graded and non-graded materials.

INTRODUCTION

Metal matrix composites reinforced with ceramic particles have received increasing attention as engineering materials being recognized as possible substitute materials of monolithic metallic materials in applications requiring high wear resistance. These composites are considered as promising candidate for tribological applications in the aerospace, aircraft and automotive industries. Also, during the last few years the need of lightweight materials became an imperative from an ecological standpoint to improve fuel efficiency in such applications. This lead to the development of aluminum based matrix composites reinforced with ceramic particles.

The tribological behavior of reinforced aluminum matrix composites has been investigated by numerous researchers and there exists a large amount of published works concerning this subject, most of them showing the better tribological performance of the composites compared to the unreinforced alloys [1-5]. Most tribological investigations of reinforced aluminum matrix composites have focused on the sliding wear of homogeneous materials against steel couterfaces, with little attention being given to the possibility of other type of counterfaces. However, some studies have shown that the reinforcing particles may not only affect the tribological performance of the composite, but they can also drastically alter the tribological response of the counterpart [1,6,7]. As a consequence, the full consideration of a sliding tribological system requires an analysis of the wear of the counterface [1,8].

Iron alloys, such as steel and cast iron, are the most common metallic materials for tribological applications, namely against aluminum matrix composites[1-8].

In this study, unlubricated pin-on-disk experiments of aluminum based materials/gray cast iron couples are performed at room temperature, 5N of normal load and constant sliding speed, 0.5 ms^{-1}. The wear resistance of aluminum matrix composites strongly increased due to the combined effect of reinforcing particles as load bearing elements and the formation of adherent tribolayers. For FGM composites there is evidences that an optimum area fraction of reinforcing SiC particles at the contact surface of FGM composites for low system wear exists. A comparative analysis is established between the functionally graded and non-graded materials.

EXPERIMENTAL

Materials

Aluminum based disks were machined from three types of materials: non reinforced Al-Si (Al12SiMg) alloy after centrifugal casting, cast SiC particulate reinforced F3S-20S aluminum matrix composite as prepared by *Duralcan* (with 20 vol.% SiC) and the same material after centrifugal casting to obtain functionally graded properties. Composition and hardness of the aluminum based materials are given in Table I. In the case of the functionally graded material a cylindrical cast dowel was produced with properties varying along the axis. A high frequency induction centrifugal casting furnace (*Titancast 700 μP Vac*, from *Linn High Therm*, Germany), equipped with a vacuum system was used to produce the samples. The sample was poured at 850°C, with a rotational speed of 418 rpm. Three cross sections were considered for tribological characterization, identified in Table I as A, B and C, located respectively at the edge, 5 mm and 20 mm from the dowel edge. Area fraction of particles in each cross-section was determined through an image analysis software (*Paqi*, CEMUP, Portugal). Gray nodular cast iron was the counterface pin material, its characteristics being presented in Table I.

Table I. Composition, Particle Fraction Area and Hardness of the Aluminum based Disks and Cast Iron Pin.

Material	Composition (wt%)		Particle Fraction Area (%)	Hardness (HV30)
Al-Si alloy	12.5% Si; 0.03% Mg; 0.14% Mn; 0.47% Fe			57
Composite (F3S-20S, *Duralcan*)	8.5-9.5% Si; 0.45-0.65% Mg; 0.2% Ti; 0.2% Fe (max); 0.2% Cu (max)		20	86
FGM composite	A	8.5-9.5% Si; 0.45-0.65% Mg; 0.2% Ti; 0.2% Fe (max); 0.2% Cu (max)	25.8	80
	B		33.4	89
	C		30.5	85
Cast iron	3.3% C; 2.0% Si; 0.5% Mn; 0.08 % S (max); 0.2% P (max)			149

A – edge cross section; B – 5 mm from the edge; C – 20 mm from the edge.

Testing Procedure
Unlubricated pin-on disk experiments were carried out in a *Plint and Partners* tribometer *model TE67 HT*. Aluminum based disks were tested at room temperature and constant sliding speed, 0.5 ms^{-1}, against gray cast iron pins. The experiments were conducted inside an acrylic chamber to isolate the rotating disc and the pin holder from the external environment. A relative humidity of 50% was kept constant inside the acrylic chamber.

A bending type force transducer was used to assess the friction coefficient, f. Tests were regularly stopped at the end of each 1 or 3 km run to measure the amount of wear of the specimens using a microbalance with an accuracy of 10 µg. The wear volume, V, was calculated from the weight loss and the density of the material.

Chemical and morphological characterization of the worn surfaces of the pins, the track surfaces of the disks, and the wear debris were performed using SEM/EDS techniques.

RESULTS
The variations of friction coefficient and volumetric wear of aluminum based disks and cast iron pins as a function of the sliding distance are shown in Figs. 1-3. In the Figs. representative micrographs showing the morphological aspect of the disk wear tracks at the end of the tests, are also shown. In these micrographs, the sliding direction of the opposite mating surface is from left to right.

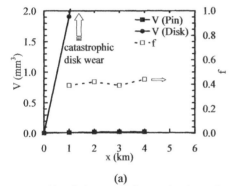

(a)   (b)

Fig. 1. (a) Friction coefficient and volumetric wear loss of Al-Si (Al12SiMg) alloy disk and cast iron pin for different sliding distances. (b) SEM micrograph of the final worn surface of the Al-Si alloy disk.

Results show important differences between the tribological response of the sliding pairs involving distinct types of aluminum based materials. When the non reinforced Al-Si alloy is tested against cast iron, the friction coefficient remains almost constant and relatively low (f≈0.40) (Fig. 1(a)), but with large amount of material being removed from the aluminum disc where catastrophic wear is rapidly attained. Conversely, almost negligible wear loss is found for the opposing cast iron pin. The final worn surface morphology of the Al-Si alloy disk (Fig. 1(b)) is characterized by large surface damage, with extensive plastic deformation and severe delamination, resulting in a rough surface appearance. EDS analysis of the disk worn track reveals only the elements corresponding to the original composition of the Al-Si alloy, while a high Al-peak is noticed in the EDS spectrum of the cast iron pin worn surface. The morphology of the loose wear debris is mainly characterized by the presence of large plate-like particles (≈500 μm) and a small portion of larger chips. Only Al and Si are found to be present in wear debris.

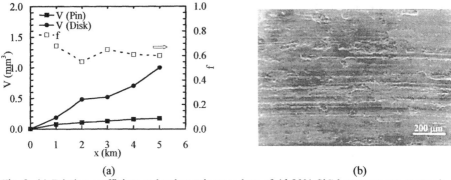

(a)   (b)

Fig. 2. (a) Friction coefficient and volumetric wear loss of Al-20% SiC homogeneous composite (F3S-20S) disk and cast iron pin for different sliding distances. (b) SEM micrograph of the worn surface of the Al-20% SiC composite disk.

For the Al-20% SiC homogeneous composite (F3S-20S) sliding against cast iron the friction coefficient slightly decreases with the sliding distance (Fig. 2(a)). Concerning wear behavior, this tribopair is characterized by a relatively high wear loss in both mating materials, particularly in the SiC reinforced Al alloy disk. The worn surface of the composite shows plastic deformation resulting in parallel, continuous and deep plowing grooves, alternately with flat smooth areas (Fig. 2(b)). Some regions of material removal by delamination are also observed. EDS analysis of the disk worn track indicated some transfer of iron from the pin counterpart, but no transfer of Al in the reverse sense was detected. The major part of the loose wear debris consisted of discrete fine wear particles (≈1 μm). Some platelet shapes up to 20 μm long and few microns thick were also found. EDS spectra of the wear debris revealed the presence of elements from both mating materials, the oxygen peak being very intensive.

Regarding contacts involving FGM Al-Si/SiC$_p$ composite materials significant differences were found. An initial running-in stage is evidenced allied to important gains of mass on the composite disks A and B (Figs. 3(a) and 3(c)) or to an intense wear loss on the disk C (Fig. 3(e)). Depending on the FGM disk, the running-in stage is characterized by stationary or increasing values of friction coefficient, but similar friction values are obtained when the sliding couple entered a steady-state regime (f≈0.50) (Fig. 3(a) 3(c) and 3(e)). Also analogous wear losses were measured for FGM disks in this stationary regime. Concerning the wear of cast iron pins, important losses of material are found, mainly during the running-in stage and for contacts with disks A an C. The final worn surfaces of FGM disks A, B and C are presented in Figs. 3(b), 3(d) and 3(f). A protrusion of smooth SiC particles above the matrix is evidenced and material removal by delamination seems to occur around these reinforcing particles. A characteristic feature of the final FGM disk worn tracks is the presence of dispersed adherent tribolayers with smooth appearance spreading along the surface in the sliding direction. EDS analysis of the tribolayer free regions inside the wear track of the disk shows only the elements corresponding to the original composition of the Al-Si/SiC$_p$ composite. If the same analysis is made on the tribolayer, the intensity of oxygen and iron peaks reveals that it is rich in iron oxides from the cast iron pin.

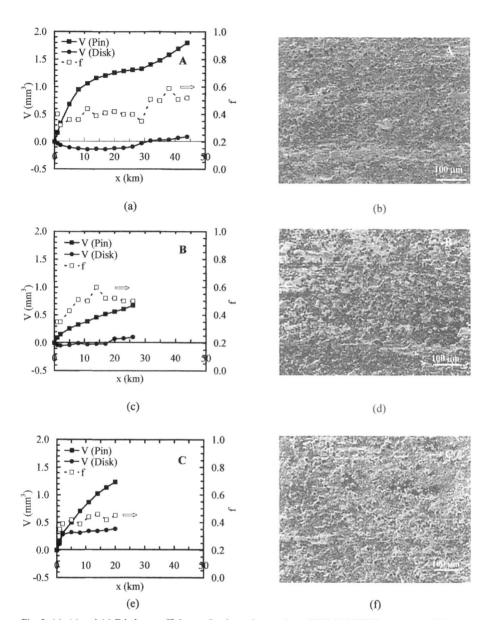

Fig. 3. (a), (c) and (e) Friction coefficient and volumetric wear loss of FGM Al-Si/SiC$_p$ composite disks and corresponding cast iron pins for different sliding distances and disk cross sections. (b), (d) and (f) SEM micrographs of the worn surfaces of FGM Al-Si/SiC$_p$ composite disks from cross sections A, B and C. **A**-disk from the edge cross section of the cast dowell; **B**-disk 5 mm from the edge; **C**-disk 20 mm from the edge.

However, no Al-peaks were detected by EDS analysis at the pin worn surface. The loose wear debris denote morphological features similar to the ones described for Al-20% SiC homogeneous composite (F3S-20S). In the EDS spectra iron and oxygen dominate, the Al-peaks being almost negligible.

DISCUSSION

The friction coefficient always exhibits a higher steady-state value when the composites are involved in the contact (f≈0.50-0.60) compared to unreinforced Al-Si alloy (f≈0.40) (Figs 1(a), 2(a), 3(a), 3(c) and 3(e)). This is attributed to the energy dissipation required for plastic deformation of the cast iron counterface during the abrasive action by SiC particles present in the composites [9]. The wear resistance of the Al-Si/SiC$_p$ composites is superior to that of unreinforced Al-Si alloy (Figs 1(a), 2(a), 3(a), 3(c) and 3(e)). which is in accordance to other studies[2-5]. However, FGM composites are characterized by superior tribological behavior when compared with the homogeneous composite.

The unreinforced Al-Si alloy does not cause wear on the cast iron counterface. By the contrary, Al- and Si-rich deposits are found on the pin surface as a consequence of catastrophic wear behavior of the Al disk (Fig. 1(a)). The occurrence of mutual transfer of Fe and Al between the two mating surfaces during the course of sliding contact is a common phenomenon, usually being attributed to the high adhesive bonding force between iron and aluminum [10]. However, in this study, the transfer of material only occurs from the least resistant Al-Si surface to the more resistant cast iron surface. As a consequence, the contact surface of the Al-Si alloy undergoes extensive plastic deformation and severe delamination (Fig. 1(b)). According to these observations, only Al and Si comprise the loose wear debris, whose morphology is characterized by large plate-like particles (≈500 μm) and a small portion of large chips. The formation of large chip-like particles is due to a macro-cutting action by the opposing gray cast iron counterface. In fact, under the influence of sliding and attendant heat generation, graphite is removed, creating large cavities on the cast iron surface which can act as cutting edges.

The high wear resistance of composite materials can be attributed to the presence of SiC particles that act as load-supporting elements. In order to remain effective load-bearing elements, the particles should maintain their structural integrity during wear, i.e. they should not be removed from the surface. Such topography is considered to be useful to prevent the softer aluminum matrix from becoming directly involved in the wear process [4]. Moreover, several important effects can arise from the presence of a tribolayer at the contact interface that may determine the tribological behavior of the mating materials. In fact, it modifies the contact pressure distribution, tends to spread the contact area and diminish the contact pressure [3]. Thus, the combined effect of reinforcing particles as load bearing elements and the presence of a protective tribolayer is decisive for the tribological response of composite materials.

Fig. 4 (a) presents a longitudinal cross-section of the Al-20% SiC homogeneous composite disk wear track where the load supporting effect of the SiC reinforcing particles at the contact surface is evidenced. Fig. 4 (b) shows that the same effect is observed for FGM composites by the presence of protruding SiC particles on the worn track. EDS analysis of the disc worn surface of the homogeneous composite indicated transfer of Fe from the pin counterpart. However, this transfer was more intense for FGM materials, being the presence of dispersed iron-reach adherent tribolayers a characteristic morphological feature of their final disk worn tracks (Figs. 3(b), 3(d) and 3(f)). Accordingly, the FGM composites cause more wear of the cast iron counterface than the homogeneous composite (Figs. 2(a), 3(a), 3(c) and 3(e)). This behavior is consistent with the SiC particles area fraction present in each surface (Table I).

Under the test conditions investigated in this work, the applied stress is lower than the fracture strength of reinforcing particles. The SiC particles serve as load-bearing elements and their abrasive action on the cast iron counterface leads to the detachment of debris, which are oxidized and transferred onto the composite surface. This phenomenon increase the wear resistance of the composites [1,2,7] and, if the process is particularly effective, it may result in important gains of mass as observed during the initial running-in stage for FGM composite disks A and B (Figs. 3(a) and 3(c)). These FGM composite disks also presented relatively low values of friction coefficient during the running-in stage. The iron oxides are known to have low friction coefficient and thus, the thick adherent tribolayers provided an *in situ* lubrication effect [4].

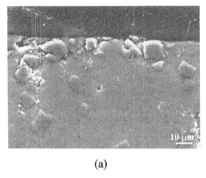 

(a)   (b)

Fig. 4. (a) Longitudinal cross-section of the Al-20% SiC homogeneous composite (F3S-20S) disk worn track showing the SiC particles as load-bearing elements. (b) Worn surface of FGM Al-Si/SiC$_p$ composite disk showing the anchoring effect of protruding SiC particles retaining adherent iron-reach tribolayers.

As mentioned before, the load supporting effect of the SiC reinforcing particles is evidenced by both homogeneous and FGM composites (Figs. 4(a) and 4(b)), but the ability to form and stabilize the iron oxide transfer layers is strongly marked in the case of FGM's. The retention of adherent iron-reach tribolayers by mechanical compaction in between particles and by anchoring effect of protruding SiC particles is well documented in Fig. 4(b) for a FGM Al-Si/SiC$_p$ composite disk worn track. As a consequence, the wear of the FGM composite is reduced, but the wear of the counterface against which the FGM composite is sliding is significantly increased (Figs. 3(a), 3(b) and 3(e)). Therefore, when both the composite and counterface wear are considered, the benefit of reinforcement on the tribological response of the whole system must be well estimated. In the steady state wear regime, very low wear losses are measured for FGM composite disks A, B and C, denoting a comparable wear resistance in this stationary regime (Figs. 3(a), 3(c) and 3(e)), which is due to the protection afforded by adherent iron-reach tribolayers similarly dispersed over the three disk wear surfaces (Figs. 3(b), 3(d) and 3(f)). However, disks A and C, which are present a lower particle area fraction, considerably increase the wear losses on the cast iron pins. This is a direct consequence of an intense abrasive action of the SiC particles against the cast iron. The stability of the iron-rich tribolayers at the disk surface appears to play an essential role on the wear mechanisms in these tribosystems. If tribolayers are removed, surface roughness increases, and a new tribolayer will be formed at expenses of the iron pin wear. In fact, the anchoring of the tribolayers by the protruding SiC particles seems to be more effective in the FGM containing more SiC particles (disk B). As a consequence, tribolayers become more stable in this system reducing the wear of the iron pin.

CONCLUSIONS

Unlubricated sliding wear tests for aluminum-based materials against gray cast iron were performed. The main conclusions resulting from this work can be listed as follows.

The unreinforced Al-Si alloy do not cause wear on the cast iron counterface but only deposits on the counterface. The contact surface of the unreinforced alloy undergoes extensive plastic deformation and severe delamination leading to catastrophic wear.

The wear resistance of the Al-Si/SiC$_p$ composites is superior to that of unreinforced Al-Si alloy, but FGM composites exhibited superior tribological behavior than the homogeneous composite. This behavior is controlled by the combined effect of reinforcing particles as load bearing elements and the formation of protective iron-reach adherent tribolayers.

FGM composites tends to cause more wear on the cast iron counterface than the homogeneous composite by an intense abrasive action of the protruding SiC particles.

Nevertheless, the best tribological system tested in this work corresponds to a FGM with 33.4% of SiC particles at the surface, where both cast iron and composite wear losses are reduced.

REFERENCES

[1] P.H. Shipway, A.R. Kennedy and A.J. Wilkes, "Sliding Wear Behaviour of Aluminium-Based Metal Matrix Composites Produced by a Novel Liquid Route", *Wear*, 216, 160-171 (1998).

[2] A.T. Alpas and J. Zhang, "Effect of SiC Particulate Reinforcement on the Dry Sliding Wear of Aluminium-Silicon Alloys (A356)", *Wear*, 155, 83-104 (1992).

[3] S.C. Sharma, B.M. Girish, R. Kamath and B.M. Satish, "Effect of SiC Particle Reinforcement on the Unlubricated Sliding Wear Behaviour of ZA-27 Alloy Composites", *Wear*, 213, 33-40 (1997).

[4] A.T. Alpas and J. Zhang, "Effect of Microstructure (Particle Size and Volume Fraction) and Counterface Material on the Sliding Wear Resistance of Particulate-Reinforced Aluminum Matrix Composites", *Metallurgical and Materials Transactions*, A, 25A, 969-983 (1994).

[5] S.Y. Yu, H. Ishii, K. Tohgo, Y.T. Cho and D. Diao, "Temperature Dependence of Sliding Wear Behavior in SiC Whisker or SiC Particulate Reinforced 6061 Aluminum Alloy Composite", *Wear*, 213, 21-28 (1997).

[6] Z.Z. Zhang, L. Zhang and Y. Mai, "The Running-in Wear of a Steel/SiC$_p$-Al Composite System", *Wear*, 194, 38-43 (1996).

[7] H. Akbulut, M. Durman and F. Yilmaz, "Dry Wear and Friction Properties of δ-Al$_2$O$_3$ Short Fiber Reinforced Al-Si (LM 13) Alloy Metal Matrix Composites", *Wear*, 215, 170-179 (1998).

[8] A.P. Sannino and H.J. Rack, "Tribological Investigation of Al-20 vol.% SiC$_p$/17-4 PH Part I: Composite Performance", *Wear*, 197, 151-159 (1996).

[9] B. Venkataraman and G. Sundararajan, "The Sliding Wear Behaviour of Al-SiC Particulate Composites-II. The Characterization of Subsurface Deformation and Correlation with Wear Behaviour", *Acta Mater.*, 44 [2] 461-473 (1996).

[10] H.C. How and T.N. Baker, "Dry Sliding Wear Behaviour of Saffil-Reinforced AA6061 Composites", *Wear*, 210, 263-272 (1997).

ACKNOWLEDGEMENTS

This work was sponsored by FCT (Portugal) under the program PRAXIS XXI (contract PRAXIS P/CTM/12301/1998).

S.J. Crnkovic gratefully acknowledges the grant from FAPESP (Brazil).

ULTRASONIC CHARACTERIZATION OF THE ELASTIC PROPERTIES OF CERAMIC-METAL GRADED COMPOSITES.

R. Marks, E. Zaretsky, N. Frage, O. Tevet, Y. Greenberg and M.P.Dariel
Department of Materials Engineering, Ben-Gurion University of the Negev, Beer-Sheva, Israel

ABSTRACT
    In graded ceramic-metal composites, the elastic properties vary in a pre-determined manner along certain directions. Gradients of elastic moduli can be established by generating spatial variations of the ceramic-to-metal ratio. The objective of the present study was to correlate composition and elastic moduli gradients in ceramic-metal composites, using sound velocity measurements.
    The elastic moduli of solids can be derived from the density and the values of the longitudinal and transverse ultrasonic wave velocity. A database of sound velocities and densities was established for homogeneous porous ceramics and cermets. Graded ceramic-to-metal composites were constructed by infiltrating with molten metal ceramic preforms that had an in-built graded porosity.
    Porosity gradients were generated in the single–phase, TiC, and in the two-phase, TiC-TiB$_2$ systems. In the TiC-TiB$_2$ system, the rate of sintering depends critically on the TiC/TiB$_2$ ratio. In the TiC system, varying the ratio of fine to coarse-sized ceramic particles can change the rate of sintering. Stacked powder layers of different mixtures of the ceramic powders were sintered at 1550°C for one hour and subsequently infiltrated with molten Al.
    The sound wave velocity was measured parallel and perpendicular to the composition gradient. The sound velocity was measured in each layer perpendicular to the composition gradient. The results and the values obtained in homogeneous samples were used to derive calculated sound velocity values parallel to the composition gradient. These values were compared to the measured sound velocities parallel to the gradient. The agreement between the two sets of values was within 3%.

INTRODUCTION
    In a graded ceramic-metal composite, material properties such as the elastic moduli vary in a pre-determined manner along one or more dimensions. In such configurations, the elastic moduli are represented by a 'field of properties' in contrast to a homogenous material, in which the elastic moduli maintain a constant value throughout the bulk of the material.
    Application of the graded material concept requires the development of appropriate characterization techniques. Only few attempts have been made [1,2] for determining the elastic moduli field in graded composites.
    The purpose of the present work was to probe the feasibility of using sound velocity values determined in homogeneous samples, in order to predict the effective values of the elastic moduli in graded materials.

In ceramic-metal composites, property gradients can be established by varying the ceramic-to-metal ratio. A varying ratio can easily be fabricated along one dimension, by sintering at a given temperature stacked layers of powder that have different rates of sintering [3]. We focused our attention on the $(TiC,TiB_2)/Al$ and TiC-Al ceramic-metal composite (CMC) systems. These systems were examined before and after infiltration with aluminum. Two-phased $TiC-TiB_2$ ceramics can be sintered to controllable and pre-determined levels of density. Molten Al wets well two-phased $TiC-TiB_2$ composites over the whole composition range and allows infiltrating the preforms and obtaining, after infiltration, composites with a pre-determined ceramic-to-metal ratio. In the same manner, one-phased TiC ceramic can be sintered to a controllable level of porosity by changing the ratio of fine and coarse powder particles, since the rate of sintering depends strongly on particle size.

The methodology consisted in establishing a database of the density and the longitudinal and transverse wave velocities [4]-[8] based on homogeneous single-phase, TiC, and two-phase, $TiC-TiB_2$, ceramic composites that had different ratios of their ceramic and metallic components. This database was used as a reference for determining the field of properties in the graded samples.

EXPERIMENTAL PROCEDURES

Different TiC and TiB_2 ceramic powder compositions (20% to 80% TiC) as well as mixtures of fine and coarse TiC particles were cold-mixed under alcohol. The mixed powders were cold pressed into samples approximately 4mm high and 20mm diameter. Graded samples were fabricated by stacking powder layers each with a different TiC/TiB_2 or TiC fine/coarse ratio.

The compacts were sintered in a vacuum furnace at 1550°C for one hour. The porosity and apparent density were measured by the liquid displacement method (in water). The compacts were ground and polished to ensure plane-parallel bases to within 0.005 mm.

The velocity of propagation of the acoustic wave was determined from direct measurement of the sound wave travel time of the longitudinal and transverse wave, generated by a 5 MHz probe, and the length of the specimen. The elastic moduli were derived from the longitudinal (V_l) and transverse (V_t) ultrasound velocity and the density of the specimen.

Since the homogenous samples were fabricated from a mixture of powders and by uni-directional cold press, there was a need to assure they were indeed homogenous and isotropic. A few measurements were taken in several regions on the sample and a slice from the sample was cut and the velocity was measured in parallel and perpendicular to the applied pressure.

In the graded preforms, the elastic properties vary along the direction of the gradient. The wave velocity was measured parallel (in the xy plane) and perpendicular (in the xz or yz planes) to the gradient axis, Fig. 1. In order to perform the perpendicular measurement a slice was cut out from the center of the graded sample in a region where the height of the layers was even. The longitudinal wave velocity within the layers (in a direction perpendicular to the gradient), was measured with a pointed, 2 mm diameter 15 MHz, probe. The layers were 3 mm thick, on the average, and as will be shown, the measured values represent the actual sound velocity in each layer.

Infiltration with molten aluminum was done in a vacuum furnace. The aluminum initially was placed on top of the sample. Infiltration due to the capillary forces took place at 1050°C. After infiltration, the sound velocity and density were measured in the homogeneous and graded samples in the same manner as in the porous samples.

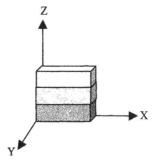

Figure 1. Gradient in the z direction, x-y plane normal to the gradient and x-z plane parallel to the gradient

The wave velocity in the graded sample, parallel to the gradient, was calculated using the sound velocities obtained in homogeneous samples or in layered samples, normal to the gradient, and the relative height of each layer, according to:

$$V = \sum_{i=1}^{n} V_i \cdot h_i$$

where n is the number of layers V_i is the velocity and h_i is the relative height of the ith layer. The Young modulus was calculated in the same way, based on the moduli of the homogenous samples and the relative height of each layer. The results were compared to the actual velocities measured along the composition gradient.

RESULTS

The measurements carried out in the homogeneous samples confirmed their homogeneity and isotropy. The results obtained from measurements of the homogeneous samples and within the layers of the graded samples were in good agreement. The agreement supports the premise that the velocities in each layer are not affected by the presence of the adjacent layers. The longitudinal and transverse sound wave velocity results, obtained in the homogeneous samples, are plotted as function of porosity and aluminum content in the porous and infiltrated samples, respectively (Figs.2 and 3). Error bars in Figs. 2 to 5 are not shown because they were of the order of the size of the data symbols.

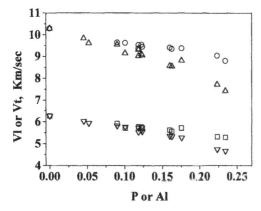

△ V_l Ceramic
○ V_l Cermet
▽ V_t Ceramic
□ V_t Cermet

Figure 2. Longitudinal and transverse wave velocity in porous TiC ceramics, and TiC-Al cermets, as a function of pore or aluminum fraction.

The Young modulus was calculated from the velocities and the measured apparent density. The results were plotted as function of the porosity or aluminum content in the porous and infiltrated samples, respectively (Figs.4 and 5).

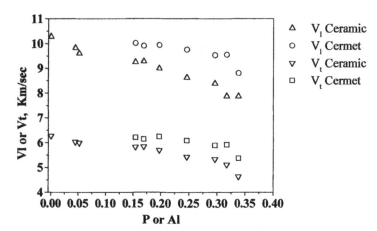

Figure 3. Longitudinal and transverse wave velocity in porous TiC-TiB$_2$ ceramics, and TiC-TiB$_2$-Al cermets, as a function of pore or aluminum volume fraction

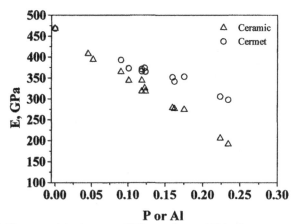

Figure 4. Young modulus in porous TiC ceramics, and TiC-Al cermets, as a function of pore or aluminum fraction.

590

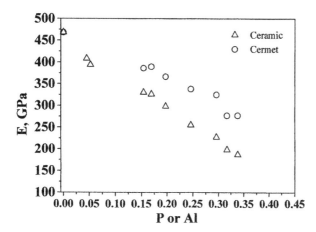

Figure 5. Young modulus in porous TiC-TiB$_2$ ceramics, and TiC-TiB$_2$ -Al cermets, as a function of pore or aluminum fraction

Sound velocity values were calculated using the data obtained from homogenous samples or using the values measured in the layers, perpendicular to the gradient axis. The results were compared to the ultrasonic wave velocity measured parallel to the gradient axis. The differences between the two sets of results are shown in Figs 6-9.

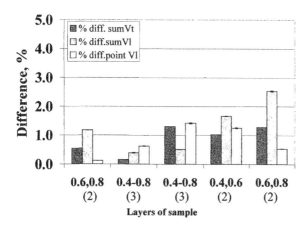

Figure 6. The difference (in percent) between calculated and measured values of the wave velocity in graded porous TiC preforms. The figures in bold on the x-axis stand for the fraction of TiC, and the numbers in parenthesis indicate the number of layers in the graded porous preforms.

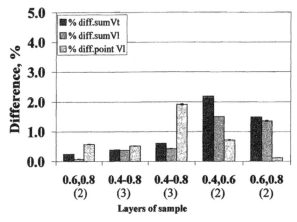

Figure 7. The difference (in percent) between calculated and measured values of the wave velocity in graded TiC-Al cermet preforms. The figures in bold on the x-axis stand for the fraction of TiC, and the numbers in parenthesis indicate the number of layers in the graded infiltrated preforms.

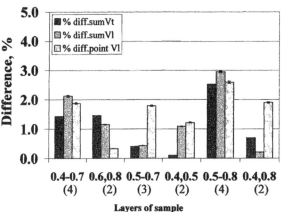

Figure 8. The difference (in percents) between calculated and measured values of the wave velocity in graded porous $TiC-TiB_2$ preforms. The figures in bold on the x-axis stand for the fraction of TiC, and the numbers in parenthesis indicate the number of layers in the graded porous preforms.

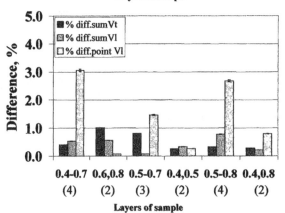

Figure 9. The difference (in percents) between calculated and measured values of the wave velocity in $TiC-TiB_2-Al$ graded cermet preforms. The figures in bold on the x-axis stand for the fraction of TiC, and the numbers in parenthesis indicate the number of layers in the graded infiltrated preforms.

The difference between the calculated and the measured sound velocities fall within a 3% range, both with regard to the values calculated on the basis of results found in homogeneous samples and in graded samples, measured normal to the gradient. Similar procedures were applied to compare calculated Young moduli, using data from homogeneous materials, to the moduli of graded samples determined on the basis of sound velocity measurements parallel to the gradient. The calculated values deviate by approximately 8% from the measured Young moduli. It is not surprising that the deviation is over a wider range than for the sound velocities, since Young moduli are determined on the basis of three independently measured quantities (V_l, V_t and the density).

No echoes from the internal interfaces between the stacked layers were detected in the graded samples. We estimate that internal flaws affect only marginally the measured sound velocities; the latter were derived from the clearly visible echoes originating from the backside of the samples.

An ultrasonic c-scan was performed on a three-layered $TiC-TiB_2$ graded preform. Fig. 10 gives a 3-D representation of the longitudinal velocity (V_l) field in the sample. One can clearly distinguish the presence of the three stacked layers in each of which the velocity values fall within rather narrow and well-defined intervals.

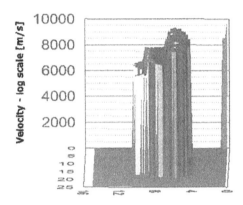

Figure 10. Ultrasonic c-scan in a sample consisting of three stacked layers. The z-axis (vertical) represents the longitudinal velocity at each point in a x-y plane. In this particular sample, a step-wise gradient of the velocity had been established.

CONCLUSIONS
1. The longitudinal and transverse wave velocities are a linear function of the porosity or aluminum fraction in both systems that were studied. A deviation from linearity appears at a high fraction (25%) of the second phase,
2. The calculated values of the wave velocity in graded samples, determined on the basis of data obtained from homogenous samples, are within 3% of the actual values measured in graded samples parallel to the composition gradient.
3. This study focused on two ceramic systems. There was no significant difference between the two systems, even though the three-phase system ($TiC-TiB_2$-pore or $TiC-TiB_2$-Al) is a more complex one.

4. The Young moduli, determined from the longitudinal and transverse wave velocity and the density of the homogenous samples, deviate by approximately 8% from values determined on the basis of the actually measured sound velocities in graded samples parallel to the gradient axis.

REFERENCES
1. P. R. Marur and H. V. Tippur "Evaluation of Elastic Properties of a Functionally Graded Composites Using an Elastic-Impact Technique", NCA **24**, Proceedings of the ASME Noise Control and Acoustics Division, P227-234 (1997).
2. Y. Fukui, K. Takashima and C. B. Ponton "Measurement of Young's Modulus and Internal Friction of an in situ Al-Al$_3$Ni Functionally Gradient Material", *J. Mater. Sci.*, **29** 2281-2288 (1994).
3. Y.Seidman, N.Frumin, N.Frage and M.P.Dariel, "Two-Phase TiC-TiB$_2$ Graded Ceramic Preforms", *6th International Symposium of Functionally Graded Materials (2000)*.
4. A. K. Maitra and K. K. Phani, "Ultrasonic Evaluation of Elastic Parameters of Sintered Powder Compacts", *J. Mater. Sci.*, **29** 4415-4419 (1994).
5. K. S. Ravichandran "Elastic Properties of Two-Phase Composites", *J. Am. Ceram. Soc.*, **77**,[5], 1178-1184 (1994).
6. J.R.Moon "Elastic Moduli of Powder Metallurgy Steel", *Powder Metallurgy*, **32** [2] 132-139 (1989).
7. L. F. Nielsen "Elastic Properties of Two-Phase Materials", Mater. Sci. Eng., 52 39-62 (1982).
8. D. N. Boccaccini and A. R. Boccaccini "Dependance of Ultrasonic Velocity on Porosity and Pore Shape in Sintered Materials", *J. Non-Destr. Eval.*, **16** [4] 187-192 (1997).

APPLICATIONS OF PHASE SHIFTED MOIRE INTERFEROMETRY

Eric D. Steffler
Idaho National Engineering
and Environmental Laboratory
PO Box 1625 MS 2218
Idaho Falls, ID 83415

ABSTRACT
Full field, high resolution displacement measurements are essential to understand complex mechanical deformation and to validate numerical results. Moiré interferometry is used to measure sub-micron displacements in regions of import such as near crack tip singularity fields, near interface deformation, shear-compression loaded fracture and mechanical response of graded materials. An overview of phase-shifted moiré interferometry (PSMI) capabilities and limitations will be presented with specific examples of recent and ongoing projects.

INTRODUCTION
Moiré interferometry is a powerful technique with a significant developmental history. The intent of this paper is to discuss current and past uses of moiré at the Idaho National Engineering and Environmental Laboratory and to showcase the applicability of the technique to various mechanics problems and to present current capabilities. For a detailed and comprehensive treatment of the subject, the reader should refer to the text by Post, Han and Ifju(1) which includes references to historical background information.

High sensitivity moiré is a technique used to create detailed contours maps of surface displacements. While interferometers can be used to measure both in and out of plane displacements, the primary focus of this discussion is the in-plane experimental problem. Application of this technique results in the full-field measurement of surface displacements. What follows is an abbreviated discussion of the most basic elements of moiré and the motivation for using the technique.

GRATINGS AND FRINGES

Fringes are produced when two sets of light and dark bands of similar spatial frequency and orientation combined to produce a secondary effect, fringes. A simple example of geometric moire is shown in Figure 1 where two identical 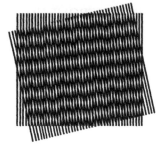 sets of lines are generated with one set rotated slight with respect to the other. The effect results in the secondary horizontal dark bands, fringes.

Figure 1. Example of geometric moiré. Secondary horizontal dark bands are fringes.

While geometrical moiré can be explained by mechanical interference, all moiré effects can be described from an optical interference perspective(2). Moiré interferometry is routinely used at the INEEL to measure displacement with a sensitivity of 2.4 fringes/micron of displacement, sensitivities as high as 1.2 fringes/micron are attainable(1).

In order to generate the desired effect, two gratings are required. A physical grating is bonded to the surface of the specimen which retains an aluminized layer to improve diffraction efficiency. The specimen is then placed in an interferometer which reproduces the same frequency constructive/destructive interference that was used to produce the physical grating. The interference pattern produced by the laser light is called the reference or virtual grating. As the specimen is physically deformed, a frequency mismatch between the physical and virtual grating occurs and generates fringes. These fringes are then collected and recorded at the image plane.

To extract the displacement information from the recorded fringe pattern, the following relation is used

$$u(x,y) = \frac{1}{f} N_x(x,y)$$

$$v(x,y) = \frac{1}{f} N_y(x,y)$$

(1)

where f is the frequency of the virtual reference grating, N_x and N_y are the number of fringes as of function of x and y respectively. The derivatives of these displacement functions will yield strain however, accuracy is lost with any kind differentiation and care must be taken. The remaining discussion will consider the displacements only. Figure 2 shows the fringe pattern that results from a simple tensile load on a flat bar and is representative of what you would observe at the image plane. Traditional moiré interferometry involves recording this fringe pattern (film or digital) to be analyzed later. For this example, the fringes would be counted manually across the specimen and used in equation (1) to extract the magnitude of displacement. For the identified gage length there are approximately 8.5 fringes, f=600 lines/mm which is equivalent to 14 microns of displacement for an approximate average strain of 1100 micro-strain. While this is considering only one line of displacement, the process is the same anywhere on the fringe pattern. The non-uniform spacing along the grating can be a result of pre-existing optical effects or non-uniform bar cross-sections. Optical effects can be accounted for by capturing an image of any fringes prior to testing which can then be subtracted from subsequent fringe patterns.

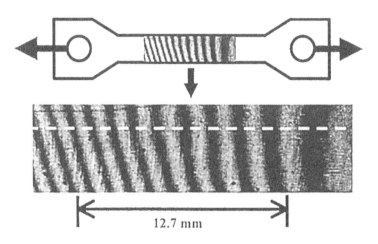

Figure 2. Simple example of a raw fringe pattern on a flat tensile bar.

PHASE-SHIFTING
 The nature of laser light is exploited to enhance available information. The interaction of the laser light that produces the constructive and destructive interference also results in the fringe pattern being a continuously changing map of light intensity. If the phase angle of one source beam is shifted by a fraction of 2π, the whole fringe pattern position will change by the same shift. A cycling shift in phase angle will result in "walking" fringes that move across the specimen with the fringe to fringe (displacement) information remaining unchanged. This

information is used to calculate the data between the peak to peak fringe separation. The current system at the INEEL uses optical fibers to deliver the laser light to the light expanding optics. A piezo-electric crystal is used to stretch one of the fiber optic cables resulting in a calibrated phase angle change in the exiting laser light. Four equal shifts in phase are imposed with digital images being acquired at each shift. These images are then used to calculate the image phase map.

Using the tensile bar example in Figure 2 the associated phase map for the same load state is shown in Figure 3a. Once the phase map is available, phase unwrapping algorithms are used to assemble the saw wave structure into a continuous gray scale representation(3,4,5,6). When frequency information and scaling are included in this process, each pixel represents a value of displacement. The images shown here were acquired on a 640 x 240, 8 bit depth (0-255 gray scale levels) pixel array resulting in over 153,000 data points for one sensitivity direction. Usually both directions of displacement are acquired for each load state.

(a)

(b)

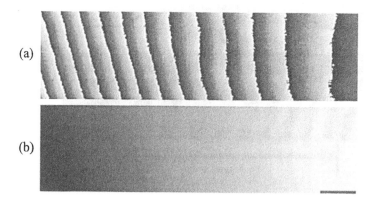

Figure 3. Computed phase map and subsequently unwrapped displacement information.

Using image analysis software, the researcher can extract lines of displacement or create complete contour maps and 3-D plots of the displacement fields to analyze results. Later versions of this acquisition hardware used full pixel arrays of 640 x 480 (also 8 bit depth) increased the image data amount to over 300,000 data points.

The Fracture Behavior Group recently upgraded the data acquisition hardware at the INEEL. Currently the large interferometer can illuminate a 90.0 mm diameter area on a specimen in both the u and v directions. An EPIX® PIXCI-D digital frame grabbing board is used in conjunction with a KODAK® ES 1.0 10 bit depth (0-1023 gray scale levels) digital camera are used to capture phase shifted images. The resulting unwrapped image contains over 1,000,000 displacement data points for each sensitivity direction at each load step. The spatial resolution is dependent on the numerical aperture of the lens, the wavelength of light and the magnification of the object. The displacement resolution is proportional to the uncertainty of fringe order, N. Visual estimations of a raw fringe patterns can be estimated as high as $1/10$ of a fringe order or 0.04 microns. Well calibrated phase shifting systems can resolve displacements as small as 0.01 microns. The reader is again referred to texts and papers that address these issues in detail(1,6).

The latest version of INEEL software stores the complete array of data for subsequent post processing. Since the human eye cannot resolve the subtle changes in even 8 bit depth gray scale representations, on screen images are only averaged approximations while the full data array is available for manipulation. For convenience, often graphs showing lines of displacement are only plotted using every 8th data point for discussion. What follows are a few examples of how moiré interferometry has been used at the INEEL with a brief description of each.

SHEAR-COMPRESSION FIELDS

Moiré was used to investigate the frictional effects of a single flaw loaded in remote shear and compression. Early attempts utilized traditional photographic data acquisition with subsequent manual fringe counting and reduction. Difficulties in analyzing complex displacement fields motivated a movement to phase-shifting data acquisition. Figure 4 shows the phase map of the near crack displacement field with data extraction lines above and below the 30 mm long flaw. The resulting contours revealed relative displacement profiles never observed at this scale using this technique. The developing displacements were surprisingly analogous to displacement profiles measured along km long strike slip faults(7). This technique continues to be applied to this geometry to further investigate any potential cross-scale insights. Prior to the use of phase shifting, the displacement resolution for this experiment was insufficient to observe this behavior. In addition, the continuous displacement information available above and below the crack face was used as boundary condition inputs for numerical models. This hybrid experimental/numerical technique resulted in additional insights to the shear stress evolution near the crack tips as a function of crack face slip(7). This particular data set was acquired using the 1st generation phase shifting software with the smaller CCD array. Results using the latest available hardware are discussed in the next section.

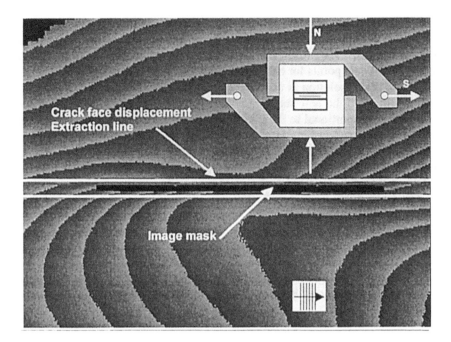

Figure 4. Photographs of shear-compressive loads on a center crack before and after secondary crack extension.

SIMULATED RESIDUAL STRESS RELIEF

Hole drilling stress relief test conducted to demonstrate capabilities. A small graphite coupon 20 x 20 x 4 mm was loaded to a compressive load of approximately 0.8 kN. The fringes generated as a result of the applied load were removed by modifying the virtual grating frequency thus simulating a piece of graphite with an unknown internal stress with a grating on its surface shown in Figure 5. A hole was introduced and the resulting fringe pattern was recorded and is shown in Figure 5b. The specimen was then additionally loaded until fractures initiated at the top and bottom of the drilled hole shown in Figure 6.

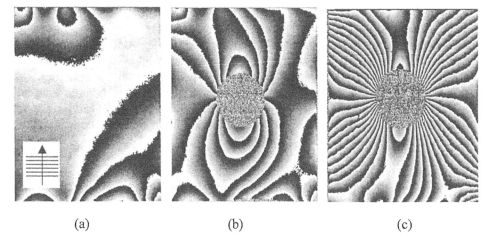

(a)   (b)   (c)

Figure 5. Graphite sample with a) simulated unknown stress, b) 6.4 mm hole drilled through specimen and c) subsequent additional load to 1.8 kN axial load.

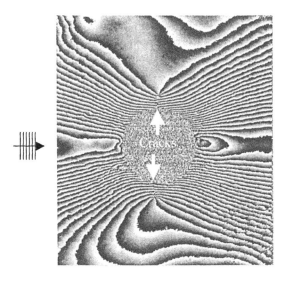

Figure 6. Increased compressive loads resulting in fracture initiation in graphite specimen.

CONCLUSIONS

Phase Shifted Moiré Interferometry has the potential to significantly enhance the understanding of mechanics problems. The technique is ideally suited where complex geometries, spatial changes in material properties and fracture behaviors need to be measured. The full field resolution and accuracy are of sufficient levels to enable direct comparisons to numerical techniques as well as supply displacement boundary conditions for complex system analyses and algorithm validation. Current capabilities at the INEEL continue to be among the best in high sensitivity moiré interferometry.

ACKNOWLEDGEMENTS

This work was supported by the U.S. Department of Energy, Office of Science, under DOE Idaho Operations Office Contract DE-AC07-99ID13727.

1 Post, D., Han, B., Ifju, P., High sensitivity moiré: experimental analysis for mechanics and materials. New York, Springer-Verlag, 1994.

2 J. Guild, "The Interference Systems of Crossed Diffraction Gratings,", Xlarendon Press, Oxford, 1956.

3 R.M. Goldstein, H.A. Zebker, and C.L. Werner. Satellite radar interferometry: Two dimensional phase unwrapping. Radio Science, 23(4):713--720, July-August 1998.

4 J.M. Huntley. Noise-immune phase unwrapping algorithm. Applied Optics, 28(15):3268--3270, August 1989.

5 J.M. Huntley. Automated fringe pattern analysis in experimental mechanics: a review. Journal of Strain Analysis, 33(2):105--125, 1998.

6 J. Mckelvie. Moiré strain analysis: an introduction, review, and critique, including related techniques and future potential. Journal of Strain Analysis, 33(2):137--151, 1998.

7 Steffler, E. D., "Frictional sliding energy partitioning of a center crack loaded in remote shear and compression.", PhD diss., Mechanical Engineering Dept., New Mexico State University, Las Cruces, NM, Aug. 1999.

DISTRIBUTION OF MACRO- AND MICRO-STRESSES IN W-CU FGM

Juergen Schreiber, Achim Neubrand[1], Thomas Wieder[1], Meinhard Stalder[2], and Nail Shamsutdinov[3]
Fraunhofer-Institute for Non-Destructive Testing, Branch Lab Dresden, Krügerstraße 22, D-01326 Dresden;
[1] Darmstadt University of Technology, Department of Material Science;
[2] University Kiel, FB Physik, Germany;
[3] Joint Institute for Nuclear Research, Laboratory of Neutron Physics, Dubna, Russia

ABSTRACT

W-Cu FGM with different gradation profiles were fabricated by electrochemical gradation and infiltration. The composites produced by this method were free of porosity and had an interpenetrating network microstructure. Residual stresses along the gradient were analyzed by neutron diffraction. The observed micro-stresses are often higher and of opposite sign than expected from the thermal expansion mismatch. These results demonstrate that non-equilibrium effects such as inhomogeneous and fast cooling, plastic Cu deformation, and micro-cracks in the W matrix can have a large effect on micro- as well as macro-stresses. After annealing of the FGM the detected distribution of macro-stresses is in qualitative agreement with FEM calculations. A model is presented to explain the unexpected experimental behavior of stress and strain during thermo-mechanical load.

INTRODUCTION

The redistribution of residual stresses avoiding abrupt interfaces between the constituents was one of the main arguments for the introduction of graded materials. Several experiments [1,2] and theoretical assessments [3,[4]] demonstrated this idea. Here we present results for a W/Cu material and discuss the effects of gradients and the manufacturing process on residual stresses. Two major processing routes for W/Cu FGMs are known: 1) sintering of appropriately mixed metal powders [5] or 2) liquid copper infiltration of electrochemically etched porous tungsten matrices [6]. Previously we reported residual stresses in different FGMs manufactured with both processing routes [7 - 9] and compared the experimental findings with the results of analytical (cf. [10,11]) and finite element (FEM) calculations [12] for thermally induced stresses. The experimental data reported in [7 - 9] surprisingly showed compressive stresses in the Cu phase which could not be explained by the thermal expansion mismatch of W and Cu. In this contribution a model is presented which includes plastic deformation of the Cu phase and micro cracks in the W phase after thermal load. Our neutron diffraction data for the as prepared, the annealed and the thermally shocked graded W-Cu samples will be analyzed in the framework of this model. Furthermore, the thermal expansion as well as mechanical compression experiments with W-Cu composites will be analyzed.

SAMPLE PREPARATION AND CHARACTERIZATION

The investigated W-Cu FGMs were prepared by electrochemical gradation of porous W preforms followed by infiltration with Cu [6]. Before the gradation process the employed preforms had a porosity of 25 % and a thickness of 6 mm. Porosity gradients were introduced into these preforms by anodic dissolution in 2N aqueous NaOH solution. Macroscopic concentration gradients were avoided by pumping the electrolyte through the porous electrode. After the electrochemical gradation step the porous preforms were thoroughly washed with distilled water and then infiltrated with electrolytic Cu at 1300°C using 5 MPa argon gas pressure. After melt infiltration the specimens consisted of a graded circular or quadratic W-Cu plate of about 6 mm thickness joined to pure Cu on the copper rich side of the FGM. Important characteristics of all investigated specimens are summarized in Table 1.

The volume content of the W and Cu phases in the prepared specimens were evaluated using an optical microscope equipped with an image analysis system (Leica Quantimet). The investigated W-Cu FGMs had an interpenetrating network microstructure and a negligible residual porosity. The W volume content changed continuously from maximum values of 75 % to values of about 20%. The size of the Cu ligaments varied from about 10 µm in the W rich region to about 30 µm in the Cu rich region (see [6]). The coarse grained samples S1 and S2 had similar microstructures except for the size of the W and Cu phase which was about ten times larger.

The thermal expansion coefficients of the graded W-Cu samples were determined as a function of W volume fraction by cutting the samples into very thin rectangular bars with a thickness of <0.6 mm. These slices were ground so that both surfaces became as parallel as possible. The final thickness of the samples was 0.3-0.4 mm, the length was 22mm and the W volume fraction within one such sample was almost constant. The samples were sandwiched between two half cylinders of alumina (length 21mm) clamped together by a tungsten wire to prevent the thin samples from bending, and heated with a rate of 5 °C/min to 1000°C in a dilatometer (DIL 402 E/7, Netzsch, Selb, Germany). After a holding time of 30min the dilatometer was cooled down to room temperature at a rate of 5 °C/min.

Tab. 1: Characteristics of specimens and processing conditions. W content varies continu-ously from 0 to 65 Vol-% W across layers of thickness A≤ 1 mm, M = 1-3 mm, and S > 3 mm.

Sample ID	Cross Section	Grain Size (µm)	Current density (mA/cm²)	Anodisation time (h)	Gradation profile
W13	square	4 µm	12.2	88	A
W17	square	4 µm	8.9	122	M
W18	square	4 µm	5.6	194	S
S1	circular	45 µm	5	389	S
S2	circular	45 µm	8.9	122	M

STRESS ANALYSIS IN FGM BY DIFFRACTION EXPERIMENTS

The neutron diffraction measurements were carried out using neutron diffractometers at three different neutron sources: i) the High Resolution Fourier Diffractometer (HRFD) at the pulsed reactor IBR 2 of the Frank Laboratory of Neutron Physics, JINR Dubna [7], ii) the High Resolution Two-Axis Diffractometer (D1A) at the High Flux Reactor of the Institute Laue-Langevin in Grenoble [9], and iii) the Two-Axis Powder Diffractometer with Multicounter (E3) at the upgraded research reactor BER II of the Hahn-Meitner Institute, Berlin [13]. By use of slit systems in the incoming as well as in the scattered neutron beam (sometimes special collimators were applied) a defined gauge volume V_g can be set, e.g. $V_g \sim 1$ mm at D1A [2] and Vg ~ 2x2x20 mm³ at E3. The Time-Of-Flight technique at HRFD was used to perform stress scans along the gradient, averaging over the cross-section with a beam width of about 1 mm and for fixed scattering geometry (cf. [7,8]).

According to the Bragg condition $\lambda = 2\,d\sin(\theta)$ (λ - wavelength, d – lattice spacing, and θ - the scattering angle) the shift of the interference peaks yields the lattice strain averaged over a certain number of grains. At D1A and E3 individual (hkl)-reflections are used with θ near 90 °. The Rietveld refinement for the full diffraction pattern at HRFD allows to take an average over different Bragg-reflections for a fixed scattering angle and to determine an averaged lattice constant $a(\alpha)$ (α = W, Cu). The corresponding lattice strain is then given by

$$\varepsilon_{33}(\alpha) \cong (a(\alpha) - a(\alpha)_0)/a(\alpha)_0 - b_{abs} = \sum_{ij} A_{3i} A_{3j} \varepsilon(\alpha)_{ij} \qquad (1)$$

and represents a projection of the strain tensor $\varepsilon(\alpha)$ onto the direction of the scattering vector (defined as the z-axis of the laboratory system). A_{3i} are the components of the direction cosine matrix. The parameters $a(\alpha)_0$ are the stress-free reference values of the lattice constants and were determined from reference samples of pure Cu and W phase. The quantity b_{abs} accounts for the absorption effects.

Using a sufficient number of sample orientations with reference to the neutron beam direction, the strain tensor $\varepsilon(\alpha)$ can be found by a least square fit procedure based on Eq. (1). Assuming isotropic behavior on a macroscopic scale the stress components averaged over the gauge volume can be computed by using Hooke's law:

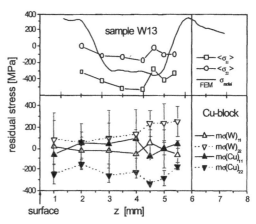

Fig.1. Residual micro-stress components perpendicular to the gradient direction of the sample W13 as a function of the depth z. FEM simulations - see [12].

$$\sigma(\alpha)_{ij} = \left(\frac{E_\alpha}{1+v_\alpha}\right)\varepsilon(\alpha)_{ij} + \left(\frac{v_\alpha E_\alpha}{(1+v_\alpha)(1-2v_\alpha)}\right)\delta_{ij}\sum_n \varepsilon(\alpha)_{nn} \cdot \qquad (2)$$

E_α and v_α are the Young's moduli and Poisson's ratio, respectively. To compare the obtained stress values with results of FEM simulations [12] the macro-stress is introduced, averaging over the phase constituents

$$<\sigma> = G_{Cu}\,\sigma(Cu) + (1-G_{Cu})\,\sigma(W) \qquad (3)$$

where G_{Cu} is a weight factor for the Cu phase. Assuming a random distribution and complete adhesion of both phases the weight factor G_{Cu} can be approximated by the Cu volume concentration $X_{Cu} = 0.25$. However, in the case of uniaxial stress it may happen that micro-cracks (local delamination of Cu and W phase or rupture of walls or stripes of the W matrix) can change the weight factor, so that $G_{Cu} \neq X_{Cu}$. In addition to the macro-stress $<\sigma>$, phase dependent micro-stresses $m\sigma(\alpha)$ are defined in the following way:

$$m\sigma(Cu) = \sigma(Cu) - <\sigma> \quad \text{and} \quad m\sigma(W) = \sigma(W) - <\sigma> \,. \qquad (4)$$

The stress fluctuations within the grains as a result of plastic deformation give no shifts of the interference peaks, however, they cause an additional line width $\Delta B(\alpha)$

Our attempt to determine residual stresses in smaller gauge volumes was accompanied by difficulties to observe Cu reflections for an arbitrary sample orientation. As already understood from the diffraction patterns obtained at the HRFD, the Cu phase is strongly textured in the investigated W-Cu FGM. For $V_g \cong 1$ mm^3 only single, crystal like spots could be found [9]. Since we could not guarantee that reflections in different spatial drections came from the same crystallite, residual stress for the Cu phase were not derived for such samples.

RESULTS AND DISCUSSION

STRESS SCAN

In Fig. 1 the experimental results are displayed for the sample W13. The residual macro-stress obtained for samples 17 and 18 (cf. [8]) deviates considerably from the stress values of the FEM simulation[1] [12]. It was found by a stress scan along the middle line of the sample using a gauge volume $V_g = 1$ mm^3 at D1A that the deviation from FEM was even more pronounced locally.

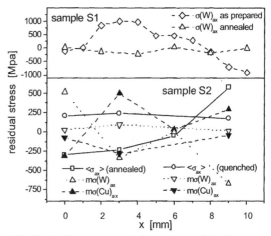

A non-equilibrium stress distribution was characteristic for most of the as prepared samples, i.e. the Cu phase has compressive stresses, while stresses in W are tensile. The deviation from the equilibrium state is the larger the faster the FGM has been cooled down from the manufacturing temperature. Annealing the S1 and S2 samples (they were tempered at 700 °C for several hours and then cooled down very slowly with cooling rates of 1-2 K per minute) a stress state was achieved close to the equilibrium with expected micro-stresses, i.e. $m\sigma_{11}(Cu) > 0$ and $m\sigma_{11}(W) < 0$. For cross-section averaged measurements, the

Fig. 2: Axial components in dependence of the distance to the centre x after different heat treatment. Above W residual stress in the as prepared and the annealed sample S1 (D1A, $V_g = 1$ mm^3, depth z = 3 mm). Below experimental results for the annealed and quenched sample S2 (E3, $V_g = 2x2x20$ mm^3, z = 1 mm).

macro-stresses did correspond to the FEM predictions qualitatively for depth values z > 2 mm. Quenching the sample S2 in water from 900°C this FGM stayed intact, however, the extreme thermo-mechanical load caused again a non-equilibrium stress state (cf. Fig. 2).

Further results for W-Cu FGM can be summarized as follows:
• The interface to the Cu block has always an important influence on the stress distribution.

---

[1] The following simplifying assumptions and approximations are used in the approach [12]:
i) consideration of a material with effective material parameters, ii) applying simple mixing rules to estimate the materials parameter from the bulk values of the constituents, iii) assuming homogeneous cooling from 1085 °C down to room temperature, and iv) effects of plastic deformation are based on fictive yield strength and hardening curves.

- Stress in Cu seems to have predominantly an upward trend towards the interface, stress in W behaves opposite.
- The observed micro-stresses are often of hydrostatic nature, which is in contradiction to the FEM simulation.
- The absolute stress values of hydrostatic stress for Cu are much larger than the bulk yield strength, which is due to the small dimensions of Cu ligaments and the strong adhesion forces.

THERMAL EXPANSION OF W-CU COMPOSITES

Using neutron diffraction, the temperature-dependent lattice strain $\varepsilon(T,\alpha) = (a(T,\alpha) - a(\alpha)_0))/a(\alpha)_0$ can be determined. $a(T,\alpha)$ is the temperature dependent lattice constant. For both the Cu and W phase the strain values $\varepsilon(T,\alpha)$ contain a stress free thermal expansion part and a contribution responsible for the residual stress. In homogenous single phase bulk materials only the first part is present. Since the diffraction method controls the elastic lattice strain in the crystallites only, the difference of the measured strain in the composite and that one in the bulk material can be attributed to the residual stress part in $\varepsilon(T,\alpha)$, i.e.

$$\sigma(T,\alpha) = (\varepsilon(T,\alpha) - \varepsilon^{bulk}(T,\alpha))\, E_\alpha(T) \,/\, (1-2\nu_\alpha(T), \qquad (5)$$

where $E_\alpha(T)$ and $\nu_\alpha(T)$ are the temperature dependent elastic constants. If we assume isotropic thermal expansion the stress is of hydrostatic nature which causes the factor $1/ (1-2\nu_\alpha(T))$ in Eq. (5). The room temperature values ($E_W^{,bulk} = 411$ GPa, $\nu_W^{,bulk} = 0.28$, $E_{Cu}^{bulk} = 145$ GPa, and $\nu_{Cu}^{bulk} = 0.34$) as well as the temperature dependence of the elastic constants were approximated by the data found for the bulk materials.

A temperature-dependent experiment was performed at HRFD with a 5x5x50 mm^3 sample of 25 vol. % Cu cut from a homogeneous composite material. Using a lamp furnace the sample was heated up to 1000 °C. In Fig. 3 the lattice strain data for both phases are shown together with the results of dilatometric measurements. The derived residual stresses based on the thermal expansion data are shown in Fig. 4. Due to the influence of temperature the sample position can be slightly

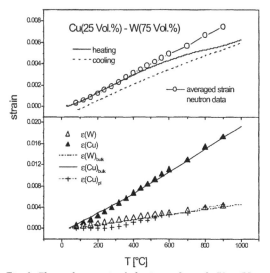

Fig. 3: Thermal expansion behaviour of sample S2 at 75 Vol.% W content in comparison with the lattice strain measured by neutron diffraction during heating. Below the plastic Cu strain is given calculated according to Eq. (8).

changed and, consequently, the diffraction peak position has to be corrected. As both, the Cu and W reflections are shifted, the relative position will not be changed. Therefore, introducing suitable phase independent strain will cause the macro-stress to vanish, i.e.

$$X_{Cu} \, \sigma(T,Cu) + (1-X_{Cu}) \, \sigma(T,Cu) = 0 \, . \tag{6}$$

The residual stress at room temperature T_R was measured separately for both phases: $\sigma(T_R, Cu) = -44.6$ MPa and $\sigma(T_R,W) = 19.1$ MPa. These values are in good agreement with the estimates above (see Fig. 4). During heating and cooling plastic deformations $\varepsilon(Cu)_{pl}$ of the Cu phase as well as micro-crack formation in the W matrix can occur. In order to get more quantitative information on these effects we have developed a macroscopic model where the plastic flow of Cu $\varepsilon(Cu)_{pl}$ is treated as a fit parameter. The local instability of the W matrix (rupture of walls and strips) is taken into account by using an effective porosity P_{eff} which lowers the elastic constant of the W matrix according to (cf. [14]) :

$$E_W(T, P_{eff}) = E_{W,}^{bulk}(T) \, (1 - 1.91 \, P_{eff} + 0.91 \, P_{eff}^2) \, , \tag{7}$$

The basic equations for evaluating the stress values in both phases are then:

$$<\sigma>(T, P_{eff}, \varepsilon(Cu)_{pl}) = X_{Cu} \, \sigma_{Cu}(T, P_{eff}, \varepsilon(Cu)_{pl}) + (1-X_{Cu}) \, \sigma_W(T, P_{eff}, \varepsilon(Cu)_{pl}$$

$$\sigma_{Cu}(T, P_{eff}, \varepsilon(Cu)_{pl}) = -E_{Cu}(T)/(1-2v_{Cu}(T)) \, [\varepsilon_g(T, P_{eff}, \varepsilon(Cu)_{pl}) + CTE(T,Cu) \, \Delta T - \varepsilon(Cu)_{pl}] \tag{8}$$

$$\sigma_W(T, P_{eff}, \varepsilon(Cu)_{pl}) = - E_W(T, P_{eff})/(1-2v_W(T)) \, [\varepsilon_g(T,P_{eff}, \varepsilon(Cu)_{pl}) + CTE(T,W) \, \Delta T]$$

Due to the mismatch of the thermal expansion coefficient CTE_α (T) both phases will be strained until reaching the equilibrium condition $<\sigma>(T, P_{eff}, \varepsilon(Cu)_{pl}) = 0$. The strain $\varepsilon_g(T, P_{eff}, \varepsilon(Cu)_{pl})$ has to be introduced in order to fulfill this condition. Furthermore, the residual stress $\sigma_{Cu}(T,Peff, \varepsilon(Cu)_{pl})$ of the Cu phase derived in the framework of the macroscopic model must coincide with $\sigma(T,Cu)$, obtained above on the basis of the neutron data (microscopic approach). Trying to solve these equations for fixed values of $P_{eff} = X_{Cu}$ only unphysical negative plastic deformations $\varepsilon(Cu)_{pl}$ follow for low temperatures $T < T^* \cong 250\ °C$. Therefore, for $T < T^*$ a solution was found for $P_{eff} > X_{Cu}$ corresponding to a lower Young's modulus of the W matrix (caused by cracking). Above T^*, $P_{eff}(T) = P_{eff}(T^*)$ was used. The onset of plastic deformation at $T > T^*$ is indicated by the increase of that part of the line width B_{Cu} caused by quadratic fluctuations of the strain within the grains. B_{Cu} is determined by the slope of the

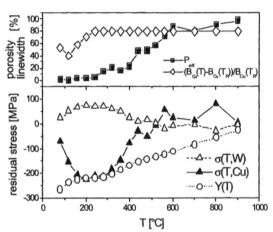

Fig. 4: Above : Effective porosity during heating and relative change of the Cu line width $B_{Cu}(T)$ caused by increasing dislocation density. Below : Estimated residual micro-stresses during the heating process for both phases of a W-Cu composite with 25 vol. % Cu. Additionally, the temperature dependent Cu yield strength $Y(T) = Y_o + Y_{bulk}(T)/Y_{bulk}(T_o)$ is drawn, where $Y_o = 350$ MPa was taken from the compression experiment.

Functionally Graded Materials 2000

whole line width as a function of the lattice displacement and is mainly determined by the dislocation density (the initial defect contribution to B_{Cu} was subtracted).

COMPRESSIVE LOAD
With the help of a special loading device equipped for measurements at HRFD compressive tests were performed with the W-Cu samples cut from a composite material of about 25 mm diameter and 25 Vol % Cu. The results for two samples with quite different initial stress state $\sigma_i(0)$ are displayed in Fig 5. From a linear fit up to 100 MPa the elastic constants were estimated as $E_{Cu} = 148.8$ GPa , $E_W = 390.7$ GPa , $v_{Cu} = 0.36$ and $v_W = 0.26$ which is in good agreement with the bulk values of these quantities.
From the analysis of the micro-stresses according to Eq. (4), it was established that the micro-stresses are mainly of hydrostatic nature. In the first sample ($\sigma_{Cu}(0) = -680$ MPa and $\sigma_W(0) = 230$ MPa) the increase of the Cu line width starts at about 350 MPa, which indicates the yield strength for Cu in the composite under compression. In the second case $\sigma_{Cu}(0) = 230$ MPa and $\sigma_W(0) = -70$ MPa) no plastic deformation of the Cu phase was detected, i.e. the Cu line width remains nearly constant with increasing load. However, at applied stresses above 100 MPa the W matrix exhibits micro-cracking and mainly the Cu phase compensates the external load.

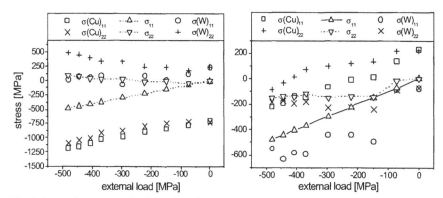

Fig. 5: Internal stresses in the Cu and W phases, averaged macro-stresses in load direction and perpendicular to it under compressive load. The two samples have different initial residual stress states

CONCLUSIONS
Macro-stresses in W-Cu FGMs were determined by neutron diffraction. A qualitative agreement of the stresses with an elasto-plastic FEM analysis could be achieved if the samples were annealed at 700°C. Very different stress states were observed after rapid cooling. This result is explained by local fracture of the W matrix and plastic deformation of the Cu phase. It was found that different gradients in W-Cu samples have only a small influence on residual stresses compared to the effects of different thermal treatments during or after the manufacturing process. This result is important, because W-Cu composites are often subjected to temperature changes in use, be it as a part of a fusion reactor wall or as a high current switch.

ACKNOWLEDGEMENT

The authors wish to thank the German Research Society (DFG) for funding these investigations under contract numbers SCHR 577/1-3 and NE599/4-1. The authors would like to thank Prof. J. Rödel and Dr. R. Jedamzik for fruitful collaboration. We gratefully acknowledge the stimulating discussions with Prof. D. Munz, Dr. Y.Y. Yang and W. Schaller. The authors are indebted to Prof. V.L. Aksenov, Prof. A.M. Balagurov, Dr. V.G. Simkin, Dr. Yu.V. Bokuchava, Dr. A. Pyzalla and Dr. Th. Pirling for support in neutron experiments.

REFERENCES

[1] Y.Itoh, M.Takahashi, H.Takano: "Design of tungsten/copper graded composites for high heat flux components", Fusion Engeneering and Design, 31 (1996) 279-289

[2] A. Neubrand, J. Rödel: „Gradient Materials: An Overview of a Novel Concept", Z. Metallkunde 88 (5) (1997) 357-371.

[3] R.L. Williamson, B.H. Rabin, J.T. Drake, „Finite Element Analysis of Thermal , Residual Stresses at Graded Ceramic-Metal Interfaces, Part I: Model Description and Geometrical Effects and Part II: Microstructure Optimisation for Residual Stress Reduction", J. Appl. Phys. 74 (2) (1993) 1321-1326.

[4] B.H. Rabin, R.L. Williamson, T.R. Watkins, X.-L. Wang, C.R. Hubbard, S. Scooner: „Characteris ation of Residual Stresses in graded Ceramic-Metal Structures: A Comparison of Diffraction Experiments and FEM-Calculation", 3rd Intern. Symp. FGM 94 (Lausanne, 1994) 209-215.

[5] M. Joensson, U. Birth, B. Kieback, "Gradient Components with High Melting Point Diffe rence", in Functionally Graded Materials 1996, Edited by I. Shiota and Y. Miyamoto, Elsevier, Amsterdam (1996) 167-172.

[6] R. Jedamzik, A. Neubrand, J. Rödel, "Functionally Graded Materials by Electrochemical Processing and Infiltration: Application to Tungsten/Copper Composites", Journal of Materials Science 35 (2000) 477-486.

[7] G. Bokuchava, N. Shamsutdinov, J. Schreiber, M. Stalder, Determination of residual stresses in WCu gradient materials, Textures and Microstructures, 33 (1999) 207-217

[8] G.Bokuchava, J. Schreiber, N.Shamsutdinov, and M.Stalder, "Residual Stress States of Graded CuW Materials", Proc. of the 5th Int. Symp. on Functionally Graded Materials, Ed.: W.A. Kaysser (ttp TransTech Publications LTD, 1998) 1018 – 1023.

[9] T. Wieder, A. Neubrand, H. Fuess, T. Pirling: "Yield stress increase in a W/Cu composite by neutron diffraction"; Journal of Materials Science Letters, 18 (1999) 1135 - 1137.

[10] R.S. Ravichandran, "Thermal Residual Stresses in Functionally Graded Material System", Mater. Sci. Eng. A201 (1995) 269-276.

[11] A.E. Giannakopulos, S. Suresh, M. Finot, M. Olsson, "Elastoplastic Analysis of Thermal Cycling: Layered Materials with Compositional Gradients", Acta Metall. Mater. 43 (4) (1995) 1335-1354.

[12] W. Schaller, Y.Y. Yang, "Thermal Stresses in a Graded Multi-Layered Joint", Forschungszentrum Karlsruhe, Wissenschaftliche Berichte FZK 6107 (1998).

[13] http://www.hmi.de/bensc/instrumentation/instrumente/e3/e3.html

[14] B.I. Ermolaev, "Thermal Expansion of Tungste n-Copper Pseudoalloys Containing up to 25-39 Vol.-% Copper", in Poroshkovaya Metallurgiya vol 3/99 (1971) 45-50.

FUNCTIONALLY GRADED MOSI₂-AL₂O₃ TUBES FOR TEMPERATURE SENSOR APPLICATIONS

M.I. Peters, R. U. Vaidya, R. G. Castro, J.J. Petrovic, K.J. Hollis and D.E. Gallegos

Los Alamos National Laboratory
Material Science and Technology Division
Los Alamos, NM 87545, USA

ABSTRACT

$MoSi_2$ and Al_2O_3 are thermodynamically stable elevated temperature materials whose thermal expansion coefficients match closely. Composites of these materials have potential for applications such as protective sheaths for high temperature sensors. $MoSi_2$-Al_2O_3 functionally graded tubes were fabricated using advanced plasma spray-forming techniques. Both continuously-graded and layered-graded tube microstructures were synthesized. The characteristics of the graded microstructures and the features of the graded mechanical properties of these tubes will be discussed.

INTRODUCTION

Using platinum coatings on alumina (Al_2O_3) sheaths for thermocouples is a widely used practice in the glass industry. Protection of the thermocouple wires and alumina (Al_2O_3) sheathing is necessary to avoid corrosion and dissolution of the temperature-sensing unit. The cost associated with providing platinum coatings on the Al_2O_3 sheath material can be prohibitively high when taking into consideration the infrastructure needed at the glass plants to maintain and secure an inventory of available platinum. There are also issues associated with improving the performance of the platinum coated Al_2O_3. The failure rate of the thermocouples can be as high as 50%. The U.S. glass industry has been in search of alternative materials that can replace platinum and still provide the durability and performance needed to survive in an extremely corrosive glass environment.

Investigations by Y.S. Park et al [1] have shown that molybdenum disilicide ($MoSi_2$) has similar performance properties in molten glass as some refractory materials that are currently being used in glass processing applications. Molybdenum disilicide is a candidate high temperature material for such applications because of its high melting temperature (2030°C), relative low density (6.24g/cm³), high thermal conductivity (52 W/mK), a brittle to ductile transition near 1000°C, and stability in a variety of corrosive and oxidative environments [2,3]. Additionally, the cost of $MoSi_2$ is significantly lower as compared to platinum coatings.

Plasma spraying has been shown to be a very effective method for producing coatings and spray formed components of $MoSi_2$ and $MoSi_2$ composites [4]. Investigations on plasma spray formed $MoSi_2$-Al_2O_3 composite gas injection tubes were shown to have enhanced high temperature thermal shock resistance when immersed in molten copper and aluminum [5]. The composite tubes outperformed high-grade graphite and SiC tubes when immersed in molten copper and had similar performance to high-density graphite and mullite when immersed in molten aluminum. Energy absorbing mechanisms such as debonding (between the $MoSi_2$ and Al_2O_3 layers) and microcracking in the Al_2O_3 layer contributed to the composites' ability to absorb thermal stresses and strain energy during the performance test (shown in Figure 1). Molybdenum disilicide and alumina are chemically compatible and have similar thermal expansion coefficients [6,7].

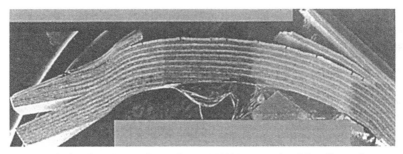

Figure 1. Four-point bend test beam after testing at 1400°C. Extensive debonding at the MoSi$_2$/Al$_2$O$_3$ interface and cracking within the Al$_2$O$_3$ was observed [5].

For thermocouple applications that require immersion of the thermocouple directly into molten glass, MoSi$_2$ coatings on Al$_2$O$_3$ protective sheaths will need to be optimized in order to perform in both a high-temperature (>1300°C) oxidizing environment (above the glass line) in addition to performing in the highly corrosive molten glass environment (below the glass line). We are currently evaluating the potential use of a graded coating of Al$_2$O$_3$ to MoSi$_2$ to enhance the performance of the MoSi$_2$ coating in molten glass. The graded microstructure of the coating will reduce the residual stresses that can develop during the spray deposition process, which can cause cracking and spallation of the coating limiting the coatings' lifetime. Preliminary results will be presented on the methodology used to produce the plasma sprayed MoSi$_2$-Al$_2$O$_3$ graded composites and on the microstructure and mechanical behavior.

EXPERIMENTAL PROCEDURES
Use of conventional plasma spraying equipment allows the flexibility of producing a variety of MoSi$_2$-Al$_2$O$_3$ microstructures, including laminate and graded structures. The plasma spraying equipment used for producing the graded structures included a Praxair Surface Technologies SG100 plasma torch and two Model 1264 powder feed hoppers. The plasma torch was mounted on a Fanuc S10 6-axis robot. A Technar DPV 2000 in-flight particle analyzer was used to measure the temperature, velocity and particle distribution of the MoSi$_2$ and Al$_2$O$_3$ particles as they exited the plasma torch. A computer control system was used to monitor and control the processing gases and the powder hoppers dispensing rate. Figure 2 shows an example of the computer control logic that was used to control the powder hopper rotation speed needed to produce the MoSi$_2$- Al$_2$O$_3$ graded structure. In this figure, pure Al$_2$O$_3$ (Powder 2) is first deposited at a powder hopper rotation speed of 0.8 rpm. After an initial period of depositing pure Al$_2$O$_3$, MoSi$_2$ (Powder 1) is gradually introduced into the plasma torch by increasing the powder hopper rotational speed for MoSi$_2$ and decreasing the rotational speed for Al$_2$O$_3$. The powder hopper speed for MoSi$_2$ subsequently reaches 0.8 rpm and the Al$_2$O$_3$ powder dispensing speed goes to zero. Pure MoSi$_2$ is then deposited on the outside diameter of the tube. Argon was used for the plasma generating gas (40 standard liters per minute, slm) and as a carrier gas for the MoSi$_2$ and Al$_2$O$_3$ powders (1 to 4 slm). A Mikron TH 4104 infrared camera was used to measure and display the temperature of the substrate. A POCO™ graphite tube (12.7mm OD, 9.5mm ID) was used as the substrate for the deposition spray trials.

To determine the mechanical behavior of the graded $MoSi_2$-Al_2O_3 structures, C-rings were machined from the material deposited on the graphite rods. C-ring samples were wire electro-discharge machined (EDM) out of the sprayed tube samples. The C-rings had an OD of 25.93 mm, an ID of 12.8mm and a width of 10.76 mm. The critical $b/(r_o$-$r_i)$ ratio was 1.64, within the required range of 1 to 4. The C-ring samples were tested in diametrical compression using a hydraulic Instron test frame (Type 1331 with an 8500 Plus controller and a 10kN load cell), at a crosshead speed of 0.125 mm/min (strain rate ~ 0.316×10^{-4} s^{-1}). Machine compliance was corrected using a standard Al_2O_3 sample of known stiffness. All of the samples were machined and tested in accordance with ASTM Standard C 1323-96. Twelve samples for each composite tube were tested. Four samples of monolithic plasma sprayed $MoSi_2$ and Al_2O_3 were tested and used for comparison. A Weibull statistical approach [8] was used to obtain the strength distributions in the coated and uncoated samples. While the standardized student-t test [8] employing the normal or gaussian frequency distribution is often used in statistical analysis, there is no theoretical or experimental justification for using it in problems involving fracture. Use of a normal distribution is often inappropriate in analyzing fracture problems with plasma-sprayed ceramic materials because of the presence of multiple flaw populations. The Weibull distribution is more appropriate (and conservative) in this scenario because it does not require that the flaw population be normally distributed.

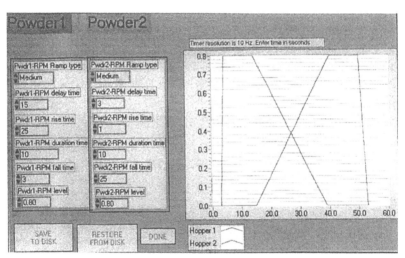

Figure 2. Computer based control system used to control the powder-hopper dispensing rate for producing the $MoSi_2$-Al_2O_3 graded structures. Powder 2 is Al_2O_3 and Powder 1 is $MoSi_2$.

RESULTS

Macrographs of two types of $MoSi_2$- Al_2O_3 graded composites produced by plasma spraying are shown in Figure 3. The white phase is the Al_2O_3 and the dark phase is $MoSi_2$. Figure 3a, shows a layered and graded microstructure where the cross-section of the sample consists of discrete individual layers that have been graded from Al_2O_3 to $MoSi_2$. Figure 3b, shows the cross-section of a continuously graded structure where pure Al_2O_3 was first deposited on a graphite rod followed by increasing amounts of $MoSi_2$ until pure $MoSi_2$ is deposited on the outside diameter of the tube. An example of a typical microstructure produced in the graded region between the MoSi$_2$ and the Al$_2$O$_3$ is given in Figure 4. A layered type of microstructure is produced when individual molten particles are flattened as they impact the substrate or previously deposited material.

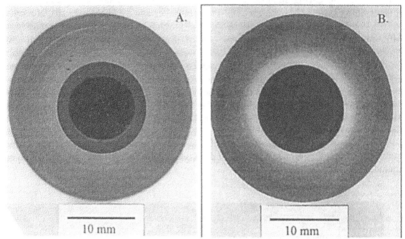

Figure 3. Macrographs of MoSi$_2$-Al$_2$O$_3$ graded tube cross-sections. a) Discrete individual layers that have been graded from Al$_2$O$_3$ to MoSi$_2$ and b) continuously graded structure from pure Al$_2$O$_3$ on the inside diameter of the tube to pure MoSi$_2$ on the outside diameter of the tube.

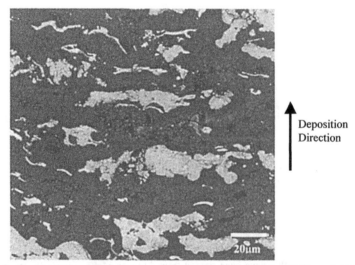

Deposition Direction

Figure 4. A typical microstructure produced in the graded regions between the MoSi$_2$ and Al$_2$O$_3$. Light areas are Al$_2$O$_3$ and dark areas are MoSi$_2$.

Residual stresses can occur in plasma sprayed deposited material as a result of quenching stresses that occur when individual particles are rapidly cooled upon impacting the substrate surface, differential thermal contraction between the deposited material and substrate and volume changes associated with any solid state phase transformation [9]. Residual stresses can give rise to deformation of the coated work piece and can result in spalling and cracking of the coating. In addition, various types of coating performance indicators such as adhesion strength, resistance to thermal shock under thermal cycling and erosion resistance are strongly influenced by the nature of the residual stresses. At elevated temperatures (> 1000°C), $MoSi_2$ goes through a brittle to ductile transformation [2] and can accommodate some of the residual stresses that occur during the deposition process and also during high temperature use of the material. Combining $MoSi_2$ with Al_2O_3 can enhance the high temperature performance of Al_2O_3 protective thermocouple sheaths by improving the poor thermal shock behavior of Al_2O_3.

THERMAL AND MECHANICAL PERFORMANCE
The mechanical behavior of the FGMs was evaluated using C-ring tests. Probability of failure at a given strength was obtained using Weibull analysis. Figure 5 shows the strength distribution plots for the layered and continuous FGMs. Both the continuous and layered FGM microstructures were found to exhibit similar mean Weibull strengths ($\sigma_f \sim 70$ MPa), which were calculated using equation 1 [8].

$$\sigma_f = a^{-1/\beta} \Gamma[1 + 1/\beta] \qquad \text{(Eq. 1)}$$

Where beta (β) is the Weibull modulus and gamma (Γ) is a function of the sample size. However, the spread of the data for the continuously graded material was smaller (hence a larger Weibull slope; 13.38 for the continuously graded samples, versus 7.635 for the layered graded samples). Interestingly, the fracture energy of the FGMs (qualitatively determined from the area under the load-displacement plot of the C-ring tests) was observed to be significantly higher (~ 3 times) than that of monolithic Al_2O_3 or $MoSi_2$. The fracture energies of the monolithic and graded coatings are listed in Table 1. We are in the process of conducting independent fracture toughness tests to validate these preliminary observations and to determine if the graded microstructures exhibit R-curve behavior.

Table 1. Fracture energies of $MoSi_2/Al_2O_3$ coatings.

Material	Fracture Energy J/m^2
Monolithic Al_2O_3	285
Monolithic $MoSi_2$	496
Continuously Graded $MoSi_2/Al_2O_3$	766
Layered & Graded $MoSi_2/Al_2O_3$	955

Fracture surfaces for the continuously graded and layered FGMs are shown in Figure 6. The fracture surface exhibited extensive microcracking and roughening in the center portion of the C-ring. We believe that the increased toughening of the composite is a direct result of microcracking. Preliminary analysis has indicated the strength and toughness of these FGM tubes to be more than acceptable for the proposed applications.

Earlier studies conducted on MoSi$_2$ indicated that the oxidation of the MoSi$_2$ was acceptable below and above the glass line. However, the high corrosion rate at the glass line was unacceptable and the resulting glass-line oxidation provided a challenge in the development of thermocouple sheaths. The answer to some of these challenges lies in understanding the mechanisms involved in the glass corrosion process. Figure 7 is a schematic of a phenomenological model that describes the oxidation mechanisms occurring above, at and below the glass line for MoSi$_2$ in a borosilicate glass [1]. Above the glass line (air environment), a SiO$_2$ coating protects the MoSi$_2$ from continued oxidation. The protective SiO$_2$ layer is disrupted in the temperature range around 500°C and pesting of the MoSi$_2$ can occur. Below the glass line, molybdenum metal, oxides, and silicides form in a silicon-depleted layer. The silicon-depleted layer consists of a mixture of phases (Mo$_5$Si$_3$, Mo$_3$Si and Mo) that are stable due to low oxygen activity [1].

The high flux of molten glass at the glass line increases the oxygen activity, resulting in increased oxidation rates. The oxidation rate is further accelerated by the dissolution of the protective SiO$_2$ layer as a result of the convective flow of glass at the glass line. Efforts are underway to produce a protective MoSi$_2$ coating to eliminate the high corrosion rate at the glass line. We are currently pursuing composite coatings of MoSi$_2$ and higher viscosity glasses to minimize glass line corrosion rates.

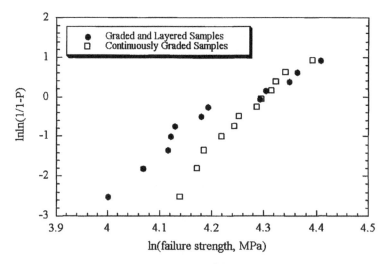

Figure 5. Results of C-ring tests performed on continuously graded and layered graded Al$_2$O$_3$-MoSi$_2$ coatings.

Functionally Graded Materials 2000

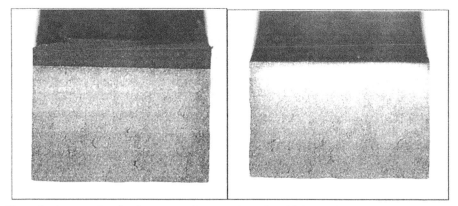

Figure 6. Macrographs of c-ring test specimens performed on Al_2O_3-$MoSi_2$ coatings sprayed on graphite substrates. a) Layered FGM b) Continuous FGM.

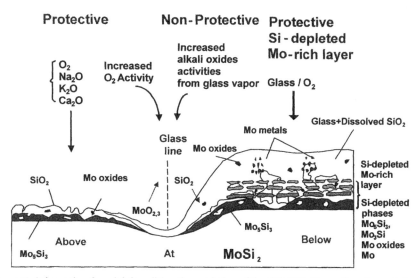

Figure 7. Schematic of model describing the oxidation of the FGM $MoSi_2$ coating above, below and at the glass line [1].

CONCLUDING REMARKS

We have demonstrated the use of conventional plasma spraying equipment in manufacturing $MoSi_2$-Al_2O_3 FGMs. The mechanical performance of these FGMs was superior to that of the monolithic materials. This was verified through C-ring tests. Although the molten glass corrosion behavior of the FGM coating is acceptable below and above the glass line, the corrosion rate at the glass line is higher than desirable. Processing of $MoSi_2$ based FGM composite coatings is ongoing and it will be determined if these new FGM coatings will reduce the glass-line corrosion.

ACKNOWLEDGEMENTS

This research has been supported by the U.S. Department of Energy, Office of Industrial Technologies, Glass Industry Program. The authors would like to acknowledge the technical support given by Richard Hoover at LANL.

REFERENCES

[1] Y.S. Park, D.P. Butt, R. Castro, J. Petrovic, W. Johnson, *Materials Science and Engineering*, **A261**, 278-283 (1999).

[2] J.J Petrovic, *Materials Research Society Bulletin*, **XVIII 35**, (1993).

[3] J.J Petrovic, and A.K. Vasudevan, *Materials Research Society Symposium Proceedings*, 322, 3 (1994).

[4] R.G. Castro, H. Kung, K.J. Hollis and A.H. Bartlett, *15th International Thermal Spray Conference*, Nice, France May 25-29, 1199-1204 (1998).

[5] A.H. Bartlett, R.G. Castro, D.P. Butt, H. Kung, Z. Zurecki and J.J. Petrovic, *Industrial Heating*, January 33-36 (1996).

[6] R.U. Vaidya, P. Rangaswamy, M.A.M. Bourke, D.P. Butt, *Acta Metallurgica*, **46**, 2047-2061 (1998).

[7] W.D. Kingery, H.K. Bowen, D.R. Uhlmann, *Introduction to Ceramics*, Second Edition, John Wiley & Sons, New York, 1976.

[8] W. A. Weibull, *Journal of Applied Mechanics*, **18** [3], 293 (1951).

[9] S. Kuroda and T.W. Clyne, *Thin Solid Films*, **200**, 49-66 (1991).

A NOVEL METHOD OF JOINING CERAMIC-METAL SYSTEMS TO REDUCE THERMALLY INDUCED STRESSES,

K. Khene, C. S. Trueman, N.D. Tinsley, M.R. Lacey and J. Huddleston
Department of Mechanical and Manufacturing Engineering
The Nottingham Trent University
Burton Street, Nottingham, NG1 4BU, England

ABSTRACT

A novel method for ceramic-metal joining, applicable for a wide range of emerging Ceramic Matrix Composites (CMC) and cermets is described. It combines singularity removal from the free surface with the introduction of a functional gradient zone, achieved by preshaping the mating parts to obtain typically a regularly repeating interpenetration of the two surfaces. This shaping is performed by CNC electrical discharge machining the CMC (generally containing conductive dispersoid nitrides, carbides or borides), by the preshaped metal part. This paper presents Finite Element Analysis predictions for a range of interface geometries, along with experimental results, illustrating the merits of the process as demonstrated through mechanical testing of finished joints.

INTRODUCTION

Ceramic-metal joints are vulnerable during processes such as brazing for, during cooling the joint may fail as a crack runs along the ceramic side of the interface. This cracking results from high differential thermal strain-induced stresses (owing to high expansion/contraction mismatch between metal and ceramic) and the high notch sensitivity of the ceramic.

The current work extends the study initiated by Suganuma et al [1] in investigating the effects and implications of interface geometry and size of components on the generation of residual stresses at ceramic-metal joint interfaces, and looking into novel ways to reduce or eliminate them and so allow the production of larger and more reliably joined parts.

This work moreover is particularly timely in that its supports current developments in the manufacture of complex ceramic shapes by electrical discharge machining (EDM) of emerging families of electroconductive ceramics [2]. EDM makes possible the manufacture of perfectly mating surfaces without design restriction or risk of machining damage that accompanies traditional diamond grinding. As electroconductive ceramics generally exhibit higher thermal expansion coefficients (CTE) than their traditional non-conductive counterparts owing to their dispersed network of conductive particles (generally carbides, nitrides and borides of titanium or zirconium) they should be preferred for joining applications for they thus generate lower residual stress levels.

INNOVATIVE CERAMIC TO METAL INTERFACES

Metal-to-ceramic joining is useful as it combines the properties of both materials to create a system which can exhibit specific properties at specific locations. The most obvious system would be to join two flat parts surface to surface. Unfortunately this simple approach represents the worst scenario, as thermally induced stresses rise sharply to values exceeding by far the strength of the ceramic part resulting in immediate failure promoted by cracks along the joint interface.

Tinsley et al [3] demonstrated by modelling, supported by experimentation conducted on joining ceramics to metals that, by using a ceramic profile, (convex especially) comprising, as optimum, a hemispherical 'domed' interface, the thermal stresses induced can be reduced by up to 74%. Thus sound and reliable joints can be produced. Their investigation however was only conducted for samples of 10mm diameter. For general engineering applications larger parts must be joined reliably.

As flat interfaces are to be avoided and domed ones are practical for only small sample size geometry, the present authors have introduced a new convex ceramic interface configuration for the joining of larger shapes. It consists in the simplest case of a half-dome profile at the free edge of the section embraced by a feather edge on the metal member. The central part of the interface is machined flat. This is termed 'the dome-flat' configuration (Figure 1). In its more complex form the flat inner section of the joint interface is replaced by single or repeated smooth interpenetration of one joint member by the other. Peak stresses generally occur at the free edge of the interface between ceramic and metal. At the free edge, the half dome convex shapes proposed here very largely eliminate the interfacial stress concentration.

All these interface designs, from simple dome through dome-flat to 'dome-complex' can be regarded as functional gradients, being a device for spreading the transition from 100% metal to 100% ceramic across the thickness of the joint and EDM is proposed here as a particularly attractive way to shape the mating interfaces when the ceramic member is made electrically conductive.

FINITE ELEMENT ANALYSIS

Predictive Finite Element Analysis [4] using a commercial package ANSYS 5.5.1* was used to identify and evaluate thermally induced stresses in regions near the joint interface between ceramic and metal. All the models considered in this study were cylindrical, therefore 2D considerations [5] were possible; this reduced the processing time. The critical regions of joint interfaces adjacent to the free surface, where thermally induced stresses peak on cooling and are most dangerous were represented on the models by the finest meshes.

In all cases eight quadratic elements were used to mesh the models. For both ceramic and metal parts, typical meshes contained 800 elements for 10mm diameter samples, 1300 elements for 20mm samples and 1600 elements for 30mm samples. The sample height for both ceramic and metal was 15mm in the models.

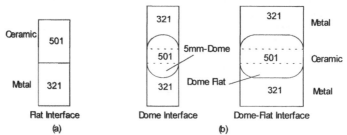

Figure 1: Simple interface configurations for AISI 321/Syalon 501 joints:
(a) Single interface, (b) double interface specimen

*Swanson Analysis System Inc. Houston, USA

The commercially available sialon ceramics, Syalon 101* and Syalon 501* were used in the modelling investigation with respective coefficients of thermal expansion of $3.04/°C \times 10^{-6}$ and $5.6/°C \times 10^{-6}$ at 20°C. Stainless Steel (AISI 321) represented the metal members of each joint. Values of its temperature-adjusted elastic modulus rise from 121.2GPa to 200GPa between 850°C and 20°C and are accompanied by values for its thermal expansion coefficient falling from $19.38/°Cx10^{-6}$ to $15.50/°Cx10^{-6}$ (values derived from [6]).

All stress modelling reported here for the ceramic-metal joints assumed non-linear elastic-plastic conditions as was reported in our earlier work [3]. The derived values of Young's Modulus and Coefficient of Thermal Expansion used in the FEA models are listed below:

Temperature (T/K)	293	523	723	923	1123
Young's Modulus (E/GPa)	200	177.6	162.9	144.7	121.2
C.T.E $(K^{-1} \times 10^{-6})$	15.50	16.65	17.80	18.61	19.38

Section 4 will show how experimentally derived results compare with the models' predictions.

RESULTS FROM THE FINITE ELEMENT ANALYSIS PREDICTIONS

Tinsley et al [3] demonstrated using non-linear elastic plastic analysis that the peak tensile residual stress induced at the free surface of a 10mm diameter flat interfaced Syalon 501-AISI 321 joint was 746 MPa. This high level of residual stress they showed subsequently by tensile testing to have limited the tensile strength of the 10 joints tested to values not exceeding 18.4 MPa.. It is to be concluded therefore that flat joint interfaces are undesirable.

In all following stress plots, SMX and SMN indicate maximum and minimum stress positions and values.

Figure 2(a) shows the principal stress (S1) distribution where the interface is domed with a 5mm-apex height. The peak value of 224MPa indicates a 69% reduction in stress in the ceramic after complete cooling in comparison with the flat system configuration with its value of 746MPa. It is further reduced by another 5% to only 192MPa when separated from the steel member by a 2mm copper soft metal interlayer (Figure 2b).

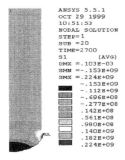

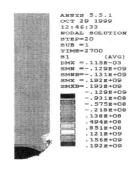

(a) Without interlayer   (b) With 2mm copper interlayer

Figure2: Convex Syalon 101 (10mm diameter) with 5mm radius dome.

Contrasting Figure 3(a) with Figure 2(a) shows the effect of increasing specimen diameter from 10 to 30mm in a dome flat configuration, whilst retaining the 5mm radius. After complete cooling over 9000s the stress has risen to 362MPa

*Syalon is a registered trade name of International Syalons Newcastle Ltd., P.O. Box26, Wallsend, Tyne & Wear, UK

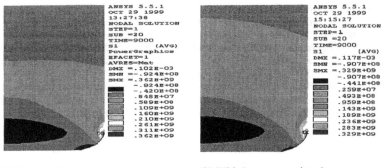

(a) Without interlayer   (b) With 2mm copper interlayer

Figure 3: Dome -Flat interfaced Convex Syalon 101 (30mm diameter).

The large increase in stress is again linked to a drop in temperature with time compounding thermal mismatch stress. Figure 3(b) shows that introducing a 2mm copper interlayer into the system gives a 9.1% reduction of the peak residual stress, decreasing it to 329MPa.

These results confirm that the introduction of a half-dome feature for the ceramic and a corresponding feather edge for the stainless steel member will reduce to a minimum the free-edge stress concentration. Thus we see a reduction in peak residual stress in the Syalon 101 by 69% and 51% for 10 and 30mm sections respectively, compared to the values for a flat interfaced 10mm joint without interlayer. Put another way, it confirms that the effect of increasing specimen diameter through 10,20 and 30 mm diameter will increase the peak residual stress and therefore lower the residual useful available strength, but for this ceramic with a tensile strength of 500MPa (approximately) that residual strength should be significant, even without the soft metal interlayer.

Figure 4(a) again shows the effect of increasing stress with increasing diameter, this time for Syalon 501; the value of 286MPa for peak stress in a sample with 20mm diameter has risen to 302MPa for 30mm diameter. Note, these values are slightly less than those calculated for identical interface shapes using Syalon 101. This is due to the fact that Syalon 501 has a larger coefficient of thermal expansion, lowering the expansion mismatch.

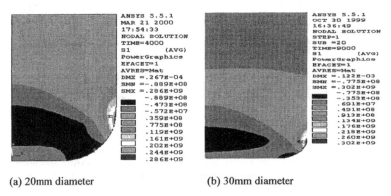

(a) 20mm diameter   (b) 30mm diameter

Figure 4: Dome -Flat interfaced Convex Syalon 501 without interlayer.

EXPERIMENTAL INVESTIGATION
Experimental procedures.

The ceramic and metal selected for the ceramic-to-metal joining study were a commercial TiN-bearing sialon, Syalon 501 and AISI 321 stainless steel. Profiling of the contoured metal interface was achieved by conventional machining involving a specially shaped knife edge tool to generate the feather edge. Electrical discharge machining, using copper electrodes, was used to profile the contoured ceramic surfaces. Final fitting of mating parts was achieved by using the shaped ceramic as an EDM tool against the AISI321 steel members of each joint. The 0.1mm thick active metal braze foils used comprised the commercial material Ticusil*, a titanium bearing silver-copper alloy.

Prior to assembly of the metal-ceramic-metal test pieces, with or without copper interlayers, any residual stresses were relieved by annealing the stainless steel at 1050°C, the copper at 600°C and the Syalon 501 at 1400°C for 30minutes for each. Fine plastic bead blasting was used to remove any recast layer from the ceramic and 800 grit emery paper was used to prepare the metal before cleaning both ultrasonically in inhibisol.

Induction heating was used to produce each joint system at 920°C for 15 minutes. Constituent parts were assembled precisely within graphite jigs accurately located within a graphite suseptor. Heating and cooling rates were 20°C per minute between 700°C - 920°C-700°C, then allowed to cool naturally, still under vacuum of 10^{-4} mbar for 1 hour. A compressive pressure of 165Pa was maintained across each joint interface throughout.

Five double flat-interfaced joints were produced (Figure 5(a)) for tensile testing (Table I) and for determining in one of them (Sample 5) the variation in hardening of the copper across the joint width after tensile testing the sample to failure. A first microhardness survey was conducted using a Leitz Miniload** hardness tester across polished cross sections cut diametrically and normal to the joint's flat interfaces close to both top and bottom edges of the copper interlayer (original annealed hardness $65H_{v100g}$), that is 75µm from the stainless steel/copper interface and 75 µm from the copper/Syalon 501 interface.

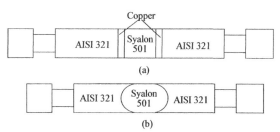

Figure 5 : Sections through (a) AISI 321/Syalon 501 double flat- interfaced joint with copper interlayers, (b) AISI 321/Syalon 501 double domed-interface joint without interlayers.

Six 10mm diameter double dome interfaced joints without interlayers were made (Figure 5(b)) and tensile tested to fracture (Table II).

*Ticusil is a registered trade name of WESGO, Inc, 477 Harbor Boulevard, Delmont, CA
**Leitz, Wetzlar, Germany.

Each brazed joint was tensile tested, ensuring axiality, using a crosshead speed of 0.5mm per minute, and the load at fracture recorded.

To investigate brazed joint quality, fractured test pieces were sectioned normal to the interface using EDM to avoid the introduction of machining damage. Standard metallographic techniques were used to polish the sections in order to gain an assessment of the degree of surface coverage by braze and of any voids' presence.

A second microhardness survey was conducted across polished cross sections cut diametrically and normal to the 10mm diameter Syalon 501-AISI 321 joint interfaces of two additional jointed samples, comprising a flat-interfaced and a 5mm-radius domed sample that were **not** to be tensile tested, to indicate the level of differential cooling-induced stress absorbed **by the stainless steel alone in the absence of a soft metal interlayer.**

All samples for microhardness testing were extracted using EDM to avoid the introduction of any additional work hardening to the joint systems under investigation.

EXPERIMENTAL RESULTS AND DISCUSSION

Table I: Tensile strengths for AISI 321-Syalon501- AISI 321 double flat-interfaced joints with copper interlayers.

Sample	1	2	3	4	5	Average
Strength (MPa)	36.46	41.63	51.84	53.16	58.86	**48.39**

These tensile strength values show relatively little spread, with brazed joints of generally good quality being produced.

The results of the first microhardness survey that was conducted within the copper interlayer after fracture for Sample 5 are shown in Figure 6.

They confirm the accepted fact that soft metal interlayers operate effectively as partial absorbers of brazing stresses in metal-to-ceramic joints. Moreover these results show, through the high level of work hardening observed here (up to 160Hv from 65Hv), that the interlayer may continue to ensure the joint's survival after brazing by a further period of work hardening under increasing 'service' stress levels until fracture can no longer be resisted.

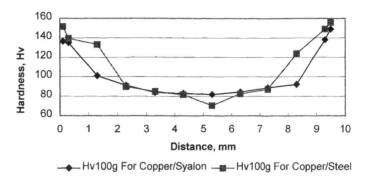

Fig 6: Comparison of microhardness of copper interlayer from fractured 10mm diameter double flat-interfaced AISI 321-Cu-Syalon501-Cu- AISI 321 joint for Specimen 5. Diamonds represent hardness values 75 μm from the copper/ ceramic interface. Squares represent hardness values 75 μm from the copper/ steel interface.

Table II: Tensile strengths for AISI 321-Syalon501- AISI 321 double dome-interfaced joints without copper interlayers.

Sample	1	2	3	4	5	6	Average
Strength (MPa)	33.14	34.47	43.09	79.15	122.63	173.67	**81.03**

The single high value of 173.7MPa achieved during the tensile testing of 6 double dome-interfaced samples (comprising 12 metal/ceramic interfaces) represents a very considerable increase on the highest value of 58.9MPa for a double flat-interface sample with soft metal interlayers. This high value of 173.7MPa indicates what can be achieved with good alignment of the braze assembly and good braze quality at each joint interface. The weaker joints in 5 of the 6 samples however were responsible for reducing the average strength of the double dome-interfaced samples to only 81.03MPa. The metallographic study of the solidified braze at all 10 of the interfaces from these weaker samples regrettably showed the presence of voids covering, typically 40% and 70% of the weakest of the two interfaces in samples 4 and 3 respectively. Clearly, great care must be exercised to ensure high brazing standards if the strength advantages to be gained by the mating of ceramic and metal interface profiles is to be realised consistently.

The second microhardness survey conducted in the AISI 321 at a distance of 100μm from the interface revealed that it had work hardened progressively from 207Hv_{100g} and 197Hv_{100g} at the centre line to 242Hv_{100g} and 254 Hv_{100g} close to the free surface of the flat- and dome-interfaced samples respectively. This clearly shows that the deliberate thinning of the stainless steel to a feather edge, coupled with greater steel/Syalon interfacial area in 10mm domed interfaced joints facilitated relief of brazing stresses greatly, as evidenced by the higher hardness differential (57points) between joint edge and centre compared with 35 points differential for the flat-interfaced joint. The hardness of the stainless steel before joining was 158Hv_{100g}.

CONCLUSION

Electrical discharge machining of newly emerging electrically conductive dispersoid-containing ceramics presents an excellent method of shaping the most complex interface configurations with minimum risk of surface damage. Furthermore these materials invariably have higher expansion coefficients than non-dispersoid-containing ceramics, so ensuring greater compatibility and therefore lower residual stresses when joined to metals.

Computer modelling predictions that the replacement of the abrupt stress concentration comprising the flat interface at the free surface of a metal-to-ceramic joint, by the incorporation of a feather edge to the metal member such as to embrace the ceramic member with its convex dome-like form, have been demonstrated experimentally to reduce very considerably the residual stress level in ceramic-to-metal joints. Tests on 10mm diameter active metal brazed metal-to-ceramic joints have demonstrated tensile strengths up to 173.7MPa. Furthermore, these joints had average strength more than double the value of double flat interfaced joints, notwithstanding the presence within the latter of stress relaxing soft copper interlayers.

The deliberately thinned feather edges, associated with the **extended** metal-to-ceramic interfacial area in dome-profile joints absorbed brazing stresses more completely than flat interfaced joints, as evidenced by the higher levels of differential strain - induced work hardening observed close to the free surfaces of the former.

Further experimental work will investigate the strengths of larger sectioned joints having dome-flat and dome-complex interface geometries, with and without soft metal interlayers. By combining appropriate profiling of interfaces with the incorporation of soft metal interlayers it is proposed that the residual strength of ceramic-to-metal joints will be further increased and that large sections will be successfully joined.

ACKNOWLEDGEMENTS

The authors are grateful to The Nottingham Trent University for the provision of facilities, to International Syalons Ltd for the supply of Syalon and to J. Wilkes, J. Shipstone, and the laboratory staff who assisted with the experimentation.

REFERENCES

1. K. Suganuma, T. Okamoto and K. Kamachi, "Influence of Shape and Size on Residual Stress in Ceramic/Metal Joining", *Journal of Material Science*, **22**, 2702-2706, 1987.
2 N.L. Mordecai, C.S. Trueman and J. Huddleston, " Die sinking technology for evaluation of spark erodable ceramics", *British Ceramic Transactions*, **95**, [4], 157-161, 1996
3. N.D. Tinsley, J. Huddleston, and M.R. Lacey, " The reduction of residual stress generated in metal-ceramic joining ", *Materials and Manufacturing Processes*, Marcel Dekker, **13**, [4], 491-504, 1998.
4. S. Kovalev, P. Miranzo and M. I. Osendi. "Finite element simulations of thermal residual stresses in joining ceramics with thin metal layers". *J. Am. Ceram. Soc.* **81**, 2342-48, 1998.
5. A. G. Foley and C. G. Winters, "The use of Finite Element Methods in the Design of Ceramic -to-Metal Joints", *British Ceramic Proceedings*, **48**, 107-122, 1991.
6. D. Peckner and I.M. Bernstein, "Handbook of Stainless Steels ", McGraw-Hill, 21.6, 1977.

HIGH THERMAL CONDUCTIVITY LOSSY DIELECTRICS USING A MULTI-LAYER APPROACH

B. Mikijelj
Ceradyne, Inc.
3169 Redhill Ave
Costa Mesa, CA 92626

J. O. Kiggans, T. N. Tiegs, P. A. Menchhofer
H. Wang, and H. T. Lin
Oak Ridge National Laboratory
P. O. Box 2008
Oak Ridge, TN 37831-6087

ABSTRACT

Lossy ceramics are used as load materials in devices that include both high power microwave sources and low power communication equipment. Desired properties of the materials are low porosity, high thermal conductivity, tailorable dielectric permittivity and loss, and good high temperature stability and mechanical properties. BeO-based composites have traditionally been used in these applications, but due to chronic beryllium disease health issues, the BeO materials are being replaced by AlN-based materials. The initial AlN substitutes did not have the desired thermal conductivities. By using a multi-layer approach, AlN-based composites having both high thermal conductivity and high dielectric loss have been produced.

INTRODUCTION

Almost every medium-to-high power vacuum microwave tube (such as traveling wave tubes and klystrons) in the U. S. employ BeO-SiC composites as lossy dielectric loads to suppress unwanted electromagnetic oscillations. Increasingly, there are economic and regulatory pressures to eliminate the use of BeO because it can cause chronic beryllium disease[1]. The development of a replacement for BeO-SiC composites would have a significant impact on commercial and military systems using microwave tubes, and possibly allow new components to be designed. These lossy dielectrics could also be of great importance for use on planar microwave circuits that are present in every day items such as cell phones. Because of the enormous world-wide commercial market for these lossy dielectrics, these materials are becoming an important topic for research.

In most of the work done to date, researchers have attempted to replace BeO directly with AlN in a dispersed structure. For example, since 1990 Ceradyne has been producing a range of AlN-SiC composites which have similar loss properties to BeO-SiC, but a lower thermal conductivity: 35-55 W/m·K for the AlN-SiC, versus 130 W/m·K for BeO-SiC[2]. Other early composites used alternative materials for the lossy phase (semiconductors such as TiC or SiC-TiC mixtures; and conductors such as Mo, W, Nb, Ta, and glassy carbon[3,4]). The most notable improvement in the thermal conductivity, while still maintaining the high loss, was in AlN proprietary composites (**Ceralloy**® 137-CA and 137-CB)[2,5] which have a thermal conductivity of 85-105 W/m·K.

This report describes efforts to fabricate composites with layered structures, containing alternating, parallel layers of thermally conductive AlN and lossy dielectric materials.

PROCEDURES

Two basic sample types were prepared for this study: hot-pressed disks made from mixed powders and hot-pressed disks fabricated from tape cast materials.

Two types of mixed powder (MP) compositions were fabricated at ORNL:

1) ALN-MP - which was composed of AlN (Grade D[*]) with 5.5 vol. % Y_2O_3 (Grade 5600[†]), and
2) ALN-SiC-MP - which was composed of AlN with 4.1 vol. % Y_2O_3, and 25 vol. % SiC (Grade A 10[*]).

These mixed-powder samples were made by blending the dry powders for 18 h in a high-density polyethylene jar with zirconia media. Powders were hot pressed at 25 MPa and 1750 – 1800 °C for 1 h in nitrogen, using a 3.8 cm ID graphite die lined with boron nitride[‡] coated Grafoil[§].

Several types of tape-laminate samples were fabricated at ORNL:

1) ALN-L, which consisted of tapes of AlN containing 5.5 vol. % Y_2O_3,- Hot pressed at ORNL;
2) ALN-SiC-L163, which consisted of alternating layers of AlN – 5.5 vol. % Y_2O_3 and layers of AlN – 2.7 vol. % Y_2O_3 – 50 vol % SiC (Grade UF-05[*]) – Hot pressed at ORNL
3) ALN-SiC-L156, which consisted of alternating layers of AlN – 5.5 vol. % Y_2O_3 and AlN - 2.7 vol. % Y_2O_3 – 50 vol % SiC (Grade A 10[*]) - Hot pressed at Ceradyne
4) ALN-SiC-L157, which consisted of alternating layers of AlN – 5.5 vol. % Y_2O_3 and AlN - 2.7 vol. % Y_2O_3 – 50 vol % SiC (Grade UF-05) - Hot pressed at Ceradyne.

Preparation of the laminate samples started with the preparation of the ceramic tapes. Tapes consisting of the four compositions listed above were made from ceramic slurries which contained the following: a) 41 – 42.7 vol % of the powder constituents of each composition, b) 40.5 – 42.5, vol. % toluene[**]–ethanol[**] (5/1 wt. ratio) mixture, c) 7.5 – 8.2 vol. % N-butyl phthalate[α], 0.38 – 0.41 vol % Hypermer PS3[δ] dispersant, and d) 7.1 – 8.5 vol % Elvacite 2010[ψ] (methyl methacrylate). The slurries were blended for 6 to 12h with zirconia media in a high-density polyethylene jar. The slurry was separated from the media and degassed for approximately 3

---

[*] H. C. Starck, New York, NY 10017
[†] Molycorp, Inc., Los Angeles, CA, 90017
[‡] ZYP Coatings, Inc., Oak Ridge, TN, 37830
[§] UCAR Carbon Company, Inc, Cleveland, OH 4101
[**] EM Science, Gibbstown, NJ, 08027
[α] Tape Casting Warehouse, Morrisville, PA 190670
[δ] ICI Surfactants, Wilmington, DE,19850
[ψ] ICI Acrylics, Wilmington, DE,19850

minutes at 0.8 bars pressure with vibration to remove trapped air. A standard tape casting process was used to produce ~ 0.04 cm thick tapes. The tapes were dried overnight at ambient temperatures. Discs, 3.6 cm or 7.3 cm in diameter, were punched from the tapes.

A number of discs were stacked on top of each of other as needed to produce a given composite specimen. A mylar film was placed on the top and bottom of the stack and the stack placed in a steel die (inner walls and punch faces coated with AZN Release Agent[††]). The die was heated to a temperature of 80 °C, and the stack was pressed at 17 MPa for 15 minutes. Each laminate, 0.5 to 2.4 - cm thick, was removed from the die and heated in a capped, tube furnace in a flowing nitrogen atmosphere to remove the binder. The binder removal cycle consisted of heating at a rate of approximately 1°C/ min to a final temperature of 500°C, with 1 h holds during the heat-up at 200°, 250°, 300°, 350°, 400°, and 500°C. Each laminate was hot-pressed using the same conditions as the mixed powder samples.

Three sets of samples were prepared from Ceradyne-supplied materials (either as mixed powders or cast tape). For both, the compositions were proprietary and are designated "Ceradyne #1 MP" and "Ceradyne #2 MP" for the samples prepared from mixed powders, and "Ceradyne #1 L", and "Ceradyne #2 L" for samples prepared from layered tapes. Ceradyne MP samples were made using methods proprietary to Ceradyne. Ceradyne laminate samples were made using methods similar to those employed for the ORNL laminate samples. Laminates consisted of approximately 75 total tape layers and were 7.6 cm diameter by 2.5 - 2.8 cm thick. The two types of laminate samples consisted of:

1) 1L - stacked layers of Ceradyne AlN composition alternating with layers of Ceradyne 1 lossy composition, and
2) 2L -stacked layers of Ceradyne AlN composition alternating with layers of Ceradyne 2 lossy composition.
3) 3L -Three alternating stacked layers: AlN/Ceradyne 1/Ceradyne 2 basic stack.

These laminates were de-bindered at Ceradyne, Inc., using a similar procedure as used at ORNL, and the laminates were hot-pressed using methods proprietary to Ceradyne.

After each sample was hot-pressed and cleaned, the density of each sample was measured using the Archimedes method. The hot-pressed discs were machined at ORNL and Ceradyne into parts for thermal diffusivity measurements using laser flash techniques. Measurements were made perpendicular and parallel to the flat punch surface (top or bottom) of the hot pressed billet. The thermal conductivity (k) of each sample was calculated using the formula, $k = \alpha\, c_p\, \rho$, where (c_p) is the heat capacity, (ρ) the density, and (α) the diffusivity. The c_p values were taken from literature sources or calculated, based on the compositions of the samples[6,7]. The dielectric constant and loss tangent of select samples were measured at Ceradyne over the 1 to 20 GHz frequency range using a Hewlett-Packard (HP) 8510 network analyzer and a HP 85070C probe. Select ORNL samples were polished and examined with a light microscope and a SEM. Select Ceradyne parts were also

---

[††] Axel Plastics Research Laboratories, Woodside, NY, 11377

machined in two orientations into "T-button" parts for standardized loss testing in the S-band by a tube manufacturer, and the "Q" factors were measured in the standard industry fixture[8].

RESULTS

All of the samples reached better than 98% of theoretical density (Table 1), which is a necessity for achieving both high thermal conductivity and for compatibility with the high vacuum environment of microwave tubes.

Thermal conductivity values for ORNL and Ceradyne parts are presented in Table 1. The "ALN-MP" and "ALN-L" parts were made as controls to test sintering conditions and to evaluate the baseline thermal conductivity of the AlN material. The thermal conductivities of 158 W/m·K and 170 W/m·K obtained for the mixed powder and the laminated sample respectively, showed that the processing conditions were adequate to produce ALN-based materials having high thermal conductivity. Typical samples used for measuring the thermal diffusivity in perpendicular and parallel orientations are shown in Figure 1.

Figure 2a shows that ALN-SiC-L had a higher thermal conductivity than the ALN-SiC-MP sample in both disc directions. The data also shows that the layered laminate sample achieved a significant increase in the thermal conductivity in the direction parallel to the layered structure. Figure 2b shows a similar trend with the Ceradyne materials. Both the Ceradyne 1 and 2 laminate samples had higher thermal conductivity values, as compared to the samples prepared from mixed powders. Once again, the thermal conductivity values in the direction parallel to the layers was slightly higher than in the perpendicular direction.

Table 1. Summary of data for fabricated samples

Sample Type	Relative density (% theoretical)	Thermal Conductivity, $\perp$ (W/m·K)	Thermal Conductivity, $\parallel$ (W/m·K)
ALN-MP	>99	158	170
ALN-L	>98	171	ND
ALN-SiC-MP	>99	76	77
ALN-SiC L163	>99	88	116
ALN-SiC L156	>99	55	66
ALN-SiC L157	>99	67	73
Ceradyne 1-MP	"	105	
Ceradyne 1-L	"	127	138
Ceradyne 2-MP	"	120	
Ceradyne 2-L	"	153	160
Ceradyne 3L	"	160	166

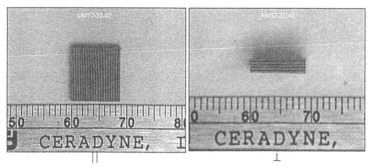

Figure 1. Thermal diffusivity samples machined in two orientations from a hot pressed layered composite of ALN-SiC L157.

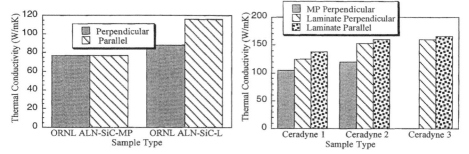

Figure 2a (left) and b (right). Thermal conductivity values for the ORNL and Ceradyne mixed powder and laminate samples, respectively.

Figure 3 and 4 show optical micrographs of ORNL mixed powder and layered samples. Figure 3a shows that the SiC grains are irregular in shape and appear to be evenly distributed throughout the matrix in ALN-SiC-MP sample. The appearance of individual layers in the ALN-SiC-L sample are shown in Figure 3b. The interface between the AlN and AlN-SiC along the boundary, and the dispersion of the AlN and SiC grains in the AlN-SiC layer of the same sample are shown in Figure 4a and 4b, respectively.

a   b

Figure 3. Optical micrographs of a) ALN-SiC-MP and b) ALN-SiC-L layers

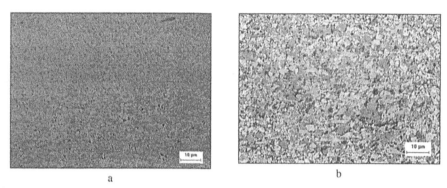

a   b

Figure 4. Optical micrographs of a) ALN-SiC-L layer interface polished samples and b) ALN-SiC-L lossy layer

Figure 5 shows T-buttons machined in two orientations (parallel or perpendicular to the layers) from the ALN-SiC L156 layered material used for standardized loss testing. Similar T-buttons were also fabricated for ALN-SiC L157 material.

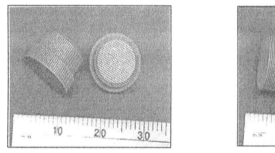

Figure 5. T-buttons (-01, left, and –02, right, orientation) machined from the ALN-SiC L156 layered material.

Figure 6 and 7 show the test fixture cavity Q factors with laminated T-buttons for the ALN-SiC L156 and ALN-SiC L157 compared to the standard BeO based lossy material.

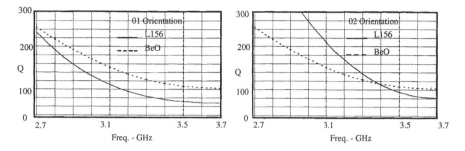

Figure 6. Q-factors for ALN-SiC L156 layered T-button samples machined in two orientations.

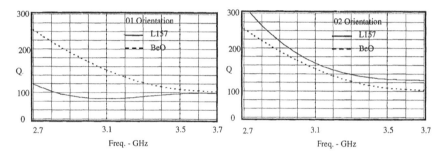

Figure 7. Q-factors for ALN-SiC L157 layered T-button samples machined in two orientations

The cavity Q factors for laminated samples show significant differences for the parallel (02) and perpendicular (01) orientations. The maximum loss (minimum Q) of both the AlN-SiC-L156 and AlN-SiC-L157 samples machined in the –01 orientation showed Qs relative to the standard BeO lossy material (dashed line which presents the maximum Q allowable over the frequency range). However, the Q varied with the position of the button in the measurement cavity. The Q factor of the AlN-SiC-L156 material, machined in the –02 orientation, was slightly higher than the maximum allowed value. For the ALN-SiC-L157 material in the –02 orientation, the Q was close to the requirement, and was very stable and independent of the button position in the cavity. Measured Q values of Ceradyne 2-L samples were too high (1200-1700), however the same dependence on the button orientation was observed as for the AlN-SiC-L156 and AlN-SiC-L157 samples. The discrepancy between the Q-values and the measured material properties for this sample is believed to be due to the variability of the material within the hot-pressed billet and the location from which the T-buttons were machined. Presented data demonstrates that the concept of layered lossy materials may be useful in replacing BeO lossy materials if the orientation dependency issue can be resolved.

CONCLUSION

The results of this study have demonstrated that the concept of layered lossy materials, consisting of alternating thermally conductive and dielectrically lossy layers, is valid. Materials manufactured using the concept have shown to have enhanced thermal conductivity values as compared to materials fabricated from dispersed powders, and in some cases have shown acceptable dielectric losses in the S-band in one of the components variations.

Due to its high orientation dependence, work is required to further refine and tune the material compositions and layer thickness', and to determine how these types of materials would be useful in the microwave tube industry.

ACKNOWLEDGMENTS

This work has been funded by the Laboratory Directed Research and Development Program of Oak Ridge National Laboratory (ORNL) managed by UT-Battelle, LLC for the U. S. Department of Energy under Contract No. DE-AC05-00OR22725 and by the DOD SBIR Contract N100014-99-M-0306 and The Office of Naval Research. Authors are grateful to Mr. Bob Tousey and Mr. Walt Gasta of CPI of Palo Alto for Q-factor measurements in their "Titanic" fixture.

REFERENCES
[1] Jeff Johnson, "We Did Wrong" C&EN, Nov. 15, 1999, p.24.
[2] Ceradyne, Inc., "Properties of Advanced Technical Ceramics for Microwave Applications," brochure 1999.
[3] B. Mikijelj and I. E. Campisi, "Structural and Electrical Properties of Artificial Dielectric Ceramics," pp. 39-47, in *Dielectric Ceramics*, Proc. Am.Cer.Soc. Pac. Coast Meeting, San Francisco 1992.
[4] I. E. Campisi, L. K. Summers, K. E. Finger, and A. M. Johnson, "Microwave Absorption by Lossy Ceramic Materials," Mat. Res. Synp. Proc., Vol. 269, pp 157-162, 1992.
[5] B. Mikijelj, "Alternative ALN Based Lossy Ceramics for High Power Applications," Abstracts. International Vacuum Electronics Conference, p364, 2000.
[6] W. D. Kingery, H. K. Bowen, and D. R. Uhlmann, *Introduction To Ceramics*, 2nd edition, edited by E. Burke, B. Chalmers, and J. Krumhansl, John Wiley and Sons, New York, 1976.
[7] CRC Handbook of Chemistry and Physics, 48th ed. edited by R. C. Weast and S. M. Selby, published by The Chemical Rubber Co, 1967.
[8] B. Mikijelj, "Alternative Lossy Dielectrics for High Power Applications," Final report for contract # N00014-99-M-0306, June, 2000.

DESIGN AND FABRICATION OF S-TYPE B_4C-SiC/C FUNCTIONALLY GRADED MATERIALS

Wei-ping Shen, Bo-zhong Wu, Jiang-tao Li and Chang-chun Ge
Lab. of Special Ceramics & Powder Metallurgy,
University of Science & Technology, Beijing, 100083,China

ABSTRACT

Homogeneous specimens of $(1-x)(80\%B_4C$-$20\%SiC)$-x C (x=0,0.2,0.4,0.6,0.8) (volume fraction φ_c) and Functionally Graded Materials were hot-pressed respectively at 2000°C and 20MPa under Ar atmosphere. The density, water absorption, linear thermal expansion, elastic modulus and flexural strength of the ceramics and commercial graphite were tested. Cracks were observed in FGMs of 6 and 11 layers that are designed by linear function. The S-type function of compositional distribution as a design method resulted in a flexural strength of 216MPa and thermal shock resistance Δ_{fc}>500°C for the 11 layered (x=0.2~1) FGM of S-type.

1. INTRODUCTION

Boron carbide, which has a high melting point, outstanding hardness, low specific weight, great resistance to chemical agents and high neutron absorption cross-section ($^{10}B_xC$,x>4), is currently used in high-technology industries, such as fast-breeders, lightweight armors and high-temperature thermoelectric conversion [1]. However, thermal conductivity of boron carbide (40-90 W/m·K) is lower than carbon (110-290 W/m·K). B_4C is a typical compound with covalent bond. It is difficult to densify B_4C by conventional sintering without sintering additives due to its low self-diffusion coefficient and micro-structural coarsening. Dense boron carbide samples have been obtained by sintering in presence of sintering additive [2].

A high flexure strength (750MPa) and fracture toughness (5.92MPa·m$^{1/2}$) were obtained for a composition of 80/20 B_4C/ SiC (volume fraction) by hot pressing at 2150°C for 30 minutes [3].

Carbon is an important additive for sintering of boron carbide. Density of >98% has been obtained for submicron boron carbide with phenolic resin (corresponding to 1%-3%C (weight fraction)) by pressureless sintering at 2150℃ [4].

Specimens with relative density of 97%-99.7% were obtained from a mixed powders of submicron boron carbide with 9%-10%SiC,1%-3%C (weight fraction) by pressureless sintering at 2000~2100℃ [5].

Functionally graded materials have been designed in which the plasma resistant materials face plasma and high heat conductive materials face a cooling medium. Thermal shock resistance is increased if it is without severe internal thermal stress due to gradual change in composition from one side to the other [7]. So it is necessary to know properties of functionally graded material layer by layer.

B_4C, SiC and C are all plasma resistant materials, but B_4C and SiC are better than C.C is a thermal shock resistant material. A FGM of B_4C, SiC and C was supposed to combine the advantages of these materials.

In this paper a new type of composition distribution profile - S type was designed and fabricated.

2. EXPERIMENTAL

Commercial powders were used as raw materials of experiments shown in table 1.

Table1.Some Characteristics of Powders

Powder	Purity %	Size
B_4C	>90	20 μ m
Si	>99	>74 μ m
C	>98.5	>74 μ m

Powders $(1 - x)$ $(80\%B_4C - 20\%SiC) - x$ C (volume fraction φ_c) $(x=0,0.2,0.4,0.6,0.8,1)$ and ethyl alcohol were mixed for 1h with cemented carbide ball in a nylon pot.

Specimens were made by hot-pressing at 2000℃,20MPa for 1h under Ar atmosphere. After hot pressing, surface grinding and polishing the specimen is 40mm in diameter and ~5mm in thickness.

The open porosity, water absorption, volume density and relative density were determined with Archimedes' method. The flexural strength and elastic modulus were measured by 3 point bending. In air thermal expansion coefficient was tested.

Flexural strength of FGM was measured with ceramic layer downward. A specimen of FGM was heated to 500℃ and kept for 0.5 h then quenched in water and its flexural strength was determined to observe thermal shock resistance.

3. RESULTS AND DISCUSSION

3.1. Open Porosity, Water Absorption, Density and Relative Density: The effects of volume fraction of carbon on open porosity, water absorption, density and relative density are shown in Fig.1~Fig.4.

The results of the experiment indicate the specimen of φ_c=20%~60% can be made with almost theoretical density by hot pressing at ~2000°C and ~20MPa. The specimen of φ_c=20% had the highest density. The density of the specimen of φ_c=0% is poor and there is laminar fracture on edge of the specimen because the temperature of hot pressing is lower. At that temperature a dense boron carbide / silicon carbide composite could not be obtained by hot pressing without carbon additive [3].

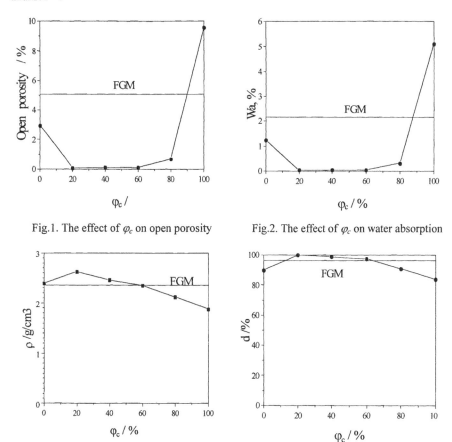

Fig.1. The effect of φ_c on open porosity

Fig.2. The effect of φ_c on water absorption

Fig.3. The effect of φ_c on volume density

Fig.4. The effect of φ_c on relative density

The water absorption of the specimen of φ_c=20,40 and 60% was lower (0.03%, 0.04% and 0.04%) than others as shown in Fig.2.

3.2. Flexural Strength and Elastic Modulus: The result of the experiment indicates that flexural strength of B$_4$C-SiC-C composites decrease gradually with increasing volume fraction of carbon φ_c (Fig.5). The flexural strength of specimen with φ_c=40% is 283.9MPa and that of specimen with φ_c=60% is 226.2MPa. Huang Qizhong et al. have attained flexural strength of 209.2MPa with a composite of 70%C - 19%B$_4$C - 11% SiC (volume fraction) [6]. The flexural strength is lower with specimens of $\varphi_c=0\%$ and 80% due to their porosity. The elastic modulus of B₄C-SiC-C composite reduces rapidly with increasing content of carbon. The elastic modulus is only 218GPa with specimen of $\varphi_c=40\%$. The elastic modulus of B₄C is 441 GPa [1] and that of SiC is 485GPa.

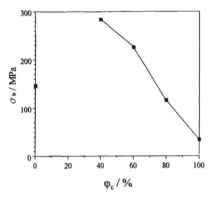

Fig.5. The effect of φ_c on flexural strength

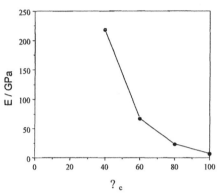

Fig.6. The effect of φ_c on elastic modulus

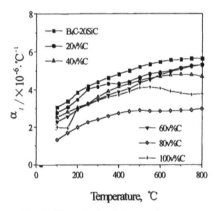

Fig.7. The effect of temperature o thermal expansion coefficient

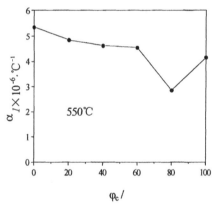

Fig.8. The effect of φ_c on thermal expansion coefficient at 550℃

3.3. Thermal Expansion Coefficient: The effect of temperature on the thermal expansion coefficients of composites with various compositions is shown in Fig.7.With increasing content of carbon the thermal expansion coefficients of B₄C-SiC-C ceramic reduce gradually, with the exemption of the φ_c =80% specimen, which has a substantially lowers thermal expansion coefficient (Fig.8). The specimen of φ_c =100% is commercial pure, dense and high strength graphite. Thermal expansion coefficients of the specimens in which φ_c are >60% decrease because of oxidation above 550℃.

3.4. Mechanism of Densification: The role of the added carbon consists in reduction of the B₂O₃ layer around the submicron B₄C particles as well as in acceleration of the process of grain

Functionally Graded Materials 2000

boundary diffusion. The role of carbon in blocking the growth of the carbide grains is also important [2].

Fig.9 shows that carbon can reduce the B_2O_3 and SiO_2 layer around the B_4C, Si and SiC particles at high temperatures. Carbon-rich boron carbide has a low viscosity and is densified more easily than the boron-rich materials [8,9].

However, the (80%B_4C-20%SiC)-x %C composites did not densify at 2000℃ and 20MPa if $x>60$ at 2000℃ and 20MPa because carbon based composites are difficult to sinter [6].

3.5 Reason of Crack Formation for the FGM Specimens and Design S-type Compositional Distribution Function: When the compositional distribution function is linear ($n=1$ in $\varphi_c=(x/\delta)^n$ as shown in Fig.11) the FGM of 6 or 11 layers have a network of cracks on their two surfaces. In addition, there are cracks between the layers in samples with 6 layers.

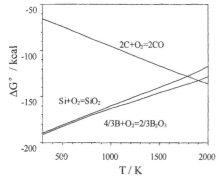

The sample possesses tensile stresses at the surface when it is cooled. This can lead to fracture.

The layer of B_4C-SiC cracks because its thermal expansion coefficient is higher and it possesses surface tension while cooling, and also the elastic modulus of B_4C-SiC end is much higher than other end (Fig.6). That is why it is necessary to reduce the composition gradient.

Fig.9. Gibbs energy of oxide as a function o temperature at 1 bar

As the flexural strength at the carbon-rich side is lower (Fig.5), it is necessary to reduce the gradient of composition in end of C, too. Unlike in ceramic/metal FGM both ends of B_4C-20%SiC/C are brittle materials. Both ends have shallow gradient of composition, the compositional distribution function becomes S-type (Equation 1, Fig.10).

The reason why cracks occurs between layers is that the number of layers is less and change in composition is larger, especially the thermal expansion coefficient changes remarkably near 80%C (Fig.8). Thus High tensile stresses are produced in the process of cooling. So it is necessary to increase number of layer.

The compositional distribution function of S-type is as Equation 1. Where φ is volume fraction of a component; x is distance from the surface; δ is thickness of specimen; n is an exponent, it is a variable that determines the shape of the curve representing the spatial distribution of the volume fraction.

For $n=2$ and first layer is $\varphi_c=20\%$ the compositional distribution function of 11 layered FGMs is shown in Fig.11.

There is no fracture occurred in 11 layered specimen with the first layer of 20v%C and S-type composition distribution. One of specimens had a flexural strength of 189MPa, another had a flexural strength of 216MPa after being kept 0.5h at 500℃ and quenched in water, so the FGM has a thermal shock resistance of $\Delta_{fc}>500℃$.

Scanning electron micrograph and compositional distribution of a S-Type specimen of 11 layers and 5mm in thickness is shown in Fig.12.

$$
\begin{cases}
\varphi = \dfrac{\left(2\dfrac{x}{\delta}\right)^{n}}{2}, & \dfrac{x}{\delta} = 0, \cdots, 0.5; \\[4mm]
\varphi = 1 - \dfrac{\left[2 \times \left(1 - \dfrac{x}{\delta}\right)\right]^{n}}{2}, & \dfrac{x}{\delta} = 0.5, \cdots, 1. \quad (n > 0)
\end{cases} \tag{1}
$$

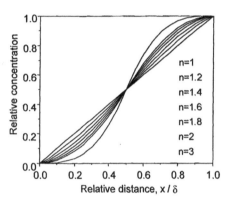

Fig.10 Diagram showing various profiles distribution obtained by varying the exponen in S-type equation.

Fig.11 S-type (n=2) compositional distributio function of 11 layers FGMs of first layer $\varphi_c=$ 20%

4. CONCLUSION

The composite of (80%B$_4$C-20%SiC)-20,40,60%C (volume fraction) can be densified by hot pressing at 2000 ℃ and 20MPa. The relative densities are ~100%, 98.8% and 97.4% respectively.

The flexural strengths are 284, 226 and 117 MPa respectively with specimens of φ_c=40,60,80v%C.

With increasing φ_c the thermal expansion coefficients of B$_4$C-SiC-C composite reduce gradually, but that of the specimen of φ_c =80% drops notably.

With increasing φ_c the elastic modulus of B$_4$C-SiC-C composite decreases rapidly.

Linear FGM (n=1) of 6 and 11 layers have cracks in the surface of the ceramic end. Specimens of 6 layers show delamination at edge.

S-type specimens of 11 layers with a first layer of 20v%C are crack-free. A flexural strength of 189MPa and a thermal shock resistance of Δ_{fc}>500℃ were obtained.

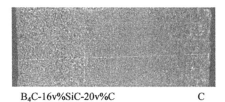

B₄C-16v%SiC-20v%C   C

11 layers, 5mm in thickness

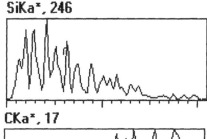

Fig.13   Scanning electron micrograph and compositional distribution of S-Type FGM

REFERENCES

[1] Thévenot F. Boron carbide-a comprehensive review. *J.Eur.Ceram.Soc.*1990; **6**:205-25

[2] Radev D D. Sintering of Boron Carbide-Based Materials. *Metall*, 1991,**51**(11): 630

[3] Niihara, Koichi. Sintered Composite Boron Carbide Body and Production Process Thereof, *European Patent Application.* EP 0 494 390 A1, 15.07.92 Bulletin 92/29

[4] Schwetz K A, Grellner W. The Influence of Carbon on the Microstructure and Mechanical properties of Sintered Boron Carbide. *J.Less.Common Met*, 1981, **82**:37.

[5] Schwetz K A, Reinmuth K, Lipp A. Polykristalline, Porenfreie Sinterkörper aus Alpha-SiC, Borcarbide und Kohlenstoff, Herstellung und Eigenschaften. Sprechsaal, 1983, **116**(12): 1063.

[6] Huang Qizhong, Yang Qiaoqin, Huang Baiyun and Lu Haibo. Effect of raw materials on properties of C-B₄C-SiC composite. *J.CENT.SOUTH UNIV. TECHNOL.*1995,**26** (2),223(Chinese)

[7] Advance of the study on the thermoelastic stress for functionally gradient material, *Material Guide*, 1998,**12**(1), 10 (Chinese)

[8] Bouchacourt, M., Etudes sur la phase carbure de bore-corr é lations propri é t é s-composition.Thése d'état. INPG, Ecole des Mines de Saint-Etienne, France,1982.

[9] Brodhag, C., bouchacourt, M. & thévenot, F., Comparison of the hot pressing kinetics of boron, boron suboxide B₆O and boron carbides. In Materials science Monographs, 16,Ceramic powders, ed. P. Vincenzini. Elsevier, Amsterdam, 1983, 881.

THERMAL SHOCK TEST ON ALUMINA/NICKEL FGM PLATE

Hideo Awaji, Hiromitu Takenaka, Yuichi Abe, Sawao Honda and Tadahiro Nishikawa
Nagoya Institute of Technology, Gokiso-cho, Showa-ku, Nagoya, Japan 466-8555

ABSTRACT

This paper presented an analysis of temperature/stress distributions in a stress-relief-type plate of functionally graded ceramic-metal-based materials and related results of thermal shock testing. The FGM disks with several gradual variations were fabricated using a powder stacking method and a pulse electric current sintering technique. Thermal shock tests on the FGM disks were performed as follows: The FGM circular disks were initially maintained at high temperature on the ceramic surface, and then cooled suddenly by cold water with constant velocity. The thermal shock resistance of the FGM was estimated using a critical temperature difference where cracks appeared on the ceramic surface. The results of the comparison of the analytical stresses and the critical temperature differences revealed that the stresses on the ceramic surface released significantly by the gradual variation of nickel across the thickness of the FGMs.

INTRODUCTION

Functionally graded materials (FGMs) have been attracting considerable attention in the field of structural ceramic applications at extremely high temperature.[1,2] These materials are composed of different materials such as ceramics and metals, with continuous and gradual variation across the thickness. Such a microstructure is expected to reduce thermal stresses and to improve the thermal shock properties of the materials. The temperature and thermal stress distributions of FGMs therefore have been analyzed intensively by many researchers.[3-5] For instance, Tanigawa et al.[1,6] analyzed one-dimensional transient thermal stress for a plate and a cylinder using a laminated composite model of numerous homogeneous layers, in relation to the

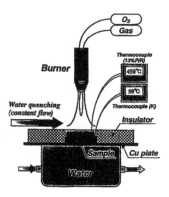

Fig. 1 Schematic representation of thermal shock equipment.

temperature dependence of the thermal and mechanical properties by an iteration technique.

The design of FGMs is now focussed on the optimization of its structure in order to generate appropriate residual stress, and also to minimize thermal stresses. In fact, thermal stresses in a structure associated with a steep temperature gradient are known to increase considerably compared with those under stable-state conditions. It is therefore necessary to estimate the transient temperature and thermal stress distributions under thermal shock environments. Thermal properties of materials are also known to have a strong temperature dependence,[6,7] which complicates the analysis of the temperature and stress distributions in structural components.

The aim of this paper is to give brief explanation of a numerical technique for analyzing one-dimensional transient temperature distributions in a ceramic-metal-based FGM plate, in terms of continuous and gradual variation of the thermal properties of the plate. Then, stress-relief-type FGMs of alumina/nickel system with different combinations of ceramics and metals were fabricated using a multilayer powder stacking method and a pulse electric current sintering technique.[8] The fabricated, circular FGM disks were examined to estimate the thermal shock resistance using a newly developed thermal shock testing equipment, as shown in Fig. 1. The FGM disks were initially exposed to high temperature on the ceramic surface using gas burner heating and to low temperature on the metallic surface, cooled using a water bucket, then the disks are suddenly cooled on the ceramic surface by cold water with a constant velocity. Thermal shock resistance was estimated using a critical temperature difference where cracks appear on the ceramic surface of the FGMs.

ANALYSIS

Temperature Distributions

Recently, we proposed a new technique for analyzing one-dimensional steady-state and transient temperature distributions in an FGM plate, in relation to both the temperature-dependent thermal properties and continuously gradual variation of the thermal properties of the FGM plate. This numerical analysis is based on a variable transformation technique[9] which is used to analyze temperature distributions in a specimen with temperature-dependent thermal properties. The variable transformation for an analysis of temperature distributions in an FGM plate is expressed as

$$G = \int_0^T \lambda^*(\xi, T)dT \qquad (1)$$

where T represents nondimensional temperature in the FGM plate, ξ the nondimensional distance, $\lambda^*(\xi, T)$ the nondimensional thermal conductivity of the FGM plate at ξ, and G the transformed variable. The variable transformation makes heat transfer equations quite simple and the transformed equation can be analyzed numerically using Crank-Nicolson's finite differential expression. The calculated G values are inverse-transformed to the normal temperature T.

We analyze the temperature distributions of an alumina/nickel FGM plate having one-dimensional symmetry, with the x axis being the direction of spatial variation shown in Fig. 2, postulating that the material is pure or 50 % nickel at x = 0 and pure alumina at x = l, and that the alumina side is maintained at the high temperature of θ_2 and the nickel-rich side at the low

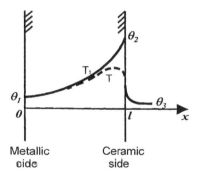

Metallic
side

Ceramic
side

Fig. 2 Temperature distributions in ceramic/metal FGM plate under thermal shock.

temperature of θ_1, then the alumina side is suddenly cooled by cold water. The unsteady-state normalized equation of heat transfer is expressed as

$$(\rho C)^*(\xi)\frac{\partial T}{\partial \tau} = \frac{\partial}{\partial \xi}\left\{\lambda^*(\xi)\frac{\partial T}{\partial \xi}\right\} \tag{2}$$

with the initial condition

$$T = T_1 \quad \text{at} \quad \tau = 0$$

and boundary conditions

$$T = 0 \quad \text{on} \quad \xi = 0,$$

$$\frac{\partial T}{\partial \xi} = \frac{\beta}{\lambda^*}(1 - T_s) \quad \text{on} \quad \xi = 1,$$

where the following nondimensional variables and material properties are used in order to normalize Eq. (2)

$$T = \frac{\theta - \theta_1}{\theta_2 - \theta_1} \quad , \quad \xi = \frac{x}{l} \quad , \quad \tau = \frac{\kappa_{ci} t}{l^2}, \quad \lambda^* = \frac{\lambda}{\lambda_{ci}}, \quad \kappa^* = \frac{\kappa}{\kappa_{ci}} \tag{3}$$

θ represents temperature in the plate, T nondimensional temperature in the FGMs, T_1 the initial temperature distribution in the FGM, T_s temperature on the ceramic surface, l the thickness of the plate, t time, ρC the heat capacity per unit volume (ρ: density, C: specific heat), λ the thermal conductivity of the FGM, τ the Fourier number, κ^* and λ^* the nondimensional thermal diffusivity and thermal conductivity of the FGM, respectively, κ_{ci} and λ_{ci} the thermal diffusivity and thermal conductivity of the monolithic ceramics (alumina) at room temperature, respectively, and $\beta = hl/\lambda_{ci}$ the Biot number. Hereafter, the material properties with the superscript * indicate the nondimensional values, the properties with the subscript c and m designate the material property of ceramics and metals, respectively, and the properties with the subscript i the values at room temperature.

In terms of G, Eq. (2) is rewritten as the simple nonlinear differential equation

$$\frac{\partial G}{\partial \tau} = \kappa^*(\xi, T)\frac{\partial^2 G}{\partial \xi^2}. \tag{4}$$

Using Crank-Nicolson's differential expression, Eq. (4) is easily analyzed numerically.

Referring to the technique proposed by Tanigawa et al.[5], the thermal stresses in the FGM plate were also analyzed using the analyzed temperature distributions. The detailed results will be shown in reference.[7]

The local volume fraction of nickel in the FGM plate in which the nickel is fully graded (namely, 0 to 100 % change of volume fraction of nickel) is described by

$$f_m(\xi) = 1 - \xi^p \qquad (5)$$

and in the case where the nickel is partially but linearly graded,

$$f_m = 0.5 - 0.5\xi \qquad (6)$$

where p represents a parameter of volume fraction. Equation (5) is shown in Fig. 3.

Table I shows the thermo-mechanical properties at room temperature of the pure alumina and nickel used.[10] The temperature dependence of the nondimensional properties of these materials is shown in reference.[6,10]

The thermal transfer coefficient, h, on the ceramic surface, during the cooling process by the cold water (300 K) of a laminar flow with the constant velocity of 1 m/s was calculated assuming that the temperature of the water will increase to 373 K during cooling. The obtained value of h was 4,600 W/(m²K).

FABRICATION OF FGM DISKS

Al₂O₃ powder with 0.25 μ m mean grain size and Ni powder with 3-7 μ m grain size were mixed homogeneously using ethanol as a suspension. The mixed powders with different volume fractions were dried and stacked in a graphite die to be an intended gradual variation and pressed at 0.2 MPa for 5 min. Because of the large difference of thermal expansion coefficient between alumina and nickel, fully graded FGMs could be fabricated only in the case where f_m = 1- ξ ^{1/5}. Therefore, the nickel content in the FGM disks was changed from 0 to 50 % in the case of a linear function of the volume fraction of nickel, as shown in Eq. (6). The conditions for the pulse electric current sintering technique were that 1150°C in vacuum at 40 MPa, and the

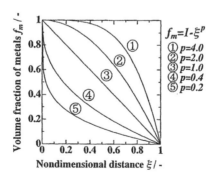

Fig. 3 Volume fraction of metals.

Table I Thermo-mechanical properties of alumina and nickel at room temperature.

Materials	Thermal Conductivity λ [W/mK]	Thermal Diffusivity κ [$\times 10^{-6}$ m^2/s]	Thermal Expansion Coefficient α [$\times 10^{-6}$ /K]	Young's Modulus E [GPa]	Poisson Ratio ν [–]
Alumina	28.2	8.3	5.4	360	0.23
Nickel	90.1	22.3	14.0	218	0.31

soaking time about 30 min.

RESULTS

Analytical Results

Figure 4 shows the nondimensional thermal stresses on the cooled alumina surface of the alumina/nickel FGM plate with two different thickness, such as 7.5 mm thick shown by bold lines and 3.5 mm thick shown by narrow lines. The dot-dashed lines indicate the stress distributions in a homogeneous alumina plate, the dotted lines the stresses in alumina/nickel FGMs with the volume fraction of nickel, $f_m = 1 - \xi^{1/5}$, and solid lines the stresses in FGMs with f_m of 0.5 - 0.5 ξ. It is recognized that a homogeneous alumina plate has the highest maximum stress, and the stresses in the FGM plates reduced significantly.

Experimental Results

The experimental results of the thermal shock testing are summarized in Fig. 5, where the

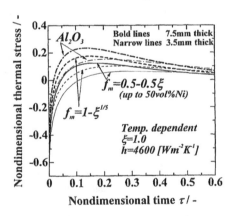

Fig. 4 Thermal stresses on the quenched surface of alumina and FGM plates.

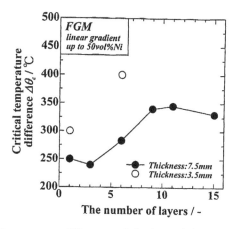

Fig. 5 Critical temperature differences of alumina and alumina/nickel FGM disks.

solid circles indicate the critical temperature differences of circular disks with 7.5 mm thick of monolithic alumina (the number of layer = 1) and FGMs with several layers, and the empty circles indicate the values of FGM disks with 3.5 thick and with the volume fraction of Ni, f_m = 0.5 - 0.5 ξ. The figure shows that the FGMs with lower number of layers have lower critical temperature difference, which suggests that there are considerable residual stresses on the layer boundaries in the FGMs with lower number of layers.

The comparison of the critical temperature differences in monolithic alumina and 10-layered FGM indicate that the thermal shock resistance is improved about 1.4 times of the monolithic alumina, whereas the analytical results shown in Fig. 4 reveals that the thermal stress in the FGM plate reduces about 1/1.9 times of the monolithic alumina plate. The reason for the discrepancy will result from the residual stresses in the FGMs[11] which we did not considered for the calculations.

CONCLUSIONS

We presented a numerical technique for analyzing one-dimensional transient temperature distributions in a plate of alumina/nickel FGM, in relation to the continuous and gradual variation of the thermal properties of the FGM. The FGM disks with several gradual variations were fabricated using a powder stacking method and a pulse electric current sintering technique. Then the thermal shock resistance was estimated using a newly developed equipment. The results of the comparison of the analytical stresses and the critical temperature differences revealed that the stress on the ceramic surface was released significantly by the gradual variation of nickels across the thickness of the FGMs.

REFERENCES

[1] A. Ghosh, Y. Miyamoto, I. Reimanis and J. J. Lannutti edited. *Functionally Graded Materials*, American Ceramic Society, Westerville, OH, 1997.

[2] Y. Miyamoto, W. A. Kaysser, B. H. Rabin, A. Kawasaki and R. G. Ford, *Functinally Graded Materials: Design, Processing and Applications*, Kluwer Academic Publishers, Norwell, MA, 1999.

[3] N.Araki, A. Makino, T. Ishiguro and J. Mihara, "An Analytical Solution of Temperature Response in Multilayered Materials for Transient Methods", *International Journal of Thermophysics*, 13, 515-538 (1992).

[4] Y. Tanigawa, "Some Basic Thermoelastic Problems for Nonhomogeneous Structural Materials", *Applied Mechanics Review*, 48, 287-300 (1995).

[5] Y. Tanigawa, M. Matsumoto and T. Akai, "Optimization of Material Composition to Minimize Thermal Stresses in Nonhomogeneous Plate Subjected to Unsteady Heat Supply", *JSME International Journal*, 40A, 84-93 (1997).

[6] H.Awaji, T. Takahashi, N. Yamamoto and T. Nishikawa, "Analysis of Temperature/Stress Distributions in Thermal Shocked Ceramic Disks in Relation to Temperature-Dependent Properties", *Journal of the Ceramic Society of Japan*, 106, 358-362 (1998).

[7] H.Awaji, H. Takenaka, S. Honda and T. Nishikawa, "Temperature Distributions in a Stress-Relief-Type Plate of Functionally Graded Material under Thermal Shock", *Journal of the Ceramic Society of Japan*, 107, 780-785 (1999).

[8] M. Omori and T. Hirai, "Fabrication of FGM Using the Function of SPS," *Functionally Graded Materials in the 21st Century*, Edited by N. Ooyama et al., March 26-28, 2000, Tsukuba.

[9] H. S. Carslaw and J. C. Jaeger, *Conduction of Heat in Solids*, 2nd ed., Oxford Univ. Press, Ely House, London 1959.

[10] Nihon-Netsubussei-Gakkai edited, *Thermophysical Properties Handbook*, Yokendo, (in Japanese) 1990.

[11] K. S. Ravichandran, "Thermal Residual Stresses in a Functionally Graded Material System", *Materials Science and Engineering, A*, 201, 269-276 (1995).

IMPACT DAMAGE IN MONOLITHIC AND FUNCTIONALLY GRADED ALUMINA

Premal Shah, Karl Jakus, and John E. Ritter
University of Massachusetts
Department of Mechanical and Industrial Engineering
Amherst, MA 01003

ABSTRACT
Previous research has shown that Hertzian cone cracking is suppressed in ceramic composites whose elastic modulus increases with depth below the surface. The objective of this research was to determine if these modulus-graded composites would also exhibit superior resistance to impact damage. Therefore, impact damage with spherical projectiles was studied in monolithic (plain) and modulus-graded alumina. The composite was fabricated by impregnating a dense, fine-grained alumina with an aluminosilicate glass having a lower elastic modulus than the alumina. This produced a modulus-graded alumina composite with decreasing glass content below the surface, thus causing the elastic modulus to monotonically increase with depth.

Small spherical chrome steel and tungsten carbide projectiles were impacted on target specimens using a particle accelerator gas-gun. The tests entailed impacting the target with increasingly greater kinetic energy until ring cracks were observed by microscopic examination. These results were compared to Hertzian impact theory and previously published quasi-static Hertzian indentation data. It was found that the modulus gradation improved the impact damage resistance when the high elastic modulus tungsten carbide projectiles were used. No improvement was observed using the lower elastic modulus chrome steel projectiles. Monolithic alumina impacted with tungsten carbide projectiles followed the predictions of impact theory and agreed reasonably well with the previously published quasi-static data. However, modulus-graded alumina impacted with tungsten carbide projectiles and both monolithic and modulus-graded alumina impacted with steel projectiles did not follow either the theory or the previous data. It is believed that the low modulus steel deforms more upon impact than the alumina specimen, thus overshadowing the effects of the modulus gradient. On the other hand, the high modulus tungsten carbide does not deform appreciably during impact, allowing the modulus gradient to play a role in reducing the impact stresses and increasing the cracking threshold.

INTRODUCTION

Contact damage resistance is an important property of structural ceramics in applications where significant contact loads are likely to be encountered. Recently a new concept has been proposed to increase the contact damage resistant of alumina ceramics.[1] According to this concept a significant increase in the cracking threshold can be realized when the elastic modulus is made to increase monotonically from the surface into the bulk of the specimen. Crack formation is suppressed because the stress field beneath the contact force is favorably altered by the modulus gradation. In practice, modulus-graded specimens have been prepared by infiltrating alumino-silicate glass into alumina substrates.[1] As the volume fraction of the lower elastic modulus glass decreased from the surface to the bulk, the effective elastic modulus increased. Time and temperature controlled the gradient and depth of infiltration. The modulus gradation suppressed Hertzian cracking in quasi-static indentation tests as compared to both the bulk glass and the alumina substrate.

When components encounter impact loading, the transient stress field generated by such impacts is likely to be different than the stress field under quasi-static loading. It is not clear if modulus gradation provides a similar increase in damage resistance under impact loading as it does in Hertzian contact. The objective of the research presented in this paper was to evaluate the performance of elastic modulus-graded alumina specimens against impact damage by small spherical objects. In this study, the projectile kinetic energy threshold for Hertzian ring crack formation was measured to assess the impact damage.

EXPERIMENTAL PROCEDURE

Aluminosilicate glass* infiltrated fine-grained (3-5 μm) alumina** discs (25 mm diameter x 4.2 mm thick) were prepared in accordance with the processing procedure outlined in Ref. 1. Namely, an alumina disc with a piece of glass (14 mm square x 4 mm thick) placed on it was heated for two hours at 1690^0C. The treatment resulted in a monotonically decreasing volume fraction of glass starting with 0.4 at the surface and ending up zero at 2 mm below the surface. The corresponding modulus gradient was estimated on the basis of rule-of-mixtures with glass having a modulus of 72 GPa and the alumina 386 GPa. The resulting modulus gradient is shown on Fig. 1. The figure shows that the elastic modulus starting from approximately 250 GPa at the surface monotonically approaches 380 GPa at 2 mm below the surface. The solid line in the figure represents best fit to the data.

---

* Code 0317, Corning Inc., Corning, NY.
** Greenleaf Technical Ceramics, East Flat Rock, NC.

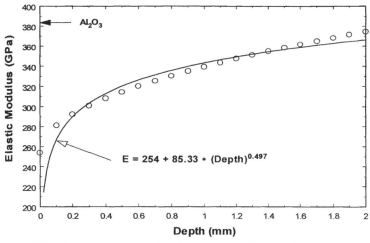

Figure 1. Elastic modulus variation in a glass infiltrated alumina specimen.

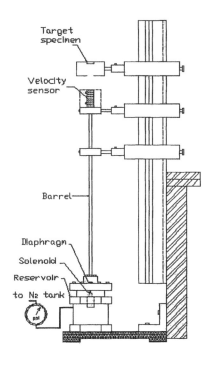

Figure 2. Particle Accelerator Gas-Gun.

The impact experiments were conducted using a particle accelerator gas-gun. Small steel and tungsten carbide spheres (2 mm and 3 mm dia.) were accelerated with high-pressure nitrogen gas against perpendicularly positioned target specimens. The velocity of the projectiles, hence their kinetic energy, was measured for each test using a time-of-flight device. A schematic of the particle accelerator is shown in Fig. 2. The tests entailed impacting the target with increasingly greater kinetic energy until ring cracks were observed by microscopic examination. In some cases the observed ring cracks were complete circles, in other cases they were only partial circles.

THEORY OF HERTZIAN CRACK FORMATION

Contact damage is a result of the concentrated stresses generated in the target material by the indenting objects, namely the indenting sphere in the case of Hertzian indentation. The stresses are compressive immediately below the indenter but they are tensile just outside the contact area. The magnitude of the stress depends on the applied load, the indenter diameter, and the material properties of the specimen and the indenter. When the tensile stress reaches a critical value a ring crack forms around the contact area. The critical load P_c that causes the formation of a ring crack in a given specimen/indenter system is given by,[2]

$$P_c = K_c^2 r \left(\frac{k}{E} \right) \phi \tag{1}$$

where K_c and E are the fracture toughness and modulus of elasticity of the specimen, respectively, r is the indenting sphere radius, ϕ is an empirical dimensionless constant, and k is a dimensionless constant given as,

$$k = (9/16)\left[\left(1 - v^2\right) + \left(1 - v'^2\right) \right]\left(\frac{E}{E'} \right) \tag{2}$$

with v Poisson's ratio (the primes refer to the indenter material).

In an impact situation the momentum of the projectile provides the indentation load. By balancing the projectile's kinetic energy with the strain energy absorbed by the specimen, a relationship between the critical load and the equivalent critical velocity V_c is obtained as,[2]

$$P_c = \left[\left(\frac{125 \, \pi^3}{48} \right)^{1/5} \left(\frac{E}{k} \right)^{2/5} \rho^{3/5} r^2 \right] V_c^{6/5} \tag{3}$$

where ρ is the density of the projectile.

The velocity V_c can be replaced by the corresponding kinetic energy U_c as,

$$V_c = \left[\frac{3 \, U_c}{2 \, \pi \rho \, r^3} \right]^{1/2} \tag{4}$$

Substituting Eq.(4) into Eq.(3) one gets the relationship between the critical load and the corresponding projectile kinetic energy as,

$$P_c = 1.54 \left(\frac{E}{k} \right)^{2/5} r^{9/5} U_c^{3/5} \tag{5}$$

Eliminating P_c between Eq.(1) and Eq.(5) and solving for r as a function of U_c, one gets,

$$U_c = 0.48 \, \phi^{5/3} \left(\frac{k}{E} \right)^{7/3} K_c^{10/3} r^{4/3} \tag{6}$$

Equation 6 shows that the critical kinetic energy depends on the fracture toughness of the specimen and the respective elastic constants of the specimen and the projectile (k). It also depends on the projectile radius to the 4/3 power.

RESULTS AND DISCUSSION

In this study modulus-graded specimens were impacted with chrome steel and tungsten carbide spheres having 1 and 1.5 mm radii (2 and 3 mm dia.). For comparison purposes monolithic alumina specimens (the substrate material used for the modulus-graded specimens) were also impacted in a similar fashion. Material properties of the specimens and the projectiles are given in Table 1.

Table 1. Material Properties (Manufacturer's or Handbook data)

Material	Elastic Modulus (GPa)	Poisson's Ratio	Density (kg/m³)
Monolithic Alumina	386	0.22	3,950
Modulus-graded Alumina	250 at the surface	~0.22	Variable
Chrome Steel	200	0.33	7,790*
Tungsten Carbide	669	0.22	14,600*

*measured in the authors' laboratory

It can be seen in the table that the chrome steel projectiles have a lower and the tungsten carbide projectiles have a higher elastic modulus than either the monolithic or the modulus-graded alumina specimens. In addition, the tungsten carbide projectiles are denser than the steel, hence have greater kinetic energy at a given diameter and velocity. The choice of these projectiles was intended to delineate the effect of the elastic modulus ratio between the specimen and the projectile, as well as the effect of projectile radius.

Table 2 gives the average critical kinetic energy required to initiate ring cracking. At least six separate test sequences were used to determine the average kinetic energy for each projectile/specimen pair. The results show that for a tungsten carbide projectile the kinetic energy required for crack initiation is about

Table 2. Average Critical Kinetic Energy for Ring Crack Initiation (Joules).

	Chrome Steel	Chrome Steel	Tungsten Carbide	Tungsten Carbide
	$r = 1$ mm	1.5 mm	1 mm	1.5 mm
Monolithic Alumina	0.0085 (0.0013)	0.0085 (0.0017)	0.0041 (0.0008)	0.0058 (0.0005)
Modulus-graded Alumina	0.0084 (0.0003)	0.0078 (0.0002)	0.0073 (0.0013)	0.0090 (0.0014)

The quantities in parenthesis represent one standard deviation.

1.5 to 1.8 times greater for modulus-graded alumina than for monolithic alumina. However, for chrome steel projectiles there is no statistical difference between the monolithic and the modulus-graded specimens. These findings imply that elastic modulus gradation is a viable method to improve the impact damage tolerance of alumina as long as the elastic modulus of the projectile is considerably higher than that of the target specimen. When the projectile is less stiff than the target specimen, modulus gradation shows no benefit.

Equation 5 gives the equivalence between quasi-static indentation load and projectile kinetic energy. This relationship allows one to compare published quasi-static indentation data to the present results. Accordingly, the data shown in Table 2 is plotted as the logarithm of the kinetic energy versus the logarithm of the projectile radius in Fig.3. The figure also shows quasi-static data converted to equivalent kinetic energy from four different invstigators.[3-6] The solid line on the figure has a slope of 4/3 as suggested by theory (Eq.(6)). It can be seen that the quasi-static data follows the theory reasonably well. Furthermore, the present tungsten carbide/monolithic alumina data also follows the theory reasonably well. However, the data for tungsten carbide/modulus-graded alumina falls outside the trend, indicating that it takes more kinetic energy to initiate cracking than theory suggests. The dependence of the kinetic energy on the projectile radius also deviates from theory. The data for modulus-graded alumina shows no radius dependence within experimental scatter.

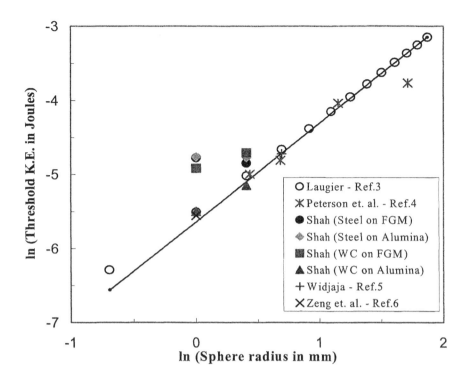

Figure 3. Average Kinetic Energy for Ring Crack formation. (Ref. 4 to7: quasi-static indentation on monolithic alumina).

The kinetic energy required to initiate cracking with a steel sphere proved to be independent of both the type of specimen and the projectile diameter. This shows that the low elastic modulus projectile deforms more on impact than the

Table 3. Average Ring Crack Diameter (μm)

	Chrome Steel	Chrome Steel	Tungsten Carbide	Tungsten Carbide
	$r = 1$ mm	1.5 mm	1 mm	1.5 mm
Monolithic Alumina	440 (10)*	448 (10)	291 (18)	328 (37)
Modulus-graded Alumina	417 (35)	443 (14)	299 (10)	323 (10)

* One standard deviation or measurement resolution, which ever is greater.

substrate. Accordingly, a larger contact area or ring crack diameter is expected for a chrome steel projectile than for a tungsten carbide one. This is indeed the case as can be seen on Table 3. These results suggest that the higher kinetic energy required for crack initiation with steel projectiles is due to the relatively large amount of strain energy absorbed by the steel spheres during impact.

CONCLUSIONS

Prior research[1] showed that elastic modulus gradation at the surface of alumina specimens could significantly increase the contact damage resistance of the material. This paper presents results for the impact damage resistance of modulus-graded alumina. The results show that modulus gradation is beneficial only when the projectile material has a substantially higher elastic modulus than the specimen.

For projectiles with elastic modulus lower than the specimen, modulus gradation is ineffective. The authors made a similar observation regarding the sharp particle impact (erosion) behavior of modulus-graded alumina.[7] Modulus gradation was found to be ineffective for reducing erosive wear.

ACKNOWLEDGEMENT

This research was supported by NSF grant #DMR-9400261. Dr. Nittin Padture of Univ. of Connecticut facilitated the preparation of the specimens.

REFERENCES
1. J. Jitcharoen, N.P. Padture, A.E. Giannakopoulos, and S. Suresh, "Hertzian Crack Supression in Ceramics with Elastic-Modulus-Graded Surfaces," *J. Am. Ceram. Soc.* **81**[9] 2301-308 (1998).
2. S. M. Wiederhorn and B. R. Lawn, "Strength Degradation of Glass Resulting from Impact with Spheres," *J. Am. Ceram. Soc.* **60**[9-10] 451-457 (1977).
3. M. T. Laugier, "Hertzian Indentation of Sintered Alumina," *J. Mat. Sci.*, **19**, 254-258 (1984).
4. I. M. Peterson, A. Pajares, B. R. Lawn, V. P. Thompson, and E. D. Rekow, "Mechanical Characterization of Dental Ceramics by Hertzian Contacts," *J. Dental Res.*, **77** [4] 427-440 (1998).
5. S. Widjaja, J. E. Ritter, and K. Jakus, "Influence of R-Curve Behavior on Strength Degradation Due to Hertzian Indentation," *J. Mat. Sci.*, **31**, 2379-2384 (1996).
6. K. Zeng, K. Breder, D. J. Rowcliffe, and C. Herrstrom, "Elastic Modulus Determination by Hertzian Indentation," *J. Mat. Sci.*, **27**, 3789-3792 (1992).
7. R.P. Panat, K. Jakus, J. E. Ritter, and P. Shah "Erosion and Strength Degradation of an Elastic Modulus Graded Alumina-Glass Composite," *Ceram. Engr. & Sci. Proc.*, **21** [3] pp635-42 (2000).

RECENT DEVELOPMENT IN THE COMPUTATIONAL MICRO-MECHANICS, THERMAL DAMAGE MODEL AND IMPACT RESPONSE OF FUNCTIONALLY GRADED MATERIALS

Qing-Jie Zhang, Peng-Cheng Zhai, Li-Sheng Liu and Run-Zhang Yuan
State Key Lab. of Advanced Technology for Materials Synthesis and Processing
Wuhan University of Technology, Wuhan, 430070, China

Shin-Ichi Moriya and Masayuki Niino
National Aerospace Lab., Kakuda Research Center, Miyagi, 981-15, Japan

ABSTRACT
The recent progress in the mechanics of functionally graded materials in the State Key Laboratory in cooperation with the Japanese National Aerospace Laboratory is reported. Emphasis is put on the computational micromechanics, thermal damage model and impact response. The computational micromechanics combines a construction technique for a two-dimensional random geometry composed of multiphases with a finite element method. The main feature of the micromechanics is its ability to exactly construct the real random and graded microstructure of FGMs and its ability to exactly determine the relationship between the microstructure and properties. The thermal damage model deals with the damage problem of FGMs under cyclic thermal shock tests. A theoretical model is developed to examine an important thermal damage phenomenon: the coupling between the damage and heat conduction under cyclic thermal shock tests. The impact mechanics considers the dynamic response of a ceramic/metal FGM under high velocity impact. The propagation characteristics of impact waves across the interface between the ceramic and ceramic/metal interface are investigated. The effect of the composite properties and the ceramic thickness on the wave reflection across the interface is emphasized.

THE COMPUTATIONAL MICROMECHANICS
Determination of the microstructure-property relationship of FGMs is a basic problem for the physics and mechanics of the materials and for the materials design. A computational micromechanical method (CMM) has been developed to

exactly determine the relationship between the microstructure and properties[1]. The CMM combines a construction technique for the real random and graded microstructure with a finite element method. Emphasis in the CMM is put on the construction of a two-dimensional random geometry composed of multiphases and on the exact solution for different physical fields in such random geometry. Therefore the feature of the CMM is its ability to exactly reconstruct the real random microstructure of a FGM sample observed with SEM and its ability to exactly solve the physical and mechanical problems with the constructed sample. The CMM has been verified by experiments and has been applied for various effective properties of FGMs and multiphase materials such as mechanical properties, transport properties and electrokinetic properties[2].

A detailed description for the CMM has been given in Refs.[1,2]. For the convenience of its use in present paper, the fulfilling procedures are briefly restated as follows.

Step 1: Construction of Random Microstructure

For an imagined composite sample with an assumed configuration and size and with either an assumed inclusion shape and volume fraction or an assumed crack shape and crack density, construct the microstructure by a computer with a random mathematical method and an automatic geometry formation technique. Inclusions/cracks in the microstructure are completely random in location and orientation. The inclusion shape can be a spheroid, which can model a sphere, disc and needle, or muti-side body, depending on requirements. The inclusion size is allowed to change between a specified maximum value (such as 100μm) and minimum one (such as 10μm). the crack shape can be elliptical or penny-shaped and the size can also be different. The microstructure constructed can be either dispersed with the inclusions separated or interwoven with the inclusions connected, depending on the inclusion volume fraction. For an actual composite

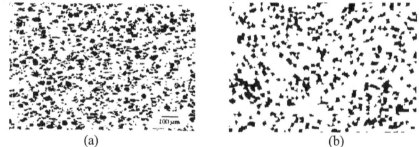

(a)   (b)

Fig. 1   The random microstructure of a SiC/Al composite:
(a) the SEM picture for a 15%SiC/85%Al composite[3];
(b) simulation for the SEM picture by CMM

sample with a known microstructure from a SEM image, by specifying the sample's configuration and size, and the inclusion shape and size as well as volume fraction, the real random microstructure can exactly be simulated.

Step 2: Finite Element Division

For the microstructure constructed, divide the highly irregular geometry with finite elements by a computer automated division technique. The element mesh in the domain near cracks and in the domain near and in aggregated or interwoven inclusions must be dense enough in order to guarantee the accuracy of the finite element analysis in the following step. The data from the divided mesh forms an input information file available for the following analysis.

Step 3: Finite Element Analysis

According to a required effective property and the corresponding test standard, impose an appropriate boundary condition and loading on the sample constructed, then find the solution of the corresponding physical field using a proper finite element code such as Super-Sap, ADINA/ADINAT,ABAQUS, NASTRAN, etc. Since for a given composite sample with the same inclusion shape, size and volume fraction the inclusions are embedded randomly in location and orientation, the effective property data calculated from the first construction of the microstructure may differ from that calculated from the second one. Therefore the effective property data for one sample should be based on calculating and averaging from several constructions of the microstructure. It has been indicated that for one sample calculating and averaging from five to six constructions of the microstructure is suitable for obtaining the accurate effective property data.

The CMM is used to examine the elastoplastic stress-strain curve of a SiC/Al composite with a random microstructure in the present paper. The SEM pictures for the microstructure for several typical SiC volume fractions and the corresponding experimental stress-strain curves have been obtained by Lloyd[3]. One of the SEM images is presented in Fig.1(a) and its simulation by CMM is presented in Fig.1(b) in the present paper. Based on the microstructure in Fig.1(b), the numerical result for the elastic-plastic stress-strain curve of the composite is calculated by CMM and given in Fig.2. In Fig.2, the corresponding experimental stress-strain curve obtained by Lloyd is also plotted for comparison. It is seen from Fig.2 that the two results are in an excellent agreement.

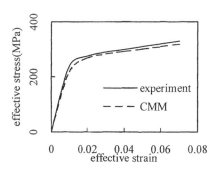

Fig.2 The effective stress-strain curve of the composite

THE THERMAL DAMAGE MODEL

Cyclic thermal shock tests for ceramic/metal FGMs have been conducted in cooperation with the Japanese National Aerospace Laboratory. It was observed from the tests that when the materials were subjected to cyclic thermal shock, microcracks inside the samples were caused and developed as the number of thermal cycles increased. Because the growing thermal damage decreased the effective thermal conductivity in the FGM interlayers, the heat conduction in subsequent thermal shock was affected. The coupling between the damage and heat conduction is an important phenomenon in the cyclic thermal shock tests and its study may lead to a new, theoretical method for thermal fatigue damage evaluation. A basic theoretical model has been proposed to examine the coupling phenomenon[4,5]. This model deals with the effective thermal conductivity of a matrix-inclusion-microcrack three-phase composite. The microcracks in the model are assumed to be randomly distributed and penny-shaped and inclusions to be spherical, the crack effect is accounted for by introducing a crack density parameter. The effective thermal conductivity of the microcracked composite is derived using a self-consistent micromechanical method (SCM) and related to the crack density parameter. The effective thermal conductivity has also been studied by the computational micromechanical method (CMM). The two results by SCM and CMM have found to be in a good agreement for metal matrix composites and metal-rich interlayers of ceramic/metal FGMs. For comparison, a typical result of a Ni-matrix/ZrO$_2$-inclusion composite containing randomly distributed, penny-shaped microcracks from Refs.[4,5] is presented in Fig.3(a). In Fig.3(a) and the subsequent 3(b), λ_1, λ_2 and λ_{ef} are the matrix thermal conductivity, inclusion thermal conductivity and the effective thermal conductivity respectively; ε is the crack density; C_2 is the inclusion volume fraction.

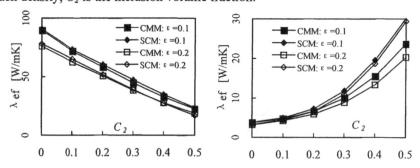

(a) Ni matrix/ZrO$_2$ inclusion ($\lambda_1 > \lambda_2$) (b) ZrO$_2$ matrix/Ni inclusion ($\lambda_1 < \lambda_2$)
Fig. 3 The effective thermal conductivity of a microcracked Ni-ZrO$_2$ composite material as a function of the inclusion volume fraction

To examine the suitability of SCM to the thermal damage of ceramic matrix composites and ceramic-rich interlayers of ceramic/metal FGMs, the effective thermal conductivity of a ZrO_2 matrix/Ni inclusion composite containing randomly distributed, penny-shaped microcracks is considered in the present paper. The effective thermal conductivity is calculated by SCM for $\varepsilon=0.1$ and $\varepsilon=0.2$ and the result is given in Fig.3(b). In Fig.3(b), the corresponding numerical result by CMM is also plotted for comparison. Unlike in Fig.3(a), it is seen in Fig.3(b) that the results by SCM and by CMM disagree as the crack density ε increases.

To explain the disagreement between the results by SCM and by CMM in Fig.3(b), the crack effect on the effective thermal conductivity needs to be investigated further. Recently, a generalized self-consistent method(GSCM) for the effective thermal conductivity of a microcracked homogeneous body has been developed to consider the interaction among the cracks in a more accurate way by one of the present authors[6]. Using GSCM in Ref.[6], the effective thermal conductivity of a homogeneous body containing randomly distributed, penny-shaped microcracks with a wide range of crack density is estimated in the present paper and the result, along with the corresponding results by SCM and CMM, is presented in Fig.4. From a comparison between SCM, GSCM and CMM in Fig.4, it is seen that SCM is inadequate to treat the interaction among the cracks as compared with GSCM and CMM as the crack density ε increases.

From Fig.3(a) and (b) and Fig.4, two points can be made: (1) Since the matrix thermal conductivity is large for metal matrix composites or metal-rich interlayers of ceramic/metal FGMs, the interaction among the cracks is less important in affecting the effective thermal conductivity and SCM is an appropriate model to examine the heat conduction behavior of the damaged composite; (2) Since the matrix thermal conductivity is relatively small for ceramic matrix composites or ceramic-rich interlayers of the FGMs, and the effect of the interaction among the cracks on the effective thermal conductivity becomes appreciable, SCM is not enough to deal with the heat conduction-thermal damage coupled problem of the composites. Therefore, a further study is needed to develop a suitable theoretical model for the effective thermal conductivity of micro-cracked ceramic matrix composites and ceramic-rich interlayers of the FGMs.

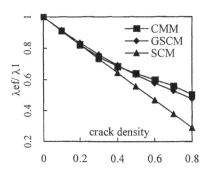

Fig.4 The effective thermal conductivity of a microcracked homogeneous body

THE IMPACT RESPONSE

The dynamics of ceramic/metal FGMs under high velocity impact is an important subject because of their potential application as new and high efficient impact-resistant materials. Recently, a theoretical study has been conducted by three of the present authors[7] to examine the propagation characteristics of impact waves across the interface between the ceramic and ceramic/metal interlayer in the FGMs. The study is based on a basic method provided by Furlong et al.[8] for predicting the reflection and refraction of spherical waves across a planar interface. Emphasis in the study is focused on the effect of the interlayer properties upon the propagation characteristics of impact waves across the interface. Two interesting results from the study are given in Fig.5(a) and (b) in the present paper which correspond to a TiB_2/Fe FGM(the volume fraction of Fe in the interlayer is allowed to change from 25% to 100%). In Fig.5(a) and (b), the dimensionless amplitude ratio is defined as a ratio of the amplitude of the reflected dilatation wave to that of the incident dilatation wave due to a normal impact on the ceramic surface; kr_0 is a parameter related to the frequency of the incident wave, the wave speed in the ceramic and the ceramic thickness. Since the incident dilatation wave due to a normal impact on the ceramic surface is a compressive one, a positive dimensionless amplitude ratio means a reflected compressive wave and a negative ratio a reflected tensile one. Therefore, if the ratio is negative and less than one, it means that the incident compressive wave is unloaded and becomes a tensile one because of the wave reflection at the interface. Moreover, for a given impact velocity and ceramic, the parameter kr_0 is related only to the ceramic thickness.

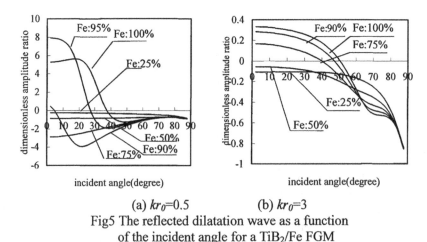

(a) $kr_0=0.5$ (b) $kr_0=3$

Fig5 The reflected dilatation wave as a function
of the incident angle for a TiB_2/Fe FGM

From Fig.5, the material properties of the interlayer and the thickness of the ceramic have a significant effect on the reflected wave. For kr_0=0.5(for an impact velocity of 1000m/s, it corresponds to a ceramic thickness of about 1mm), when the volume fraction of Fe is less than 90%, the reflected wave strength increases as the volume fraction increases and the back of the ceramic is subjected to a tensile wave within a large range of the incident angle. When the volume fraction is greater than 90%, a significant change in the reflected wave takes place and a ring area which undergoes a tensile wave is formed on the ceramic back. The tensile ring diminishes as the volume fraction increases. For kr_0=3(for an impact velocity of 1000m/s, which corresponds to a ceramic thickness of about 6mm), the reflected wave strength increases as the volume fraction increases within the whole range of the volume fraction considered. Since the absolute value of the dimensionless amplitude ratio is less than one for kr_0=3, the whole back of the ceramic is subjected to a compressive wave.

To further examine the effect of the ceramic thickness on the reflected wave, the result of the reflected wave under different kr_0 values and a given volume fraction of Fe is presented in Fig.6. Since kr_0 is related only with the ceramic thickness in a linear form for a given impact velocity and ceramic, Fig.6 gives a clear picture on the variation of the reflected wave with the ceramic thickness. From Fig.6, there exists a critical ceramic thickness for the tensile ring. When the ceramic thickness is relatively thin, a tensile ring exists on the back of the ceramic. When the ceramic is relatively thick, the tensile ring vanishes and the whole back of the ceramic bears a compressive wave. The critical ceramic thickness corresponds to a value of kr_0 which is between 0.5 and 0.6.

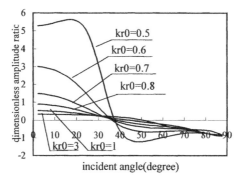

Fig.6 The reflected dilatation wave as a function of incident angle
for different kr_0 values (Fe=100%)

ACKNOWLEDGMENT
This work was supported by the National Natural Science Foundation of China, the Ministry of Education of China and the Chinese "863" High-Technology Program.

REFERENCES
[1] P.C.Zhai, Q.J.Zhang and R.Z.Yuan, "Random microstructure finite element method and its verification for effective properties of composite materials", Journal of Wuhan University of Technology (Mater. Sci. Ed.), 15(1), 6-12, 2000.

[2] P.C.Zhai, Q.J.Zhang and R.Z.Yuan, "Random microstructure finite element method for effective nonlinear properties of composite materials", Journal of Wuhan University of Technology (Mater. Sci. Ed.), 15(2), 1-6, 2000.

[3] D.J.Lloyd, "Particle reinforced aluminum and magnesium matrix composite", Int. Mater. Rev., 39(1), 1-23, 1994

[4] Q.J.Zhang, P.C.Zhai and R.Z.Yuan, "A thermal damage-heat conduction coupled model of ceramic-metal functionally graded materials", Materials Science Forum, Vols.308-311, 1030-1034, 1999.

[5] Q.J.Zhang, P.C.Zhai and Y.Li, " A theory of effective thermal conductivity for matrix-inclusion-microcrack three-phase heterogeneous materials based on micromechanics", Acta Mechanica Solida Sinica, 13(2), 179-187, 2000

[6] Y.Li and Q.J.Zhang, " A generalized self-consistent method for the effective conductivity of cracked body", Acta Mechanica Solida Sinica(in Chinese), Vol.20,S.Issue, 26-30, 1999

[7] L.S.Liu, P.C.Zhai and Q.J.Zhang, "The propagation characteristics of impact waves across a planar interface between a ceramic and ceramic/metal composite", submitted to Int. J. Impact Eng., 2000

[8] J.R.Furlong, C.F.Westbury and E.A.Phillips, "A method for predicting the reflection and refraction of spherical waves across a planar interface", J. Appl. Phys., 76(1), 24-32, 1994

GENERATING QUASI-ISENTROPIC COMPRESSION WAVES VIA LAYERED FLIER-PLATE MATERIALS

L.M. Zhang, C.B. Wang, Q. Shen and J.G. Li
State Key Laboratory of Advanced Technology for Materials Synthesis and Processing
Wuhan University of Technology
Wuhan 430070, China

J.S.Hua
Laboratory for Shock Waves and Detonation Physics
Engineering Physics Academic of China
Mianyang 621900, China

ABSTRACT
Recent studies indicate that a kind of Layered Flier-Plate (LFP) materials can be used to generate quasi-isentropic compression waves in target materials as opposed to conventional shock loading. In this study several different plates were joined together to form such LFP materials with a parabolic distribution of density along the thickness direction. The impact experiments of the LFP materials on the targets were carried out on a light gas gun, and the histories of the free surface velocity of the targets were measured by VISAR. The results showed that wave profiles with a stepwise front were created, i.e. quasi-isentropic compression waves had been successfully generated via LFP materials.

1.INTRODUCTION

Shock loading[1] technique offers an important way to study the behavior and dynamic response of materials under extremely high pressure. As its characteristics of high loading rate and great heat dissipation, however, this technique usually causes the rising of entropy and temperature in materials[2], which limits the further compression to materials and therefore higher dynamic pressure ($10^2 \sim 10^3$ GPa) may not be easily obtained.

Recent studies[3-6] indicate that a new kind of Layered Flier-Plate (LFP) materials with graded density can be used to generate quasi-isentropic compression waves in materials. Such quasi-isentropic compression waves have slowly increasing fronts compared with general shock loading waves, and thus the temperature increase is much lower. Therefore, the kinetic energy of the LFP materials can be thoroughly converted into the compressive energy of the target materials, which results in a much higher pressure than conventional shock loading technique.

In this paper, such kind of LFP materials with a parabolic distribution of density along the thickness direction was prepared. The impact experiments of the fabricated LFP materials on the targets were then carried out on a light gas gun, and the results showed that quasi-isentropic compression waves had been successfully generated via the LFP materials.

2.EXPERIMENTAL

During a high density range, five metallic plates----tungsten alloy (93W), oxygen free copper (Cu), titanium-aluminum alloy (TC$_4$), aluminum (Al), magnesium-aluminum alloy (Mg) and a piece of plastic material polyethylene (PE) were chosen to fabricate this kind of LFP materials. Their main compositions and densities are shown in Table I . The methods used for the fabrication were diffusion bonding and glue bonding.

Table I . Main compositions and densities of the experimental materials

Material	Main composition (mass %)	Density ($\times 10^3$ kg/m^3)
93W	W 92.95, Ni 4.20, Fe 2.45	17.64
Cu	Cu	8.93
TC$_4$	Ti 89.15, Al 5.5, V 4.5	4.45
Al	Al	2.70
Mg	Al 3~4, Zn 0.2~0.8, Mn 0.15~0.5, Mg Bal.	1.77
PE		0.92

All the experimental materials were made into round plates which were 32 mm in diameter and whose thickness accorded with the design. Then they were put into a mold layer by layer, and heated in a furnace under preset temperature, pressure and time. Two types of LFP materials, that is LFP- I and LFP- II, were fabricated in this paper. Their compositions and fabrication methods can be seen in Table II . The density of both these two LFP materials changed gradually through the thickness in a parabolic distribution, which can be controlled by adjusting the thickness of the experimental materials. Fig.1 shows the density distribution of the LFP- II material which had a total thickness of 6.5 mm.

Table II. The compositions and fabrication methods of the LFP materials used in this paper

LFP type	Composition	Fabrication method	Parameter
LFP- I	93W/Cu/TC$_4$/Al/Mg/PE	Diffusion bonding of 93W/Cu/TC$_4$	800°C-10MPa-10min
		Glue bonding of TC$_4$/Al/Mg/PE	60°C-10MPa-30min
LFP- II	93W/Cu/TC$_4$/Al/Mg	Diffusion bonding of 93W/Cu/TC$_4$	800°C-10MPa-10min
		Glue bonding of TC$_4$/Al/Mg	60°C-10MPa-30min

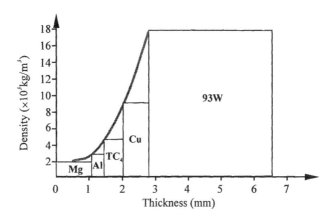

Fig.1　The change of density of the LFP- II material along the thickness direction

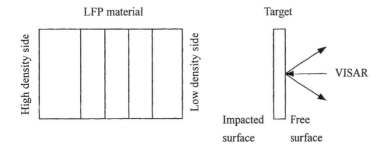

Fig.2　Schematic diagram of the impact of the LFP materials on the targets

Quasi-isentropic compression experiments were conducted on a light gas gun, which propelled the LFP materials to high velocities (500~2100 m/s) and then impacted the targets. The experimental system is illustrated in Fig.2. 93W alloys with the same thickness of 2.2 mm were used as the target materials, whose free surface velocity was measured by VISAR (Velocity Interferometric System for Any Reflector) to investigate the dynamic pressure produced in materials.

3.RESULTS AND DISCUSSION

At first, the targets were impacted by two LFP- I materials respectively at the velocity of 500 m/s. The variation of the free surface velocity of the targets with the impact time is shown in Fig.3. It can be seen that wave profiles which had a small initial velocity jump followed by a stepwise increase were produced and the rise time was extended to 2.5 μs too, obviously differing from the conventional shock loading waves. It showed that quasi-isentropic compression waves had been successfully generated via the LFP materials.

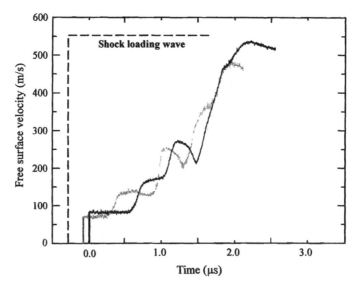

Fig.3   Free surface velocity histories of the targets impacted by the LFP- I materials
(Impact velocity was 500 m/s)

The impact process of the LFP- I materials on the targets was then computed (only the result was given here and more detailed process will be published in other papers). Fig.4 shows the results of experiment and calculation obtained under the same impact conditions. From the comparison, it can be seen that both the front's rise time and the peak velocity of the calculated wave profile, which mainly characterize the quasi-isentropic compression waves, were in accord with those of the experiment. Therefore, the design and fabrication of the LFP materials might be guided by the theoretical calculation.

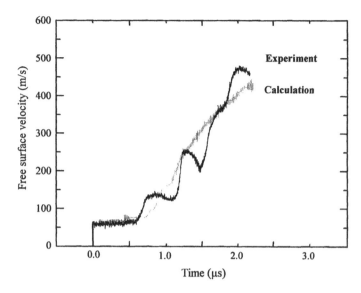

Fig.4   The results of calculation and experiment of the LFP- I materials
(Impact velocity was 500 m/s)

As mentioned above, the LFP materials were made of different plates layer by layer. In the impact process, accordingly, the compression of the LFP materials to the targets was step-by-step, and then the shock waves produced in materials were much smaller as compared to a conventional shock loading process. As a result, the successive interaction of these smaller shock waves led to the formation of the wave profiles with a stepwise front, and the free surface velocity of the targets was increased much slowly too just as seen in Fig.3. In other words, the process of the quasi-isentropic compression was the overlap of a series of small shock loading waves.

In addition, velocity minima are apparent in the wave profiles shown in Fig.3. This resulted from the complex wave interactions such as reflection and transmission between the interfaces, since the LFP materials consisted of many interfaces. Furthermore, according to PE which was a composite material of the LFP- I material, the transit speed of shock waves in PE was so slow that the rarefactive unloading waves reflected from the free surface might lower the intensity of the shock loading waves, which also caused velocity decreases. In order to reduce the negative effect of the PE material, in the following experiments the LFP-II material, i.e. 93W/Cu/TC₄/Al/Mg system LFP was then used.

Fig.5 shows the free surface velocity histories of the targets impacted by two LFP-II materials at the velocity of 600 m/s and 2100 m/s respectively. From the wave profiles, we can find that the wave's fronts were both extended even at different impact velocities, which indicates that quasi-isentropic compression has been realized.

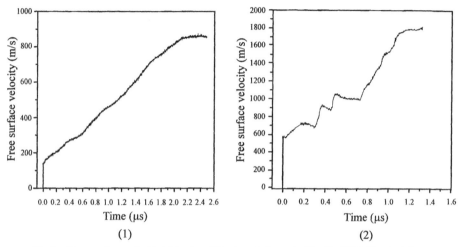

Fig.5   Free surface velocity histories of the targets impacted by the LFP-II materials
(1) Impact velocity was 600 m/s   (2) Impact velocity was 2100 m/s

On the other hand, the differences between the two wave profiles were obvious too. That is, the higher the impact velocity, the smaller the extended wave's front and the higher the peak free surface velocity. For example, the rise time of the wave front generated at the impact velocity of 600 m/s had been extended to 2.5 μs, while the peak free surface velocity was only 860 m/s. Conversely, the peak free surface velocity of the wave produced at 2100 m/s had reached 1800 m/s and the corresponding calculated pressure value in the targets was nearly 80GPa, but the front's rise time was merely 1.3 μs. All these resulted from the higher loading rate of the waves produced in the targets when the LFP materials impacted the targets at a higher velocity. By adjusting the impact velocity, therefore, the profile of the generated quasi-isentropic compression waves can be controlled.

As indicated in some studies[4], wave profiles can also be improved by adjusting the design of LFP and target materials including their compositions and dimensions. Thereby, more effective quasi-isentropic compression and higher dynamic pressure can be produced.

4.CONCLUSIONS

1) Generating quasi-isentropic compression waves by the impact of LFP materials on the targets is feasible.

2) The wave profiles with a stepwise front have been created via the LFP materials fabricated in this paper, and the rise time also extended to 1.3~2.5µs.

3) Different wave profiles which have different extended wave fronts and peak free surface velocity, were obtained at different impact velocities.

ACKNOWLEDGEMENTS

This work was supported by the National Science Foundation of China (No. 59771028).

REFERENCES

[1] F.Q. Jing, "Ultrahigh Dynamic Pressure Techniques (I)", *Explosion and Shock (in Chinese)*, 4[3] 1-9 (1984)

[2] S.H. Huang, F. Ding, F.Q. JING, Y.B. Dong and Z.R. Li, "Dynamic Quasi-Isentropic Compression of Oxygen Free Copper"; pp. 313-316 in *Shock Compression of Condensed Matter-1989*. Edited by S.C. Schmidt, J.N. Johnson and L.W. Davison. Elsevier Science Publishers B.V., Amsterdam, 1990

[3] L.M. Barker, "High-Pressure Quasi-Isentropic Impact Experiments"; pp. 217-224 in *Shock Waves in Condensed Matter-1983*. Edited by J.R.Asay, R.A.Graham and G.K.Straub. Elsevier Science Publishers B.V., Amsterdam, 1984

[4] L.M. Barker and D.D. Scott, "Development of a High-Pressure Quasi-Isentropic Plane Wave Generating Capability". SAND 84-0432

[5] L.C. Chhabildas, J.R. Asay and L.M. Barker, "Shear Strength of Tungsten under Shock- and Quasi-Isentropic Loading to 250 GPa". SAND 88-0306, UC-704

[6] L.C. Chhabildas and L.M. Barker, "Dynamic Quasi-Isentropic Compression of Tungsten"; pp. 111-114 in *Shock Waves in Condensed Matter-1987*. Edited by S.C. Schmidt and J.N. Johnson. Elsevier Science Publishers B. V., Amsterdam, 1988

ANISOTROPY OF WEAR RESISTANCE IN Al-Al$_3$Ti FGMs FABRICATED BY A CENTRIFUGAL METHOD

Yoshimi WATANABE and Hiroyuki ERYU
Department of Functional Machinery and Mechanics, Shinshu University
3-15-1 Tokida, Ueda 386-8567, Japan

Yasuyoshi FUKUI
Department of Mechanical Engineering, Kagoshima University
1-21-40 Korimoto, Kagoshima 890-0065, Japan

ABSTRACT

Al-Al$_3$Ti FGMs were fabricated by a centrifugal method with three different G numbers, *i.e.*, G=10, 30 and 50. Three types of wear specimens were prepared taking into account the Al$_3$Ti platelet morphology in the thick-walled FGM ring; the Al$_3$Ti platelets were arranged with their platelet planes nearly normal to the radial direction as a result of the applied centrifugal force. A significant anisotropic wear resistance was noted at the ring's outer region whereas anisotropic wear resistance was not observed at the ring's inner region. Based on experimental results, the origin of anisotropic wear resistance in Al-Al$_3$Ti FGMs is discussed.

INTRODUCTION

Properties of aluminum alloys may be improved through the addition of aluminum-transition metal intermetallic phases [1]. Compared to most other aluminum-rich intermetallic phases, Al$_3$Ti is attractive due to its higher melting point (about 1620K) and relatively low density (about 3300 kg/m^3). Moreover, titanium has low diffusivity and solubility in aluminum, hence Al$_3$Ti may be expected to exhibit low coarsening rates at elevated temperature. In addition, the Young's modulus of Al$_3$Ti phase is about 220 GPa [2], and the presence of an Al$_3$Ti phase is very effective in increasing the stiffness of aluminum alloys [1].

The authors have proposed a centrifugal method as a fabrication method for functionally graded materials (FGMs) [3-6]. A centrifugal force applied to a mixture of molten metal and dispersed materials, such as ceramic powder or intermetallic compounds, leads to formation of the desired composition gradient. The gradient is controlled mainly by the difference in density between the matrix and dispersed material. It was found that Al-Al$_3$Ti FGMs fabricated by the

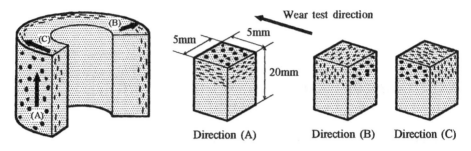

Fig. 1   Schematic representation of the arrangement of Al₃Ti platelets in the Al-Al₃Ti FGM ring. Three types of wear specimens were prepared taking into account the morphology of the Al₃Ti platelets in the thick-walled FGM ring, $i.e$, directions (A), (B), and (C).

centrifugal method contain anisotropically oriented Al₃Ti platelets [7] and exhibit anisotropic wear resistance [8].   **Figure 1** shows a schematic representation of the arrangement of Al₃Ti platelets. However, since wear resistances were previously measured for only one alignment, the relationship between degree of orientation and anisotropic wear resistance is still an open question. In a recent study, the orientation of Al₃Ti platelets was found to be gradually distributed in the Al-Al₃Ti FGM as well as the mean volume fraction of particles [9].   Therefore, the positional dependence of wear resistance is expected to occur in Al-Al₃Ti FGMs.

With the above in mind, Al-based FGMs with different Al₃Ti platelet orientations were fabricated by the centrifugal method with different G numbers.   Three types of wear specimens were prepared taking into account the Al₃Ti platelet morphology in the thick-walled FGM ring as shown in Fig. 1.   Based on experimental results, the origin of anisotropic wear resistance in Al-Al₃Ti FGMs is discussed.

EXPERIMENTAL METHODS

Al-Al₃Ti FGMs were fabricated by the centrifugal method (solid particle technique) [6] in a similar fashion to that of previous studies.   As the details of Al-Al₃Ti FGM fabrication can be found elsewhere [7-9], only the essential points are summarized below.   The mother alloy ingot was a commercial aluminum alloy containing 5mass% Ti.   The ingot was melted under an argon gas atmosphere at a solid/liquid co-existing temperature.   The applied G numbers were 10, 30 and 50, where G indicates the centrifugal force in units of gravity.

A block-on-disc type wear apparatus was used for the wear tests.   Specimens for wear testing were cut into blocks of size 5 mm x 5 mm x 20 mm.   The Al-Al₃Ti FGMs were machined from the outer or inner region of the FGM thick-walled ring, and the specimens were used without any additional heat treatment.   The anisotropic wear resistance in the Al-Al₃Ti FGMs was measured in three directions based on the expectation that most of Al₃Ti platelets were arranged with their platelet planes aligned approximately perpendicular to the radial direction.   The test directions were along the longitudinal direction on the ring outer surface (A), along the radial direction on the radial plane (B), and along the hoop direction on the radial plane (C). In direction (A), therefore, the worn plane nearly coincides with the Al_3Ti platelet planes, as shown in Fig. 1. In directions (B) and (C), on the other hand, the worn planes coincided with the platelets' two edge planes, and the wear directions are along the thickness direction and the longitudinal direction of the Al_3Ti platelets, respectively (Fig. 1). The counter-face disc for the wear test was S45C steel heat-treated to a hardness of 185 Hv. The relative sliding speed and sliding distance were 1 m/s and 1 km, respectively. All wear tests were performed at room temperature in open air without the use of any lubricants. In this study, the initial loading stress was 1.0 MPa and kept constant during the wear test. The morphology of the Al_3Ti platelet-particles in the Al matrix before wear testing was observed using an optical microscope.

RESULTS AND DISCUSSION

Distribution of Microstructures

Figure 2 shows the distribution of Al_3Ti platelets in FGMs fabricated by the centrifugal method. Here, the abscissa represents the position in the thickness direction of the ring, normalized by the thickness, i.e., 0.0 and 1.0 correspond to the inner and outer surfaces, respectively. It is seen that the volume fraction of Al_3Ti platelets increases towards the ring's outer edge. In addition, it was also noted that a steeper distribution of Al_3Ti platelets existed in the larger G number specimens. In this way, platelet-particles dispersed FGMs may be fabricated by the centrifugal method, similar to the case in which spherical particles are present.

The positional dependence on the orientation along the radial direction is also shown in Fig. 2. Here, the ordinate is Herman's orientation parameter, fp [10, 11], which ranges between values of $fp = 0$ (for a random distribution of planes) and $fp = 1$ (for perfect alignment). It is clearly seen that the orientation parameter increases with the normalized thickness. Thus, the orientation of Al_3Ti platelets is gradually distributed in the Al-Al_3Ti FGM as well as the mean volume fraction of particles. It was also noted that a steeper gradient distribution of orientation parameter formed for a larger G number specimen. Therefore, the positional dependence of wear properties might be expected in Al-Al_3Ti FGMs.

Wear Resistance

The wear volumes of FGMs fabricated for G=10, 30, and 50 are shown in Figs. 3 (a), (b), and (c), respectively. The result of a pure Al specimen made by the same process is also shown in these figures for comparison. The wear volumes in the Al-Al_3Ti FGMs are much smaller than that of pure Al. Also seen from these figures is that the wear volume at the ring's outer region is smaller than that at the inner region. The positional dependence of wear residence is emphasized with an increase in G number.

Figure 4 shows the mean wear volume as a function of volume fraction of Al_3Ti platelet-particles. It is seen that the mean wear volume decreases with increasing volume fraction

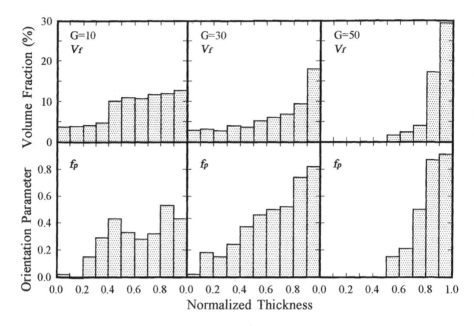

Fig. 2 Distributions of volume fraction (a) and the Herman's orientation parameter (b) for Al$_3$Ti platelets in the FGMs. The abscissa in this figure is the position in the thickness direction normalized by the thickness; *i.e.*, 0.0 is the inner surface and 1.0 is the outer surface of the ring.

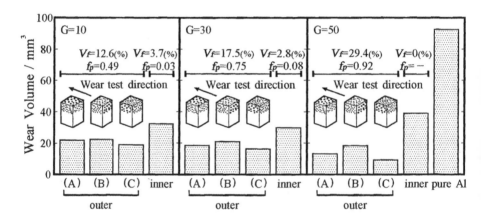

Figs. 3 (a), (b) and (c) Wear volumes of the FGMs fabricated for G=10, 30 and 50, respectively. The result of a pure Al specimen made by the same process is also shown in this figure for comparison.

of Al_3Ti platelet-particles. Thus, the wear resistance is significantly improved by introducing Al_3Ti platelets-particles into the Al matrix.

In the case of the ring's outer region, it is worth mentioning here that a notable anisotropy exists in the wear volume among the three platelet orientations tested. Namely, direction (B) shows the lowest wear resistance among the three orientations. In contrast to this, directions (A) and (C) show relatively better wear resistance and the wear volume of direction (C) is slightly smaller than that of direction (A). Moreover, greater anisotropy of wear resistance is found for larger orientation parameter specimens. Although not presented here, anisotropic wear resistance was not observed at the ring's inner region.

In order to express the anisotropic wear resistance quantitatively, the wear volume ratio (wear volume of direction (B) / wear volume of direction (A) or (C)) was calculated. We have

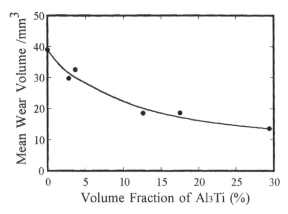

Fig. 4 Relation between mean wear volume and volume fraction of Al_3Ti platelet-particles

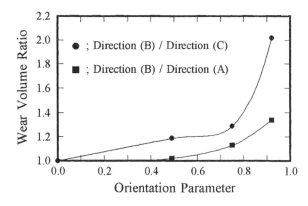

Fig. 5 Wear volume ratio (wear volume of direction (B) / wear volume of direction (A) or (C)) plotted against the orientation parameter

considered that the larger the value of wear volume ratio, the larger the anisotropy of wear resistance. **Figure 5** shows the wear volume ratio plotted against orientation parameter. As can be seen, since the wear volume ratio increases with increasing orientation parameter, we can conclude that the anisotropy of wear resistance is emphasized with an increase in the orientation of the Al_3Ti platelets. The origin of anisotropic wear resistance in $Al-Al_3Ti$ FGMs will be discussed later.

Cross-section near the Worn Surface

In order to investigate microstructural features near the worn surface, cross-sections of specimens parallel to the sliding direction were observed and are shown in **Fig. 6**. Damaged layers containing broken Al_3Ti platelets are found in (a) and (b). Particularly, in the case of direction (B), the wear tests led to fracture and bending of Al_3Ti platelets along the wear test direction. This result is similar to that of Si particles in the damaged layer reported for the case of Al-Si alloy worn tests [12]. In contrast to this, damaged Al_3Ti platelets were not observed for direction (C) (Fig. 6 (c)).

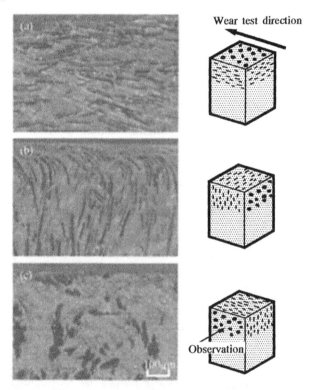

Figs. 6 (a), (b) and (c)   Cross-sections parallel to the sliding directions (A), (B) and (C), respectively.

Origin of the Anisotropic Wear Resistance

The present Al-Al$_3$Ti FGMs exhibit anisotropic wear resistance as shown in Fig. 3. The main difference between directions (A), (B), and (C) is the orientation of the Al$_3$Ti platelets in the Al matrix. In direction (A) the wear plane is nearly parallel to the Al$_3$Ti platelet face planes whilst in directions (B) and (C) the wear planes are parallel to the platelet edge planes along the thickness and longitudinal directions, respectively. It was shown by transmission electron microscopy (TEM) that the Al$_3$Ti platelets are a few micrometers thick [13] and that Al$_3$Ti is brittle in nature. Judging from these facts, the shear flow stress for forming a damaged region in direction (B) should be lower than that in directions (A) and (C). This is supported by the evidence that Al$_3$Ti platelets for direction (B) break easily during the wear test (Figure 6 (b)), whereas the platelets for directions (A) and (C) are more difficult to break. Thus, direction (B) shows a lower wear resistance whilst directions (A) and (C) show better wear resistance.

Based on the longitudinal orientation of the platelets in both directions (A) and (C), it could be expected that their wear properties would be similar. This, however, was not the case. Direction (A) exhibits lower wear resistance compared to direction (C). The volume fractions of Al$_3$Ti in both specimens are the same and both are tested along the longitudinal direction of the Al$_3$Ti platelets, however, the wear plane of the Al$_3$Ti platelets in the two specimens differs, i.e., the platelet face for direction (A) and the edge plane for direction (C). The Al matrix for direction (A) is easier to deform around the worn surface compared to direction (C). Thus, the wear rate of direction (A) is greater than that of direction (C).

CONCLUSIONS

(1) Al-Al$_3$Ti FGMs showed superior wear resistance compared to a pure Al specimen fabricated by the same process.

(2) Anisotropic wear resistance was observed to be dependent on the test direction relative to the Al$_3$Ti platelet orientation. Specimens tested along the Al$_3$Ti platelets thickness direction show the poorest wear resistance among the three orientations due to the ease in which the Al$_3$Ti platelets broke.

(3) A greater anisotropy in wear resistance was found for specimens with larger orientation parameters.

(4) It is possible to improve the wear resistance of Al-Al$_3$Ti FGM so that improved wear properties could be achieved by considering the morphology and distribution of Al$_3$Ti particles.

Acknowledgments

One of the authors (YW) gratefully acknowledges financial support from the Tokuyama Science Foundation. This work was supported by a Grant-in-Aid for COE Research (10CE2003) by the Ministry of Education, Science, Sports and Culture of Japan.

REFERENCES

[1] S. H. Wang and P. W. Kao, "The Strengthening Effect of Al_3Ti in High Temperature Deformation of Al- Al_3Ti Composites," *Acta mater.*, **46** [8] 2675-82 (1998).

[2] M. Nakamura and K. Kimura, "Elastic Constants of $TiAl_3$ and $ZrAl_3$ Single Crystals," *J. Mater. Sci.*, **26** [8] 2208-14 (1991).

[3] Y. Fukui, ""Fundamental Investigation of Functionally Gradient Material Manufacturing System using Centrifugal Force," *JSME Int. J. Series III*, 34 [1] 144-48 (1991).

[4] Y. Fukui and Y. Watanabe, "Analysis of Thermal Residual Stress in a Thick-Walled Ring of Duralcan-Base Al-SiC Functionally Graded Material," *Metall. Mater. Trans. A*, **27A** [12] 4145-51 (1996).

[5] Y. Watanabe, N. Yamanaka and Y. Fukui, "Control of Composition Gradient in a Metal-Ceramic Functionally Graded Material Manufactured by the Centrifugal Method," *Composites Part A*, **29A** [5-6] 595-601 (1998).

[6] Y. Watanabe and Y. Fukui, "Fabrication of Functionally-Graded Aluminum Materials by the Centrifugal Method," *Aluminum Trans.*, 2 [2] 195-209 (2000).

[7] Y. Watanabe, N. Yamanaka and Y. Fukui, "Orientation of Al_3Ti Platelets in Al-Al_3Ti Functionally Graded Material Manufactured by Centrifugal Method," *Z. Metallkd.*, **88** [9] 717-21 (1997).

[8] Y. Watanabe, N. Yamanaka and Y. Fukui, "Wear Behavior of Al- Al_3Ti Composite Manufactured by a Centrifugal Method," *Metall. Mater. Trans. A*, **30A** [12] 3253-3261 (1999).

[9] Y. Watanabe, H. Eryu and K. Matsuura, "Evaluation of Three-Dimensional Orientation of Al_3Ti Platelet in Al based FGMs Fabricated by a Centrifugal Casting Technique," *Acta Mater.*, **49** [5] 775-783, (2001).

[10] S. H. McGee and R. L. McCullough, "Characterization of Fiber Orientation in Short-fiber Composites," *J. Appl. Phys.*, **55** [5] 1394-1403 (1984).

[11] L. M. Gonzalez, F. L. Cumbrera, F. Sanchez-Bajo and A. Pajares, "Measurement of Fiber Orientation in Short-fiber Composites," *Acta metall. Mater.*, **42** [3] 689-94 (1994).

[12] J. W. Liou, L. H. Chen and T. S. Lui, "The Concept of Effective Hardness in the Abrasion of Coarse Two-phase Materials with Hard Second-phase Particles," *J. Mater. Sci.*, **30** [1] 258-62 (1995).

[13] K. Yamashita, A. Sato, Y. Watanabe, N. Yamanaka and Y. Fukui, "Crystallization of Al_3Ti in Al-5wt%Ti by Centrifugal Casting and the Strengthening," International Symposia on Advanced Materials and Technology for the 21st Century (JIM' 95 Fall Annual Meeting in Hawaii), abstracts, 6, (1995).

FABRICATION AND THERMOMECHANICAL PROPERTIES OF MoSi₂-Mo FUNCTIONALLY GRADED MATERIALS

Jae-Ho Jeon, Yoo-Dong Hahn
Department of Materials Engineering, Korea Institute of Machinery and Materials, 66 Sangnam-Dong, Changwon 641-010, Korea

Zhong-Da Yin
Materials Science and Engineering School, Harbin Institute of Technology, Harbin 150001, P.R. China

ABSTRACT
For the application of Mo as a structural material in an oxidizing atmosphere, an oxidation-resistant surface layer should be introduced because Mo has poor resistance to oxidation over 700K. When $MoSi_2$ was used as the surface layer, cracks would form due to the difference of thermal expansion coefficient between $MoSi_2$ and Mo. In order to overcome this problem, a functionally graded $MoSi_2$-Mo with varying amounts of these two materials was fabricated by hot pressing at 1650°C for 4h at a pressure of 30MPa. The $MoSi_2$-Mo FGM shows excellent heat-resistant and thermal-shock properties when it was evaluated by gas burner heating and acoustic emission test.

INTRODUCTION
The concept of functionally graded materials (FGMs) was originated in the research field of thermal barrier materials.[1,2] Continuous changes in the composition, grain size, porosity, etc., of these material result in gradients in such properties as mechanical strength and thermal conductivity. When ceramic/metal functionally graded materials are fabricated by powder metallurgy, sintering temperatures are usually limited by the melting temperature of the metal. Therefore the ceramic side of the FGMs usually has a poor relative density which is expected to results in inferior properties. In order to improve the density of ceramic side by increasing the sintering temperature, in this paper, Mo metal was chosen as a metal side because Mo has a relatively high melting temperature of 2893K.

$MoSi_2$-Mo composites are strong candidates for gas-fired burner tubes for glass batch melting.[3-6]   These tubes must simultaneously survive molten glass corrosion and a combustion gas environment.   Mo showed the lowest recession rates during corrosion tests in glass melts.[3]   Mo, unfortunately, oxidizes to MoO_3 at high temperatures in a combustion product environment.[3] $MoSi_2$ showed excellent oxidation resistance by forming a continuous SiO_2 layer which prohibits oxidation of the metallic interior.[3]   A simple inner coating of $MoSi_2$ on a Mo tube was rejected due to the coefficient of thermal expansion mismatch between coating and the substrate.[3]   We developed a $MoSi_2$-Mo FGM by hot-pressing and investigated a correlation between the thermo-mechanical properties and the phase distribution in this material system.

EXPERIMENTAL

The powders used for this research were 99.95% pure molybdenum metal (Aldrich, USA) and 99.5% molybdenum disilicide (Cerac, USA).   Both powders have the same average particle size of <45µm.   Because of low room-temperature fracture toughness and ductility of molybdenum disilicide, it was reinforced with a small amount of partially stablized zirconia (doped with 3mol% yitria) powder.[7-9]   Four powder mixtures (80Mo-20$MoSi_2$, 60Mo-40$MoSi_2$, 40Mo-60$MoSi_2$, 20Mo-80$MoSi_2$, vol.%) were wet milled for 24 hours in a polyethylene bottle with ethyl alcohol and zirconia balls.   The dried slurry was crushed in an agate bowl and sieved to 125µm.

The prepared powders were stacked sequentially with stepwise changes in mixing ratio and compacted in a steel die of 16mm diameter.   Then the green compacts were hot pressed at 1650°C under a pressure of 30 MPa for 4 hours in argon atmosphere.   In order to evaluate mechanical properties of each layer, the individual plates of 30mm diameter with various mixing ratios corresponding to each layer of the FGM were fabricated as well under the same conditions.   X-ray diffraction (XRD) was employed to provide phase identification in the individual plates.   Samples for microstructural inspection, phase analysis, and mechanical tests were cut with a diamond saw, and their surfaces were ground and polished. A scanning electron microscope equipped with an energy dispersive spectrometer was used for microstructural and compositional analysis.   The Vickers hardness and fracture toughness were measured by indenting with a 5 kg load.

Heat-resistant and thermal shock property of $MoSi_2$-Mo FGMs were evaluated by gas burner heating test using C_2H_2/O_2 combustion flame and acoustic emission (AE) test.   A schematic diagram of the experimental set-up for gas burner heating and acoustic emission monitoring is shown in Fig. 1.   The ceramic surface was heated with the burner flame while the bottom surface was cooled with flowing water.   The ceramic surface was quickly heated from room temperature to 1600°C, held for 500 sec and then it was cooled in air.   The temperature of the ceramic surface was measured by a radiation thermometer.

Three K-type thermocouples were used for measuring the temperature of the metal side. AE monitoring was carried out using MISTRAS-2001 Analyzer produced by Physical Acoustics Corporation (PAC). The AE signals were counted during air cooling because severe noise was included due to burner flame during heating.

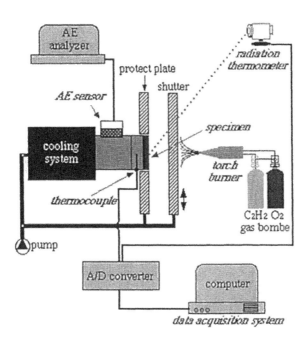

Fig. 1. The schematic drawing of experimental set-up for gas burner heating and acoustic emission monitoring

RESULTS AND DISCUSSION

Phase Distribution

The XRD results of individual layers are shown in Fig. 2. The reaction between Mo and $MoSi_2$ produced Mo_3Si and Mo_5Si_3. Mo_3Si phases begin to appear in the 80Mo-20MoSi$_2$ layer, and is a dominent phase in the 60Mo-40MoSi$_2$ layer. The 40Mo-60MoSi$_2$ layer consisted of Mo_3Si and Mo_5Si_3 with a little amount of Mo. Remnant $MoSi_2$ phase was present along with Mo_3Si and Mo_5Si_3 in the 20Mo-80MoSi$_2$ layer. The sixth layer is strictly $MoSi_2$.

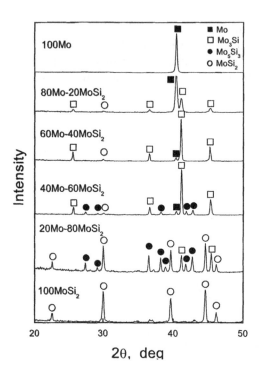

Fig. 2. XRD patterns of individual layers in $MoSi_2$-Mo functionally graded material.

Mechanical Properties

Figure 3 illustrates the relationship between Vickers hardness and composition in the $MoSi_2$-Mo system. The hardness increases with increasing $MoSi_2$ content until 60vol%, and then decreases with further increases in $MoSi_2$ content. These hardness test results indicate the hardness of reaction products Mo_3Si and Mo_5Si_3 is higher than that of $MoSi_2$.

Figure 4 displays the variation of fracture toughness with composition, which reveals a minimum fracture strength around 40Mo-60MoSi$_2$. The varying trend of the fracture strength is just opposite from the distribution of Vickers hardness, that is, the higher hardness corresponds to the lower fracture toughness. Without assessments of crack behavior and fracture surface, it is difficult to explain the fracture toughness variation with the composition. The fracture toughness measurement results, however, indicate reaction products, Mo₃Si and Mo₅Si₃, are more brittle than Mo and MoSi₂.

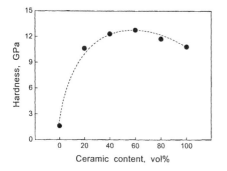

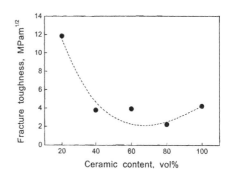

Fig.3 Variation of Vickers hardness with composition in the MoSi₂-Mo system.

Fig. 4. Variation of fracture toughness with composition in the MoSi₂-Mo system.

Heat Resistance and Thermal Shock Property

Figure 5 shows the results of gas burner heating and AE test. The temperatures of the ceramic surface and the metal side were indicated by Ts and Tb, respectively. The temperature history of this MoSi₂-Mo FGM reveals

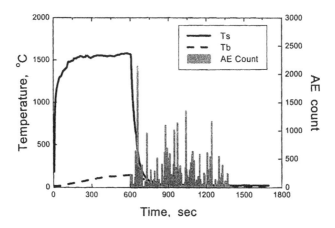

Fig. 5. AE behavior and temperature history by burner heating in MoSi₂-Mo

functionally graded material.
excellent heat-resistant property because the temperature difference between the ceramic and metal side was over 1250°C. The total AE counts during cooling shows 46503 and the maximum AE count is 2354. The AE is thought to closely relate with the initiation and propagation of microcracks.

Figure 6 is a cross-sectional SEM micrograph of the $MoSi_2$-Mo FGM after thermal shock test. The left-hand side is Mo and the opposite is $MoSi_2$ and ▼ denotes the interface between each layer. Several microcracks were found in the $60Mo$-$40MoSi_2$ layer and a few cracks in the $40Mo$-$60MoSi_2$ layer. These microcracks are thought to form during cooling as suggested by the AE behavior in Fig. 5. The reason for the formation of microcracks in the central area of the FGM seems to be the brittle behavior of Mo_3Si and Mo_5Si_3, as shown in Fig. 4, but additional studies are required to quantify the effect.

Mo ▼   ▼   ▼   ▼   ▼ $MoSi_2$

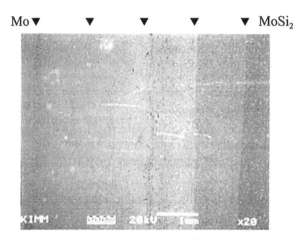

Fig. 6. A SEM micrograph of a cross-section of the $MoSi_2$-Mo FGM after thermal shock test.

CONCLUSIONS

A functionally graded $MoSi_2$-Mo composite was fabricated by a hot-pressing process. Reaction products, Mo_3Si and Mo_5Si_3, were found to form in the intermediate compositional layers after sintering. The hardness and fracture toughness of individual layers of this FGM were strongly dependent on the reaction products that formed during sintering. The $MoSi_2$-Mo FGM showed excellent heat-resistant and thermal-shock properties when it was evaluated by gas burner heating and acoustic emission tests. The reason for the formation of microcracks in $60Mo$-$40MoSi_2$ and $40Mo$-$60MoSi_2$ layer during thermal shock test are thought to be the brittleness of the reaction products. Additional studies are required to quantify the effect of reaction products on the thermo-mechanical properties of this FGM.

ACKNOWLEDGMENTS

This study has been done as a part of the Korea-China cooperative research program sponsored by the MOST of Korea. The authors would like to thank professor J.-K. Lim and Dr J.-H. Song for their measurement of the heat-resistance and thermal shock properties of the FGM.

REFERENCES

[1]A. Mortensen and S. Suresh, "Functionally Graded Metals and Metal-Ceramic Composites: Part 1 Processing," *Int. Mat. Rev.*, **40**[6] 239-65 (1995).

[2]A. Neubrand and J. Rödel, "Gradient Materials: An Overview of a Novel Concept," *Z. Metallkd*, **88**[5] 358-71 (1997).

[3]G. Agawal, W.-Y. Lin and R. F. Speyer, "Fabrication and Characterization of a Functionally Gradient Mo-MoSi$_2$ Composite," *Mat. Res. Soc. Symp. Proc.*, **322** 297-302 (1994).

[4]W.-Y. Lin, J.-Y. Hsu and R. F. Speyer, "Stability of Molybdenum Disilicide in Combustion Gas Environment," *J. Am. Ceram. Soc.*, **77**[5] 1162-68 (1994).

[5]S. Kamakshi, J.-Y. Hsu and R. F. Speyer, "Molten Glass Corrosion Resistance of Immersed Combustion-Heating Tube Materials in Soda-Lime-Silicate Glass," *J. Am. Ceram. Soc.*, **77**[6] 1613-23 (1994).

[6]S. Kamakshi, J.-Y. Hsu and R. F. Speyer, "Molten Glass Corrosion Resistance of Immersed Combustion-Heating Tube Materials in E-Glass," *J. Am. Ceram. Soc.*, **78**[7] 1940-46 (1995).

[7]J. J. Petrovic and R. E. Honnell, "Partially Stablized ZrO$_2$ Particle-MoSi$_2$ Matrix Composites," *J. Mat. Sci.*, **25** 4453-56 (1990).

[8]W. Soboyejo, D. Brooks and L.-C. Chen, "Transformation Toughening and Fracture Behavior of Molybdenum Disilicide Composite Reinforced with Partially Stabilized Zirconia," *J. Am. Ceram. Soc.*, **78**[6] 1481-88 (1995).

[9]J. J. Petrovic, A. K. Bhattacharya, R. E. Honnell and T. E. Mitchell, "ZrO$_2$ and ZrO$_2$-SiC Particle Reinforced MoSi$_2$ Matrix Composites," *Mat. Sci. and Eng.*, **A155** 259-66 (1992).

MICROSTRUCTURE EFFECTS OF PSZ/NI FGM UPON MATERIAL PROPERTIES AND THERMAL STRESS CONDITIONS

S. Moriya[1], P.C. Zhai[2], Q.J. Zhang[2] and M. Niino[1]

[1] National Aerospace Laboratory, 1 Koganezawa, Kimigaya, Kakuda, Miyagi 981-1525, Japan

[2] State Key Lab of Advanced Technology, Wuhan University of Technology, Wuhan 430070, China

ABSTRACT

Microstructure of PSZ/NiCr FGM fabricated by LPPS (Low Pressure Plasma Splay) method shows a pile of PSZ layers and NiCr layers. On the other hand, a PSZ/NiCr FGM fabricated by a sintering method or an EF (Electro Forming) method has particles dispersion structure. In this paper, finite element analysis was conducted by using microstructure models to study the microstructure effects of PSZ/NiCr FGM upon effective material properties and thermal stress conditions.

INTRODUCTION

PSZ/NiCr FGM thermal barrier coating has been developed to increase performance of a high temperature component such as a combustion chamber [1]. And high resistance to a vertical crack was demonstrated as a result of relaxation of a circumferential stress. However, "swelling" of the PSZ topcoat was frequently observed as shown in Fig.1. This phenomenon is the most serious problem to enhance the life and the reliability of TBC (Thermal Barrier Coating). It was pointed out that a delamination of a topcoat was induced by the combination of a corrugated interface and a large circumferential compressive stress [2]. The corrugated

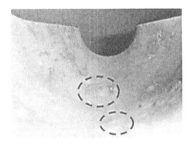

Fig.1 Swelling of TBC

interface and the large compressive stress are typical of a plasma spray coating method and a combustion chamber, respectively. Using FEM numerical analysis, it was demonstrated in [3] that a tensile stress induced by a heating of the ceramic-metal interface was locally normal to the interface. Therefore, it is necessary to understand an effect of a microstructure, such as shapes of an inclusion or interface, upon a thermal stress condition to enhance durability of TBC.

In this paper, FEM analysis was performed on microstructure models simulating a piled up structure as fabricated by a LPPS method to study an influence of a microstructure on effective material properties. FEM analysis was also conducted on a particles dispersion structure as fabricated by an EF (Electro Forming) method or a sintering method to study an influence of a microstructure on thermal stress conditions.

ANALYSIS MODEL AND PROCEDURE

(1) Effective material property analysis

The microstructure of PSZ/NiCr FGM whose composition profile is PSZ/NiCr=50/50 vol.% fabricated by a LPPS method is shown in Fig.2. Porosity was estimated from the measured density to be 9.2 vol.%. It shows that the PSZ layers and the NiCr layers were piled up irregularly. In such case, it is difficult to evaluate material properties by a mixture rule. In this paper, material properties such as an elastic modulus, a thermal expansion coefficient and a thermal conductivity were calculated to study the effect of microstructures upon the effective material properties by using microstructure simulated models. Numerical

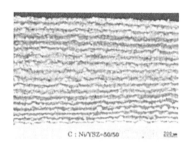

Fig.2 A microstructure of PSZ/NiCr FGM fabricated by a LPPS method

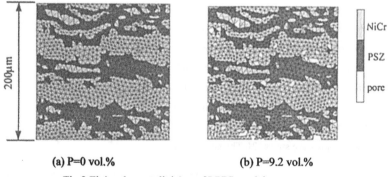

(a) P=0 vol.%   (b) P=9.2 vol.%

Fig.3 Finite element division of LPPS model

analysis was performed by using FEM code MARC. Fig.3 shows finite element divisions of the models simulating the microstructure fabricated by a LPPS method: (a) excluding pores (P=0%) and (b) including pores (P=9.2%) (later, this model is indicated as "LPPS model"). An effective elastic modulus was derived from a slope of stress-strain relationship. Both the effective elastic modulus and the effective thermal expansion coefficient were calculated in parallel and horizontal directions. An effective thermal conductivity was calculated from small temperature differences between the top and the bottom and a heat flux.

Calculations were performed on particles uniformly dispersed microstructure models simulating a material fabricated by an EF method and a sintering method to make comparison with above LPPS models. The finite element division was same as indicated in Fig.3, but the each material component was distributed uniformly (later, this model is indicated as "particles dispersion model").

(2) Thermal stress analysis

An influence of the FGM microstructure on a thermal stress was studied for a particles dispersion material with varying a diameter of particles. These analyses were conducted to study the combination effects of a thermal compressive stress and a microstructure of a FGM layer upon the delamination of a coating layer. Fig.4 shows the finite element divisions of the models used in the thermal stress analyses. A FGM layer whose volume contents of PSZ and NiCr were 75 and 25 vol.% respectively was inserted between a PSZ layer and a NiCr layer. The NiCr particles were arranged uniformly in the PSZ matrix. The diameters of NiCr particles were (a) 0.04

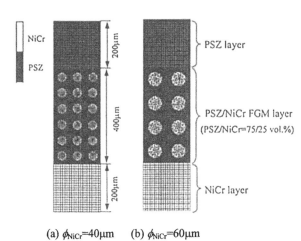

(a) ϕ_{NiCr}=40μm   (b) ϕ_{NiCr}=60μm

Fig.4 Finite element division of particle dispersion model

Table 1 Thermal boundary conditions

(i) Heated Side
-Heat Flux : q=4MW/m2
(ii) Cooled Side
-Heat Transfer Co. : h=3.75x10⁴W/m²K
-Coolant Temp.   : T_c=340K

and (b) 0.06 mm. Thickness of the PSZ layer, the FGM layer and the NiCr layer were 0.2, 0.4 and 0.2 mm respectively. Both sides were fixed in the horizontal direction to simulate the restrained condition of TBC of a combustion chamber. Thermal boundary conditions were listed in Table 1.

RESULT AND DISCUSSION

1. Influence of microstructure on effective material property

(a) Effective elastic modulus

Fig.5 shows calculated elastic moduli of the LPPS models in the parallel and the vertical directions (P=0 and 9.2%), the particles dispersion models (P=0%) and the experimental results as a function of temperature. In the case of P=0%, the moduli of the LPPS model in both the parallel and the vertical directions, and of the particles dispersion model were quite in good agreement. On the other hand, in the case of P=9.2%, the moduli of the LPPS model in the parallel direction were slightly larger than the moduli in the vertical direction. It is considered that the pores were mainly distributed in the NiCr phases and the whole stiffness of the model was dominated mainly by the PSZ layers, so the stiffness in the horizontal direction that corresponded to the direction of the PSZ layers was higher than the stiffness in the vertical direction.

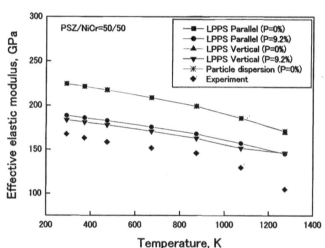

Fig.5 Effective elastic modulus of LPPS model

(b) Effective thermal expansion coefficient

Fig.6 indicates the calculated effective thermal expansion coefficients of LPPS models in the parallel and the vertical directions (P=0 and 9.2%) and the particle dispersion model (P=0%), and the experimental results as a function of temperature. The calculated results were almost same excepting the result of the LPPS model with pores (P=9.2%) in the parallel direction. The

Functionally Graded Materials 2000

coefficient of the LPPS model with pores in the parallel direction was lower than other models by 2%. It is considered that the reason is same as mentioned in the above section.

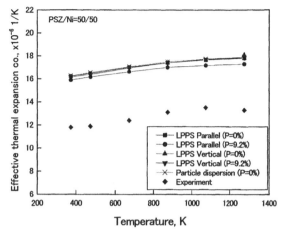

Fig.6 Effective thermal expansion coefficient of LPPS model

(c) Effective thermal conductivity

Fig.7 indicates the calculated effective thermal conductivities of the LPPS models in the vertical direction (P=0 and 9.2%), the particle dispersion model (P=0%), and the experimental results. As can be seen from Fig.7, a microstructure has significant effect upon the effective thermal conductivity. In the case of the LPPS models, the effective thermal conductivity of P=9.2% was smaller than that of P=0% by 14%. On the other hand, comparing the LPPS model and the particle

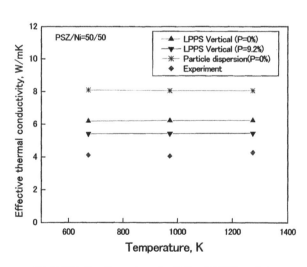

Fig.7 Effective thermal conductivity of LPPS model

dispersion model of P=0%, the effective thermal conductivity of the latter was higher than the former one by 30%. Therefore, the morphology of the dispersion phase and the porosity should be taken into consideration in the thermal analysis.

2. Influence of microstructure on thermal stress condition

Thermal stress analyses were performed on the particle dispersion models with changing the diameter of particles such as 40 and 60μm. Fig.8 shows the temperature change at the heated surface and the cooled surface as a function of time. Temperature changes of both models were almost same. This means that the particle diameter has no significant effect upon the effective thermal conductivity, if the composition

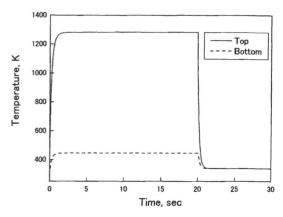

Fig.8　Analyzed temperature of particle dispersion model as a function of time

profile in a macro scale is same. Maximum temperature difference between the top and the bottom surfaces was 835K during the heating period. Fig.9 shows the contour of maximum principal stress of the upper half of FGM layer at the end of the heating period. As can be seen this figure, the maximum vertical tensile stress, namely the stress component causing the delamination of the top

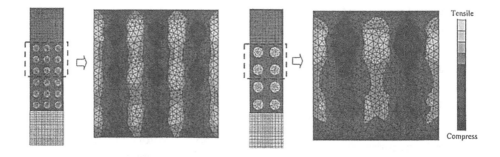

(a) Particle diameter of 40μm　　　　　　(b) Particle diameter of 60μm

Fig.9 Contour of maximum principal stress of particle dispersion model

coat, was induced near the interface between the NiCr particle and the PSZ matrix. Fig.10 shows the ratio of the maximum vertical stress to the absolute value of the maximum compressive stress at the heated surface as a function of time. As can be seen in this figure, the smaller particles dispersed microstructure model can reduce the maximum stress during a heating period.

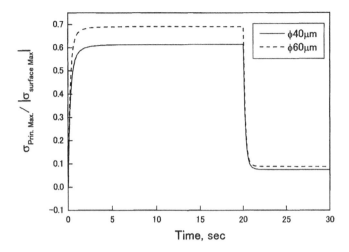

Fig. 10   Ratio of the maximum vertical stress to the absolute value of the maximum compressive stress at the heated surface as a function of time

It has been demonstrated by a numerical analysis that FGM TBC of a combustion chamber can relax the circumferential tensile stress caused by the thermal expansion mismatch during a cooling period [3]. However, TBC of a combustion chamber has a significant problem with the tensile stress that is locally normal to the interface and causes a delamination during a heating period. Taking into account above analysis results, it is possible to design FGM TBC layer with increased delamination resistance by an engineering of the interface microstructure.

CONCLUDING REMARK

In this paper, numerical analysis was conducted by using PSZ/NiCr FGM microstructure models to study the influence of the microstructure on effective material properties and thermal stress conditions. It was demonstrated that a significant effect of the microstructure upon a thermal conductivity and possibility of the increased delamination resistance by an engineering of the interface microstructure.

ACKNOWLEDGMENT

This work was supported by the National Science Foundation of China under the Grant 19772037.

REFERENCE

[1] Y.Kuroda, M.Tadano, A.Moro, Y.Kawamata and N.Shimoda, "Durability and High Altitude Performance Tests of Regeneratively Cooled Thrust Engine Made of ZrO_2/Ni Functionally Graded Materials", Proceedings of the 4th Int. Symposium on Functionally Graded Materials, pp. 469-474, October 21-24, 1996

[2] D.M.Nissley, "Thermal Barrier Coating Life Modeling in Aircraft Gas Turbine Engines", Proceedings of Thermal Barrier Coating Workshop, pp.265-281, March 27-19, 1995

[3] S.Moriya, Y.Kuroda, M.Sato, M.Tadano, A.Moro and M.Niino, "Research on the Application of PSZ/Ni FGM Thermal Barrier Coating to the Combustion Chamber (Damage Conditions of TBC and its Mechanism)", Proceedings of the 5th Int. Symposium on Functionally Graded Materials, pp.410-415, October 26-29, 1998

QUASI-ISENTROPIC COMPRESSION CHARACTERISTICS OF W-MO-TI GRADED DENSITY FLIER-PLATE MATERIALS

Q.Shen, J.G.Li, C.B.Wang and L.M.Zhang
State Key Lab. of Materials Synthesis and Processing, Wuhan University of Technology, Wuhan 430070, P.R.China;

H.Tan
Laboratory for Shock Waves and Detonation Physics, Institute of Fluid Physics, Engineering Physics Academic of China, Mianyang 621900, P.R.China

ABSTRACT
Quasi-isentropic compression techniques via graded density flier-plate materials, allow investigations of material properties in a high-pressure, low-temperature regime which is inaccessible by conventional shock wave experiments. In this study, quasi-isentropic experiments were performed with a light gas gun and W-Mo-Ti graded density flier-plate materials. The histories of the free surface velocity in targets were measured by VISAR. Effects of density distributions in the flier-plate materials on the wave profiles are discussed. It is found that wave profiles with a smoothly-rising front were created inside the targets, and the maximum pressure peak reached 167 GPa. The flier-plates with parabolic distribution of density can produce a better quasi-isentropic compressive wave profile.

INTRODUCTION
It is necessary to improve our understanding of how materials behave and response at extreme pressures, especially in the research fields of astrophysics, geophysics and nuclear engineering. Normally, such extreme pressures are mainly obtained from explosions and dynamic high-pressure techniques. According to the flier-plate impact method, a homogeneous flier-plate can be accelerated up to 7-8

km/s and can only reach pressures of several hundred GPa[1-2], but generates a shock wave as the pressure increases abruptly. The acutely increasing temperature leads to shock-induced melting, even shock-induced vaporization. Yet deeper compression can not be achieved, and from which the properties of materials are affected simultaneously by pressure and temperature.

Since 1980 researches[3-8] have shown that if the density of a flier-plate changes gradually from a low value to a high value in its flying direction, a long risetime wave profile inside the target will be created. That is, the target can be loaded step by step and the entropy rise is much smaller than that of general shock loading. Therefore, the variations of material properties are then affected only by pressure, thus the contributions of pressure and temperature on material properties may be resolved. This is of great importance for studying the constitutive relations and static equations at extreme conditions.

In the present paper, quasi-isentropic compression experiments using flyer-plates with graded density were carried out on a two-stage light gas gun. The effects of density distribution along the thickness direction of the flier-plate on the compressive wave profile are discussed.

EXPERIMENTS AND MEASUREMENTS

W-Mo-Ti graded density flier-plates were prepared by powder metallurgy method [9]. Density values of 4.5×10^3 kg/m^3 to 17.0×10^3 kg/m^3 were well controlled by inserting ten transient layers with graded compositions between Ti and W. Flier-plates with linear, parabolic and cubic density distributions were

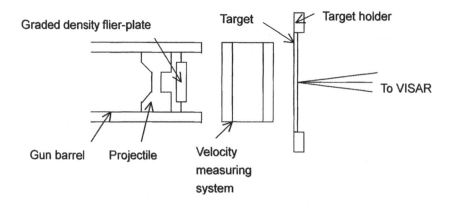

Figure 1 Schematic diagram of impact experimental system

The size of each flier-plate was 32 mm in diameter and 3.66 mm in height. The impact experiments were carried out with a two-stage light gas gun, shown schematicly in figure 1. The graded density flier-plate were cemented to a polycarbonate projectile. A set of magnet coils was installed in front of the barrel to measure the impact velocities. The laser alignment method was performed to adjust the target holder for assurance that the flier-plate would impact normally with the target and the average impact tilt would be less than 0.09°. Tungsten alloy sized 2.2 mm in thickness was used as the target. The compression of the target was monitored with VISAR system[8] and the velocity profile on the rear surface of the target was measured.

RESULTS AND DISCUSSIONS

Three flier-plates with different density distributions were launched at nearly the same velocity of 2.61 km/s. Figure 2 to figure 4 show the VISAR-measured wave profiles, the histories of the free surface velocities. Compared with the traditional shock wave profile, the graded density flier-plate produces an initial velocity jump followed by a smoothly-rising profile to the peak velocity amplitude, which identifies that dynamic quasi-isentropic compression in the target is created[3].

The initial jump was caused by the density of Ti alloy at the impact surface of the flier-plate. It means that a small amplitude shock wave firstly entered the target. Then, due to the varying density distribution in the thickness direction, the target would be subjected to a continuous shock by the transient layers with gradually increasing densities. According to the conservation of momentum and mass for wave propagation, the target would be compressed continuously, or step by step. The wave profile could be seen as the piling up of a series of small shock waves, which were generated by the transient layers. Because the density difference between adjacent layers was so small, a smoothly-rising front was exhibited. This processing could be illustrated as the curve marked with Q, as shown in figure 5. It was found that Q-curve was located between the H-curve for Hugoniot shock situation and S-curve for isentropic compression, but more close to the S-curve. That is, when a wave profile with extended front instead of a traditional shock wave profile is introduced into the target, the entropy rising will be much smaller.

As seen from the profiles in figure 2 to figure 4, the following were of interest. First, all of the wave front rise-times were about the same, 0.8µs, which shows

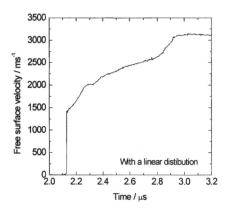

Figure 2 The VISAR-measured wave profile of flier-plate with a linear distribution of impedance

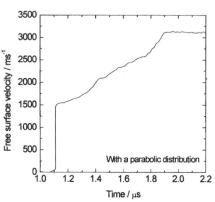

Figure 3 The VISAR-measured wave profile of flier-plate with a parabolic distribution of impedance

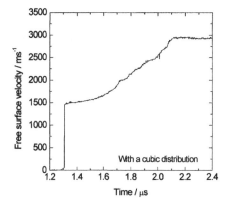

Figure 4 The VISAR-measured wave profile of flier-plate with a cubic distribution of impedance

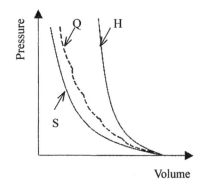

Figure 5 The Schematic process of generating a quasi-isentropic wave profile

that the rise-time is independent of the density distribution. Secondly, the wave profile produced by the flier-plate with linear density distribution has a slightly convex front, but that of the other two are concave. Thirdly, with the exponent of the density distribution function increasing, the velocity peaks of the wave profiles decrease. It can be seen that, the linear gradient reaches 3155 m/s, and the parabolic gradient reaches 3150 m/s, while that of the cubic gradient is only 2950 m/s.

Theoretical analysis in reference [3] pointed out that a wave with concave rising front could achieve a better quasi-isentropic compressive effect. Combined with our experimental results, it can be easily deduced that the flier-plate with a parabolic density distribution generates a better quasi-isentropic wave profile. Therefore, the design for the graded density flier-plate must conform to the parabolic function relationship between the density distribution and the thickness.

Consequently, according to the free surface velocity double law, the particle velocity u in the target is equal to half of the free surface velocity u_{fs}, that is,

$$u = \frac{1}{2} u_{fs} \qquad (1)$$

And the wave propagation velocity D, was determined by the following formula,

$$D = C_0 + \lambda u \qquad (2)$$

where C_0 is the sound velocity of the target and λ is the shock constant. For impacting a stationary target, the impact pressure P can be calculated from the equation,

$$P = \rho_0 D u \qquad (3)$$

ρ_0 is the density of the target. Then, inserting the corresponding values of ρ_0, D, u, the impact pressure can be calculated. The results for the pressure peaks of each profile are 167.2 GPa, 166.8 GPa and 153 GPa respectively, at the velocity peak. If a homogeneous flier-plate of tungsten alloy were used to impact the same target at the same impact velocity, the pressure peak would only reach 131 GPa, which is smaller than those of the graded density flier-plates. That is to say, the graded density flier-plate can generate quasi-isentropic compression which allows more of the kinetic energy of the flier-plate to be converted into the compressive energy of the target, thus achieving a higher pressure. In addition, such graded density flier-plates can also be used to increase the velocities of impact experiments.

CONCLUSIONS

1) It is feasible to generate quasi-isentropic compressions and gain higher pressures via graded density flier-plate materials.

2) Wave profiles with smoothly-rising fronts were created in the target, which is significantly different from traditional shock waves.

3) The maximum pressure peak reached 167 GPa, and the flier-plate with

parabolic density distribution could produce a better quasi-isentropic wave profile.

ACKNOWLEDGEMENT
This work was supported by the National Natural Science Foundation of China(Grant No.59771028) and the Doctor Foundation of the Education Ministry of China(Grant No.1999049702)

REFERENCES
[1]J.Q.Fu, "Dynamic high pressure techniques(in Chinese)," *Physics*, **15** [5] 305-10 (1986).
[2]J.Q.Fu et al, "Introduction to experimental static equations(in Chinese)," Science Press, Beijing, 1986.
[3]L.M.Barker and D.D.Scott, "Development of a high-pressure quasi-isentropic plane wave generating capability," *SAND* 84-0432 (1984).
[4]L.C.Chhabildas, J.R.Asay and L.M.Barker, "Shear strength of tungsten under shock- and quasi-isentropic loading to 250GPa," *SAND* 88-0306 (1988).
[5]L.C.Chhabildas and L.M.Barker, "Dynamic quasi-isentropic compression of tungsten"; pp.111-114 in *Shock Waves in Condensed Matter 1987*, edited by S.C.Schmid and N.C.Holmes, Elsevier Science Publishers B.V. 1988.
[6]L.C.Chhabildas, J.R.Asay and L.M.Barker, "Dynamic quasi-isentropic loading of tungsten," *SAND* 89-0975C (1989).
[7]L.C.Chhabildas, L.N.Kmetyk, W.D.Reihart and C.A.Hall, "Enhanced hypervelocity launcher--capabilities to 16km/s," *International Journal of Impact Engineering*,**17** 183-94 (1995).
[8]F.Ding, S.H.Huang, F.Q.Jing, Y.B.Dong and Z.R.Li, "Quasi-isentropic compression characteristics of oxygen free copper," *Chinese Journal of High Pressure Physics*, **4** [2] 150-5 (1990).
[9]Q.Shen, L.M.Zhang, H.P.Xiong, J.S.Hua and H.Tan, "Fabrication of W-Mo-Ti system flier-plate with graded impedance for generating quasi-isentropic compression," *Chinese Science Bulletin*, **45** [15] 1421-4 (2000).

THERMAL CYCLING BEHAVIOUR OF A Cu/Al$_2$O$_3$ - FUNCTIONALLY GRADED MATERIAL

A. Neubrand
University of Technology Darmstadt
Department of Materials Science
Petersenstr. 23
64287 Darmstadt
Germany

A. Kawasaki
Tohoku University
Department of Materials Processing
Materials Systems and Design
Sendai 980-8579
Japan

Y.Y. Yang
Institut für Materialforschung
Forschungszentrum Karlsruhe
Postfach 3640
DE-76021 Karlsruhe
Germany

ABSTRACT

Cu/Al$_2$O$_3$ FGMs were produced by infiltration of Cu into porous graded Al$_2$O$_3$, and their thermal cycling behavior was investigated for two different gradation functions by heating with a H$_2$/O$_2$ flame from the Al$_2$O$_3$ side. At low heat fluxes, cracks perpendicular to the surface were initiated. Critical heat fluxes for surface crack initiation were 1.5-2 MW/m^2 with only slight differences between different gradation profiles. At higher heat fluxes additional horizontal cracks were formed at the interface between the ceramic coating and the Cu/Al$_2$O$_3$ composite. The formation of this type of cracks was delayed for a joint with a linearly graded Cu/Al$_2$O$_3$ interlayer compared to a three- layer joint.

INTRODUCTION

Hasselman et. al predicted as early as 1978 that thermal stresses in a hollow tube subjected to outward heat flow can be substantially reduced by a properly designed thermal conductivity gradient[1]. It took almost ten years until more systematic theoretical studies addressing the issue of thermal stresses in graded materials were carried out. Many of these earlier papers dealt with unconstrained plates or beams. For this kind of geometry , a considerable reduction of stresses near the interface was predicted by elastic[2] or elastoplastic[3] analyses if the gradient extends over a large part of the plate's or beam's thickness. As a consequence, a reduced plastic strain accumulation during thermal cycling is expected. More recently it has been demonstrated by FEM that a concave porosity gradient in a porous SiC tube subjected to outward heat flow reduces the thermal stresses below the local strength during transient heating and in the steady state[4]. For components containing only thin graded layers, e.g. in graded coatings or joints, the overall force balance of the component is only slightly affected by the small graded zone. The stress situation outside the graded region is therefore practically unchanged. However, for a graded interface the stress concentration at the free edge of the interface is generally reduced[5]. In thermal cycling experiments, this should result in an increased resistance of such a graded joint to spallation.

Experimental studies on the performance of FGMs during thermal cycling is mainly restricted to thermal barrier coatings. In 1992 Henning et. al. demonstrated qualitatively the superior thermal shock resistance of a piston crown with a functionally graded thermal barrier coating[6]. PSZ thermal barrier coatings are in use for high temperature components of gas turbines. It has been demonstrated that the cyclic spallation life at a given temperature difference across the coating is considerably increased in a graded versus a duplex PSZ coating[7].

Whereas most experimental investigations on the thermal cycling behavior of FGMs focussed on thermal barriers, the present paper investigates Cu/Al_2O_3, which is a potential heat sink material. This material combination may also serve in dies for wire drawing combining a hard surface with a high heat conductivity. Symmetric $Cu/Al_2O_3/Cu$ FGMs which are electrically insulating parallel to the gradient direction are suitable for compliant pads in thermoelectric conversion units[8]. Such symmetric $Cu/Al_2O_3/Cu$ FGMs have been produced by a spark plasma sintering process[9] by one of the authors. In the present study the cyclic thermal fracture behavior of Cu/Al_2O_3 FGMs produced by an infiltration process is studied and typical failure mechanisms are identified.

EXPERIMENTAL PROCEDURES

Cu/Al_2O_3 - FGMs were produced by infiltration of alumina preforms with graded porosity. The graded alumina preforms were fabricated by replication of compressed polyurethane foams as described earlier[10]. Infiltration of the preforms with molten Cu was carried out at 1350°C and 10 MPa pressure. Detailed information on the entire process can be found in Ref [11]. Cylindrical specimens of 14 mm diameter and a total length of 22 mm were prepared. One face of the cylinder consisted of an alumina layer whereas the bulk of the cylinder was Cu. In one sample group (B) the alumina top layer (1mm thickness) was bonded to the copper substrate by a linearly graded layer of 3.4 mm thickness consisting of eight single layers of equal thickness and increasing Cu volume fraction (Fig.1, left). In the other sample group (U) the alumina top layer (1.34 mm thick) was bonded to the copper substrate by a 3.4 mm thick alumina/copper composite layer containing 30vol. % Cu (Fig. 1, right).

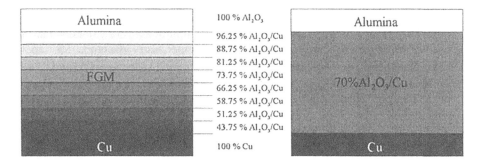

Fig. 1 Structure of graded specimen (B, on the left) and the three-layered specimen (U, on the right)

The alumina surface was cyclically heated in burner rig test system as described in Ref. 7. The back of the copper substrate was kept at 10°C during the whole experiment. Three thermocouples at a distance of 1, 4, and 7 mm from the bottom of the graded zone were used to monitor the temperature in the copper substrate, and the average heat flux was calculated from the measured temperature gradient. The surface temperature was determined using an infrared camera assuming an emissivity of alumina of 0.71 in the temperature range above 300°C. An acoustic emission sensor connected to the substrate holder was used to detect crack initiation and crack growth during thermal cycling.

A sequence of the burner rig test consisted in heating the sample to a given temperature, holding for three minutes and cooling down to room temperature. After three minutes the cycle was repeated with an increased sample surface temperature. Samples of both groups were cycled until first signs of spallation could be detected (acoustic emissions accompanied by a strong increase in thermal resistance). For some samples thermal cycling was stopped at earlier stages of damage for microstructural observation.

RESULTS AND DISCUSSION

A typical set of acoustic emissions recorded during the thermal shock experiment is reproduced in Fig. 2. In this figure the temperature difference ΔT is the difference between the average temperature recorded at the surface of the specimen and the temperature at the bottom of the graded layer which was calculated from the temperatures recorded by the thermocouples. For both types of sample, a similar acoustic emission pattern is observed : No acoustic emissions were observed below a critical value of ΔT. Above this critical value a small number of very strong acoustic emissions was observed during heating of the specimen, typically after the specimen reached thermal equilibrium. Immediately after these first emissions in heating, a larger number of acoustic emissions was observed during cooling, however the sound amplitudes were smaller than those of the heating cycle. The average critical temperature difference for the first acoustic emission activity was 207 K ± 9 K for graded specimens (B) and 198 K ± 36 K for the ungraded specimens (U). Hence, there was no significant difference in the damage threshold for the two types of design. The heat flux at this critical temperature was 1.74 ± 0.18 and 2.10 ± 0.18 MW/m², for type B and U, respectively. Obviously, type U specimens had a thermal resistance which was about 15 to 20 % lower than type B samples. Therefore, the critical heat flux was slightly higher for this type of samples at a comparable temperature difference. A few cycles after the first acoustic emissions were recorded, surface cracks could be observed in the radial direction of the samples. Apparently, they originated in the outer rim of the sample during heating. After a few more cycles they had propagated to the center of the sample.

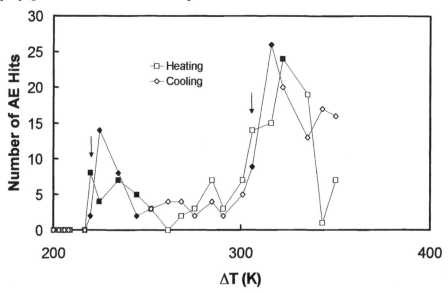

Fig. 2 Number of acoustic emissions per cycle in a B type specimen as a function of ΔT. Full data points reflect strong emissions exceeding 100 dB amplitude.

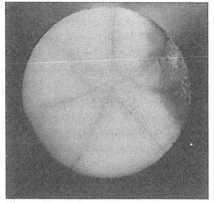

Fig.3a Top View ΔT 214 °C

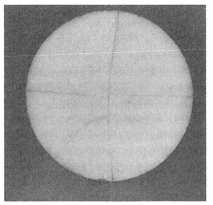

Fig 3b Top view ΔT 266°C

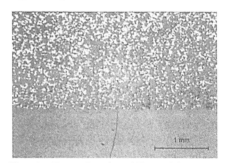

Fig. 4a Cross Section ΔT 214°C

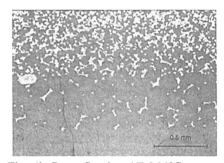

Fig. 4b Cross Section ΔT 266°C

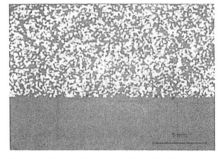

Fig. 5a Cross Section ΔT 303 K

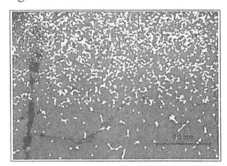

Fig. 5b Cross Section ΔT 373 K

Micrographs of Type U specimens at different stages of damage

Micrographs of Type B specimens at different stages of damage

Typically the acoustic emission activity decreased slightly and no strong emissions (caused by crack initiation) were observed in the propagation regime (near $\Delta T = 250K$ in Fig. 2). Specimen surfaces display a radial crack pattern in this stage (Fig. 3). The fact that the radial cracks do not meet exactly in the center of the sample also indicates crack growth from the rim towards the center of the specimen. No significant difference in the crack density was observed for sample group B compared to group U. However, there were differences in the cross section between these two samples : In sample group U the surface cracks penetrated only down to the interface between the alumina layer and the alumina/copper composite (Fig. 4a). At higher magnifications strong crack branching and ductile bridging are observed in the region where the crack reaches the interface. As a result the crack is arrested near the interface. In contrast, the crack is not detained at the ceramic/FGM interface in the graded sample (Fig. 4b). Although some ductile bridges are observed the crack penetrates about 2 mm into the graded layer. The crack is thus only arrested at a copper volume fraction which is similar to that in the type U sample (about 30%, the exact fraction depends on the maximum temperature difference). The total crack depth of the surface cracks was thus clearly longer in the graded sample B compared to sample U where the metal content increased discontinuously.

In contrast to earlier observations in TBCs where surface cracks were initiated during the cooling cycle [7], the surface cracks described above are initiated during the heating cycle. In the TBCs tensile stresses occur during cooling as a result of inelastic relaxation of compressive stresses at elevated temperature. In the investigated heat sink materials the maximum temperature at the surface of the samples when the cracking occurred was below 500°C and thus much too low for inelastic stress recovery. Hence, the tensile stresses must be of different origin. In the experimental arrangement the central part of the specimen surface is subjected to a much higher heat flux than the outer rim - thus a temperature gradient in the radial direction exists during the heating cycle. An elastoplastic FEM analysis including residual stresses was carried out assuming that a heat flux of 9.2 MW/cm^2 was concentrated in an area of 7 mm diameter in the center of the specimen surface. This corresponds to an average heat flux of 2.3 MW/m^2 across the entire specimen cross section. In the center of the surface of sample B a maximum tangential and radial compressive stress of 815 MPa was predicted. The calculated tensile stress at the outer edge the sample was 372 MPa and the maximum tensile stress of 451 MPa occurred at the surface, about 3mm away from the edge. The height of the tensile stresses did not depend very much on the exact distribution of heat flux as long as the rim of the sample was not heated. For sample U the maximum tangential and radial compressive stress at the same heat flux was 1040 MPa and the maximum tangential tensile stress at the surface was only 270 MPa. However, higher stresses occurred at the outer face of the specimen near the interface of the alumina top layer and the Cu/Al_2O_3 composite. The maximum tensile stress in this position was 395 MPa. As alumina has a flexural strength of about 400 MPa both types of specimen fail under a heat flux of about 2 MW/m², albeit the exact location of the crack initiation should be at the surface for the graded sample B and below the surface for sample U. This failure mechanism is also supported by the observation that the cracks initiate at the rim and grow only during the next cooling cycles towards the center of the sample. The growth of the initiated cracks towards the center of the surface occurs during the cooling cycle - indeed FEM predicts low tangential tensile stresses (below 200 MPa) at the whole sample surface in the uncracked state at the end of the cooling cycle. Similar tensile stresses probably exist also in the cracked sample and may cause crack propagation during cooling. This is supported by the fact that acoustic emissions during cooling are observed immediately after the first emissions during heating occurred. The crack patterns in Fig. 3 also suggest that the cracks have grown from the rim to the center of the sample.

At higher temperature differences, a second increase in acoustic emission activity with strong emissions was observed both in samples of type U and B (Fig.2). The critical temperature difference at which this emission occurred was about 270 K for samples of type U and 315 K for samples of type B. The surface of the samples did not show any significant changes after the onset of this second acoustic emission maximum - however the cross section of the samples showed a number of cracks parallel to the surface near the ceramic/composite interface (Fig.5a). Such horizontal cracks were also discovered in sample group B, but it appears that the crack density was much smaller as only very few cracks of this type could be found. The cracks propagated through the ceramic rich part of the graded region - typically the cracks were deflected towards the metal rich part of the gradient where they stopped (Fig. 5b). Obviously the graded structure has an improved resistance towards the formation of cracks parallel to the surface. The advantage of the graded structure B can be explained by a reduction of the stress concentration at pores or other defects in the interface region[12]. This improved performance of the graded structure may be even more pronounced in high cycle fatigue experiments as delamination cracks are deflected towards the metal rich region where they eventually are arrested.

CONCLUSION

It has been demonstrated that Cu/Al_2O_3 bimaterials can be exposed to heat fluxes of 1.5 MW/m² without being damaged. At higher heat fluxes surface cracks are initiated due to a lateral temperature gradient at the sample surface. This type of damage is not affected by a gradation. Cracks parallel to the surface were observed at higher heat fluxes. The initiation of these cracks was significantly delayed in the linearly graded material compared to a material with one interlayer.

The increased resistance towards horizontal crack formation found in this work is in agreement with earlier findings for graded PSZ/NiCrAlY - TBC's[7] and appears thus to be a general trend observed in thermal cycling of FGMs.

REFERENCES

[1] D.P.H. Hasselman and G.E. Youngblood, "Enhanced Thermal Stress Resistance of Structural Ceramics with Thermal Conductivity Gradient", J. Amer. Ceram. Soc. 61 49-52 (1978)

[2] K.S. Ravichandran, "Thermal Residual Stresses in a Functionally Graded Material System", Mater. Sci. & Eng. A201 269-276 (1995)

[3] A. E. Giannakopoulos, S. Suresh, M. Finot and M. Olsson, "Elastoplastic Analysis of Thermal Cycling : Layered Materials with Compositional Gradients", Acta Metall. Mater. 43, 1335-1354 (1995)

[4] M. Dröschel, R. Oberacker, M.J. Hoffmann, W. Schaller, Y.Y. Yang and D. Munz, "Silicon Carbide Evaporator Tubes with Porosity Gradient Designed by Finite Element Calculations", in Functionally Graded Materials 1998, W.R. Kaysser (ed.), Trans Tech Publications, Switzerland, p.820-826

[5] Y.Y. Yang, "Stress Analysis in a joint with a functionally graded material under a thermal loading by using the mellin transform method", Int. J. Solids Structures 35, 1261-1287 (1998)

[6] W. Henning, C. Melzer, S. Mielke, "Ceramic Gradient Materials for Components in Combustion Engines", Metall 5, 436-439 (1992), in German

[7] A. Kawasaki and R. Watanabe, "Cyclic Thermal Fracture Behaviour and Spallation Life of PSZ/NiCrAlY Functionally Graded Thermal Barrier Coatings", in Functionally Graded Materials 1998, W.R. Kaysser (ed.), Trans Tech Publications, Switzerland, p.402-409

[8] M. Kambe, in Functionally Graded Materials 1998, "Design of High Energy Density Thermoelectric Energy Conversion Unit by Using FGM Compliant Pads"; in Functionally Graded M aterials 1998, W.R. Kaysser (ed.), Trans Tech Publications, Switzerland, p.653-658

[9] A.Ohtsuka; A. Kawasaki; R. Watanabe; U.T. Young, "Fabrication of Cu/Al$_2$O$_3$/Cu Symmetrical Functionally Graded Material by Spark Plasma Sintering Process", J.of the Japan Society of Powder and Powder Metallurgy, 45[3] 220-224 (1998)

[10] F.R. Cichocki Jr, K. P. Trumble, J. Rödel, "Tailored Porosity Gradients via Colloidal Infiltration of Compression Molded Sponges", J.Am.Ceram.Soc. 81 (1998), 1661-64

[11] A. Neubrand and A. Kawasaki, "A new manufacturing process for Al$_2$O$_3$/Cu FGMs", Proc. of the National Symposium on FGM '99, Sendai

[12] Y.Y. Yang, D. Munz, W. Schaller, "Effect of the stress jump at the interface of a joint on the failure behavior", Int. J. Fract. 87, L113-L188 (1997)

Fracture Mechanics Modeling

FRACTURE MECHANICS OF VISCOELASTIC FUNCTIONALLY GRADED MATERIALS

Glaucio H. Paulino and Z.-H. Jin
Department of Civil and Environmental Engineering
University of Illinois at Urbana-Champaign
Newmark Laboratory
205 North Mathews Avenue
Urbana, IL 61801, USA

ABSTRACT

In this paper, the basic equations of viscoelasticity in functionally graded materials (FGMs) are formulated. The "correspondence principle" is revisited and established for a class of FGMs where the relaxation moduli for shear and dilatation $\mu(x, t)$ and $K(x, t)$ take separable forms in space and time, i.e. $\mu(x, t) = \mu_0 \tilde{\mu}(x) f(t)$ and $K(x, t) = K_0 \tilde{K}(x) g(t)$, respectively, in which x is the position vector, t is the time, μ_0 and K_0 are materials constants, and $\tilde{\mu}(x)$, $\tilde{K}(x)$, $f(t)$ and $g(t)$ are nondimensional functions. The "correspondence principle" states that the Laplace transforms of the nonhomogeneous viscoelastic variables can be obtained from the nonhomogeneous elastic variables by replacing μ_0 and K_0 with $\mu_0 p \bar{f}(p)$ and $K_0 p \bar{g}(p)$, respectively, where $\bar{f}(p)$ and $\bar{g}(p)$ are the Laplace transforms of $f(t)$ and $g(t)$, respectively, and p is the transform variable. The final nonhomogeneous viscoelastic solution is realized by inverting the transformed solution. The "correspondence principle" is then applied to a crack in a viscoelastic FGM layer sandwiched between two dissimilar homogeneous viscoelastic layers under antiplane shear conditions. Results for stress intensity factors, including their time evolution, are presented considering the power law material model.

INTRODUCTION

One of the primary application areas of functionally graded materials (FGMs) is high temperature technology. Materials will exhibit creep and stress relaxation behavior at high temperatures. Viscoelasticity offers a basis for the study of phenomenological behavior of creep and stress relaxation. The elastic-viscoelastic correspondence principle (or elastic-viscoelastic analogy) is probably one of the most useful tools in viscoelasticity because the Laplace transform of the viscoelastic solution can be directly obtained from the corresponding elastic solution. In this paper, the basic equations of viscoelasticity in FGMs are formulated. The correspondence principle is revisited and established for a class of FGMs where the relaxation moduli for shear and dilatation $\mu(x, t)$ and $K(x, t)$ take the forms $\mu(x,t) = \mu_0\tilde{\mu}(x)f(t)$ and $K(x,t) = K_0\tilde{K}(x)g(t)$, respectively, where x is the position vector, t is the time, μ_0 and K_0 are materials constants, and $\tilde{\mu}(x)$, $\tilde{K}(x)$, $f(t)$ and $g(t)$ are nondimensional functions. The correspondence principle is then applied to a crack in a viscoelastic FGM layer sandwiched between two dissimilar homogeneous viscoelastic layers under antiplane shear conditions. An elastic crack problem of the composite structure is first solved and the correspondence principle is used to obtain stress intensity factors (SIFs) for the viscoelastic system.

BASIC EQUATIONS

The basic equations of quasi-static viscoelasticity of FGMs are

$$\sigma_{ij,j} = 0, \qquad \text{(equilibrium equation)} \tag{1}$$

$$\varepsilon_{ij} = (u_{i,j} + u_{j,i})/2, \qquad \text{(strain-displacement relation)} \tag{2}$$

$$s_{ij} = 2\int_0^t \mu(x, t-\tau)\frac{de_{ij}}{d\tau}d\tau, \quad \sigma_{kk} = 3\int_0^t K(x, t-\tau)\frac{d\varepsilon_{kk}}{d\tau}d\tau, \quad \text{(constitutive law)} \tag{3}$$

$$s_{ij} = \sigma_{ij} - \sigma_{kk}\delta_{ij}/3, \quad e_{ij} = \varepsilon_{ij} - \varepsilon_{kk}\delta_{ij}/3, \tag{4}$$

in which σ_{ij} are stresses, ε_{ij} are strains, s_{ij} and e_{ij} are deviatoric components of stress and strain tensors, u_i are displacements, δ_{ij} is the Kronecker delta, $x = (x_1, x_2, x_3)$, $\mu(x, t)$ and $K(x, t)$ are the appropriate relaxation functions, t is the time, the Latin indices have the range 1, 2, 3, and repeated indices imply the summation convention. Note that the relaxation functions also depend on spatial positions, whereas in homogeneous viscoelasticity, they are only functions of time, i.e. $\mu \equiv \mu(t)$ and $K \equiv K(t)$ (Christensen[1]).

The boundary conditions are given by

$$\sigma_{ij}n_j = S_i, \qquad \text{on } B_\sigma, \tag{5}$$

$$u_i = \Delta_i, \qquad \text{on } B_u, \tag{6}$$

where n_j are the components of the unit outward normal to the boundary of the body, S_i are the tractions prescribed on B_σ, and Δ_i are the prescribed displacements on B_u. Notice that B_σ and B_u are required to remain constant with time.

CORRESPONDENCE PRINCIPLE, REVISITED

In general, the correspondence principle of homogeneous viscoelasticity does not hold for FGMs. To circumvent this problem, here we consider a special classss of FGMs in which the relaxation functions have the following general form (Paulino and Jin[2])

$$\mu(x, t) = \mu_0\tilde{\mu}(x)f(t),$$

$$K(x, t) = K_0\tilde{K}(x)g(t). \tag{7}$$

The constitutive law (3) is then reduced to

$$s_{ij} = 2\mu_0\tilde{\mu}(x)\int_0^t f(t-\tau)\frac{de_{ij}}{d\tau}d\tau, \quad \sigma_{kk} = 3K_0\tilde{K}(x)\int_0^t g(t-\tau)\frac{d\varepsilon_{kk}}{d\tau}d\tau. \tag{8}$$

By assuming the material is initially at rest, the Laplace transforms of the basic equations (1), (2), (8), and the boundary conditions (5) and (6) are obtained as

$$\bar{\sigma}_{ij,j} = 0, \tag{9}$$

$$\bar{\varepsilon}_{ij} = (\bar{u}_{i,j} + \bar{u}_{j,i})/2, \tag{10}$$

$$\bar{s}_{ij} = 2\mu_0\tilde{\mu}(\boldsymbol{x})p\bar{f}(p)\bar{e}_{ij}, \tag{11}$$

$$\bar{\sigma}_{kk} = 3K_0\tilde{K}(\boldsymbol{x})p\bar{g}(p)\bar{\varepsilon}_{kk}, \tag{12}$$

$$\bar{\sigma}_{ij}n_j = \bar{\tilde{S}}_i, \quad \text{on } B_\sigma, \tag{13}$$

$$\bar{u}_i = \bar{\Delta}_i, \quad \text{on } B_u, \tag{14}$$

where a bar over a variable represents its Laplace transform, and p is the transform variable. It is seen that the set of equations (9) - (12), and conditions (13) and (14) have a form identical to those of nonhomogeneous elasticity with shear modulus $\mu = \mu_0\tilde{\mu}(\boldsymbol{x})$ and bulk modulus $K = K_0\tilde{K}(\boldsymbol{x})$ provided that the transformed viscoelastic variables are associated with the corresponding elastic variables, and $\mu_0 p\bar{f}(p)$ and $K_0 p\bar{g}(p)$ are associated with μ_0 and K_0, respectively. Therefore, the *correspondence principle* in homogeneous viscoelasticity still holds for the FGM with the material properties given in equation (7), i.e., *the Laplace transformed nonhomogeneous viscoelastic solution can be obtained directly from the solution of the corresponding nonhomogeneous elastic problem by replacing μ_0 and K_0 with $\mu_0 p\bar{f}(p)$ and $K_0 p\bar{g}(p)$, respectively. The final solution is realized upon inverting the transformed solution.*

Among the various models for graded viscoelastic materials are the *standard linear solid* defined by

$$\mu(\boldsymbol{x}, t) = \mu_\infty(\boldsymbol{x}) + [\mu_e(\boldsymbol{x}) - \mu_\infty(\boldsymbol{x})]\exp\left[-t/t_\mu(\boldsymbol{x})\right],$$

$$K(\boldsymbol{x}, t) = K_\infty(\boldsymbol{x}) + [K_e(\boldsymbol{x}) - K_\infty(\boldsymbol{x})]\exp\left[-t/t_K(\boldsymbol{x})\right], \tag{15}$$

the *power law model*

$$\mu(\boldsymbol{x}, t) = \mu_e(\boldsymbol{x})\left[t_\mu(\boldsymbol{x})/t\right]^q, \quad K(\boldsymbol{x}, t) = K_e(\boldsymbol{x})\left[t_K(\boldsymbol{x})/t\right]^q, \quad 0 < q < 1, \tag{16}$$

and the *Maxwell material*

$$\mu(\boldsymbol{x}, t) = \mu_e(\boldsymbol{x})\exp\left[-t/t_\mu(\boldsymbol{x})\right], \quad K(\boldsymbol{x}, t) = K_e(\boldsymbol{x})\exp\left[-t/t_K(\boldsymbol{x})\right], \tag{17}$$

where $t_\mu(\boldsymbol{x})$ and $t_K(\boldsymbol{x})$ are the relaxation times in shear and bulk moduli, respectively, and q is a material constant.

The discussion below indicates the revisions needed in the general models so that the correspondence principle holds.

• *Standard Linear Solid* (15). If the relaxation times t_μ and t_K are constant, if $\mu_e(\boldsymbol{x})$ and $\mu_\infty(\boldsymbol{x})$ have the same functional form, and if $K_e(\boldsymbol{x})$ and $K_\infty(\boldsymbol{x})$ have the same functional form, then the standard linear solid satisfies assumption (7).

• *Power Law Model* (16). If the relaxation times t_μ and t_K are independent of spatial position, then the assumption (7) is readily satisfied. Moreover, even if the relaxation times depend on the spatial position in (16), the correspondence principle may still be applied with some revision, which consists of taking the corresponding nonhomogeneous elastic material with the following properties

$$\mu = \mu_e(\boldsymbol{x})[t_\mu(\boldsymbol{x})]^q, \quad K = K_e(\boldsymbol{x})[t_K(\boldsymbol{x})]^q, \tag{18}$$

instead of $\mu = \mu_e(\boldsymbol{x})$ and $K = K_e(\boldsymbol{x})$.

• *Maxwell Material* (17). If the relaxation times t_μ and t_K are independent of spatial position, then (7) is satisfied.

A CRACK IN A LAYERED FGM SYSTEM UNDER ANTIPLANE SHEAR

Under antiplane shear conditions, the only nonvanishing field variables are

$$u_3(\boldsymbol{x}, t) = w(x, y, t),$$
$$\sigma_{31}(\boldsymbol{x}, t) = \tau_x(x, y, t), \quad \sigma_{32}(\boldsymbol{x}, t) = \tau_y(x, y, t),$$
$$2\varepsilon_{31}(\boldsymbol{x}, t) = \gamma_x(x, y, t), \quad 2\varepsilon_{32}(\boldsymbol{x}, t) = \gamma_y(x, y, t),$$

where $\boldsymbol{x} = (x_1, x_2) = (x, y)$. The basic equations of mechanics satisfied by these variables are

$$\partial \tau_x / \partial x + \partial \tau_y / \partial y = 0, \tag{19}$$
$$\gamma_x = \partial w / \partial x, \quad \gamma_y = \partial w / \partial y, \tag{20}$$
$$\tau_x = \int_0^t \mu(x, y, t - \tau) \frac{d\gamma_x}{d\tau} d\tau, \quad \tau_y = \int_0^t \mu(x, y, t - \tau) \frac{d\gamma_y}{d\tau} d\tau, \tag{21}$$

where $\mu(x, y, t)$ is the shear relaxation function.

In the present study of the crack problem, a *power law material model* is employed. The shear relaxation modulus for the FGM is assumed as

$$\mu = \mu_0 \exp(\beta\, y/h) \left[t_0 \exp(\delta y/h)/t\right]^q = \mu_0 \exp[(\beta + \delta q)\, y/h]\, (t_0/t)^q, \tag{22}$$

where $\mu_0, t_0, \beta, \delta, q$ are material constants and h is a scale length (see Figure 1).

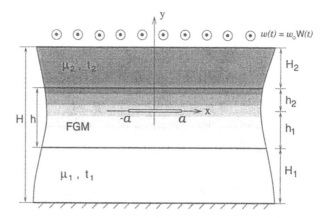

Figure 1: A viscoelastic FGM layer sandwiched between two dissimilar homogeneous viscoelastic layers occupies the region $|x| < \infty$ and $-H_1 - h_1 \leq y \leq H_2 + h_2$ with a crack at $|x| \leq a$ and $y = 0$. The lower boundary of the system $(y = -H_1 - h_1)$ is fixed and the upper boundary $(y = H_2 + h_2)$ is subjected to the uniform antiplane displacement $w_0 W(t)$.

An infinitely long viscoelastic FGM system containing a crack of length $2a$ is shown in Figure 1. The FGM layer is sandwiched between two dissimilar homogeneous viscoelastic layers. The system is fixed along the lower boundary $(y = -H_1 - h_1)$ and is displaced $w(t) = w_0 W(t)$ along the upper boundary $(y = H_2 + h_2)$, where w_0 is a constant, $W(t)$ is a nondimensional function of time t, H_1 and H_2 are the thicknesses of the bottom and top layers, respectively, and h_1 and h_2 (define the location of the crack) satisfy $h = h_1 + h_2$ in which h is the thickness of the FGM layer. *Notice that the model of Figure 1 allows solution of various coating problems* (see Paulino and Jin[3]). It is assumed that the crack lies on the x−axis from $−a$ to a and is of infinite extent in the $z−$ direction (normal to the $x − y$ plane). The crack surfaces remain traction free. The boundary conditions of the crack problem, therefore, are

$$w = 0, \qquad y = -(H_1 + h_1), \ |x| < \infty, \tag{23}$$

$$w = w_0 W(t), \qquad y = (H_2 + h_2), \ |x| < \infty, \tag{24}$$

$$\tau_y(x, -h_1^+) = \tau_y(x, -h_1^-), \quad w(x, -h_1^+) = w(x, -h_1^-), \qquad |x| < \infty, \tag{25}$$

$$\tau_y(x, h_2^+) = \tau_y(x, h_2^-), \quad w(x, h_2^+) = w(x, h_2^-), \qquad |x| < \infty, \tag{26}$$

$$\tau_y = 0, \qquad y = 0, \ |x| \le a, \tag{27}$$

$$\tau_y(x, 0^+) = \tau_y(x, 0^-), \quad w(x, 0^+) = w(x, 0^-), \qquad a < |x| < \infty. \tag{28}$$

The shear relaxation modulus of the FGM layer is given in (22). The relaxation functions for the two homogeneous viscoelastic layers are assumed as follows

$$\mu = \mu_i (t_i/t)^q = \mu_i (t_i/t_0)^q (t_0/t)^q, \qquad i = 1, 2, \tag{29}$$

where $\mu_i(i = 1, 2)$ are characteristic moduli, $t_i(i = 1, 2)$ are characteristic relaxation times, $i = 1$ represents the bottom layer and $i = 2$ the top layer.

By considering *continuity of shear relaxation modulus across the interfaces between the homogeneous layers and the FGM layer*, the constants β, δ, μ_0 and t_0 in (22) can be expressed by the material properties in the homogeneous layers as follows

$$\beta = \ln(\mu_2/\mu_1), \quad \delta = \ln(t_2/t_1), \quad \mu_0 = \mu_1(\mu_2/\mu_1)^{h_1/h}, \quad t_0 = t_1(t_2/t_1)^{h_1/h}. \tag{30}$$

According to the correspondence principle established previously, one can first consider a nonhomogeneous elastic material with the following shear modulus

$$\mu = \mu_0(\mu_i/\mu_0)(t_i/t_0)^q : \qquad \text{bottom } (i{=}1) \text{ and top } (i{=}2) \text{ layers} \tag{31}$$

$$\mu = \mu_0 \exp[(\beta + q\delta)(y/h)] : \qquad \text{FGM layer.} \tag{32}$$

Then the viscoelastic solution can be obtained by the correspondence principle.

The displacements $w(x, y)$ are harmonic functions in the two homogeneous layers, i.e.

$$\nabla^2 w = 0, \tag{33}$$

and are governed by the following equation in the nonhomogeneous (FGM) layer

$$\nabla^2 w + (\tilde{\beta}/h)(\partial w/\partial y) = 0 \qquad \text{with} \quad \tilde{\beta} = \beta + q\delta. \tag{34}$$

By using the Fourier transform method (see, for example, Erdogan et al.[4]), the boundary value problem (described by (23) to (28), (33) and (34)) can be reduced to the following singular integral equation

$$\int_{-1}^{1} [1/(s - r) + k(r, s)] \varphi(s) ds = -2\pi \tau_0/\mu_0, \quad |r| \le 1, \tag{35}$$

where τ_0 is given by

$$\tau_0 = \mu_0 \tilde{\beta} w_0 / \{h[\exp(\tilde{\beta} h_1/h) - \exp(\tilde{\beta} h_2/h) + (H_1/h)(\tilde{\beta}/\tilde{\mu}_1) + (H_2/h)(\tilde{\beta}/\tilde{\mu}_2)]\}, \qquad (36)$$

$\tilde{\mu}_i$ are constants given by

$$\tilde{\mu}_i = (\mu_i/\mu_0)\,(t_i/t_0)^q \qquad i = 1, 2, \qquad (37)$$

the unknown density $\varphi(r)$ is the slope function

$$\varphi(x) = \partial[w(x, 0^+) - w(x, 0^-)]/\partial x, \qquad (38)$$

the nondimensional coordinates r and s are

$$r = x/a, \qquad s = x'/a, \qquad (39)$$

respectively, and the Fredholm kernel $k(r, s)$ is given by Paulino and Jin[3].
The function $\varphi(r)$ can be further expressed as (see Erdogan et al.[4])

$$\varphi(r) = \psi(r)/\sqrt{1 - r^2}, \qquad (40)$$

where $\psi(r)$ is continuous for $r \in [-1, 1]$. When $\varphi(r)$ is normalized by w_0/H, the elastic Mode III stress intensity factor (SIF) , K_{III}^e, is obtained as

$$K_{III}^e = -\mu_0(w_0/H)\sqrt{\pi a}\,\psi(1)/2. \qquad (41)$$

Here, $H = H_1 + H_2 + h$, is the total thickness of the FGM system (see Figure 1). It is noted that $\psi(1)$ depends on several material and geometric parameters including μ_2/μ_1, t_2/t_1, q, h/a, $h1/h$, H_1/h and H_2/h.
The SIF for the viscoelastic FGM system can be obtained using the correspondence principle between the elastic and the Laplace transformed viscoelastic equations as follows

$$K_{III} = -\mu_0(w_0/H)\sqrt{\pi a}\,\psi(1)F(t)/2, \qquad (42)$$

where the time-dependent portion of the SIF is given by

$$F(t) = \mathcal{L}^{-1}\left[t_0^q\,\Gamma(1 - q)\,p^q\,\bar{W}(p)\right], \qquad (43)$$

in which p is the Laplace transform variable, $\mathcal{L}^{-1}$ represents the inverse Laplace transform, $\bar{W}(p)$ is the Laplace transform of $W(t)$, and $\Gamma(\cdot)$ is the Gamma function.
For the Heaviside step function loading, $W(t) = H(t)$, where $H(t)$ is the Heaviside step function. In this case, the function $F(t)$ reduces to

$$F(t) = (t_0/t)^q. \qquad (44)$$

For the exponentially decaying loading, $W(t) = \exp(-t/t_L)$ where t_L is a positive constant measuring the time variation of loads. In this case, the function $F(t)$ becomes

$$F(t) = \left\{(t_0/t)^q - \frac{1}{t_L}\int_0^t (t_0/\tau)^q \exp[-(t - \tau)/t_L]d\tau\right\}. \qquad (45)$$

NUMERICAL RESULTS

By observing equation (42), one verifies that the SIF is a multiplification of three parts. The *first part* is a dimensional base, $\mu_0(w_0/H)\sqrt{\pi a}$; the *second part* is a geometrical and material nonhomogeneity correction factor, $-\psi(1)/2$, which can be obtained from the numerical solution of the singular integral equation (35); and the *third part* is the time evolution of SIF, $F(t)$, which is obtained analytically from the inverse Laplace transform. The numerical solution of the integral equation (35) is obtained by a Chebyshev polynomial expansion method and a collocation technique (see, for example, Erdogan *et al.*[4]).

Figures 2-4 show the normalized SIF (see (42)), $-\psi(1)/2$, versus the nondimensional crack length $2a/h$ considering various modulus ratios μ_2/μ_1 and relaxation time ratios t_2/t_1. The geometric parameters are taken as $H_1 = H_2 = h$ and $h_1 = 0.5h$, i.e. the crack is located at the center of the layered structure. The solution is valid for both Heaviside step function loading and exponentially decaying loading. In general, the normalized SIF decreases with increasing $2a/h$. This occurs because the normalization includes the parameter $1/\sqrt{a}$ in the denominator (cf. (42)). The SIF is always lower than that of the corresponding homogeneous material ($\mu_2 = \mu_1$ and $t_2 = t_1$). Further SIF results can be found in Paulino and Jin[3].

Figure 5 illustrates the time evolution of the normalized SIF, $F(t)$, considering both Heaviside step function loading and exponentially decaying loading (see(44) and (45)). It is evident that, under the fixed displacement condition, SIFs decrease monotonically with increasing time. By observing the plots in Figure 5, one notices that, for time-dependent loading, the SIF can become negative as the ratio t_L/t_0 decreases, which occurs, for example, for $t_L/t_0 = 1.0$. This happens because of stress relaxation for long-time behavior. It is noted that a negative SIF is allowed without violating the crack face traction free conditions. The crack faces are not closed, but just slide in the opposite direction.

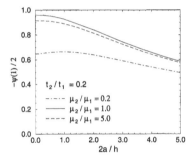 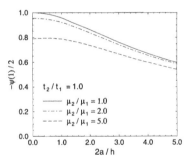

Figure 2: Normalized mode III SIF vs nondimensional crack length $2a/h$ for various shear modulus ratios μ_2/μ_1 and $t_2/t_1 = 0.2$.

Figure 3: Normalized mode III SIF vs nondimensional crack length $2a/h$ for various shear modulus ratios μ_2/μ_1 and $t_2/t_1 = 1.0$.

CONCLUSIONS

The correspondence principle is revisited and established for a class of FGMs where the relaxation functions for shear and dilatation take separable forms in space and time, i.e. $G_1(x,t)/2 = \mu(x,t) = \mu_0\tilde{\mu}(x)f(t)$ and $G_2(x,t)/3 = K(x,t) = K_0\tilde{K}(x)g(t)$, respectively. The correspondence principle states that the Laplace transforms of the nonhomogeneous

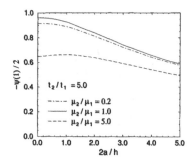

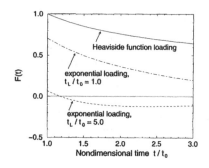

Figure 4: Normalized mode III SIF vs nondimensional crack length $2a/h$ for various shear modulus ratios μ_2/μ_1 and $t_2/t_1 = 5.0$.

Figure 5: Time variation of normalized mode III SIF.

viscoelastic variables can be obtained from the nonhomogeneous elastic variables by replacing μ_0 and K_0 with $\mu_0 p \bar{f}(p)$ and $K_0 p \bar{g}(p)$, respectively, where $\bar{f}(p)$ and $\bar{g}(p)$ are the Laplace transforms of $f(t)$ and $g(t)$, respectively, and p is the transform variable. The final nonhomogeneous viscoelastic solution is realized by inverting the transformed solution. Equivalently, if the creep functions $J_1(x,t)$ and $J_2(x,t)$ have separable forms in space and time, then the correspondence principle (as employed here) is also directly applicable. Following this theoretical development, the correspondence principle is then used to study a crack in a viscoelastic FGM layer embedded between two homogeneous viscoelastic layers under antiplane shear conditions. The elastic FGM crack problem is solved first and the correspondence principle between the elastic and the Laplace transformed viscoelastic equations is used to obtain SIFs for viscoelastic FGMs with position-dependent relaxation time.

ACKOWLEDGMENTS

We would like to acknowledge the support from the National Science Foundation (NSF) under grant No. CMS-9996378 (Mechanics & Materials Program).

REFERENCES

[1] R. M. Christensen, *Theory of Viscoelasticity*. Academic Press, New York, 1971.

[2] G. H. Paulino, and Jin, Z.-H., "Correspondence Principle in viscoelastic Functionally Graded Materials," *ASME Journal of Applied Mechanics*, **68**, 129-132 (2001).

[3] G. H. Paulino, and Jin, Z.-H., "A Crack in a Viscoelastic Functionally Graded Material Layer Embedded between Two Dissimilar Homogeneous Viscoelastic Layers – Antiplane Shear Analysis," *International Journal of Fracture*, 2001 (in press).

[4] F. Erdogan, G. D. Gupta, and T. S. Cook, "Numerical Solution of Singular Integral Equations"; pp. 368-425, in *Mechanics of Fracture*, Vol. 1, Edited by G. C. Sih. Noordhoff, Leyden, 1973.

STRESS INTENSITY FACTORS FOR A CRACK ARBITRARILY ORIENTED IN A FUNCTIONALLY GRADED LAYER

Sami El-Borgi
Associate Professor
Applied Mechanics Research Laboratory
Polytechnic School of Tunisia
La Marsa 2078, Tunisia

Fazil Erdogan
G. Whitney Snyder Professor
Dept. of Mech. Eng. and Mechanics
Lehigh University
Bethlehem, PA 18015, USA

Lotfi Hidri
Graduate Student
Applied Mechanics Research Laboratory
Polytechnic School of Tunisia
La Marsa 2078, Tunisia

ABSTRACT

In this paper, we consider an infinitely long elastic layer made of a Functionally Graded Material (FGM) with an embedded center crack subjected to arbitrary crack surface tractions. The material property grading is assumed to be exponential. Both the direction of material property variation and crack orientation are assumed to be arbitrary. The medium is modeled as a nonhomogeneous elastic solid with an isotropic stress-strain law under plane strain or generalized plane stress conditions. Fourier transforms are used to convert the coupled Navier's equations into a system of singular integral equations with the crack surface displacements as density functions. The integral equations are solved numerically to yield the displacement field in the medium. The primary objective of the paper is to study the effect of the nonhomogeneity parameters and the direction of material property variation on the crack tip stress intensity factors.

INTRODUCTION

Recently the materials research community has been exploring the possibility of using new concepts in coating design, such as Functionally Graded Materials (FGMs), as an alternative to the conventional homogeneous coatings. FGMs have promised attractive applications in a wide variety of thermal shielding problems such as high temperature chambers, furnace liners, turbines, micro-electronics and space structures [1]. Other important potential applications include their use as interfacial zones to improve the bonding strength and to reduce the residual and thermal stresses in bonded dissimilar materials [2] and as wear resistant layers in such components as gears, ball and roller bearings, cams and machine tools [3].

In designing components involving FGMs, an important aspect of the problem is the question of mechanical failure, specifically the fracture failure. Experimental observations of cracking in FGMs were reported in recent tests especially in the ceramic rich part of the material [4]. The fatigue and fracture characterization of materials and the related analysis require the solution of certain standard crack problems.

From the viewpoint of applied mechanics, FGMs can be mathematically modeled as nonhomogeneous materials. During the past two decades, a number of researchers considered various crack problems in nonhomogeneous media [5,6]. These studies among others provided the basis for the fracture mechanics research on FGMs. [3] gave a brief discussion of the elementary concepts of fracture mechanics in nonhomogeneous materials and identified a number of typical problem areas relating to the fracture of FGMs. [7] showed theoretically that for an FGM the stress distribution near the crack tip is of the same form as for a homogeneous material (at least asymptotically) provided the model used for the material gradient is a continuously differentiable function. Linear elastic fracture mechanics theory could, therefore, be applied to FGMs and the notion of stress intensity factors could be defined. [8] studied the problem of interface fracture between an FGM coating and a metallic substrate of finite thickness subject to mechanical loading. [9] considered the axisymmetric mixed-mode crack problem in an infinite FGM medium with elastic properties varying in the axial direction only. The same problem was studied by [10] under the assumption of plane-strain or generalized plane-stress conditions with a material gradient arbitrarily oriented with respect to the crack axis. [11] considered the plane stress/strain problem of an FGM layer with an embedded or a surface crack perpendicular to the boundaries and subject to mechanical crack surface tractions.

In this paper, we consider an infinitely long elastic FGM layer of finite thickness with an internal crack parallel to the horizontal boundary subjected to mechanical stresses. The problem is modeled as a nonhomogeneous elastic medium with an isotropic stress-strain law under the assumption of plane strain or generalized plane stress conditions. The material properties are assumed to vary exponentially in a direction arbitrarily oriented with respect to the crack axis. This problem consists of solving the Navier's equations of elasticity subject to the appropriate boundary conditions. Fourier transforms are used to convert the two coupled partial differential equations into a two coupled singular integral equations which are then solved numerically to yield, the displacement field in the layer. The main purpose of this paper is to study the effect of the nonhomogeneity parameters and the direction of material property variation on the crack tip stress intensity factors.

FORMULATION OF THE PROBLEM

The problem under consideration is described in Figure (1). The medium is unbounded in the x-direction which is the horizontal axis. A crack of length 2a is located at the center of the medium and oriented along the x-direction. The medium is subjected to mechanical stresses applied as crack surface tractions which can be expressed in terms of the external mechanical loads. The material gradient varies along the x'-direction which is arbitrarily oriented by an angle θ with respect to the crack axis as shown in Figure (1). For this graded medium, the Poisson's ratio ν is assumed to be a constant because the effect of its variation on the crack-tip stress intensity factors was shown to be negligible [5,11].

The shear modulus in the FGM layer depends on the x'-coordinate only (or on both x and y coordinates) and is modeled by an exponential function as follows:

$$\mu = \mu_0 \exp(\beta x') = \mu_0 \exp(\beta_1 x + \beta_2 y) \tag{1}$$

where μ_0 is a material constant, β is a nonhomogeneity parameter and the parameters β_1 and β_2 are related to β using the following equations: $\beta_1 = \beta \cos\theta$ and $\beta_2 = \beta \sin\theta$

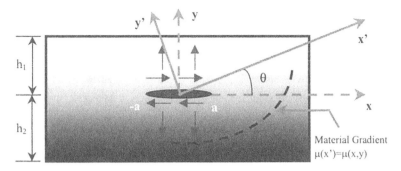

Figure 1. Notation for a plane crack in an FGM layer

The primary unknowns of this problem are the displacements u and v along the x and y-axis. The basic equations of the plane problem for nonhomogeneous isotropic elastic bodies consist of the equilibrium equations (ignoring body forces), the strain-displacement relationships and the linear elastic stress-strain law given respectively by:

$$\frac{\partial \sigma_{xx}}{\partial x} + \frac{\partial \sigma_{xy}}{\partial y} = 0 \qquad\qquad \frac{\partial \sigma_{xy}}{\partial x} + \frac{\partial \sigma_{yy}}{\partial y} = 0 \qquad (2)$$

$$\varepsilon_{xx} = \frac{\partial u}{\partial x} \qquad \varepsilon_{yy} = \frac{\partial v}{\partial y} \qquad \varepsilon_{xy} = \frac{1}{2}\left(\frac{\partial u}{\partial y} + \frac{\partial v}{\partial x}\right) \qquad (3)$$

$$\sigma_{xx} = \frac{\mu}{\kappa-1}\left[(1+\kappa)\varepsilon_{xx} + (3-\kappa)\varepsilon_{yy}\right] \qquad \sigma_{yy} = \frac{\mu}{\kappa-1}\left[(3-\kappa)\varepsilon_{xx} + (1+\kappa)\varepsilon_{yy}\right] \qquad \sigma_{xy} = 2\mu\varepsilon_{xy} \qquad (4)$$

where $\kappa = 3 - 4\nu$ for plane strain and $\kappa = \dfrac{3-\nu}{1+\nu}$ for generalized plane stress.

Substituting equations (3) into (4), inserting the resulting expressions into equation (2) and using (1) yields the following Navier's equations:

$$(\kappa+1)\frac{\partial^2 u}{\partial x^2} + (\kappa-1)\frac{\partial^2 u}{\partial y^2} + 2\frac{\partial^2 v}{\partial x \partial y} + \beta_1(\kappa+1)\frac{\partial u}{\partial x} + \beta_2(\kappa-1)\frac{\partial u}{\partial y} + \beta_1(3-\kappa)\frac{\partial v}{\partial y} + \beta_2(\kappa-1)\frac{\partial v}{\partial x} = 0 \qquad (5a)$$

$$(\kappa-1)\frac{\partial^2 v}{\partial x^2} + (\kappa+1)\frac{\partial^2 v}{\partial y^2} + 2\frac{\partial^2 u}{\partial x \partial y} + \beta_2(3-\kappa)\frac{\partial u}{\partial x} + \beta_1(\kappa-1)\frac{\partial u}{\partial y} + \beta_1(\kappa-1)\frac{\partial v}{\partial x} + \beta_2(\kappa+1)\frac{\partial v}{\partial y} = 0 \qquad (5b)$$

Navier's equations (5a) and (5b) are subject to the following boundary conditions:

$$\sigma_{xy}(x,0^+) = \sigma_{xy}(x,0^-) = \omega_1(x) \qquad\qquad |x| \le a \qquad (6)$$

$$\sigma_{yy}(x,0^+) = \sigma_{yy}(x,0^-) = \omega_2(x) \qquad\qquad |x| \le a \qquad (7)$$

$$\sigma_{xy}(x,y) = 0 \qquad\qquad y - h_1, |x| < +\infty \qquad (8)$$

$$\sigma_{yy}(x,y) = 0 \qquad\qquad y = h_1, |x| < +\infty \qquad (9)$$

$$\sigma_{xy}(x,y) = 0 \qquad\qquad y = -h_2, |x| < +\infty \qquad (10)$$

$$\sigma_{yy}(x,y) = 0 \qquad\qquad y = -h_2, |x| < +\infty \qquad (11)$$

$$\sigma_{xy}(x,0^+) = \sigma_{xy}(x,0^-) \qquad\qquad |x| > a \qquad (12)$$

$$\sigma_{yy}(x,0^+) = \sigma_{yy}(x,0^-) \qquad\qquad |x| > a \qquad (13)$$

$$u(x,0^+) = u(x,0^-) \qquad\qquad |x| > a \qquad (14)$$

$$v(x,0^+) = v(x,0^-) \qquad\qquad |x| > a \qquad (15)$$

Equations (6) and (7) describe, respectively, the applied tangential and normal crack surface tractions which can be expressed in terms of the external mechanical loads. Equations (8) to (11) represent the free stress conditions on the planes $y = h_1$ and $y = -h_2$. Equations (12) to (15) describe the continuity conditions of the displacement and stress fields along the crack axis outside the crack.

DISPLACEMENT FIELD

Equations (5a) and (5b) are solved using the Fourier transform and its inverse to yield the expressions of the displacement field above and below the crack:

$$u(x,y) = \int_{-\infty}^{+\infty}\!\left[\left\{C_1 e^{m_1 y} + C_2 e^{m_2 y} + C_3 e^{m_3 y} + C_4 e^{m_4 y}\right\}e^{-ix\lambda}d\lambda\right] \qquad 0 < y \le h_1 \qquad (16)$$

$$v(x,y) = \int_{-\infty}^{+\infty}\!\left[\left\{C_1 k_1 e^{m_1 y} + C_2 k_2 e^{m_2 y} + C_3 k_3 e^{m_3 y} + C_4 k_4 e^{m_4 y}\right\}e^{-ix\lambda}d\lambda\right] \qquad 0 < y \le h_1 \qquad (17)$$

$$u(x,y) = \int_{-\infty}^{+\infty}\!\left[\left\{C_5 e^{m_1 y} + C_6 e^{m_2 y} + C_7 e^{m_3 y} + C_8 e^{m_4 y}\right\}e^{-ix\lambda}d\lambda\right] \qquad -h_2 \le y < 0 \qquad (18)$$

$$v(x,y) = \int_{-\infty}^{+\infty}\!\left[\left\{C_5 k_1 e^{m_1 y} + C_6 k_2 e^{m_2 y} + C_7 k_3 e^{m_3 y} + C_8 k_4 e^{m_4 y}\right\}e^{-ix\lambda}d\lambda\right] \qquad -h_2 \le y < 0 \qquad (19)$$

where m_1, m_2, m_3 and m_4 are the roots of the characteristic polynomial associated with the partial differential equations (5a) and (5b), C_j (j=1,...,8) are unknown functions determined from the boundary conditions and k_j (j=1,...,4) are known constants given by:

$$k_j = \frac{[2i\lambda - \beta_1(3-\kappa)]m_j + i(1+\kappa)\lambda\beta_2}{(\kappa-1)m_j^2 + (\kappa-1)\beta_2 m_j - (\kappa+1)\lambda(\lambda+i\beta_1)}$$

Introducing the unknown density functions $\psi_1(x) = \frac{\partial}{\partial x}[u(x,0+) - u(x,0-)]$ and $\psi_2(x) = \frac{\partial}{\partial x}[v(x,0+) - v(x,0-)]$ and applying all boundary conditions yield two singular integral equations of the form

$$\int_{-a}^{+a}\left\{\left[\frac{1}{t-x} + k_{11}(x,t)\right]\psi_1(t) + [k_{12}(x,t)]\psi_2(t)\right\}dt = \frac{2\pi}{\mu_0\exp(\beta_1 x)}\omega_1(x) \qquad |x| \le a \qquad (20a)$$

$$\int_{-a}^{+a}\left\{[k_{21}(x,t)]\psi_1(t) + \left[\frac{1}{t-x} + k_{22}(x,t)\right]\psi_2(t)\right\}dt = \frac{2\pi(\kappa-1)}{\mu_0\exp(\beta_1 x)}\omega_2(x) \qquad |x| \le a \qquad (20b)$$

where the Fredholm kernels $k_{11}(x,t)$, $k_{12}(x,t)$, $k_{21}(x,t)$ and $k_{22}(x,t)$ depend on the parameters β_1 and β_2.

It was shown by [12] that the solutions of equations (20a) and (20b) may be expressed as $\psi_j(t) = \Psi_j(t)/\sqrt{1-t^2}$ (j=1,2) in which $\Psi_1(t)$ and $\Psi_2(t)$ are bounded and continuous functions in the interval [-a,a]. A suitable choice for these functions is to express them as a truncated series of Chebychev polynomials of the first kind. Using this solution and a suitable collocation method, the singular integral equations (20a) and (20b) are converted into a linear algebraic system whose unknowns are the coefficients of the series in the Chebychev polynomial. This system is solved numerically leading to the displacement field in the layer.

STRESS INTENSITY FACTORS

The stress intensity factors at both crack tips can be estimated from the expressions of the stress field as follows:

$$k_1(a) = \lim_{x \to a} \sqrt{2(x-a)}\,\sigma_{yy}(x,0) \qquad k_1(-a) = \lim_{x \to -a} \sqrt{2(-x-a)}\,\sigma_{yy}(x,0) \qquad (21a)$$

$$k_2(a) = \lim_{x \to a} \sqrt{2(x-a)}\,\sigma_{xy}(x,0) \qquad k_2(-a) = \lim_{x \to -a} \sqrt{2(-x-a)}\,\sigma_{xy}(x,0) \qquad (21b)$$

RESULTS AND DISCUSSION

The main objective of this study is to investigate the influence of the dimensionless material nonhomogeneity parameter βa, normalized dimensions h_1/a and h_2/a, the direction of material property grading θ and the loading conditions on the stress intensity factors. Some sample results on plane strain problems are shown in Figures 2-9. The external loads used in the examples are constant crack surface tractions $\sigma_{yy}(x,0) = -\sigma_0$; $\sigma_{xy}(x,0) = 0$ and $\sigma_{yy}(x,0) = 0$; $\sigma_{xy}(x,0) = -\tau_0$ or constant strain $\varepsilon_{yy}(x,h_1) = \varepsilon_{yy}(x,-h_2) = \varepsilon_0$ giving the crack surface tractions $\sigma_{yy}(x,0) = -\dfrac{8\mu_0}{1+\kappa}\varepsilon_0 e^{\beta\cos\theta}\cos^2\theta$; $\sigma_{xy}(x,0) = -\dfrac{8\mu_0}{1+\kappa}\varepsilon_0 e^{\beta\cos\theta}\sin\theta\cos\theta$ in which $|x| \le a$.

Figure 2 shows the effect of βa and the dimensions on the normalized mode I stress intensity factor k_1 at the crack tips $x = \pm a$ in an FGM layer under crack surface pressure $\sigma_{yy}(x,y) = -\sigma_0$. Since $\theta = 0$ and $h_1 = h_2$, $y = 0$ is a plane of symmetry and consequently $k_2(\pm a) = 0$. The results obtained for the limiting case of the homogeneous medium $\beta = 0$ as well as for $h_1 = h_2 = \infty$ agree rather well with that for of previous studies [5,10]. Note that for $\beta < 0$, $k_1(-a) > k_0 > k_1(a)$ in all cases, where k_0 is the corresponding homogeneous strip result and the material stiffness $\mu(x)$ is a monotonically decreasing function of x, that is, k_1 at the stiffer end of the crack $x = -a$ is greater than that at the less stiff end. This somewhat paradoxical result can be explained by observing that k_1 is proportional to the local values of $\mu(x)$ and $\dfrac{\partial v}{\partial x}(x,0)$ at the crack tips and for $\beta < 0$, $\dfrac{\partial v}{\partial x}$ at $x = -a$ is greater and $\dfrac{\partial v}{\partial x}$ at $x = +a$ is smaller than the local homogeneous values [5,6]. Figure 3 shows the results corresponding to constant strain loading $\varepsilon_{yy}(x,h_1) = \varepsilon_{yy}(x,-h_2) = \varepsilon_0$. The limiting cases of these results for $\beta = 0$ and $h_1 = h_2 = \infty$ agree well with the previous solutions [5,10].

The normalized stress intensity factors for constant strain loading are seen to deviate from the homogeneous results to a greater extent than the constant load results.

The stress intensity factors in the semi-infinite FGM loaded by crack surface pressure σ_0 as a function of the crack location h_1/a are shown in Figure 4. Note that as the crack approaches the free surface, that is, as $h_1/a \to 0$ the stress intensity factors become unbounded. In the other limiting case, for $h_1/a \to \infty$, k_1 and k_2 approach asymptotically the results obtained previously for the infinite medium [10]. For $\beta \neq 0$ the problem has no symmetry. Again for $h_1/a \to 0$ the stress intensity factors k_1 and k_2 become unbounded and as $h_1/a \to \infty$ they approach the known infinite plane results [10]. The corresponding results for constant strain loading (ε_0) are given in Figure 5.

The effect of the direction of the material property grading θ on the stress intensity factors is shown in Figures 6 and 7 for crack surface traction $\sigma_{yy} = -\sigma_0$, $\sigma_{xy} = 0$ and in Figures 8 and 9 for tractions $\sigma_{yy} = 0$, $\sigma_{xy} = -\tau_0$. As pointed out in [10], an interesting feature of these results is that in the case of constant pressure loading the angle θ corresponding to the maximum stress intensity factors is usually neither zero nor $\pi/2$.

As a conclusion it may be stated that deviations of the stress intensity factors in FGMs from the corresponding homogeneous values monotonically increase with the increasing material non-homogeneity parameter and the deviation could be considerably more pronounced for constant strain loading ε_0 than for constant traction σ_0. Also, generally the stress intensity factor at the crack tip embedded in a part of the medium with greater stiffness would be greater than that embedded in the less stiff part.

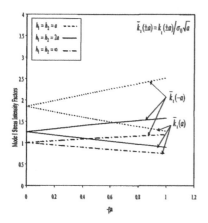

Figure 2. Effect of material nonhomogeneity and thickness on the mode I stress intensity factor in an FGM layer. $\theta = 0, \nu = 0.3$, loading: $\sigma_{yy}(x,0) = -\sigma_0, \sigma_{xy}(x,0) = 0, |x| < a$

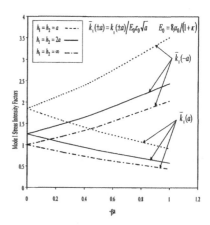

Figure 3. Effect of material nonhomogeneity and thickness on the mode I stress intensity factor in an FGM layer. $\theta = 0, \nu = 0.3$, constant strain loading: $\varepsilon_{yy}(x, \pm h_1) = \varepsilon_0$

Functionally Graded Materials 2000

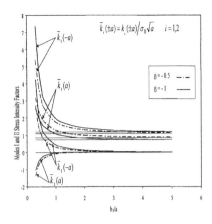

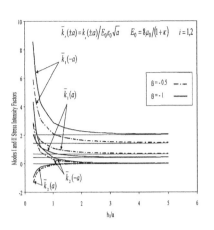

Figure 4. Effect of crack location h_1/a on the stress intensity factors in a semi-infinite graded medium. $\theta = 0$, $v = 0.3$, $h_2 = \infty$ (100a) loading: $\sigma_{yy}(x,0) = -\sigma_0$, $\sigma_{xy}(x,0) = 0$, $|x| < a$

Figure 5. Effect of crack location h_1/a on the stress intensity factors in a semi-infinite graded medium. $\theta = 0$, $v = 0.3$, $h_2 = \infty$ (100a) constant strain loading: $\varepsilon_{yy}(x,h_1) = \varepsilon_{yy}(x,-h_2) = \varepsilon_0$

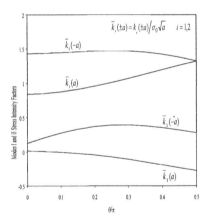

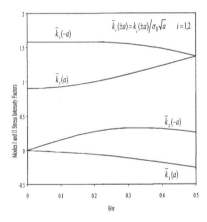

Figure 6. Effect of material property gradient direction θ on the stress intensity factors $v = 0.3$, $h_1 = 2a$, $h_2 = \infty$ (100a), $\beta a = -1$ loading: $\sigma_{yy}(x,0) = -\sigma_0$, $\sigma_{xy}(x,0) = 0$, $|x| < a$

Figure 7. Effect of material property gradient direction θ on the stress intensity factors $v = 0.3$, $h_1 = h_2 = 2a$, $\beta a = -1$ loading: $\sigma_{yy}(x,0) = -\sigma_0$, $\sigma_{xy}(x,0) = 0$, $|x| < a$

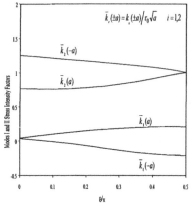

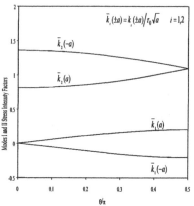

Figure 8. Effect of material property gradient direction θ on the stress intensity factors $\nu = 0.3$, $h_1 = 2a$, $h_2 = \infty$ (100a), $\beta a = -1$ loading: $\sigma_{yy}(x,0)=0$, $\sigma_{xy}(x,0) = -\tau_0$, $|x| < a$

Figure 9. Effect of material property gradient direction θ on the stress intensity factors $\nu = 0.3$, $h_1 = h_2 = 2a$, $\beta a = -1$ loading: $\sigma_{yy}(x,0)=0$, $\sigma_{xy}(x,0) = -\tau_0$, $|x| < a$

REFERENCES

[1] J. Holt, M. Koizumi, T. Hirai and Z.A. Munir, (Editors), "Functionally Gradient Materials," Ceramic Transactions, 34, The American Ceramic Society, Ohio, 1992.

[2] K. Kerrihara, K. Sasaki, K. and M. Kawarada, "Adhesion Improvement of Diamond Films"; pp. 65-90 in Proceedings of the First International Symposium on Functionally Graded Materials. Edited by Yamanouchi et al., Sendai, Japan, 1990.

[3] F. Erdogan, "Fracture mechanics of functionally graded materials," Composites Engineering, 5 [7] 753-770 (1995).

[4] J.J. Lanutti, "Functionally graded materials: properties, potential and design guidelines," Composites Engineering, 4, 81-94 (1994).

[5] F. Delale and F. Erdogan, "The crack problem for a nonhomogeneous plane," ASME Journal of Applied Mechanics, 50, 609-614 (1983).

[6] F. Erdogan, "The crack problem for bonded nonhomogeneous materials under antiplane shear loading," ASME Journal of Applied Mechanics, 52, 823-828 (1985).

[7] Z-H. Jin and N. Noda, "Crack-tip singular fields in nonhomogeneous materials," ASME Journal of Applied Mechanics, 61, 738-740 (1994).

[8] Y.F. Chen and F. Erdogan, "The interface crack problem for a nonhomogeneous coating bonded to a homogeneous substrate," Journal of Mechanics and Physics of Solids, 44, 771-787 (1996).

[9] M. Ozturk and F. Erdogan, "The axisymmetric crack problem in a nonhomogeneous medium," ASME Journal of Applied Mechanics, 44, 631-636 (1993).

[10] N. Konda and F. Erdogan, "The Mixed Mode Crack Problem in a Nonhomogeneous Elastic Medium," Engineering Fracture Mechanics, 47 [4] 533-545 (1994).

[11] F. Erdogan, F. and B.H. Wu, "The surface crack problem for a plate with functionally graded properties," ASME Journal of Applied Mechanics, 64, 449-456 (1997).

[12] F. Erdogan, G.D. Gupta and T.S. Cook, "Numerical solution of singular integral equations"; pp. 368-425 in Mechanics of Fracture, Edited by G.C. Sih. Norrdhoff, Leyden, 1973.

GRADIENT ELASTICITY THEORY FOR MODE I CRACK IN FUNCTIONALLY GRADED MATERIALS

Youn-Sha Chan
Department of Mathematics
University of California, Davis
Davis, CA 95616, USA

Glaucio H. Paulino
Department of Civil and Environmental Engineering
University of Illinois at Urbana-Champaign
Urbana, IL 61801, USA

Albert C. Fannjiang
Department of Mathematics
University of California, Davis
Davis, CA 95616, USA

ABSTRACT

High-order gradients play an important role in modeling fracture phenomena. Thus, an anisotropic strain-gradient elasticity theory with two characteristic lengths, ℓ and ℓ', associated to volumetric and surface elastic strain energy, respectively, is applied to fracture mechanics. The mode I fracture problem on a half-space is modeled and solved by means of a system of coupled hypersingular integrodifferential equations, which are discretized by means of the collocation method. Numerical results include displacement profiles and the stress intensity factors (SIFs) corresponding to various material parameters.

INTRODUCTION

Size (or scale) effects have been attracting intense attention in many recent applications in which various material length scales are involved and microstructure is considered. While there is no explicit microstructure information present in classical linear elastic fracture mechanics (LEFM), strain-gradient elasticity enhances the classical theory by introducing a characteristic length, which accounts for size effect and microstructure of the material. In this paper, strain-gradient elasticity with both volumetric and surface energy terms is applied to functionally graded materials (FGMs) in which the shear modulus assumes an exponential space variation. The mode I crack problem, formulated as a mixed-valued boundary condition problem, is solved by using the method of integral equations. It leads to a system of coupled hypersingular integrodifferential equations. The collocation method is employed in the numerical approximation of the governing coupled integral equations. Crack displacement profiles are obtained, and the (modified) stress intensity factors (SIFs) are derived and computed.

CONSTITUTIVE EQUATIONS OF GRADIENT ELASTICITY

The constitutive equations of gradient elasticity in mode I problems for a functionally graded material are (Paulino et al.[1], Exadaktylos[2]):

$$
\begin{cases}
\sigma_{zz} = \sigma_{yz} = 0, \quad \sigma_{xy} = \sigma_{yx} = 2G(x,y)\epsilon_{xy} - 2G(x,y)\ell^2\nabla^2\epsilon_{xy}, \\[4pt]
\sigma_{zz} = \lambda(x,y)(\epsilon_{xx} + \epsilon_{yy}) - \lambda(x,y)\ell^2\nabla^2(\epsilon_{xx} + \epsilon_{yy}), \\[4pt]
\sigma_{xx} = [\lambda(x,y) + 2G(x,y)]\epsilon_{xx} + \lambda(x,y)\epsilon_{yy} - \\
\qquad [\lambda(x,y) + 2G(x,y)]\ell^2\nabla^2\epsilon_{xx} - \lambda(x,y)\ell^2\nabla^2\epsilon_{yy}, \\[4pt]
\sigma_{yy} = [\lambda(x,y) + 2G(x,y)]\epsilon_{yy} + \lambda(x,y)\epsilon_{xx} - \\
\qquad [\lambda(x,y) + 2G(x,y)]\ell^2\nabla^2\epsilon_{yy} - \lambda(x,y)\ell^2\nabla^2\epsilon_{xx};
\end{cases}
\tag{1}
$$

$$
\begin{cases}
\mu_{xxx} = [\lambda(x,y) + 2G(x,y)]\ell^2\frac{\partial}{\partial x}\epsilon_{xx} + \lambda(x,y)\ell^2\frac{\partial}{\partial x}\epsilon_{yy}, \\[4pt]
\mu_{xyy} = [\lambda(x,y) + 2G(x,y)]\ell^2\frac{\partial}{\partial x}\epsilon_{yy} + \lambda(x,y)\ell^2\frac{\partial}{\partial x}\epsilon_{xx}, \\[4pt]
\mu_{yxx} = -[\lambda(x,y) + 2G(x,y)]\ell'\epsilon_{xx} - \lambda(x,y)\ell'\epsilon_{yy} + \\
\qquad [\lambda(x,y) + 2G(x,y)]\ell^2\frac{\partial}{\partial y}\epsilon_{xx} + \lambda(x,y)\ell^2\frac{\partial}{\partial y}\epsilon_{yy}, \\[4pt]
\mu_{yyy} = -[\lambda(x,y) + 2G(x,y)]\ell'\epsilon_{yy} - \lambda(x,y)\ell'\epsilon_{xx} + \\
\qquad [\lambda(x,y) + 2G(x,y)]\ell^2\frac{\partial}{\partial y}\epsilon_{yy} + \lambda(x,y)\ell^2\frac{\partial}{\partial y}\epsilon_{xx}, \\[4pt]
\mu_{xxy} = \mu_{xyx} = 2G(x,y)\ell^2\frac{\partial}{\partial x}\epsilon_{xy}, \\[4pt]
\mu_{yyx} = \mu_{yxy} = -2G(x,y)\ell'\epsilon_{xy} + 2G(x,y)\ell^2\frac{\partial}{\partial y}\epsilon_{xy}.
\end{cases}
\tag{2}
$$

Here, the shear modulus varies in both (x,y)-direction and it assumes the particular form $G \equiv G(x,y) = G_0 e^{\beta x + \gamma y}$, where, G_0, β, and γ are material constants; $\lambda \equiv \lambda(x,y) = [(3 - \kappa)/(\kappa - 1)]G(x,y)$, is also a function of (x,y), where $\kappa = 3 - 4\nu$ as plane strain is considered in this paper; ℓ is the characteristic length of the material responsible for volumetric strain-gradient terms, and ℓ' is responsible for surface strain-gradient terms; σ_{ij} is the stress tensor, and μ_{ijk} is the couple-stress tensor.

MIXED-VALUED BOUNDARY CONDITIONS PROBLEM

By imposing the equilibrium equations

$$
\frac{\partial \sigma_{xx}}{\partial x} + \frac{\partial \sigma_{xy}}{\partial y} = 0 \quad \text{and} \quad \frac{\partial \sigma_{xy}}{\partial x} + \frac{\partial \sigma_{yy}}{\partial y} = 0,
\tag{3}
$$

the following system of partial differential equations (PDEs) in terms of the x and y components of the displacement vector, u and v, respectively, is obtained:

$$
(1 - \ell^2\nabla^2)\left[(\kappa + 1)\frac{\partial^2 u}{\partial x^2} + (\kappa - 1)\frac{\partial^2 u}{\partial y^2} + 2\frac{\partial^2 v}{\partial x \partial y} + \beta(\kappa + 1)\frac{\partial u}{\partial x} + \right.
$$
$$
\left. \gamma(\kappa - 1)\frac{\partial u}{\partial y} + \gamma(\kappa - 1)\frac{\partial v}{\partial x} + \beta(3 - \kappa)\frac{\partial v}{\partial y}\right] = 0,
\tag{4}
$$

$$
(1 - \ell^2\nabla^2)\left[(\kappa - 1)\frac{\partial^2 v}{\partial x^2} + (\kappa + 1)\frac{\partial^2 v}{\partial y^2} + 2\frac{\partial^2 u}{\partial x \partial y} + \gamma(3 - \kappa)\frac{\partial u}{\partial x} + \right.
$$
$$
\left. \beta(\kappa - 1)\frac{\partial u}{\partial y} + \beta(\kappa - 1)\frac{\partial v}{\partial x} + \gamma(\kappa + 1)\frac{\partial v}{\partial y}\right] = 0.
\tag{5}
$$

Note that as the volumetric strain-gradient parameter $\ell \to 0$, the governing system of PDEs (4) and (5) becomes the classical one (Konda and Erdogan[3]). In this paper we consider the case that the shear modulus is a function of x only (see Figure 1), *i.e.* $G \equiv G(x) = G_0 e^{\beta x}$. Thus all the terms in equations (4) and (5) with coefficient γ in front will not appear.

Based on the principle of virtual work, the following mixed-valued boundary conditions are derived (Paulino *et al.*[1], Exadaktylos[2]):

$$\left\{ \begin{array}{lll} \sigma_{xy}(x,0^+) = 0 \,, & \mu_{yyx}(x,0^+) = 0 \,, & -\infty < x < \infty \\ \sigma_{yy}(x,0^+) = p(x) \,, & \mu_{yyy}(x,0^+) = 0 \,, & |x| < a \\ v(x,0^+) = 0 \,, & \frac{\partial}{\partial y}v(x,0^+) = 0 \,, & |x| > a \end{array} \right. \tag{6}$$

where $p(x)$ is the traction on crack surface $(-a,\ a)$ (see Figure 1), and $\partial v(x,0^+)/\partial y = 0$ can be interpreted as a rigid-substrate type of boundary condition. Thus, the mode I crack problem has been formulated as a mixed-valued boundary condition problem – governing PDEs (4) and (5) with boundary conditions (6).

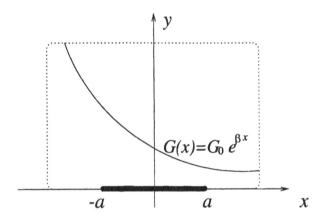

Figure 1: Geometry of the problem.

FOURIER TRANSFORM

Let the Fourier transform be defined by

$$\mathcal{F}(f)(\xi) = \hat{f}(\xi) = \frac{1}{\sqrt{2\pi}} \int_{-\infty}^{\infty} f(x) e^{ix\xi} dx \,, \tag{7}$$

then by the inversion formula for the Fourier transform,

$$\mathcal{F}^{-1}(\hat{f})(x) = f(x) = \frac{1}{\sqrt{2\pi}} \int_{-\infty}^{\infty} \hat{f}(\xi) e^{-ix\xi} d\xi \,. \tag{8}$$

Now let us assume that

$$u(x,y) = \frac{1}{\sqrt{2\pi}} \int_{-\infty}^{\infty} \hat{u}(\xi,y) e^{-ix\xi} d\xi \quad \text{and} \quad v(x,y) = \frac{1}{\sqrt{2\pi}} \int_{-\infty}^{\infty} \hat{v}(\xi,y) e^{-ix\xi} d\xi \,, \tag{9}$$

i.e. $u(x,y)$ and $v(x,y)$ are the inverse Fourier transform of the functions $\hat{u}(\xi,y)$ and $\hat{v}(\xi,y)$, respectively. Considering each term in equations (4) and (5), combined with equation (9), one obtains the following system of ordinary differential equations (ODEs) in terms of $\hat{u}(\xi,y)$ and $\hat{v}(\xi,y)$:

$$-\ell^2(\kappa-1)\frac{d^4\hat{u}}{dy^4} + \left[(\kappa-1) + 2\ell^2\kappa\xi^2 + i\ell^2\beta(\kappa+1)\xi\right]\frac{d^2\hat{u}}{dy^2} +$$

$$\left[(\kappa+1)\xi^2 + i\beta(\kappa+1)\xi + i\ell^2\beta(\kappa+1)\xi^3 + \ell^2(\kappa+1)\xi^4\right]\hat{u} +$$

$$\left[2i\ell^2\xi - \ell^2\beta(3-\kappa)\right]\frac{d^3\hat{v}}{dy^3} - \left[2i\xi - \beta(3-\kappa) + 2i\ell^2\xi^3 - \ell^2\beta(3-\kappa)\xi^2\right]\frac{d\hat{v}}{dy} = 0 , \quad (10)$$

and

$$\left[2i\ell^2\xi - \ell^2\beta(\kappa-1)\right]\frac{d^3\hat{u}}{dy^3} - \left[2i\xi - \beta(\kappa-1) + 2i\ell^2\xi^3 - \ell^2\beta(\kappa-1)\xi^2\right]\frac{d\hat{u}}{dy} -$$

$$\ell^2(\kappa+1)\frac{d^4\hat{v}}{dy^4} - \left[(\kappa+1) + 2\ell^2\kappa\xi^2 + i\ell^2\beta(\kappa-1)\xi\right]\frac{d^2\hat{v}}{dy^2} -$$

$$(\kappa-1)\left[i\beta\xi + \xi^2 + i\ell^2\beta\xi^3 + \ell^2\xi^4\right]\hat{v} = 0 . \quad (11)$$

The corresponding characteristic equation to the system of ODEs (10) and (11) can be found by solving the eigenvalues of an 8×8 matrix (see Paulino *et al.*[1]), and one can show that there are six roots, three with non-positive real parts, and three with non-negative real parts. Due to the far-field conditions

$$u(x,y) , \ v(x,y) \to 0 \ \text{as} \ \sqrt{x^2+y^2} \to \infty .$$

Taking into account the symmetry of the problem, the mode I problem is simplified by considering the upper-half plane ($y \geq 0$). Thus only the three roots with non-positive real parts are adopted, *i.e.*

$$\lambda_1(\xi) = -\frac{\beta\sqrt{\frac{3-\kappa}{\kappa+1}}}{2} - \frac{\sqrt{\frac{3-\kappa}{\kappa+1}\beta^2 + 4\xi^2} + i\beta\xi}{2} , \quad (12)$$

$$\lambda_2(\xi) = \frac{\beta\sqrt{\frac{3-\kappa}{\kappa+1}}}{2} - \frac{\sqrt{\frac{3-\kappa}{\kappa+1}\beta^2 + 4\xi^2} + i\beta\xi}{2} , \quad (13)$$

$$\lambda_3(\xi) = -\sqrt{\frac{\ell^2\xi^2 + 1}{\ell^2}} . \quad (14)$$

Then the general solutions of $\hat{v}(\xi,y)$ and $\hat{u}(\xi,y)$ take the following forms:

$$\hat{v}(\xi,y) = c_1(\xi)e^{\lambda_1(\xi)y} + c_2(\xi)e^{\lambda_2(\xi)y} + c_3(\xi)e^{\lambda_3(\xi)y}$$
$$\hat{u}(\xi,y) = d_1(\xi)e^{\lambda_1(\xi)y} + d_2(\xi)e^{\lambda_2(\xi)y} + d_3(\xi)e^{\lambda_3(\xi)y} . \quad (15)$$

According to the system of ODEs (10) and (11), one may find that

$$\hat{u}(\xi,y) = \eta_1(\xi)c_1(\xi)e^{\lambda_1(\xi)y} + \eta_2(\xi)c_2(\xi)e^{\lambda_2(\xi)y} + d_3(\xi)e^{\lambda_3(\xi)y} , \quad (16)$$

in which

$$\eta_1(\xi) = \frac{\lambda_1(\xi) + \lambda_2(\xi) - \beta\kappa\sqrt{\frac{3-\kappa}{\kappa+1}}}{2\,i\,\xi - \beta(\kappa-1)} , \quad (17)$$

$$\eta_2(\xi) = \frac{\lambda_1(\xi) + \lambda_2(\xi) + \beta\kappa\sqrt{\frac{3-\kappa}{\kappa+1}}}{2\,i\,\xi - \beta(\kappa - 1)}\,. \tag{18}$$

There are four unknown coefficients $c_1(\xi)$, $c_2(\xi)$, $c_3(\xi)$, and $d_3(\xi)$ to be determined by the boundary conditions (6).

FREDHOLM INTEGRAL EQUATIONS

By inverting the Fourier transform, one can express $v(x,y)$ and $u(x,y)$ as

$$v(x,y) = \frac{1}{\sqrt{2\pi}} \int_{-\infty}^{\infty} \left[c_1(\xi)e^{\lambda_1(\xi)y} + c_2(\xi)e^{\lambda_2(\xi)y} + c_3(\xi)e^{\lambda_3(\xi)y} \right] e^{-ix\xi} d\xi\,, \tag{19}$$

$$u(x,y) = \frac{1}{\sqrt{2\pi}} \int_{-\infty}^{\infty} \left[\eta_1(\xi)c_1(\xi)e^{\lambda_1(\xi)y} + \eta_2(\xi)c_2(\xi)e^{\lambda_2(\xi)y} + d_3(\xi)e^{\lambda_3(\xi)y} \right] e^{-ix\xi} d\xi\,. \tag{20}$$

By choosing the density functions

$$\phi(x) = \frac{\partial}{\partial x}v(x,0) \quad \text{and} \quad \psi(x) = \frac{\partial}{\partial y}v(x,0)\,, \tag{21}$$

and using the boundary conditions (6), one may obtain a system of integral equations in the limit form (Paulino $et\ al.$[1]):

$$\begin{cases} \lim_{y \to 0} \frac{1}{2\pi} \int_{-a}^{a} \int_{-\infty}^{\infty} \left[\mathcal{K}_{11}(\xi,y)\phi(t) + \mathcal{K}_{12}(\xi,y)\psi(t) \right] e^{i(t-x)\xi}\, d\xi\, dt = \frac{p(x)}{G(x)}, & |x| < a \\ \lim_{y \to 0} \frac{1}{2\pi} \int_{-a}^{a} \int_{-\infty}^{\infty} \left[\mathcal{K}_{21}(\xi,y)\phi(t) + \mathcal{K}_{22}(\xi,y)\psi(t) \right] e^{i(t-x)\xi}\, d\xi\, dt = 0, & |x| < a. \end{cases} \tag{22}$$

After performing asymptotic analysis and splitting the singularity out of the four kernels $\mathcal{K}_{ij}$, one reaches a system of hypersingular integrodifferential equations (Paulino $et\ al.$[1]):

$$\fint_{-a}^{a} \left[\frac{-16\ell^2}{\kappa+2} \frac{\phi(t)}{(t-x)^3} - \frac{4\ell^2\beta(\kappa+4)}{(\kappa+2)^2} \frac{\phi(t)}{(t-x)^2} \right] dt + \int_{-a}^{a} \frac{A_1}{t-x}\phi(t)dt + \int_{-a}^{a} K_{11}(x,t)\phi(t)dt$$
$$+ \int_{-a}^{a} K_{12}(x,t)\psi(t)dt + \frac{8\ell^2}{\pi(\kappa+2)}\psi''(x) + \frac{4\ell^2\beta(\kappa+4)}{\pi(\kappa+2)^2}\psi'(x) + B_1\psi(x) = \frac{\pi p(x)}{G(x)}, |x| < a, \tag{23}$$

$$\fint_{-a}^{a} \frac{2\ell^2(\kappa^2+\kappa+4)}{(\kappa+2)(\kappa-1)} \frac{\psi(t)}{(t-x)^2}dt + \int_{-a}^{a} D_2\frac{\psi(t)}{t-x}\,dt + \int_{-a}^{a} K_{22}(x,t)\,\psi(t)\,dt + D_1\psi(x)$$
$$+ \fint_{-a}^{a} \frac{2\ell'\kappa(\kappa-3)}{(\kappa+2)(\kappa-1)} \frac{\phi(t)}{t-x}dt + \int_{-a}^{a} K_{21}(x,t)\phi(t)dt + C_2\phi'(x) + C_1\phi(x) = 0,\ |x| < a\,, \tag{24}$$

where the six coefficients A_1, B_1, C_1, C_2, D_1, and D_2, the four kernels K_{ij} ($i,j = 1,2$), and the additional crack tip conditions, are described in detail by Paulino $et\ al.$[1].

NUMERICAL SOLUTION

By choosing the dimensionless variables

$$r = x/a\,, \quad s = t/a\,, \quad \tilde{\ell} = \ell/a\,, \quad \text{and} \quad \tilde{\ell}' = \ell'/a\,,$$

where a is the half crack length, the hypersingular integrodifferential equations (23) and (24) can be treated only on the normalized interval $(-1, 1)$. Based on the analysis of the asymptotics (Shi et al.[4]), the normalized density functions $\Phi(s)$ and $\Psi(s)$ can be expanded by the Chebyshev polynomials of the second kind $U_n(s)$:

$$\Phi(s) = \sqrt{1-s^2} \sum_{n=0}^{\infty} A_n U_n(s) , \quad \Psi(s) = \sqrt{1-s^2} \sum_{n=0}^{\infty} B_n U_n(s) . \tag{25}$$

Substituting the representation of $\Phi(s)$ and $\Psi(s)$ into the normalized version of integral equations (23) and (24), one may form a linear system of equations in terms of the Cauchy singular and hypersingular integrals. Evaluating the singular and hypersingular integrals (Kaya and Erdogan[5], Chan et al.[6]), computing the derivatives of the density functions, and numerically approximating (Gaussian quadrature) the non-singular integrals, one may discretize the integral equations (23) and (24) into an algebraic linear system of equations with unknowns A_n and B_n (see, for example, Erdogan and Gupta[7]).

STRESS INTENSITY FACTORS (SIFs)

The classical definition of SIFs in linear elastic fracture mechanics may not hold for strain-gradient elasticity because $\sigma_{yy}(x,0)$ has a stronger singularity. Thus, the SIFs will be redefined as (Paulino et al.[1]):

$$\ell K_I(a) = \lim_{x \to a^+} 2\sqrt{2\pi(x-a)}(x-a)\,\sigma_{yy}(x,0) , \tag{26}$$

$$\ell K_I(-a) = \lim_{x \to -a^-} 2\sqrt{2\pi(-x-a)}(-x-a)\,\sigma_{yy}(x,0) . \tag{27}$$

Therefore, the following formulas for the mode I SIFs in the strain-gradient elasticity may be derived:

$$\begin{aligned} \ell\, K_I(a) &= \lim_{x \to a^+} 2\sqrt{2\pi(x-a)}(x-a)\sigma_{yz}(x,0) \\ &= \lim_{r \to 1^+} 2\sqrt{2\pi(ar-a)}(ar-a)\sigma_{yz}(ar,0) \\ &= \frac{2a\sqrt{2\pi a}G_0}{\kappa+2} \lim_{r \to 1^+} e^{\beta ar}\sqrt{(r-1)}(r-1)\left(\frac{\ell}{a}\right)^2\left[\frac{1}{\pi}\int_{-1}^{1}\frac{-16\Phi(s)}{(s-r)^3}ds + 8\Psi''(r)\right] \end{aligned} \tag{28}$$

After cancellation of the common terms, equation (28) becomes (Paulino et al.[1]):

$$K_I(a) = -\frac{8\sqrt{\pi a}\,G_0}{\kappa+2}(\ell/a)e^{\beta a}\sum_{n=0}^{N}(n+1)(A_n - B_n) . \tag{29}$$

Similarly,

$$K_I(-a) = \frac{8\sqrt{\pi a}\,G_0}{\kappa+2}(\ell/a)e^{-\beta a}\sum_{n=0}^{N}(-1)^n(n+1)(A_n - B_n) . \tag{30}$$

Formulas (29) and (30) will be used to obtain numerical results for SIFs, in which N is the number of collocation points.

NUMERICAL RESULTS

Numerical solutions for the crack opening displacements and approximations to the stress intensity factors (SIFs) are reported in this paper. The (modified) stress intensity factors for various values of the material gradation parameter are given in Table 1. As βa increases, the SIF at the left tip decreases, and the SIF at the right tip increases. Figures 2 and 3 show the trend of the influence of the material nonhomogeneity (determined by the parameter β) on the crack opening displacements. If $\beta > 0$, then the crack profiles tilt to the left; otherwise ($\beta < 0$), the crack profiles tilt to the right. This trend is consistent with the classical results of Delale and Erdogan[8]. Figures 4 and 5 show the effect of the gradient parameters ℓ and ℓ' on the crack profiles, respectively. Figure 4 shows that the crack profiles increases as ℓ decreases. An analogous effect for the parameter ℓ' is shown in Figure 5. Notice that the negative values of ℓ' allow a more compliant fracture behavior than the positive ones. This is an important feature of the present gradient elasticity theory based on the Casal's continuum. The numerical outcome for the gradient parameters ℓ and ℓ' are consistent with the results from gradient elasticity in Mode III crack problem (Paulino et al.[9], Fannjiang et al.[10]). Other observations regarding fracture behavior in gradient elastic medium with functionally graded materials can be found in Paulino et al.[1].

Table 1: Normalized SIFs for a mode I crack under uniform loading, $p(x) = -p_0$. ($\ell/a = 0.1$, $\ell'/a = 0.01$, $\nu = 0.3$)

βa	$\frac{K_I(-a)}{p_0\sqrt{\pi a}}$	$\frac{K_I(a)}{p_0\sqrt{\pi a}}$
0.01	0.3166	1.0861
0.10	0.3078	1.1376
0.25	0.2864	1.2060
0.50	0.2535	1.3263
0.75	0.2246	1.4579
1.00	0.1996	1.6058

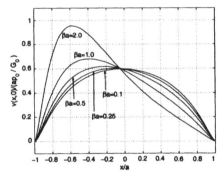

Figure 2 : Normalized crack profiles for $\beta a > 0$, $\tilde{\ell} = 0.10$, $\tilde{\ell}' = 0.01$; $\nu = 0.3$.

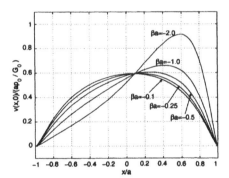

Figure 3 : Normalized crack profiles for $\beta a < 0$, $\tilde{\ell} = 0.10$, $\tilde{\ell}' = 0.01$; $\nu = 0.3$.

CONCLUSION

This paper has provided theoretical and numerical approaches for modeling mode I crack problems in FGMs using strain gradient elasticity that involves both volumetric and surface energy terms. The crack-tip singularity is not affected by the nonhomogeneity of the material, however, it is changed by the strain gradient elasticity (Casal's continuum) that has been used in modeling crack problems.

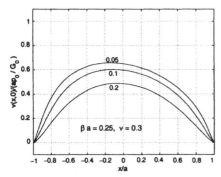

Figure 4: Normalized crack profiles for $\tilde{\ell}' = 0.01$; $\tilde{\ell} = 0.05, 0.1, 0.2$.

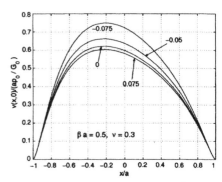

Figure 5: Normalized crack profiles for $\tilde{\ell} = 0.1$; $\tilde{\ell}' = 0.075, 0, -0.05, -0.075$.

ACKNOWLEDGMENTS

We would like to acknowledge the support from the National Science Foundation (NSF) under grants CMS-9996378 (previously CMS-9713798) and DMS-9971322.

REFERENCES

[1] G.H. Paulino, Y.-S. Chan and A.C. Fannjiang, "Gradient Elasticity Theory for a Mode I Crack in Both Homogeneous and Nonhomogeneous Materials," *To be Submitted for Journal Publication* (2001).

[2] G. Exadaktylos, "Gradient Elasticity with Surface Energy: Mode-I Crack Problem," *International Journal of Solids and Structures*, **35** [5] 421-456 (1998).

[3] N. Konda and F. Erdogan, "The Mixed Mode Crack Problem in a Nonhomogeneous Elastic Medium," *Engineering Fracture Mechanics*, **47** [4] 533-545 (1994).

[4] M.X. Shi, Y. Huang and K.C. Hwang, "Fracture in a Higher-order Elastic Continuum," *Journal of the Mechanics and Physics of Solids*, **48** [12] 2513-2538 (2000).

[5] A.C. Kaya and F. Erdogan, "On the Solution of Integral Equations with Strongly Singular Kernels," *Quarterly of Applied Mathematics*, **45** [1] 105-122 (1987).

[6] Y.-S. Chan, A.C. Fannjiang and G.H. Paulino, "Integral Equations with Hypersingular Kernels – Theory and Application to Fracture Mechanics," *International Journal of Engineering Science* (2002) (in press).

[7] F. Erdogan and G.D. Gupta, "On the Numerical Solution of Singular Integral Equations," *Quarterly of Applied Mathematics*, **30** 525-534 (1972).

[8] F. Delale and F. Erdogan, "The Crack Problem for a Nonhomogeneous Plane," *ASME Journal of Applied Mechanics*, **50** [5] 609-614 (1983).

[9] G.H. Paulino, A.C. Fannjiang and Y.-S. Chan, "Gradient Elasticity Theory for Mode III Fracture in Functionally Graded Materials – Part I. Crack Perpendicular to the Material Gradient," *ASME Journal of Applied Mechanics* (2002) (in press).

[10] A.C. Fannjiang, Y.-S. Chan and G.H. Paulino, "Strain Gradient Elasticity for Mode III cracks: A Hypersingular Integrodifferential Equation Approach" *SIAM Journal on Applied Mathematics* (2002) (in press).

CRACK AND CONTACT PROBLEMS IN FUNCTIONALLY GRADED MATERIALS

F. Erdogan and S. Dag
Department of Mechanical Engineering and Mechanics
Lehigh University
19 Memorial Drive West
Bethlehem, PA 18015

ABSTRACT

In this article the failure of FGM substrates due to sliding contact is considered. The first stage of the failure process involves crack initiation on the surface of FGM substrates or FGM coatings bonded to homogeneous substrates caused by the tensile stress due to friction forces. The second stage is the crack propagation under repeated loading. After a brief review of the related contact problem leading to the calculation of complete stress state on the surface, the coupled crack-contact problem for an FGM substrate loaded by a rigid stamp is considered and the stress intensity factors are calculated.

INTRODUCTION

Many of the present and potential applications of FGMs involve *contact problems*. These are mostly the load transfer problems between two deformable solids, generally in the presence of friction. In such applications the concept of FGMs appears to be ideally suitable to improve the surface properties and wear resistance of the components that are in contact. From the standpoint of failure mechanics an important aspect of contact problems is surface cracking which is caused by friction forces and which invariably leads to fretting fatigue. The main objective of this study is to investigate the problem of contact mechanics and the associated fracture phenomenon in FGM substrates or in FGM coatings bonded to homogeneous substrates subjected to repeated friction forces. The physical problem is the initiation and propagation of surface cracks under repeated sliding contact. The examination of crack initiation requires the determination of the complete stress state on the surfaces of contacting solids outside as well as within the contact region. This, in turn, requires the calculation of the in-plane stress $\sigma_{yy}(0, y)$ in addition to contact stresses σ_{xx} and σ_{xy}, $x = 0$ being the tangent plane to contacting surfaces (Figure 1). After crack initiation, the study of the subsequent subcritical crack propagation requires the evaluation of stress intensity factors. In this study it is assumed that the contacting solids are in relative motion and contact stresses are related by $\sigma_{xy} = \eta \sigma_{xx}$, η being the coefficient of friction. In this article after briefly reviewing the sliding contact problem and presenting some sample results for stresses on the surface, the main topic of the study namely the coupled crack/contact problem will be considered. Studies in contact mechanics originated with Hertz [1]. A thorough description of the underlying solid mechanics problems in homogenous materials may be found for example, in [?] The elasticity problem for an FGM half plane acted upon by a frictionless rigid stamp was considered in [3]. The corresponding axisymmetric rigid stamp problem for a semi infinite graded half space was studied in [4,5]. In this study after a brief description of the contact problems in FGMs, the crack/contact problem for a half plane will be formulated and some sample results showing the influence of the material nonhomogeneity parameter, the coefficient of friction, the stamp geometry and the relative dimensions on the stress intensity factors are presented.

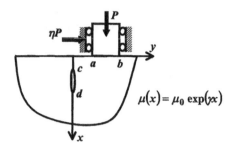

Figure 1. The basic geometry of crack-contact problem.

ON THE FORMULATION AND SOLUTION OF SLIDING CRACK-CONTACT PROBLEMS FOR FGMs

The general contact problem for a rigid stamp acting on an FGM subspace containing a crack is described in Figure 1. Initially it is assumed that the relative locations of the stamp and the crack are arbitrary and the crack is embedded in FGM, that is $b > a > 0$ and $h > d > c > 0$ where h is the thickness of the FGM layer if there is a homogenous substrate ($h < x < \infty$) or $h = \infty$ if the graded medium is infinitely thick. The shear modulus of FGM is assumed to be $\mu(x) = \mu_0 \exp(\gamma x)$. In the coupled problem shown in Figure 1 the unknown functions are the crack surface displacements and contact stresses defined by

$$f_1(x) = \frac{2\mu_0}{\kappa + 1} \frac{\partial}{\partial x}(v(x, 0^+) - v(x, 0^-)), \quad c < x < d, \tag{1}$$

$$f_2(x) = \frac{2\mu_0}{\kappa + 1} \frac{\partial}{\partial x}(u(x, 0^+) - u(x, 0^-)), \quad c < x < d, \tag{2}$$

$$f_3(y) = \sigma_{xx}(0, y) = \frac{1}{\eta}\sigma_{xy}(0, y), \quad a < y < b, \tag{3}$$

where η is the coefficient of friction and $\kappa = 3 - 4\nu$ for plane strain and $\kappa = (3 - \nu)/(1 + \nu)$ for generalized plane stress conditions, ν being the Poisson's ratio. The input functions are the crack surface tractions and stamp profile given by

$$\sigma_{yy}(x, 0) = 0, \quad \sigma_{xy}(x, 0) = 0, \quad c < x < d, \tag{4}$$

$$\frac{4\mu_0}{\kappa + 1} \frac{\partial}{\partial y}u(0, y) = f(y), \quad a < y < b. \tag{5}$$

By using equations of elasticity and the definitions given by (1) - (5), the mixed boundary value problem described in Figure 1 may be reduced to the following system of integral equations:

$$\int_c^d f_1(t)dt \int_0^\infty k_{11}(x,t,\rho)d\rho + \int_a^b f_3(t)dt \int_0^\infty k_{13}(x,t,\rho)d\rho = \sigma_{yy}(x,0), \quad c < x < d, \quad (6)$$

$$\int_c^d f_2(t)dt \int_0^\infty k_{22}(x,t,\rho)d\rho + \int_a^b f_3(t)dt \int_0^\infty k_{23}(x,t,\rho)d\rho = \sigma_{xy}(x,0), \quad c < x < d, \quad (7)$$

$$\int_c^d f_1(t)dt \int_0^\infty k_{31}(y,t,\rho)d\rho + \int_c^d f_2(t)dt \int_0^\infty k_{32}(y,t,\rho)d\rho + \int_a^b f_3(t)dt \int_0^\infty k_{33}(y,t,\rho)d\rho$$

$$= f(y), a < y < b, \quad (8)$$

$$\int_c^d f_1(t)dt = 0, \quad \int_c^d f_2(t)dt = 0, \quad \int_a^b f_3(t)dt = -P. \quad (9)$$

The singular nature of the integral equations and consequently the singular behavior of the unknown functions f_1, f_2, f_3 are determined by examining the singularities of the kernels given by the infinite integrals in (6)-(8). With the exception of one case that will be discussed below, the integrands k_{ij} are bounded and continuous for $\rho < \infty$ and integrable near $\rho = 0$. The singular nature of the kernels is, therefore, determined by examining the asymptotic behavior of the integrals as $\rho \to \infty$. Designating the asymptotic values of k_{ij} by k_{ij}^∞ it can be shown that for $c > 0$

$$\int_0^\infty k_{11}^\infty(x,t,\rho)d\rho = \int_0^\infty k_{22}^\infty(x,t,\rho)d\rho = \frac{1}{\pi}\frac{1}{t-x}, \quad c < (x,t) < d, \quad (10)$$

$$\int_0^\infty k_{33}^\infty(y,t,\rho)d\rho = -\frac{1}{\pi}\frac{1}{t-y} - \eta\frac{\kappa-1}{\kappa+1}\delta(t-y), \quad a < (y,t) < b, \quad (11)$$

and all other integrals involving k_{ij}, $(i \neq j)$ and $k_{ii} - k_{ii}^\infty$ are bounded. For $c = 0$ and $a > 0$ we have the standard edge crack and k_{11}^∞ and k_{22}^∞ would give the well-known generalized instead of simple Cauchy kernels [6].

In the problem under consideration there are two cases that may be of considerable theoretical interest. The first is the limiting case of $a = c = 0$. In this case by defining

$$f_1(t) = t^\alpha(d-t)^\delta g_1(t), \quad f_2(t) = t^\alpha(d-t)^\delta g_2(t), \quad f_3(t) = t^\alpha(b-t)^\beta g_3(t), \quad (12)$$

and using a function-theoretic method, the condition of boundedness of $\sigma_{yy}(x,0)$, $\sigma_{xy}(x,0)$, $0 < x < d$ and $f(y)$, $0 < y < b$ would give the following characteristic equations to determine δ, β and α

$$\cot(\pi\delta) = 0, \quad \cot(\pi\beta) + \eta\frac{\kappa-1}{\kappa+1} = 0, \quad (13)$$

$$\eta\left(4\alpha^2 + 10\alpha + 5 + (\kappa-1)\cos(\pi\alpha) + \kappa(2\alpha+3)\right) + (\kappa+1)\sin(\pi\alpha) = 0. \quad (14)$$

One may note that these results are independent of μ_0 and the material nonhomogeneity constant γ and dependent on η and κ only, meaning that the stress singularities for FGMs and homogeneous materials are identical. Generally the contact stresses are concentrated toward the trailing end of the stamp. For $c > 0$ it may easily be shown that [7]

$$f_1(x) = (x - c)^\theta (d - x)^\delta g_1(x), \quad f_2(x) = (x - c)^\theta (d - x)^\delta g_2(x), \quad \cot(\pi\theta) = 0, \quad \cot(\pi\delta) = 0,$$

$$f_3(y) = (y - a)^\omega (b - y)^\beta g_3(y), \quad \cot(\pi\omega) = \eta \frac{\kappa - 1}{\kappa + 1} = -\cot(\pi\beta), \quad \omega < 0, \quad \beta < 0. \quad (15)$$

Note that for $\eta > 0$ the stress singularity at $y = a$ is greater than that at $y = b$. For a flat stamp the singularities α, ω and β are shown in Figure 2 as functions of the friction coefficient η. From the standpoint of cracking $\eta > 0$ is the physically meaningful case for which α is real and, for high values of η, can be greater than the corresponding uncracked value ω. This somewhat unusual result given by (14) has also been verified independently by using Mellin Transforms.

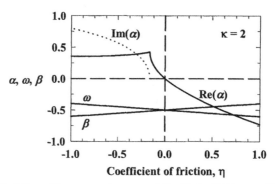

Figure 2. Variation of the exponents α, w and β with the friction coefficient η.

The second interesting case is concerned with an uncracked FGM half plane having a shear modulus $\mu(x) = \mu_0 \exp(\gamma x)$ and loaded by a rigid stamp. By letting $c = d$, the only remaining relevant term in (6) - (8) would be that involving k_{33} which can be expressed as

$$k_{33}(y, t, \rho) = \frac{1}{2\pi} \left[K_1(\rho) \sin(\rho(y - t)) + \eta K_2(\rho) \cos(\rho(y - t)) \right] \quad (16)$$

It can be shown that for $\gamma = 0$, $k_{33} = k_{33}^\infty$ (see (11)) and for $\gamma \neq 0$, as $\rho \to 0$ the functions K_1 and K_2 have the following asymptotic behavior

$$K_1(\rho) = a_1 \rho + a_3 \rho^3 + O(\rho^5),$$

$$a_1 = \frac{2(\kappa - 1)}{\gamma(\kappa + 1)\mu_0}, \qquad a_3 = \frac{8(\kappa - 3)(\kappa(1 + \text{sign}(\gamma)) - 4)}{\gamma^3(\kappa + 1)^2(1 + \text{sign}(\gamma))\mu_0}, \quad (17)$$

$$K_2(\rho) = b_2 \rho^2 + b_4 \rho^4 + O(\rho^6),$$

$$b_2 = \frac{4(2 - \kappa)}{\gamma^2(1 + \kappa)\mu_0}, \qquad b_4 = -\frac{16(\kappa^2(1 + \text{sign}(\gamma)) - \kappa(7 + 9\text{sign}(\gamma)) + 10 + 16\text{sign}(\gamma))}{\gamma^4(\kappa + 1)^2(1 + \text{sign}(\gamma))\mu_0}, \quad (18)$$

These results indicate that for $\gamma < 0$, $\text{sign}(\gamma) = -1$ and terms such as a_3 and b_4 become unbounded. Consequently, for a graded medium with an exponentially decaying stiffness the contact problem is not a well-posed problem. Physically the analogous problem is a homogeneous infinite strip of finite thickness under an unbalanced transverse force P which has no solution (see [3] for explanation). Thus, for FGM half planes with or without a crack if $\gamma < 0$ the contact problem has no solution. On the other hand if the medium consists of an FGM coating ($\gamma > 0$ or $\gamma < 0$) bonded to a homogeneous half plane, the analogous problem is contact problem for a homogeneous medium which has a closed form solution, hence the problem is well-posed. From (13)-(15) it may be observed that the characteristic roots δ, θ, β, ω and α are multiple-valued. The particular values of these exponents within the acceptable range (-1,+1) are determined from the index of the integral equations [9]. For the crack $\delta = -1/2$ and $\theta = -1/2$ for $c > 0$ and $\theta = 0$ for $c = 0$. For $a = c = 0$ and $\eta > 0$ the dominant and acceptable root of (14) is real and $\alpha < 0$. In the general stamp problem at an end point a (or b) ω (or β) is positive if the contact is smooth and negative if the stamp has a sharp corner [7].

Once the exponents δ, θ, β, ω and α are determined the weight functions $w_i(r)$ and the form of the solution of the integral equations may be obtained by normalizing the intervals $a < t < b$ and $c < t < d$ to $-1 < r < 1$ and by expressing the unknown functions as (see (1) - (3) and (6) - (8))

$$f_i(t) = w_i(r)\sum_0^\infty C_{in}P_n^{(\delta,\theta)}(r), \quad i = 1,2, \quad w_1(r) = w_2(r) = (1-r)^\delta(1+r)^\theta,$$

$$f_3(t) = w_3(r)\sum_0^\infty C_{3n}P_n^{(\beta,w)}(r), \quad w_3(r) = (1-r)^\beta(1+r)^\omega, \tag{19}$$

where P_n are the Jacobi polynomials associated with the weight functions w_j and C_{in} are unknown coefficients ($j = 1,2,3$). In the general form given by (19) it is assumed that $c > 0$ (and $-\infty < a < \infty$). In the special cases of ($c = a = 0$) and ($c = 0$, $a > 0$) we have ($\theta = \alpha$, $\omega = \alpha$) and ($\theta = 0$), respectively. The integral equations are solved by truncating the series in (19) and by using a suitable numerical method. After solving the integral equations, the quantities of physical interest, namely the stress intensity factors and the in-plane stress on the surface $\sigma_{yy}(0,y)$ may be obtained from

$$k_i(d) = -\lim_{x\to d} e^{\gamma x}\sqrt{2(d-x)}f_i(x), \quad k_i(c) = \lim_{x\to c} e^{\gamma x}\sqrt{2(x-c)}f_i(x), \tag{20}$$

$$\sigma_{yy}(0,y) = \sum_1^2 \int_c^d h_j(y,t)f_j(t)dt + \int_a^b h_3(y,t)f_3(t)dt, \tag{21}$$

where k_1 and k_2 are the modes I and II stress intensity factors and h_1, h_2, h_3 are the known kernels associated with the in-plane stress component $\sigma_{yy}(0,y)$. From the standpoint of crack initiation the critical point on the surface is the trailing end ($y = a$, Figure 1) of the contact region where the cleavage stress $\sigma_{\theta\theta}(r,\theta)$ is positive and may be obtained from

$$\sigma_{\theta\theta}(r,\theta) = \sigma_{xx}\sin^2(\theta) + \sigma_{yy}\cos^2(\theta) - \sigma_{xy}\sin(\theta)\cos(\theta), \tag{22}$$

In the notation of Figure 1, from (22) it may be shown that $\theta_{cr} = 0$, $\sigma_{\theta\theta cr} = \sigma_{yy}(0,a)$.

SAMPLE RESULTS AND DISCUSSION

The sliding contact problem for an FGM coating bonded to a homogeneous substrate is solved for four typical stamp profiles, namely the flat, circular, semi-circular and triangular [7]. In addition to contact stresses, the primary interest in these studies was, in the calculation of in-plane surface stresses which control the crack initiation. Figure 3 shows some sample results obtained from a flat and a cylindrical stamp. In this figure the stiffness ratio $\Gamma_3 = 1$ corresponds to a homogeneous half plane for which the solution exists in closed form. The stiffness of the FGM coating is given by $\mu(y) = \mu_{30}\exp(\gamma y)$, $\gamma = -(\log\Gamma_3)/h_3$. Note that the peak values of the in-plane stress on the surface, $\sigma_{xx}(x,0)$ increase with increasing Γ_3, where $\Gamma_3 = \mu_4/\mu_{30}$, $\sigma_{xx}(x,0)$ is tensile at and near the trailing end, and at the trailing end it is bounded for cylindrical stamp and has a singularity for the flat stamp, indicating that the trailing end of the contact region is a likely location of crack initiation. Extensive results for the FGM coated elastic solids (e.g., gear teeth, bearings), as well as the rigid stamps are given in [7].

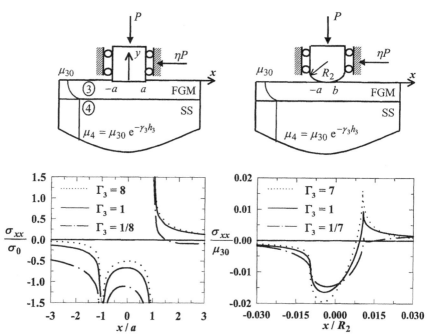

Figure 3. Stamp geometry and the in-plane stress σ_{xx} on the surface of an FGM coating; $\Gamma_3 = \mu_4/\mu_{30}$, $\eta = 0.3$; $\sigma_0 = P/2a$, $a/h_3 = 0.1$ for flat stamp, $(a+b)/R_2 = 0.01$, $R_2/h_3 = 50$ for cylindrical stamp

Figure 4 shows the stress intensity factors in a homogeneous half plane containing a surface crack and loaded by a moving flat stamp ($c = 0$, Figure 4, top row). The lines are calculated by using the fundamental functions given by (15) (with $c = 0$, $a > 0$), whereas the full circles for $a/d = 0$ are determined independently by using the fundamental functions given by (12). The trends that may be observed regarding the influence of η and a/d are self-explanatory. Note that in the absence of friction $k_1 < 0$, $k_2 > 0$ for all values of a/d and k_1 and k_2 vanish as $a/d \to \infty$ for

Functionally Graded Materials 2000

all values of η. Similar results for a homogeneous half plane with a surface crack loaded by a moving cylindrical stamp are shown in the bottom row of Figure 4.

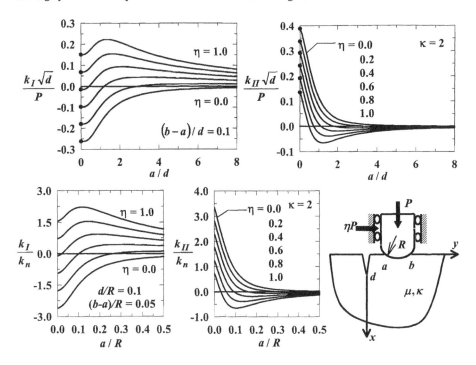

Figure 4. Stress intensity factors in a homogeneous half plane with a surface crack loaded by a flat stamp stamp, Figure 1 $c = 0$ (top row) and a cylindrical stamp, (bottom row), $k_n = (P/R)\sqrt{d}$.

Some sample results giving the stress intensity factors for an FGM half plane containing a surface crack and subjected to a sliding flat rigid stamp are shown in Figure 5 (Figure 1, $c = 0$). In Figures 5a and 5b the full circles are obtained from the homogeneous plane solution, whereas the corresponding lines are given by the FGM plane solution with $\gamma d = 0.0001$. The figures clearly show the strong influence of stamp location a/d and the material nonhomogeneity parameter γd on the stress intensity factors.

CONCLUDING REMARKS

In homogeneous substrates with FGM coatings, the peak value of in-plane surface stress increases with increasing stiffness ratio $\Gamma_3 = \mu_s/\mu_0$ where μ_s and μ_0 are the shear moduli of the substrate and FGM surface, respectively. The sliding contact problem for an FGM half plane with exponentially decaying stiffness has no solution. In crack/contact problems the stress intensity factors decrease as the stamp-to-crack distance increases. The coefficient of friction, the material nonhomogeneity parameter and the stamp-to-crack distance have a significant influence on the signs as well as the magnitudes of the stress intensity factors.

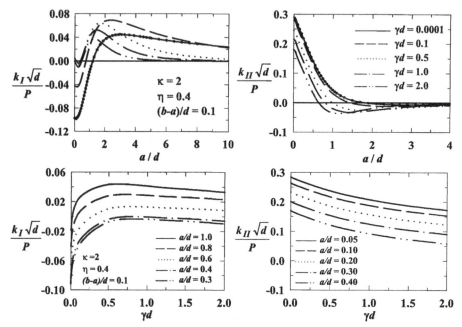

Figure 5. SIFs in an FGM half plane with a surface crack loaded by a rigid stamp (Figure 1 $c = 0$)

ACKNOWLEDGMENTS

This work was partially supported by AFOSR under the grant F49620-98-1-0028. The senior author also gratefully acknowledges the support provided by the Alexander von Humboldt Foundation during his stay at the FZK in Karlsruhe, Germany.

REFERENCES

[1] H. Hertz, "Uber die Beruhrung fester elastischer Korper," *J. reine und augewandte Mathematik*, **92** 156-171 1882.

[2] K. L. Johnson, Contact Mechanics, Cambridge University Press, 1985.

[3] I. Bakirtas, "The problem of a rigid punch on a nonhomogeneous elastic half space," *Int. J. Engng. Sci.*, **18** 597-610 1980.

[4] A. E. Giannakopoulos and S. Suresh, "Indentation of solids with gradients in elastic properties: Part II Axisymmetric indenters," *Int. J. Solids Structures*, **34** 2393-2428 1997.

[5] S. Suresh, A. E. Giannakopoulos and J. Alcala, "Spherical indentation of compositionally graded materials: Theory and experiments," *Acta Mater.*, **45** 1307-1321 1997.

[6] F. Erdogan and B. H. Wu, "Surface crack problem for a plate with functionally graded properties," *ASME Journal of Applied Mechanics*, **64** 449-456 1997.

[7] M. A. Guler, Contact Mechanics of FGM Coatings, Ph.D. Dissertation, Lehigh University, ME-MECH Department, Jan. 2001.

[8] M. Ratwani and F. Erdogan, "On the plane contact problem for a frictionless elastic layer," *Int. J. Solids Structures*, **9** 921-936 1973.

[9] F. Erdogan, "Mixed boundary value problems in mechanics"; pp. 1-86 in *Mechanics-Today*, **4**. Edited by S. Nemat-Nasser. Pergamon Press Inc. 1978.

TRANSIENT THERMAL STRESS ANALYSIS OF CRACKED FUNCTIONALLY GRADED MATERIALS

Z.-H. Jin and Glaucio H. Paulino
Department of Civil and Environmental Engineering
University of Illinois at Urbana-Champaign
Urbana, IL 61801, USA

ABSTRACT

An edge crack in a strip of a functionally graded material (FGM) is studied under transient thermal loading conditions. The FGM is assumed having constant Young's modulus and Poisson's ratio, but the thermal properties of the material vary along the thickness direction of the strip. Thus the material is elastically homogeneous but thermally non-homogeneous. This kind of FGMs include some ceramic/ceramic FGMs such as TiC/SiC and $MoSi_2/Al_2O_3$, and also some ceramic/metal FGMs such as zirconia/nickel and zirconia/steel. A multi-layered material model is used to solve the temperature field. By using the Laplace transform and an asymptotic analysis, an analytical first order temperature solution for short times is obtained. Thermal stress intensity factors (TSIFs) are calculated for a TiC/SiC FGM with various volume fraction profiles of the constituent materials. It is found that the TSIF could be reduced if the thermally shocked cracked edge of the FGM strip is pure TiC.

INTRODUCTION

The knowledge of thermal fracture behavior of functionally graded materials (FGMs) is important in order to evaluate their integrity. The existing analytical studies in this aspect have been mainly related to thermal stress intensity factors (TSIFs) for FGMs with specific material properties. For example, by assuming exponential variations in both elastic and thermal properties, Noda and Jin[1] and Erdogan and Wu[2] computed TSIFs for cracks in thermally loaded FGM strips. In this paper, an edge crack in an FGM strip under transient thermal loading conditions is studied. The FGM is assumed having constant Young's modulus and Poisson's ratio, but the thermal properties of the material vary along the thickness direction of the strip. A multi-layered material model is used to solve the temperature field. By using Laplace transform and asymptotic analysis, an analytical first order temperature solution for short times is obtained. Thus numerical inversion of the Laplace transform is not needed in the present work. Thermal stresses and TSIFs are calculated for a TiC/SiC FGM, and the effect of the volume fraction profiles of the constituent materials on thermal stresses and TSIFs is discussed.

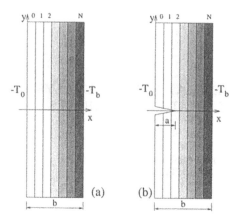

Figure 1: An FGM strip occupying the region $0 \leq X \leq b$ and $|Y| < \infty$. The bounding surfaces of the strip are subjected to uniform thermal loads T_0 and T_b. (a) a layered material; (b) an edge crack in the layered material.

TEMPERATURE FIELD

Consider an infinite strip of thickness b with a crack of length a as shown in Figure 1. The strip is initially at a constant temperature. Without loss of generality, the initial constant temperature can be assumed as zero. The surfaces $X = 0$ and $X = b$ of the strip are suddenly cooled down to temperatures $-T_0$ and $-T_b$, respectively. Since the heat will flow only in the $X-$ direction, the initial and boundary conditions for the temperature field are

$$T = 0, \quad t = 0; \qquad T = -T_0, \quad X = 0; \qquad T = -T_b, \quad X = b. \tag{1}$$

where T denotes the temperature field and t denotes time. Here an idealized thermal shock boundary condition is assumed, i.e., the heat transfer coefficient on the surfaces of the FGM strip is infinitely large. This corresponds to the most severe thermal stress induced in the strip.

The heat flow is controlled by the following conduction equation

$$\frac{\partial}{\partial X}\left[k(X)\frac{\partial T}{\partial X}\right] = \rho(X)c(X)\frac{\partial T}{\partial t}. \tag{2}$$

where $k(X)$ is the thermal conductivity, $\rho(X)$ the mass density and $c(X)$ the specific heat. Tanigawa et al.[3] used a laminated material model to solve (2). They modeled the FGM by a laminated composite and each lamina was assumed as a homogeneous layer. With this model, they were able to obtain the temperature field. The disadvantage of their analysis method is that to ensure convergence of the series solution, one has to numerically determine a sufficiently large number of the roots (eigenvalues) of a transcendental equation and the transcendental function is a determinant whose order is twice the number of layers, which may become very large if the FGM strip is to be reasonably modeled by a layered material. This is particularly true for small times at which the series solution converges very slowly. In this study, we also use a discrete model, i.e., the FGM strip is divided into many layers in the $X-$ direction, say $N + 1$ layers, as shown in Figure 1, and in each layer, the material properties are assumed as constants. However, we first determine the temperatures at the interfaces between layers and represent intra-layer temperatures by those interface temperatures without solving an eigenvalue problem. Further, an asymptotic temperature solution for small times is obtained here.

Since the layers are homogeneous, the temperature in the nth layer is well known as follows (Ozisik[4])

$$
\begin{aligned}
T(x^*, \tau) &= (1 - x^*)T_n(\tau) + x^* T_{n+1}(\tau) \\
&\quad -2\sum_{\ell=1}^{\infty} \frac{\sin(\ell\pi x^*)}{\ell\pi} \left[T_n(\tau) - \beta_{\ell n} \int_0^{\tau} \exp(-\beta_{\ell n}(\tau - \tau'))T_n(\tau')d\tau' \right] \\
&\quad +2\sum_{\ell=1}^{\infty}(-1)^{\ell} \frac{\sin(\ell\pi x^*)}{\ell\pi} \left[T_{n+1}(\tau) - \beta_{\ell n} \int_0^{\tau} \exp(-\beta_{\ell n}(\tau - \tau'))T_{n+1}(\tau')d\tau' \right]
\end{aligned}
$$

$$0 < x^* < 1, \qquad n = 0, 1, ..., N, \tag{3}$$

where $T_n(\tau)$ and $T_{n+1}(\tau)$ are the temperatures at the boundaries of the nth layer, x^* is the nondimensional local coordinate given by

$$x^* = (X - X_n)/h_n, \qquad h_n = X_{n+1} - X_n, \tag{4}$$

with X_n and X_{n+1} being the $X-$ coordinates of the two boundaries of the nth layer, τ is the nondimensional time

$$\tau = t\kappa_0/b^2, \tag{5}$$

and $\beta_{\ell n}$ is a constant defined by

$$\beta_{\ell n} = (\kappa_n/\kappa_0)(b/h_n)^2(\ell\pi)^2, \tag{6}$$

where κ_n is the thermal diffusivity of the layer given by

$$\kappa_n = k_n/(\rho_n c_n), \qquad n = 0, 1, ..., N, \tag{7}$$

in which k_n, ρ_n and c_n are the thermal conductivity, mass density and specific heat of the nth layer, respectively (note that k_n and κ_n are distinct quantities). Here it is understood that $X_0 = 0$ and $X_{N+1} = b$, and $T_0(\tau) = -T_0$ and $T_{N+1}(\tau) = -T_b$. The N unkown interface temperatures, $T_n(t)$, $(n = 1, 2, ..., N)$, are to be determined by the heat flux continuity conditions (e.g., Ozisik[4]; Gray and Paulino[5]) at the interfaces between the layers

$$k_{n-1}\frac{\partial T}{\partial X}\Big|_{X \to X_n^-} = k_n \frac{\partial T}{\partial X}\Big|_{X \to X_n^+}, \qquad n = 1, 2, ..., N. \tag{8}$$

This will result in a system of Volterra integral equations of $T_n(\tau)$. These equations are not appropriate to numerically determine the unknowns $T_n(\tau)$ because the series involved converge slowly and, more importantly, the convergence is dependent on the yet to be determined $T_n(\tau)$. To overcome this difficulty, we will study the problem in the Laplace transformed plane and try to first obtain the Laplace transforms of $T_n(\tau)$. The Laplace transforms of the equations to determine $T_n(\tau)$ are

$$
\begin{aligned}
&\frac{k_{n-1}}{k_n} \left\{ \bar{T}_n(s) \left[1 + 2\sum_{\ell=1}^{\infty} \frac{s}{s + \beta_{\ell(n-1)}} \right] - \bar{T}_{n-1}(s) \left[1 + 2\sum_{\ell=1}^{\infty}(-1)^{\ell} \frac{s}{s + \beta_{\ell(n-1)}} \right] \right\} \\
&= \frac{h_{n-1}}{h_n} \left\{ \bar{T}_{n+1}(s) \left[1 + 2\sum_{\ell=1}^{\infty}(-1)^{\ell} \frac{s}{s + \beta_{\ell n}} \right] - \bar{T}_n(s) \left[1 + 2\sum_{\ell=1}^{\infty} \frac{s}{s + \beta_{\ell n}} \right] \right\}
\end{aligned}
$$

$$n = 1, 2, ..., N, \tag{9}$$

in which $\bar{T}_n(s)$ is the Laplace transform of $T_n(\tau)$. It is noted that $\bar{T}_0(s) = -T_0/s$ and $\bar{T}_{N+1}(s) = -T_b/s$ since $T_0(\tau) = -T_0$ and $T_{N+1}(\tau) = -T_b$. After $\bar{T}_n(s)$ are solved from (9), we may use inverse Laplace transform to get $T_n(\tau)$ in the time domain. In general, numerical inversion has to be invoked. However, there are difficulties with the numerical inverse Laplace transform since the numerical algorithms are generally not stable. Here we propose an asymptotic solution of the interface temperatures $T_n(\tau)$ for small values of time (τ). Thus the equations (9) are first solved for large values of s. The solutions $\bar{T}_n(s)$ are then inverted analytically. The normalized interface temperatures $T_n(\tau)$ (normalized by T_0) for short time (τ) are finally obtained as follows

$$T_n(\tau) = T_n^{(1)}(\tau) + \left(\frac{T_b}{T_0}\right) T_n^{(2)}(\tau), \qquad T_n^{(1)}(\tau) = L_n^{(0)} \mathrm{erfc}\left(\frac{1}{2\sqrt{\tau}} \sum_{i=1}^{n} \frac{1}{\gamma_{i-1}}\right),$$

$$T_n^{(2)}(\tau) = P_n^{(0)} \mathrm{erfc}\left(\frac{1}{2\sqrt{\tau}} \sum_{i=n}^{N} \frac{1}{\gamma_i}\right), \qquad n = 1, 2, ..., N, \tag{10}$$

where $L_n^{(0)}$ and $P_n^{(0)}$ are known constants which can be found in Jin and Paulino[6], erfc($\cdot$) is the complementary error function, and γ_i is given by

$$\gamma_i = (b/h_i) \sqrt{\kappa_i/\kappa_0}, \qquad i = 0, 1, ..., N. \tag{11}$$

After obtaining the interface temperatures, the intra-layer temperatures (normalized by T_0) can be calculated by substituting (10) into (3). Usually the series in (3) converge slowly. As an alternative, a linear interpolation may be used to obtain the temperatures within the layers with satisfactory accuracy if a large number of layers is chosen. The intra-layer temperature is then obtained approximately as

$$T(x^*, \tau) = (1 - x^*)T_n(\tau) + x^* T_{n+1}(\tau), \qquad n = 1, 2, ..., N. \tag{12}$$

The above equation will be used to study thermal stresses and TSIFs in the FGM strip.

THERMAL STRESS

In the following study of thermal stresses in the FGM strip and TSIFs at the tip of an edge crack, as shown in Figure 1, a special kind of FGM is considered in which the Young's modulus and Poisson's ratio are constant. This assumption limits the applications of the present analysis to a certain extent, however, there do exist FGM systems for which Young's modulus may be approximately assumed as constant. Examples include TiC/SiC and zirconia/nickel FGMs. The Young's moduli of these materials may not change significantly because the basic constitutents have similar Young's modulus.

The FGM strip is assumed to undergo plane strain deformations and is free from constraints at the far ends (see Figure 1). The only nonzero in-plane stress σ_{YY} is given by (Jin and Batra[7])

$$\sigma_{YY}^T(X, \tau) = -\frac{E\alpha(X)}{1 - \nu} T(X, \tau) + \frac{E}{(1 - \nu^2)A_0}[(A_{22} - XA_{21}) \int_0^b \frac{E\alpha(X')}{1 - \nu} T(X', \tau)dX'$$

$$- (A_{12} - XA_{11}) \int_0^b \frac{E\alpha(X')}{1 - \nu} T(X', \tau)X'dX'], \tag{13}$$

where E is Young's modulus, ν is Poisson's ratio, $\alpha(X)$ is the coefficient of thermal expansion, the superscript T in σ_{YY}^T stands for thermal stress, $A_{ij}(i, j = 1, 2)$ and A_0 are constants given by Jin and Batra[7], and the temperature $T(X, \tau)$ is given by (10) and (12).

THERMAL STRESS INTENSITY FACTOR (TSIF)

Consider an edge crack in the FGM strip shown in Figure 1(b). The boundary conditions for the thermal crack problem are

$$\sigma_{XX} = \sigma_{XY} = 0, \qquad X = 0 \text{ and } X = b; \; Y \geq 0, \tag{14}$$

$$\sigma_{XY} = 0, \qquad 0 \leq X \leq b, \; Y = 0, \tag{15}$$

$$v = 0, \qquad a < X \leq b, \; Y = 0, \tag{16}$$

$$\sigma_{YY} = -\sigma_{YY}^T(X, \tau), \qquad 0 \leq X \leq a, \; Y = 0, \tag{17}$$

where $\sigma_{YY}^T(X, \tau)$ is given in (13), v is the displacement in Y-direction, a and b are the crack length and the strip thickness, respectively. By using Fourier transform and integral equation methods, the above boundary value problem is reduced to the following singular integral equation

$$\int_{-1}^{1} \left[\frac{1}{s-r} + K(r,s) \right] \phi(s,\tau)ds = -\frac{2\pi(1-\nu^2)}{E}\sigma_{YY}^T(X, \tau), \quad |r| \leq 1, \tag{18}$$

where the unknown density function $\phi(r,\tau)$ is

$$\phi(X,\tau) = \partial v(X,0,\tau)/\partial X, \tag{19}$$

with the notation $v \equiv v(X, Y, \tau)$, and $r = 2X/a - 1$, $s = 2X'/a - 1$. The kernel $K(r, s)$ is singular only at $(r, s) = (-1, -1)$ and can be found in Gupta and Erdogan[8]. The function $\phi(r,\tau)$ can be further expressed as

$$\phi(r,\tau) = \psi(r,\tau)/\sqrt{1-r}, \tag{20}$$

where $\psi(r,\tau)$ is continuous and bounded for $r \in [-1, 1]$. When $\phi(r,\tau)$ is normalized by $(1 + \nu)\alpha_0 T_0$, the normalized TSIF, K^*, at the crack-tip is obtained as

$$K^* = \frac{(1-\nu)K_I}{E\alpha_0 T_0\sqrt{\pi b}} = -\frac{1}{2}\sqrt{\frac{a}{b}} \; \psi(1,\tau), \tag{21}$$

where K_I denotes the mode I TSIF.

NUMERICAL RESULTS AND DISCUSSION

In the following numerical calculations, a TiC/SiC FGM is considered. The material properties of the titanium carbide (TiC) and the silicon carbide (SiC) are listed in Table I (Munz and Fett[9]; Sand et al.[10]). This is a ceramic/ceramic FGM with potential applications in areas such as cutting tools and turbines. We will assume that the thermally shocked edge $X = 0$ is pure TiC (as the matrix phase) and the opposite edge is pure SiC (as the particulate phase). We also only consider the case of $T_b/T_0 = 0$, i.e. $T_b = 0$ and $T_0 \neq 0$. This represents a severe thermal shock load on the strip, which is important for engineering applications.

Table I   Material Properties of TiC and SiC

Materials	Coefficient of thermal expansion $(10^{-6}K^{-1})$	Thermal conductivity $(Wm^{-1}K^{-1})$	Mass density $(g\;cm^{-3})$	Specific heat $(Jg^{-1}K^{-1})$
TiC	7.0	20	4.9	0.7
SiC	4.0	60	3.2	1.0

The FGM is assumed as a two phase composite material with graded volume fractions of its constituent phases. Reiter and Dvorak[11] have argued that the micromechanics models for homogeneous composite materials may be utilized for FGMs with reasonable accuracy. Thus the FGM thermal conductivity, coefficient of thermal expansion, specific heat and mass density are calculated from conventional micromechanics models (Christensen[12]). The thermal conductivity of the FGM, $k(X)$, is

$$k(X) = k_m \{1 + [V_i(X)(k_i - k_m)]/[k_m + (k_i - k_m)(1 - V_i(X))/3]\}, \qquad (22)$$

where the subscripts i and m stand for the inclusion and matrix properties, respectively. The mass density, $\rho(X)$, the coefficient of thermal expansion, $\alpha(X)$, and the specific heat, $c(X)$, of the FGM are described by a rule of mixtures, i.e.

$$F(X) = V_i(X)F_i + [(1 - V_i(X)]F_m, \qquad (23)$$

where $F(X)$ stands for any of the quantities $\rho(X)$, $\alpha(X)$ or $c(X)$. The particle volume fraction, $V_i(X)$, is assumed in the form of power function, i.e.,

$$V_i(X) = (X/b)^p. \qquad (24)$$

Thus $X = 0$ corresponds to pure matrix phase and $X = b$ is pure inclusion material.

The temperature in the FGM strip is calculated from the asymptotic solution (10) and (12). To obtain an idea to what extent the temperature can be approximated by (12), the solution is first applied to a homogeneous strip where each layer has identical material properties. The results are plotted against the complete temperature solution (normalized by T_0) for a homogeneous strip (Ozisik[4]). Figure 2 shows the normalized temperatures at different nondimensional times (τ) for a 30 layer model where the layers have equal thickness. It is seen from Figure 2 that the asymptotic solution and the complete solution are almost identical in the entire range of the strip for nondimensional times up to $\tau = 0.05$. Those solutions also agree well with each other in the entire strip for times up to $\tau = 0.10$. For times up to $\tau = 0.15$, the solutions are in good agreement in the region of $X/b < 0.6$. It will be seen that the thermal stress in the strip and the TSIF reach their maximum values before the normalized time of $\tau = 0.10$. Hence, the asymptotic solution (10) and (12) offer a reliable basis to obtain the maximum thermal stress and the maximum TSIF.

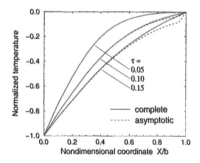

Figure 2: Temperature distribution in a homogeneous strip: asymptotic solution versus complete solution.

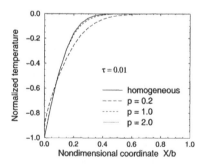

Figure 3: Temperature distribution in the FGM strip for $\tau = 0.01$.

Figure 3 shows normalized temperatures at $\tau = 0.01$ in both the homogeneous strip and FGM strip for volume fraction profiles $p = 0.2$, 1.0 and 2.0 (see (24)). The layered model consists of 45 layers with more layers intensively deployed near the edge $X = 0$ as the compositional profile of the FGM varies dramatically near the edge $X = 0$ in the case of $p = 0.2$. The 45 layer model is used in all calculations of temperatures, thermal stresses and TSIFs for the FGM strip. Notice that the temperature remains at the initial value $(T = 0)$ in the region $0.7 < X/b \leq 1$. The temperature starts dropping around the middle of the strip to the boundary value of -1 at the thermally shocked edge $X = 0$.

The normalized thermal stresses in the FGM strip are calculated from (13) with the asymptotic temperature solution (12). Figure 4 shows the normalized thermal stress at $\tau = 0.01$ in both the homogeneous strip and the FGM strip for the volume fraction profiles $p = 0.2$, 1.0 and 2.0 (see (24)). As in the homogeneous case, tensile stresses develop in the edge regions of the FGM strip and compressive stresses develop in the middle portion. It is observed that the thermal stress reaches the peak value at $X = 0$. The thermal stress in the FGM strip for $p = 0.2$ decreases from its peak value sharply near the thermally shocked edge $X = 0$ with an increase in X. The normalized peak thermal stresses (at $X = 0$) in the FGM strip for $p = 0.2$ are 0.9587, 0.8778, 0.6763 and 0.4484 for times $\tau = 0.0001$, 0.001, 0.01 and 0.1, respectively. The corresponding peak values in the homogeneous strip are 0.9549, 0.8632, 0.6087 and 0.1569, respectively. The stress plot for the times $\tau = 0.0001$, 0.001 and 0.1 are not provided here, but they show similar trends to those in Figure 4.

Figure 5 shows the normalized TSIF versus nondimensional time τ for an edge crack of length $a/b = 0.1$ in both the homogeneous strip and the FGM strip for the volume fraction profiles $p = 0.2$, 1.0 and 2.0 (see (24)). Some relevant observations can be made from this figure. First, the TSIF for the crack in the FGM strip varies with time and crack length in a similar way to that of the TSIF for a homogeneous strip, i.e., for a given normalized crack length a/b, the TSIF increases with time, reaches a peak value at a particular time (that increases with the crack length), and then decreases with further increase of time. There exists a critical crack length at which the peak TSIF reaches a maximum. The time at which the FGM TSIF reaches the peak decreases with decreasing power index p of the volume fraction of SiC. Second, the TSIF for the FGM is lower than that for the homogeneous strip for short times, but may be higher than that for the homogeneous strip for extended times. However, the peak TSIF for the FGM is lower than that for the homogeneous strip. More detailed results can be found in Jin and Paulino[6].

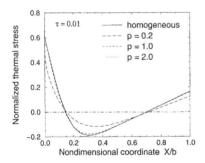

Figure 4: Thermal stresses in the FGM strip for $\tau = 0.01$.

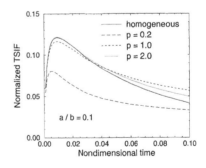

Figure 5: Thermal stress intensity factor for the FGM strip with $a/b = 0.1$.

CONCLUSIONS

A multi-layered material model is employed to solve the temperature field in a strip of a functionally graded material subjected to transient thermal loading conditions. The FGM is assumed having constant Young's modulus and Poisson's ratio, but the thermal properties of the material vary along the thickness direction of the strip. This kind of FGMs include some ceramic/ceramic FGMs such as TiC/SiC and $MoSi_2/Al_2O_3$, and also some ceramic/metal FGMs such as zirconia/nickel and zirconia/steel. By using Laplace transform and an asymptotic analysis, an analytical first order temperature solution for short times is obtained. For a homogeneous strip, the asymptotic solution for temperature agrees well with the complete solution for nondimensional times up to about $\tau = 0.10$, and so do the thermal stress and the TSIF of an edge crack. It is noted that the peak TSIF occurs at times less than $\tau = 0.1$. The thermal stresses and the TSIFs of edge cracks are calculated for a TiC/SiC FGM with various volume fraction profiles of SiC represented by the power index p. It is found that the peak TSIF decreases with a decrease in p if the thermally shocked cracked edge of the FGM strip is pure TiC.

ACKNOWLEDGMENTS

We would like to acknowledge the support from the National Science Foundation (NSF) under grant No. CMS-9996378 (Mechanics & Materials Program).

REFERENCES

[1]N. Noda and Z.-H. Jin, "Thermal Stress Intensity Factors for a Crack in a Strip of a Functionall Gradient Material," *International Journal of Solids and Structures*, **30**, 1039-1056 (1993).

[2]F. Erdogan and B. H. Wu, "Crack Problems in FGM Layers Under Thermal Stresses," *Journal of Thermal Stresses*, **19**, 237-265 (1996).

[3]Y. Tanigawa, T. Akai, R. Kawamura and N. Oka, "Transient Heat Conduction and Thermal Stress Problems of a Nonhomogeneous Plate with Temperature-Dependent Material Properties," *Journal of Thermal Stresses*, **19**, 77-102 (1996).

[4]M. N. Ozisik, *Heat Conduction*, John Wiley & Sons, New York, 1980.

[5]L. J. Gray and G. H. Paulino, "Symmetric Galerkin Boundary Integral Formulation for Interface and Multi-zone Problems," *International Journal for Numerical Methods in Engineering*, **40**, 3085-3101 (1997).

[6]Z.-H. Jin and G. H. Paulino, "Transient Thermal Stress Analysis of an Edge Crack in a Functionally Graded Material," *International Journal of Fracture*, **107**, 73-98 (2001).

[7]Z.-H. Jin and R. C. Batra, "Stress Intensity Relaxation at the Tip of an Edge Crack in a Functionally Graded Material," *Journal of Thermal Stresses*, **19**, 317-339 (1996).

[8]G. D. Gupta and F. Erdogan, "The Problem of Edge Cracks in an Infinite Strip," *ASME Journal of Applied Mechanics*, **41**, 1001-1006 (1974).

[9]D. Munz and T. Fett, *Ceramics*, Springer, Berlin, 1999.

[10]Ch. Sand, J. Adler and R. Lenk, "A New Concept for Manufacturing Sintered Materials with a Three Dimensional Composition Gradient Using a Silicon Carbide - Titanium Carbide Composite," pp. 65-70, *Materials Science Forum*, **308-311**, Edited by W. A. Kaysser, Trans Tech Publications, Switzerland, 1999.

[11]T. Reiter and G. J. Dvorak, "Micromechanical Models for Graded Composite Materials: II. Thermomechanical Loading," *Journal of the Mechanics and Physics of Solids*, **46**, 1655-1673 (1998).

[12]R. M. Christensen, *Mechanics of Composite Materials*, John Wiley & Sons, New York, 1979.

THERMO-MECHANICAL STRESS INTENSITY FACTORS FOR A PARTIALLY INSULATED CRACK IN A FUNCTIONALLY GRADED MEDIUM

Sami El-Borgi
Associate Professor of Structural Engineering
Applied Mechanics Research Laboratory
Ecole Polytechnique de Tunisie
B.P. 743, La Marsa 2078, Tunisia

Fazil Erdogan
G. Whitney Snyder Professor
Dept. of Mech. Eng. and Mechanics
Lehigh University
Bethlehem, PA 18015, USA

Houda Hila
Graduate Student
Applied Mechanics Research Laboratory
Ecole Polytechnique de Tunisie
B.P. 743, La Marsa 2078, Tunisia

Hichem Smaoui
Professor of Structural Engineering
Ecole Nationale d'Ingénieurs de Tunis
B.P. 37, Le Belvédère
Tunis 1002, Tunisia

ABSTRACT

In high-temperature applications, the materials research community has been recently exploring the possibility of using new nonhomogeneous coatings made of Functionally Graded Materials (FGMs) as an alternative to the conventional homogeneous ceramic coatings. In designing components involving FGMs, an important aspect of the problem is the fracture failure. In this paper, we consider an infinite medium made of a Functionally Graded Material (FGM) with a partially insulated crack subjected to a steady-state heat flux at infinity as well as mechanical stresses. The problem is modeled as a nonhomogeneous elastic medium with an isotropic stress-strain law under the assumption of plane strain or generalized plane stress conditions. The heat conduction and the plane elasticity partial differential equations are converted analytically into singular integral equations which are solved numerically. The equivalent thermal stresses to be applied as crack surface tractions are also obtained. The main objective of the paper is to study the effect of the material nonhomogeneity parameters and the partial crack surface insulation on the crack tip stress intensity factors for the purpose of gaining better understanding on the behavior and design of graded materials.

INTRODUCTION

In high-temperature applications, the potential of using homogeneous materials appears to be limited and in recent years the new trends in material design seem to evolve toward coating the main load-bearing component by a heat-resistive layer, generally a ceramic. Because of the relatively high mismatch in thermal expansion coefficients, the bonded structure generally gives rise to very high residual and thermal stresses. As a result, the composite medium becomes very vulnerable to cracking, debonding and spallation [1].

Recently the materials research community has been exploring the possibility of new concepts in coating design, such as Functionally Graded Materials (FGMs), as an alternative to the conventional homogeneous coatings. FGMs have promised attractive applications in a wide variety of thermal shielding problems such as high temperature chambers, furnace liners, turbines, microelectronics and space structures [2].

In designing components involving FGMs, an important aspect of the problem is the question of mechanical failure, specifically the fracture failure. Experimental observations of cracking in FGMs were reported in recent tests especially in the ceramic rich part of the material [3]. Fatigue and fracture characterization of materials and related analysis require the solution of certain standard crack problems. Most of the crack problems solved over the past two decades on nonhomogeneous materials [4,5,6] provide the basis for the fracture mechanics research on FGMs.

In [1] a brief discussion is given on the application of elementary concepts of fracture mechanics in nonhomogeneous materials and a number of typical problem areas are identified which relate to the fracture of FGMs. A number of crack problems in FGMs were solved accounting only for mechanical loading [7,8] while other studies considered thermal loading as well. For example, [9] studied the crack problem for an infinite FGM medium subject to a steady-state heat flux over the crack surfaces by assuming continuously varying thermal properties. [10] considered the problem of an edge crack in a semi-infinite nonhomogeneous medium subject to a steady-state heat flux. The case of an internal fully insulated crack parallel to the boundary of a semi-infinite FGM medium subject to a steady-state heat flux applied at the free surface was studied by [11]. This problem was later extended to the case of transient heat flux by [12]. [13] studied the problem of an internal crack in an FGM layer subject to a thermal gradient parallel to both the crack axis and the material gradient axis.

In this paper, we consider an infinite medium made of a Functionally Graded Material with a partially insulated center crack subjected to a steady-state heat flux at infinity as well as mechanical stresses. The problem is modeled as a nonhomogeneous elastic medium with an isotropic stress-strain law under the assumption of plane strain or generalized plane stress conditions. The material properties are assumed to vary exponentially in the direction perpendicular to the plane of the crack. Assuming no coupling between thermal and mechanical effects, this problem consists of solving first the heat conduction equation and then the Navier's equations of elasticity subject to the appropriate thermal and mechanical boundary conditions. Fourier transforms are used to convert the three partial differential equations into a system of singular integral equations which are then solved numerically to yield, the temperature and the displacement fields in the medium. The equivalent thermal stresses to be applied as crack surface tractions are also obtained. Finally, we can estimate the parameters that govern the crack growth such as the crack tip stress intensity factors under both thermal and mechanical loading.

FORMULATION OF THE PROBLEM

The problem under consideration is described in Figure (1). The medium is unbounded in both the x and y directions which are respectively the horizontal and vertical axes. The partially insulated crack of length 2a is located at the center of the medium and oriented along the x-axis. The material gradient is directed along the y-direction. For this graded medium, the Poisson's ratio v is assumed to be a constant because the effect of its variation on the crack-tip stress intensity factors was shown to be negligible [5,10] and the remaining thermo-mechanical properties depend on the y-coordinate only and are modeled as follows:

$$k = k_0 \exp(\delta y) \qquad \mu = \mu_0 \exp(\beta y) \qquad \alpha = \alpha_0 \exp(\gamma y) \qquad (1)$$

In equation (1), k, μ and α are, respectively, the heat conductivity, the shear modulus and the thermal expansion coefficient. k_0, μ_0 and α_0 correspond to the values of k, μ and α along the crack axis, and δ, β, γ are material constants.

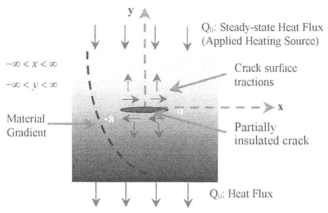

Figure 1. Crack problem geometry and applied loading

Two types of loading are considered: (i) a steady-state heat flux applied in y direction away from the crack region; and (ii) mechanical stresses applied as crack surface tractions which can be expressed in terms of the external mechanical loads. The primary unknowns of this problem are the temperature T and the displacements u and v along the x and y-axes. The basic equations of the plane problem for nonhomogeneous isotropic elastic bodies consist of the steady-state heat equations (without heat generation), the equilibrium equations (ignoring body forces), the strain-displacement relationships and the linear elastic stress-strain law given respectively by:

$$\frac{\partial}{\partial x}\left(k\frac{\partial T}{\partial x}\right)+\frac{\partial}{\partial y}\left(k\frac{\partial T}{\partial y}\right)=0 \tag{2}$$

$$\frac{\partial\sigma_{xx}}{\partial x}+\frac{\partial\sigma_{xy}}{\partial y}=0 \qquad \frac{\partial\sigma_{xy}}{\partial x}+\frac{\partial\sigma_{yy}}{\partial y}=0 \tag{3}$$

$$\varepsilon_{xx}=\frac{\partial u}{\partial x} \qquad \varepsilon_{yy}=\frac{\partial v}{\partial y} \qquad \varepsilon_{xy}=\frac{1}{2}\left(\frac{\partial u}{\partial y}+\frac{\partial v}{\partial x}\right) \tag{4}$$

$$\sigma_{xx}=\frac{\mu}{\kappa-1}\left[(1+\kappa)(\varepsilon_{xx}-\alpha T)+(3-\kappa)(\varepsilon_{yy}-\alpha T)\right] \quad \sigma_{yy}=\frac{\mu}{\kappa-1}\left[(3-\kappa)(\varepsilon_{xx}-\alpha T)+(1+\kappa)(\varepsilon_{yy}-\alpha T)\right] \quad \sigma_{xy}=2\mu\varepsilon_{xy} \tag{5}$$

where $\kappa=3-4\nu$ for plane strain and $\kappa=(3-\nu)/(1+\nu)$ for generalized plane stress.

Substituting the expression of the thermal conductivity k given by (1) into the heat equation (2) gives the following partial differential equation:

$$\frac{\partial^2 T}{\partial x^2}+\frac{\partial^2 T}{\partial y^2}+\delta\frac{\partial T}{\partial y}=0 \tag{6}$$

Substituting equations (4) into (5), inserting the resulting expressions into equation (3) and using (1) yields the following Navier's equations:

$$(\kappa+1)\frac{\partial^2 u}{\partial x^2}+(\kappa-1)\frac{\partial^2 u}{\partial y^2}+2\frac{\partial^2 v}{\partial x\partial y}+\beta(\kappa-1)\frac{\partial u}{\partial y}+\beta(\kappa-1)\frac{\partial v}{\partial x}=4\alpha_0 e^{yy}\frac{\partial T}{\partial x} \tag{7a}$$

$$(\kappa - 1)\frac{\partial^2 v}{\partial x^2} + (\kappa + 1)\frac{\partial^2 v}{\partial y^2} + 2\frac{\partial^2 u}{\partial x \partial y} + \beta(3 - \kappa)\frac{\partial u}{\partial x} + \beta(\kappa + 1)\frac{\partial v}{\partial y} = 4\alpha_0 e^{\gamma y}\left[(\beta + \gamma)T + \frac{\partial T}{\partial y}\right] \quad (7b)$$

The heat equation (6) is subject to the following thermal boundary conditions:

$$k\frac{\partial T}{\partial y} = -Q_0 \qquad\qquad y \to +\infty, |x| \le +\infty \qquad\qquad (8)$$

$$k\frac{\partial T}{\partial y} = -Q_0 \qquad\qquad y \to -\infty, |x| \le +\infty \qquad\qquad (9)$$

$$k\frac{\partial T}{\partial y} = k^*Q_c \qquad\qquad y = 0, |x| \le a \qquad\qquad (10)$$

$$T(x,0^+) = T(x,0^-) \qquad\qquad |x| > a \qquad\qquad (11)$$

$$\frac{\partial T}{\partial y}(x,0^+) = \frac{\partial T}{\partial y}(x,0^-) \qquad\qquad |x| > a \qquad\qquad (12)$$

Equations (7a) and (7b) are subject to the following mechanical boundary conditions:

$$\sigma_{xy}(x,0^+) = \sigma_{xy}(x,0^-) = \omega_1^M(x) \qquad |x| \le a \qquad\qquad (13)$$

$$\sigma_{yy}(x,0^+) = \sigma_{yy}(x,0^-) = \omega_2^M(x) \qquad |x| \le a \qquad\qquad (14)$$

$$\sigma_{yy}(x,y) = \sigma_{xy}(x,y) = 0 \qquad y \to \pm\infty, |x| \prec +\infty \qquad\qquad (15)$$

$$\sigma_{xy}(x,0^+) = \sigma_{xy}(x,0^-) \qquad |x| > a \qquad\qquad (16)$$

$$\sigma_{yy}(x,0^+) = \sigma_{yy}(x,0^-) \qquad |x| > a \qquad\qquad (17)$$

$$u(x,0^+) = u(x,0^-) \qquad |x| > a \qquad\qquad (18)$$

$$v(x,0^+) = v(x,0^-) \qquad |x| > a \qquad\qquad (19)$$

Equations (8) and (9) represent, respectively, the applied heat flux away from the crack region. Equation (10) describes the partial insulation of the crack surfaces which is modeled by assuming that the crack allows some heat flux Q_y that is only a certain percentage of the flux Q_c corresponding to the perfect conduction case. In this equation, k^* is a heat conductivity index assumed to be a constant, $0 \le k^* \le 1$. The limiting values $k^* = 0$ and $k^* = 1$ represent, respectively, perfect insulation and perfect conduction along the crack surfaces. Equations (11) and (12) represent the continuity of the temperature field and the heat flux along the crack axis and outside the crack.

Equations (13) and (14) describe, respectively, the applied tangential and normal crack surface tractions which can be expressed in terms of the external mechanical loads. Equation (15) represents the regularity conditions at $y \to \pm\infty$. Equations (16) to (19) describe the continuity conditions of the displacement and stress fields along the crack axis and outside the crack.

TEMPERATURE FIELD

The thermo-mechanical coupling being assumed negligible, we solve first the heat equation (6) using the Fourier transform and its inverse to yield the expressions for the perturbed temperature field above and below the crack:

$$T(x,y) = \int_{-\infty}^{+\infty}\left[\left\{A_1 e^{n_1 y} + A_2 e^{n_2 y}\right\}e^{-ix\lambda}d\lambda\right] \quad y \succ 0 \tag{20a}$$

$$T(x,y) = \int_{-\infty}^{+\infty}\left[\left\{A_3 e^{n_1 y} + A_4 e^{n_2 y}\right\}e^{-ix\lambda}d\lambda\right] \quad y \prec 0 \tag{20a}$$

where n_1 and n_2 are the roots of the characteristic polynomial associated with the partial differential equation (6) and A_1, A_2, A_3 and A_4 are unknown functions determined from the thermal boundary conditions.

Introducing the unknown density function

$$\psi(x) = \frac{\partial}{\partial x}\left[T(x,0^+) - T(x,0^-)\right] \tag{21}$$

and applying all thermal boundary conditions yields a singular integral equation, in which the only unknown is the function ψ, and which after extracting the Cauchy singularity from its kernel takes the following form:

$$\int_{-a}^{+a}\left[\frac{1}{t-x} + k(x,t)\right]\psi(t)dt = 2\pi\left(k^* - 1\right) \quad -a < x < a \tag{22}$$

where the known function $k(x,t)$ is a Fredholm kernel that depends on the nonhomogeneity parameter δ.

It was shown by [14] that the solution of equation (22) may be expressed as $\psi(t) = \Psi(t)/\sqrt{1-t^2}$ in which $\Psi(t)$ is a bounded and continuous function in the interval [-a, a]. A suitable choice for $\Psi(t)$ is to express it as a truncated series of Chebychev polynomials of the first kind. Using this solution as well as a suitable collocation method, the singular integral equation (22) is converted into a linear algebraic system whose unknowns are the coefficients of the series in the Chebychev polynomial. This system is solved numerically leading to the temperature distribution in the medium.

DISPLACEMENT FIELD

Substituting the expressions of the temperature field into Navier's equations (7a) and (7b), and using the same approach as for the thermal problem, we obtain a system of two coupled singular integral equations:

$$\int_{-a}^{+a}\left\{\left[\frac{1}{t-x} + k_{11}(x,t)\right]\psi_1(t) + \left[k_{12}(x,t)\right]\psi_2(t)\right\}dt = \frac{2\pi}{\mu_1}\left[\omega_1^M(x) + \omega_1^T(x)\right] \quad -a < x < a \tag{23a}$$

$$\int_{-a}^{+a}\left\{\left[k_{21}(x,t)\right]\psi_1(t) + \left[\frac{1}{t-x} + k_{22}(x,t)\right]\psi_2(t)\right\}dt = 2\pi\frac{\kappa-1}{\mu_1}\left[\omega_2^M(x) + \omega_2^T(x)\right] \quad -a < x < a \tag{23b}$$

where the unknown density functions ψ_1 and ψ_2 are given by:

$$\psi_1(x) = \frac{\partial}{\partial x}\left[u(x,0+) - u(x,0-)\right] \qquad \psi_2(x) = \frac{\partial}{\partial x}\left[v(x,0+) - v(x,0-)\right] \tag{24}$$

In equations (23a) and (23b), the temperature dependent terms are brought to the right hand side resulting in the expressions of the equivalent tangential and normal thermal stresses applied as crack surface tractions: $\omega_1^T(x)$ and $\omega_2^T(x)$ which depend on the nonhomogeneity parameters δ, β, γ. The fredholm kernels $k_{11}(x,t)$, $k_{12}(x,t)$, $k_{21}(x,t)$ and $k_{22}(x,t)$ depend on the nonhomogeneity parameter β. The singular integral equations (23a) and (23b) are solved numerically in the same manner as the thermal singular integral equation (22) to yield the x and y components of the displacement field in the medium as well as the stress field.

STRESS INTENSITY FACTORS

The stress intensity factors at both crack tips can be estimated from the expressions of the stress field as follows:

$$k_1(a) = \lim_{x \to a} \sqrt{2(x-a)}\sigma_{yy}(x,0) \qquad k_1(-a) = \lim_{x \to -a} \sqrt{2(-x-a)}\sigma_{yy}(x,0) \qquad (25a)$$

$$k_2(a) = \lim_{x \to a} \sqrt{2(x-a)}\sigma_{xy}(x,0) \qquad k_2(-a) = \lim_{x \to -a} \sqrt{2(-x-a)}\sigma_{xy}(x,0) \qquad (25b)$$

RESULTS AND DISCUSSION

Typical results for the problem considered are shown in Figures (2) to (4). Figures (2a) and (2b) illustrate the temperature distribution along the crack axis (y=0+ and y=0-) for, respectively, different crack insulation factors and various values of the nonhomogeneity parameter δ. The horizontal axis represents the crack axis normalized with respect to half the crack length a (ie, the crack is located in the interval [-1,1]). As expected, the temperature jump across the crack becomes more pronounced as the crack surfaces become more insulated (Figure (2.a)). This jump increases for increasing values of the nonhomogeneity parameter δ (Figure (2.b)). This result is different from [9] where it was assumed for the same medium that the heat flux is applied from within the crack and concluded that the temperature jump decreases with increasing values of the nonhomogeneity parameter δ. For the FGM medium, the temperature distribution is not symmetric with respect to the x-axis. This asymmetry can be explained by noting that the crack being partially insulated is playing the role of a heat barrier in addition to the fact that the heating source is located in the upper half of the medium. The temperature distribution was obtained in closed-form for the homogeneous medium and was plotted in Figures (2a) and (2b) in order to compare with the FGM medium.

The distribution of the equivalent thermal tangential and normal crack surface stresses are plotted respectively in Figures (3a) and (3b) for fixed values of δ and γ and different values of β. To our knowledge, the distribution of these crack surface stresses were not plotted in any of the previous studies. Closed-form expressions of these stresses were derived for the case of the homogeneous medium in order to provide a basis of comparison with the FGM medium. Figures (3a) and (3b) indicate that the distribution of the tangential crack surface stress is linear and the distribution of the normal stress is zero for the case of the homogeneous medium. It is also clear from these figures that the distribution of the crack surface stresses is rather sensitive to the choice of the nonhomogeneity parameter β.

The effect of the parameter β on the opening and sliding mode stress intensity factors is shown respectively in Figures (4a) and (4b) for a fixed value of δ and different values of γ. Figure (4a) indicates that the mode I stress intensity factor is relatively sensitive to the choice of the nonhomogeneity parameter β. The mode II stress intensity factor was almost insensitive to this parameter β. These results were also reported by [11]. In Figure (4a), it should be pointed out that a negative value of the mode I stress intensity factor is physically meaningless. The negative value of k_1 indicates that the crack surfaces may be in contact. To avoid this problem, a contact algorithm similar to the one developed by [13] should be utilized in the future.

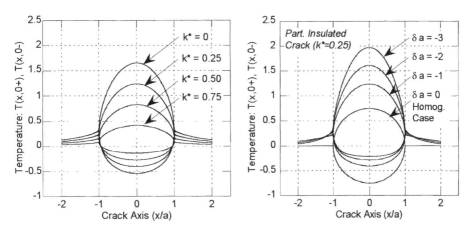

(a) Effect of varying crack insulation factor   (b) Effect of varying the parameter δ

Figure (2): Temperature distribution around the crack

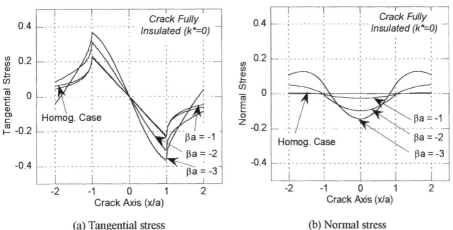

(a) Tangential stress   (b) Normal stress

Figure (3): Effect of β on equivalent thermal crack surface stresses

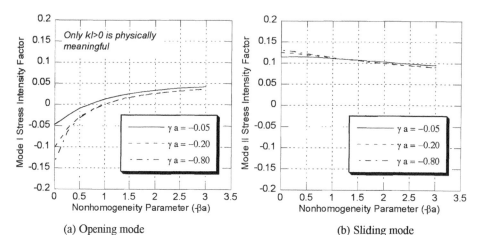

(a) Opening mode   (b) Sliding mode

Figure (4): Effect of β on thermal stress intensity factors

REFERENCES

[1] F. Erdogan, "Fracture mechanics of functionally graded materials," *Composites Engineering*, **5** [7] 753-770 (1995).

[2] J. Holt, M. Koizumi, T. Hirai and Z.A. Munir, (Editors), "Functionally Gradient Materials," Ceramic Transactions, **34**, The American Ceramic Society, Ohio, 1992.

[3] J.J. Lanutti, "Functionally graded materials: properties, potential and design guidelines," *Composites Engineering*, **4**, 81-94 (1994).

[4] R.S. Dhaliwal and B.M. Singh, "On the theory of elasticity of a non-homogeneous medium," *Journal of Elasticity*, **8**, 211-219 (1978).

[5] F. Delale and F. Erdogan, "The crack problem for a nonhomogeneous plane," *ASME Journal of Applied Mechanics*, **50**, 609-614 (1983).

[6] F. Erdogan, "The crack problem for bonded nonhomogeneous materials under antiplane shear loading," *ASME Journal of Applied Mechanics*, **52**, 823-828 (1985).

[7] M. Ozturk and F. Erdogan, "The axisymmetric crack problem in a nonhomogeneous medium," *ASME Journal of Applied Mechanics*, **44**, 631-636 (1993).

[8] F. Erdogan and B.H. Wu, "The surface crack problem for a plate with functionally graded properties," *ASME Journal of Applied Mechanics*, **64**, 449-456 (1997).

[9] N. Noda and Z-H. Jin, "Steady thermal stresses in an infinite nonhomogeneous elastic solid containing a crack," *Journal of Thermal Stresses*, **16**, 181-196 (1993).

[10] Z-H. Jin and N. Noda, "Edge crack in a nonhomogeneous half plane under thermal loading," *Journal of Thermal Stresses*, **17**, 591-599 (1994).

[11] Z-H. Jin and N. Noda, "An internal crack parallel to the boundary of a nonhomogeneous half plane under thermal loading," *Int. Journal of Engineering Sciences*, **31** [5] 793-806 (1993).

[12] Z-H. Jin and N. Noda, "Transient thermal stress intensity factors for a crack in a semi-infinite plate of a functionally gradient material," *International Journal of Solids Structures*, **31** [2] 203-218 (1994).

[13] F. Erdogan and B.H. Wu, "Crack problems in FGM layers under thermal stresses," *Journal of Thermal Stresses*, **19**, 237-265 (1996).

[14] F. Erdogan, G.D. Gupta and T.S. Cook, "Numerical solution of singular integral equations"; pp. 368-425 in *Mechanics of Fracture*, Edited by G.C. Sih. Norrdhoff, Leyden, (1973).

TORSIONAL PROBLEM OF TRIANGULAR BAR WITH FUNCTIONALLY GRADED STRUCTURE

Shigeyasu Amada
Gunma University
1-5-1, Tenjin, Kiryu,
Gunam 376-8515
Japan

Yasushi Terauchi
Katsura Machinery Ltd.
2-27-3, Bunkyo, Maebashi
Gunma 371-0801
Japan

ABSTRACT
Hemp palm branches have a triangular cross section which is very exceptional and a functionally graded structure for the reinforced fiber distribution. The fiber distributes densely in the outer surface layer and sparsely in the core region. This is modeled by a three layered composite bar with a triangular cross section. A torsional problem of this bar is solved using conformal mapping and complex torsional function approach and an analytical solution is derived in terms of complex power series. The computed maximum shear stress is reduced compared to the stress in the homogeneous bar. This reduction rate increases with the difference of both the shear modulus of the core and outer layer. Under the twisting moment the induced shear stress distribution is similar to the fiber distribution with the functionally graded structure. It was concluded that this structure like the Hemp palm branch is more adapted to the environmental loading than a homogeneous structure.

1. INTRODUCTION
Plants and animals have a superior ability called adaptive growth which changes their body by themselves to survive from the struggle of existence, and eventually it determines their shape and characteristics [1] to adapt to their environment. These factors lead to various superior structural properties[2]. A biologist Wainwright insists that the main frame of every plants and animals has a circular cross section in his book[3]. It is reasonable for them when subjected to loading from various directions and is the most rational shape to adapt to the environments. But there are some exceptions. One of those is the Hemp palm branch the cross section of which is triangular. Fig.1(a) shows Hemp palm tree, (b) branch and leaf, (c) cross section of branch. It is easily recognized that the branch with 80~100 cm in length is very slender as compared with the large leaf in a fan shape. A large leaf withstands the a heavy loading subjected to the leaf surface. To keep

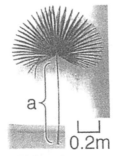

(a) Hemp palm tree   (b) Branch and leaf   (c) Cross section

Fig.1 Hemp palm tree

its surface toward the sun's direction, the branch must maintain the leaf's position. To support the loading hemp palm branches has avoid a circular shape and adapt a triangular cross section as shown in Fig.1(c), which is an isosceles triangle. This change of cross section could reduced the bending stress about 34%[4].

The branch is a composite material reinforced by strong fibers. Furthermore it has adapted a functionally graded structure of the fiber distribution. This structure provides a high bending strength[4]. The fibers distribute densely in the outer surface layer and sparsely in the core region. This distribution contributes not only to a bending but also to twisting.

There are some studies on torsional problems for inhomogeneous bars. An inhomogeneous bar with a fan shape made of layers was studied based on a polar coordinate system[5] and a composite of the laminated plate was investigated by FEM[6]. St. Venant's principle of a rectangular bar was discussed[7]. There exists no study on torsional problem of an inhomogeneous bar with triangular cross section.

This study focuses a torsion of the Hemp palm branch with a triangular cross section and functionally graded structure of the fiber distribution. Using conformal mapping approach, analytical solution is derived in terms of complex torsional function. Finally, we present how the Hemp palm branch reduces torsional stress and what kind of the role the functionally graded structure of the fiber distribution plays.

2. SHAPE AND STRUCTURE

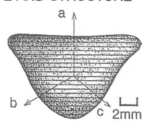

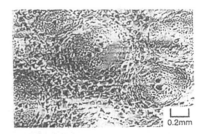

(a) Cross section of branch   (b) Magnified cross section of fiber

Fig.2 Cross section of branch and fiber

Fig.2(a) shows the cross section of the branch where the distributed solid dots correspond to fibers called vascular bundle sheath as shown in Fig.2(b). A vascular bundle is made up of two kinds of tubes, sieve tubes(transporting nutrition) and vessels(transporting water), for which the outside region is protected by fine strong cells.

The width of the branch cross section is given with non-dimensional length in Fig.3(a) and (b), and its aspect ratio (width/height) becomes approximately constant at 1.36 as shown in Fig.3.(c).

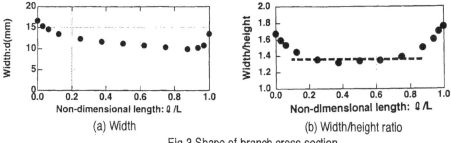

(a) Width   (b) Width/height ratio

Fig.3 Shape of branch cross section

Using image analysis of the photograph of the cross section, the volume fraction V_f of fibers was evaluated and the results are shown in Fig.4(a) in the a(vertical)-direction, and (b) in the b -and c-directions. The layer number one corresponds to the outer surface and layer number seven indicates the center core. The volume fraction is 60~80% in the outer layer and 20% in the core region.

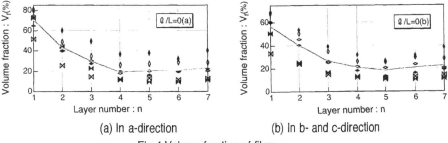

(a) In a-direction   (b) In b- and c-direction

Fig.4 Volume fraction of fiber

Fig.5 shows the distribution of the estimated shear modulus with respect to V_f based on the bamboo data[8]. It changes considerably with the volume fraction of fibers.

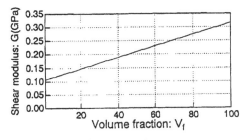

Fig.5 Shear modulus

3. FUNDAMENTAL EQUATIONS

A bar consisting of several different materials is analyzed. The outer layer material is denoted by S_o and its outer periphery by C_o as shown in Fig.6. The inside materials are denoted by S_j and their periphery by $C_j(j=1,2,\cdots,m)$. The X-Y-Z coordinate system Is glued on the cross section of the bar as shown in Fig.6. The displacements in the X, Y and Z directions are given by

$$U_j = -azy$$

$$V_j = azx \tag{1}$$

$$W_j = a\varphi_j(x,y)\cdots(j=0,1,\cdots,m)$$

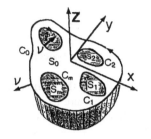

where a is the twisting angle, φ_j is the torsional function in the j-th region. The fundamental equation for the torsional problem is written by

Fig.6 Configuration and coordinate

$$\nabla^2\varphi_j = 0 \tag{2}$$

where ∇^2 is the Laplace operator. The complex torsional function $f(t)$ is introduced as follows,

$$f(t) = \varphi + i\psi,\cdots t = x + iy \tag{3}$$

where i is the imaginary number and ψ is the conjugate function ψ of φ.
The mapping function from the t-plane to the ξ plane is represented by

$$t = x + iy = \omega(\xi) \tag{4}$$

where $\xi = \rho e^{i\theta}$. The following relation is easily derived,

$$x^2 + y^2 = \omega(\xi)\overline{\omega(\xi)} \tag{5}$$

where $\overline{\omega(\xi)}$ stands for the conjugate function of $\omega(\xi)$.

The boundary conditions at the interface $\rho = \rho_i$ for the displacement becomes

$$\frac{\partial\psi_o}{\partial\rho} = \frac{\partial\psi_i}{\partial\rho} \qquad \text{at } \rho = \rho_i \tag{6}$$

The stress continuity condition is given by

$$\mu_o\psi_o - \mu_i\psi_i = \frac{\mu_o - \mu_i}{2}\omega(\xi)\overline{\omega(\xi)} + const. \qquad (i=1,2,\cdots m) \tag{7}$$

and there is no stress action on the outer surface, that is

$$\psi_o = \frac{x^2 + y^2}{2} + const. \qquad \text{at } \rho = \rho_0 \tag{8}$$

The torsional function which satisfies eq.(2) can be expressed by the Laurent series expansion of $f_0(\xi)$ in the outer layer and by the Taylor series expansion of $f_i(\xi)$ in the inner layers as follows,

$$f_o(\xi) = \sum_{k=1}^{\infty}\left(a_{3k}^{(0)} + ib_{3k}^{(0)}\right)\xi^{3k} + \sum_{k=1}^{\infty}\left(a_{-3k}^{(0)} + ib_{-3k}^{(0)}\right)\xi^{-3k} + a_0^{(0)} + ib_0^{(0)}$$

$$f_i(\xi) = \sum_{k=0}^{\infty}\left(a_{3k}^{(i)} + ib_{3k}^{(i)}\right)\xi^{3k} + a_0^{(i)} + ib_0^{(i)}\cdots(i=1,2,\cdots,m) \tag{9}$$

where a_j and b_j are real unknown constants to be determined later.

Let Im[f] be denoted by the imaginary part of the complex function f. Since we have the relations $\psi_0 = \mathrm{Im}\left[f_{t_0}(\xi)\right]$ and $\psi_i = \mathrm{Im}\left[f_i(\xi)\right]$, the conjugate function ψ of φ is given by

$$\psi_o = b_o^{(0)} + \sum_{k=1}^{\infty}\left[\left(a_{3k}^{(0)}\rho^{3k} - a_{-3k}^{(0)}\rho^{-3k}\right)\sin 3k\theta + \left(b_{3k}^{(0)}\rho^{3k} + b_{-3k}^{(0)}\rho^{-3k}\right)\cos 3k\theta\right]$$

$$\psi_i = b_o^{(i)} + \sum_{k=1}^{\infty}\left(a_{3k}^{(i)}\rho^{3k}\sin 3k\theta + b_{3k}^{(i)}\rho^{3k}\cos 3k\theta\right)$$

(10)

4. MAPPING FUNCTIONS

Although the Hemp palm branch has an isosceles triangle, the cross section of the bar is assumed to be an equiltaeral triangle. Using a Schwarz-Christoffel formula[9], a conformal mapping function $\omega(\xi)$ which maps an equilateral triangle to a circle is given by

$$\omega(\xi) = \int_0^{\xi}(1-\xi)^{-2/3}d\xi = A\sum_{k=1}^{\infty}a_{3k-2}\xi^{3k-2}$$

(11)

where A is the magnification factor. For the case of A=1.0, the mapping function can be represented by

$$\omega(\xi) = \sum_{k=1}^{\infty}a_{3k-2}\rho^{3k-2}\left[\cos(3k-2)\theta + i\sin(3k-2)\theta\right]$$

(12)

The configuration of t-plane from ξ-plane calculated by eq.(12) with k=40 is given in Fig.7.

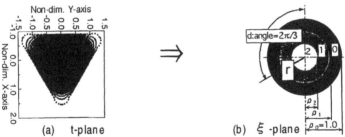

(a) t-plane   (b) ξ-plane

Fig.7 Mapping from triangle to circle

5. 3-LAYERED BAR

It is assumed that the triangular bar with the functionally graded structure is made of the three layers. The boundary conditions (6)-(8) can determine the unknown constants a_j and b_j in eq.(10). The torsional functions f0 ~ f2 become

$$f_0(\xi) = i\sum_{k=1}^{\infty}\left(D_{3k}\xi^{3k}\right) + i\sum_{k=1}^{\infty}\left(P_{3k}\xi^{-3k}\right) \qquad \rho_1 \le \rho \le \rho_0$$

(13)

$$f_1(\xi) = i\sum_{k=1}^{\infty}\left(E_{3k}\xi^{3k}\right) + i\sum_{k=1}^{\infty}\left(Q_{3k}\xi^{-3k}\right) \qquad \rho_2 \le \rho \le \rho_1$$

(14)

$$f_2(\xi) = i\sum_{k=1}^{\infty}\left(F_{3k}\xi^{3k}\right) \qquad 0 \le \rho \le \rho_2$$

(15)

where the coefficients D_{3k}, E_{3k}, F_{3k}, Q_{3k} and P_{3k} are given by

$$D_{3k} = \frac{\alpha_{3k}}{2\rho_o^{3k}} - \frac{P_{3k}}{\rho_o^{6k}}$$

$$E_{3k} = \frac{\alpha_{3k}}{2\rho_o^{3k}} - \frac{\mu_0 - \mu_1}{2\mu_1}\left(\frac{\beta_{3k}}{2\rho_1^{3k}} - \frac{\alpha_{3k}}{2\rho_0^{3k}}\right)$$

$$-\left\{\frac{1}{\rho_0^{6k}} + \frac{1}{\rho_1^{6k}} - \frac{1}{2\mu_1}\left[\mu_0\left(\frac{1}{\rho_1^{6k}} - \frac{1}{\rho_0^{6k}}\right) + \mu_1\left(\frac{1}{\rho_1^{6k}} + \frac{1}{\rho_0^{6k}}\right)\right]\right\}P_{3k}$$

$$F_{3k} = \frac{\alpha_{3k}}{2\rho_o^{3k}} + \frac{\rho_1^{6k}(\mu_0 - \mu_1)}{2\mu_1}\left(\frac{1}{\rho_2^{6k}} - \frac{1}{\rho_1^{6k}}\right)\left(\frac{\beta_{3k}}{2\rho_1^{3k}} - \frac{\alpha_{3k}}{2\rho_0^{3k}}\right)$$

$$-\left\{\frac{1}{\rho_1^{6k}} + \frac{1}{\rho_0^{6k}} + \left(\frac{1}{\rho_2^{6k}} - \frac{1}{\rho_1^{6k}}\right)\left[\mu_0\left(\frac{1}{\rho_1^{6k}} - \frac{1}{\rho_0^{6k}}\right) + \mu_1\left(\frac{1}{\rho_1^{6k}} + \frac{1}{\rho_0^{6k}}\right)\right]\frac{\rho_1^{6k}}{2\mu_1}\right\}P_{3k}$$

$$Q_{3k} = -\frac{\rho_1^{6k}(\mu_0 - \mu_1)}{2\mu_1}\left(\frac{\beta_{3k}}{2\rho_1^{3k}} - \frac{\alpha_{3k}}{2\rho_0^{3k}}\right) - (\mu_2 - \mu_1)\left(\frac{1}{\rho_1^{6k}} + \frac{1}{\rho_0^{6k}}\right)P_{3k}$$

$$-\frac{\rho_1^{6k}}{2\mu_1}\left[\mu_0\left(\frac{1}{\rho_1^{6k}} - \frac{1}{\rho_0^{6k}}\right) + \mu_1\left(\frac{1}{\rho_1^{6k}} + \frac{1}{\rho_0^{6k}}\right)\right]\left[\mu_2\left(\frac{1}{\rho_2^{6k}} - \frac{1}{\rho_1^{6k}}\right) + \mu_1\left(\frac{1}{\rho_2^{6k}} + \frac{1}{\rho_1^{6k}}\right)\right]P_{3k}$$

$$P_{3k} = U_{3k}/V_{3k}$$

and

$$U_{3k} = \frac{(\mu_1 - \mu_2)\gamma_{3k}}{2\rho_2^{3k}} - \frac{(\mu_1 - \mu_2)\alpha_{3k}}{2\rho_0^{3k}} + \frac{\rho_1^{6k}}{2\mu_1}\left[\frac{(\mu_0 - \mu_1)\beta_{3k}}{2\rho_1^{3k}} - \frac{(\mu_0 - \mu_1)\alpha_{3k}}{2\rho_0^{3k}}\right]$$

$$\times\left[\mu_2\left(\frac{1}{\rho_2^{6k}} - \frac{1}{\rho_1^{6k}}\right) + \mu_1\left(\frac{1}{\rho_2^{6k}} + \frac{1}{\rho_1^{6k}}\right)\right]$$

$$V_{3k} = (\mu_2 - \mu_1)\left(\frac{1}{\rho_1^{6k}} + \frac{1}{\rho_0^{6k}}\right) + \frac{\rho_1^{6k}}{2\mu_1}\left[\mu_0\left(\frac{1}{\rho_1^{6k}} - \frac{1}{\rho_0^{6k}}\right) + \mu_1\left(\frac{1}{\rho_1^{6k}} + \frac{1}{\rho_0^{6k}}\right)\right]$$

$$\times\left[\mu_2\left(\frac{1}{\rho_2^{6k}} - \frac{1}{\rho_1^{6k}}\right) + \mu_1\left(\frac{1}{\rho_2^{6k}} + \frac{1}{\rho_1^{6k}}\right)\right]$$

6. NUMERICAL RESULTS

The non-dimensional outer radius in the ξ-plane is $\rho_0 = 1.0$, and the ones of the 1st and 2nd layer take $\rho_1 = 0.7$ and $\rho_2 = 0.4$, respectively. The numerical calculations were carried out under the specified torque for the fixed shear modulus $\mu_2 = 0.12$ GPa, in the core layer and variable modulus μ_1 and μ_2 in the middle and outer layers, respectively. Three kinds of the bars are analyzed, those are (i) the homogeneous fiber distribution, (ii) the uniform fiber distribution in arithmetic average of the fiber volume fraction, (iii) the 3-layers (functionally graded structure : FGM). Shear stress is normalized by the maximum shear stress in the homogeneous bar, which is denoted by $\bar{\tau}$. The calculated shear stress for the FGM bar with $(\mu_1, \mu_2) = (0.22, 0.26)$GPa is represented in 3-dimensional configuration as shown in Fig.8. A large stress takes

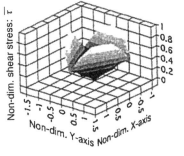

Fig.8 Shear stress distribution

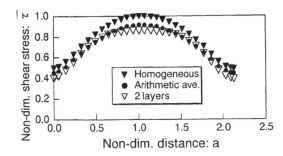

Fig.9 Shear stress along side

place on the sides of the triangular cross section and it decays sharply toward the core.

Fig.9 shows the shear stress distribution along the side of the triangle, where the solid and white triangle marks indicate the homogeneous and the FGM bars, respectively, and the solid circles stand for the bar with the uniform distribution of the fiber volume fraction in arithmetic average. The maximum stress occurs on the middle point of the side and approaching the corners the shear stress decreases to one half. It is easily recognized that the generated stress is the lowest for the FGM bar. Fig.10 gives the shear stress distribution along radius which corresponds to the line connecting the middle point of the side and the center of the triangle. The stresses in both the homogeneous bar and the one with the uniform fiber distribution vary approximately in a linear form. On the other hand, the FGM bar shows the lower distribution as compared with other two bars. The maximum shear stress reduces by about 15% as compared with the homogeneous bar. So that, the FGM structure leads to a considerable advantage in the stress distribution.

The shear modulus distribution influences the stress value in an interesting may.

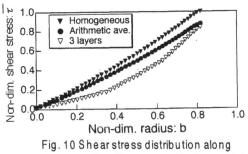

Fig. 10 Shear stress distribution along radius(line perpendicular to side)

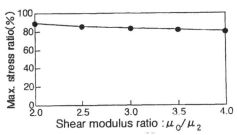

Fig.11 Max. shear stress ratio with μ_0 / μ_2

Defining the maximum shear stress ratio which the maximum shear stress in the FGM bar is divided by the one in the homogeneous bar, its variation is shown in Fig.11 with respect to the shear modulus ratio $\mu_0/\mu_2(\mu_2 = 1.0GPa)$. The ratio decreases linearly with μ_0/μ_2. It was concluded that the high shear modulus in the outer layer can effectively reduce the

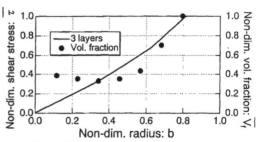

Fig. 12 Shear stress and fiber distributions

maximum shear stress. Fig.12 shows the shear stress distribution marked by a solid line and volume fraction of fibers marked by solid circles along radius. It is easily recognized that both the distributions provide a good agreement except at the core region. Judging from this result, the object of the functionally graded structure of fibers is similar to the shear stress distribution.

7. CONCLUSIONS

The Hemp palm branch was modeled by a bar with the equilateral triangular cross section and with the functionally graded structure of the fiber distribution. The following conclusions were obtained:

(1) The functionally graded structure of the fiber distribution can reduce the maximum shear stress about 15% as compared with the homogeneous bar.

(2) The maximum shear stress decreases linearly as the ratio μ_0/μ_2 increases.

(3) The fiber distribution in the functionally graded structure is approximately similar to the shear stress distribution under a twisting, which is the object of the functionally graded structure of the fibers in the Hemp palm branch.

Acknowledgment: This study was financially supported by the Ministry of Education of the Japanese Government and carried out under the project " Physics and Chemistry of Functionally Graded Structure".

REFERENCES

1) F.R.Paturi, " Generale Ingenieure der Natur", Econ Verlag(1974)

2) K.J.Niklas,"Plant Biomechanics: An Engineering Approach to Plant Form and Function", The University of Chicago Press(1992)

3) S.P. Wainwright, "Axis and Circumference: Cylindrical Shape of Plants and Animals", Harvard University Press(1988)

4) S.Amada and Y.Terauchi, " Mechanical Characteristics of Hemp Palm branch Wit Triangular Cross-section", *Trans Japan Soc. Mech. Eng.*, Ser.A, Vol.65(1999),2418

5) C.Y.Wang, " Torsion of a Compound Bar Bounded by Cylindrical Polar Coordinates" , *J. Appl. Math .*, Vol.48, Pt.3(1995),389

6) J.C.Fish, " Torsion and Twisting of Symmetric Composite Laminates" , *AIAA-92-2425-CP*(1992),736-744

7) N.K.Akhmedov and Y.A.Ustinov, "On St.Venant's Principle in the torsion Problem for a Laminated Cylinder", *Prikl Math. Mekhhan.*, USSR, Vol.52(1988),207

8) S. Amada and Y.Terauchi, " Torsion of Triangular Bar" , *Trans. Japan Soc. Mech. Eng.*, Ser.A, Vol.66(2000),1806-1811

9) R.V.Churchill, Complex Variables and Applications, McGraw-Hill, 2nd Edt.(1960)

Fracture Characterization

FRACTURE IN NOTCHED PARTS WITH A MICROSTRUCTURAL GRADIENT

F. Bohner, J.K. Gregory
Institute of Materials in Mechanical Engineering
Technical University of Munich
D-85747 Garching, GERMANY

ABSTRACT

Notched parts are subjected to loading or stress gradients, hence, it makes sense to tailor the local strength to this profile by generating an appropriate microstructural gradient. This was achieved in the age-hardenable alloys Ti-2.5Cu and 2017 Al by thermomechanical surface treatment. While the macroscopic yield stress in graded notched specimens can be reliably evaluated and reasonably well predicted using a simple local strength approach, an appropriate choice of specimen geometry is apparently critical to adequately assess fracture toughness. In contrast to the tensile behavior, no effect of the gradation on fracture properties is observed.

INTRODUCTION

It has been common industrial practice for many years to alter the surface properties of mechanically loaded components either by a mechanical surface treatment such as shot peening or a thermochemical surface treatment such as carburizing. The goal is generally to improve fatigue behavior, i.e., resistance to cyclic loading. Parts which have been treated in this manner can essentially be regarded as graded materials.

A newer method for developing graded microstructures iun commercial structural alloys using economical nethods is is known as thermomechanical surface treatment, or for the special case of age-hardenable alloys, selective or preferential surface aging [1,2]. In certain metallic materials of this type, the precipitation reaction which results in age-hardening occurs more rapidly and leads to higher strength levels if cold work is applied prior to aging. While this phenomenon is usually exploited industrially by forming after solution treating and before aging, conventional applications involve cold working the entire part. The novelty of thermomechanical surface treatment resides in the fact that only the surface layer, or "skin" is cold worked by a mechanical surface treatment such as shot peening, causing a gradient in the degree of cold work and hence in dislocation density. Upon aging, a precipitate structure forms which reflects the gradient in cold work. Furthermore, the gradient in dislocation density is retained to a certain extent. Both of these features give rise to a gradient in mechanical properties which can be quantified using microhardness measurements.

An example of the microhardness gradients which can be obtained in readily available commmercial alloys is shown in Fig. 1 for a heat-treatable aluminum alloy. Shot peening was used to generate a plastically deformed surface layer, hence the Almen intensity was used to quantify the shot peening process. This parameter measures the deflection in a standardized metal strip after shot peening. The higher the Almen intensity, the more severe the shot peening conditions, which in turn result in a higher surface hardness upon aging.

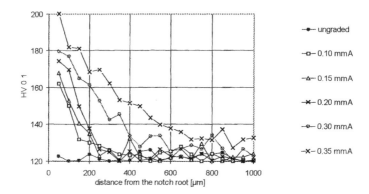

Figure 1: Microhardness profiles in 2017 Al after thermomechanical surface treatment using the Almen intensities indicated [8]

Most of the available literature on the effects of mechanical or thermochemical treatments on surface layer properties focus on those aspects which affect fatigue [3,4]. Despite the fact that such parts may also be subjected to monotonic loading, comparatively little work has been done to assess the influence of a change in surface layer properties on tensile behavior in notched parts [5,6,7]. Recent work on graded materials has shown that a simple approach based on comparing the local stress resulting from a notch stress field with a local yield stress adequately predicts a macroscopic yield stress [8], while finite element modelling can be successfully applied to predict the flow curve of notched parts with microstructural stress gradients [9]. In both cases, the microhardness profiles were successfully used to assess the local value of a property which can be measured macroscopically, for example yield stress, work hardening coefficient, etc. via an emprical correlation.

Very little information is available on the fracture properties of graded materials. A comparison of the fracture toughness of a high-strength beta-titanium alloy which has a high toughness in the solution heat treated condition and a low toughness in the age hardened condition showed that grading the solution treated material such that the skin was selectively surface aged had not detrimental effect on fracture toughness [10]. The present work examines further the possibility of assessing fracture toughness in notched parts in the presence of a microstructural gradient, using an approach analogous to that used to predict the yield stress.

MATERIAL AND EXPERIMENTAL PROCEDURE

The age-hardening alloys 2017 Al (nominal composition: 4Cu,0.7Mg, 0.7Mn, 0.5Si, 0.25(Ti+Zr), balance Al) and Ti-2.5Cu (actual composition: 2.1Cu, balance Ti) were chosen for this investigation because they are amenable to thermomechanical surface treatment. The processing parameters used to develop the microstructural gradients are given in Table I.

Fracture toughness K_{1c} was measured on conventionally treated material using the procedure described in ASTM E399 and standard compact tension specimens. Vickers microhardness was measured on the same specimens in order to obtain an empirical correlation between microhardness and fracture toughness.

Table I: Parameters used to generate graded microstructures

material	solution heat treatment	Almen intensity	aging treatment
2017 Al	1 h 496 °C / water quench	0.10 mmA	6 h 170 °C
2017 Al	1 h 496 °C / water quench	0.35 mmA	6 h 170 °C
Ti-2.5Cu	1 h 805 °C / water quench	0.25 mmA	24 h 400 °C
Ti-2.5Cu	1 h 805 °C / water quench	0.50 mmA	24 h 400 °C
Ti-2.5Cu	1 h 805 °C / water quench	0.25 mmA	8 h 400 °C + 8 h 450 °C

Cylindrical specimens with the same geometry used in previous work on assessment of tensile properties of graded material [8,9] were used to assess the toughness of material with a microstructural gradient (Figure 2). These dimensions were chosen to ensure that the scale of the gradation corresponded with the depth of the stress field from the notch, which is generally equal to the notch root radius. Cracks having various lengths were introduced into the notch cross-section by fatigue. Specimens were subsequently fractured and F, the load necessary for fracture, recorded. A value for the stress intensity K_I was calculated according to [11]

$$K_I = \frac{F}{D^{3/2}}\left[-1.27 + 1.72\frac{D}{d}\right]$$

(1)

where the parameters d and D are defined in Fig. 2. The minimum diameter d took the fatigue pre-crack into account.

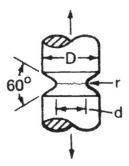

Fig. 2. Schematic geometry of notched cylindrical specimens for assessment of fracture toughness K_I. The notch is semi-circular with a radius r=1 mm, D= 8 mm, d=6 mm (without the fatigue pre-crack).

RESULTS

Figure 3 shows the results of the conventional fracture toughness measurements plotted versus yield stress. (K_Q is plotted instead of K_{Ic} owing to the fact that the thickness criterion was not met.) The aluminum alloy exhibits what is considered normal behavior in that an increasing yield stress is associated with a decreasing fracture toughness. It should be pointed out that while this is the most commonly observed behavior, there is no theoretical justification for this dependence. The fact that this trend does not necessarily hold for titanium alloys has been reported [12], and in the present case the opposite trend is observed. For the aluminum alloy, increasing cold work leads to an increase in yield stress and a decrease in fracture toughness, while for the titanium alloy, cold work increases both yield stress and fracture toughness.

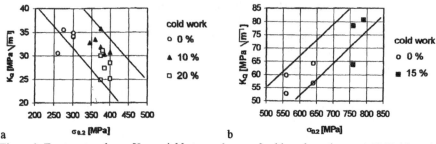

Figure 3: Fracture toughness K_Q vs yield stress, degree of cold work as shown a) 2017 Al, various aging times at 170 °C or 190 °C b) Ti-2.5Cu aged 24 h 400 °C or 8 h 400 °C + 8 h 450 °C.

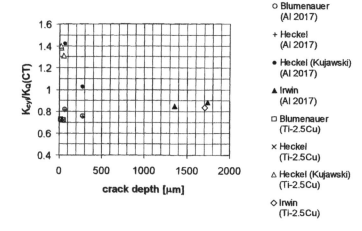

Figure 4: Results of fracture toughness tests in pre-cracked cylindrically notched specimens with an ungraded microstructure.

Before the toughness in cylindrically notched specimens with a graded microstructure was assessed, the values provided by equation 1 were compared with similar calculations [13] both without and with correction factors for the stress field of the notch [14] on ungraded samples. For the purpose of verifying the calculation, the K_Q-values obtained on the standard C(T)-specimens are considered to be the correct absolute values. It should be noted that when the thickness criterion is violated as in the present case for the data in Fig. 3, the K_Q-values are in all probability higher than the actual K_{Ic}-values. Figure 4 shows the results normalized as $K_{cyl}/K_{Q(CT)}$ obtained from ungraded specimens using equation 1 for circumferential cracks with depths of up to 500 μm. For crack depths greater than 1000 μm, the crack shape was semi-elliptical rather than circumferential, hence the standard calculation according to Irwin ($K_I \approx 1.26\sigma\sqrt{a}$)[15] is more appropriate. Two results are evident which are relevant for the calculations on graded samples.

Firstly, there is no difference in result between [11] and [13]. Secondly, the use of a notch correction factor for the stress intensity as suggested in [14] leads to values which depend on crack depth. In the case of ungraded material, no crack depth dependence should be present. Equation 1 [11] yields values which are independent of crack depth in the region of interest, namely, where the material is graded, hence it is considered adequate for assessing fracture toughness in the graded specimens.

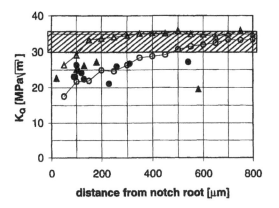

Figure 5: Fracture toughness values for 2017 Al aged 6 h 170 °C as determined on cylindrically notched specimens (solid symbols), standard C(T)-specimens (shaded scatterband) and as predicted from microhardness measurements (open symbols).
▲△, Almen intensity 0.10 mmA ●○, Almen intensity 0.35 mmA

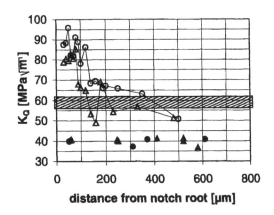

Figure 6: Fracture toughness values for Ti-2.5Cu aged 24 h 400 °C as determined on cylindrically notched specimens (solid symbols), standard C(T)-specimens (shaded scatterband) and as predicted from microhardness measurements (open symbols). ▲△, Almen intensity 0.25 mmA ●○, Almen intensity 0.50 mmA

Figures 5-7 show the fracture toughness values determined experimentally on graded samples of 2017 Al, Ti-2.5Cu aged 24 h 400 °C and 8 h 400 °C + 8 h 450 °C respectively. Also shown on these diagrams are the scatterbands for the K_Q-data from the C(T)-specimens. At depths greater than about 500 μm, i.e, outside of the graded region, the data should be identical, but are not because the calculations for the cylindrical specimens are somewhat low. Emprical predictions based on the correlation between hardness measurements and K_Q-values are also shown on these diagrams. Based on the results from the C(T)-specimens, the higher degree of cold work in the skin of the aluminum alloy should lead to a decrease in fracture toughness when the crack depth is within the graded region, while the reverse is the case for the titanium alloy. However, in no case does the graded microstructure result in a toughness value different from that of the bulk.

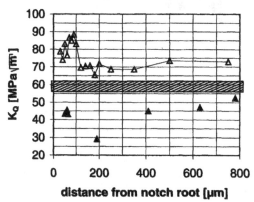

Figure 7: Fracture toughness values for Ti-2.5Cu aged 8 h 400 °C + 8 h 450 °C as determined on cylindrically notched specimens (solid symbols), standard C(T)-specimens (shaded scatterband) and as predicted from microhardness measurements (open symbols).
▲Δ, Almen intensity 0.25 mmA

DISCUSSION

Although the present results strongly suggest that a gradation has no effect on fracture toughness, supporting the results in [10], it should be pointed out that the absolute values of strength and toughness may in fact prevent an experimental assessment of the actual toughness values for the non-standard geometry used. If it is assumed that failure must occur when the ultimate tensile strength is reached, independent of any crack propagation effects, it can be shown that the toughness is in fact too high to be measured in the present case. Using the values for the ultimate tensile strength given in Table II, an upper limit for the maximum toughness which can be measured can be calculated using equation 1. These values are also shown in Table II. It can be seen that these upper limits are lower than the K_Q-values obtained for the C(T)-specimens. If indeed the prediction for the aluminum alloy in the notch region is correct, toughness values in the bulk are too high to be measured. If however the prediction for the titanium alloy is correct, then the values in the notch region are too high to be measured. This type of specimen geometry has been previously used to assess fracture toughness [10,16], and investigation of the fracture surfaces confirmed that failure had indeed occurred by crack propagation and not by tensile fracture. This failure mechanism was presumably has enhanced by the multiaxial stress state caused by the notch. In contrast to the present work, the absolute values of the fracture toughness calculated for notched cylindrical specimens were consistently only about half those of the C(T)-specimens [10,16].

An additional consideration is the fact that while fracture mechanics presumes to describe the stress state and fracture resistance at a crack tip, the inevitable local deformation at the crack tip in fact means that material a certain distance away from the crack tip can still influence crack propagation behavior. Using the relation [17]

$$r_p{}^* = \frac{K_I{}^2}{2\pi\sigma_y{}^2} \tag{2}$$

to calculate the size of the plastic zone $r_p{}^*$ using the yield strength σ_y for unnotched tensile specimens at the maximum K-values which can be achieved for infinitesmally small cracks, it can be seen that the calculated plastic zone size has the same dimensions as the depth of the gradation. Hence, even a substantial difference in local fracture properties will not be measurable if it is contained within the plastic zone.

Table II: Strength and toughness parameters for conventionally treated material

Material / aging treatment	2017 Al / 6 h 170 °C	Ti-2.5Cu / 24 h 400 °C	Ti-2,5Cu / 8 h 400 °C + 8 h 450 °C
notch tensile strength (MPa)	530	940	1070
upper limit for K (MPa√m)	20-21	37-38	42-43
unnotched yield strength (MPa)	375 (20 % cold work)	790 (15 % cold work)	760 (15 % cold work)
plastic zone size at upper limit for K (μm)	480	360	500

CONCLUSIONS

1) Fracture toughness values measured on cylindrically notched specimens without a gradient in microstructure consistently underestimate the values from standard C(T)-specimens by 20 %. This is most likely related to the fact that the standard specimens violated the thickness criterion, which tends to yield values which are too high. However, the assessed values are independent of crack depth.

2) The fracture toughness values obtained on cylindrical specimens with a graded micro-stucture did not exhibit any effect of the gradation, i.e., the values were independent of crack depth. This observation could be a result of the fact that the expected fracture toiughness values may be too high to be measured. This is the case when the notch tensile strength is the limiting factor rather than the fracture toughness. A further hindrance to assessing any effect of a gradation may be caused when the size of the plastic zone is comparable to the depth of the gradation. Because the material in this region ahead of the crack tip contributes to fracture behavior, it may only be possible to determine gradation effects when the plastic zone is much smaller than the graded region.

ACKNOWLEDGMENTS

Funding for this work was provided by the Deutsche Forschungsgemeinschaft under contract no. Gr 974/4 within the framework of a focussed effort on graded materials (SPP 322).

REFERENCES

1 J.K. Gregory, C. Müller and L. Wagner: Preferential Surface Aging: Novel Methods for Improving Fatigue of Mechanically Loaded Parts (in German) *Metall*, **47**, Heft 10 (1993) 915-919.

2 L. Wagner and J.K. Gregory: Improve the Fatigue Life of Titanium Alloys - Part II, *Advanced Materials and Processes*, **145** (1994) 50HH-50JJ.

3 K.H. Kloos, E. Velten: Calculation of the Endurance Limit of Plasma-Nitrided Samples with a Component-like Geometry with Respect to the Hardness and Residual Stress Profile (in German) *Konstruktion* **36** (1984) 181-188.

4 B. Winderlich: The Concept of a Local Endurance Limit and its Application to Martensitic Surface Layers, Particularly after Age-Hardening (in German) *Materialwissenschaften und Werkstofftechnik* **21** (1990) 378-389.

5 J.D. Lubahn: On the Applicability of Notch Tensile Test Data to Strength Criteria in Engineering Design *Trans. ASME* (1957) 111-115.

6 W. Backfisch, E. Macherauch: Notch Tensile Tests on Quenched and Tempered 32 NiCrMo 14 5 (in German) *Arch. Eisenhüttenwesen* **50** (1979) 167-170.

7 C. Müller, U. Holzwarth, J.K. Gregory: Influence of Nitriding on Microstructure and Fatigue Behaviour of a Solute-Rich Titanium Alloy, *Fatigue Fract. Engng. Mater. Struct.* **20** (1997) 1665-1676.

8 F. Bohner, J.K. Gregory: Mechanical Behavior of a Graded Aluminum Alloy, Materials Science Forum 308-311 Trans-Tech Publications, Switzerland (1999) 313-318.

9 F. Bohner, J.K. Gregory, U. Weber, S. Schmauder: Yielding in Notch Tensile Specimens with Graded Microstructures, *Mechanics of Materials* **31** (1999) 627-636.

10 A. Berg, J. Kiese, L. Wagner: Microstructural Gradients in Ti-3Al-8V-6Cr-4Mo-4Zr for Excellent HCF-Strength and Toughness, *Mater. Sci. Engg. A* **243** (1998) 146.

11 H. Blumenauer: Technical Fracture Mechanics (in German) VEB Verlag, Leipzig (1987)

12 C. Q. Bowles: Fracture and Structure in ASM Handbook, Vol. 19, Fatigue and Fracture (1996) American Society for Materials, Materials Park, OH 5-14

13 K. Heckel: Introduction to the Technical Application of Fracture Mechanics (in German) C. Hanser Verlag Munich/Vienna (1983).

14 D. Kujawski: Estimations of Stress Intensity Factors for Small Cracks at Notches, *Fatigue Fract. Engng. Mater. Struct.* **14** (1991) 953-965.

15 G.R. Irwin: Trans. ASME *J. Appl. Mech.* **24** (1957) 361.

16 J.K. Gregory, L. Wagner, C. Müller: Notch Fatigue Behavior in Aged Ti-38-644 after Mechanical Surface Treatments, Beta Titanium Alloys, A. Vassel, D. Eylon, Y. Combres, eds,, SFMM, Paris (1994) 229-235.

17 G. R. Irwin: Plastic Zone near a Crack and Fracture Toughness, Proc. 7th. Sagamore Conf. (1960) IV-63.

A WEIGHT FUNCTION ANALYSIS OF R-CURVE BEHAVIOR IN GRADIENT ALUMINA-ZIRCONIA COMPOSITES

Robert Moon and Mark Hoffman
The University of New South Wales
NSW 2052, Australia

Jürgen Rödel
University of Technology, Darmstadt
D-64287 Darmstadt, Germany

Jon Hilden, William Blanton, Keith Bowman, and Kevin Trumble
Purdue University
West Lafayette, IN 47907, USA

ABSTRACT
    The single-edge-V-notched-beam (SEVNB) testing geometry was used to measure the crack growth resistance (R-curve) behavior of gradient multilayer alumina-zirconia composites. Fracture mechanics weight function analysis was applied to estimate the change in the apparent fracture toughness as a function of notch position, resulting from residual thermal stress distributions. These results were then used to differentiate the influence of residual stress from other closure stresses attributed to crack bridging on the measured R-curve behavior.

INTRODUCTION
    The measured apparent fracture toughness, K_R, and crack growth resistance behavior, which depend on both microstructure and residual stress, become functions of position within multilayered or gradient composites [1-4]. The ability to differentiate between microstructure-related toughening mechanisms (i.e. crack bridging, kinking, etc.) from residual stress based mechanisms is difficult [2]. Moon et al. [4,5] demonstrated that the weight function analysis could be used to differentiate the influence of a macro residual stress distribution (stepwise change in residual stress) from that of other microstructural mechanisms on the measured R-curve behavior of a layered alumina-zirconia composite. Experimental results showed that the macro residual stress distribution significantly influenced the measured R-curve behavior. This paper is a continuation of the work completed in reference [5].

EXPERIMENTAL PROCEDURE
Sample Preparation
    Multi-layered alumina-zirconia composites were produced by sequential centrifugal consolidation [6] of flocculated [2] aqueous alumina-zirconia slurries. Further details of the centrifugal procedure can be found in references [3] and [4]. Composite green bodies were dried at room temperature and then fired at 1600°C for 4 h in air. Specimens (4 mm x 3 mm x 35 mm) had a V-notch cut across the 3 mm x 35 mm face, perpendicular to the length of the bend bar, as described in previous work [7]. V-notch tip radii were measured to be 5 to 10 μm using optical microscopy. The 4 mm x 35 mm side-surfaces were additionally polished to 15 μm diamond abrasives in order to facilitate observation of crack growth.

Specimen Characterization

The specimens produced for this study (Table 1) consisted of ~6 layers of 76 vol% alumina + 19 vol% zirconia + 5 vol% platelike alumina. Additional information concerning the ceramic powders used in this study may be found in references [3,4,8]. A macro-view of a bend bar side surface is shown in Figure 1. During centrifugation, the larger platelike alumina particles preferentially settled, resulting in a gradient in the alumina/zirconia volume fraction, porosity, Young's modulus, coefficient of thermal expansion (CTE), and residual stress across each layer (Figure 2). The composition profile was determined using the ASTM E562 standard point counting method (900 points per location). The composite was tested in two different orientations with respect to the gradient within the layers. In orientation 1 (samples 1 and 2), the cracks were extended in the particle settling direction and in orientation 2 (samples 3 and 4) the cracks were extended opposite to the particle settling direction.

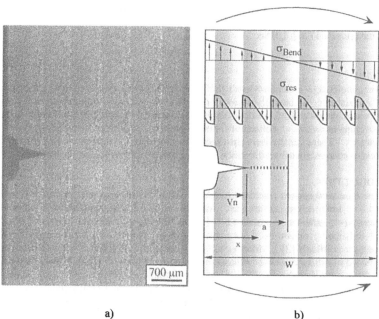

a)   b)

Figure 1   a) Optical micrograph showing the layer stacking and gradient microstructure with respect to the V-notch, and b) a schematic of the two stress components considered within the bend bar: bending ($\sigma_{Bend}(x)$), and residual stress ($\sigma_{res}(x)$).

Table I. List of samples tested.

Sample	Layer Thickness (μm)	Testing Orientation	Sample Cross-Section (mm) B	W	V-Notch Depth (μm)
1	~700	1	2.97	4.02	920
2	~700	1	2.96	4.03	935
3	~700	2	2.97	4.03	1045
4	~700	2	3.03	4.02	1340

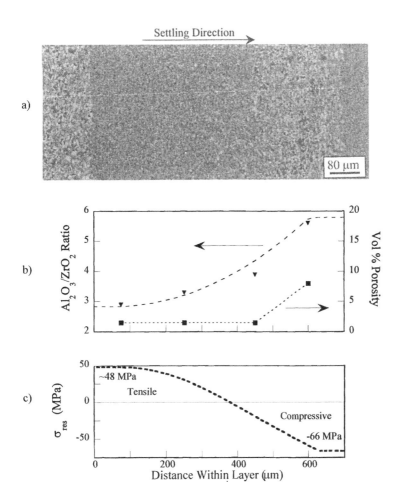

Figure 2   The gradient properties within each layer: a) optical micrograph, b) measured alumina-zirconia ratio (▼) and porosity (■), and c) the estimated residual stress distribution. Note that in b) the points mark actual data while the dashed curves are the estimated profiles used.

Mechanical Testing

Direct observations of crack initiation and extension were made *in situ* on the bend bar side surfaces using a specialized four-point bend fixture mounted on the stage of an optical microscope [3,9]. Specimens were tested under displacement-control where subcritical crack initiation and further crack extensions (~10 μm increments) were achieved using an incremental loading technique described in reference [3]. The load and crack length plus notch depth (on both sides of the bend bars) were measured, and then used to calculate the apparent fracture toughness [3].

EXPERIMENTAL ANALYSIS

Weight Function Analysis

Bueckner [10] showed that the stress intensity factor for an edge crack of depth a can be calculated by intergrating over the crack length the product of a weight function, $h(x,a)$ and any stress distribution, $\sigma(x)$, acting normal to the proposed fracture plane:

$$K_w = \int_0^a h(x,a)\sigma(x)dx \tag{1}$$

where x is the distance along the crack measured from the surface. The weight function, for a specific crack-component configuration, used in this study was developed by Fett and Munz [11,12] for an SENB sample and notch geometry.

Stress Distribution

Two independent stress distributions were considered in this study: an applied bending stress distribution, $\sigma_{Bend}(x)$, and a residual thermal stress distribution, $\sigma_{res}(x)$. A standard bending-stress formula was used for $\sigma_{Bend}(x)$, while $\sigma_{res}(x)$ was determined from the thermal expansion mismatch strains. Figure 1b shows schematically the two stress distributions acting within the sample.

The residual thermal stress distribution (Figure 2c) was determined by considering thermal strains in a single layer of the specimen. The layer was composed of several volume elements of width dx, length L_o, and depth B, each having uniform Young's modulus, $E(x)$, Poisson's ratio, $v(x)$, and CTE, $\alpha(x)$, as estimated by the Voigt Rule-of-Mixtures model for composite properties (Table II). If each element were allowed to cool separately by an amount ΔT, then each element, would have a different stress-free final length $(\alpha(x)\Delta T+1)L_o$. However, since each volume element is physically constrained by neighboring elements, each must reach the same final length, L_f. Thus, some elements are driven into a state of plane tension while others into plane compression. It was assumed that the stress distribution developing within the volume elements were alleviated via creep for temperatures above ~1025°C, thus, during cooling to room temperature, the temperature range over which the residual stress develops was $\Delta T = $ ~1000°C. The incremental load on each volume element is given by:

$$dF = \left[\frac{L_f - (\alpha(x)\Delta T+1)L_o}{(\alpha(x)\Delta T+1)L_o}\right]\frac{BE(x)}{1-v(x)}dx \tag{2}$$

where the term in square brackets defines the residual strain, $\varepsilon(x)$. The sum of all incremental loads was assumed to be zero, and the ratio L_f/L_o may be solved through the relation:

$$\int_0^{Layerthickness}\left[\frac{L_f/L_o - (\alpha(x)\Delta T+1)}{(\alpha(x)\Delta T+1)}\right]\frac{BE(x)}{1-v(x)}dx = 0 \tag{3}$$

Obtaining a L_f/L_o value of 0.99112, the residual stress in each volume element was determined from:

$$\sigma_{res}(x) = \frac{\varepsilon(x)E(x)}{1-v(x)} \tag{4}$$

Table II.   Material Properties of Composite Constituents

Layer (vol%)	E (GPa)	ν	α (10^{-6}/°C)
100 Al$_2$O$_3$	380	0.19	8.4
100 t-ZrO$_2$	205	0.3	11.5

Stress Intensity Factor Equations

The weight function analysis was used to calculate the stress intensity factors associated with each stress distribution, assuming that they act independently. The stress intensity factor, K_{Bend}, resulting from the applied bending loading condition, is as follows:

$$K_{Bend} = \int_0^a h(x,a)\sigma_{Bend}(x)dx \tag{5}$$

The stress intensity factor, K_{res}, resulting from the residual stress distribution acting along the total flaw length, a, is as follows:

$$K_{res} = \int_0^a h(x,a)\left(-\sigma_{res}(x)\right)dx \tag{6}$$

Note that the residual stress sign convention was reversed (i.e., - $\sigma_{res}(x)$), so that tensile stresses act to reduce K_{res} and compressive stresses act to increase K_{res}. For K_{res} <0, the residual stress distribution acting along a, reduces the measured stress intensity as compared to the zero residual stress situation, while K_{res} >0 increases the measured stress intensity factor.

The stress intensity factor, K_{br}, resulting from the bridging stress acting along the crack length extending from the V-notch was not calculated. An independent bridging stress function, $\sigma_{br}(a-Vn)$, could not be derived because the bridging stresses not only vary as a function of crack length but they also vary as a function of crack position within the gradient microstructure.

Apparent Stress Intensity Calculations

The stress intensity factors resulting from the two stress distributions were calculated separately as a function of total flaw length, a. The $K_{res}(a)$ was obtained by solving K_{res} for several values of a while the $K_{Bend}(a)$ profile was estimated by setting Eqn (5) equal to the critical stress intensity factor of 3.6 MPa·m$^{\frac{1}{2}}$, which was estimated from experimental results of monolithic samples having the 80 vol% alumina + 20vol% zirconia composition [5].

Using the principle of superposition, the two stress intensity factors were summed, producing an apparent stress intensity factor that accounts for the influences of the residual stress distribution, $K_{RR}(a)$, as shown below:

$$K_{RR}(a) = K_{Bend}(a) + K_{res}(a) \tag{7}$$

The resulting K_{RR} vs. a profile predicts what the measured apparent stress intensity, K_R, is for crack initiation from a notch tip at various depths, a, and should not be mistaken as an R-curve. The R-curve behavior of a material only reflects toughening influences associated with an increasing crack length.

RESULTS AND DISCUSSION

The microstructure and R-curves for the gradient samples tested in orientation 1 (samples 1 and 2) and in orientation 2 (samples 3 and 4) are shown in Figures 3 and 4, respectively. The plots consist of three curves: two measured R-curves and the calculated K_{RR} profile for the given testing orientation. The superposition of several R-curves measured from a layer or gradient sample can be inappropriate due to the influences caused by starting the R-curve measurement at different locations within the microstructure. For samples 1 and 2 the V-notch position (R-curve starting position) within a given layer were nearly identical, whereas the sample 4 V-notch was ~20 µm closer to the bottom of the next layer as compared to sample 3. To line up the layer interface for both samples, 20 µm was subtracted from the measured crack length for each data point of the sample 4 R-curve.

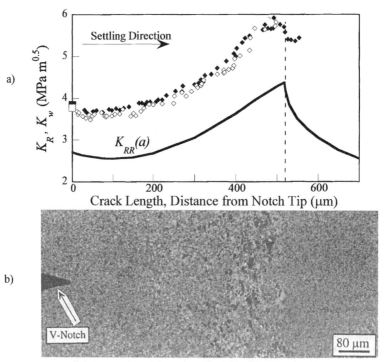

Figure 3   Orientation 1: a) The measured sample 1 (◆) and 2 (◇) R-curves with the K_{RR} profile b) A representative microstructure through which the crack extended. The (■ and □) symbols on the K_R-axis represents the K_R for crack initiation from the V-notch tip.

The similarity of the two measured R-curves for each testing orientation demonstrates the consistency of the R-curve measurement technique. Additionally, the gradient microstructure resulted in an asymmetry in the measured R-curves and K_{RR} profiles from orientations 1 and 2. The orientation 1 R-curves have a gradual increase in slope until the end of the platelike particle region was reached, resulting from the gradual increase in residual compressive stress and bridging stresses within the platelike particle region of the layer. The orientation 2 R-curves initially have a near flat R-curve until the crack impinged on the bottom of the next layer, the resulting steeper rise in K_R was due to the immediate increase in the residual compressive stress and bridging stresses.

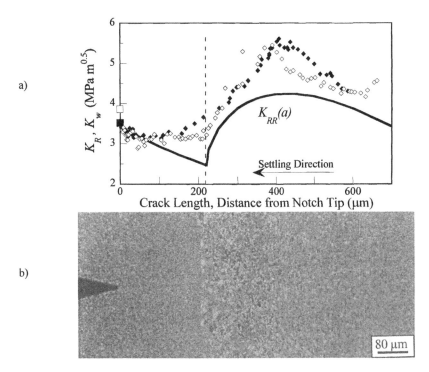

a)

b)

Figure 4   Orientation 2: a) The measured sample 3 (◆) and 4 (◇) R-curves with the K_{RR} profile. b) A representative microstructure through which the crack extended. The (■ and □) symbols on the K_R-axis represent the K_R for crack initiation from the V-notch tip.

The calculated K_{RR} profiles demonstrated that the residual stress distribution influenced the apparent stress intensity as a function of position and that this variation was dependent on the testing orientation (1 or 2). The study by Moon et al. [5] showed that the K_{RR} profile directly influenced the measured R-curve in specimens containing step-wise changes in residual stress. Additionally, by including the influence of the bridging stresses acting along the crack length to the K_{RR} calculations, the resulting stress intensity profile (K_{RRB}-including bridging) closely estimated the experimentally measured R-curve. In the current study, the K_{RR} profiles deviated from the measured R-curves in part because the bridging stresses were not taken into account. However, the K_{RR} profile still shows how the residual stress distribution influenced the measured R-curve where, in general, the measured R-curve follows the K_{RR} profile, in orientation 1 more so than in orientation 2. Notice that where the K_{RR} profile is increasing, the measured R-curve increases, and where the K_{RR} profile decreases, the measured R-curve decreases.

It should be noted that for orientation 1 the K_{RR} profile underestimated the measured R-curve by ~1 MPa·m$^{1/2}$ over the entire extension of the crack. This is unexpected since toughening by bridging is small at short crack lengths, thus for distances <10 μm from the notch tip both curves should have similar K_R values (Figure 3). This difference may have resulted from $\sigma_{Bend}(x)$ not being accurately estimated due to the variation of modulus (~24 GPa) across each layer. However, this irregularity was not as apparent in samples 3 and 4, which were tested in orientation 2. Additionally, incorrect estimates of the elastic modulus and coefficient of thermal expansion across each layer may have caused other deviations between the measured R-curve and the K_{RR} profiles.

SUMMARY

Results have shown that the gradient microstructure and residual stress distribution both have a significant influence on the measured R-curve behavior. However, without an accurate estimate of $\sigma_{Bend}(x)$ and $\sigma_{br}(a-Vn)$, the weight function analysis could not predict the measured R-curve behavior of the samples tested. This research was supported by the Army Research Office MURI grant no. DAAH04-96-1-0331.

REFERENCES

[1] R. Lakshminarayanan, D. Shetty, and R. Culter, "Toughening of Layered Ceramic Compoites with Residual Surface Compression," J. Am. Ceram. Soc., 79 [1] 79-87 (1996).

[2] D. Marshall, J. Ratto, and F. Lange, "Enhanced Fracture Toughness in Layered Microcomposites of Ce-ZrO$_2$ and Al$_2$O$_3$," J. Am. Ceram. Soc., 74 [12] 2979-87 (1991).

[3] R. Moon, K. Bowman, K. Trumble, and J. Rödel, "A Comparison of R-Curves from SEVNB and SCF Fracture Toughness Test Methods on Multilayered Alumina-Zirconia Composites," J. Am. Ceram. Soc., 83 [2] 445-47 (2000).

[4] R. J. Moon, "Static Fracture Behavior of Multilayer Alumina-Zirconia Composites," PhD Thesis, Purdue University (May 2000).

[5] R. Moon, M. Hoffman, J. Hilden, K. Bowman, K. Trumble, and J. Rödel, "A Weight Function Analysis on the R-curve Behavior of Multilayered Alumina-Zirconia Composites," unpublished work.

[6] J. C. Chang, B. V. Velamakanni, F. F. Lange, and D. S. Pearson, "Centrifugal Consolidation of Al$_2$O$_3$ and Al$_2$O$_3$/ZrO$_2$ Composite Slurries Vs Interparticle Potentials: Particle Packing and Mass Segregation," J. Am. Ceram. Soc., 74 [9] 2201-204 (1991).

[7] J. Kübler, "Fracture Toughness Using The SEVNB Method: Preliminary Results," Ceram. Eng. & Sci. Proc., 18 [4] 155-162 (1997).

[8] R. Moon, K. Bowman, K. Trumble, and J. Rödel, "Fracture Resistance Curve Behavior of Multilayered Alumina-Zirconia Composites Produced by Centrifugation," submitted to Acta Materialia (August 2000).

[9] M. Stech and J. Rödel, "Method for measuring short-crack R-curves without calibration parameters: case studies on Alumina and Alumina/Aluminum composites", J. Am. Ceram. Soc., 79 [2] 291-97 (1996).

[10] H. F. Bueckner, "A Novel Principle for the Computation of Stress Intensity Factors," Z. Angew Math Mech., 50 529-46 (1970).

[11] T. Fett and D. Munz, "Influence of Crack-Surface Interactions on Stress Intensity Factor in Ceramics," J. Mater. Sci. Lett., 9, 1403-406 (1990).

[12] T. Fett and D. Munz, "Determination of Fracture Toughness at High Temperature After Subcritical Crack Extension," J. Am. Ceram. Soc., 75 [11] 3133-36 (1992).

FRACTURE TOUGHNESS AND R-CURVE BEHAVIOR OF Al$_2$O$_3$/Al FGMs

T.-J. Chung, A. Neubrand and J. Rödel
University of Technology Darmstadt
Department of Materials Science
Petersenstr. 23
64287 Darmstadt
Germany

T. Fett
Institut für Materialforschung
Forschungszentrum Karlsruhe
Postfach 3640
76021 Karlsruhe
Germany

ABSTRACT
    Quantitative determination of fracture toughness and R-Curve behavior in FGMs is still an issue as a precise determination of the stress intensity factor in FGMs is not straightforward due to the spatial variation in elastic properties. In the present work it is demonstrated that the weight function method can be adapted to FGMs. Fully dense Al$_2$O$_3$/Al FGM compact tension specimens with different well controlled gradients in the direction of crack propagation were prepared by an infiltration technique, and experimental data for the R-curve behavior of the different Al$_2$O$_3$/Al FGMs were determined.

INTRODUCTION
    Systematic experimental investigations on fracture toughness and R-curve behavior of FGMs are still a rarity. In many cases, reported fracture toughness values were determined by the indentation method [1,2,3]. The simplicity of this method is attractive, but the validity of the obtained results is questionable for metal/ceramic composites in general and metal/ceramic FGMs in particular. The SENB method is a viable alternative and has been used to determine K$_{Ic}$ of metal/ceramic FGMs [4,5]. The stress intensity factor for a FGM differs from that in a homogeneous material due to the gradient in Young's modulus [6]. Experimental investigations of crack propagation and R-curve behavior of FGMs thus require special methods for the determination of the stress intensity. Butcher et. al. [7] determined stress intensity factors for a compact tension (CT) specimen with linear gradation of the Young's modulus in the direction perpendicular to the crack. For a graded epoxy/glass composite they determined the stress intensity factors experimentally by an optical method and by finite element analysis (evaluation of the contour integral around the crack tip). The results obtained with these two methods agreed very well. In the present work, Al/Al$_2$O$_3$ FGMs with continuous gradients were fabricated by a foam

replication method, and the R-curve behavior of FGMs with concave, linear and convex gradation profiles was determined experimentally. The stress intensity factors and the weight function for CT specimens with a gradient parallel to the crack were calculated for these gradation profiles and the results were compared with those for the homogeneous case.

THEORY
Stress intensity factors

R-curves were determined with standard compact tension (CT) specimens with dimensions $W = 28$ mm, $c = W/4$ and thickness $B = 3$ mm (for the geometric data see insert in Fig. 1a). In the case of a homogeneous material, the stress intensity factor under the load P, is for a crack length a [8]

$$K_{hom} = \frac{P}{B\sqrt{W}} \frac{(2+\alpha)(0.886 + 4.64\alpha - 13.32\alpha^2 + 14.72\alpha^3 - 5.6\alpha^4)}{(1-\alpha)^{3/2}} \tag{1}$$

$$\text{where } \alpha = a/W$$

In terms of the total crack length $a+c$ and the total specimen width $W+c$ the relative crack length reads :

$$\beta = \frac{a+c}{W+c} = \tfrac{4}{5}(\alpha + \tfrac{1}{4}) \tag{2}$$

The graded material addressed in this contribution exhibits a change of Young's modulus in the region $x_0 \le x \le W$ according to :

$$E(x) = E(x_0) + [E(W) - E(x_0)]\left(\frac{x - x_0}{W - x_0}\right)^n \tag{3}$$

Strictly, Eq.(1) does not hold for graded materials. For our material composition (max. 30 vol.% Al) and the extension of the graded region $(0.5 \le \beta \le 1)$ we obtained the ratio of the stress intensity factor for graded material K_{grad} and homogeneous material K_{hom} as :

$$\frac{K_{grad}}{K_{hom}} = \begin{cases} 1.8834 - 2.8637\beta + 3.2071\beta^2 - 1.2264\beta^3 & \text{for} \quad n = 1/3 \\ 1.022 + 0.249\beta - 0.2713\beta^2 & \text{for} \quad n = 1 \\ 0.8204 + 0.3424\beta + 0.4768\beta^2 - 0.6399\beta^3 & \text{for} \quad n = 3 \end{cases} \tag{4}$$

Weight function:

In the next step of evaluation, it is planned to determine the relation between the bridging stresses σ_{br} and the crack face separation δ. In order to evaluate the bridging relation $\sigma_{br} = f(\delta)$, the fracture mechanics weight function is necessary for the actual transition function. In previous studies [9,10] it has been found that the general weight function procedure is applicable also for graded materials and no changes are necessary compared with the procedure for homogeneous materials.

By use of the Rice equation [11], the representation of stress intensity factors by the weight function and the separation of the total stresses into the applied stresses in the uncracked specimen along the prospective crack line and the crack bridging stresses in the cracked body :

$$h = \frac{E'}{K_I}\frac{\partial \delta}{\partial a} \quad , \quad K_I = \int_0^a \sigma(x)h(x,a)dx \quad , \text{ and } \sigma = \sigma_{appl}(x) + \sigma_{br}(x), \tag{5}$$

we obtain for the crack opening displacement δ the integral equation :

$$\Rightarrow \delta(x) = \int_x^a \frac{h(x,a')}{E'(a')}\left[\int_0^{a'} h(x',a')[\sigma_{appl}(x') + \sigma_{br}(x')]dx' \right] da' \tag{6}$$

where σ_{br} is again a function of the total crack opening displacement. The solution of the integral equation can be determined by several methods. The simplest one is iterative approximation (see e.g. [12]). Equation (6) is nearly identical with the relation for homogeneous materials. The only difference is the occurrence of E' under the integral sign.
A weight function with three terms reads [10]

$$h = \sqrt{\frac{2}{\pi(a+c)}}\left[\frac{1}{\sqrt{1-\rho}} + D_1\sqrt{1-\rho} + D_2(1-\rho)^{3/2} \right], \quad \rho = \frac{x+c}{a+c} \text{ where} \tag{7}$$

$$D_1 = \frac{A_0 + A_1\beta + A_2\beta^2 + A_3\beta^3}{(1-\beta)^{3/2}} \text{ and } D_2 = \frac{B_0 + B_1\beta + B_2\beta^2 + B_3\beta^3}{(1-\beta)^2} \tag{8}$$

with the coefficients compiled in Table 1. The weight functions are shown in Fig. 1 for several relative crack sizes, for different values of n and for homogeneous material (dashed lines).

n	A_0	A_1	A_2	A_3	B_0	B_1	B_2	B_3
1/3	1.5193	2.450	-2.757	1.000	0.4808	-0.0868	1.660	-1.2166
1	1.5729	0.9567	1.6357	-2.1667	0.1222	1.2117	0.2214	-0.6666
3	1.4716	-1.2203	7.9763	-6.2476	0.2485	0.23665	2.1143	-1.6666

Table 1 Coefficients for the weight function representation according to eqs.(5) and (6).

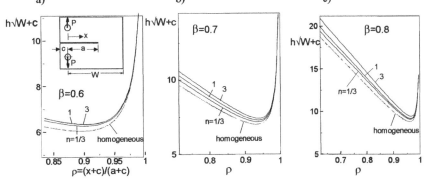

Fig. 1 Weight function for different values of the .exponent of Young's modulus n. a) relative crack size a) $\beta = 0.6$, b) $\beta = 0.7$ c) $\beta = 0.8$

EXPERIMENTAL PROCEDURES

Al/Al$_2$O$_3$ FGMs with compact tension geometry were prepared using the GMFC (Gradient Materials by foam compaction) replication process first described in [13]. The process is outlined in Fig.2. Soft Polyurethane foam (Bulpren S90, Eurofoam, Troisdorf, Germany) with a relative density of 2.35% and a pore size corresponding to 87 ppi (pores per inch) was used as a starting material to produce graded preforms. Different FGMs with gradation exponents n=1/3, 1 and 3 (cf. Eq. 3) were produced. For this purpose wedges with different shapes were cut from PU foams using a hot wire (Item 1 in Fig.2b). The wedges were compacted to a square plate by enclosing them in a die of uniform thickness at 200°C for 30 min (Item 2 in Fig.2b). Compaction occurred only in the thickness direction. As the Poisson's ratio of the PU foams was close to zero, the porosity gradients of the retained PU foams were almost identical with the intended gradation functions. The reproducibility of the gradients was high and deviations in porosity were typically less than 2% (Fig.3). The compacted foams were placed on a plaster of Paris mold and infiltrated with alumina slips in the next production step. As the pore size of the polyurethane foam (several tens of microns depending on compaction ratio) is much larger than the particle size of the alumina a green body was produced inside the pore network of the foam during slip casting (Item 3 in Fig.2b). After carefully drying the green bodies the polyurethane foam was burned out by heating to 400°C at 0.4K/min, to 800°C at 0.7K/min and to 1050°C at 2K/min. After sintering at 1450°C for 60 min a plate with a porosity gradient in the length direction was obtained (Item 4 in Fig.2b). The plates were infiltrated at 1050°C with molten Al applying 10 MPa argon gas pressure for 2h. The Al phase of the obtained Al/Al$_2$O$_3$ FGM (Item 5 in Fig.2b) was a replica of the struts of the PU foam - it had identical microstructure and volume content (Fig.4). Finally the FGM plate was cut and ground to the shape of a CT specimen (Item 6 in Fig.2b). The specimen was polished and a crack was initiated from a Vickers indentation starter crack. The specimen was mounted into a piezoelectric in situ loading device which can be used in an optical or electron microscope [14]. R-curves were determined by observation of crack extension in a microscope and Eqs.(1) and (4) were used for the calculation of stress intensity factors.

Fig.2
a)) Outline of the GMFC process.
b) Graded Al/Al$_2$O$_3$- compact tension specimen at consecutive processing stages. The numbers correspond to Fig. 2a.

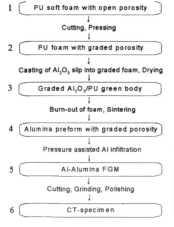

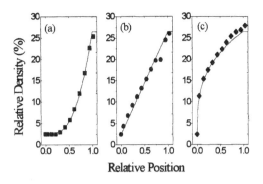

Fig. 3 Gradation profiles of PU foams. Full line - intended density gradients. Data points: Observed density gradients.
a) gradation exponent n=3
b) gradation exponent n=1
c) gradation exponent n=1/3

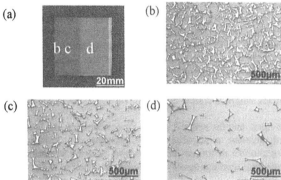

Fig. 4 Microstructure of Al/Al$_2$O$_3$ FGM plate.
a) overview
b) microstructure at 31% metal content
c) microstructure at 15% metal content
d) microstructure at 4% metal content.

RESULTS AND DISCUSSION

Homogeneous Al/Al$_2$O$_3$ Composites

The R-curve of a composite containing 13 vol.% Al is shown in displays three regimes : I) At short crack lengths below 0.5 mm K_R increases rapidly to values of 5 MPa√m. II) For crack lengths between 0.5 and 3.5 mm K_R a slower linear increase to 8 MPa√m is observed. III) At even longer cracks the plateau of the R-curve is reached at K_R values of about 8.2 MPa√m. The three regimes may be explained qualitatively as follows : In region I) both the number of Al ligaments and the bridging stress exerted by a single ligament $p(\delta)$ increase. Additionally there may be a small contribution of bridging alumina grains at very short crack lengths (and crack opening displacements). In region II) the number of Al ligaments bridging the crack still increases, but $p(\delta)$ no longer increases for those ligaments which span large crack openings. The latter effect is due to the onset of yielding of the metal. Finally the plateau is reached if the number of bridging ligaments becomes constant as the creation of new ligaments near the crack tip is compensated by failure of ligaments at larger crack length (and crack opening displacement).

Graded Al/Al$_2$O$_3$ Composites

For our case of Al/Al$_2$O$_3$ with a graded metal content the determination of stress intensity factors according to (Eq.1) produces very similar results as compared to Eq.(4) for all three gradients (Figs.6, 7 and 8). This is explained by the relatively small ratio of maximum/minimum Young's modulus $E(x_0)/E(W)$ along the gradient which is only about 4/3 for the FGMs presented here. This relatively small change in Young's modulus does not strongly affect the weight function and therefore, K_{hom} and K_{grad} differ only by 10% or less. Obviously it is possible to use K_{hom} for graded CT specimens if deviations of 10% from the actual K value can be tolerated. This agrees well with stress intensity values determined for a graded Ti alloy/ZrO$_2$ CT specimen with exponential gradation function [6]. The Young's modulus of the Ti alloy (110MPa) was similar to that of ZrO$_2$ (150 MPa) and the stress intensity factors calculated for the homogeneous and graded case differed by 10% or less. It should be pointed out, however, that the errors in the determination of K values using Eq.(1) increase as the ratio of Young's moduli $E(x_0)/E(W)$ deviates more and more from 1. For ratios below ½ or above 2 Eq.(1) should not be used irrespective of the value of the gradation exponent n.

The R-curve for FGMs with n=1/3 (Fig.6) resembles the distribution in metal content. For this kind of FGM the effects of rising number density of ligaments and increasing closure stresses $p(\delta)$ are most prominent at low metal contents, and consequently K_R reaches substantial values exceeding 10 MPa√m even for short crack lengths. As the metal content increases only very slowly for large crack lengths a plateau-like behavior is observed for long cracks.

For the FGM specimen with n=1, K_R rises steeply for short crack lengths of less than 1mm (Fig.7). For higher crack extensions, the R-curve rises continuously and the R-curve shows no plateau. This behavior can be explained by the continuously rising metal content of the samples which is 3% at Δa=0mm and about 25% at Δa=14mm. As the size of the metal ligaments is independent of position (they are replicated struts of the PU foam which have an average thickness of about 40μm) the ligaments have the same $p(\delta)$-function for all positions along the gradient and the increase of K_R with increasing crack length is caused by the increasing number density of the metal ligaments.

Although the metal content for short crack lengths below 3 mm is almost constant, the R-curve of FGMs with n=3 also shows an early increase of K_R (Fig.8). The explanation is similar to the homogeneous CT sample : increasing number of bridges and increasing bridging stresses exerted by the ligaments. The fact that the absolute ligament density is low reflects itself in a low K_R (<4 MPa√m) in this region. No plateau is found as the metal content increases very rapidly for long crack extensions.

Finally, K_R values observed for different gradation functions were compared for identical metal content (Table 2). The graded samples have a higher fracture toughness at a given metal content irrespective of the gradation function. There may be two reasons for this behavior : The gradation in metal ligament density or residual stresses caused by the gradient. Unfortunately, the bridging stresses as a function of crack opening displacement are not known at present which precludes a calculation of expected K_R values for different gradients. However, it is expected that the graded materials have lower bridging stresses as the crack propagates always from low to high metal content in the FGMs. Thus, it is likely that residual macrostresses produced during cooling of the FGMs from the infiltration temperature are responsible for the increased fracture toughness of the graded specimens. Such effects of residual stresses on the fracture toughness of FGMs can be very pronounced [5,15]. In the Al$_2$O$_3$/Al FGMs, the metal rich part has a higher thermal expansion coefficient and residual stresses are expected to be tensile near the end of the sample and compressive near the crack tip. These residual stresses near the crack tip add to the applied and bridging stresses and decrease the effective stress intensity at the crack tip of the FGM. Obviously, in the homogeneous material no residual macrostresses should exist which can explain its lower fracture toughness.

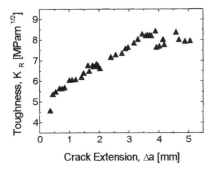

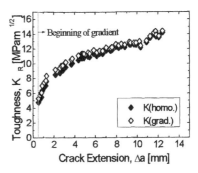

Fig.5 R-curve of homogeneous CT specimen containing 13 Vol.% Al.

Fig. 6 R-curve of CT specimen with gradation exponent n=1/3.

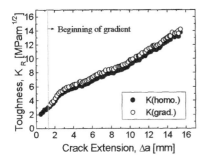

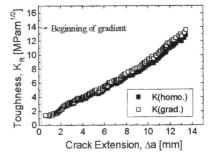

Fig.7 R-curve of CT specimen with gradation exponent n=1.

Fig 8 R-curve of CT specimen with gradient exponent n=3.

n	homogeneous	1/3	1	3
K_R (MPa$\sqrt{m}$)	8.0	9.4	9.3	12.7

Table 2 K_R at 13% metal content for homogeneous samples and FGMs with different gradation exponents n.

CONCLUSIONS

Weight functions and stress intensity factors for graded CT specimens have been determined. For moderate differences in Young's modulus, stress intensity factors in graded samples were similar to those in homogeneous samples. R-curve behavior of the graded samples was strongly affected by the graded metal distribution and residual stresses in the FGM samples.

ACKNOWLEDGEMENTS

Financial support of this project by the Deutsche Forschungsgemeinschaft is gratefully acknowledged.

REFERENCES

[1] K. Hirano, Application of a Ductile Metallic Phase Toughening Mechanism to Ceramic/Metals, Functionally Graded Materials, Proceedings of the 3rd International Symposium on FGM, Lausanne 1995, pp. 301-306

[2] A. Kawasaki und R. Watanabe, Effect of Gradient Microstructure on Thermal Shock Crack Extension in Metal/Ceramic Functionally Graded Materials, Functionally Graded Materials 1996 (I. Shiota and Y. Miyamoto eds.), Proceedings of 4th Int. Symp. on Structural and Functional Gradient Materials, Oct.20-24, 1996, Tsukuba, Japan, pp. 143-148

[3] Z. Li, K. Tanihata und Y. Miyamoto, Elaboration of Symmetric Functionally Gradient Materials of the Al_2O_3/TiC/Ni/TiC/Al_2O_3 System, Proceedings of the 3rd International Symposium on FGM, Lausanne 1995, pp. 109-114

[4] C.-Y. Lin, H.B. McShane und R.D. Rawlings, Fracture Behaviour of Silicon Carbide/Aluminium-2124 Alloy Functionally Graded Materials, Proceedings of the 3rd International Symposium on FGM, Lausanne 1995, pp. 327-332

[5] J.S. Lin and Y. Miyamoto, Internal Stress Behaviour of Symmetric Al_2O_3/TiC/Ni FGMs, Materials Science Forum 308-311, W.A. Kaysser (ed.), Trans Tech Publications, Switzerland 1999, pp.855-860

[6] Z.-H. Jin and R.C. Batra, R-Curve and Strength of a functionally graded material, Materials Science and Engineering A242, 70-76 (1998)

[7] R.J. Butcher, C.-E. Rousseau, and H.V. Tippur, "A Functionally Graded Particular Composite : Preparation, Measurements and Failure Analysis", Acta Mater. 47, 259-268 (1999)

[8] J.E. Srawley, Wide range stress intensity factor expressions for ASTM E399 standard fracture toughness specimens, Int. J. Fract. Mech. 12, 475-476 (1976)

[9] T. Fett, D. Munz, Y.Y. Yang, Applicability of the extended Petroski-Achenbach weight function procedure to graded materials, *Engineering Fracture Mechanics* 65, 393-403 (2000)

[10] T. Fett, D. Munz, Y.Y. Yang, Direct adjustment procedure for weight function of graded materials, Fat. Fract. Engng. Mater. Struct. 23,191-198 (2000)

[11] J.R. Rice, Some remarks on elastic crack-tip stress fields, Int. J. Solids and Structures 8, 751-758 (1972)

[12] D. Munz, T Fett, CERAMICS, Springer Verlag, Heidelberg, 1999

[13] F.R. Cichocki Jr, K. P. Trumble, J. Rödel, Tailored Porosity Gradients via Colloidal Infiltration of Compression Molded Sponges, J.Am.Ceram.Soc. 81, 1661-64 (1998)

[14] J. Rödel, J.F. Kelly, B. Lawn, In Situ Measurements of Bridged Crack Interfaces in the Scanning Electron Microscope, J. Am. Ceram. Soc. 73[11], 3313-18 (1990)

[15] R. Moon, M. Hoffman, J. Hilden, K. Bowman, K.P. Trumble and J. Rödel, A weight function analysis on gradient alumina-zirconia composites, these proceedings

FRACTURE IN DUCTILE/BRITTLE GRADED COMPOSITES

Jesus Chapa, Keith Rozenburg and Ivar Reimanis
Metallurgical and Materials Engineering Department
Colorado School of Mines
Golden, Colorado 80401

Eric D. Steffler
Idaho National Engineering and
Environmental Laboratory
Metals and Ceramics Group
PO Box 1625 MS 2218
Idaho Falls, ID 83415

ABSTRACT

Experiments in conjunction with finite element analysis (FEA) were used to better understand the crack propagation in ductile/brittle graded composites. Graded Cu/W composites were fabricated by stacking discrete layers of powders and hot pressing. Notches were cut perpendicular to the gradient on different composition layers. Experiments show that the gradient in material properties shifts the phase angle of loading, which causes the crack to deflect. Moiré interferometry and FEA are used to assess the relative influence of elastic-plastic gradient and thermal residual stresses on the phase angle. FEA and experimental results suggest that the elastic-plastic gradient has the largest effect on shifting the phase angle.

INTRODUCTION

Past theoretical and experimental work has shown that a crack kinks out of plane under mixed-mode loading. Crack kinking under mixed-mode loading of brittle materials has been described by the maximum principal stress criterion [1], minimum strain energy density criterion [2] and maximum energy release rate criterion [3,4,5]. These different criteria predict the kink angle to increase proportional to the applied shear. Analytical studies [6] show that a crack in a linear-elastic graded material loaded under remote $K^R_{II}=0$ load kinks out of plane due to shear introduced by the elastic gradient resulting in local $K^L_{II} \neq 0$ at the tip of the un-kinked crack. Gu and Asaro [6] obtained the crack kinking angle using the Cotterell and Rice [4] analytical solution for which the mode II stress intensity factor at the tip of the kink K^K_{II} vanishes. Recent numerical studies [7] show that accurate predictions of crack kink angle must include the effect of the T-stress in addition to the local phase angle effect.

Thermal residual stresses resulting from processing or heat treatment could have an effect on the local phase angle of the un-kinked crack, ultimately affecting the crack kink angle. Plasticity would be expected to alter shear stress distributions, and therefore should also be an important parameter in governing the kink angle. A theory for predicting the crack path for elastic-plastic graded composites with residual stresses does not exist.

The present study builds on the approach of crack kinking based on the criteria mentioned above; however, the influence of a plasticity gradient and thermal residual stresses introduces further challenges. Systematic fracture studies, combined with moiré interferometry techniques and FEA are used in this work to better understand the fracture behavior for cracks perpendicular to the gradient in graded composites.

EXPERIMENTAL

Dense Cu/W composites were hot pressed by powder metallurgy methods. High purity (99.9%) Cu powder, with a particle diameter between 1 and 5 μm, was purchased from Atlantic Equipment Engineering in Bergenfield, NJ. High purity (99.95%) W powder with an average particle size of 0.5 μm was purchased from Goodfellow Cambridge Ltd., in Huntington, England. The powders of Cu and W were weighed to produce a certain volume percentage per interlayer, assuming zero porosity. The powders were then mixed and ball milled in a polycarbonate bottle with four to five cleaned alumina balls for 24 hrs. Longer mixing times tends to result in agglomeration of the copper powder. A graphite die was prepared by spraying the interior walls with boron nitride aerosol paint. The die was then baked in an oven at 75°C for 1 hour prior to filling with powders to inhibit carbon diffusion into the sample. Before the powder was poured in the die, graphoil was used to cover all the interior surfaces, facilitating sample extraction after fabrication. Samples of specific compositions were then poured by hand into a 1 inch diameter graphite die for hot pressing. After each composition was poured in, the powder was leveled using a clean, graphite die punch. The samples were then hot pressed in a Thermal Technologies vacuum hot press. A vacuum of 10^{-4} torr was maintained at all times. A very small pressure was applied at room temperature, prior to heating to insure that the hot press ram made contact with the die assembly. The die was then heated at 20°C/min, with a 30 min soak at 200°C and a 15 hr soak at 985°C. A load of 40 MPa was applied at the start of the run. Cooling was accomplished with the load still applied at a rate of 20 °C/min, until about 500 °C at which point the rate slowed substantially. The load was removed at about 400 °C.

The hot pressed cylinders consisted of 60 vol. % W composite material on the ends of the cylinder, and monolithic Cu in the middle, as shown in Figure 1. The ends and the cylinder center were separated by graded layers of composition 20 and then 40 vol. % W. All layers were either 2 mm or 4 mm in thickness except the 60 vol. % W layers which were varied to insure that the overall length of the cylinder was about 25 mm. Based on FEA modeling studies, the magnitude of the residual stresses would be expected to be different for the two types of composites produced here [8]. The composition range stopped at 60 %W because sintering of higher composition W interlayers could only be accomplished above the melting point of Cu, and the higher temperatures resulted in substantial loss of copper during hot pressing. Bars approximately 25 mm x 4 mm x 8 mm in the orientation shown in Figure 2 were cut from the cylinders using a diamond cutting wheel. The diamond saw was used to cut notches approximately 4 mm deep, symmetrically within either the 20 % W or 40 % W layer, as illustrated schematically in Figure 2. The notch width was approximately 0.2 mm. It was assumed in the fracture experiments that the damage induced from the saw blade was distributed well around the notch, and that crack nucleation would not be controlled by the location of a flaw. This assumption was validated by conducting several tests and observing highly repeatable results.

The bend bar specimens were loaded under bending in a four point bending fixture an Instron uniaxial tester, such that the notch tip experienced the highest tensile stress. The specimens were loaded until a crack propagated from the notch. In most of the cases, the crack ran unstably until the specimen fractured. In some of the specimens for which the notch was in the 20 %W layer, the crack ran towards the 0 %W/20%W interface and arrested. Some four point bending experiments were conducted in an optical moiré interferometer so that displacement fields could be determined while the sample was being loaded, as described elsewhere [9]. The crack deflection angles were measured by scanning electron microscopy. The crack deflection angle is defined as the angle away from the direction of the notch. Thus, the crack deflection angle would be zero for an isotropic four point bend bar; experiments on 20% W and 40% W composites made by hot pressing have indeed confirmed this. Numerical simulations and experiments have confirmed that the presence of the deep notch on one side of the sample does not induce shear stresses at the base of the notch.

Figure 1. Hot pressed cylinder and test bar of a typically Cu/W graded composite with 2 mm interlayers. The center layer is 0% W and the subsequent layers on either side are 20 %W/40% W/60% W.

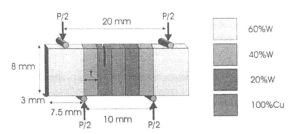

Figure 2. Schematic of four point bend bar geometry used in fracture tests. The interlayer thickness, t, varied from 2 mm to 4 mm in the present study.

FINITE ELEMENT MODELING
    The finite element code ABAQUS was employed to evaluate displacement and stress fields at the base of the notch. Plane strain quadratic elements were used, and the mesh around the notch tip was refined to 0.05 mm. The average number of nodes used was 4250. Specimens were heated to 985°C and then cooled to 25°C to simulate the composite fabrication. Simple rules of mixtures have been used to estimate the material properties of the different layers. Table 1 shows the properties of Cu and W. Due to a difference in thermal expansion coefficient between Cu and W, the cooling induced thermal residual stresses. Several simulations were carried out without the initial heating/cooling cycle, so that specimens devoid of thermal residual stress could be simulated for the purposes of examining the effect of residual stress on crack deflection.

Table I. Materials properties used in the FE model. W was modeled as an elastic solid. The copper yield stress varied with temperature [10].

Material	Coefficient of Thermal Expansion (x 10^{-6} /°C)	Elastic Modulus (GPa)	Yield Stress (MPa)
Cu	17	130	42-210
W	4.5	408	NA

RESULTS

The crack deflection angle depends on the notch location in the graded composite, as shown in the scanning electron micrographs in Figure 3 for four samples fractured in four point bending. In Figures 3a), b) and c), the crack is deflected towards the copper-rich part of the graded composite, to the right of the figure. The crack deflection angle is smaller when the crack resides in the 40 %W interlayer (figures 3b) and d)) compared with when the crack resides in the 20 %W interlayer (Figures 3a) and c)). Furthermore, the crack deflection angle decreases when the interlayer thickness increases. The average crack deflection angles measured from three to five samples for each condition are summarized in Table 2.

Figure 4a) reveals a distortion in the local displacement field around the loaded notch tip predicted by finite element analysis. The contour lines represent the in-plane, horizontal displacement for the specimen geometry of Figure 2. Each contour color is approximately 5.5 μm. It is noted that an isotropic bar in bending would exhibit a symmetric displacement field around the notch tip, even when the notch is not centered. The shift in displacement field from a symmetric pattern corresponds qualitatively to the shift observed in optical Moiré interferometry experiments on Figure 4b).

Because the maximum principal stress is frequently an appropriate crack propagation criterion for brittle materials, its location in the notch vicinity was determined from the FE model, as shown in Figure 5. Figure 5a) and 5b) show the inclination of maximum principal stress for the 20 %W and 40 %W interlayers in the FE model which did not include the thermal residual stress. The maximum principal stress is inclined toward the direction of the pure Cu, similar to the crack deflection angles measured in Figures 3a) and c), and observed in the FEA predicted displacement field in Figure 4, but the magnitudes are very different. When thermal residual stress is added to the FE model, then the angle of inclination of maximum stress changes only slightly, as shown in Figure 5c) and 5d) for the 20 %W and 40 %W interlayers.

Table II. Average crack deflection angles measured in the scanning electron microscope, in degrees.

	Interlayer Thickness	
Interlayer Composition	2 mm	4 mm
20 %W	50	33
40 %W	25	0

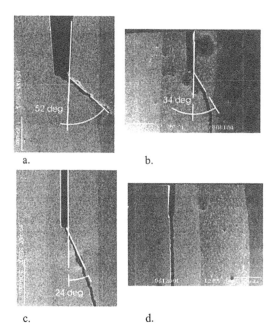

a.   b.

c.   d.

Figure 3. Scanning electron micrographs showing a crack in a) 2 mm thick 20 %W interlayer, b) 4 mm thick 20 %W interlayer, c) 2 mm thick 40 %W interlayer, and d) 4 mm thick 40 %W interlayer. In all cases increasing copper composition is to the right. The test was conducted in the conditions indicated in Figure 2.

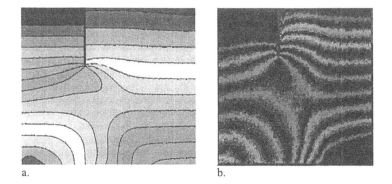

a.   b.

Figure 4. a) FEA predicted displacement contours, and b) Moiré displacement contours around notch in the 2 mm thick 20 %W interlayer under an applied bending load. The pure Cu layer is to the right of each figure . The range of displacement covered in one contour color is approximately 5.5 µm.

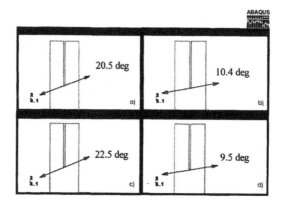

Figure 5. FEA predicted direction of highest maximum principal stress at notch base. a) and b) are the 20 % W and 40 % W interlayers respectively, modeled without the presence of thermal residual stress. c) and d) are the 20 % W and 40 %W interlayers respectively, modeled including the presence of thermal residual stress.

DISCUSSION

It is clear that the elastic mismatch between the various composite layers has a large influence on the stress field which the beam experiences in bending, since the loading at low loads is mostly, if not entirely, elastic. Systematic loading/unloading studies must be conducted to establish precisely what the relative role of elastic mismatch and plasticity is in influencing the displacement fields. However, the notion that this is primarily an elastic effect is supported by the FEA results, which show that a significant asymmetric shift in displacement field occurs under fully elastic conditions. The fact that the FE model shows a similar displacement field pattern to that observed using optical Moiré interferometry provides confidence in the FE model.

As shown in Figure 5, FEA predicts that the maximum principal tensile stress around the notch tip in a composite beam under load is inclined towards the copper rich layers. It is also apparent from the FEA results (e.g., Figure 5) that thermal residual stresses have a very small influence on shifting the notch tip stress fields. The fact that they have a small influence on the tip stress fields suggests that thermal residual stresses do not significantly alter the crack propagation direction (Figure 3); however, the authors are conducting further studies to better understand the influence of thermal residual stress. The different crack deflection angles observed in Figure 3 are most likely due to an alteration of the local stress field because of the elastic mismatch between layers. Plastic mismatch may also contribute to shifting the local stress field, but a significant asymmetry in the displacement field was noted under mostly elastic conditions, indicating that the plastic mismatch is not as important as the elastic mismatch. The possibility that stresses are applied asymmetrically in the sample simply due to a geometrical effect of the sample was considered. It was considered that such a geometrical effect would be due to asymmetric residual stress relaxation during notching, or due simply to the presence of a compliant layer (pure Cu) on one side of the notch (a 'hinge' effect). Optical moiré interferometry experiments which involved sequential material removal in Ni/Al₂O₃ graded composites have shown that machining a notch relieves stress, but does not induce shear stresses at the crack tip, even when the notch is not placed symmetrically in the sample [11]. Finally, if the presence of a hinge-like, compliant Cu layer on one side of the notch induced shear stresses during bending, then it would be expected that an increase in the thickness of the pure Cu layer would result in an increase in the crack deflection angle. However, the opposite was observed: thicker interlayers resulted in a decrease in crack deflection angle (cf. Figure 3). Thus, crack deflection is not an artifact of the specimen geometry.

A comparison of Figure 5 and 3 reveals that the maximum principal tensile stress (MPTS) criterion for fracture does not accurately predict the direction of crack propagation. In fact one might expect a priori that a fracture criterion for brittle materials may not apply to a composite which likely behaves in a ductile manner. Before dismissing the application of a MPTS criterion, it is noted that the FE result predicts crack nucleation, not propagation. To predict crack propagation beyond nucleation, it would be necessary to re-evaluate the MPTS once the crack has nucleated and examine whether the crack continues along the same direction. In other words, the FEA-predicted MPTS theory as currently used only applies to a kinking crack. Furthermore, the experimental observations do not reveal the crack kinking angle, the angle predicted by the MPTS criterion, unless that angle and the macroscopic deflection angle are identical. In order to measure the crack kinking angle, it would be necessary to make a higher resolution measurement and to start with a sharp crack. To summarize, a rigorous validation of the MPTS theory in predicting crack propagation has not been completed, though it appears from these initial results that the MPTS criterion does not work well.

An alternative crack propagation criterion was evaluated, particularly since the materials with high Cu volume fraction are expected to be relatively ductile. Crack propagation in a ductile material typically occurs at a location which experiences a minimum in the strain energy density ahead of a notch. For a round notch the crack nucleation corresponds to the location of maximum strain energy density on the notch surface. The strain energy density around the notch in the 2 mm thick 20 %W interlayer was extracted from the FE model and plotted as shown in Figure 6a). The angle at which the maximum strain energy density occurs is about 21°. As shown in the corresponding micrograph in Figure 6b), the crack seems to nucleate close to the location of maximum strain energy density, suggesting that the minimum strain energy density theory may be better suited in predicting the crack path.

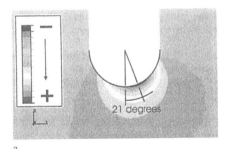 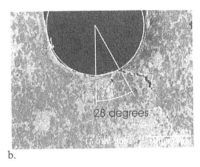

a.   b.

Figure 6. a) FEA-predicted contour plot of the strain energy density around the notch tip in the 2 mm thick 20 %W interlayer. b) Scanning electron microscope image of a notch in the 2 mm thick, 20 %W interlayer showing the initial stages of crack propagation. In both cases, the pure copper layer is to the right of the picture.

The application of the minimum strain energy density theory is currently being examined. Ultimately, the development of a predictive model will require combining crack nucleation and propagation.

CONCLUSIONS

Experimental and FEA results suggest that elastic mismatch in material properties are the principal reason the applied stress field around the notch in the 20 and 40% W layers in Cu/W composites is asymmetric. It is likely that this asymmetric stress field causes crack deflection at a prescribed angle, as observed experimentally (Figure 3).

ACKNOWLEDGEMENTS

The authors would like to acknowledge the U.S. Department of Energy, Office of Basic Energy Sciences for funding this research under contract DE-FG03-96ER45575. JC further acknowledges the Mexican National Council for Science and Technology (CONACYT) and Fulbright for the scholarship support to conduct graduate studies at the Colorado School of Mines.

REFERENCES

1. Erdogan, F. and Sih, G.C., "On the crack extension in plates under plane loading and transverse shear", J. Basic Eng., 85 (1963) 519-527.
2. Sih, G.C., "Strain energy density factor applied to mixed mode crack problems", Int. J. Fracture, 10 (1974) 305-322.
3. K. Hayashi and S. Nemat-Nasser, "Energy-Release Rate and Crack Kinking under Combined Loading", J. Appl. Mech. 48, (1981) 520-524.
4. B. Cotterell and J. R. Rice, "Slightly Curved or Kinked Cracks", Int. J. Fract. Vol. 16, (1980) 155-169.
5. He, M. Y., and Hutchinson, J.W., "Kinking of a crack out of an interface", J. Appl. Mech., 56, (1989) 270-8.
6. Gu and Asaro, "Cracks in Functionally Graded Materials", Int. J. Solids Structures, Vol 34, No. 1, (1997) 1-17.
7. Becker, Jr. T.L., Cannon, R.M. and Ritchie, R.O., "Finite crack kinking and T-Stresses in functionally graded materials", Submitted to Int. J. Solids and Structures (2000).
8. R. D. Torres, G. G. W. Mustoe, I. E. Reimanis and J. J. Moore, "Evaluation of Residual Stresses Developed in a Functionally Graded Material Using the Finite Element Technique", Processing and Fabrication of Advanced Materials IV, Edited by T. S. Srivatsan and J. J. Moore, The mineral, Metals & Materials Society (1996) 431-438.
9. I. Reimanis, J. Chapa, A. N. Winter, W. Windes and E. Steffler, "Fracture and Deformation in Ductile/Brittle Joints with Graded Structures", to appear in proceedings from the International Workshop on Functionally Graded Materials in the 21st Century: Trends and Forecasts, March 27-31, 2000, Epochal Tsukuba, Japan (2000).
10. J. Chapa and I. E. Reimanis, unpublished work.
11. A. N. Winter, Ph.D. Thesis, Colorado School of Mines (1999). Synthesis of Ni-Al2O3 Composites with Compositionally Graded Microstructures; Deformation Analysis with Phase Shifted Moiré Interferometry,

Functionally Graded Materials 2000

AUTHOR AND KEYWORD INDEX

Printed and bound by CPI Group (UK) Ltd, Croydon, CR0 4YY

16/04/2025